FIELD GUIDE

Birds of Western Africa

FIELD GUIDE TO THE

Birds of Western Africa

Nik Borrow and Ron Demey

CHRISTOPHER HELM
LONDON

Published 2004 by Christopher Helm, an imprint of A&C Black Publishers Ltd.,
38 Soho Square, London W1D 3HB

Reprinted with corrections 2008

www.acblack.com

ISBN 978-0-7136-6692-2

A CIP catalogue record for this book is available from the British Library.

Commissioning Editor: Nigel Redman

Designer: Julie Dando, Fluke Art, Cornwall

Produced by Phoenix Offset / The Hanway Press Ltd
Printed in China

10 9 8 7 6 5 4 3 2

CONTENTS

ACKNOWLEDGEMENTS

Since the publication of *Birds of Western Africa* (2001) we have benefited from the constructive comments of a large number of individuals including Thierry Bara, Clive Barlow, Peter Bijlmakers, Chris Bowden, Joost Brouwer, Iain Campbell, Claude Chappuis, Patrice Christy, William S. Clark, Nigel Cleere, Mary Crickmore, Richard Cruse, Tim Dodman, Andrew Dunn, Geoffrey Field, Lincoln Fishpool, Cornelis Hazevoet, Paul Herroelen, Ian Hinze, Olivier Lachenaud, Michel Louette, Bruno Portier, Bill and Rowena Quantrill, Hugo Rainey, Adam Riley, Stephen Rumsey, Jean-Marc Thiollay, Alan Tye and Tim Wacher.

Mark Adams, Peter Colston and Robert Prys-Jones, at the Natural History Museum in Tring, UK, and Michel Louette, at the Royal Museum for Central Africa, Tervuren, Belgium, are thanked for continuing to facilitate access to skins. Michel Louette is also thanked for stimulating discussions and generous help in various ways.

A special word of thanks is due to Robert Dowsett and Françoise Dowsett-Lemaire for their repeated generosity in providing us with invaluable help, advice, and stimulating discussions.

BirdLife International kindly allowed the use of some of its maps.

We thank Nigel Redman, the managing editor, for his assistance in the production of this book and Julie Dando, of Fluke Art, for the production and design, and her patience with the endless updating of the maps.

It is with pleasure that we take this opportunity to repeat our expressions of thanks made in *Birds of Western Africa* and to include a number of additional individuals.

NB would like to thank his mother for continued support and the many Birdquest clients, friends and companions in the field, including, in particular, Mark Andrews, George Angehr, Ombrou Antoine, Bamenda Highlands Forest Project, Lawrence Bangura, Mark Beaman, Christian Boix, Sering Bojang, Callan Cohen, Joachim Dibakou, Tony Disley, Enongene Louis Epie, Kingsley Epie, Brian Finch, Germain Gagné, Alan Greensmith, Ndong Bass Innocent, Solomon Jallow, Ekpe Kennedy, Korup National Park Project, Pedro Leitão, Norberto de Lima Vidal, Nomo Guirobo Luc, Pierfrancesco Micheloni, Michael Mills, Jerome Mokoko, Mount Kupe Forest Project, Idrissa Ndiaye, Edward Njie, Aoudou Oumarou, Robert Payne, Gerhard Radl, Sio, Zo Beugre Sylvain, Wandifa Touray and Mark Van Beirs.

RD thanks his companions in the field in various African countries, in particular Jane and Peter Chandley, Jean-Michel Borie, Thierry Bara, Martine Cmok, Ian Davidson, Marc Languy, Gondo Manh, Michel Nicole, Hugo Rainey, Alain Rousseau, Jan Van de Voorde and the members of the field teams in Burkina Faso, Cameroon, Ethiopia, Guinea, Ivory Coast and Nigeria. RD also wishes to thank Saliou Diallo (Guinée Ecologie) for his friendship and assistance in organising field work in Guinea, Lincoln Fishpool (BirdLife International) for his companionship in the field and help in various ways, Jennifer McCullough and Leeanne Alonso (Conservation International) for their invitations to participate in Rapid Assessment Program expeditions, and Francis Lauginie (Afrique Nature) for his superb hospitality and much practical assistance.

Once again, RD is particularly grateful to Rita Swinnen for her keen and unwavering support, her patience, and her indefatigable optimism and stimulating company in the field – without her input this book could not have been completed.

INTRODUCTION

AREA COVERED

This work illustrates and briefly describes all species definitely recorded from western Africa, as well as some that have been claimed but whose occurrence requires proof.

Western Africa, as defined here and subsequently defined as 'the region', comprises the 23 countries, south of the Sahara, from Mauritania in the northwest, to Chad and the Central African Republic in the east, and Congo-Brazzaville in the southeast, including the Cape Verde and Gulf of Guinea islands. The term 'region' is used in a general sense and does not indicate an avifaunal/biogeographical region or subregion. For ease of reference, all range states considered to comprise western Africa are covered in their entirety; thus parts of northern Mauritania, Niger and Chad that are generally considered within the Western Palearctic are included. The four principal Gulf of Guinea islands comprise Bioko (formerly Fernando Po), Príncipe, São Tomé and Annobón (formerly Pagalú). Bioko and Annobón form part of Equatorial Guinea.

Since the publication of our *Birds of Western Africa* (2001, hereafter abbreviated as *BOWA*) six additional species have been documented from the region: Great Blue Heron *Ardea herodias* (Hazevoet 2003), Ring-necked Duck *Aythya collaris* (Hazevoet 2003), Sociable Lapwing *Vanellus gregarius* (Messemaker 2004), Greater Yellowlegs *Tringa melanoleuca* (Hazevoet 2003), Moussier's Redstart *Phoenicurus moussieri* (Salewski *et al.* 2003) and Yellow-browed Warbler *Phylloscopus inornatus* (Cruse 2004). Three species are no longer considered to occur: Nyanza Swift *Apus niansae* (Herroelen 2003), Variable Indigobird *Vidua funerea* and Eurasian Rock Bunting *Emberiza cia* (Fry & Keith 2004). Additions to the list since 2004 are detailed on page 487. In total, 1306 species (including some potential species) are treated.

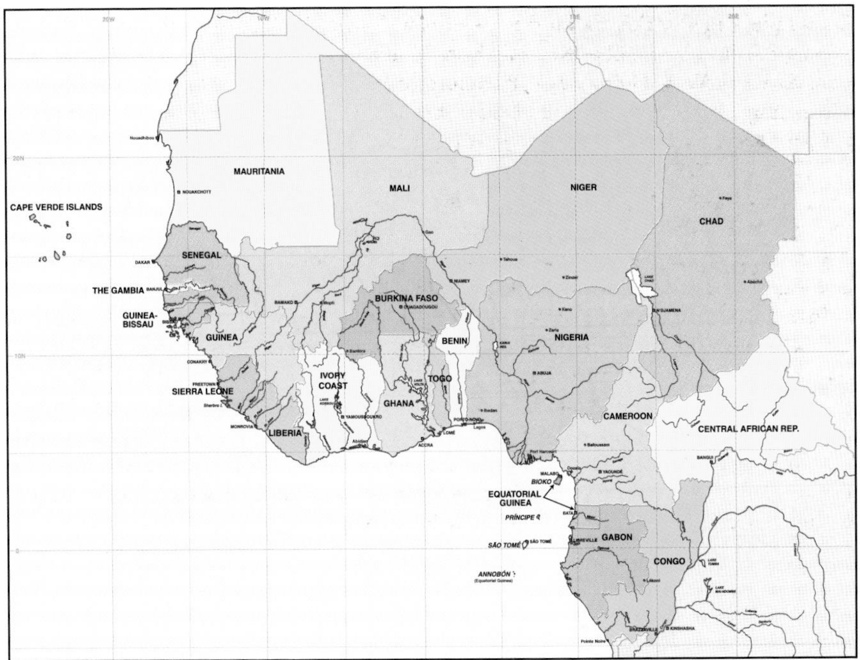

Figure 1. Western Africa.

NOMENCLATURE

Taxonomy and scientific names

In general, we have followed *The Birds of Africa*, vols 1–7 (Brown *et al.* 1982; Urban *et al.* 1986, 1997; Keith *et al.* 1992; Fry *et al.* 1988, 2000; Fry & Keith 2004; hereafter abbreviated as *BoA*), although in some cases we have preferred Dowsett & Forbes-Watson (1993), del Hoyo *et al.* (1992–2003), Dickinson (2003) or other recent authors, where these adopt what we consider a more advanced or consistent view.

In a few cases, where a taxon has been variably treated as a subspecies or a species by different authors, the taxon has been treated under a separate English name, but to indicate that this does not imply an undisputed taxonomic decision, the specific name under which it is also often treated is placed within parentheses. This has been done where there is strong evidence to suggest that the taxon may preferably be treated as a separate species, e.g. *Bostrychia (olivacea) bocagei*, *Bubo (bubo) ascalaphus* and *Saxicola (torquatus) rubicola*, or in the, rather few, cases where evidence suggests that forms treated as full species by *BoA* may actually best be 'lumped' again, e.g. *Indicator (minor) conirostris*, *Acrocephalus (scirpaceus) baeticatus* and *Eremomela (icteropygialis) salvadorii*.

We have been rather more 'liberal' in treating forms separately than in *BOWA*, thus following a current trend. Name changes are explained below.

Changes in the scientific name resulting from the correction of the gender

Scientific names have been corrected following David & Gosselin (2002a,b). These are, in alphabetical order:

*Alopochen aegyptiac***a**, *Amaurornis flavirostr***a**, *Ammomanes cinctur***a**, *Bleda canicapill***us**, *Bleda eximi***us**, *Bleda notat***us**, *Bleda syndactyl***us**, *Butorides striat***a**, *Chlidonias hybrid***a**, *Cinnyris coccinigast***rus**, *Cinnyris ose***a**, *Cisticola fulvicapill***a**, *Columba livia gymnocycl***a**, *Delichon urbic***um**, *Euplectes macrour***a**, *Halcyon senegalensis fuscopile***us**, *Heliolais erythropter***us**, *Nigrita canicapill***us**, *Nigrita fusconot***us**, *Numida meleagris galeat***us**, *Phalaropus fulicari***us**, *Ploceus nigriment***us**, *Saxicola torquat***us**, *Saxicola torquat***us** *moptan***us**, *Schoenicola platyur***us**, *Tchagra senegal***us**, *Tigriornis leucoloph***a**, *Treron calv***us**, *Turdoides fulv***a**, *Turdoides reinwardtii stictilaem***a**, *Turdoides tenebros***a**, *Turnix hottentott***us**, *Turnix sylvatic***us**, *Urolais epichlor***us**.

New scientific names resulting from taxonomic changes

Name used in *BOWA* (2001)	Name used in this book	References
Diomedea melanophris	*Thalassarche melanophris*	Knox *et al.* (2002), Sangster *et al.* (2002), Dickinson (2003)
Diomedea chlororhynchos	*Thalassarche chlororhynchos*	Knox *et al.* (2002), Sangster *et al.* (2002), Dickinson (2003)
Phoenicopterus roseus	*Phoenicopterus ruber*	Knox *et al.* (2002).
Eupodotis savilei	*Lophotis savilei*	del Hoyo *et al.* (1996), Dickinson (2003)
Eupodotis melanogaster	*Lissotis melanogaster*	del Hoyo *et al.* (1996), Dickinson (2003)
Catharacta skua	*Stercorarius skua*	Dickinson (2003), Sangster *et al.* (2004)
Catharacta maccormicki	*Stercorarius maccormicki*	Dickinson (2003), Sangster *et al.* (2004)

Caprimulgus binotatus	*Veles binotatus*	Cleere (2001), Dickinson (2003)
Cyanomitra obscura	*Cyanomitra olivacea*	Dickinson (2003)
Passer motitensis	*Passer cordofanicus*	*BoA*
Ploceus superciliosus	*Pachyphantes superciliosus*	*BoA*
Ortygospiza locustella	*Paludipasser locustella*	*BoA*
Amandava subflava	*Sporaeginthus subflavus*	*BoA*
Lonchura cucullata	*Spermestes cucullatus*	*BoA*
Lonchura bicolor	*Spermestes bicolor*	*BoA*
Lonchura fringilloides	*Spermestes fringilloides*	*BoA*
Lonchura cantans	*Euodice cantans*	*BoA*
Miliaria calandra	*Emberiza calandra*	Dickinson (2003), Sangster *et al.* (2004)

African Masked Weaver *Ploceus velatus*, treated as a single species in *BOWA* (2001) is now considered to consist of two species, Vitelline Masked Weaver *P. vitellinus* and Southern Masked Weaver *P. velatus* (*BoA*).

English names

With the aim of establishing a standardised world list of English names, many novel bird names have been coined in recent years, which has often resulted in confusion and frustration for users of ornithological works. In order to avoid further complication, we have principally followed *BoA*, supplemented by Dowsett & Forbes-Watson (1993), even if we do not personally favour the name chosen. Where these sources offer alternatives, we have chosen the name most commonly used in western Africa. Where Palearctic bird names, as listed by Beaman (1994), differ from these, based on convincing rationale, we have followed the latter. If none of these three sources proposes a name (e.g. in certain cases arising from taxonomic uncertainty), or if none of the proposed names is in frequent usage in western African ornithological literature, we have used other sources.

We have not, however, followed *BoA*'s controversial spelling of some names. Debate principally centres on the use of hyphenation and capitalisation. We agree with Beaman (1994) that the current fashion of introducing hyphens into bird names is both inconsistent and ugly, and have therefore used hyphens only when we considered it essential. A clear and consistent explanation of the use of hyphens is given by Inskipp *et al.* (1996), and their reasoning, which is briefly reproduced hereafter, has been largely followed here. They state that hyphens in group names should be used in two circumstances:

1. to link two nouns in apposition
2. between an adjective and a noun, but only in cases to avoid a misleading impression of the species' relationships.

Truly apposed nouns appear to be rare in bird names and most qualifying words within a group name that seem to be nouns are actually adjectival nouns, i.e. nouns functioning as adjectives that, for construction purposes, should be treated as ordinary adjectives and, therefore, should not take a hyphen. Thus, a snake eagle is a 'snake-eating eagle', not a 'snake-eagle' (the latter implies an unlikely intermediate between a snake and an eagle!). Adjectival nouns are abbreviations and can be identified as such if it makes sense to add a suffix such as -like, -sized, -billed, -eating, -nesting, -loving, -dwelling, or -driven. Examples: Tiger Heron, Night Heron, Cuckoo Hawk, Bat Hawk, Fish Eagle, Stone Partridge, Water Rail, Wood Dove, Grass Owl, Bush Lark, Hoopoe Lark, Sand Martin, Cliff Swallow, Ant Thrush, Scrub Robin, Robin Chat, Ground Thrush, Reed Warbler, Swamp Warbler, Wren Warbler, Woodland Warbler, Forest Flycatcher, Hill Babbler, Thrush Babbler, Sparrow Weaver, etc.

Names such as Scops Owl or Turtle Dove are more obscure but their qualifiers will probably also be demonstrated to be adjectival. Thus, they function in the same way as a combination of an overt adjective with a noun, as in Crested Tern, Green Pigeon, Crested Flycatcher, Penduline Tit and Glossy Starling.

Hyphens are placed between nouns and adjectives (or adjectival nouns) only in the interests of matching English with scientific classification. Painted Snipes and Cuckoo Shrikes are not, respectively, snipes and shrikes. The two components of the name are therefore better hyphenated, with the initial letter of the second word being in lower case: Painted-snipe and Cuckoo-shrike. (Hyphenated constructions with the initial letter of the second word being capitalised, e.g. 'Cuckoo-Shrike', are not used in this book.) It is, however, impossible to apply this rule consistently within a family and in some cases it appears preferable to leave well-established names unhyphenated, e.g. European Honey Buzzard (Beaman 1994, *contra* Inskipp *et al.* 1996).

The use of a terminal 's' in possessives is retained in all cases, including those of personal names ending in an 's' sound, following Dowsett & Forbes-Watson (1993) and other recent authors; thus Ayres's and Bates's.

New English names

In a few cases we have replaced the English name of a species used in *BOWA* (2001) by the one used in *BoA*'s last volume. *Glaucidium sjostedti* (previously Chestnut-backed Owlet) has become Sjöstedt's Owlet in order to avoid confusion with the similarly named *G. castanonotum*. *Telophorus viridis* (previously Gorgeous Bush-shrike) has been given its old and widely accepted name again, Perrin's Bush-shrike, in order to distinguish it from extralimital forms that have been lumped with it. In all these cases the previously used name has been put within parentheses.

Name used in *BOWA* (2001)	**Name used in this book**	**References**
Short-toed Eagle	Short-toed Snake Eagle	del Hoyo *et al.* (1999)
Chestnut-backed Owlet	Sjöstedt's Owlet	del Hoyo *et al.* (1999), Dickinson (2003)
Gorgeous Bush-shrike	Perrin's Bush-shrike	Bannerman (1939, 1953), Mackworth-Praed & Grant (1973)
Preuss's Golden-backed Weaver	Preuss's Weaver	*BoA*
Black-winged Red Bishop	Black-winged Bishop	*BoA*
Marsh Widowbird	Hartlaub's Marsh Widowbird	*BoA*
Grey-crowned Negrofinch	Grey-headed Negrofinch	*BoA*
Red-headed Antpecker	Woodhouse's Antpecker	*BoA*
Little Oliveback	Shelley's Oliveback	*BoA*
White-cheeked Oliveback	Grey-headed Oliveback	*BoA*
Reichenow's Firefinch	Chad Firefinch	*BoA*
African Quailfinch	Black-faced Quailfinch	*BoA*
Parasitic Weaver	Cuckoo Finch	*BoA*
Cut-throat	Cut-throat Finch	*BoA*
Baka Indigobird	Barka Indigobird	*BoA*

THE PLATE CAPTIONS

Measurements

To give an indication of the bird's relative size, **length** (in cm) is mentioned after the species name. Wingspan (**WS**) and tail-streamer length are mentioned where relevant. Measurements are taken from authoritative sources, complemented by our own mensural data obtained from specimens. Some measurements from *BOWA* (2001) have been corrected. Total-length measurements represent the length of museum skins stretched out on their backs and measured from bill tip to tail tip. It should be noted that direct comparison of these measurements can be quite misleading because they do not take into account other aspects of physiognomy, such as relative bill, neck and tail length, bulk, etc. It is therefore most useful only when comparing related species.

Status categories

The status of each species is denoted by the following letters (in bold):

R Resident — a species that resides within its range throughout the year and breeds; the opposite of a migrant.

M Intra-African migrant — a species that breeds in one part of Africa and spends the post-breeding season in a different area, or appears only seasonally in another part.

P Palearctic migrant — a species that breeds in the Palearctic region (Europe, N Africa and part of Asia) and spends the northern winter in sub-Saharan Africa.

V Vagrant — a species outside its normal range.

If more than one category is applicable, the commoner is placed first.

It should be appreciated that it is sometimes difficult to distinguish between a genuine vagrant and a species with only a few records. Some 'vagrants' may prove to be more or less regular, if rare or scarce, with increased observer coverage.

Abundance categories

Our aim has been to be as user-friendly as possible. Therefore we use only five, easily understood, abundance categories and one qualifier (**l** = local):

c common — invariably encountered singly or in significant numbers within its normal habitat

f fairly common — usually, but not invariably, encountered within its normal habitat (= 'not uncommon' in *BOWA* 2001, or 'frequent' in several other works)

u uncommon — relatively frequently, but not regularly, encountered within its normal habitat

s scarce — only irregularly and infrequently encountered within its normal habitat

r rare — rarely encountered, often implying fewer than *c.* 10 records.

The category of the most frequently occurring abundance is placed first; e.g. 'f/s' means that the species is more often (= in more countries) fairly common than scarce; 'u/lc' means 'uncommon to locally common'.

Threatened species

Species included in *Threatened Birds of the World* (BirdLife International 2000) and its most recent update (BirdLife International 2004) are indicated by one of the following letters (in bold):

CR	Threatened (Critical)	species facing an extremely high risk of extinction in the wild in the immediate future.
EN	Threatened (Endangered)	species facing a very high risk of extinction in the wild in the near future.
VU	Threatened (Vulnerable)	species facing a high risk of extinction in the wild in the medium-term future.
NT	Near Threatened	species coming very close to qualifying as threatened.
DD	Data Deficient	species for which there is inadequate information to make an assessment of its risk of extinction.

For a comprehensive discussion of the above categories, see BirdLife International (2000).

Identification features

The most important distinguishing identification features are succinctly given. We have attempted to include as much information as space permitted.

The term 'immature' (used generally for a non-adult bird) has normally not been used to indicate juvenile plumage (the first plumage of true, non-downy, feathers). For young passerines of Palearctic origin wintering in our area, 'immature' indicates 'first-winter' plumage. In some cases 'immature' is used when the available information prevents greater precision. The terms 'winter' and 'summer' refer to northern hemisphere seasons.

If the plumage illustrated is not specifically named, it is the adult plumage. For the sake of brevity, subspecies are indicated by their subspecific name only, when preceded by 'Ad' or 'Juv' (e.g. 'Ad *gabonensis*').

The terms 'small', 'medium-sized' and 'large' indicate only the relative size of a species compared with its close relatives. For the sake of brevity, and when not indicated otherwise (or obviously different), the term 'top of head' includes forehead and crown, 'throat' includes the chin, and 'legs' includes legs and feet. 'Above' generally encompasses head and upperparts, and sometimes also tail; 'below' the entire underparts; in some cases, certain parts are excluded, but this should be obvious from the context. 'Flight feathers' include primaries and secondaries (but not tail feathers). 'Wing feathers' include flight feathers and wing-coverts. 'Upperparts' often includes scapulars and wing-coverts (if concolorous). Study of the illustrations should preclude any confusion.

Colour names have deliberately been kept as simple as possible and most will be readily understood. Where two colours are combined (compound colours), the last named is dominant; in other words, they should be interpreted as the second colour tinged with the first; e.g. reddish-brown is brown tinged red. Where the suffix -ish is added to a colour, this indicates a weaker or less distinct shade of that colour.

Frequently used colours include buff (pale or dull yellowish with a brownish or beige tinge), chestnut (dark reddish-brown), olive (dull yellowish-green, like the fruit), and horn (pale brownish-yellow; used solely for bills). Dusky (dirty greyish or brownish) is often used for rather indistinct darkish markings. The term dark (opposite pale) is used for dark plumage markings that do not possess any obvious colour.

Habitat and behaviour (▲)

The habitat in which the species normally occurs is briefly indicated. Some typical behaviour relevant to identification is mentioned wherever space permitted to do so. If a species is endemic to the region, this has been mentioned.

Voice (✤)

Only the most characteristic vocalisations are (briefly) given. In some cases, where the species is mainly silent or the vocalisations are unimportant for identification purposes, these have been omitted when space was limited. Transcriptions of calls and songs are placed in *italics*. References to Claude Chappuis' (2000) outstanding collection of 15 CDs of *African Bird Sounds* – an invaluable tool – are presented within square brackets, with CD and track number. It should be noted that these CDs inevitably contain certain errors (see for example Françoise Dowsett-Lemaire's thorough review in *Bull. Afr. Bird Club* 9: 74–78, 2002). Discrepancies between Chappuis' accompanying booklet and this book are therefore due to the fact that references of vocalisations known to us as having been misidentified have been omitted or placed under the correct species.

As is widely acknowledged, transcribing bird sounds in such a way that they can be unambiguously interpreted by others is almost impossible. Phonetic renditions and verbal descriptions are necessarily highly subjective and open to misinterpretation, but there appears to be no convenient alternative. The user of this book is therefore advised to listen to sound recordings and compare them with the transcriptions, in order to understand the authors' interpretation.

In the transcriptions of vocalisations the following conventions apply:

single vowels are pronounced short (thus *a* as in 'apple', *e* as in 'extra', *i* as in 'it', *u* as in 'full')
ee as in 'see', 'be'
iiiii is higher pitched than *eeee*
k as in 'cat' (c is not used for the hard 'k' sound)
ch as in 'check'
sh as in 'sheep'

CAPITAL LETTERS indicate that the component in question is considerably louder than the others.

Pauses between notes or syllables are denoted as follows (after Lewington *et al.* 1991):
see-see very short pause
see see normal pause, as in ordinary conversation
see, see longer pause (at least *c.* 1 second)
see...see pause of more than 2 seconds

THE PLATES

All species recorded or claimed from the region are illustrated in colour. Our aim has been to illustrate as many distinct plumages as space would permit. Thus, distinctive male, female and immature plumages are depicted, subject to the availability of representative specimens or personal field notes. This is also the case with races that are sufficiently distinctive to be separated in the field. Only in a few cases, however, has the juvenile plumage of passerines been illustrated, as this appears unnecessary and potentially confusing. Indeed, this plumage is usually acquired for only a short period and identification is normally facilitated by the presence of the parents.

Wherever possible or desirable, care has been taken to respect family groupings—this has resulted in some plates containing more illustrations than others. On each plate, the species are arranged in order of resemblance. The order of the plates generally follows taxonomic sequence. Species and distinctive subspecies that occur exclusively on Cape Verde and the Gulf of Guinea islands are grouped on the final four plates (145–148).

The majority of the plates are those of *BOWA* (2001), but ten are entirely new and a significant number have been amended.

THE MAPS

Distribution maps are provided for all species except those occurring exclusively in the Cape Verde and Gulf of Guinea islands. They are placed (generally alternately) either immediately after or immediately before each double-page spread of plate and caption text. The maps are based on published data up to March 2004. A few unpublished records have been included, when considered reliable. The maps reflect the known or inferred distribution of a species in areas of suitable habitat within a broadly defined range. As locality data remain scant for many species in our region, the maps should not necessarily be taken as providing a true reflection of actual distributions and must therefore be used with caution and common sense. Some records from Mali and Mauritania (Lamarche 1980–1981, 1988), included in the distribution maps in *BOWA* (2001), appear to be either erroneous or highly unlikely and have been omitted or indicated with a '?'.

Key to the maps:

☐	resident	☐	mainly resident but partially migratory or erratic within range	☐	breeding visitor (intra-African migrant)
☐	non-breeding visitor (main range)	☐	non-breeding visitor (sparse occurrence)	X	vagrant or isolated record

AUTHORS' NOTE

Many interesting observations remain buried in personal notebooks or unpublished reports. It is therefore recommended that relevant data be documented and submitted to a refereed journal for publication, e.g. *Malimbus* or *Bulletin of the African Bird Club* (see page 24).

In a work of this scope some errors and omissions appear inevitable, despite the care with which museum specimens were examined and data collected. Future fieldwork will also certainly add to our knowledge of the region's avifauna. The authors (c/o the publishers, Christopher Helm) would therefore be pleased to receive any information which updates and corrects that presented in this book.

CLIMATE, TOPOGRAPHY AND MAIN HABITATS

CLIMATE

The principal feature of the region's climate is the alternate wet and dry seasons, which are governed by the movement of the Intertropical Convergence Zone (ITCZ), a zone of low pressure towards which blow winds from the northern and southern high pressure belts. Southwesterly winds from the Atlantic Ocean are warm and moist, while northeasterly winds from the Sahara, known as Harmattan, are hot, dry and dusty. The ITCZ annually oscillates north and south following, with a lag of 1–2 months, the position of the sun. At the northern summer solstice it lies near the Tropic of Cancer and wet maritime winds produce a rainy season north of the equator. When the ITCZ moves south and the sun is over the Tropic of Capricorn, most of the region comes under the influence of the continental, hot, dry Harmattan. Overall, the dry season lengthens and rainfall decreases with increasing latitude. In the south, the dry season generally extends from December to March, while in the north it lengthens to at least seven months, from October to April. This pattern can, however, exhibit considerable annual variation, especially in the north where the wet season is short (July–September) and rainfall is erratic in timing, quantity and distribution. In the south, from Ivory Coast to Nigeria, the rains decrease in August, resulting in a 'short dry season'.

In the part of our region south of the equator, namely Gabon and Congo, the weather pattern is the reverse of that to the north, with a dry season from mid-May to September and a long rainy season from October to mid-May. Rains decrease during a short dry spell in December–January. Here also, monthly rainfall varies greatly from year to year.

Maximum temperatures and temperature ranges increase with latitude. In the forest zone temperatures vary little, averaging around 27°C throughout the year, while in the desert they may range from around 0°C to more than 45°C.

TOPOGRAPHY

Most of the western African mainland is, unlike eastern and southern Africa, largely flat and low lying, with altitudes rarely exceeding 400 m except for granite inselbergs that can reach 700 m.

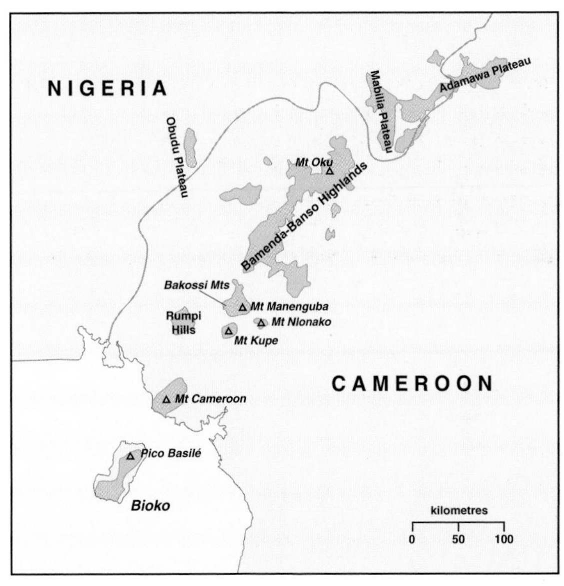

Figure 2. The Cameroon highlands

Notable montane or hilly areas include the Fouta Djalon (Guinea), the Loma Mts (Sierra Leone), Mt Nimba (1752 m, Liberia/Guinea/Ivory Coast), the Jos Plateau (C Nigeria), the Aïr Mts (Niger), and the Tibesti and Ennedi Mts (N Chad). The interior of Gabon and Congo is occupied by a vast plateau ranging at 300–1000 m.

The most important, however, is the Cameroon highlands, a chain of mountains in western Cameroon running southwest and extending across the border into SE Nigeria (Mambilla and Obudu Plateaux) (see fig. 2). It contains the highest peak in W Africa, Mt Cameroon (4095 m), which is volcanic and still active. Other mountains include Mt Rata (1768 m) in the Rumpi Hills, Mt Kupe (2064 m), Mt Nlonako (1825 m), Mt Manenguba (2411 m) in the Bakossi Mts, and Mt Oku (3011 m) in the Bamenda Highlands.

The Gulf of Guinea islands are part of the line of volcanoes, extending from the Cameroon highlands southwest into the Atlantic Ocean. Bioko lies on the continental shelf just *c.* 30 km off the coast of Cameroon and was probably linked to the African mainland in the past. Pico Basilé (3011 m) is still active. Príncipe, São Tomé and Annobón, however, are true volcanic islands surrounded by deep seas. This isolation has led to a species-poor avifauna with a high degree of endemism.

The oceanic Cape Verde Islands are situated 460–830 km off the coast of Senegal and consist of ten islands and several islets of volcanic origin (see fig. 3). The mountainous islands, with peaks reaching over 1000 m (and up to 2800 m on Fogo, the only active volcano), receive annual rain or at least some precipitation, while those with low relief receive only little and irregular rain (Hazevoet 1995).

The region has five or six main river systems. The 4030 km-long Niger River, with its main tributary the Bénoué (or Benue), drains the majority of the region. It arises in the high-rainfall Fouta Djalon hills of Guinea, only a few hundred kilometres from the ocean, flows through every climatic and vegetation zone of the region and forms a huge inland delta, which considerably delays flooding downstream such that maximum water levels here occur well into the dry season, coinciding with the northern hemisphere winter and important for migrant waterbirds. The other main river systems are the Senegal River, the Ogooué, the Congo River and its tributary the Ubangi, and the Chari (sole affluent of Lake Chad).

There are few large natural lakes in the region, the largest being Lake Chad, which exhibits considerable fluctuations in water levels.

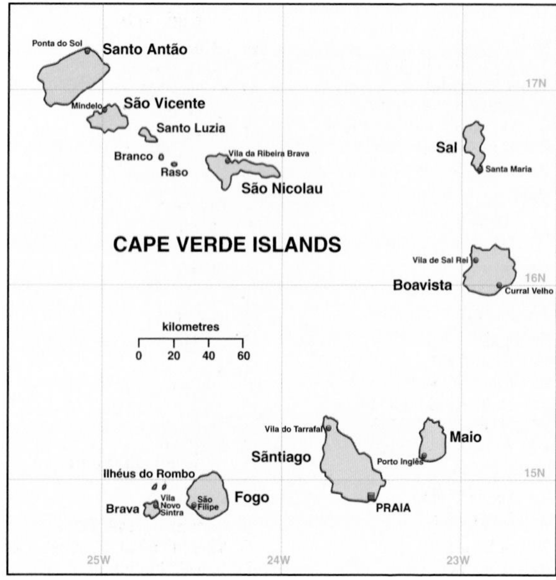

Figure 3. Cape Verde Islands

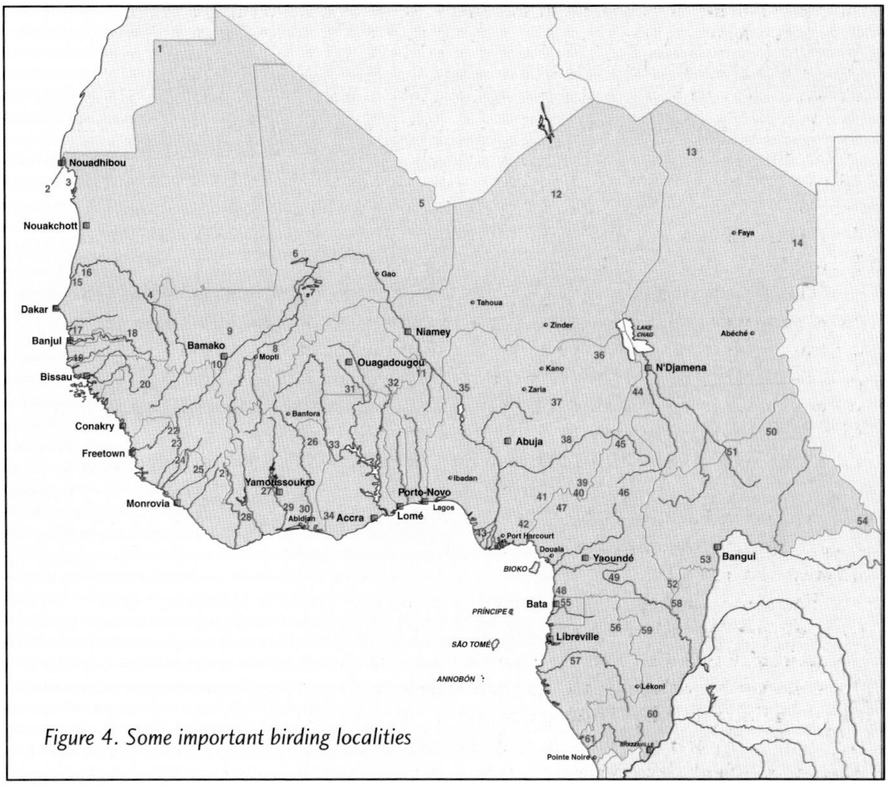

Figure 4. Some important birding localities

Mauritania
1 Zemmour
2 Baie de l'Etoile
3 Banc d'Arguin NP
4 Guidimaka

Mali
5 Adrar des Iforhas
6 Lac Faguibine
7 Central Niger delta
8 Bandiagara escarpment
9 Boucle du Baoulé NP
10 Mandingues Mts

Niger
11 'W' NP
12 Aïr Mts

Chad
13 Tibesti
14 Ennedi

Senegal
15 Djoudj NP
16 Lac de Guier
17 Delta du Saloum NP
18 Niokolo-Koba NP
19 Basse Casamance

Guinea
20 Fouta Djalon
21 Mt Nimba

Sierra Leone
22 Loma Mts
23 Tingi Mts
24 Gola Forest

Liberia
21 Mt Nimba
25 Wonegizi Mts

Ivory Coast
21 Mt Nimba
26 Comoé NP
27 Marahoué NP
28 Taï NP
29 Lamto
30 Yapo Forest

Burkina Faso
11 'W' NP
31 Po NP
32 Arli NP

Ghana
33 Mole NP
34 Bia NP

Benin
11 'W' NP

Nigeria
35 Kainji NP
36 Hadejia-Nguru wetlands
37 Jos Plateau
38 Wase Rock
39 Gashaka-Gumti NP
40 Mambilla Plateau; Gotel Mts
41 Obudu Plateau
42 Cross River NP; Oban Hills
43 Lower Niger delta

Cameroon
44 Waza NP
45 Bénoué NP
46 Adamawa Plateau
47 Bamenda Highlands
48 Campo-Ma'an NP
49 Dja Game Reserve

Central African Republic
50 Manovo-Gounda-St Floris NP
51 Bamingui-Bangoran NP
52 Dzanga-Sanga Forest
53 Lobaye Préfecture
54 Ouossi R. (Baroua and Zémio area)

Equatorial Guinea
55 Mt Alen

Gabon
56 Makokou
57 Lopé NP

Congo
58 Nouabalé-Ndoki NP
59 Odzala NP
60 Léfini Reserve (Bateke Plateau)
61 Conkouati Faunal Reserve

NP = National Park

MAIN HABITATS

Western Africa contains a broad range of habitats, from rainforest to desert. They are arranged in a series of parallel latitudinal bands orientated west–east, reflecting the decreasing northward rainfall gradient. Except locally, the lack of relief means there is little disturbance to the zonal arrangement of both climatic and vegetation belts.

The sea coast is generally flat and sandy, with some intertidal mud and sand flats. Africa's largest intertidal mudflats occur in the region, in N Mauritania (Banc d'Arguin area, 46,000 ha) and along the coast of Guinea-Bissau, Guinea and Sierra Leone (284,300 ha). Smaller mudflats, which seasonally hold significant aggregations of waders, are the Senegal delta and Sine-Saloum, both in Senegal.

Brackish creeks and lagoons may occur behind the narrow coastal belt; these may be very large locally (e.g. in Ivory Coast). Mudflats and lagoons typically are bordered by mangroves. Open swamps and swamp forest may be found further inland.

Africa's two major lowland rainforest blocks occur in the region, often referred to as the Upper Guinea and Lower Guinea forests, separated, from E Ghana to Benin, by the Dahomey Gap, where savanna reaches the coast. These forests, except within the Dahomey Gap, originally covered most of the area between Sierra Leone and SE Guinea to SW CAR and Congo. Much of this area is now deforested and replaced by 'derived savanna', a mosaic of cultivation, farmbush and secondary forest, which permits savanna species to penetrate the forest zone. Of the remaining forest, very little is true primary: most has been modified by man. As the degree of secondary modification may be hard to gauge, the term 'high forest' has been proposed. This forest zone may be subdivided into the moister evergreen forests in the south and the drier semi-deciduous forests further north, but the change is progressive.

The savanna zone, immediately north of the forest, is usually subdivided, on the basis of the density of the woodland and species composition, into two: the Guinea savanna in the

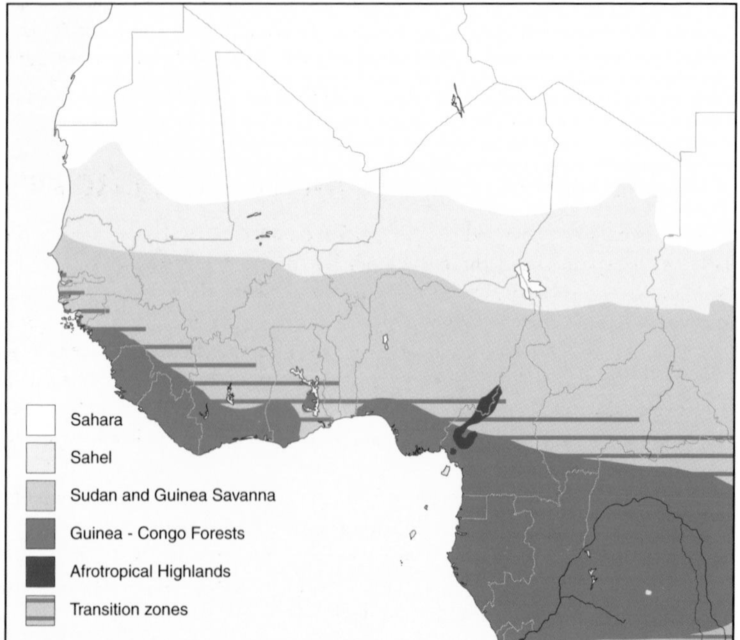

Figure 5. The vegetation zones of western Africa

south and the drier Sudan savanna further north. The zone is characterised by various types of wooded grassland, in which there is a gradual decrease of tree cover and tree height, reflecting the decline in rainfall, northward. The other dominant feature of the vegetation here is the grasses that grow under and between the trees. In Guinea savanna they are tall (2–3 m), dense perennial species which, with increasing latitude, give way to smaller species in which annuals become increasingly common. Watercourses are bordered by gallery forest of various widths. Some dense forest patches may remain; these permit forest species to penetrate deep into the savanna zone. Bare, granitic inselbergs are a typical feature of the landscape, especially in the northern Guinea savanna. A feature of these savannas, especially in the south, are the fires that regularly rage through them in the late dry season (December–February). Although a natural phenomenon to which taxa of the region are well adapted, these fires now are almost all deliberately started by man.

In the southeast of our region, in C Congo and Gabon, a northward extension of the vast woodlands of southern Africa penetrates, as a wedge, into the forest zone. This gives rise to the presence of bird species and subspecies of southern affinities.

The Sahel zone, north of the savanna zone, is characterised by thorn scrub (with *Acacia* spp. and *Ziziphus* spp.), sparse, mostly annual, grasses, and very low and often erratic rainfall (mean 100–600 mm). The most important wetlands of the region stretch across this semi-arid belt and include the Senegal delta (with Djoudj National Park), the central Niger delta in Mali, the Hadejia–Nguru wetlands in N Nigeria, and the Logone floodplain in N Cameroon.

The northern border of the region is formed by the Sahara desert, which consists of arid landscapes with sandy, stony or rocky substrates and sparse plant cover, except in depressions, wadis and oases, where water is retained.

Afrotropical highlands with montane forest and montane grassland occur only in W Cameroon and Bioko. Montane forest differs from lowland forest in tree species composition and is also relatively rich in endemic bird species. Local variations in rainfall result in differences in the altitude at which it is found. A combination of high rainfall and reduced temperature causes montane forest to occur at relatively low elevations on Mt Cameroon and Bioko (generally above 800 m). Further inland, montane forest is only found higher, e.g. above 1200 m on Mt Kupe and at 2000–2950 m in the Bamenda Highlands (Stattersfield *et al.* 1998).

RESTRICTED-RANGE SPECIES AND ENDEMIC BIRD AREAS

BirdLife International has analysed the distribution patterns of birds with restricted ranges, defined as landbird species which have or have had a total global breeding range of less than 50,000 km^2 throughout historical times (i.e. post-1800) (Stattersfield *et al.* 1998). The results demonstrate that these restricted-range species tend to cluster, often on islands or in isolated patches of a particular habitat. Regions where two or more species of restricted range share completely or partially overlapping distributions are termed Endemic Bird Areas (EBAs). Where the distributions of such species only partially overlap, the total area of the EBA may be considerably larger than 50,000 km^2. Areas where only one restricted-range species occurs are termed Secondary Areas (SAs).

Most EBAs also have one or more globally threatened bird species and are important for restricted-range species from other wildlife groups. They are therefore clearly priorities for conservation action. EBAs vary considerably in size (from a few square kilometres to more than 100,000 km^2) and in the numbers of restricted-range species that they support (from two to 80), and thus their relative importance also varies.

In western Africa, a total of 87 restricted-range species occur in seven EBAs and four SAs. The region's EBAs comprise the Cape Verde Islands (4 restricted-range species), Annobón

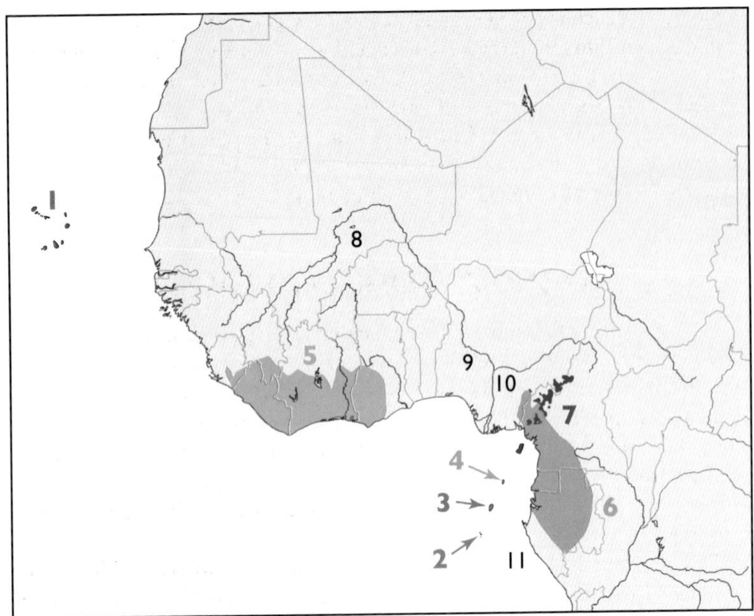

Figure 6. Endemic Bird Areas of western Africa

Endemic Bird Areas
1 Cape Verde Islands
2 Annobón
3 São Tomé
4 Príncipe
5 Upper Guinea forests
6 Cameroon and Gabon lowlands
7 Cameroon mountains

Secondary Areas
8 Upper Niger valley
9 South-west Nigeria
10 Lower Niger valley
11 Gabon-Cabinda coast

Source: BirdLife International

(3), São Tomé (21), Príncipe (11), the Upper Guinea forests (15), the Cameroon and Gabon lowlands (6), and the Cameroon mountains (29). The region's Secondary Areas (all with one restricted-range species) are the Upper Niger valley, South-west Nigeria, Lower Niger valley and Gabon–Cabinda coast. Six restricted-range species are shared between EBAs.

Tropical lowland and montane forests are the predominant habitats, and in the Upper Guinea forests and Cameroon mountains, in particular, it is estimated that there has been major (>50%) loss of these key habitat types. The Cameroon mountains are particularly important for the absolute numbers of restricted-range species occurring (the third largest total in the African region) and for the proportion of these that are threatened (12 species). São Tomé and the Upper Guinea forests also rate very highly in terms of their biological importance and threat levels.

Examples of highly threatened (according to BirdLife International 2000) restricted-range species include: Raso Lark *Alauda razae* (confined to one minute island in the Cape Verdes and at risk from introduced predators); Dwarf Olive Ibis *Bostrychia (olivacea) bocagei*, São Tomé Fiscal *Lanius newtoni* and São Tomé Grosbeak *Neospiza concolor* (all from São Tomé and assumed to have tiny populations in a small area of primary forest); Rufous Fishing Owl *Scotopelia ussheri* (a rarely seen species with a small, fragmented population seriously threatened by habitat loss); Liberian Greenbul *Phyllastrephus leucolepis* (a poorly known species that appears to have an extremely small, severely fragmented range and is inferred to be declining owing to habitat destruction); Bannerman's Turaco *Tauraco bannermani* (which is only likely to survive if the Kilum-Ijim forest, the largest remaining montane forest area in the Cameroon mountains, is preserved) and Mount Kupe Bush-shrike *Malaconotus kupeensis* (which has a very small population with a small range and suffers from habitat loss).

The BirdLife International African Partnership has identified potentially conservable sites termed Important Bird Areas throughout the continent, including suites of sites within EBAs that seek to protect the unique species within these areas.

TAXONOMY: SOME DEFINITIONS

Taxonomy is the science of classification and naming of living organisms. The classification of organisms attempts to reflect relationships between them and works on a hierarchical system by which an organism is placed in categories of decreasing level, from 'Kingdom' to 'Species' and 'Subspecies'. These categories are known as taxa (singular: taxon; hence 'taxonomy'). The main categories useful to the field ornithologist are the following.

Vertebrates are divided into **classes**: mammals, birds, reptiles, amphibians and fish. Birds are vertebrates characterised by the possession of feathers and belong to the class Aves, which is the scientific name for 'birds'.

The class Aves contains approximately 29 large groups called **orders**, with names ending in '-iformes'. Birds of the order Passeriformes (passerines, or songbirds) are placed together because they share certain morphological characters. Other orders include, for example, Ciconiiformes (storks, herons and ibises), Strigiformes (owls) and Piciformes (woodpeckers, honeyguides and barbets).

Orders are divided into *c.* 180 **families**, with names ending in '-idae', following the same logic of shared characters. Examples of families within the order of Passeriformes are: Hirundinidae (swallows and martins), Pycnonotidae (bulbuls) and Ploceidae (weavers).

Within families, birds are clustered into more than 2000 **genera** (singular: genus). Examples of different genera within the family Sylviidae are: *Acrocephalus*, *Prinia*, *Apalis*, *Phylloscopus*, *Cisticola*, etc.

A genus comprises one or several **species**. For example, four species belonging to the genus *Prinia* occur in western Africa: *Prinia subflava*, *Prinia fluviatilis*, *Prinia bairdii* and *Prinia molleri*. According to the biological species concept, which is followed in this book, a species can be defined as a population, or group of populations, of actually or potentially interbreeding individuals, reproductively isolated from all other such populations. Members of a species should be able to interbreed freely and produce fertile offspring.

The scientific name of a species is based on an internationally accepted, binomial system and consists of a two-part, latinised name, which is conventionally written in *italics*. The first part is the generic name (with first letter capitalised), the second the specific name (all lower case).

Finally, a species may be divided into different **subspecies** (also called races). Subspecies are groups of similar-looking individuals, slightly different from other groups, but belonging to the same species. Members of subspecies can or could still interbreed with other members of the same subspecies. A subspecies' name is added as a third part to the species' scientific name, e.g. *Prinia subflava pallescens*. The first population to be described becomes the **nominate subspecies** and carries the same subspecific and specific names, e.g. *Prinia subflava subflava*. If there is no doubt as to which species or genus is involved, the name can be shortened to *Prinia s. subflava* or *P. s. subflava*.

ORGANISATIONS

International

BirdLife International

Wellbrook Court, Girton Road, Cambridge CB3 0NA, UK.
Tel. +44 (0)1223 277318 Fax. +44 (0)1223 277200
E-mail: birdlife@birdlife.org.uk Website: http://www.birdlife.org

BirdLife International (formerly the International Council for Bird Preservation) strives to conserve birds, their habitats and global biodiversity, and is the leading authority on the status of the world's birds and the urgent problems that confront them. BirdLife aims to:

- prevent the extinction of any bird species
- maintain and where possible improve the conservation status of all bird species
- conserve and where appropriate improve and enlarge sites and habitats important for birds
- help, through birds, to conserve biodiversity and to improve the quality of people's lives
- integrate bird conservation into sustaining people's livelihoods

Publications: *World Birdwatch* magazine (quarterly) and, with Cambridge University Press, *Bird Conservation International* (quarterly).

African Bird Club

c/o BirdLife International, Wellbrook Court, Girton Road, Cambridge CB3 0NA, UK.
E-mail: info@africanbirdclub.org Website: http://www.africanbirdclub.org

The African Bird Club aims to:

- provide a worldwide focus for African ornithology
- encourage an interest in the conservation of the birds of the region
- liaise with and promote the work of existing regional societies
- encourage observers to visit lesser known areas of the region
- encourage observers to actively search for globally threatened and near-threatened species
- run the ABC Conservation Programme

Publication: *Bulletin of the African Bird Club* (bi-annual).

West African Ornithological Society / Société d'Ornithologie de l'Ouest Africain

Secretary: R. E. Sharland, 1 Fisher's Heron, East Mills, Fordingbridge, Hampshire, SP6 2JR, UK.

The Society aims to promote scientific interest in the birds of West Africa and to further the region's ornithology, principally through publication of its journal.

Publication: *Malimbus* (bi-annual).

National

Burkina Faso: Fondation des Amis de la Nature (Naturama)
01 BP 6133, Ouagadougou 01, Burkina Faso
E-mail: naturama@fasonet.bf

Cameroon: Cameroon Biodiversity Conservation Society
P O Box 3055, Messa, Yaoundé, Cameroon
E-mail: coc@iccnet.cm

Ghana:	Ghana Wildlife Society P O Box 13252, Accra, Ghana E-mail: wildsoc@ighmail.com
Guinea:	Guinée Ecologie BP 3266, Conakry, Guinée E-mail: dmsaliou@afribone.net.gn
Nigeria:	Nigerian Conservation Foundation P O Box 74638, Victoria Island, Lagos, Nigeria E-mail: ncf@hyperia.com
Sierra Leone:	Conservation Society of Sierra Leone P O Box 1292, Freetown, Sierra Leone E-mail: cssl@sierratel.sl

BIRD TOPOGRAPHY

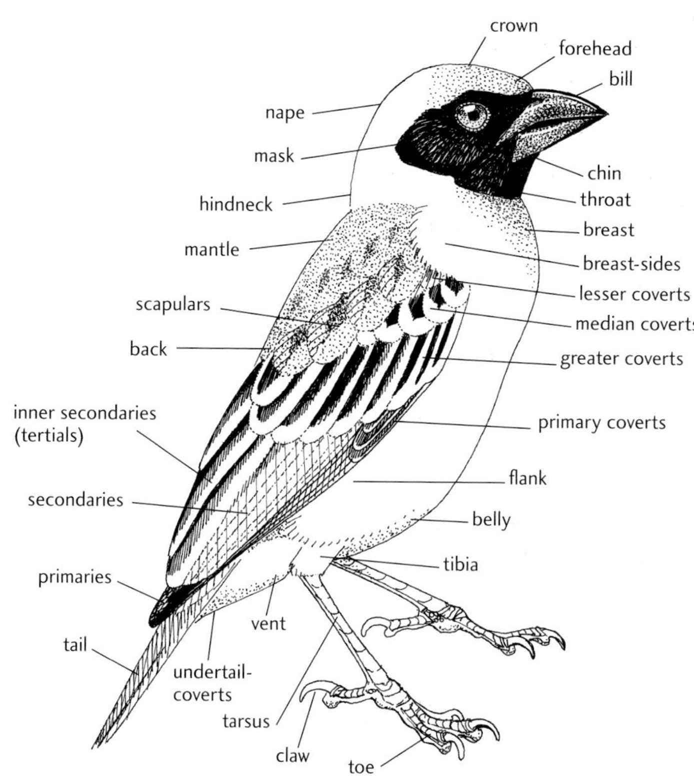

Vitelline Masked Weaver *Ploceus vitellinus*

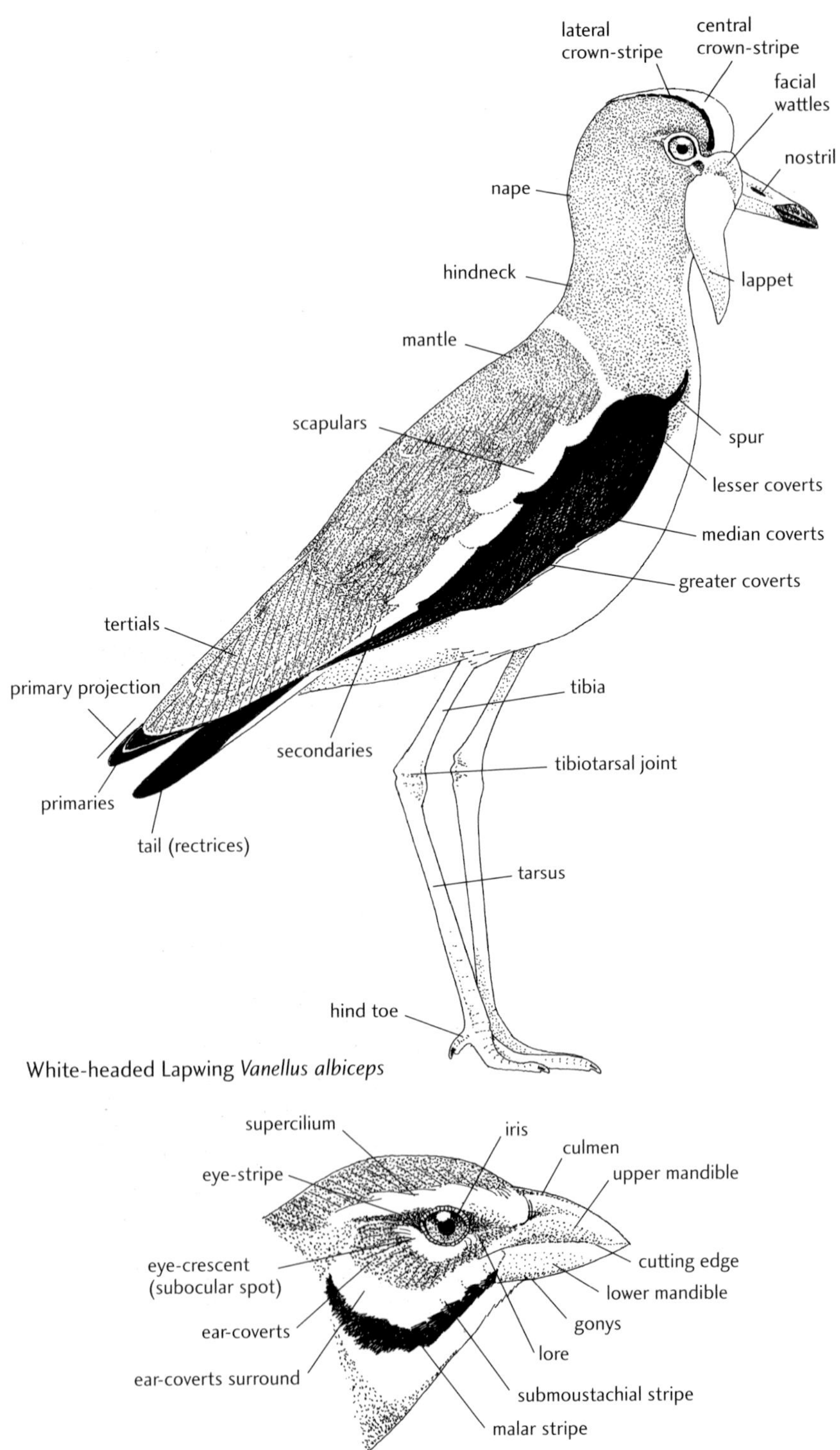

White-headed Lapwing *Vanellus albiceps*

Chestnut-crowned Sparrow Weaver *Plocepasser superciliosus*

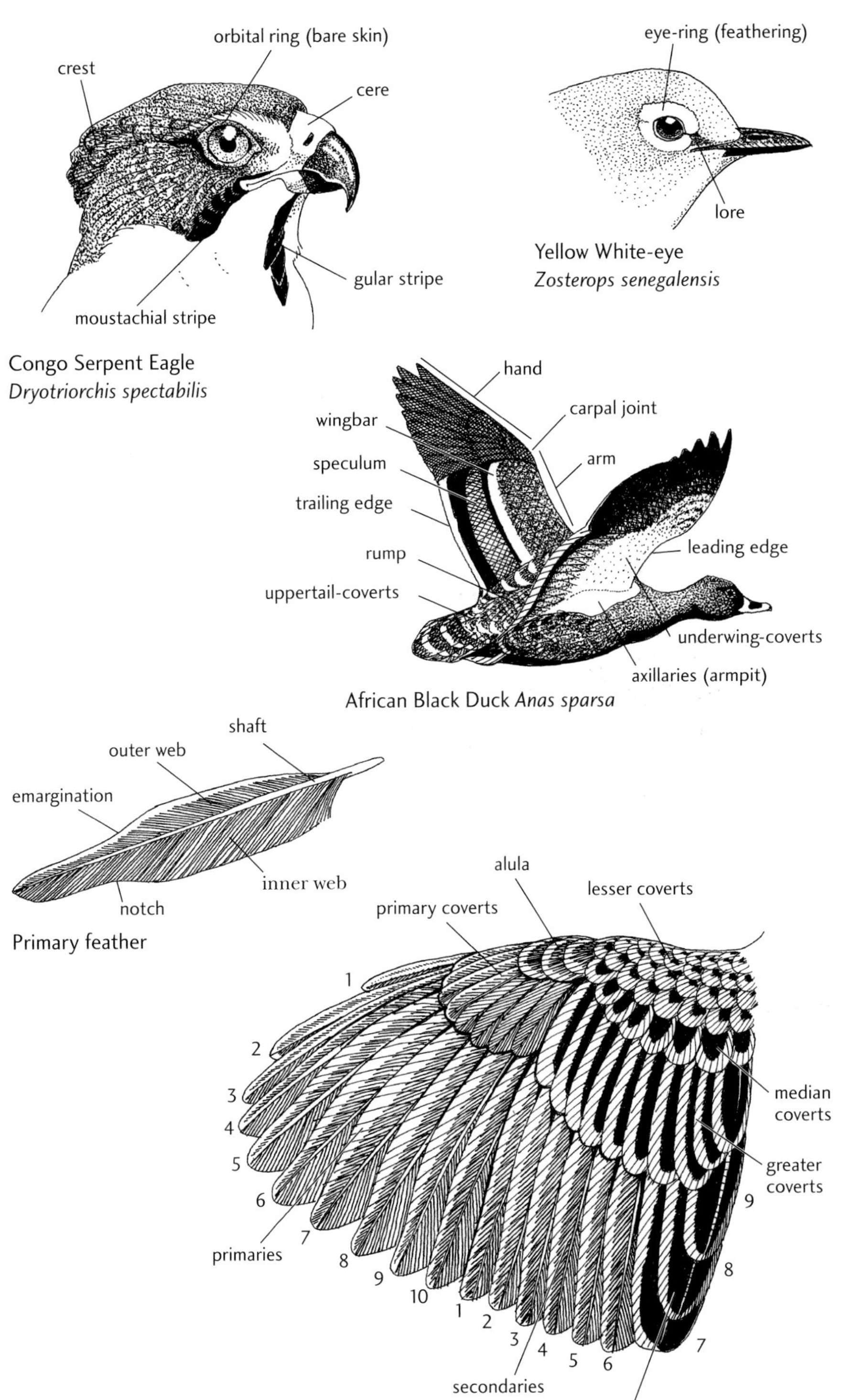

Congo Serpent Eagle *Dryotriorchis spectabilis*

Yellow White-eye *Zosterops senegalensis*

African Black Duck *Anas sparsa*

Primary feather

Zitting Cisticola *Cisticola juncidis*

GLOSSARY

adult: a bird in final plumage (plumage no longer changing with age).
aerial: air-frequenting (e.g. an aerial feeder).
antiphonal: referring to precisely timed alternating singing or calling by two birds, usually a mated pair (see also 'duetting').
aquatic: living in water.
arboreal: living in trees.
arm: inner part of wing, between the body and the carpal joint (wing-coverts and secondaries).
axillaries: the feathers at the junction of the underwing and the body; the 'armpit' of a bird.
bare parts: those parts not covered by feathers (including bill, cere, eyes, wattles, orbital ring and feet).
bib: a contrasting, usually dark area, on the throat and/or upper breast.
brood parasite: a species that lays its eggs in the nest of another (the 'host') and plays no parental role in raising its young, e.g. cuckoos, honeyguides and indigobirds.
call: brief vocalisation, used mainly to maintain contact with conspecifics or to alert to danger. Often consists of a single, simple note (cf. song).
cap: a contrasting patch on the top of the head.
carpal: the bend of the wing between 'hand' and 'arm', or carpal joint.
carpal bar: a contrasting, dark bar on the upperwing, running diagonally from the carpal joint towards the body. Exhibited in certain plumages by some gull species.
carpal joint: the joint at the bend of the wing.
carpal patch: a contrasting mark or area at or near the carpal joint.
casque: an enlargement on the upper surface of the bill, in front of the head, as on hornbills.
cere: bare and often brightly coloured skin at the base of the upper mandible, containing the nostrils.
cheek: loosely applied to the area on the side of the head.
collar: a band of contrasting colour on the neck.
colony: an assemblage of breeding birds within a discrete area (hence colonial species).
congeneric: belonging to the same genus (hence congeners).
conspecific: belonging to the same species.
crepuscular: active at dawn and dusk.
crest: a tuft of elongate feathers on the head.
cryptic: colours or markings aiding concealment (as in nightjars).
culmen: the ridge of the upper mandible.
dambo: seasonally wet grassland on acid soils.
deciduous: a tree that loses its leaves during set periods of the year, remaining leafless for some time.
dimorphic: having two distinct morphs or forms.
dissonant: not in harmony, harsh toned.
distal: furthest from centre of body or point of attachment; opposite of proximal.
disyllabic: consisting of two syllables.
diurnal: active during the day / in daylight.
duetting: male and female of a pair singing simultaneously or antiphonally in response to each other.
eclipse plumage: a female-like plumage acquired by males of some species (e.g. ducks and some sunbirds) during or following breeding.
edgings/edges: in relation to feather patterns, indicates outer feather margins. Edgings can result in distinct paler or darker panels of colour on wings or tail.

endemic: native or confined to a particular area (e.g. turacos are endemic to Africa, Bannerman's Turaco is endemic to montane forests of SW Cameroon).
epiphyte: a plant that uses another for support, not for nutrients.
eye-crescent: a crescent above or below the eye; often occurring both above and below (forming broken eye-ring).
eye-ring: a ring of tiny feathers surrounding the orbital ring.
eye-stripe: a usually dark stripe that extends back from the bill over the lores and through the eye (post-ocular stripe).
face: anthropomorphic term, encompassing the forehead, lores, supercilium, ear-coverts and upper throat, or any combination of these.
first-winter: an immature plumage stage acquired after juvenile plumage and recognised by the presence of retained, and more worn, juvenile feathers in wings and tail (refers to winter season in northern hemisphere and is therefore used for Palearctic migrants).
flight feathers: main wing feathers (primaries and secondaries), but excluding tail feathers.
forest: a continuous stand of trees at least 10 m tall, their crowns interlocking.
forest outliers: patches of forest away from main forested areas.
form: a general term for distinguishable entities, including species, subspecies, morphs, etc.
fringes: in relation to feather tracts, indicates complete feather margins (compare edgings). Contrasting fringes can result in a scaly appearance to body feathers or wing-coverts.
frugivorous: fruit-eating (hence frugivore, fruit-eater).
gallery forest: forest along rivers.
gape: the fleshy interior and corners of the bill.
genus (plural **genera**): a taxonomic category between family and species, representing one or more species with a common ancestor (phylogenetic origin); the genus name forms the first part of the two-part scientific name.
gliding: flight on a direct course without, or between, wingbeats.
gorget: a distinctively coloured or streaked band across the throat or upper breast ('necklace').
graduated: referring to a tail in which the central rectrices are longest, the others becoming progressively shorter toward the sides.
granivorous: feeding on grains or seeds (hence granivore, seed-eater).
grassland: land covered by grasses and other herbs, either with woody plants, or the latter not covering more than 10% of the ground.
gregarious: commonly assembling in groups.
gular: related to the throat. A gular pouch is a loose area of skin extending from the throat (e.g. on pelicans). A gular stripe is a stripe extending on the centre of the throat (usually narrow and dark).
hand: the outer part of the wing, between carpal joint and tip.
herbaceous: of or similar to herbs.
Holarctic: biogeographical region that includes the Palearctic and Nearctic regions, i.e. most of the temperate zone of the northern hemisphere.
hood: a contrasting area covering all or most of the head and neck.
hybrid: the product of a cross between individuals of unlike genetic constitution; usually used for a cross between individuals of different species.
immature: a general term for a non-adult bird.
inner wing: inner part of the wing; also called arm.
insectivorous: insect-eating (hence insectivore, insect-eater).
inselberg: an isolated hill or mountain rising abruptly from its surroundings.
iridescent: the glossy or 'metallic' effect of changing colours caused by reflected light from specially structured feathers, e.g. on glossy starlings and sunbirds.

iris (plural **irides**): the round, coloured membrane surrounding the pupil of the eye.
juvenile: a bird in its first feathered, non-downy, plumage.
jizz: an overall impression of the appearance a bird in the field based on a combination of characters.
leading edge: the front edge of the wing. Generally referred to when it is marked with a contrasting (dark or pale) band.
littoral: situated near a (sea) shore.
local: occurring within a small or restricted area.
mangrove: open or closed stands of salt-adapted evergreen trees or bushes occurring on shores between high- and low-water mark.
mask: a dark area of plumage surrounding the eye, usually extending from the base of the bill to the ear-coverts.
melanistic: a blackish morph.
midwing panel: a contrasting pale area in the middle of the inner wing on some gull species.
migratory: making regular geographic movements.
mirror: a subterminal white spot on the wingtip of a gull. Not to be confused with the white primary tips (these, if present, are always visible on a gull at rest; mirrors are usually not visible on the closed wing).
monotypic: a biological group having a single representative. E.g. 'monotypic genus' (a genus with a single species), 'monotypic species' (a species with no subspecies). See polytypic.
montane: growing or living in mountainous areas.
morph: a normal but distinct plumage variant which is not related to sex, age or season.
morphological: pertaining to form and structure.
mottled: plumage marked with coarse spots or irregular blotches.
moult: the process of replacement of old feathers by new. Moult may be complete (all head, body, wing and tail feathers replaced during the same period) or partial (involving the renewal of all or most contour feathers, except those of wings and tail). In smaller species moult is usually repeated during a set period each year; in larger species (e.g. raptors) it may be spread over several years with overlapping moults.
Nearctic: biogeographical region comprising N America south to the tropics.
nocturnal: active at night.
nomadic: referring to a wandering or erratically occurring species with no fixed territory when not breeding.
nominate (**race** or **subspecies**): the first described and named form of a species, typified by its subspecific name being the same as the specific e.g. *Sylvietta virens virens*.
non-passerines: all orders of birds except the passerines.
notched: referring to a tail in which the central feathers are slightly shorter than the outer ones, forming a very shallow fork or notch.
nuchal: relating to the hindneck or nape (used with reference to a crest, patch or collar).
orbital ring: ring of bare skin immediately surrounding the eye (not to be confused with eye-ring).
outer wing: outer part of the wing; also called hand.
Palearctic: biogeographical region that includes Europe, N Africa, the Middle East and N Asia south to the Himalayas and Yangtze River.
partial migration: migration by part of a population.
passage migrant: a migrant that occurs regularly but only briefly at a locality during migration to and from its breeding and wintering grounds.
passerines: members of the large order Passeriformes, often referred to as 'perching

birds' or 'songbirds', characterised by perching with three toes pointing forward and one toe back. Includes all species from broadbills onward.

pectoral tufts: coloured tufts at each side of the breast (often concealed under the wings on perched birds).

pelagic: of the open sea. A pelagic species spends most of its life at sea, far from land.

pied: patterned black and white.

polygamous: an animal that has two or more mates of the opposite sex at the same time.

post-ocular stripe: a short, usually pale, stripe, which extends back from the eye. Sometimes reduced to a post-ocular spot.

primary forest: forest in a virgin or undisturbed state.

primary projection: the distance that the tips of the primaries project beyond the tertials on a closed wing.

proximal: nearest to centre of body or to point of attachment; opposite of distal.

race: synonymous with subspecies.

rainforest: closed-canopy forest in areas of high rainfall.

range: geographical area in which a taxon is distributed.

raptor: bird of prey (generally refers to diurnal species, not owls).

rectrices (singular **rectrix**): main tail feathers.

rictal bristles: sparse, though often prominent, bristles at the base of the bill.

remiges (singular **remex**): flight feathers (primaries and secondaries).

resident: a species or population that occurs year-round in the same area and breeds there, even though some individuals may not remain in the same area throughout the year.

riparian: bordering a river (or water).

riverine: living or growing on a river bank.

roost: a sleeping or resting place.

saddle: generally used to indicate a part of the upperparts contrasting in colour with the rest of the upper surface (e.g. in juvenile marsh terns).

Sahel: semi-arid zone between savanna and desert, characterised by scattered, thorny vegetation.

savanna: habitat dominated by grasses with a varying proportion of trees and shrubs.

secondary forest: forest regenerating after a greater or lesser degree of disturbance, often by selective logging or agriculture. It is characterised by a lack of large mature trees and a significant proportion of coloniser species.

secondary bar: a contrasting dark bar on the secondaries, as in immature plumages of some gull species.

sedentary: a species that remains in the same site throughout the year, individuals wandering no more than a few kilometres at the most.

shoulder patch: an area of contrastingly coloured wing-coverts.

soaring: circling flight (often in thermal of warm air).

song: a more complex pattern of vocalisations, mainly uttered in the breeding season (cf. call).

speculum: a usually iridescent panel on the secondaries in dabbling ducks.

stratum (plural **strata**): level (used to indicate levels within forest, e.g. middle strata = mid-levels).

streamer: an exceptionally long, slender tail feather, as in some bee-eaters, paradise flycatchers and sunbirds.

square: referring to a tail in which all feathers are of equal length, forming a straight border.

subadult: an imprecise term indicating a bird in nearly adult plumage; often used when precise age is difficult to establish, e.g. in large raptors.

subspecies: a population of a given species that differs more or less obviously in appearance from one or more other populations of the same species. Subspecies are assumed to be (at least theoretically) capable of interbreeding but to be relatively isolated from each other, with interbreeding limited to areas of contact. The border between species and subspecies is often arbitrary. The nominate subspecies is the one first named: its scientific name has the second and the third terms identical, e.g. *Trochocercus nitens nitens*.

subterminal: near the end. A subterminal band is a contrasting (dark or pale) band, usually broad, situated near the tip of a feather or feather tract (used particularly in reference to the tail).

supercilium: a usually pale stripe, which extends from the base of the upper mandible, above the lores and the eye (extending behind the eye; compare supraloral stripe).

supraloral stripe: a short, usually pale, stripe, which extends from the base of the upper mandible, over the lores, to just above or in front of the eye (not extending behind the eye; if it does it becomes a supercilium).

taxon (plural **taxa**): a named form (this unit of biological classification can refer to a group of organisms of any taxonomic rank, e.g. family, genus, species, subspecies, etc.)

taxonomy: the science of classification and naming of all life forms.

terminal: at the end. A terminal band is a contrasting (dark or pale) band, usually broad, situated at the tip of a feather or feather tract (used particularly in reference to the tail).

terrestrial: living or occurring mainly on the ground.

thicket: a closed stand of bushes and climbers usually between 3 and 7 m tall.

top of head: includes forehead and crown.

trailing edge: the rear edge of the wing, often contrasting with the rest of the wing (cf. leading edge).

understorey: the lowest stratum in forest or woodland.

vent: the area around the cloaca (anal opening), just behind the legs (not to be confused with the undertail-coverts).

vermiculations: narrow, often wavy, bars that generally create an overall effect and are visible only at close range.

vestigial: a feature that is very much reduced (almost absent).

wattle: naked, fleshy, usually brightly coloured, skin on head (e.g. around eye), base of bill or throat (e.g. as in wattle-eyes, African Wattled Plover and Black-casqued Hornbill).

window: a contrasting pale panel in the outer wing of some raptors and gulls.

wingbar: generally a narrow and well-defined dark or pale bar on the upperwing, and often referring to a band formed by pale tips to the greater or median coverts (or both, as in 'double wingbar').

wing feathers: includes flight feathers and wing-coverts.

wing panel: a contrasting, usually pale band on the wing; broader and generally more diffuse than a wingbar (often formed by pale edges to the remiges or coverts). At rest, usually refers to a panel on the secondary coverts, secondaries and/or tertials; in flight, usually refers to a panel on the primary bases.

wooded grassland: land covered by grasses and other herbs, with woody plants covering 10–40% of the ground.

woodland: an open stand of trees at least 8 m tall with a canopy cover of 40% or more. The field layer is usually dominated by grasses.

zygodactyl: form of foot with the outer and inner toes pointing backward, the middle toes forward, as in turacos, cuckoos, parrots, owls and woodpeckers.

ABBREVIATIONS AND SYMBOLS

Status

R Resident
M Intra-African migrant
P Palearctic migrant (including a few species of Nearctic origin)
V Vagrant
* Cape Verde only
+ also Cape Verde

Abundance categories

c common
f fairly common (= frequent; the category 'not uncommon' in *BOWA* 2001)
u uncommon
s scarce
r rare
l local (e.g. 'lc' = locally common)
* indicates abundance in Cape Verde only (if very different from elsewhere)

Threat categories

CR Critically Endangered
EN Endangered
VU Vulnerable
NT Near Threatened
DD Data Deficient

ad adult
c. circa (about)
C central
CAR Central African Republic
esp. especially
imm immature
incl. including
juv juvenile
Mt Mount, Mountain
N, NE, E, SE, S, SW, W, NW compass directions (northern, northeastern, eastern, etc.)
R. River
WS wingspan

▲ habitat and behaviour
✣ voice

[CD2:67] CD and track number in *African Bird Sounds* (Chappuis 2000)

PLATE 1: OSTRICH, GREBES, DARTER, CORMORANTS AND PELICANS

1 GREAT CRESTED GREBE *Podiceps cristatus* 46–51 cm V
1a Ad non-breeding Long, mainly white neck; blackish crown; long, sharp-pointed bill. **1b Ad breeding** Double-horned crest; chestnut and black tippets. ▲ Various aquatic habitats. Palearctic vagrant. ✤ [CD1:10b]

2 BLACK-NECKED GREBE *Podiceps nigricollis* 28–34 cm V/P r
2a Ad non-breeding Steep forehead; black cap to below eye; slightly upturned bill. **2b Ad breeding** Black neck; loose golden ear-tufts. ▲ Various aquatic habitats. ✤ [CD1:11]

3 LITTLE GREBE *Tachybaptus ruficollis* 25–29 cm R/P? lc
3a Ad breeding Dumpy; bright chestnut cheeks and foreneck; yellowish gape patch. **3b Ad non-breeding** Duller; pale brown cheeks and foreneck; contrasting dark cap. ▲ Various aquatic habitats. ✤ Loud, descending, whinnying trill *bi-i-i-i-i-i...* uttered frequently and year-round, often in duet. [CD1:10a]

4 OSTRICH *Struthio camelus* R r/s
4a Ad male Huge (height 210–275 cm); black-and-white plumage. **4b Ad female** Smaller (height 175–190 cm); dull grey-brown plumage. **4c Chick at *c.* 4 weeks** Buffish, spiky down tipped black; black spots on neck. ▲ Open plains, thorn scrub, semi-desert. ✤ [CD5:1a]

5 AFRICAN DARTER *Anhinga rufa* *c.* 80 cm; WS 120 cm R s/u
Slender, kinked neck; pointed, dagger-shaped bill. **5a Ad male** Mainly black; chestnut foreneck; white stripe on head- and neck-sides. **5b Ad female** Duller and browner than male. **5c Juv** As female, but even browner and paler; no white neck-stripe. **5d In flight** Long, graduated tail (often held in fan shape when soaring). ▲ Various aquatic habitats. Swims with only head and neck above water. ✤ [CD5:4a]

6 LONG-TAILED CORMORANT *Phalacrocorax africanus* 51–56 cm; WS 85 cm R c
6a Ad breeding Black; long tail; short crest on forehead. **6b Ad non-breeding** Duller; no crest. Some brownish like immature, but eye red. **6c Juv/imm** Mainly dull brownish above, pale brownish to buffish-white below. **6d In flight** Outstretched neck; long tail. ▲ Various aquatic habitats. Swims low in water, with neck erect and bill held up at angle; often only head and neck above surface. Stands upright, wings often held open to dry. ✤ [CD5:3b]

7 GREAT CORMORANT *Phalacrocorax carbo* 80–100 cm; WS 130–160 cm R/V lc/r
7a Ad *lucidus* breeding (White-breasted Cormorant) (R) Large; white throat and breast; white thigh patch. Non-breeding adult duller; no thigh patch. **7b Juv** Brownish above, whitish below. **7c Ad *maroccanus* breeding** (V) White throat, black breast. Breeds Morocco; possibly moves south outside breeding season. **7d In flight** Thick, outstretched neck. ▲ Aquatic habitats. Flies in line or V formation. ✤ [CD1:15]

8 GREAT WHITE PELICAN *Pelecanus onocrotalus*
140–175 cm; WS 270–360 cm R/M/V* s/lc
8a Ad Huge; mainly white; massive bill with yellow pouch. **8b Imm** Mottled greyish-brown above; dirty white below. **8c–d Ad in flight** Strongly contrasting black-and-white wings. **8e–f Imm in flight** Note underwing pattern (brown on coverts progressively lost with age); contrasting white rump. ▲ Large aquatic habitats. Flight strong; often soars. Flies in V-shaped or curved skeins, often at great height. ✤ [CD1:18]

9 PINK-BACKED PELICAN *Pelecanus rufescens* 125–132 cm; WS 265–290 cm R r/lc
9a Ad Whitish with pale grey cast; pale yellowish bill and pouch. **9b Imm** Mottled brownish above; bare parts dull coloured. **9c–d Ad in flight** Less contrasting wing pattern than 8c–d. **9e–f Imm in flight** Underwing pattern as adult; contrasting white rump. ▲ Large aquatic habitats. ✤ [CD5:4b]

Maps on page 36

1b
1a
2a
2b
3b
3a
4b
4a
4c
6d
7d
5c
5b
5a
6c
6b
6a
7b
7a
7c
8f
8e
8d
8c
9e
9f
9c
9d
8b
8a
9b
9a

PLATE 1: OSTRICH, GREBES, DARTER, CORMORANTS AND PELICANS

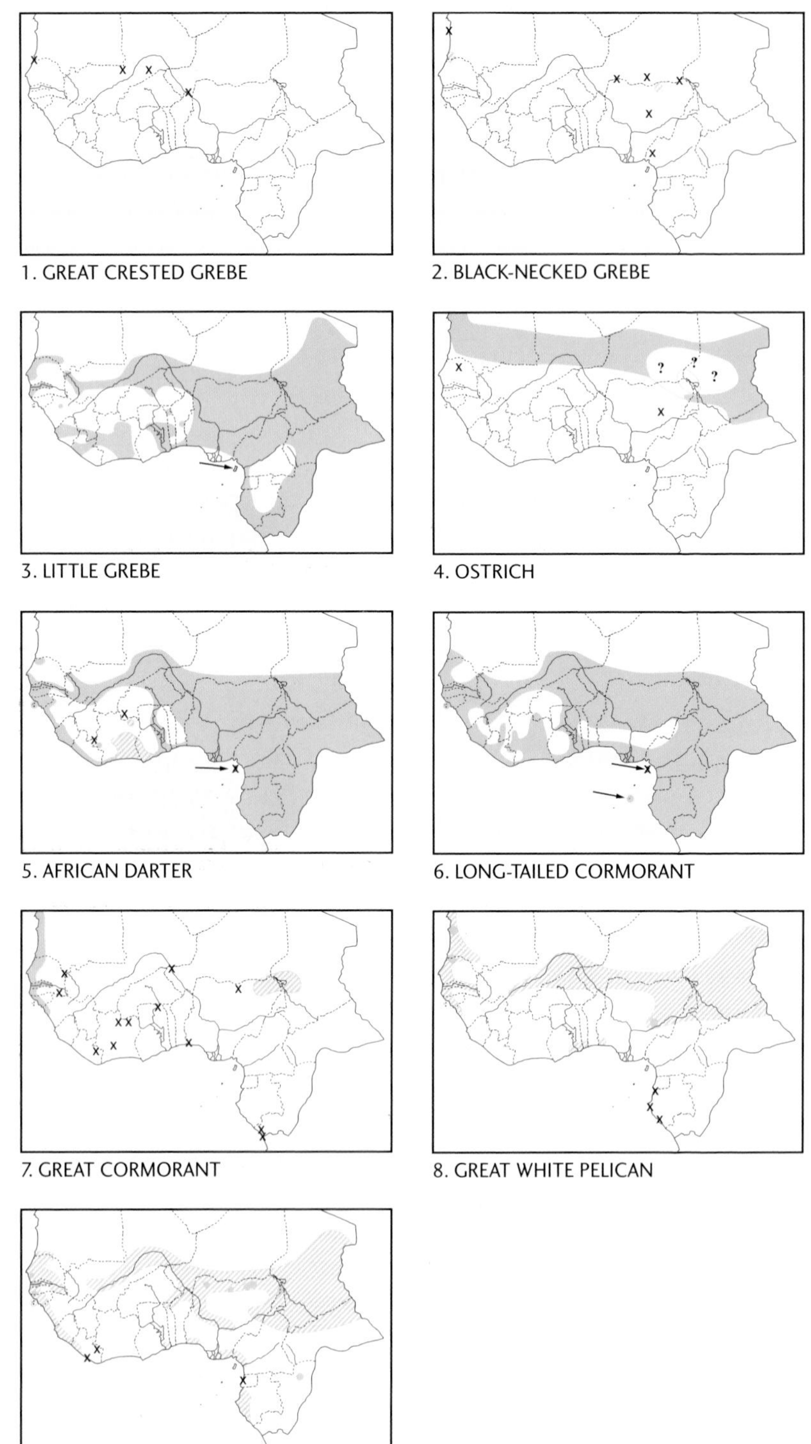

1. GREAT CRESTED GREBE

2. BLACK-NECKED GREBE

3. LITTLE GREBE

4. OSTRICH

5. AFRICAN DARTER

6. LONG-TAILED CORMORANT

7. GREAT CORMORANT

8. GREAT WHITE PELICAN

9. PINK-BACKED PELICAN

Plate on page 34

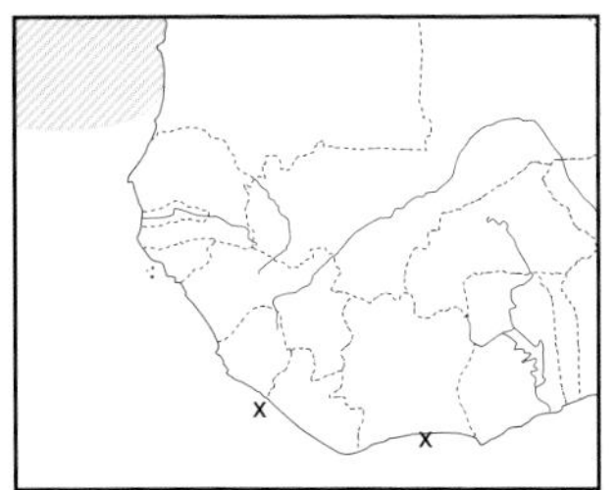

1. FEA'S PETREL
2. ZINO'S PETREL

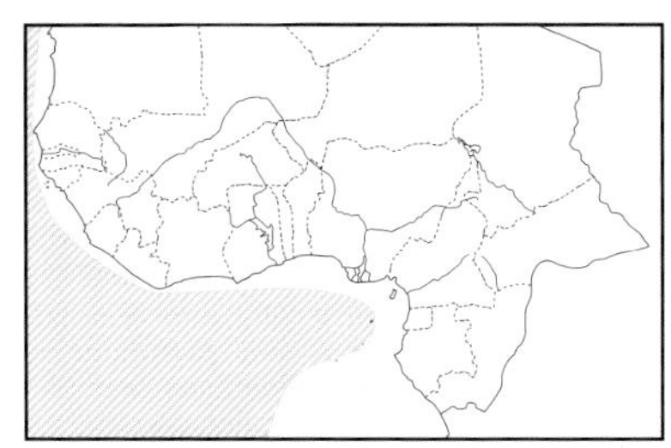

3. BULWER'S PETREL

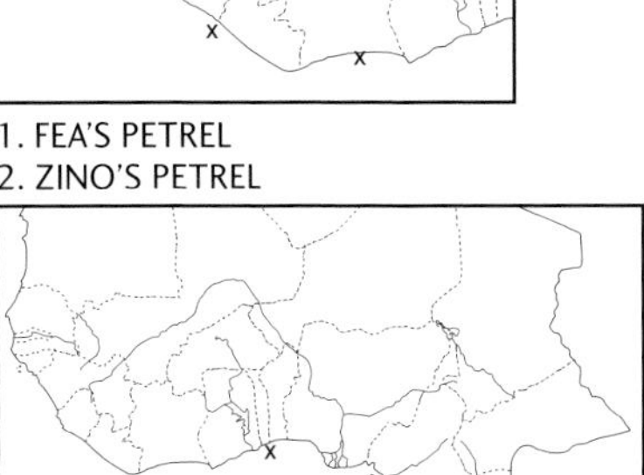

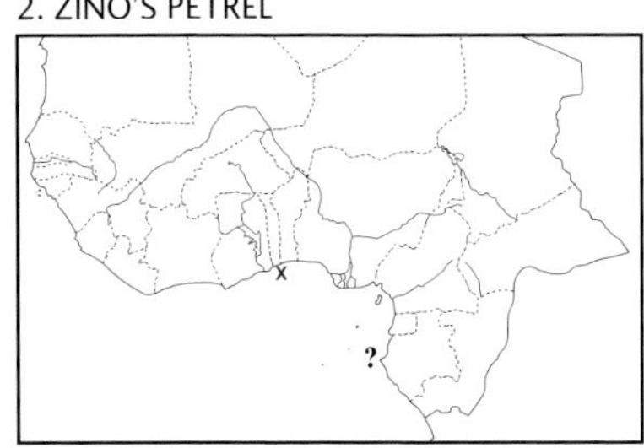

4. CAPE PETREL

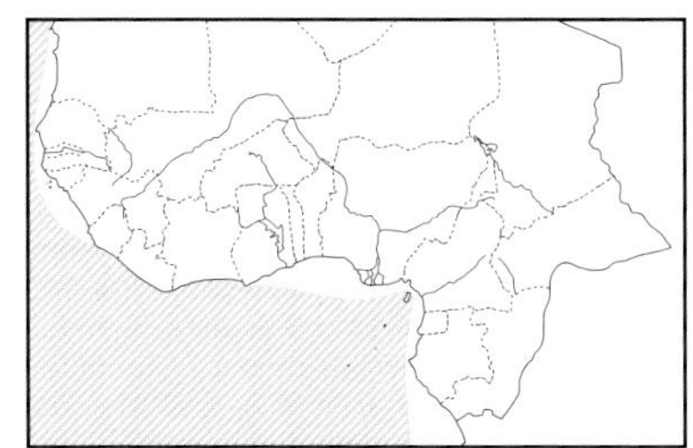

5. CORY'S SHEARWATER

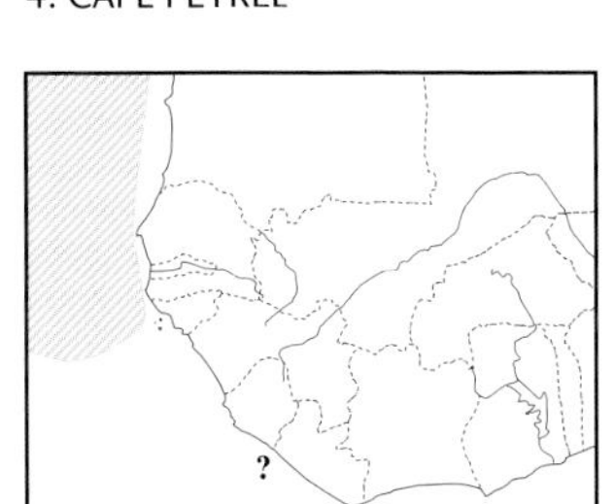

6. CAPE VERDE SHEARWATER

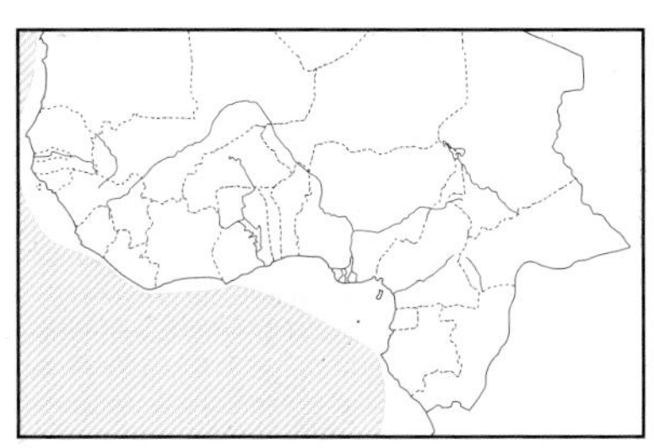

7. GREAT SHEARWATER

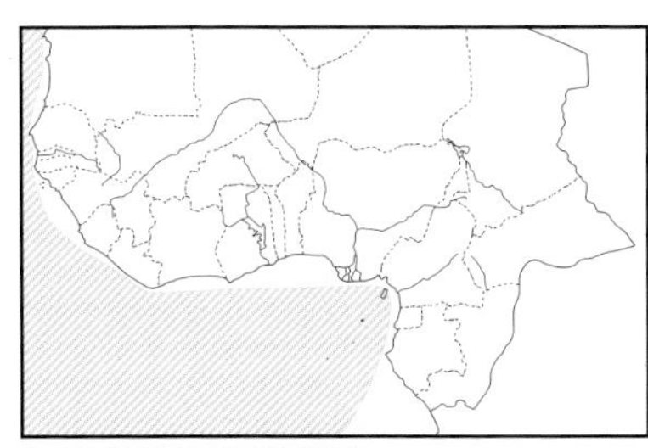

8. SOOTY SHEARWATER

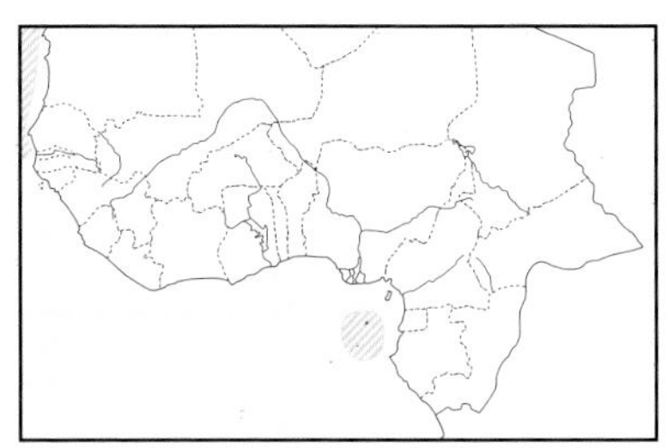

9. LITTLE SHEARWATER

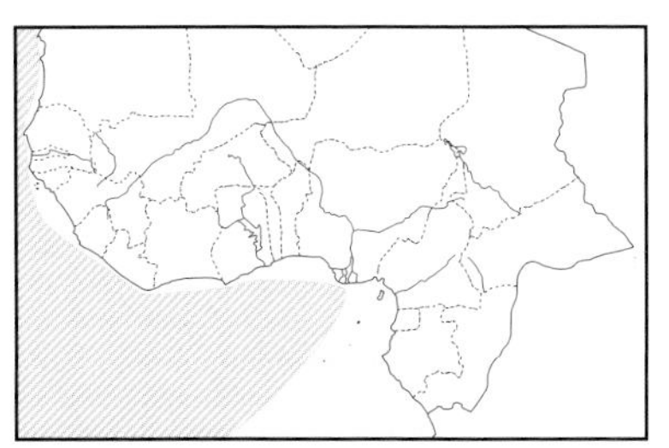

10. MANX SHEARWATER

PLATE 2: PETRELS AND SHEARWATERS

Medium-sized to large pelagic seabirds with long narrow wings. Fly low over waves on stiff wings, long glides alternating with a few fast flaps. Spend most of their lives at sea and are normally only seen on land at breeding colonies or as storm-driven waifs. Normally silent away from breeding colonies. Occurrence in W Africa inadequately known.

1 FEA'S PETREL *Pterodroma feae* 35 cm; WS 84–91 cm **R*/V? NT**
1a Dark upperwings contrasting with greyer mantle; pale grey tail; dark mark behind eye. **1b** Dark underwings; white underparts with dusky 'shawl' extending on neck-sides. Erratic, zigzag flight; towering in strong winds. **1c Head** Bill thicker than 2; crown darker. ▲ Breeds Cape Verde. Non-breeding visitor offshore south to 9°S; rare. ✤ [CD1:1a]

2 ZINO'S PETREL *Pterodroma madeira* 33 cm; WS 78–83 cm **V? EN**
Extremely similar to 1; slightly smaller; bill slimmer; forehead and crown paler. ▲ Possible non-breeding visitor (breeds Madeira). ✤ [CD1:1b]

3 BULWER'S PETREL *Bulweria bulwerii* 27 cm; WS 68–73 cm **R*/M s**
Entirely blackish with pale brown band on upperwing-coverts; long pointed tail. Flight buoyant and erratic, circling low over waves. ▲ Breeds Cape Verde. Non-breeding visitor offshore Mauritania–Gulf of Guinea. ✤ [CD1:2a]

4 CAPE PETREL *Daption capense* 39 cm; WS 81–91 cm **V**
4a–b Striking black-and-white plumage above and below. Flight 5-8 quick beats on stiff wings followed by glide. ▲ Vagrant from Southern Ocean, recorded off Togo (Gabon?).

5 CORY'S SHEARWATER *Calonectris diomedea* 45–56 cm; WS 100–125 cm **P+ r/f**
C. d. borealis Large. Dull brownish upperparts contrast with white underparts; bowed wings; pale bill. Rather heavier-looking than nominate; underwing pattern different (see Plate 5: 1a-c). Flight appears lazy, several deep wingbeats followed by long, low glide. ▲ Non-breeding visitor, Mauritania–Gulf of Guinea. ✤ [CD1:2b, 3a]

6 CAPE VERDE SHEARWATER *Calonectris edwardsii* 40cm; WS 90–110 cm **R*/M r/f NT**
As 5 but slimmer; head smaller, more angular; tail relatively longer; head and upperparts slightly darker and greyer brown; bill slimmer, grey with dark subterminal band. Underwing pattern and flight as 5 (see Plate 5:1d). ▲ Breeds Cape Verde. Non-breeding visitor, Mauritania–Gulf of Guinea. ✤ [CD1:3b]

7 GREAT SHEARWATER *Puffinus gravis* 43–51 cm; WS 105–122 cm **M+/V r/f**
7a–b Black cap and brown upperparts contrast with white collar and white uppertail-coverts. White underwing with dark band on axillaries; dark belly patch. Flight strong and powerful, stiff wingbeats interspersed by glides. ▲ Non-breeding visitor offshore Mauritania–Gabon; rare. Also Cape Verde waters; fairly common. ✤ [CD1:4a]

8 SOOTY SHEARWATER *Puffinus griseus* 40–51 cm; WS 94–109 cm **M/V+ u/r NT**
8a–b Entirely dark; silvery underwing-coverts. Long, slim body; long swept-back wings. Flight strong, direct with stiff-winged flaps followed by descending glides; soars in high winds. ▲ Non-breeding visitor offshore, throughout. ✤ [CD1:6a]

9 LITTLE SHEARWATER *Puffinus assimilis* 25–30 cm; WS 58–67 cm **R+/M s?**
9a *P. a. baroli* (Breeds Atlantic islands.) Small; black above; white below; face mainly white. **9b *P. a. boydi*** (Breeds Cape Verde.) As 9a but face-sides and undertail-coverts dark. Stiff-winged flight usually low, consisting of several rapid wingbeats followed by brief glide. ▲ Probably scarce but regular offshore to at least 9°S. ✤ [CD1:5b]

10 MANX SHEARWATER *Puffinus puffinus* 30–35 cm; WS 76–82 cm **P s**
10a–b Black above; white below; head-sides black. Typical flight a series of rapid stiff-winged beats followed by shearing over waves, banking from side to side. ▲ Non-breeding visitor offshore, mostly Mauritania–Liberia. ✤ [CD1:4b]

1c
2
1b
1a
4b
4a
3
6
7a
7b
5
9a
8a
8b
10a
9b
10b

PLATE 3: STORM-PETRELS AND TROPICBIRDS

Storm-petrels. Small, swallow-like pelagic seabirds. Largely dark with white rump. Solitary or gregarious at sea. Identification often difficult; flight action is an important clue. Breed colonially in burrows or crevices on oceanic islands. Occurrence in W Africa inadequately known.

1 EUROPEAN STORM-PETREL *Hydrobates pelagicus* 15–16 cm; WS 38–42 cm **P f/r**
1a–b Small. Wings and tail rounded; conspicuous white band on underwing; white rump patch. Fluttering flight. ▲ Regular offshore Mauritania–Ivory Coast. Rare, Gulf of Guinea. ✤ [CD1:7b]

2 WILSON'S STORM-PETREL *Oceanites oceanicus* 15–19 cm, WS 38–42 cm **M/V u/s?**
2a–b Small. Distinct pale diagonal band on upperwing; underwing all dark. Feet project beyond square tail. Wings typically held rather straight when gliding. Flies above waves with legs dangling when foraging. ▲ Probably uncommon/scarce offshore. ✤ [CD1:6b]

3 LEACH'S STORM-PETREL *Oceanodroma leucorhoa* 19–22 cm; WS 45–48 cm **P/V**
3a–b Relatively large. Prominent pale diagonal band on upperwing; rump divided in centre or smudged grey. Tail forked; feet do not project. Buoyant, bounding flight. Wings held angled when gliding. ▲ Probably regular offshore. ✤ [CD1:8b]

4 MADEIRAN STORM-PETREL *Oceanodroma castro* 19–21 cm; WS 43–46 cm **R+/M/V f?**
4a–b Relatively large. Tail less forked than 3; more white on rump. Compare also 2. Buoyant and zigzagging flight. ▲ Breeds Cape Verde (locally frequent) and possibly São Tomé; probably frequent offshore. ✤ [CD1:8a]

5 WHITE-FACED STORM-PETREL *Pelagodroma marina* 20–21 cm; WS 41–43 cm **R*/M**
5a–b Grey above; white below; distinctive head pattern. Very long legs project beyond slightly forked tail. Flies close to waves with legs dangling. ▲ Breeds Cape Verde; probably regular offshore Mauritania to at least Liberia. ✤ [CD1:7a]

Tropicbirds. Medium-sized, graceful, highly aerial seabirds. Buoyant, rather tern-like, purposeful flight, fluttering wingbeats alternated with long glides. Feed by hovering, then plunging vertically. Local. Breed colonially on islands.

6 WHITE-TAILED TROPICBIRD *Phaethon lepturus* 40 cm; WS 92 cm **R/V lc**
6a Ad Pure white upperparts; diagnostic black diagonal bar on upperwing; long tail streamers (33–40 cm); yellowish-orange bill. **6b Juv/imm** Note coarse barring on upperparts and extensive white on upperwing; pale yellow, black-tipped bill. ▲ Breeds São Tomé (Príncipe?), Annobón; locally common. Offshore vagrant elsewhere. ✤ Similar to 7, but rather higher pitched and faster: *kirrik-kirrik-kirrik*. [CD1:13]

7 RED-BILLED TROPICBIRD *Phaethon aethereus* 48 cm; WS 103 cm **R+/V ls**
7a Ad Barred upperparts; long tail streamers (46–56 cm); red bill. **7b Juv/imm** Closely barred upperparts appear grey at distance; inner secondaries and upperwing-coverts barred; black nuchal collar; black-tipped yellowish bill. ▲ Breeds Senegal and Cape Verde; scarce and local. Offshore vagrant elsewhere. ✤ Shrill *keek* or *karreek*. Also loud, piercing calls, only likely to be heard at breeding colonies. [CD1:12]

Maps on page 42

1a
1b
2a
2b
3a
3b
4a
4b
5a
5b
6a
6b
7a
7b

PLATE 3: STORM-PETRELS AND TROPICBIRDS

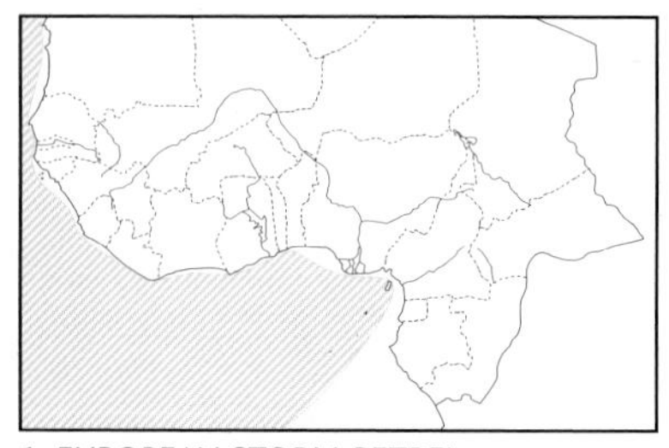
1. EUROPEAN STORM-PETREL

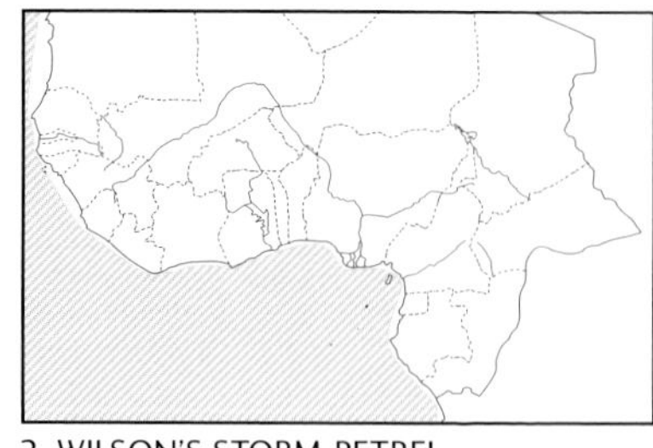
2. WILSON'S STORM-PETREL

3. LEACH'S STORM-PETREL

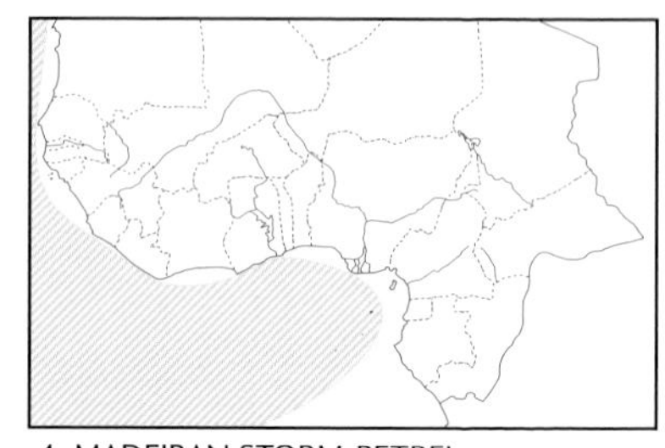
4. MADEIRAN STORM-PETREL

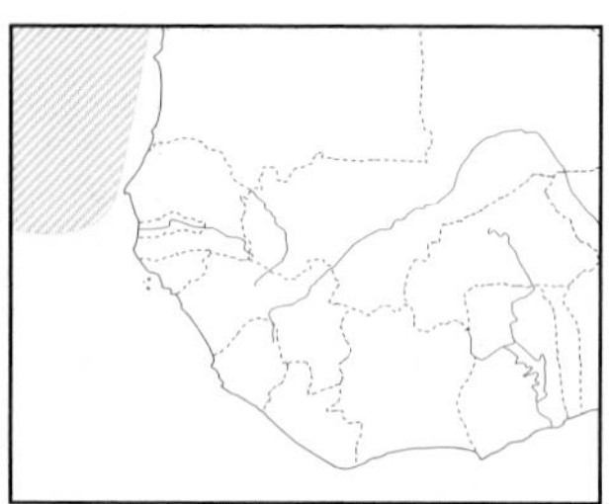
5. WHITE-FACED STORM-PETREL

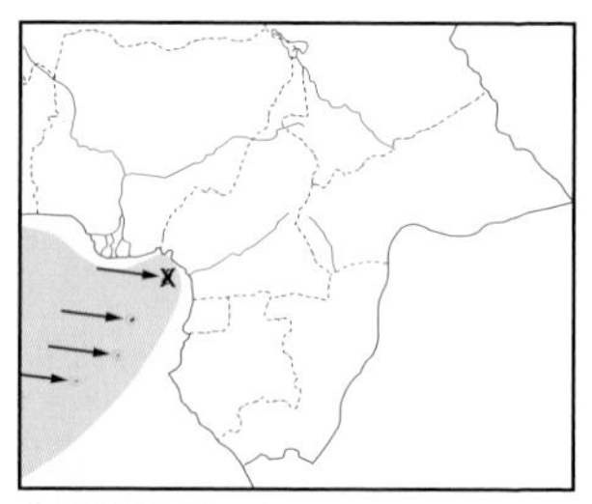
6. WHITE-TAILED TROPICBIRD

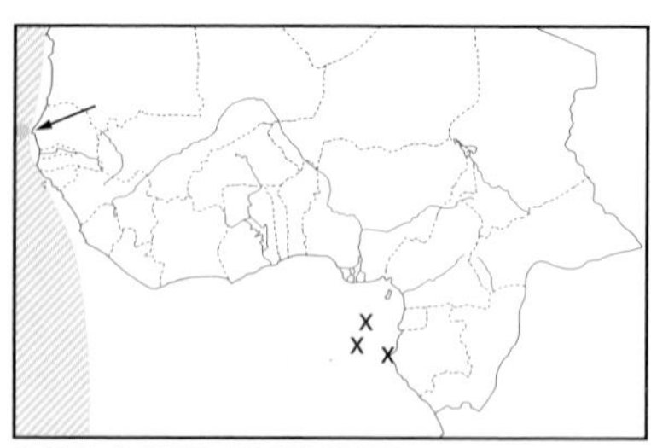
7. RED-BILLED TROPICBIRD

Plate on page 40

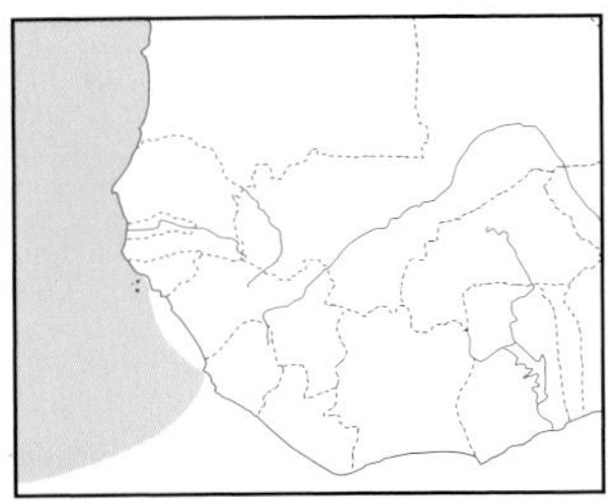
1. NORTHERN GANNET

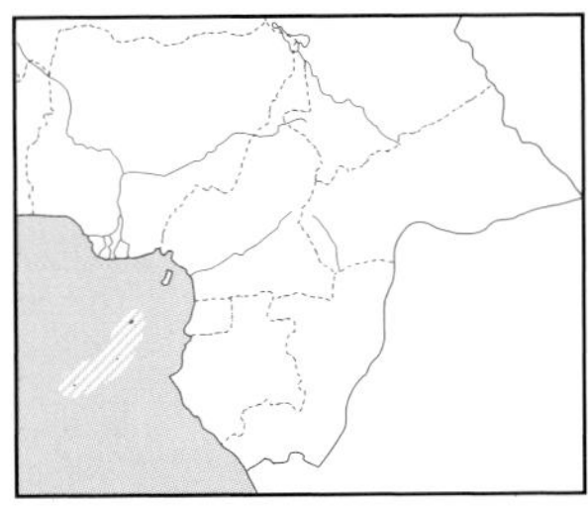
2. CAPE GANNET

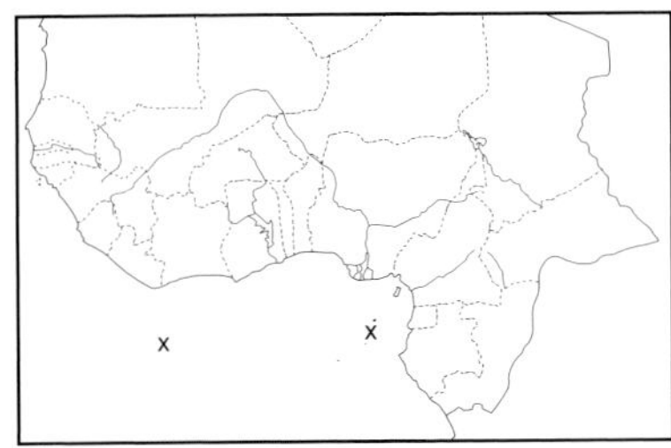

3. MASKED BOOBY

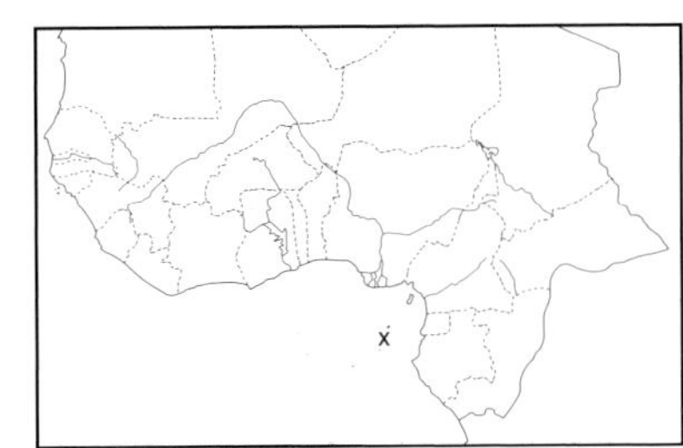

4. RED-FOOTED BOOBY

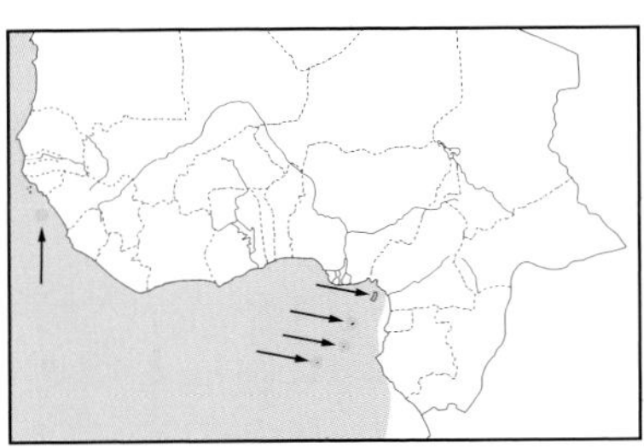
5. BROWN BOOBY

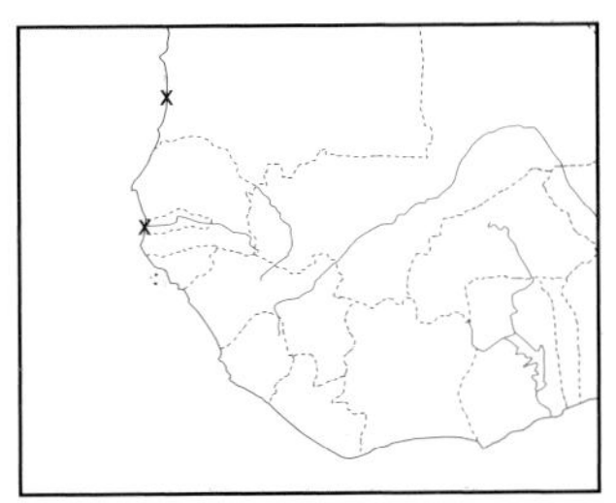

6. MAGNIFICENT FRIGATEBIRD

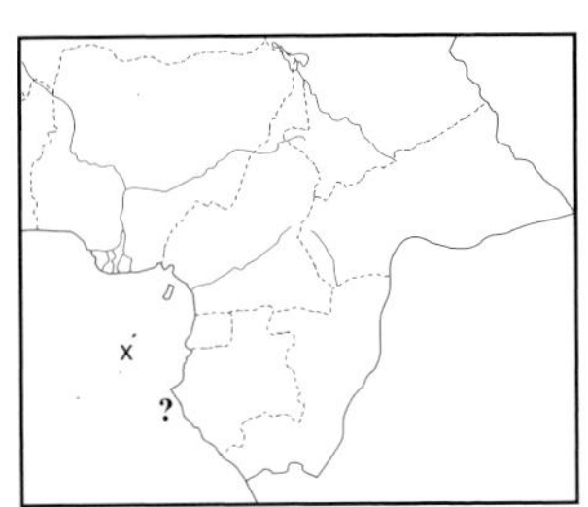

7. ASCENSION FRIGATEBIRD

PLATE 4: GANNETS, BOOBIES AND FRIGATEBIRDS

Gannets and boobies. Large, conspicuous seabirds with stout, tapering bills, cigar-shaped bodies, long, narrow wings and wedge-shaped tails. Flight steady and purposeful with shallow flaps followed by a glide. Flocks often fly in single file. Feed principally by plunge-diving from relatively great height. Breed colonially. In W Africa only likely to be encountered offshore, unless sick or storm-driven.

1 NORTHERN GANNET *Sula bassana* 87–100 cm; WS 165–180 cm P/V* u
1a Ad All-white secondaries and tail; yellowish crown and nape. **1b Ad (head)** Short gular stripe. **1c Second-year** White feathers appear on secondaries and tail. Adult plumage acquired over 4 years. **1d Juv** Dark brown; uppertail-coverts white. Indistinguishable from juvenile Cape Gannet. ▲ Non-breeding visitor to at least 10°N. ✤ [CD1:14]

2 CAPE GANNET *Sula capensis* 84–94 cm; WS 165–180 cm M s VU
2a Ad Black secondaries and tail; yellowish crown and nape. Compare 1. **2b Ad (head)** Long gular stripe. **2c Second-year** Like 1c but secondaries and tail all black. ▲ Non-breeding visitor from southern African waters to Gulf of Guinea. ✤ [CD5:2a]

3 MASKED BOOBY *Sula dactylatra* 81–92 cm; WS 152 cm V
3a Ad All-black secondaries and tail; white crown; small black mask. **3b Juv** All dark above with white collar. **3c Imm** Dark head contrasts with white breast and belly; white collar broader and more conspicuous than in juvenile. ▲ Vagrant, Gulf of Guinea. Nearest breeding colonies on Ascension.

4 RED-FOOTED BOOBY *Sula sula* 66–77 cm; WS 91–101 cm V+
4a Ad white morph Black secondaries; all-white tail; red feet. **4b Ad brown morph** Entirely dull brown above and below; red feet. **4c Juv (all morphs)** As 4b, but with dark bill and yellowish-grey feet. ▲ Vagrant, Cape Verde and Gulf of Guinea. Nearest breeding colony on Ascension.

5 BROWN BOOBY *Sula leucogaster* 64–74 cm; WS 132–150 cm R+/M/V lc/u
5a Ad (perched) All brown above; white breast and belly. **5b Ad** Distinctive underwing pattern; brown of head extending onto breast. **5c Juv** White parts of adult dusky-brown. ▲ Breeds Cape Verde, Alcatraz and Gulf of Guinea Is; locally common. Uncommon elsewhere. ✤ [CD5:2b]

Frigatebirds. Large seabirds with extremely long, angular wings and very long, deeply forked tails. Very buoyant flight; soar effortlessly for long periods. Forage in flight by swooping down on prey or by harassing other seabirds, forcing them to disgorge their catch. Readily take offal. Identification often difficult; plumages extremely variable, differing according to age (period needed to acquire adult plumage unknown, probably 4–6 years), sex and colour morph. Note in particular amount of white on underparts and occurrence of 'spurs' on underwing.

6 MAGNIFICENT FRIGATEBIRD *Fregata magnificens* 89–114 cm; WS 217–244 cm R*/V r
6a-b Ad male Entirely black; red gular pouch (inflated during display). **6c Ad female** All-dark underwings; white breast. **6d Imm** All-dark underwing; broken breast-band. ▲ Breeds Cape Verde; rare. Offshore Mauritania and Gambia; rare. ✤ [CD1:20]

7 ASCENSION FRIGATEBIRD *Fregata aquila* 89–96 cm; WS 196–201 cm V VU
7a Ad male As 6a, but somewhat smaller. Typical female very similar. **7b Ad female pale morph** Note narrow white 'spur' on axillaries. **7c Imm** White on underwing-coverts; complete brown breast-band. ▲ Vagrant, Gulf of Guinea.

Maps on page 43

1a
1c
1d
2c
2a
1b
4c
4b
3b
3a
4a
2b
3c
5c
5a
5b
6a
6c
7a
7b
6d
7c
6b

PLATE 5: VAGRANT SEABIRDS

1 **CORY'S SHEARWATER** *Calonectris diomedea* 45–56 cm; WS 100–125 cm **P r/f**
1a *C. d. diomedea* (Scopoli's Shearwater) Slightly smaller than *borealis* (Plate 2:5), with slimmer bill, rather paler head and mantle. **Underwing patterns: 1b *C. d. borealis*** Primaries wholly dark, creating solid, well defined, broad black band on outer wing and rounded white underwing panel. **1c *C. d. diomedea*** White wedges on inner webs of otherwise black outer primaries, resulting in narrower black band on outer wing and more angled white underwing panel. ▲ Non-breeding visitor, Mauritania–Gulf of Guinea.
1d Cape Verde Shearwater *C. edwardsii* Underwing pattern as 1b.

2 **NORTHERN FULMAR** *Fulmarus glacialis* 43–52 cm; WS 101–117 cm **V**
Stocky; head and underparts typically white; upperparts pale grey. Flies on stiff, straight wings. ▲ Palearctic vagrant, offshore Senegal.

3 **ATLANTIC YELLOW-NOSED ALBATROSS** *Thallasarche chlororhynchos*
71–81 cm; WS 178–205 cm **V EN**
3a Ad Underwing white with narrow black border; head pale grey; bill black with orange-yellow culmen and pinkish-orange tip. **3b Juv** Head white; bill all black. ▲ Potential vagrant from Southern Ocean. See 4.

4 **BLACK-BROWED ALBATROSS** *Thallasarche melanophris*
80-95 cm; WS 213-246 cm **V EN**
4a Ad from below Underwing white with broad black band on leading edge; small dark 'brow'; bill orange-yellow. **4b Ad from above** Dark 'saddle' and upperwings; white rump; grey tail. **4c Juv** Underwing largely dark, with only ill-defined whitish stripe along centre. Bill greyish-brown with blackish tip. ▲ Vagrant from Southern Ocean. Albatross seen offshore Congo (Aug 1988) was presumably this species or 3.

5 **SOUTHERN GIANT PETREL** *Macronectes giganteus* 86-99 cm; WS 185-210 cm **V VU**
Very large (like small albatross). Massive, pale bill. **5a Ad dark morph** Greyish-brown with whitish head. **5b Ad white morph** All white with some black specks. Flight laboured, stiff wingbeats interspersed with short glides. ▲ Vagrant/rare visitor from Southern Ocean, offshore Congo.

6 **CAPE CORMORANT** *Phalacrocorax capensis* 61–64 cm; WS 109 cm **V NT**
6a Ad breeding Medium-sized; black; yellow-orange gular area. **6b Ad non-breeding/imm** Dull brown, underparts paler; yellowish-brown gular area. ▲ Vagrant from southern African coasts, north to Gabon. ✤ [CD5:3a]

7 **JACKASS (AFRICAN) PENGUIN** *Spheniscus demersus* *c.* 60 cm **V VU**
Stocky, flightless seabird, with flipper-like wings, and short legs and tail. **7a Ad** Black face and throat surrounded by white; black band across breast down flanks. **7b Juv** Sooty-grey head and upperparts. ▲ S African vagrant, Gabon and Congo. Floats with head held high and most of body submerged. On land, walks upright with waddling gait.

8 **BALEARIC SHEARWATER** *Puffinus mauretanicus* 30–40 cm; WS 76–93 cm **V CR**
8a-b Resembles Manx Shearwater (Plate 2:10), but brownish above; dirty greyish-buff below; dark axillaries and vent. ▲ Vagrant from W Mediterranean, recorded off Mauritania. ✤ [CD1: 5a]

9 **BLACK-BELLIED STORM-PETREL** *Fregetta tropica* 20 cm; WS 46 cm **V**
9a-b White rump; white belly with black stripe on centre; white underwing-coverts. ▲ Vagrant from Southern Ocean, Gulf of Guinea. Claimed off Liberia and Sierra Leone. ✤ [CD5:1b]

10 **WHITE-BELLIED STORM-PETREL** *Fregetta grallaria* 20 cm; WS 46 cm **V?**
Similar to 9; belly all white; underwing with more white, dark leading edge narrower. ▲ Potential vagrant (single record north of Cape Verde).

Maps on page 48

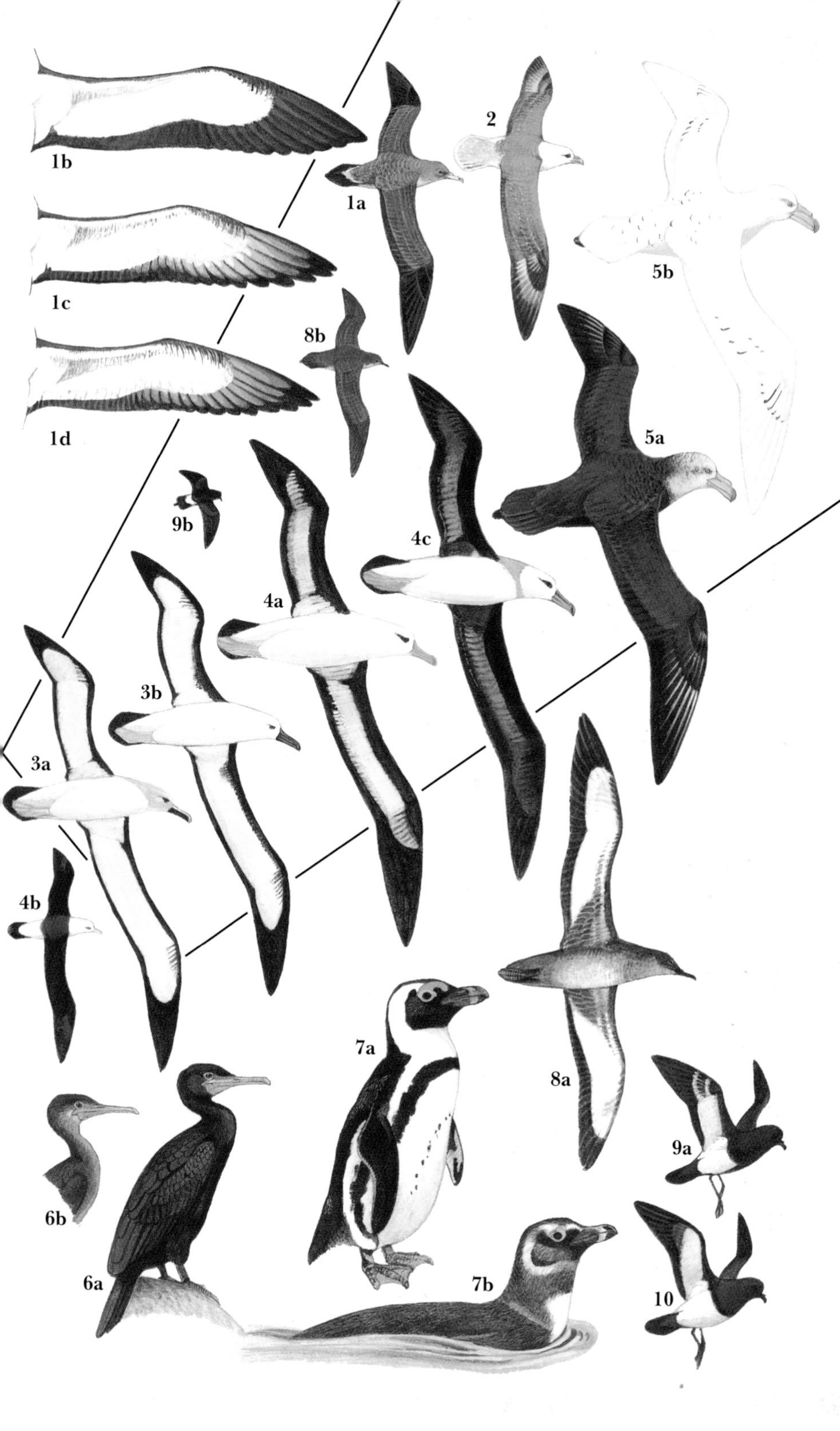

1b
1c
1d
1a
2
5b
8b
5a
9b
4c
4a
3b
3a
4b
7a
8a
9a
6b
6a
7b
10

PLATE 5: VAGRANT SEABIRDS

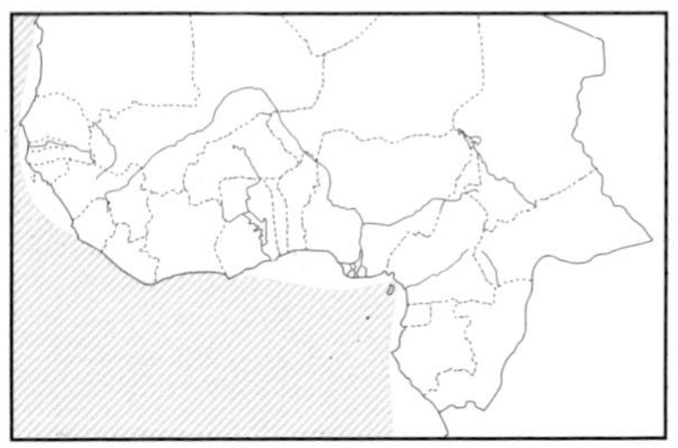

1. CORY'S SHEARWATER

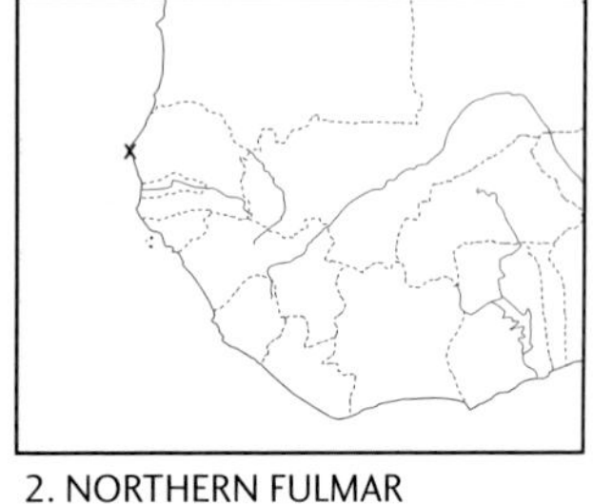

2. NORTHERN FULMAR

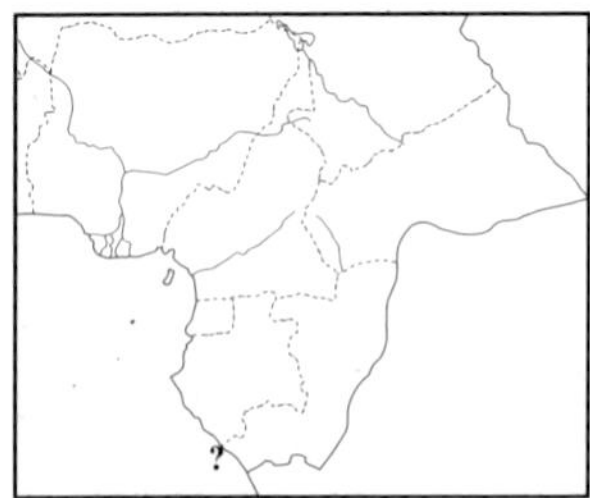

3/4. ATLANTIC YELLOW-NOSED/ BLACK-BROWED ALBATROSS

5. SOUTHERN GIANT PETREL

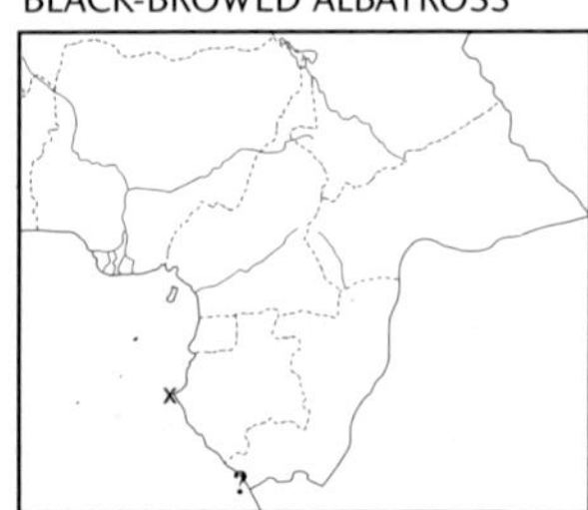

6. CAPE CORMORANT

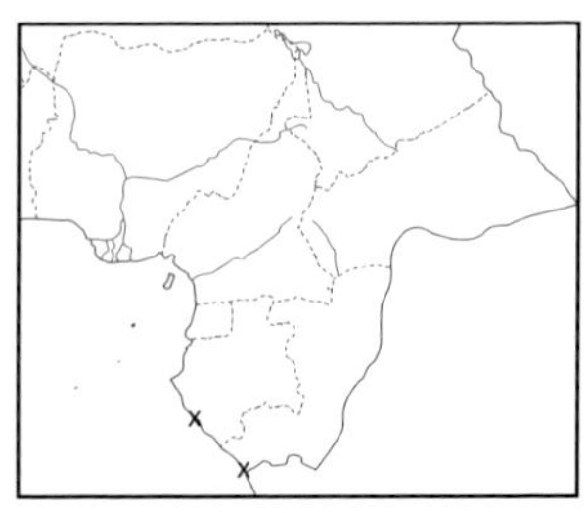

7. JACKASS PENGUIN

8. BALEARIC SHEARWATER

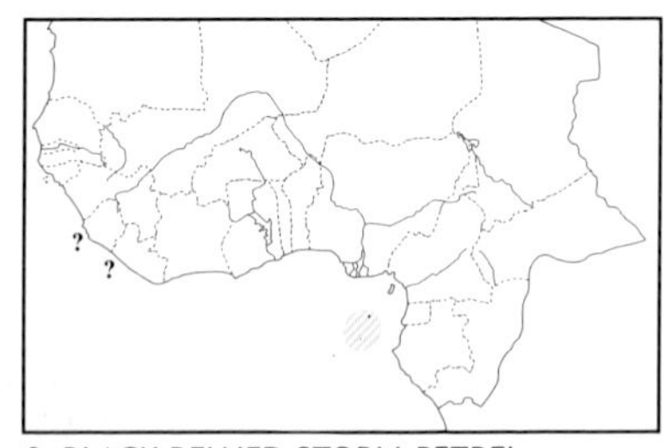

9. BLACK-BELLIED STORM-PETREL

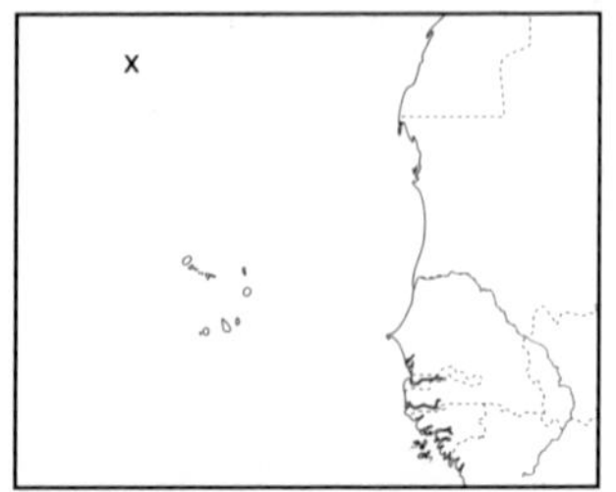

10. WHITE-BELLIED STORM-PETREL

Plate on page 46

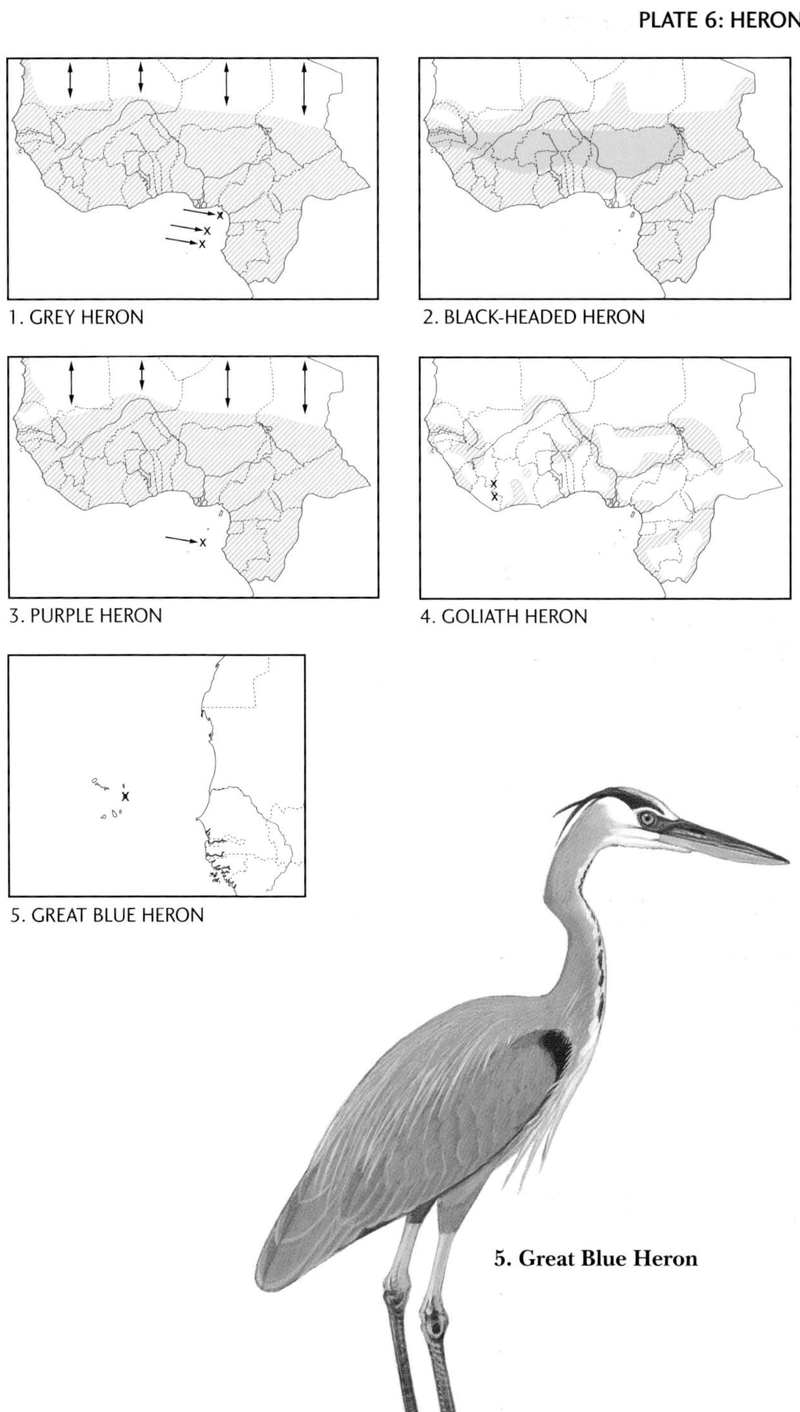

Plate on page 50

Slender, medium-sized to large wading birds with long necks and legs and long, straight, pointed bills. Flight strong with regular wingbeats and neck retracted, forming an S (unlike storks, spoonbills, ibises and cranes). Most species usually found in or near water.

1 GREY HERON *Ardea cinerea* 90–100 cm; WS 155–175 cm **R/P+ c/f**
1a Ad *cinerea* White crown and head-sides; black crown-sides; narrow black nape plumes; yellow bill. Local race *monicae* (Banc d'Arguin islands, Mauritania) very pale. **1b Juv** Grey foreneck and crown. **1c In flight** Underwing wholly dark grey. ▲ Aquatic habitats. Mostly solitary away from breeding colonies, but sometimes in small groups. Feeds by waiting motionless in shallow water. Also wades and walks through shallows. ✤ Loud, croaking *fraank* or *kraak*. [CD1:30,31]

2 BLACK-HEADED HERON *Ardea melanocephala* 92–96 cm; WS 150–160 cm **R/M u/c**
2a Ad Black crown, head-sides and hindneck. White throat contrasts with black-and-white streaked foreneck. **2b Juv** Dark crown and head-sides contrast with white foreneck. **2c In flight** White underwing-coverts contrast with dark grey flight feathers. ▲ Grassland, farmland; less often aquatic habitats. Solitary and largely terrestrial; often in association with man. Feeds by walking slowly over open areas. ✤ Raucous, nasal *kuark*. [CD5:8b]

3 PURPLE HERON *Ardea purpurea* 78–90 cm; WS 120–150 cm **R/P/V* u/c**
3a Ad Rufous-buff head-sides and neck; snake-like neck streaked black; dark grey back; purple-brown shoulders; narrow, yellowish bill. **3b Juv** Sandy-brown above; indistinctly streaked neck; yellow legs. **3c Ad in flight** Distinctive, deeply pouched appearance of neck; large feet; brownish underwing-coverts. **3d Juv/imm in flight** Paler than adult. ▲ Aquatic habitats. Usually solitary away from breeding colonies. Skulks in dense aquatic vegetation. Feeds by standing motionless with body almost horizontal, catching fish with rapid strike of head. ✤ Loud, harsh *kaark*, similar to, but higher pitched than 1. [CD1:29]

CAPE VERDE PURPLE HERON *Ardea (purpurea) bournei* See Plate 145:7.

4 GOLIATH HERON *Ardea goliath* 135–150 cm; WS 210–230 cm **R s/u**
4a Ad Massive size; rufous-brown head and hindneck; dark grey back; strong, blackish bill. **4b Juv** Paler and browner than adult; more white on head-sides and neck; black legs. **4c In flight** Brown underwing-coverts contrast with dark grey flight feathers. ▲ Aquatic habitats. Mostly solitary. Feeds by standing motionless for long periods, occasionally by walking and wading. ✤ Far-carrying, raucous *kwaaark*. [CD5:9a]

5 GREAT BLUE HERON *Ardea herodias* 110–125 cm; WS 175–195 cm **V***
Larger than 1; sides of head and neck tinged pinkish-grey; small shoulder patch and thighs chestnut. Juvenile as 1b, but browner above, more buff below. ▲ N American vagrant, Cape Verde. ✤ Low, harsh *kraak*. See illustration on page 49.

1b
1a
2b
2a
4c
4a
4b
1c
2c
3b
3a
3c
3d

1 BLACK-CROWNED NIGHT HERON *Nycticorax nycticorax* 56–65 cm; WS 90–100 cm **R/P/V* u/lc**
1a Ad breeding Stocky; black crown and back; white cheeks and neck; grey wings; two long white nuchal plumes (absent in non-breeding). **1b Juv** Dark brown streaked buff with conspicuous buff spots on wings. **1c Ad in flight** Feet project clearly beyond tail. **1d Juv in flight** Note absence of white on back and distinct feet projection. ▲ Aquatic habitats. Gregarious; largely nocturnal. ✤ Distinctive, low, harsh *kwok* in flight. [CD1:23]

2 WHITE-BACKED NIGHT HERON *Gorsachius leuconotus* 50–55 cm **R r/lc**
2a Ad Black head; dark upperparts; rufous neck and breast; white throat. **2b Juv** Darker than 1b with fewer wing spots and blackish forehead; huge eye. **2c Ad in flight** Inconspicuous white patch on back; feet project slightly beyond tail. **2d Juv in flight** Note presence of white scapulars and slight feet projection. ▲ Aquatic habitats in forest zone and wooded savanna. Nocturnal; secretive. ✤ Usually silent. Croaking alarm note. [CD5:6a]

3 SQUACCO HERON *Ardeola ralloides* 42–47 cm; WS 80–92 cm **R/P/V* f**
3a Ad breeding Wholly warm buff; crown and nape streaked black; bright blue bill tipped black. **3b Juv** Dull greyish-brown above; streaked below; white wings. Non-breeding adult similar but paler above, less streaked below. **3c Ad in flight** White wings contrast with buff upperparts. ▲ Aquatic habitats. Often solitary; occasionally in loose groups. ✤ [CD1:24]

4 WESTERN REEF EGRET *Egretta gularis* 55–65 cm; WS 86–104 cm **R/M+ f/lc**
4a Ad dark morph Slate-black to pale grey; white throat; greenish-black legs; yellowish feet and lower legs. **4b Ad white morph** Very similar to 7; note rather heavier bill with slightly drooped tip and pale base; yellowish feet and lower legs. **4c Ad dark morph in flight** Variable white patch on primary-coverts. ▲ Aquatic habitats, mainly coastal. ✤ [CD1:26]

5 BLACK HERON *Egretta ardesiaca* 48–55 cm; WS 90–95 cm **R/M/V* u/lc**
Entirely black; orange-yellow feet; shaggy plumes on hindneck, mantle and breast. ▲ Aquatic habitats. Distinctive 'canopy-feeding' posture (see illustration on page 54). ✤ [CD5:7b]

6 CATTLE EGRET *Bubulcus ibis* 48–56 cm; WS 90–96 cm **R/M+ c**
Small; mainly white; distinct jowl. **6a Ad breeding** Crown, mantle and breast buff; bill and legs reddish. **6b Ad non-breeding** All white; bill yellowish; legs and feet greenish-yellow to greyish. **6c In flight** Stocky; shortish bill. Rapid, shallow wingbeats. ▲ Various open habitats; also rubbish dumps and around human habitation. Gregarious and tame; often with domestic livestock. ✤ Short, gruff *kok* or *kwok*. [CD1:25]

7 LITTLE EGRET *Egretta garzetta* 55–65 cm; WS 88–95 cm **R+/M/P c**
All white; slender, black bill; black legs; sharply demarcated yellow feet. Compare 4b. **7a Ad breeding** Plumes on nape, mantle and breast; yellow lores. Non-breeding adult lacks plumes; lores grey-green. **7b Juv** Bare parts dull coloured; no plumes. ▲ Aquatic habitats. ✤ [CD1:27]

8 INTERMEDIATE EGRET *Egretta intermedia* 65–72 cm; WS 105–115 cm **R/M/V* u/c**
All white; bill yellow; gape line ends below eye; legs and feet black with yellowish tibia. **Ad breeding** Extensive plumes hang over back and tail. ▲ Aquatic habitats. ✤ [CD5:8a]

9 GREAT EGRET *Egretta alba* 85–100 cm; WS 140–170 cm **R/M/P/V* c/f**
Large; all white; long, kinked neck; gape line ends behind eye. **9a Ad *melanorhynchos* breeding** Bill, legs and feet black; lores emerald-green; long plumes hang over back and tail. *E. a. alba* (Palearctic vagrant, CAR, N Mali) has tibia and rear edge of tarsus yellowish. **9b Ad *melanorhynchos* non-breeding** Bill all yellow; no long plumes. ▲ Aquatic habitats. ✤ Loud, deep, croaking *krraak*. [CD1:28]

Maps on page 54

1a
1b
2b
2a
2c
2d
1c
1d
3c
3b
3a
4c
4a
5
6a
6b
6c
9a
9b
8
4b
7b
7a

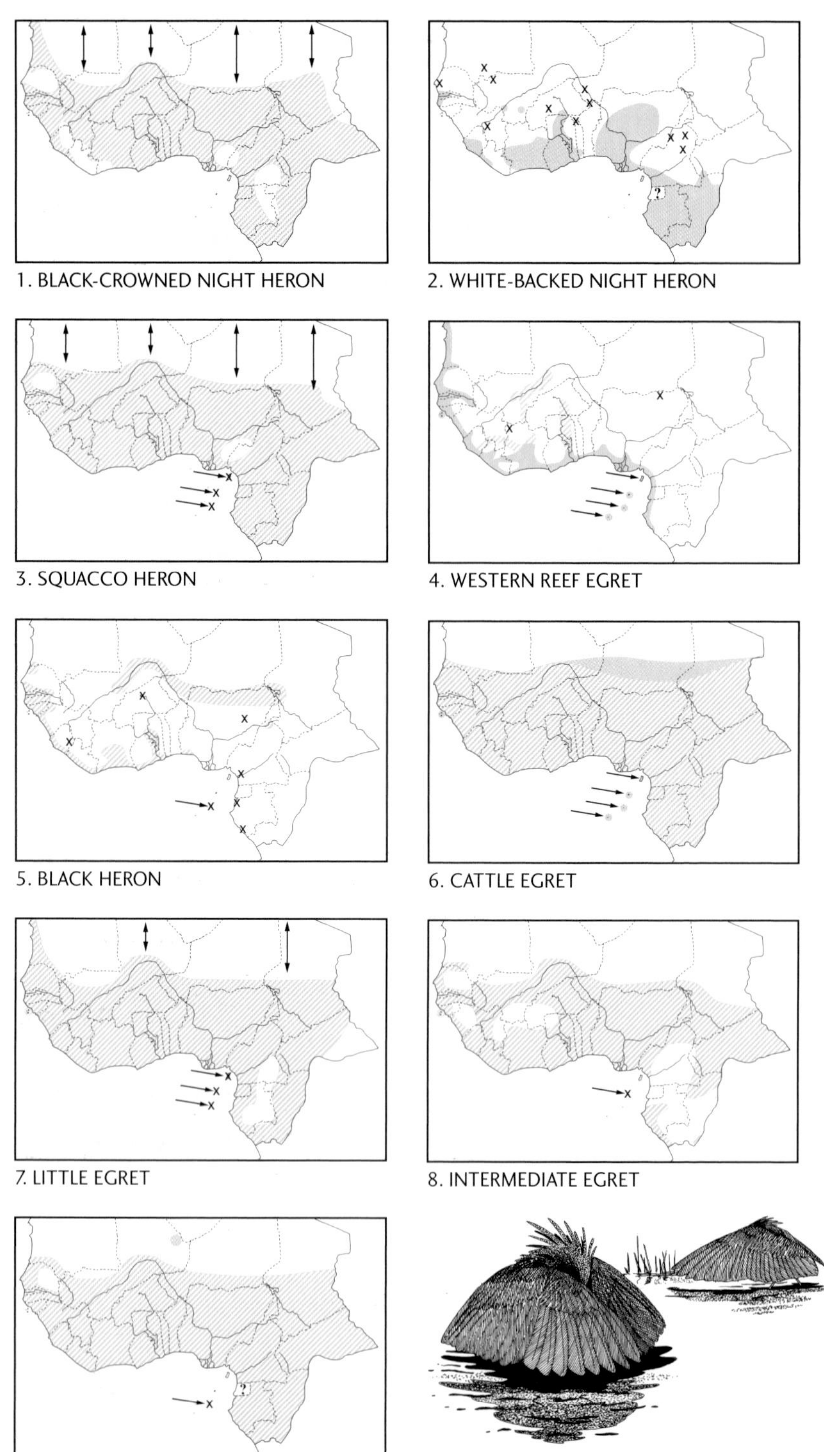

1. BLACK-CROWNED NIGHT HERON

2. WHITE-BACKED NIGHT HERON

3. SQUACCO HERON

4. WESTERN REEF EGRET

5. BLACK HERON

6. CATTLE EGRET

7. LITTLE EGRET

8. INTERMEDIATE EGRET

9. GREAT EGRET

Black Herons *Egretta ardesiaca*

Plate on page 52

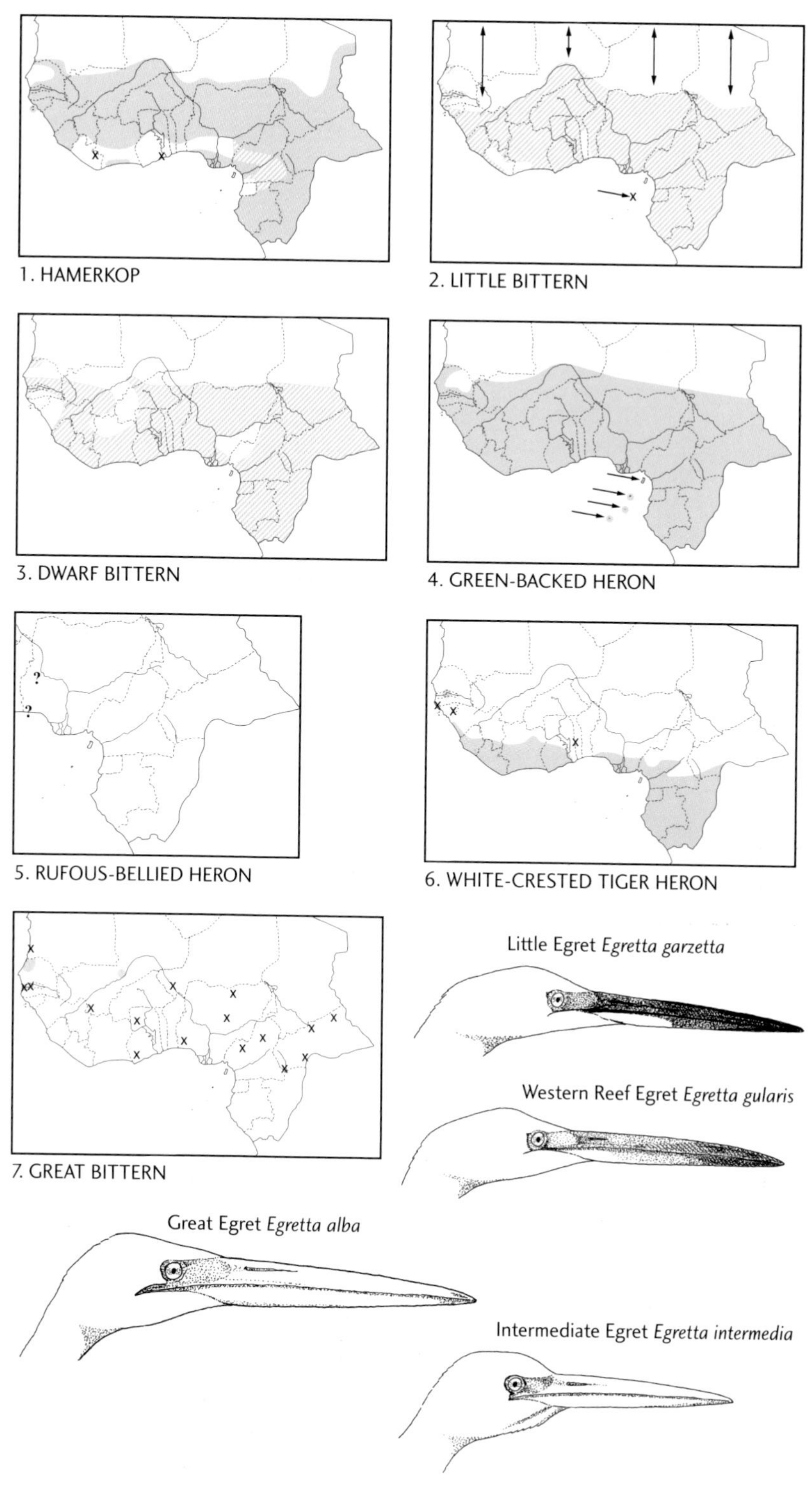

1. HAMERKOP

2. LITTLE BITTERN

3. DWARF BITTERN

4. GREEN-BACKED HERON

5. RUFOUS-BELLIED HERON

6. WHITE-CRESTED TIGER HERON

7. GREAT BITTERN

Plate on page 56

PLATE 8: HAMERKOP, HERONS AND BITTERNS

Hamerkop. Unique waterbird. Builds huge domed nest in fork of large tree.

1 HAMERKOP *Scopus umbretta* 50–56 cm; WS 85–94 cm **R u/c**
1a Entirely dark brown; characteristic head shape. **1b In flight** Broad wings. Flight rather buoyant and owl-like; often rather jerky wingbeats. Glides with neck extended. Can appear raptor-like when soaring. ▲ Aquatic habitats in open woodland; also drier areas. ✤ Distinctive nasal trumpeting *yip-purr, yip-yip-yip-purr-purr-yip-yip* uttered when in groups or during display. Flight call a high-pitched nasal *yip* or *wek*. [CD5:9b]

Herons (continued).

2 LITTLE BITTERN *Ixobrychus minutus* 27–38 cm; WS 40–58 cm **R/P/V* u**
2a Ad male *payesii* (R) Small; black crown and back contrast with rich chestnut head-sides and neck. **2b Ad male *minutus*** (P) Much paler head-sides and neck than 2a. **2c Ad female** Duller than male; upperwing-coverts browner, less contrasting. **2d Juv** Streaked, pale buffy upperparts. **2e Ad male in flight** Conspicuous pale wing panel (upperwing-coverts). ▲ Reed-beds, marshes, areas of rank vegetation near water, occasionally mangroves. 'Freezes' in erect posture with bill 'sky-pointing' when threatened. ✤ Usually silent. Hard *ker-ek* and sharp, rapid *kekekekek*. Breeding male utters deep croaking *hogh* or *woof* at *c.* 2-second intervals. [CD1:22]

3 DWARF BITTERN *Ixobrychus sturmii* 25–30 cm; WS 50 cm **M u/s**
3a Ad Very small; wholly slate-grey above; buffish underparts (darkest on belly) heavily streaked black; bill blackish with variable amount of yellow at base and below. **3b Juv** Upperpart feathers edged buff. **3c In flight** Small; dark. ▲ Vegetated streams and ponds, marshes, reedbeds, seasonally inundated areas, mangroves. 'Freezes' in typical bittern stance when alarmed. ✤ Occasionally a loud croak when disturbed. [CD5:5a]

4 GREEN-BACKED HERON *Butorides striata* 40–48 cm; WS 60–70 cm **R c**
4a Ad Small; dark crown and upperparts; grey head-sides and underparts. Greenish tinge to upperparts only visible at close range or in good light. **4b Juv** Brown upperparts spotted buffish. **4c Ad in flight** Dark; conspicuous yellow or orange legs. ▲ Heavily vegetated margins of rivers, streams, lakes and pools, mangroves; occasionally floodplains. In tidal habitats often far from cover at low tide. Usually adopts hunched stance when feeding, with body almost horizontal and legs bent. ✤ Loud, explosive *kyah* when flushed. [CD5:7a]

5 RUFOUS-BELLIED HERON *Ardeola rufiventris* 38 cm **V?**
5a Ad male Black, with maroon wings, belly and tail; heavy, two-toned bill. Female duller; buffish stripe on throat; bill duskier. **5b Juv** Streaked buffish-brown on head-sides, neck and upper breast. **5c Ad in flight** Dark; legs and feet bright yellow. ▲ Aquatic habitats; S & E African species; doubtful claims from S Nigeria.

6 WHITE-CRESTED TIGER HERON *Tigriornis leucolopha* 66–80 cm **R u/r**
Cryptically patterned, barred black and buffish-brown plumage; long, dagger-shaped bill. **6a Ad** White crest (half concealed by nape feathers when not raised; often hard to see). **6b Juv** More strongly barred than adult. **6c In flight** Dark; primaries white tipped. ▲ Shaded forest streams; also mangroves. Secretive; largely nocturnal. 'Freezes' in bittern-like posture when alarmed. ✤ Far-carrying, single or double, low moaning note, regularly repeated in slow tempo, usually for brief periods before dawn and after sunset. [CD5:5b]

7 GREAT BITTERN *Botaurus stellaris* 70–80 cm; WS 125–135 cm **P/V r**
Stocky and thick-necked with brown, cryptically patterned plumage. Heavier looking than 6; streaked (not barred); lacks white crest. **7a Ad** Black crown and malar stripe. Heavy, dagger-shaped yellowish bill. **7b In flight** Broad, rounded wings strongly bowed downwards. ▲ Dense reedbeds, marshes, swamps. Active by day and night. 'Freezes' in upright posture, bill 'sky-pointing' when threatened. Very secretive. ✤ Silent in W Africa. [CD1:21]

 Maps on page 55

1a
1b
2b
2a
2c
2d
2e
3a
3b
3c
4a
4b
4c
5a
5b
5c
6a
6b
6c
7a
7b

PLATE 9: STORKS

Large to very large wading and terrestrial birds with long bills, necks and legs. Fly with neck outstretched, except Marabou Stork, and legs trailing. Normal flight action consists of soaring and gliding, often at great heights, alternated with slow wingbeats. Gregarious or solitary. Most species normally silent.

1 YELLOW-BILLED STORK *Mycteria ibis* 95–105 cm; WS 155–165 cm **M/R u/r**
1a Ad Mainly white; yellow bill; red facial skin and legs. **1b Juv** Mainly greyish-brown; dull-coloured bare parts. **1c Ad in flight** Black flight feathers and tail. Compare 3c (red bill; white tail). ▲ Aquatic habitats, mainly in savanna belt. Gregarious, but seldom in large flocks. Often associated with other waterbirds such as herons, storks and pelicans. ✤ [CD5:10a]

2 AFRICAN OPENBILL STORK *Anastomus lamelligerus* 80–94 cm **M/R lu/r**
2a Ad All dark; glossy plumage; distinctly shaped bill with gap between mandibles. **2b Juv** Dull brown; buff-edged feathers on upperparts; bill almost straight. ▲ Aquatic habitats, mainly fresh water. Gregarious. Feeds entirely on snails and freshwater mussels. ✤ [CD5:10b]

3 WHITE STORK *Ciconia ciconia* 100–120 cm; WS 155–165 cm **P u/lc**
3a Ad Mainly white with black on wings; red bill and legs. **3b Juv** Bill dusky. **3c Ad in flight** Black flight feathers; white tail. Compare 1c (yellow bill; black tail). ▲ Grassland, open savanna, wetlands. Gregarious. Mainly Sahel; rare/scarce further south. ✤ [CD1:33]

4 BLACK STORK *Ciconia nigra* 95–100 cm; WS 145–155 cm **P/V r/s**
4a Ad Black, with white breast and belly; red bill and legs. **4b Juv** Sooty-brown; dull-coloured bare parts. **4c–d Ad in flight** All-black upperparts. Compare 5 (smaller; white rump and lower back; yellowish-green bill and legs; shorter legs projecting less in flight. Immatures more alike, but white rump and dusky-greenish legs with pink joints diagnostic of Abdim's). ▲ Aquatic and dry habitats. Solitary or in pairs; occasionally small parties. Associates with other storks. ✤ [CD1:32]

5 ABDIM'S STORK *Ciconia abdimii* 75–81 cm **M lc/r**
5a Ad Mainly black, with white breast and belly; greenish bill with red tip; dull green legs with pink joints. **5b Juv** Duller. **5c–d Ad in flight** White lower back and rump. Compare 4. ▲ Grassland, cultivated areas. Gregarious. Congregates at swarms of locusts, outbreaks of army worm caterpillars and grass fires. Occurrence strongly seasonal, breeding in wet season (May–Aug), and moving south at onset of dry season, to arrive in southern tropics early in rains. ✤ [CD5:11a]

6 WOOLLY-NECKED STORK *Ciconia episcopus* *c.* 90 cm **R/M u/r**
6a Ad Dark; white head and neck; bill blackish tipped reddish; legs black. **6b Juv** Lacks white forehead. **6c Ad in flight** White head, neck, belly and projecting undertail-coverts. ▲ Various habitats, usually near water. Mainly solitary, but may gather in small flocks. ✤ [CD5:11b]

7 SADDLE-BILLED STORK *Ephippiorhynchus senegalensis*
145–150 cm; WS 240–270 cm **R/M r/u**
7a Ad male Very large; black and white; red, black and yellow bill; dark eyes; yellow wattles. **7b Ad female** Yellow eyes; typically no wattles. **7c Juv** Greyish; bill dusky. **7d Ad in flight** White flight feathers. ▲ Aquatic habitats. Singly or in pairs. Partially nomadic, moving in response to local conditions. Normally shy and wary. ✤ [CD5:12a]

8 MARABOU STORK *Leptoptilos crumeniferus* *c.* 150 cm; WS 230–285 cm **R/M r/lf**
8a Ad Huge; bare pinkish head and neck; massive bill. **8b Juv** Upperparts dark brown. **8c Ad in flight** Black wings contrast with white underparts; head tucked into shoulders. ▲ Open habitats, terrestrial and aquatic. Often associated with man at fishing villages, slaughterhouses and rubbish dumps. Generally gregarious. ✤ [CD5:12b]

Maps on page 60

1a
1b
2a
2b
3c
1c
4c
4d
3a
5c
5d
3b
6c
4a
4b
5a
5b
6a
6b
7d
7a
8a
8b
8c
7c
7b

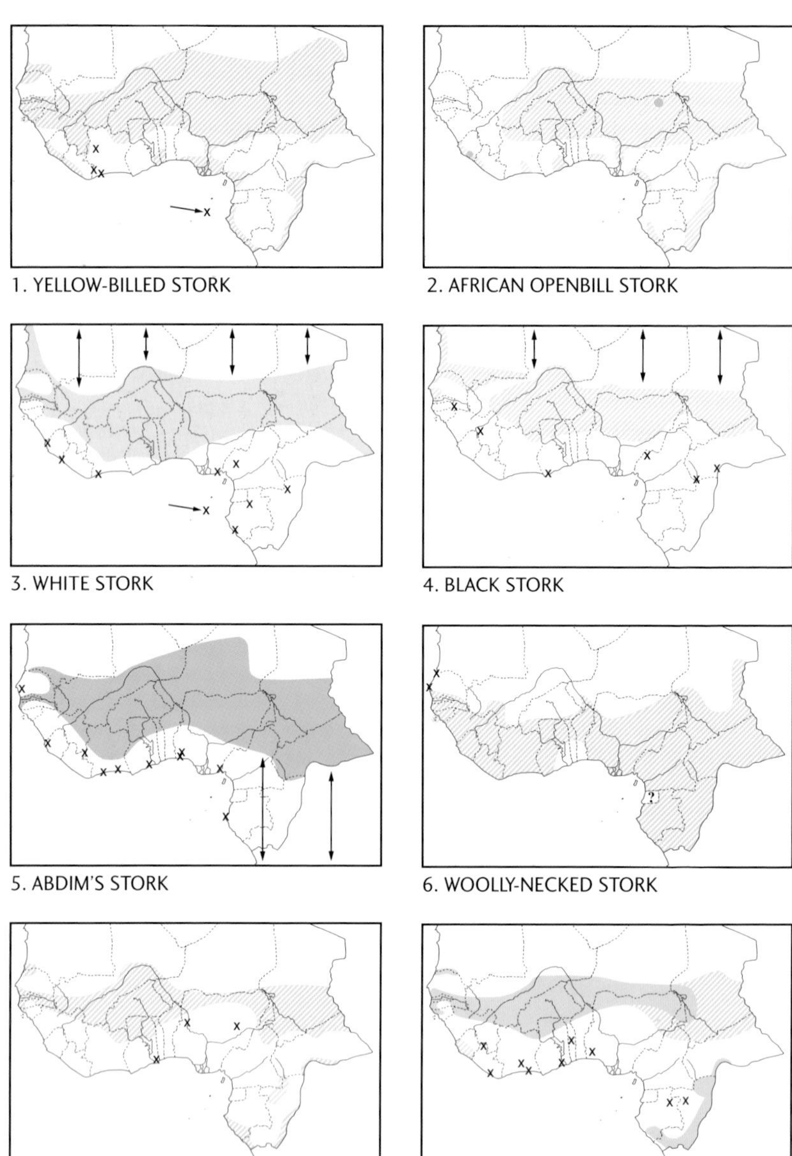

1. YELLOW-BILLED STORK

2. AFRICAN OPENBILL STORK

3. WHITE STORK

4. BLACK STORK

5. ABDIM'S STORK

6. WOOLLY-NECKED STORK

7. SADDLE-BILLED STORK

8. MARABOU STORK

Plate on page 58

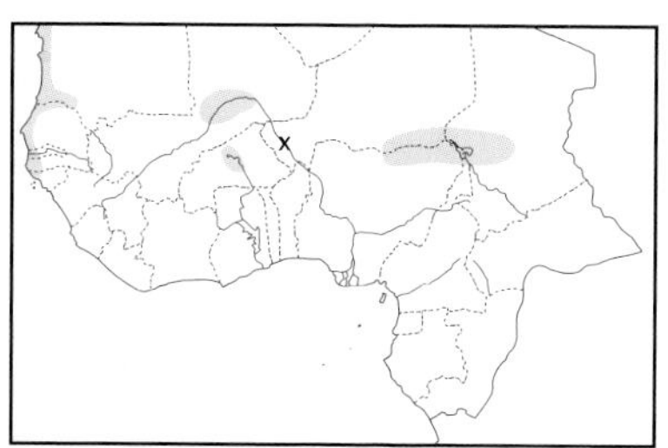

1. EUROPEAN SPOONBILL

2. AFRICAN SPOONBILL

3. SHOEBILL

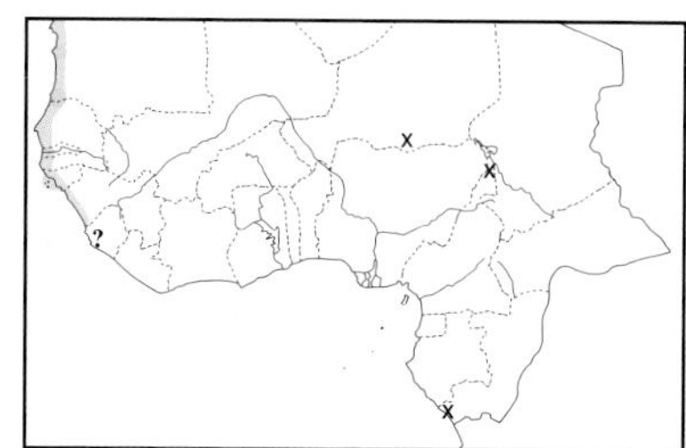

4. GREATER FLAMINGO

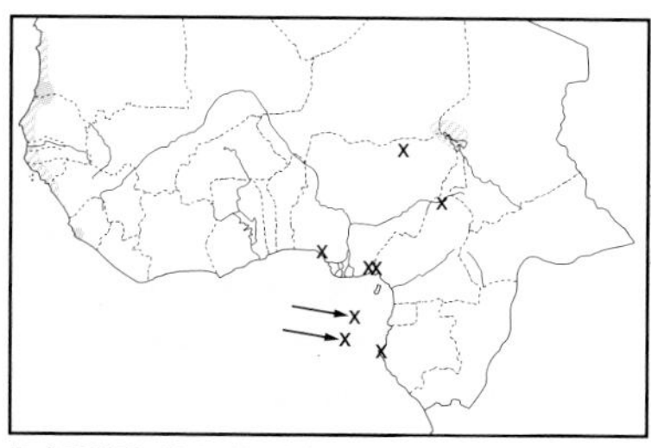

5. LESSER FLAMINGO

Plate on page 62

Spoonbills. Large, white wading birds with characteristic spoon-shaped bill and long necks and legs. Fly with neck outstretched; sometimes soar. Gregarious or solitary. Feed by sweeping bill from side to side in water. Mainly silent away from breeding colonies.

1 EUROPEAN SPOONBILL *Platalea leucorodia* 80–90 cm; WS 115–130 cm **R/P r/lc**
1a Ad *leucorodia* breeding (P; uncommon/rare) Bill black with yellow tip; legs black; long, loose nuchal crest; yellowish breast-band. Non-breeding adult lacks crest and breast-band. **1b Ad *balsaci* breeding** (R) Bill all black. Breeds Mauritania (Banc d'Arguin; locally common). **1c Juv** Bill dull pinkish, legs yellowish-brown. **1d Juv in flight** Outstretched neck; primaries tipped black (all white in adult). ▲ Mainly coastal, in shallow water of e.g. lagoons and estuaries. Occasional inland at lakes and marshes. Flocks tend to fly in line or V formation. ✤ [CD1:36]

2 AFRICAN SPOONBILL *Platalea alba* *c.* 90 cm **R u/s**
2a Ad breeding Bare face red; bill grey with red margins; legs red; loose, fluffy nuchal crest. Non-breeding adult lacks crest; bare-part coloration somewhat duller. **2b Juv** Bill yellowish; legs black; no red on face. **2c Juv in flight** Note amount of black on wingtips and primary-coverts. ▲ Aquatic habitats (as 1). ✤ [CD5:15b]

Shoebill. Very large, long-legged wading bird with unique, slightly shoe-shaped bill. Normally silent.

3 SHOEBILL *Balaeniceps rex* 110–140 cm **M r VU**
3a Ad All grey; distinctively shaped, huge pale bill. Juvenile similar, but darker and browner grey; bill smaller and pinkish. **3b In flight** Neck usually tucked in. At great distance could be confused with Goliath Heron, but huge swollen bill and lack of any rufous in plumage distinctive. Soaring Marabou Stork has white on belly and axillaries, and differently shaped bill. ▲ Extensive freshwater swamps, esp. with papyrus. Generally shy. Frequently soars. Flight on take-off laborious, with deep, heavy wingbeats. ✤ [CD5:13a]

Flamingos. Large wading birds with extremely long necks and legs, pink plumage and unique down-curved bills adapted for filter-feeding. Fly in lines or V formations with neck outstretched. Highly gregarious and quite noisy. Occur on shallow brackish, alkaline or saline lakes and lagoons.

4 GREATER FLAMINGO *Phoenicopterus roseus*
127–140 cm; WS 140–165 cm **R/P/M lc/r**
4a Ad Pale pink plumage (appearing largely white at distance); pink bill with contrasting black tip. **4b Juv** Dull greyish plumage; note rather large bill. In subsequent plumages, first becomes whiter, then progressively more pink. Adult plumage attained in 3–4 years. **4c–d Ad in flight** Scarlet lesser and median coverts contrast strongly with black flight feathers. ▲ Coastal aquatic habitats. Wades with head submerged when feeding, or swims and up-ends in deeper water. ✤ Goose-like *hank-hank* and a nasal *gnaaaa*. [CD1:37]

5 LESSER FLAMINGO *Phoeniconaias minor* 80–90 cm; WS 95–100 cm **M/V r NT**
5a Ad Rose-pink plumage; dark crimson bill (appearing entirely dark at distance); much smaller than 4a. **5b Juv** Brownish-grey; bill smaller than 4b. Subsequent plumages as 4. **5c–d Ad in flight** Lesser and median coverts contrast less than in 4c–d. ▲ Aquatic habitats, mainly coastal. Breeds SW Mauritania. Normally highly gregarious, but in W Africa also observed singly and in small groups. Occasionally associates with 4. ✤ High-pitched *chissik* or *kwirrik*. Feeding flocks utter constant murmuring *murr-err*. [CD5:16a]

 Maps on page 61

1b
1a
1c
3b
3a
2a
2b
1d
4c
2c
4d
4a
4b
5a
5b
5d
5c

Medium-sized terrestrial and wading birds with rather long necks and legs and slender decurved bills. Fly with strong wingbeats and neck outstretched. Feed in dry habitats, on forest floor or in shallow wet areas by probing in soft mud.

1 GLOSSY IBIS *Plegadis falcinellus* 55–65 cm; WS 80–95 cm **R/P/V+ f/r**
1a Ad breeding Dark plumage; rich purplish-chestnut glossed with green on wings; slender build (other dark ibises less slender). **1b Ad non-breeding** Duller; head and neck with variable amount of fine white streaking. **1c Juv** Sooty-brown; head and throat variably mottled white. **1d In flight** Legs project well beyond tail. ▲ Freshwater habitats incl. marshes, floodplains, inundated areas, margins of rivers and lakes. Occasionally coastal lagoons, estuaries. Gregarious; usually in small groups. Often associated with herons, egrets and storks. Usually flies in line, sometimes in V formation. ✤ Mostly silent. Occasionally a harsh, low *graa-graa-graa* in flight. [CD1:34]

2 HADADA IBIS *Bostrychia hagedash* 76–89 cm **R f/u**
2a Ad Larger and heavier than 1; white malar stripe; metallic green gloss on wing-coverts; dark bill with red on upper mandible. Juvenile duller. **2b In flight** Legs do not project beyond tail. ▲ Along streams and rivers in open woodland, also marshes, mangroves, forest edge; occasionally near human habitation. ✤ Vocal. Distinctive, far-carrying, nasal *haa! haa-de-dah!* frequently uttered in flight. Also a loud *HAAA!* when alarmed. [CD5:13b]

3 OLIVE IBIS *Bostrychia olivacea* 65–75 cm **R r**
3a Ad All dark; relatively short, coral-red bill; loose nuchal crest (often hard to see, esp. in flight); wing-coverts metallic coppery and purple-blue. Larger than 4. Juvenile duller; crest shorter. **3b In flight** Legs do not project beyond tail; red bill conspicuous. Compare 2b and 4b. ▲ Streams and rivers in lowland forest; occasionally away from water and in mangroves. Shy and rarely seen. Flies above forest at dawn and dusk, calling. ✤ Loud, resonant, disyllabic *HAH-hah* or *k-HA-haw* with stress on first syllable, usually uttered in flight. From perch also a single *haaw!* [CD5:14a]

DWARF OLIVE IBIS *Bostrychia (olivacea) bocagei* See Plate 146:11.

4 SPOT-BREASTED IBIS *Bostrychia rara* 47–55 cm **R r/lc**
4a Ad Small; dark; red bill; crest (hard to see); bright turquoise patches in front of and behind eye, and in a line below base of bill. Juvenile duller; crest and bill shorter. **4b In flight** Legs do not project beyond tail. Compare 3b. ▲ Forest streams, swamp forest. ✤ Vocal. Fairly loud, raucous and nasal *k-HAH! k-HAH!* or *ah-HAW ah-HAW* and *ah-HAH-hah* with stress on 2nd syllable; uttered in flight and usually frequently repeated; higher pitched than 3. Also a nasal *haw*. Usually calls at dusk and during night. [CD5:14b]

5 NORTHERN BALD IBIS *Geronticus eremita* 70–80 cm; WS *c.* 130 cm **P/V r CR**
5a Ad Unmistakable. All dark; bare reddish head; reddish bill and legs; shaggy crest. **5b Juv** Head feathered; no crest; duller plumage. In 2nd year nuchal crest and violet gloss on coverts begin to appear, but head still largely feathered. Full adult bare-part coloration probably attained during 3rd year. **5c In flight** Legs do not project beyond tail. ▲ Dry rocky areas; also open fields, lagoons. Very rare. ✤ [CD1:35]

6 SACRED IBIS *Threskiornis aethiopica* 65–82 cm; WS *c.* 120 cm **R/M u/s**
6a Ad Unmistakable. White with bare black head and neck, and black ornamental plumes at rear. **6b Juv** Head feathered, blackish mottled white. In subsequent plumages ornamental scapulars develop, black on primaries and greater primary-coverts gradually decreases. Full adult plumage with completely bare head and neck attained in 3–4 years. **6c In flight** White wings with narrow black edge. ▲ Margins of rivers and lakes, marshes, floodplains, grasslands, cultivation, rubbish dumps; occasionally estuaries, lagoons, sea coast. Gregarious. Often very tolerant of man. Flocks tend to fly in V formation. ✤ [CD5:15a].

 Maps on page 66

1a
1b
1c
1d
3b
4b
2a
3a
4a
5a
5b
2b
5c
6a
6b
6c

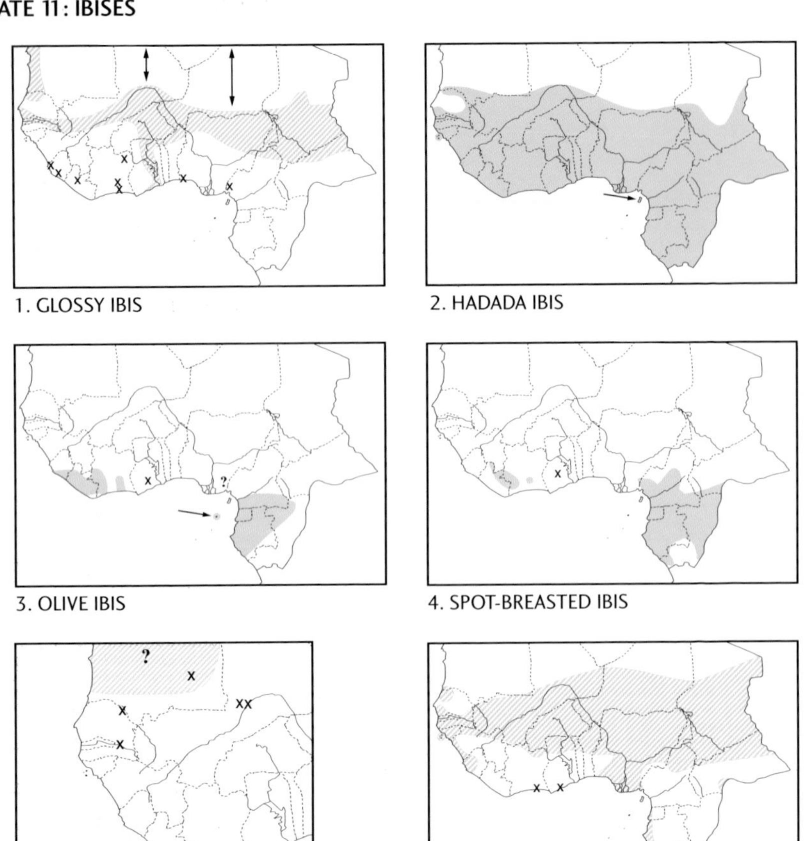

1. GLOSSY IBIS

2. HADADA IBIS

3. OLIVE IBIS

4. SPOT-BREASTED IBIS

5. NORTHERN BALD IBIS

6. SACRED IBIS

Plate on page 64

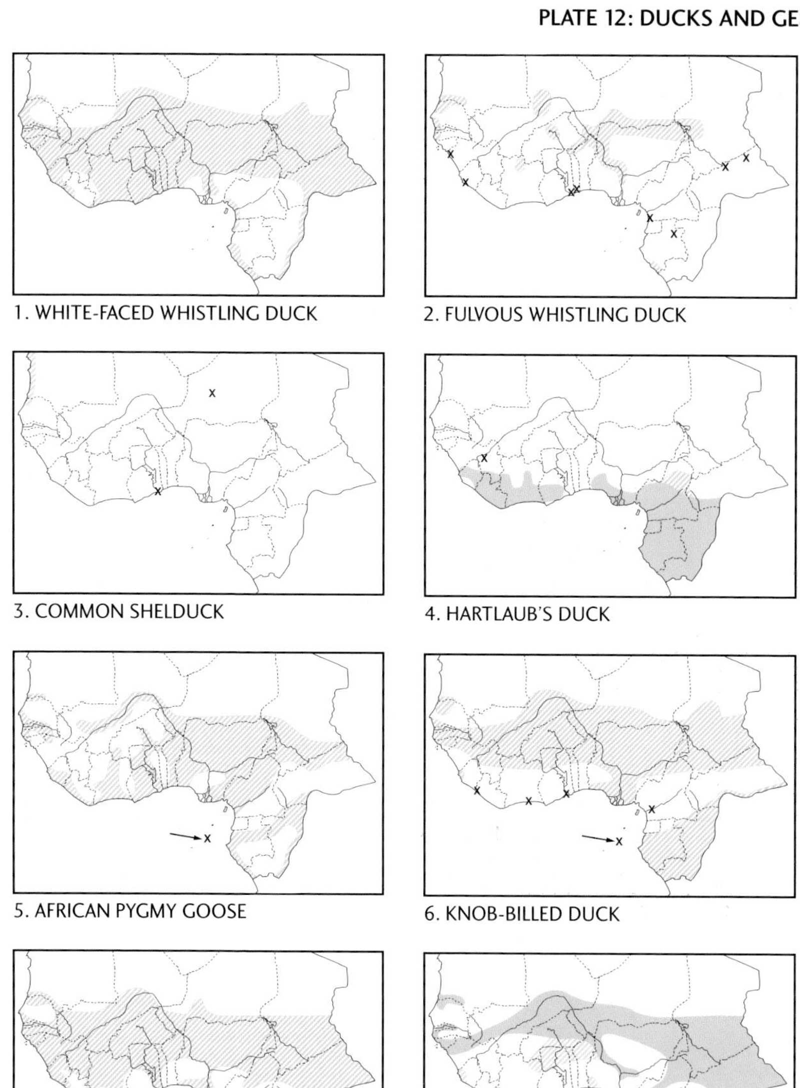

1. WHITE-FACED WHISTLING DUCK

2. FULVOUS WHISTLING DUCK

3. COMMON SHELDUCK

4. HARTLAUB'S DUCK

5. AFRICAN PYGMY GOOSE

6. KNOB-BILLED DUCK

7. SPUR-WINGED GOOSE

8. EGYPTIAN GOOSE

PLATE 12: DUCKS AND GEESE

Medium-sized to large waterbirds with plump bodies, short, robust legs, webbed feet and flat bills rounded at tip. Flight fast and direct with neck outstretched. Often gregarious.

1 WHITE-FACED WHISTLING DUCK *Dendrocygna viduata* 43–48 cm R/M lc
1a Ad Dark; contrasting white face. Upright stance, with long neck and legs. **1b Juv** (plumage retained for only a few weeks) Duller; face buffish. **1c Ad in flight** Wholly dark; chestnut forewing. ▲ Freshwater wetlands. Gregarious. ✤ Vocal, esp. in flight. Characteristic, clear, sibilant 3-note whistle *swee-swee-sweeoo* or *swee-whee-wheew*, usually repeated. [CD5:17a]

2 FULVOUS WHISTLING DUCK *Dendrocygna bicolor* 45–53 cm R/M lc
2a Ad Mainly fulvous-brown; bold white flank streaks. Shape and stance as 1. **2b Ad in flight** Blackish wings; fulvous body; white, U-shaped rump. ▲ Freshwater wetlands. Gregarious. Flocks often mix with 1. ✤ Clear, slightly nasal whistle *k-wheew* or *ksweeoo*, in flight and usually repeated. [CD5:16b]

3 COMMON SHELDUCK *Tadorna tadorna* 58–71 cm P/V r
3a Ad male Mainly black and white with rufous breast-band; red bill with knob. **3b Ad female** Duller; often some white at bill base; no knob on bill. **3c In flight** Heavy; white with black head and black flight feathers. ▲ Coastal and freshwater wetlands. ✤ Male utters low whistling calls, female a rapid *ga-ga-ga-ga-ga-ga-gak*. [CD1:48]

4 HARTLAUB'S DUCK *Pteronetta hartlaubii* 56–58 cm R u/lf
4a Ad male Chestnut body; black head; some white on forehead (extent variable). **4b Ad female** Duller; usually no white on head; some pink on bill. Juvenile duller still. **4c In flight** Blue-grey forewing. ▲ Forested streams, lakes and ponds. ✤ Low, fast *kakakakarrr* in flight, and *whit-whit-whit*. [CD5:18b]

5 AFRICAN PYGMY GOOSE *Nettapus auritus* 30–33 cm R f/s
5a Ad male Small; white face; green neck-sides; yellow bill; rusty-orange underparts. **5b Ad female** Duller; head-sides mottled greyish. **5c In flight** Dark above; white patch on inner secondaries. ▲ Freshwater wetlands rich in emergent vegetation, particularly water-lilies. ✤ Male utters soft, rather melodious, whistled *kewheep* and *khep-khep-kheew*, usually repeated. Female a soft quack. [CD5:19b]

6 KNOB-BILLED DUCK *Sarkidiornis melanotos* 56–76 cm M u/lc
6a Ad male Head and underparts mainly white; upperparts black, strongly glossed; knob on bill (reduced in non-breeding). **6b Ad female** Smaller; no knob on bill. **6c Juv** Dark brown above; buffish-brown below. **6d Ad male in flight** Black wings and upperparts; white underparts. ▲ Freshwater wetlands. ✤ Mainly silent. Calls varied; usually in display. [CD5:19a]

7 SPUR-WINGED GOOSE *Plectropterus gambensis* 75–100 cm R u/lc
7a Ad male Huge; glossy black-and-white plumage; pinkish bill and legs; red frontal knob. **7b Ad female** Smaller and duller; frontal knob reduced or absent. **7c Ad male in flight** White forewing above and below. ▲ Freshwater wetlands. ✤ Mainly silent. Male occasionally utters soft, wheezy *cheweh* or *cherwit*. [CD5:18a]

8 EGYPTIAN GOOSE *Alopochen aegyptiaca* 63–73 cm R/M/V lf
8a Ad Large; pink bill and legs; dark eye patch. **8b Juv** Duller; bill and legs greyish. **8c Ad in flight** White wing-coverts above and below. ▲ Wetlands. Readily perches in trees. ✤ Hoarse *taash taash...* in flight. Male utters harsh, wheezy hiss when displaying. Female gives harsh, nasal, slightly trumpeting cackle *honk-haah-haah-haah*. [CD1:46]

Maps on page 67

1a
1b
1c
2a
2b
3a
3b
3c
4a
4b
4c
5a
5b
5c
6a
6b
6c
6d
7a
7b
7c
8a
8b
8c

Dabbling ducks (*Anas*) feed in water by up-ending or skimming the surface with bill. Rise vertically from the water without any foot-pattering. Upperwing pattern and speculum colour often important identification marks. **Diving ducks** (*Aythya*; see Plates 14–15) feed by diving. Swim under water and run along the surface to take off. In Palearctic species of both groups males adopt female-like 'eclipse' plumage soon after breeding, from May/June until October/November. Females and juveniles of African species usually very similar to males.

1 AFRICAN BLACK DUCK *Anas sparsa* 48–57 cm **R r**
1a Ad Blackish-brown; buffish spots on upperparts; pinkish bill with dusky saddle. **1b Juv** Duller; buff spots reduced; bill greyish. **1c In flight** Dark blue speculum bordered white; barred tail; white underwing-coverts. ▲ Rocky, forested streams and lakes. ✤ Mainly silent. Male utters soft, wheezy *wheep*; female a loud quack, sometimes repeated. [CD5:21b]

2 YELLOW-BILLED DUCK *Anas undulata* 51–58 cm **R? ls**
2a Dark plumage; yellow bill. **2b In flight** Dark green speculum bordered white; white underwing-coverts. ▲ Freshwater wetlands. ✤ Male utters low whistles; female a series of quacks. [CD5:21a]

3 GADWALL *Anas strepera* 46–55 cm **P/V r**
3a Ad male Mainly greyish; black rear end; lead-grey bill. **3b Ad female** Bill with orange sides and dark culmen. Compare 4b. **3c Ad male in flight** Small white speculum; white belly; white underwing. **3d Ad female in flight** Well-defined white belly. Compare 4d. ▲ Freshwater wetlands. ✤ Mainly silent. Female gives loud quack, similar to 4b but slightly higher pitched and harder. [CD1:50]

4 MALLARD *Anas platyrhynchos* 50–65 cm **P/V r**
4a Ad male Green head; narrow white collar; purplish-brown breast; yellow bill. **4b Ad female** Pale brown marked darker; bill orange with irregular dark area on culmen. Compare 3b. **4c Ad male in flight** Dark blue speculum bordered white. **4d Ad female in flight** Brown belly. Compare 3d. ▲ Freshwater wetlands. ✤ Male has low, nasal *vrreb*; female a loud quack. [CD1:52]

5 NORTHERN SHOVELER *Anas clypeata* 44–52 cm **P c/s**
5a Ad male Large, spatulate bill; dark green head; white breast; chestnut flanks. **5b Ad female** Bill shape diagnostic. **5c Ad male in flight** Bluish forewing; front-heavy appearance. **5d Ad female in flight** Greyish-blue forewing. ▲ Freshwater and coastal wetlands. Local concentrations. ✤ Mostly silent. Male utters gruff, nasal *took-took*; female a variety of quacks. [CD1:55]

6 NORTHERN PINTAIL *Anas acuta* 51–56 cm **P c/s**
6a Ad male Brown head; white stripe on side of long neck; long, finely pointed tail. **6b Ad female** Rather plain buff-brown head; slim grey bill. **6c Ad male in flight** Long neck and tail; dark green and bronze speculum bordered white at rear. **6d Ad female in flight** Brownish speculum bordered white; greyish underwing. ▲ Freshwater and coastal wetlands. ✤ Male has clear, melodious, mellow *krrup*, reminiscent of Common Teal; female a low quack and Eurasian Wigeon-like growl. [CD1:53]

7 EURASIAN WIGEON *Anas penelope* 45–51 cm **P/V s/u**
7a Ad male Chestnut head; yellow forehead and crown; pinkish breast. **7b Ad female** Rather uniform rusty-brown to brownish-grey. **7c Ad male in flight** White forewing. **7d Ad female in flight** White belly; short pointed tail; grey underwing and axillaries. ▲ Freshwater and coastal wetlands. ✤ Male utters distinctive melodious, whistling *swheeoooh*; female a growling *krrr*, sometimes repeated. [CD1:49]

Maps on page 72

1c
1b
2a
2b
1a
3b
3a
3d
3c
4b
4a
4d
4c
5c
5b
5a
5d
6b
6a
6c
6d
7c
7b
7a
7d

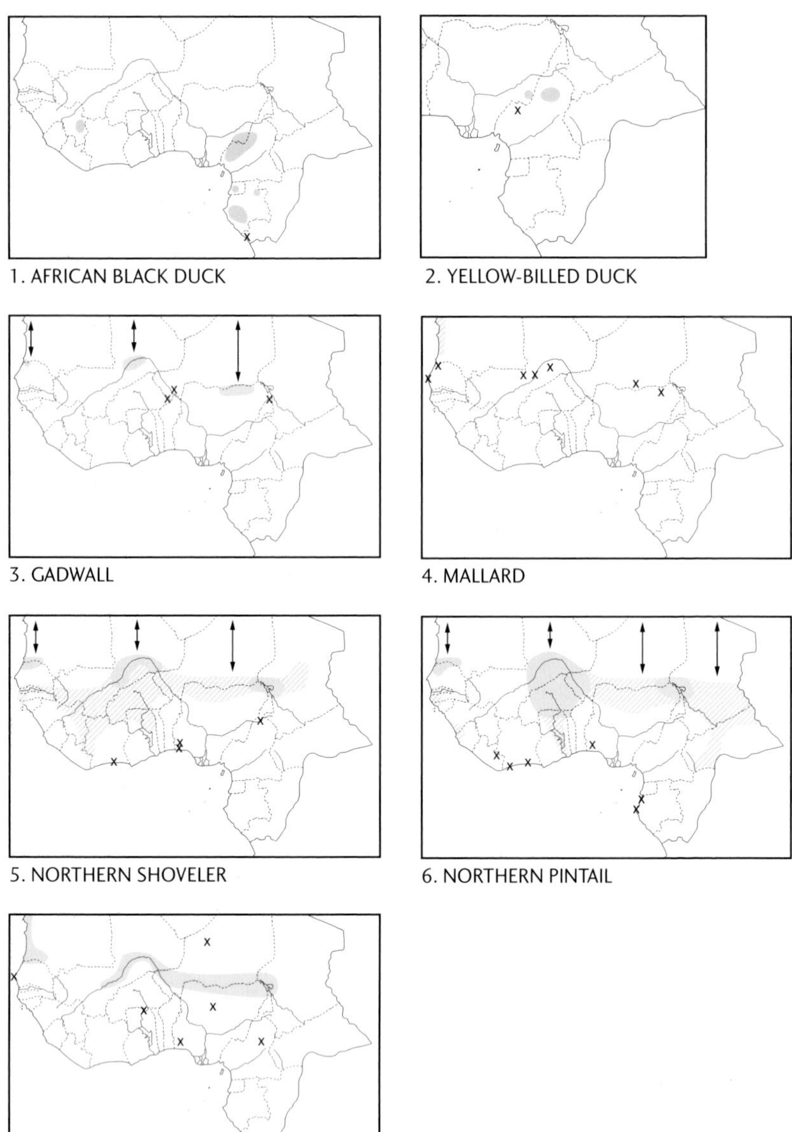

1. AFRICAN BLACK DUCK

2. YELLOW-BILLED DUCK

3. GADWALL

4. MALLARD

5. NORTHERN SHOVELER

6. NORTHERN PINTAIL

7. EURASIAN WIGEON

Plate on page 70

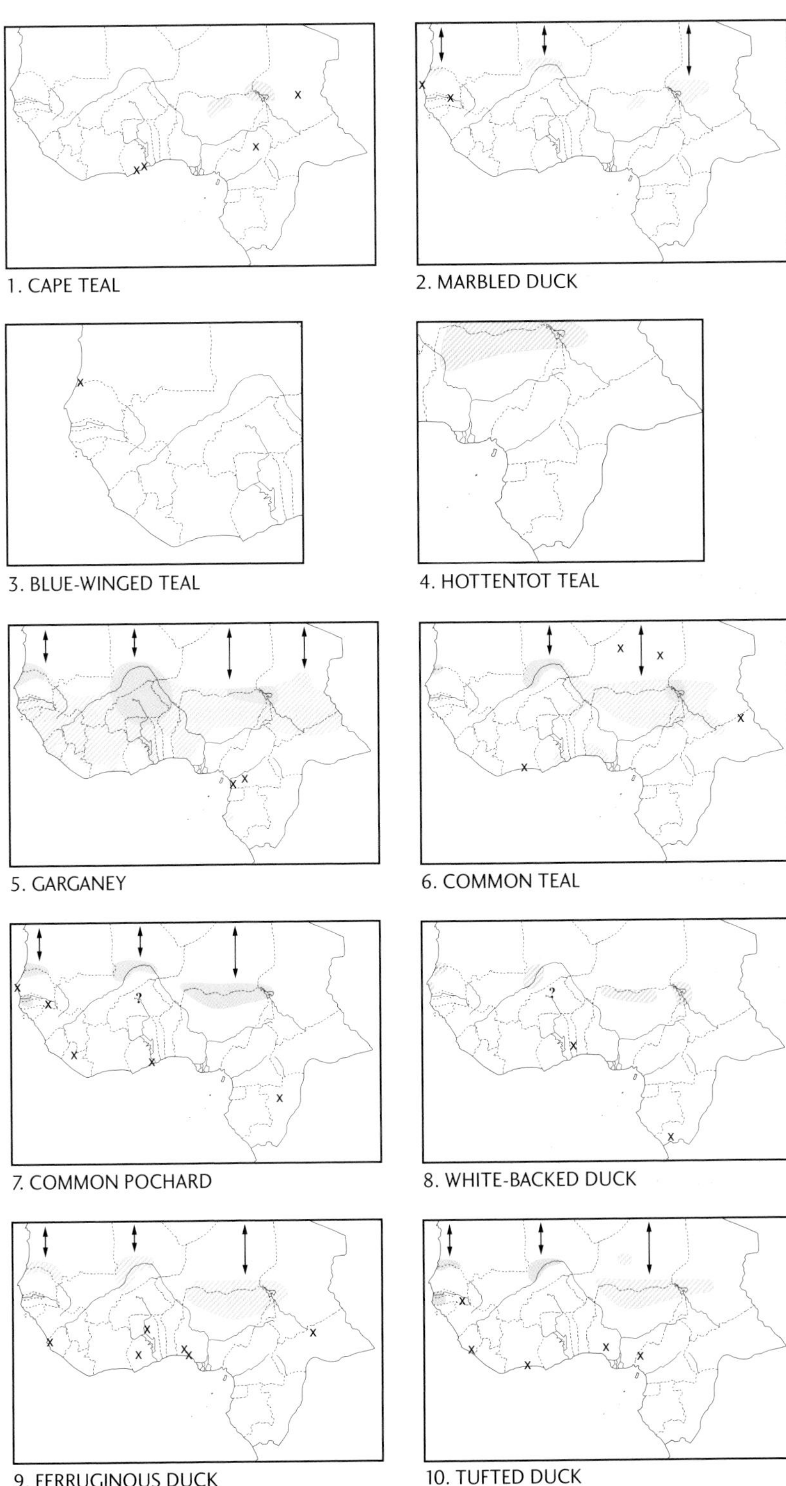

1. CAPE TEAL

2. MARBLED DUCK

3. BLUE-WINGED TEAL

4. HOTTENTOT TEAL

5. GARGANEY

6. COMMON TEAL

7. COMMON POCHARD

8. WHITE-BACKED DUCK

9. FERRUGINOUS DUCK

10. TUFTED DUCK

Plate on page 74

1 CAPE TEAL *Anas capensis* 44–48 cm R/M/V s/u
1a Very pale; pink bill. **1b In flight** Green speculum broadly bordered white. ▲ Freshwater wetlands. ✤ Usually silent. [CD5:20b]

2 MARBLED DUCK *Marmaronetta angustirostris* 39–42 cm P/V+ lf VU
2a Pale grey-brown; dark eye patch; large buffish spots on upperparts. **2b In flight** Pale secondaries; no speculum nor wingbars; white underwing. ▲ Freshwater and coastal wetlands. ✤ Mostly silent. [CD1:56]

3 BLUE-WINGED TEAL *Anas discors* 37–41 cm V+
3a Ad male Dull violet-blue head with large white crescent. **3b Ad female** Head pattern as 5b but more indistinct; prominent pale loral spot. **3c Ad male in flight** Blue forewing; no white trailing edge. **3d Ad female in flight** As male but duller. ▲ Freshwater and coastal wetlands. N American vagrant. ✤ Mostly silent.

4 HOTTENTOT TEAL *Anas hottentota* 30–35 cm R u
4a Very small; dark crown; buff head-sides; blue bill. **4b In flight** Green speculum; white trailing edge. ▲ Freshwater wetlands. ✤ Fast, harsh, nasal *kekekekekeh*. Male also utters metallic clicking notes in display. [CD5:22b]

5 GARGANEY *Anas querquedula* 37–41 cm P/V* c/u
5a Ad male White stripe from eye to nape; dark brown breast; pale grey flanks. **5b Ad female** Pale supercilium; dark crown and eye-stripe; whitish loral spot. **5c Ad male in flight** Pale grey-blue forewing. **5d Ad female in flight** Pale greyish forewing; white trailing edge. ▲ Freshwater and coastal wetlands. Local concentrations. ✤ Male has distinctive dry rattling call like scratching fingernail along comb; female a harsh, nasal *kheh* and soft quack. [CD1:54]

6 COMMON TEAL *Anas crecca* 34–38 cm P/V+ lc/r
6a Ad male Chestnut head; green band from eye to nape; yellow on rear end. **6b Ad female** Rather plain face; dark crown and eye-stripe; white stripe at sides of undertail-coverts. Compare 5b. **6c Ad male in flight** Green speculum bordered white (broader in front). **6d Ad female in flight** Wing pattern similar to male; forewing brown. ▲ Freshwater wetlands. ✤ Male has clear, liquid *kreek*, frequently uttered. Female gives sharp nasal quack and low growling *trrr*. [CD1:51]

7 COMMON POCHARD *Aythya ferina* 42–49 cm P/V u/r
7a Ad male Chestnut head; black breast and rear end; pale grey body. **7b Ad female** Brownish; peaked crown; sloping forehead; longish bill. **7c Ad male in flight** Pale grey band on flight feathers. **7d Ad female in flight** Grey wing band; greyish underwing and belly. ▲ Freshwater and coastal wetlands. ✤ Mainly silent. [CD1:58]

8 WHITE-BACKED DUCK *Thalassornis leuconotus* 38–40 cm R/V lu/r
8a Buff-and-brown plumage; darkish top of head; white loral patch. Note shape. **8b In flight** White lower back. ▲ Freshwater wetlands. ✤ Clear, sharp whistle *whit-wee* or *si-wee-wheet*, not unlike whistling duck, when swimming and in flight. [CD5:17b]

9 FERRUGINOUS DUCK *Aythya nyroca* 38–42 cm P/V+ u/r NT
9a Ad male Mainly dark chestnut; white undertail-coverts; white eye. **9b Ad female** Duller; eye dark. Compare 10b. **9c Ad male in flight** Broad white band on flight feathers; white belly. ▲ Freshwater and coastal wetlands. ✤ Mainly silent. Male utters soft *wheeoo*. [CD1:59]

10 TUFTED DUCK *Aythya fuligula* 40–47 cm P/V+ lc/r
10a Ad male Glossy black; white flanks; drooping crest; yellow eye. **10b Ad female** Mainly dark brown; flanks paler; short tuft; yellow eye. **10c Ad male in flight** White band on flight feathers. **10d Ad female in flight** Pattern similar to male. ▲ Freshwater and coastal wetlands. ✤ Mainly silent. Female has growling *krrr*. [CD1:60]

Maps on page 73

1a
2b
2a
3d
1b
4b
3a
3c
4a
5c
3b
5d
6b
6a
5b
6d
6c
5a
7a
7b
8b
8a
9c
10d
10c
7c
7d
10b
9b
10a
9a

PLATE 15: RARE AND VAGRANT DUCKS AND GEESE

1 **BLACK-BELLIED WHISTLING DUCK** *Dendrocygna autumnalis* 48–53 cm V
Pinkish-red bill; dark cap; whitish eye-ring. ▲ Neotropical vagrant, Gambia. ✤ Whistled *wee-cheew* or *weechew-weeweewheew*..., mainly in flight.

2 **RUDDY SHELDUCK** *Tadorna ferruginea* 61–67 cm V?
2a Ad male Plain rusty-orange; narrow blackish neck collar (indistinct or absent in eclipse). In flight, white wing-coverts, black flight feathers. **2b Ad female** No collar. ▲ Palearctic vagrant claimed from N Mauritania. ✤ Main call a loud honking *aangh*. [CD1:47]

3 **BEAN GOOSE** *Anser fabalis* 66–88 cm V
Large; brownish-grey; head and longish neck darker; legs bright orange. ▲ Palearctic vagrant, Mali. ✤ Varied, nasal, quite lively cackle *kayakak* or *kayak*. [CD1:41]

4 **GREATER WHITE-FRONTED GOOSE** *Anser albifrons* 65–78 cm V
Brownish-grey; white blaze on forehead; blackish bars on belly; pink bill. ▲ Palearctic vagrant, N Mauritania, Niger, N Nigeria. ✤ Disyllabic *kow-yoo* or *kyo-kyok* suggestive of yelping dogs.

5 **BRENT GOOSE** *Branta bernicla* 55–62 cm V
Dark; head, thick, rather short neck and bill black; white vent. ▲ Palearctic vagrant, N Mauritania, S Senegal. ✤ Low, guttural *hrot*. [CD1:44]

6 **AMERICAN WIGEON** *Anas americana* 45–56 cm V
6a Ad male Crown cream-white; dark metallic green band from eye to nape. **6b Ad female** As Eurasian Wigeon (Plate 13:7b) but head, neck and upperparts greyer. **6c Ad male in flight** White wing panel; white belly. Compare Eurasian Wigeon (Plate 12:7c). **6d Ad female in flight** Greyish underwing; white axillaries. ▲ N American vagrant, N Senegal. ✤ Similar to Eurasian Wigeon but male's whistle more throaty and disyllabic *wheeoh-woh*. [CD5:20a]

7 **LESSER SCAUP** *Aythya affinis* 38–45 cm V*
7a Ad male Peaked hindcrown; finely vermiculated grey upperparts. **7b Ad female** White ring at base of bill. Compare Tufted Duck (Plate 14:10b). **7c Ad male in flight** White band on upperside of secondaries; pale grey inner primaries; whitish underwing. **7d Ad female in flight** Pattern similar to male. ▲ N American vagrant, Cape Verde.

8 **COMMON SCOTER** *Melanitta nigra* 44–45 cm P r/s
8a Ad male Compact; black; yellow patch on knobbed black bill. **8b Ad female** Dark sooty-brown; pale cheeks and foreneck. **8c Ad male in flight** Uniformly dark; paler flight feathers. ▲ Open sea and estuaries. Mauritania.

9 **RING-NECKED DUCK** *Aythya collaris* 37–46 cm V*
High, steep forehead; peaked, rounded hindcrown; small indentation between crown and nape; bill (adult) three-coloured: slate-grey with broad black tip and white subterminal band. **9a Ad male** Like Tufted Duck (Plate 14:10a), but flanks pale grey with vertical white 'spur' at front; flank panel more S-curved; tail somewhat longer and often raised; white line at bill base. **9b Ad female** Note head shape and bill pattern. Diffuse pale patch at bill base; pale eye-ring (often extending as line behind eye); dark eye. **9c Ad male in flight** & **9d Ad female in flight** Like Tufted Duck (Plate 14:10c–d) but wingbar grey (not white). ▲ N American vagrant, Cape Verde.

9c

9d

9b

9a

Maps on page 78

1
2a
2b
3
4
5
6a
6b
6c
6d
7a
7b
7c
7d
8a
8b
8c

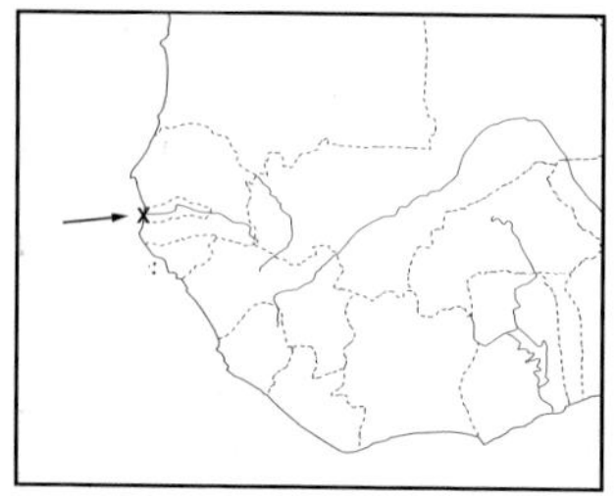

1. BLACK-BELLIED WHISTLING DUCK

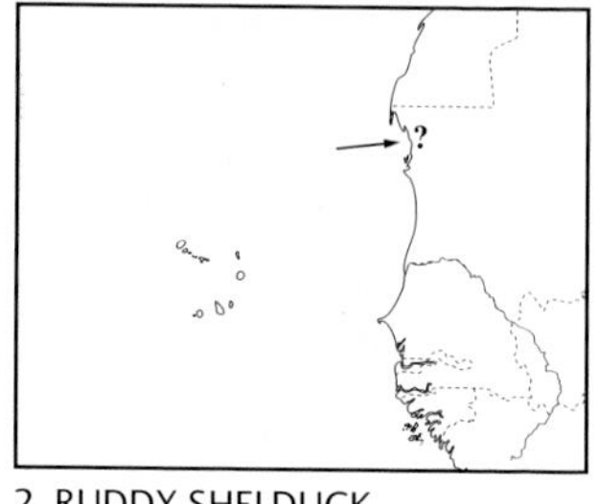

2. RUDDY SHELDUCK

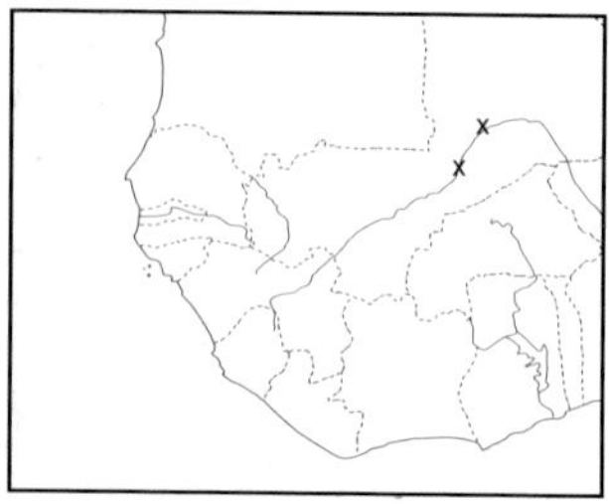

3. BEAN GOOSE

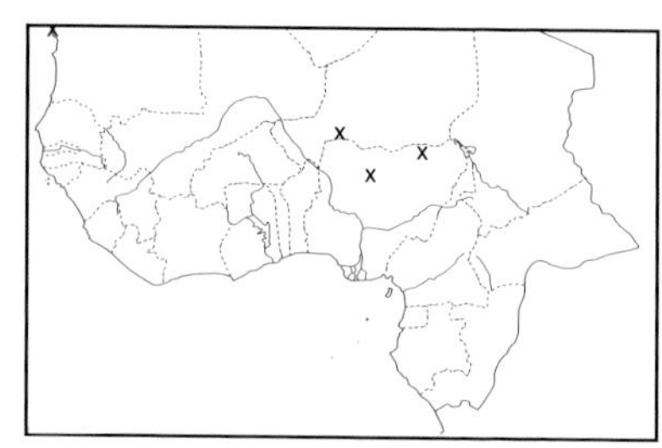

4. GREATER WHITE-FRONTED GOOSE

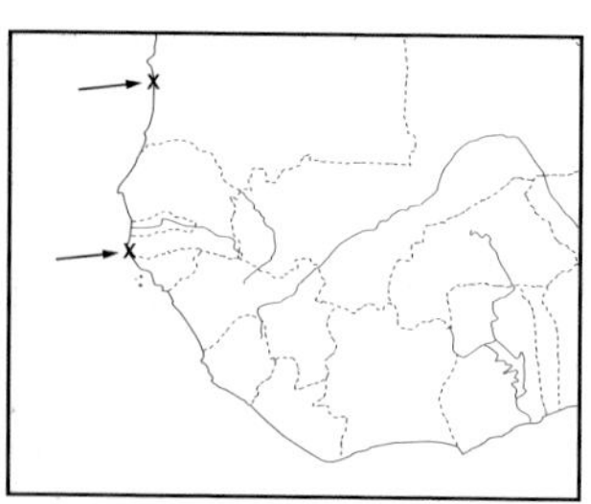

5. BRENT GOOSE

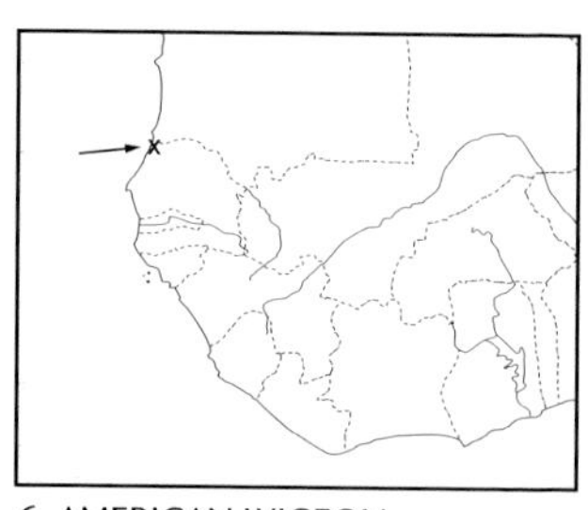

6. AMERICAN WIGEON

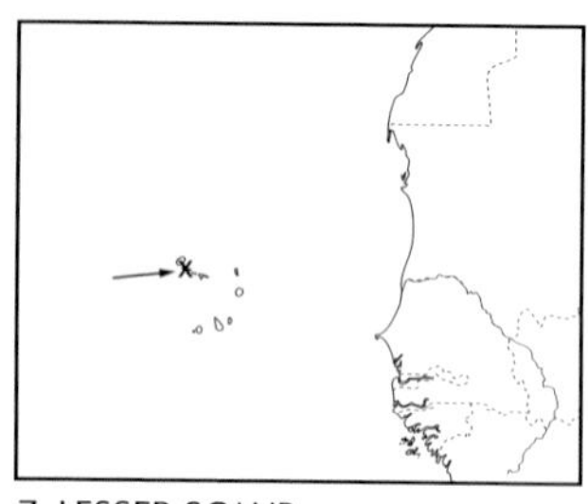

7. LESSER SCAUP

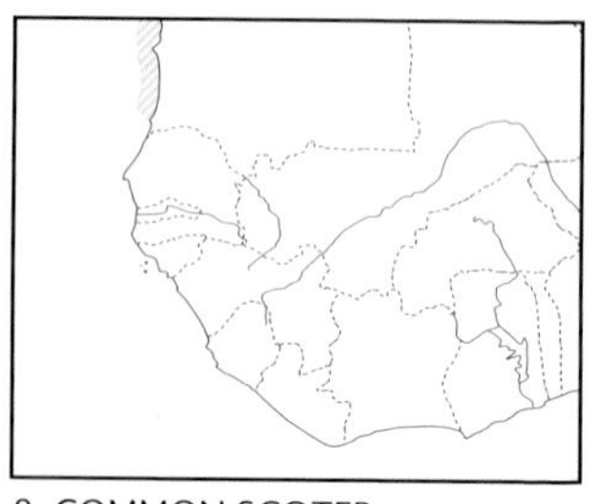

8. COMMON SCOTER

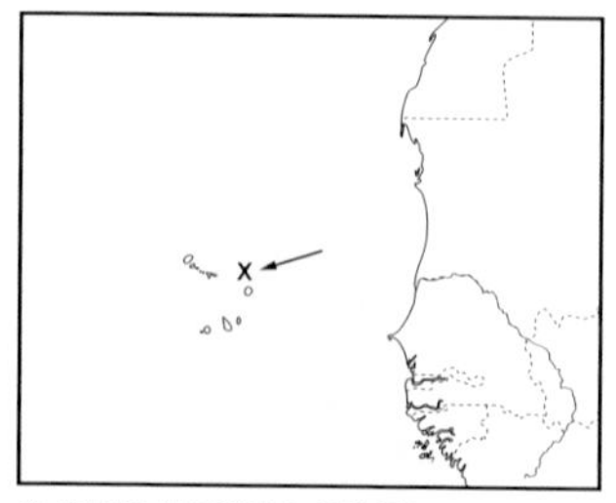

9. RING-NECKED DUCK

Plate on page 76

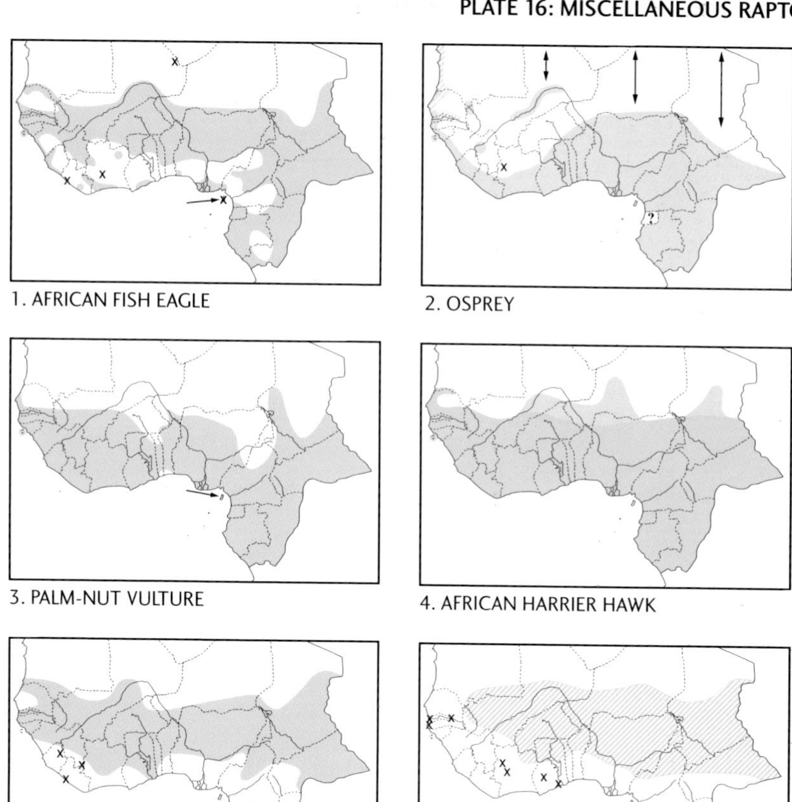

1. AFRICAN FISH EAGLE

2. OSPREY

3. PALM-NUT VULTURE

4. AFRICAN HARRIER HAWK

5. BATELEUR

6. SECRETARY BIRD

Plate on page 80

Accipitridae (plates 16–26) is a large and diverse-looking assemblage of diurnal raptors. Its members characteristically have hooked bills and powerful talons. Exhibit a wide variety of wing shapes, tail shapes and flight actions, according to habitat and hunting techniques. Sexes more or less alike; females almost always larger. Normal flight consists of regular wingbeats alternated with gliding. Almost all species carnivorous. Conspicuous aerial displays used by many. Identification often problematic; correct assessment of jizz (combination of structure, proportions, wing attitudes, and flight actions) essential.

1 AFRICAN FISH EAGLE *Haliaeetus vocifer* 63–73 cm; WS 190–240 cm R r/lc
1a Ad White head, mantle and breast; chestnut lower underparts. **1b Juv** Mainly mottled brown; whitish breast streaked brown. **1c Ad in flight** White head and breast; white tail; chestnut underwing-coverts and belly. Soars on broad flat wings. **1d Juv in flight** Pale 'window' in primaries; whitish tail with black terminal band. ▲ Large rivers, lagoons, lakes, reservoirs. ✤ Vocal. Far-carrying, ringing *WEEE-AH kleeuw kleeuw kleew*, uttered by both sexes at rest and in flight; male higher pitched than female. Often duets. [CD5:29]

2 OSPREY *Pandion haliaetus* 52–61 cm; WS 145–173 cm P/R* u/lc
2a Ad Broad dark stripe through eye; short nuchal crest. **2b In flight** Long, narrow wings with long 'hand'; black carpal patches. ▲ Large rivers, lagoons, lakes, etc. Exclusively preys on fish caught by plunge-diving. ✤ Usually silent in Africa. [CD1:69]

3 PALM-NUT VULTURE *Gypohierax angolensis* *c.* 60 cm; WS 140–150 cm R f/lc
3a Ad Black-and-white plumage; bare reddish-pink face; heavy yellowish-horn bill. **3b Juv** Drab brown; dull yellowish face; darkish bill; whitish legs. **3c–d Ad in flight** Black secondaries, wingtips and tail base. **3e Juv in flight** Darkish; broad rounded wings; short rounded tail. ▲ Various habitats, often near water. Relishes nuts of oil and raphia palms; also feeds on various small animals and scavenges. ✤ Usually silent. [CD5:30]

4 AFRICAN HARRIER HAWK *Polyboroides typus* 60–68 cm; WS *c.* 160 cm R c
4a Ad Mainly grey; small head with bare, yellow to orange face; finely barred underparts; long bare legs. **4b Juv/imm** Dark brown to tawny, variably mottled, streaked and barred. **4c Juv** Dark brown individual. **4d Ad in flight from above** Broad wings with broad black trailing edge and tips; black tail with white bar. **4e Ad in flight from below** Narrow white line along black trailing edge and tips. **4f Juv in flight** Remiges and rectrices more or less distinctly barred; greater coverts tipped dark. ▲ Forest, wooded savanna. Flight with slow, buoyant wingbeats alternated with glides; frequently soars. Undulating display flight. ✤ Plaintive whistling *sueeeee* and high-pitched *hueeeup-hueeeup-hueeeup-*... [CD5:40]

5 BATELEUR *Terathopius ecaudatus* 55–70 cm; WS 170–187 cm R u/lc
5a Ad male Jet-black head and body; chestnut back; grey shoulders; red face and feet. **5b Ad male** Cream-backed form (uncommon). **5c Ad female** As male but with grey panel on secondaries. **5d Juv** Entirely brown. Note large head, long wings, invisible tail, dark eye. **5e Ad male in flight from above** Black, bow-shaped wings; grey shoulders; chestnut back. **5f Ad male in flight from below** Black secondaries contrast with white coverts. **5g Ad female in flight** Black trailing edge to largely white underwing. **5h Juv in flight** All brown; unique silhouette similar to adult, but tail slightly longer. ▲ Savanna. Soars with wings held in marked V, frequently canting. Rarely flaps once airborne. ✤ Far-carrying, barking *kow-aw* and soft *ko-ko-ko-ko-koaaagh*. Usually silent. [CD5:38]

6 SECRETARY BIRD *Sagittarius serpentarius* 125–150 cm; WS 212 cm R/M r/u
6a Ad Unmistakable. Very long legs; plume-like feathers on nape; bare reddish face. Juvenile has facial skin yellow; central tail feathers shorter. **6b In flight** Black remiges contrast with pale grey wing-coverts; very long central rectrices project beyond long legs. ▲ Open savanna. Terrestrial. Strides slowly across grassland, searching for prey. Also soars high, like vulture. ✤ Usually silent. [CD5:66]

 Maps on page 79

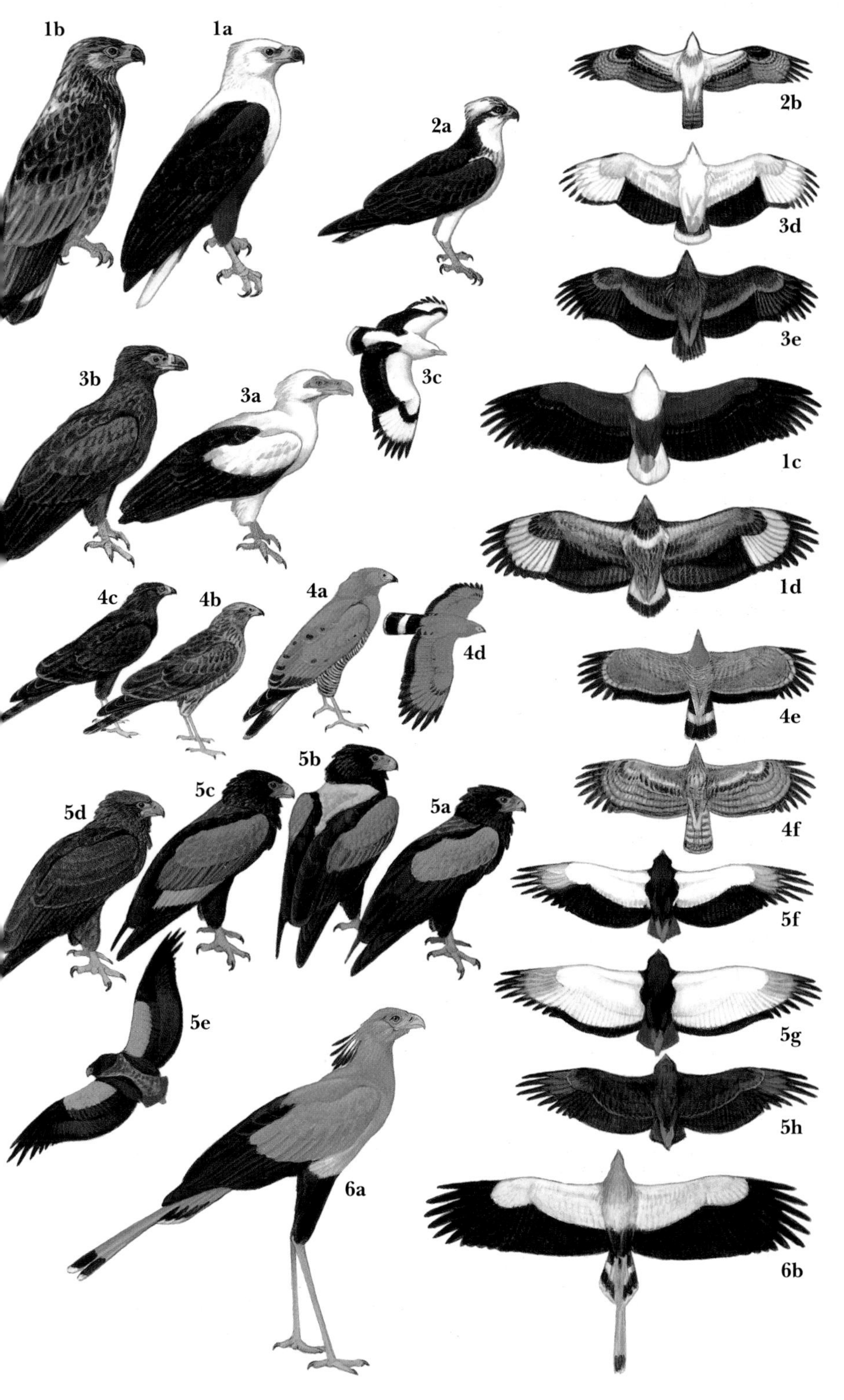

1b
1a
2a
2b
3d
3e
3b
3a
3c
1c
1d
4c
4b
4a
4d
4e
4f
5b
5c
5d
5a
5f
5g
5e
5h
6a
6b

Medium-sized to huge with long, broad, strongly 'fingered' wings, and usually unfeathered head and neck. Larger species have long neck with ruff at base, and short tail. Carrion eaters. Given to soaring, often at great height. In flight head is tucked between shoulders and appears relatively small. Usually silent.

1 HOODED VULTURE *Necrosyrtes monachus* 65–75 cm; WS 170–182 cm R lc
1a Ad Dark brown; bare face and neck pink (red in excitement); long, slender bill. **1b Juv** Bare face and neck pale bluish. **1c Ad in flight** Uniformly dark; broad wings; short, rounded tail; slim bill. ▲ Various habitats in Sahel and savanna belts; patchily in forest zone. Often near habitation, scavenging on offal and refuse. ✤ [CD5:31]

2 EGYPTIAN VULTURE *Neophron percnopterus* 55–75 cm; WS 155–175 cm R+/P u/f
2a Ad Mainly white; bare yellow face; long, narrow bill; long feathers at back of head. **2b Juv** Dark brown; greyish face and legs. **2c Ad in flight** White, wedge-shaped tail; black remiges contrast with white coverts. **2d Juv in flight** Distinctive silhouette similar to adult. ▲ Desert, arid savanna. Soars with wings held flat; when gliding, wings slightly arched. ✤ [CD1:76]

3 WHITE-HEADED VULTURE *Trigonoceps occipitalis* 78–85 cm; WS 202–230 cm R u
3a Ad female White to pink face and neck; red bill; blue cere; peaked hindcrown; white inner secondaries (dark in male). **3b Juv** Mainly dark brown; bare parts duller. **3c Ad male in flight** White line on greater coverts; white belly. **3d Ad female in flight** As male but inner secondaries white. Juvenile in flight entirely dark with contrasting white band on underwing as adult. ▲ Dry to lightly wooded savanna.

4 AFRICAN WHITE-BACKED VULTURE *Gyps africanus* 80–98 cm; WS 212–218 cm R u
4a Ad Very large; mainly brown; bare blackish face; black bill; white back and rump. **4b Juv** Darker and streaky; brownish ruff; no white on back and rump. **4c Ad in flight** Black remiges contrast with whitish coverts. From above: white back. **4d Juv in flight** Dark brown; white line along leading edge of wing; no white on back. ▲ Lightly wooded savanna. Soars with wings held in shallow V. ✤ [CD5:32]

5 RÜPPELL'S GRIFFON VULTURE *Gyps rueppellii* 85–107 cm; WS 220–250 cm R/M f
5a Ad Grey-brown feathers broadly tipped buffish; yellowish-horn bill; pale yellow eye. **5b Juv** Plain above, streaky below; blackish bill; dark eye. **5c Ad in flight** Dark; white bar parallel to leading edge of wing; two white lines on coverts. **5d Juv in flight** Very similar to (and often indistinguishable from) 4d when very young. ▲ Dry, open savannas. Also open habitats in mountainous areas, Cameroon. ✤ [CD5:33]

6 EURASIAN GRIFFON VULTURE *Gyps fulvus* 95–110 cm; WS 230–280 cm P u/r
6a Ad Huge; mainly sandy-brown; whitish ruff; yellowish bill and eye. **6b Juv** Slightly darker and streaky; pale brownish ruff; dark bill and eye. **6c Ad in flight** Blackish remiges; pale brownish underwing-coverts with 1–2 pale lines. **6d Juv in flight** Underwing-coverts paler and plainer than adult. ▲ Sahel, desert. Soars with wings held in shallow V. ✤ [CD1:77]

7 LAPPET-FACED VULTURE *Torgos tracheliotus* 98–115 cm; WS 255–290 cm R u VU
7a Ad Huge; mainly blackish; underparts streaked white; bare pink head and neck; massive yellowish-horn bill. **7b Juv** Duller and more uniformly darkish. **7c Ad in flight** Distinct white bar along leading edge of wing; white thighs. **7d Juv in flight** Wholly dark. Compare distant birds with 1c. ▲ Dry country. Soars with wings held flat or slightly arched. ✤ [CD1:79]

 Maps on page 84

1b
1a
2b
2a
1c
2c
2d
3a
3b
3d
3c
4c
4b
4a
4d
5c
5d
5b
5a
6c
6d
7c
7d
7a
7b
6a
6b

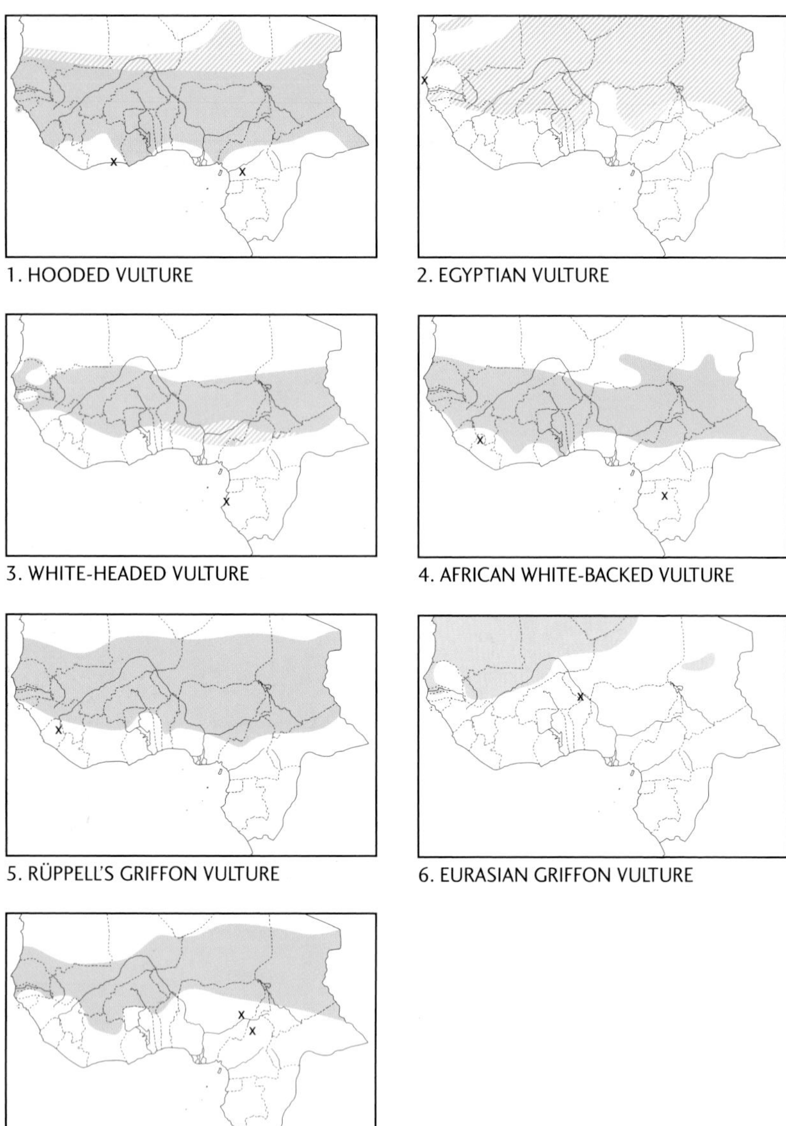

1. HOODED VULTURE

2. EGYPTIAN VULTURE

3. WHITE-HEADED VULTURE

4. AFRICAN WHITE-BACKED VULTURE

5. RÜPPELL'S GRIFFON VULTURE

6. EURASIAN GRIFFON VULTURE

7. LAPPET-FACED VULTURE

Plate on page 82

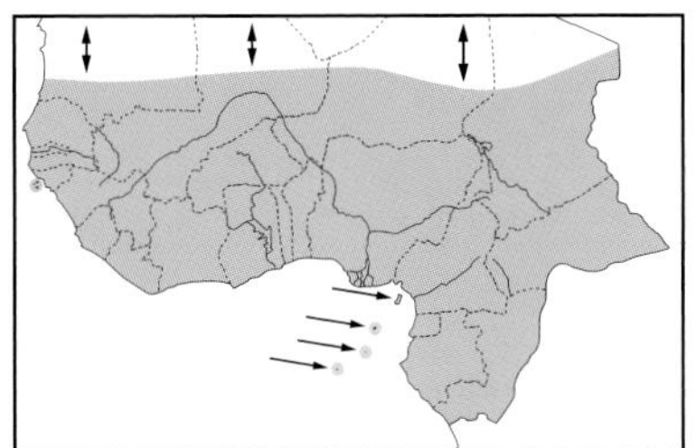
1. BLACK KITE

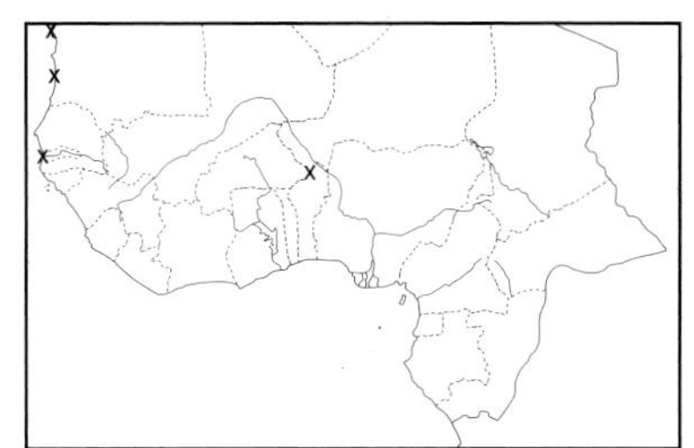
2. RED KITE

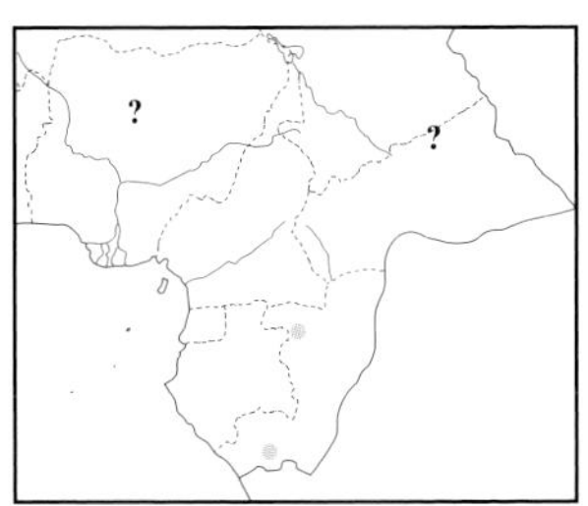
3. AFRICAN MARSH HARRIER

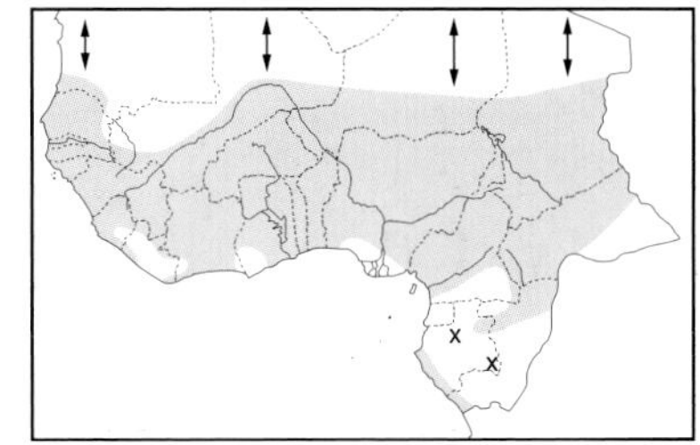
4. EURASIAN MARSH HARRIER

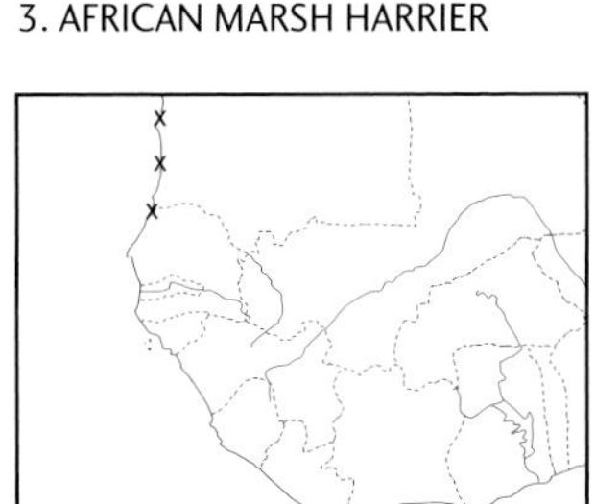
5. HEN HARRIER

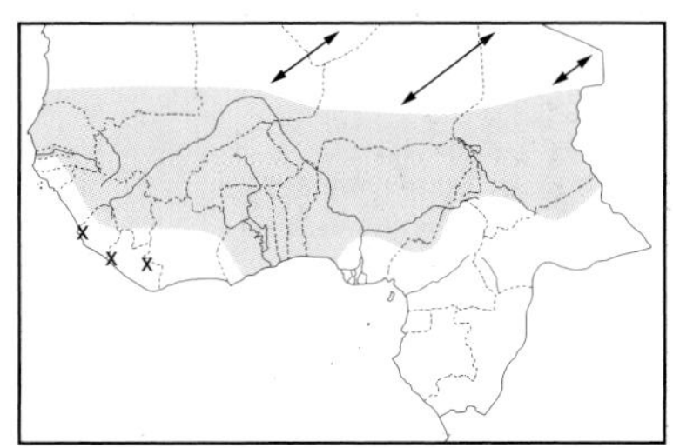
6. PALLID HARRIER

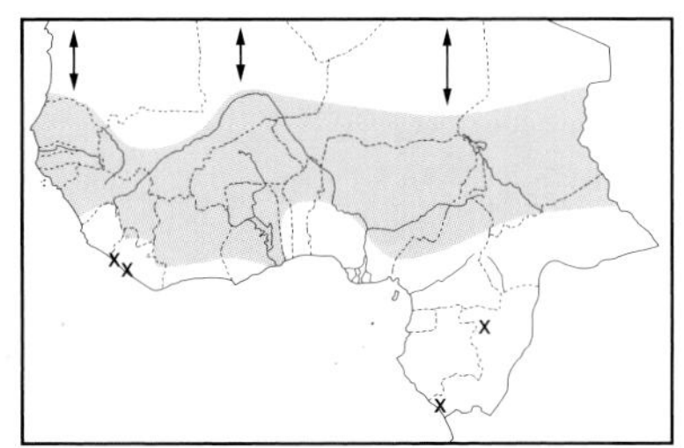
7. MONTAGU'S HARRIER

Plate on page 86

Kites. Medium-sized raptors with graceful, buoyant flight and, in most species, forked tail.

1 BLACK KITE *Milvus migrans* 50–60 cm; WS 130–155 cm **M/P/R* c**
1a Ad *parasitus* (Yellow-billed Kite; R/M) Entirely chocolate-brown; yellow bill; long, slightly forked tail. **1b Ad *migrans*** (P) Pale greyish head streaked black; black bill. **1c Juv *parasitus*** Dark bill. **1d Ad *parasitus* in flight** Long, narrow wings angled back at carpal joint; constantly twisting tail (sometimes appearing square or only slightly notched). Nominate has more contrasting wing. ▲ Various habitats. Commonest African raptor. ✤ Plaintive, tremulous *keeeey-aarrrr*. [CD1:72; 5:28]

2 RED KITE *Milvus milvus* 55–72 cm; WS 140–180 cm **V NT**
2a Ad Mainly rufous streaked black; pale greyish head; long, deeply forked reddish tail. **2b Ad in flight** Large whitish 'window' on primaries below. ▲ Palearctic vagrant. ✤ Usually silent. [CD1:73]

CAPE VERDE KITE *Milvus (milvus) fasciicauda* See Plate 145:8.

Harriers. Medium-sized and slender with long wings and tails. Females mainly brownish. Flight buoyant with wings characteristically held in shallow V. Hunt low, quartering the ground and dropping on prey. Frequent open country, cultivated areas and marshes. Usually silent.

3 AFRICAN MARSH HARRIER *Circus ranivorus* 44–50 cm; WS 110 cm **R u**
3a Ad Mainly brown variably streaked darker; lower underparts more rufous. **3b Juv** Darker and plainer; variable whitish areas on nape, throat and breast. **3c Ad in flight** Whitish leading edge to wing; barred tail; no white on uppertail-coverts. **3d Juv in flight** Dark brown with irregular whitish breast-band. ▲ Open habitats. ✤ [CD5:41]

4 EURASIAN MARSH HARRIER *Circus aeruginosus* 42–56 cm; WS 110–140 cm **P+ f/lc**
4a Ad male Mainly brown, with buff-brown head and breast streaked darker. **4b Ad female** Mainly dark brown; creamy crown, nape and throat. **4c Ad male in flight** Tricoloured wings with large black tips; plain, blue-grey tail. **4d Ad female in flight** Leading edge of wing usually creamy; often pale patch on breast. **4e Ad male dark morph** Blackish-brown; whitish band on underwing. Rare. ▲ Open habitats. ✤ [CD1:84]

5 HEN HARRIER *Circus cyaneus* 43–55 cm; WS 97–121 cm **P?**
5a Ad male Mainly blue-grey; lower underparts white. **5b Ad female** Buffish underparts heavly streaked brown. **5c Juv** As female; underparts usually darker and washed rufous. **5d Ad male in flight** Large black wingtips; white uppertail-coverts; dusky trailing edge to underwing. **5e Ad female in flight** White uppertail-coverts; three well-defined dark bands on underwing. ▲ Open habitats. Claimed from Mauritania (rare). ✤ [CD1:81]

6 PALLID HARRIER *Circus macrourus* 40–50 cm; WS 95–120 cm **P u/s NT**
6a Ad male Pale grey above, mainly white below. **6b Ad female** Narrow pale collar; less white above eye than 7b. **6c Juv** Plain rufous below; black cheek patch reaching base of bill; pale collar. **6d Ad male in flight** Very pale; black wedge at wingtips. **6e Ad female in flight** White uppertail-coverts; upper- and inner underwing darker than 7e. **6f Juv in flight** Lacks dark tips to primaries below (present in 7f). ▲ Open habitats. ✤ [CD1:82]

7 MONTAGU'S HARRIER *Circus pygargus* 38–50 cm; WS 96–120 cm **P/V+ s/lf**
7a Ad male Mainly blue-grey; belly white variably streaked chestnut. **7b Ad female** Facial pattern less distinct than 6b; more white around eye. **7c Juv** Usually deeper rufous below than 6c; cheek patch smaller; no pale collar. **7d Ad male in flight** Wings with black tips and one black bar above, two below. **7e Ad female in flight** Dark bar on base of secondaries above; underwing more uniform than 6e, with three well-defined dark bands and broad pale band along dark trailing edge. **7f Juv in flight** Dark tips to primaries. **7g Ad male dark morph** Blackish; tail greyish. Rare. **7h Ad female dark morph** Blackish-brown; pale greyish on underwing; tail banded. Rare. ▲ Open habitats. ✤ [CD1:83]

1c
1b
1a
2a
1d
2b
3b
3a
4b
4a
4e
5c
5b
5a
3d
3c
4d
4c
6c
6b
6a
5e
5d
7h
7c
7b
6e
6d
6f
7e
7g
7a
7f
7d

PLATE 19: SNAKE EAGLES

Rather large and eagle-like, with relatively large heads and large yellow eyes (producing somewhat owl-like appearance), long, broad wings, bare tarsi and short toes. Given to soaring and perching for long periods. Drop on their prey (mainly snakes and other reptiles) and kill it on ground with their powerful feet. Occur in open country and woodland.

1 SHORT-TOED SNAKE EAGLE *Circaetus gallicus* 59–69 cm; WS 162–195 cm **P u**
1a Ad male (typical) As female but breast-band broken. Some have brown parts darker, creating dark hood, and more distinct markings below; a few are very pale, without dark hood and almost lacking markings below. **1b Ad female (typical)** Grey-brown above; upper breast mottled brown; white underparts irregularly blotched and barred brown. **1c Juv** Head and breast more rufous-brown. **1d Ad female (typical) in flight** White underwing-coverts with dark lines. **1e Ad/juv (pale) in flight** Underwing pattern more indistinct. ▲ Open savanna woodland. Frequently hovers or hangs motionless in wind. ✤ Usually silent in Africa. [CD1:80]

2 BEAUDOUIN'S SNAKE EAGLE *Circaetus beaudouini* 62–69 cm; WS c. 170 cm **R u**
2a Ad male As female but more white on breast. **2b Ad female** Darker brown than 1b; white underparts with narrower bars. **2c Juv** Dark brown above, more rufous-brown below. **2d Ad female in flight** Underwing-coverts white. **2e Juv in flight** Brownish underwing-coverts; indistinct dusky bars on remiges. Compare 4c. **2f Ad male hovering.** ▲ Open savanna woodland. Frequently hovers or hangs motionless in wind. ✤ Clear *kee-u*, uttered in series. Usually silent. [CD5:34]

3 BLACK-BREASTED SNAKE EAGLE *Circaetus pectoralis* 63–68 cm; WS 178 cm **M r**
3a Ad Blackish above and on breast, pure white below. **3b Imm** Underparts with bars and blotches. **3c Juv** Entirely rufous-brown, darker on wings. **3d Ad in flight** White underwing; dark bars on remiges. **3e Juv in flight** Rufous-brown underwing-coverts; indistinct dusky bars on remiges. ▲ Open savanna woodland. Frequently hovers or hangs motionless in wind. ✤ Usually silent. [CD5:35]

4 BROWN SNAKE EAGLE *Circaetus cinereus* 66–75 cm, WS 200 cm **R u**
4a Ad Entirely dark brown. Compare 2c and 3c. **4b Imm** As adult or streaked white on head; may have white feathers below. **4c Ad in flight** Unmarked whitish remiges; dark brown body and underwing-coverts. ▲ Savanna woodland. Perches very upright and for long periods in prominent vantage point. Rarely hovers or hunts in flight. ✤ Usually silent. Guttural *khok-khok-khok-khaw* in soaring display. [CD5:36]

5 WESTERN BANDED SNAKE EAGLE *Circaetus cinerascens* 55–60 cm; WS 114 cm **R u**
5a Ad male Grey-brown; broad white band on tail. **5b Ad female** As male or all brown as illustrated. **5c Juv** Whitish head and underparts. **5d Ad in flight** Underwing white; four dark bars on remiges; distinctive tail pattern. **5e Juv in flight** Body whitish; wing and tail pattern as adult but duller. ▲ Savanna woodland. Perches unobtrusively for long periods on large tree, often in riverine woodland. ✤ Vocal. Loud *kho-kho-kho-kho-*... descending in pitch, in aerial display or from perch. Also a plaintive, nasal *ko-ah ko-ah*... and *ko-ah ko-koaaah* from perch. [CD5:37]

Maps on page 90

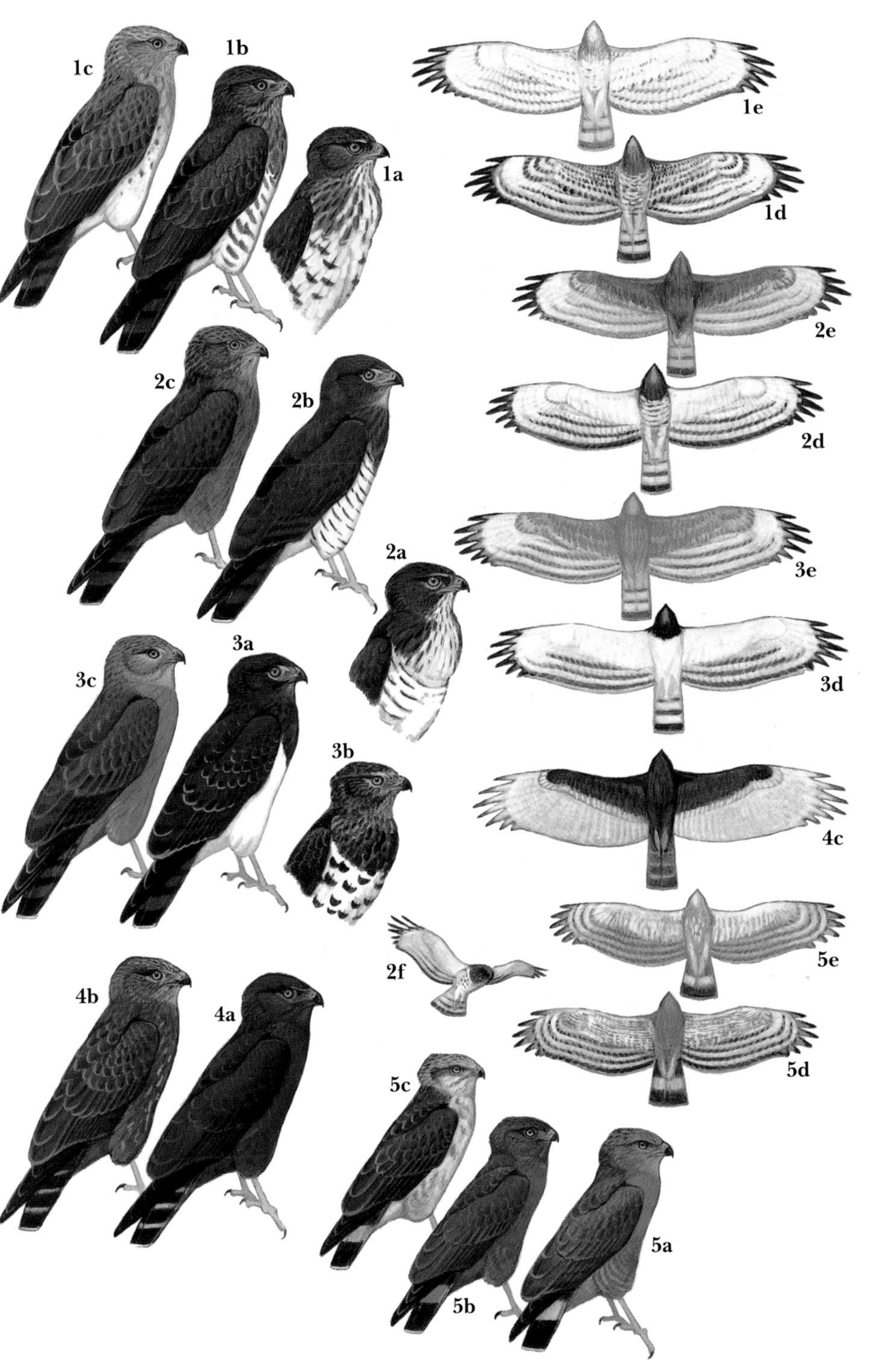
1c
1b
1e
1a
1d
2e
2c
2b
2d
2a
3e
3a
3c
3d
3b
4c
5e
2f
4b
4a
5d
5c
5a
5b

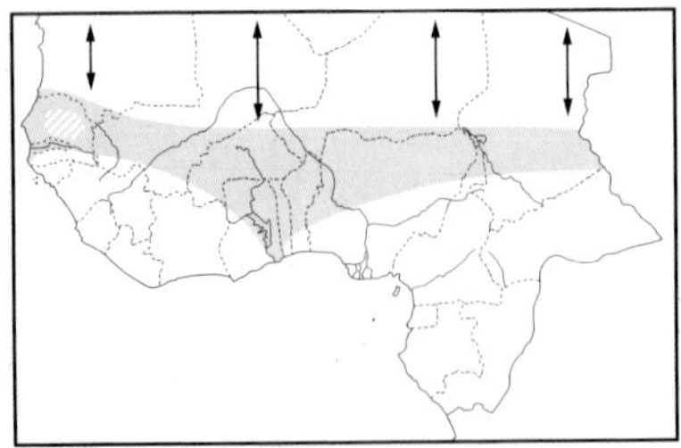

1. SHORT-TOED SNAKE EAGLE

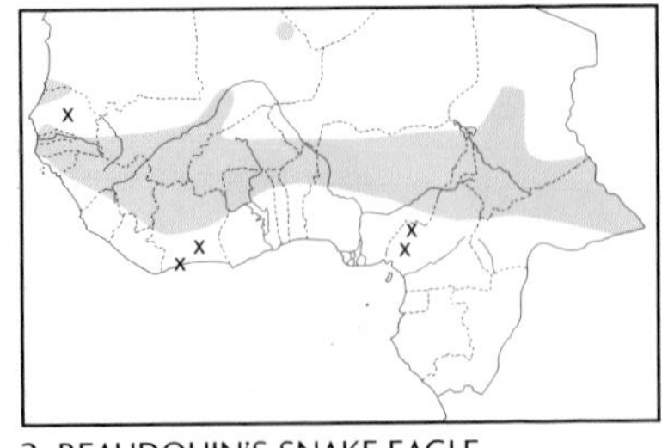

2. BEAUDOUIN'S SNAKE EAGLE

3. BLACK-BREASTED SNAKE EAGLE

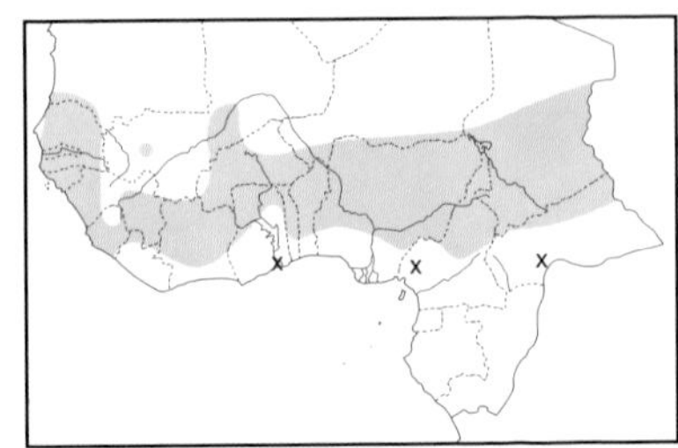

4. BROWN SNAKE EAGLE

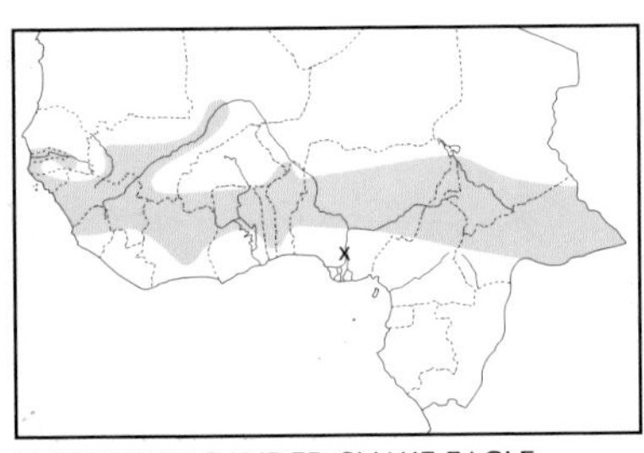

5. WESTERN BANDED SNAKE EAGLE

Plate on page 88

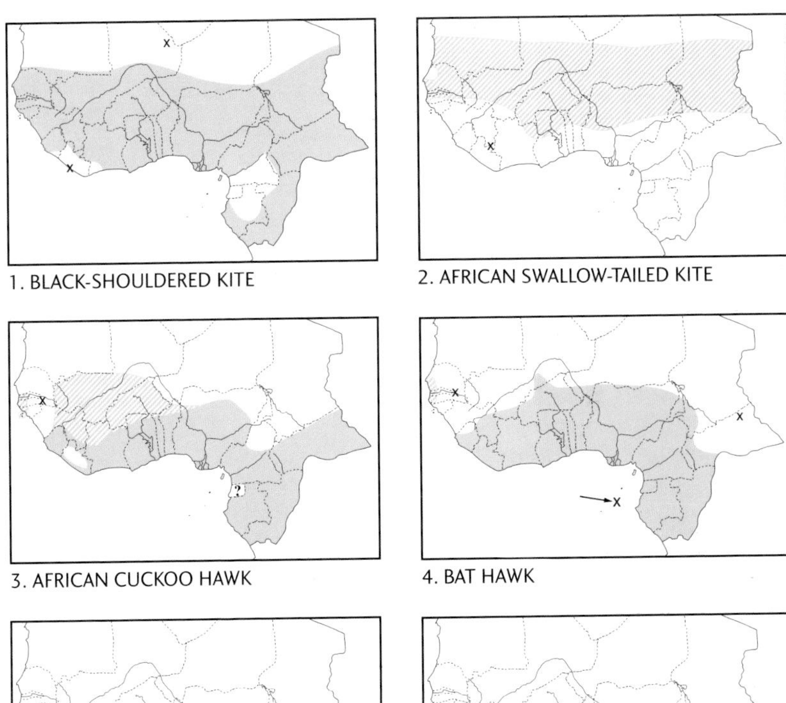

1. BLACK-SHOULDERED KITE

2. AFRICAN SWALLOW-TAILED KITE

3. AFRICAN CUCKOO HAWK

4. BAT HAWK

5. CONGO SERPENT EAGLE

6. LONG-TAILED HAWK

Plate on page 92

PLATE 20: SMALL KITES, ATYPICAL HAWKS AND SERPENT EAGLE

1 BLACK-SHOULDERED KITE *Elanus caeruleus* 31–35 cm; WS 75–85 cm **R c/u**
1a Ad Smallish; blue-grey above; dark eye patch; black shoulders; white below. **1b Juv** Upperparts tinged brownish, feathers edged white; rufous wash on breast. **1c–d Ad in flight** Long, pointed, broad-based wings; short tail; black shoulders; all white below with black primaries. ▲ Various open habitats. Hovers frequently. ✤ Mostly silent. [CD1:71]

2 AFRICAN SWALLOW-TAILED KITE *Chelictinia riocourii* 35–38 cm; WS 90 cm **M u/c**
2a Ad Smallish; grey and white; long, deeply forked tail. **2b Juv** Short tail; feathers of upperparts edged rufous; pale rufous wash on breast. **2c Ad in flight** Slender, tern-like; long pointed wings with black carpal patches. **2d Juv in flight** Almost square tail. ▲ Dry, open country. Often in small flocks. ✤ Usually silent. [CD5:27]

3 AFRICAN CUCKOO HAWK *Aviceda cuculoides* 40 cm; WS 91 cm **R u/f**
3a Ad Dark grey above and on breast; broad chestnut bars on white underparts. **3b Juv** Dark brown above; whitish supercilium; white below with brown spots. **3c Ad in flight** Chestnut underwing-coverts; longish wings and tail. **3d Juv in flight** Buffish underwing-coverts; remiges barred like adult. ▲ Woodland, forest clearings. Usually unobtrusive. Pairs perform undulating and tumbling display flights, with much calling. ✤ Usually silent. Loud, plaintive *peee-uuw*; also sharp *wheet-wheet*. [CD5:25]

4 BAT HAWK *Macheiramphus alcinus* 45 cm; WS 110 cm **R lu**
4a Ad Blackish-brown; dark stripe down white throat; short occipital crest. **4b Juv** Underparts with variable amount of white. **4c Ad in flight** Long, broad and pointed wings; longish, square tail. ▲ Forest edge, wooded savanna; often near rivers. Crepuscular. ✤ Rather hoarse, high-pitched *kwheet-kwheet-wheet-wheet-*... and *kwik-kwikik-kwik-kwikik-*... [CD5:26]

5 CONGO SERPENT EAGLE *Dryotriorchis spectabilis* 51 cm; WS 85–96 cm **R s/f**
5a Ad *spectabilis* (Upper Guinea forest) Rather large head; long tail; bare legs; dark throat-stripe, bold spots on breast, more bar-like markings on belly. **5b Ad *batesi*** (Lower Guinea forest) Markings on underparts restricted to flanks. Compare adult Cassin's Hawk Eagle (Plate 26:4a). **5c Juv** Head mottled whitish; underparts white boldly spotted blackish. **5d Ad *spectabilis* in flight** Short wings with barred remiges; long, rounded tail. **5e Ad *batesi* in flight** Whiter below. **5f Juv in flight** Pale head. ▲ Forest interior and edges. ✤ Vocal. Series of nasal *kow* or *klow* (=*klah*) notes. Also a drawn-out, plaintive *kloooooow*. [CD5:39]

6 LONG-TAILED HAWK *Urotriorchis macrourus* 57–73 cm; WS 90 cm **R s/f**
6a Ad typical form Slate grey above, mainly chestnut below, long black tail. **6b Ad grey form** Underparts grey. Rare. **6c Juv** Brown above; white variably marked dark brown below. **6d–e Ad in flight** Very long tail tipped and barred white; white uppertail-coverts; chestnut underwing-coverts. **6f Juv in flight** Long tail distinctive. ▲ Forest interior; occasionally near tracks and clearings. Mainly at mid-levels, also lower. ✤ Two drawn-out, plaintive whistles *teeu-ieeew teeu-ieeeew*, the second slightly higher pitched, with variations. Also soft *klee-klee-klee-klee-*.... [CD5:52]

Maps on page 91

1b
1a
2b
2a
1c
1d
2d
2c
3b
3a
4b
4a
4c
3d
3c
5c
5a
5b
5d
5f
5e
6c
6b
6a
6f
6e
6d

PLATE 21: SPARROWHAWKS, GOSHAWKS AND LIZARD BUZZARD

1 SHIKRA *Accipiter badius* 28–30 cm; WS 58–60 cm **R/M c**
1a Ad Small; blue-grey above, white below with fine rufous barring; red/orange eye. **1b Juv** Brown above; dark throat-stripe; blotched and barred below; tail barred. **1c–d Ad in flight** Plain grey above (no white on rump or tail); very pale below; wingtips dusky. **1e Juv in flight from below** Indistinctly barred on wings and tail. ▲ Savanna, woodland and other open habitats. Soars frequently. Southward movement in dry season. ✤ Vocal. Fast, sharp *ki-ki-ki-ki-*... Also *kiwik-kiwik-kiwik-kiwik-*... (male), and plaintive *keeu-keeu-keeu-*... (female). [CD5:46]

2 GABAR GOSHAWK *Micronisus gabar* 28–36 cm; WS 60 cm **R f/s**
2a Ad Smallish; grey breast; cere and legs pinkish-red. 7a is much larger. **2b Ad melanistic form** All black; tail faintly barred white. **2c Juv** Pale, streaky head; cere and legs yellowish to orange. **2d–e Ad in flight** White uppertail-coverts; narrowly barred underwing; four black bars on tail. **2f Ad melanistic form in flight** White barring on primaries above. **2g Juv in flight** Underwing-coverts narrowly barred rufous-brown. ▲ Wooded savanna, thornbush. ✤ Rapid, high-pitched *kik-kik-kik-*... or *kwk-kwik-kwik-*...; also more melodious *kwee-kwee-kwee-*... [CD5:42]

3 OVAMBO SPARROWHAWK *Accipiter ovampensis* 33–40 cm; WS 67 cm **M u/r**
3a Ad Underparts, incl. throat, narrowly barred grey; central tail feathers with white shafts between dark bands. Compare 2a. **3b Ad melanistic form** All black; paler tail bars; remiges barred dark below. **3c Juv rufous-breasted form** Whitish supercilium; dark ear-coverts; variably streaked underparts. **3d Juv pale-breasted form** As 3c but rufous areas whitish. **3e–f Ad in flight** Three faint white spots on uppertail; finely barred underwing. ▲ Woodland. Fast, dashing flight; also soars. ✤ Usually silent except when breeding. High-pitched *keep-keep-keep-*... [CD5:49]

4 LIZARD BUZZARD *Kaupifalco monogrammicus* 35–37 cm; WS 79 cm **R f**
4a Ad Stocky; black throat-stripe; grey breast; pinkish-red cere and legs. Juvenile has bare parts duller; upperpart feathers edged buff. **4b–c Ad in flight** Black tail with broad white band; white uppertail-coverts; underwing appears white from distance. ▲ Various types of woodland and cultivation. Perches for long periods. ✤ Far-carrying, melodious *KLEEUUu-kluklukluklu*. [CD5:54]

5 EURASIAN SPARROWHAWK *Accipiter nisus* 28–41 cm; WS 58–80 cm **P/V s**
5a Ad male Narrowly barred rufous below; rufous ear-coverts. **5b Ad female** Much larger, browner; white supercilium. **5c Juv** Browner than 5b; barring more irregular. **5d Ad male in flight** No white on uppertail-coverts; tail square with sharp corners. **5e Ad female in flight** As male but larger; flight steadier. ▲ Open country, woodland. ✤ Usually silent on migration. [CD1:86]

6 LEVANT SPARROWHAWK *Accipiter brevipes* 29–39 cm; WS 63–80 cm **V?**
6a Ad male Grey cheeks; dark eye; tail unbarred above, narrowly barred below. **6b Ad female** Only slightly larger; browner; dark throat-stripe. **6c Juv** Boldly spotted below. **6d Ad male in flight** Very pale below; wings tipped black above and below. Compare 1d. **6e Ad female in flight** Body, wings and tail narrowly barred below. ▲ Palearctic vagrant claimed from N Cameroon. ✤ Usually silent on migration.

7 DARK CHANTING GOSHAWK *Melierax metabates* 38–48 cm; WS 95–110 cm **R f**
7a Ad Long pinkish-red legs; pinkish-red cere; long, rounded tail. Compare 2a. **7b Juv** Brown above, variably marked brown and white below; long bare legs. **7c–d Ad in flight** Black tips to long broad wings; uppertail-coverts finely barred grey; black central rectrices. **7e–f Juv in flight** Note shape of wings and tail; uppertail-coverts narrowly barred grey-brown. ▲ Open woodland, thornbush. Perches prominently. ✤ Usually silent except when breeding. Melodious call, a loud, accelerating, piping *wheeeow-whew-whew-whew* or *kleeu-kleeu-klu-klu-klu* by both sexes, from perch or in flight. [CD1:85]

Maps on page 96

1b
1a
2b
2c
1c
1e
1d
2a
2f
2d
3b
2g
3e
2e
3c
3d
3f
3a
5c
4c
4a
4b
5d
5b
5a
5e
6c
6d
6b
6e
6a
7b
7f
7a
7e
7c
7d

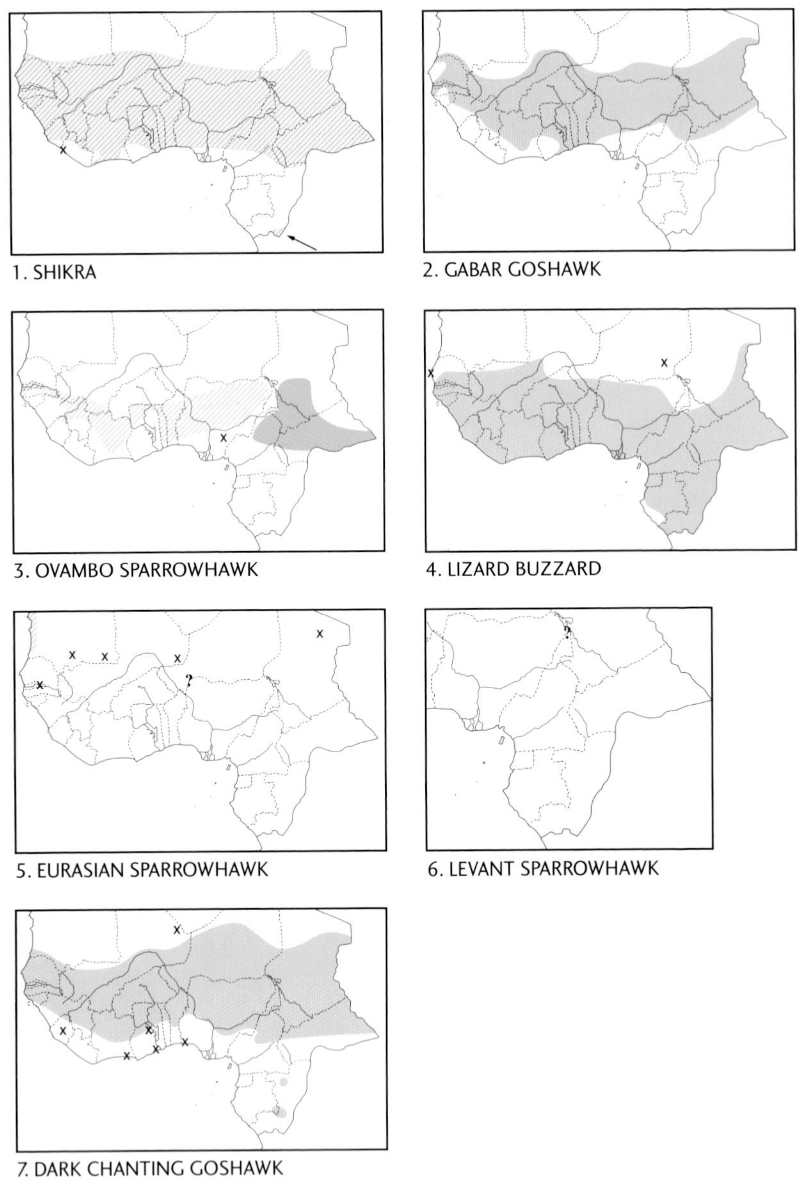

1. SHIKRA

2. GABAR GOSHAWK

3. OVAMBO SPARROWHAWK

4. LIZARD BUZZARD

5. EURASIAN SPARROWHAWK

6. LEVANT SPARROWHAWK

7. DARK CHANTING GOSHAWK

Plate on page 94

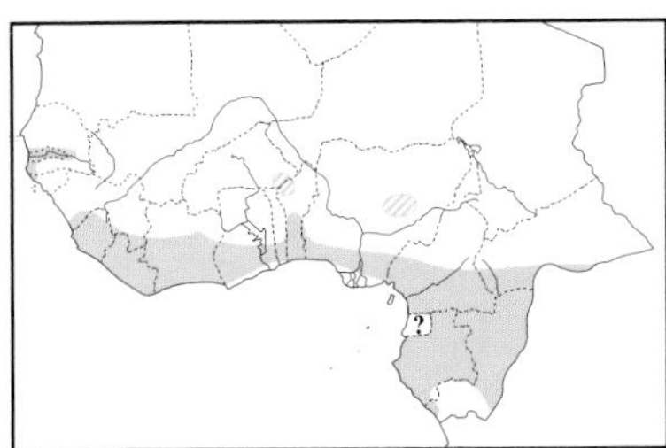

2. RED-THIGHED SPARROWHAWK

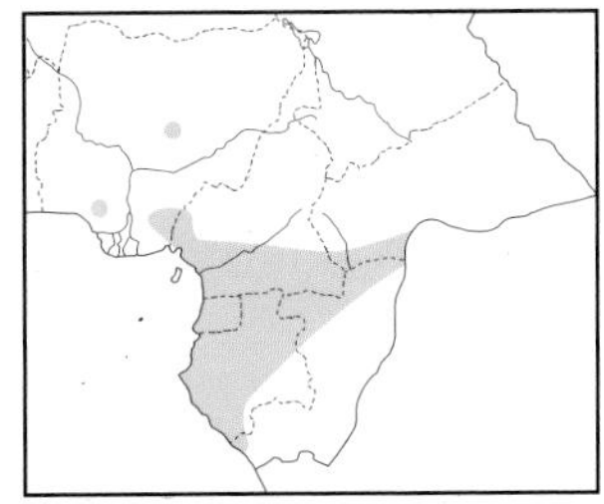

3. CHESTNUT-FLANKED SPARROWHAWK

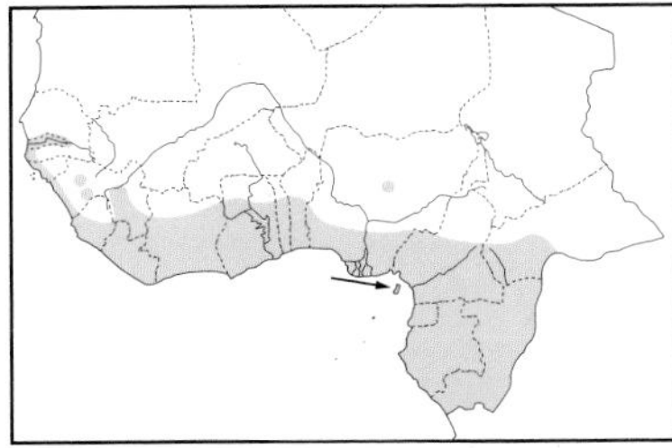

4. AFRICAN GOSHAWK

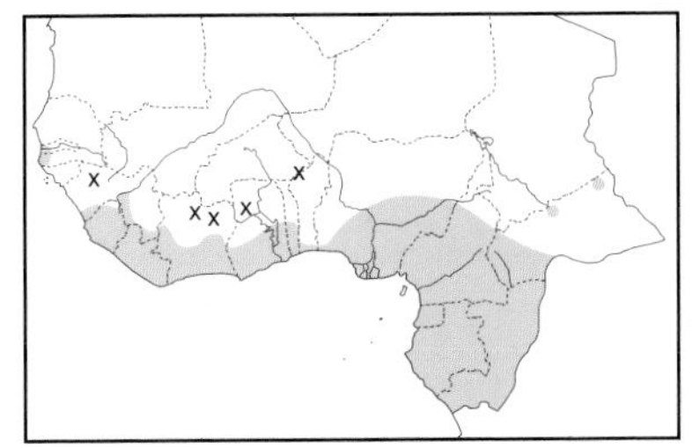

5. BLACK SPARROWHAWK

Plate on page 98

PLATE 22: SPARROWHAWKS AND GOSHAWKS II

Small to medium-sized, with rather short, rounded wings and long tails. Tarsi bare and long. Large size difference between sexes; females much larger. Swift and agile. Hunt by dashing on their prey (mainly birds and small mammals), following a stealthy approach behind cover.

1 LITTLE SPARROWHAWK *Accipiter minullus* 23–27 cm; WS 39 cm
1a Ad Very small; dark grey above; narrowly barred chestnut below. **1b Juv** Brown above; heavily spotted and barred brown below. **1c–d Ad in flight** White band on uppertail-coverts; two white spots on uppertail. Below: finely barred remiges; dark bars on white tail. ▲ Woodland. No certain records from our region. Claims possibly due to confusion with 2. Resident in E Africa and south of equator. ✤ [CD5:48]

2 RED-THIGHED SPARROWHAWK *Accipiter erythropus* 23–28 cm; WS 40 cm **R u/s**
2a Ad *erythropus* (Upper Guinea forest) Very small; blackish above; grey-white tinged pinkish below. **2b Ad *zenkeri*** (Lower Guinea forest) Breast and belly deep rufous. **2c Juv *erythropus*** Brown above; barred brown below. **2d Juv *zenkeri*** Sparsely spotted and barred below. **2e–f Ad *erythropus* in flight** White band on uppertail-coverts; three white spots on uppertail partially hidden by dark central tail feathers; underwings and undertail barred blackish. **2g–h Ad *zenkeri* in flight** Two unbroken white tail spots above; below as 2e but breast and belly deep rufous. ▲ Forest. ✤ Usually silent. High-pitched *kik-kik-kik-kik-*... or *ki-ki-ki-ki-kiw-kiw.* [CD5:47]

3 CHESTNUT-FLANKED SPARROWHAWK *Accipiter castanilius*
28–36 cm; WS 60 cm **R u/s**
3a Ad Blackish-grey above (incl. rump); barred dark below; breast-sides, flanks and thighs bright rufous; toes dusky. **3b Juv** Probably indistinguishable from 4b except for dusky toes. **3c–d Ad in flight** Three white spots in centre of blackish tail above; underwing whitish with dark barring on remiges. ▲ Forest. ✤ Unknown.

4 AFRICAN GOSHAWK *Accipiter tachiro* 36–48 cm; WS 70 cm **R c**
4a Ad *macroscelides* (Red-chested Goshawk; from W Cameroon westwards) Narrowly barred orange-chestnut below; three white spots in centre of blackish tail. **4b Juv *macroscelides*** Dark throat-stripe; spotted breast; barred flanks and thighs. **4c Ad *toussenelii*** (from SE Cameroon east and south) Paler and plainer below than 4a. **4d Juv *toussenelii*** Whiter and less marked below than 4b. **4e Ad hybrid *toussenelii* x *canescens*. 4f–g Ad *macroscelides* in flight** Three white spots in centre of blackish tail above; underwing-coverts barred orange. **4h Ad *toussenelii* in flight** Underwing-coverts white. ▲ Forest and outliers in savanna. ✤ Series of sharp, abrupt *kwit!* notes, in (usually high) display flight over territory or from perch, esp. at dawn. *A. t. toussenelii* has slightly longer *kewit!* Resembles call of Velvet-mantled Drongo. Juvenile has plaintive, nasal *we-aaaaauw.* [CD5:43,44]

5 BLACK SPARROWHAWK *Accipiter melanoleucus* 46–55 cm; WS 102 cm **R f/r**
5a Ad typical form Large; black above; white below; black patches on flanks and thighs. Compare adult Cassin's Hawk Eagle (Plate 26:4a,d). **5b Juv rufous-breasted form** Compare juvenile Ayres's Hawk Eagle (Plate 26:3b). **5c Juv white-breasted form** As 5b but white ground colour to underparts. **5d Ad in flight from above** Black. **5e Ad female in flight from below** Mainly white; irregular blackish patches on flanks. Male smaller. **5f Ad melanistic form in flight from below** Black breast and underwing-coverts, throat white. Rare. **5g Juv male rufous-breasted form in flight from below** Rufous underwing-coverts. ▲ Various types of forest. ✤ Usually silent; vocal when breeding. Call *kyip* or *klee-ep* (male) and deeper *chep* (female); duet an alternating series. [CD5:51]

Maps on page 97

1b
1a
2c
2a
1c
1d
2e
2f
2g
2h
2b
2d
3c
3d
4g
4e
3b
4b
3a
4f
4h
5g
4a
4d
4c
5e
5f
5d
5b
5c
5a

1 **EUROPEAN HONEY BUZZARD** *Pernis apivorus* 51–60 cm; WS 113–150 cm P f/s
1a Ad (typical) Small-looking grey head; slender neck; yellow eye. **1b Ad (dark)**. **1c Juv (pale)**. **1d Ad (typical) in flight** Well-protruding head; tail longish with rounded corners, dark terminal band and two narrower bars near base; dark carpal patch. **1e Ad (dark) in flight**. **1f Juv (pale) in flight** Flight and tail feathers with 4–5 more evenly spread bars. ▲ Forest, wooded savanna. Soars on flat wings, slightly bowed when gliding. ✤ Usually silent in Africa. [CD1:70]

2 **GRASSHOPPER BUZZARD** *Butastur rufipennis* 41–44 cm; WS 102 cm M c
2a Ad Slender; wings, tail and legs long; plumage mainly brown and rufous. **2b Juv** More rufous than adult; feathers of upperparts edged rufous. **2c Ad in flight from above** Conspicuous rufous wing patches diagnostic. **2d Ad in flight from below** Well-defined blackish trailing edge; pale rufous remiges. ▲ Open country; south to northern edge of forest in dry season. Perches conspicuously and for long periods atop small trees and bushes. Often near bush fires. Flight low and buoyant, rather 'harrier-like', also soars. ✤ Loud *ki-ki-ki-ki-kee* when breeding. Otherwise silent. [CD5:53]

***Buteo* buzzards**. Medium-sized, with broad-winged, eagle-like shape, but generally smaller and with only moderately 'fingered' wings. Tarsi bare. Plumages very variable, sometimes creating identification problems. Mostly seen soaring, wings held in shallow V, or perching on vantage points. Prey mainly upon mammals and reptiles caught on ground. Occur in woodland and open country.

3 **RED-NECKED BUZZARD** *Buteo auguralis* 40-50 cm; WS 95 cm R/M f/s
3a Ad Chestnut head-sides and nape; dark brown throat and breast. **3b Juv** White below with black blotches on breast and flanks. **3c Ad in flight** Rufous tail with black subterminal band. **3d Juv in flight** Brown tail barred black. ▲ Woodland, forest clearings and edges, cultivated areas. Moves north with rains. ✤ Loud, drawn-out, mewing *peeee-ah*. [CD5:56]

4 **COMMON BUZZARD** *Buteo buteo* 45–58 cm; WS 110–132 cm P r/s
B. b. vulpinus (Steppe Buzzard) **4a Ad (pale)** Pale rufous-brown underparts and tail. **4b Ad (foxy)** Uniform dark chestnut underparts and tail. **4c Ad (dark)** Mainly blackish. **4d Ad (pale) in flight** Broad rounded wings and tail without distinct terminal band. **4e Ad (foxy) in flight** White remiges with narrow barring; black trailing edge and fingers; small carpal patch; rufous tail with subterminal band. **4f Ad (dark) in flight** Mainly blackish; rare. *B. b. buteo* is very dark to pale brown (less rufous); usually has diffuse pale breast-band; underwing pattern less contrasting; barring on flight and tail feathers broader and less distinct. ▲ Wooded habitats. Distribution of two races unclear; both claimed from Liberia; those from Senegal and from Ivory Coast eastwards attributed to *vulpinus*. ✤ Usually silent in Africa. [CD1:88]

CAPE VERDE BUZZARD *Buteo (buteo) bannermani* See Plate 145:9.

5 **LONG-LEGGED BUZZARD** *Buteo rufinus* 50–65 cm; WS 115–163 cm P s/u
5a Ad (typical) Pale head and breast; dark rufous-brown belly and thighs. Compare 4a. **5b Ad (dark)** Mainly blackish; tail broadly barred. **5c Juv** Tail narrowly barred. **5d Ad (typical) in flight** Large black carpal patch and trailing edge; plain orange tail paler at base. **5e Ad (dark) in flight** Barred tail with broad subterminal band. **5f Juv in flight** No distinct trailing edge to wing; faintly barred buff-grey tail. ▲ Sahel. ✤ Usually silent. [CD1:90]

6 **AUGUR BUZZARD** *Buteo augur* 55–60 cm; WS 132 cm V?
6a Ad (typical) Black above, white below; short rufous tail. **6b Ad (dark)** Entirely black with rufous tail. **6c Juv** Brown above; dark-streaked throat; brown tail barred dark. **6d Ad (typical) in flight** Very broad wings; very short rufous tail. **6e Ad (dark) in flight** Black underwing-coverts; distinctive silhouette. **6f Juv in flight** Duller than adult; silhouette distinctive. ▲ E African species. One (doubtful) claim from N Cameroon. ✤ [CD5:57]

Maps on page 102

1b
1c
1a
2b
2a
2c
1f
1d
1e
2d
3b
3a
3d
3c
4c
4b
4a
4d
4e
4f
5b
5c
5a
5f
5d
5e
6c
6b
6a
6f
6d
6e

PLATE 23: BUZZARDS AND HONEY BUZZARD

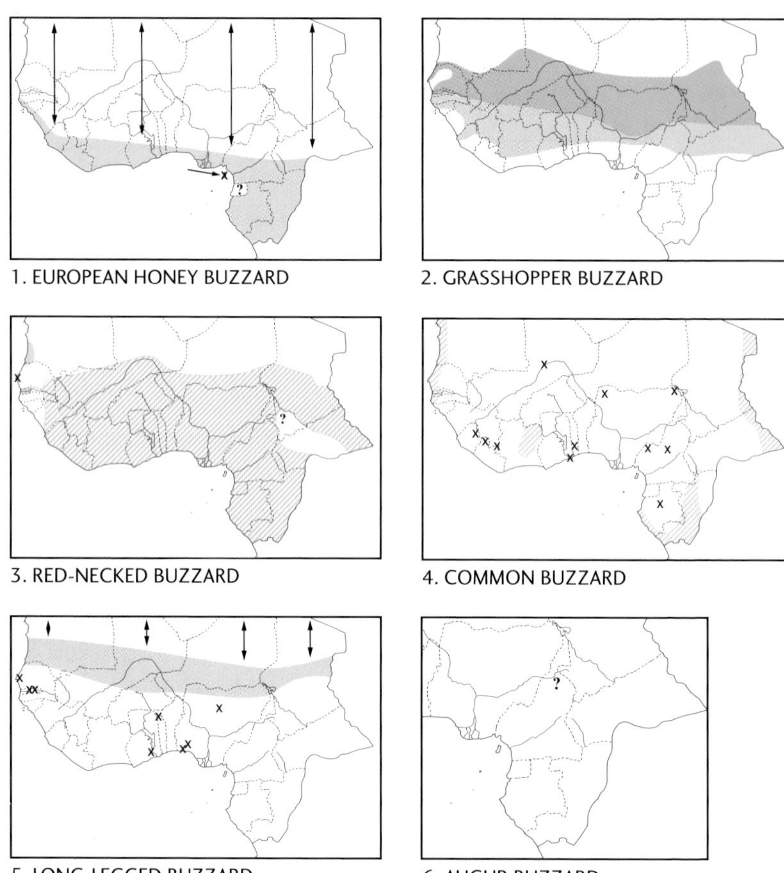

1. EUROPEAN HONEY BUZZARD

2. GRASSHOPPER BUZZARD

3. RED-NECKED BUZZARD

4. COMMON BUZZARD

5. LONG-LEGGED BUZZARD

6. AUGUR BUZZARD

Plate on page 100

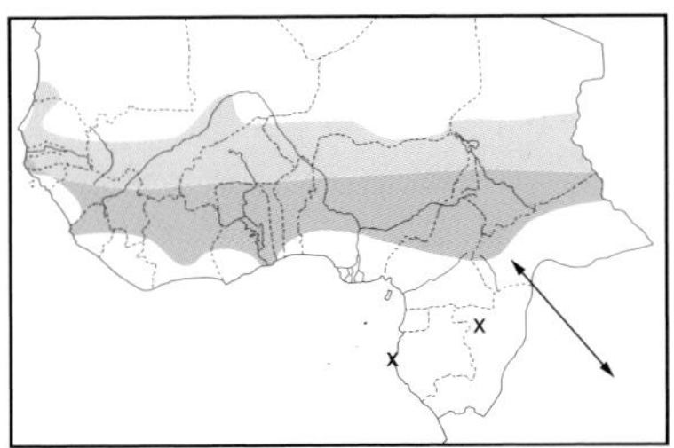

1. WAHLBERG'S EAGLE

2. TAWNY EAGLE

3. STEPPE EAGLE

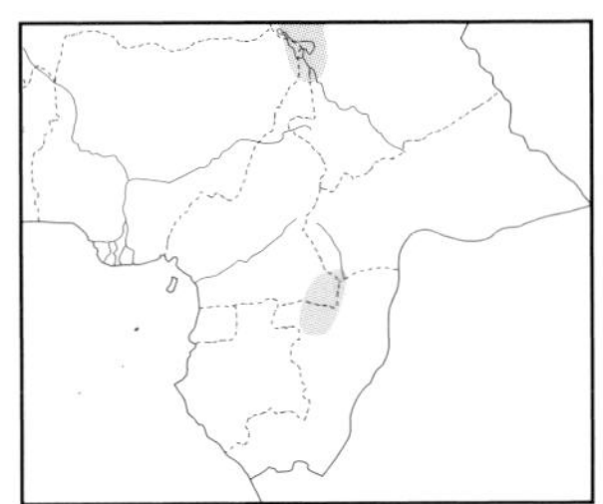

4. LESSER SPOTTED EAGLE

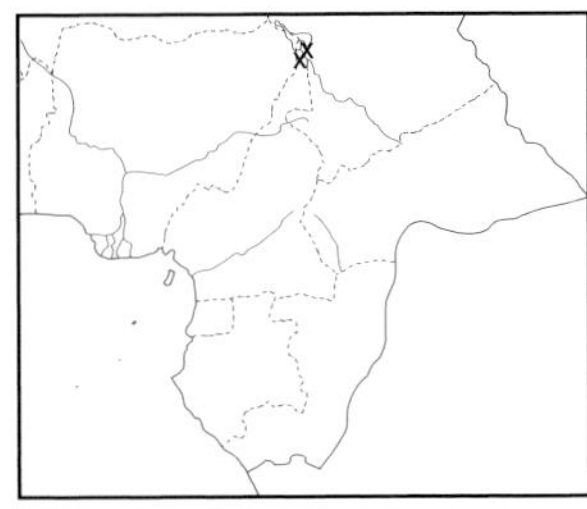

5. GREATER SPOTTED EAGLE

Medium-sized to very large with broad 'fingered' wings. Tarsi feathered. Given to soaring. Prey on large and small mammals, birds, and reptiles; some eat carrion and insects. In general fiercer and more active than buzzards. Occur in open country, woodland and forest.

1 WAHLBERG'S EAGLE *Aquila wahlbergi* 55–61 cm; WS 141 cm R/M f
1a Ad (dark) Dark brown; slight occipital crest; dark eye. Juvenile similar. **1b Ad (pale)** White to buffy head and underparts. Uncommon. **1c–d Ad (dark) in flight** Long, parallel wings held flat or slightly arched; longish, square tail usually held closed. **1e–f Ad (pale) in flight** White underwing-coverts. Note characteristic silhouette. ▲ Savanna. ✤ Usually silent. Distinctive, clear and loud *kleeeee-ay*. Also fast *kyip-kyip-kyip-kyip-*... [CD5:58]

2 TAWNY EAGLE *Aquila rapax* 62–75 cm; WS 165–185 cm R/M f/s
2a Ad (dark) Yellow gape extending to below centre of eye. **2b Ad (pale)** Tawny to pale buffish. Appearance often rather ragged. **2c Juv** Typically paler than adult; white-tipped greater coverts and trailing edge. **2d–e Ad (typical) in flight** Creamy rump patch; usually pale 'window' on inner primaries below. **2f Ad (dark) in flight**. **2g–h Juv in flight** Narrow pale line on greater upperwing-coverts; narrow white trailing edge to wing and tail. ▲ Wooded habitats. Moves south in dry season. ✤ Usually silent. Barking *kyow* and guttural *kwork* in display. [CD1:91]

3 STEPPE EAGLE *Aquila nipalensis* 62–81 cm; WS 163–200 cm V
3a Ad Long yellow gape extending to below rear of eye; oval nostrils; pale nape patch. **3b Juv** Typically paler than adult; white-tipped greater coverts, trailing edge and tail. **3c–d Ad in flight** Whitish primary patch on upperwing; small white patch on back. Compare 4 and 5. Below: dark trailing edge to wings and tail. **3e–f Juv in flight** Pale line on greater upperwing-coverts; pale U on uppertail-coverts; pale band on central underwing; white trailing edge to wings and tail. ▲ Palearctic vagrant. Feeds on insects (esp. termites) on ground; also other small animals and carrion. ✤ Generally silent in Africa.

4 LESSER SPOTTED EAGLE *Aquila pomarina* 57–64 cm; WS 145–170 cm P r
4a Ad Dark brown; narrow 'trousers'; yellowish eye. Compare 1a and 5a. **4b Juv** Pale nape patch; pale tips to greater coverts and trailing edge of wing. **4c–d Ad in flight** Small, distinct primary patch on upperwing; pale U on uppertail-coverts; underwing-coverts typically paler than flight feathers; two whitish carpal crescents. **4e–f Juv in flight** Primary patch and U on uppertail-coverts more distinct than 4c; pale tips to greater underwing-coverts and trailing edge of wing (soon abraded). ▲ In Africa principally feeds on termites, grasshoppers and other insects taken on ground. ✤ Usually silent. Barking *kow-kow-kow*. [CD1:92]

5 GREATER SPOTTED EAGLE *Aquila clanga* 60–70 cm; WS 155–180 cm V VU
5a Ad Darker, more uniform and heavier than 4a; 'trousers' more bushy; eye dark. **5b Juv** Darker and more spotted than 4b. **5c Juv pale morph** ('*fulvescens*') Pale buffish or yellowish-brown to rufous. Rare. **5d–e Ad in flight** Diffuse primary patch on upperwing; underwing-coverts typically as dark or slightly darker than flight feathers; single carpal crescent; tail relatively shorter than in 4. **5f–g Juv in flight** Spotted upperwing-coverts; white trailing edge; pale vent. **5h Juv pale morph in flight** Pale carpal crescent; note silhouette. ▲ Palearctic vagrant. Feeds on termites, grasshoppers and other insects; also small mammals, lizards, amphibians and birds. ✤ Usually silent.

 Maps on page 103

1a
1b
1c
1d
1e
1f
2a
2b
2c
2d
2e
2f
2g
2h
3a
3b
3c
3d
3e
3f
4a
4b
4c
4d
4e
4f
5a
5b
5c
5d
5e
5f
5g
5h

1 **EASTERN IMPERIAL EAGLE** *Aquila heliaca* 72–85 cm; WS 180–220 cm **V VU**
1a Ad Blackish-brown; pale tawny crown and nape; white 'braces'. **1b Juv** Pale and streaky. **1c–d Ad in flight** Long parallel wings held flat when soaring; tail held closed; pale top of head; broad black terminal band on grey tail. **1e–f Juv in flight** Two pale lines on upperwing-coverts; pale wedge on innermost primaries. ▲ Palearctic vagrant. ✤ Silent on migration.

2 **GOLDEN EAGLE** *Aquila chrysaetos* 75–90 cm; WS 190–225 cm **P/R s/r**
2a Ad Rufous-yellow crown and nape. **2b Juv** Darker than adult; no 'golden' nape; white tail with broad black terminal band. **2c–d Ad in flight** Longish tail; long wings held in shallow V; wings with slight S-curved trailing edge; diffuse pale panel on upperwing-coverts. **2e–f Juv in flight** Broad black terminal tail band; variably distinct wing patches. ▲ Gorges, open country. Feeds mainly on medium-sized mammals, also large birds, reptiles and carrion. ✤ Usually silent. [CD1:94]

3 **VERREAUX'S EAGLE** *Aquila verreauxii* 80–90 cm; WS 190–210 cm **R r**
3a Ad Jet-black with narrow white V on back. **3b Juv** Tawny-rufous crown and mantle. **3c–d Ad in flight** White Y on back; white 'windows' in primaries. Soars with wings held in shallow V; distinctive silhouette with leaf-shaped wings; longish, often partially spread tail. **3e–f Juv in flight** Pale 'windows' in primaries; pale crescent at tail base. ▲ Crags, inselbergs, gorges. Feeds mainly on hyrax. ✤ Usually silent. [CD5:59]

4 **CROWNED EAGLE** *Stephanoaetus coronatus* 80–99 cm; WS 163–209 cm **R u/r**
4a Ad Very large and powerful; boldly barred and blotched black, rufous and buff below; short occipital crest. **4b Juv** Head and underparts mainly white. **4c–d Ad in flight** Broad rounded wings; long tail; rufous underwing-coverts; broadly barred tail. **4e–f Juv in flight** Pale rufous-buff underwing-coverts; 3–4 bars on remiges. ▲ Forest and outliers in savanna. Preys on monkeys and other mammals to size of duiker. ✤ Vocal. Far-carrying, melodious *kewee-kewee-kewee-...* (male) and deeper *kowi-kowi-kowi-...* (female), in undulating display flight, often at great height, esp. in hot midday hours. [CD5:64]

5 **MARTIAL EAGLE** *Polemaetus bellicosus* 78–86 cm; WS 195–260 cm **R u**
5a Ad Very large and powerful; dark brown above and on breast; white below spotted dark brown; short occipital crest. Compare snake eagles (Plate 19:1–3). **5b Juv** Grey above; white below. **5c–d Ad in flight** Long broad dark wings; short tail; contrastingly white lower underparts. Soars with wings held flat and primary tips upturned. **5e–f Juv in flight** Whitish body and underwing-coverts; finely grey-barred remiges. ▲ Savanna. ✤ Usually silent. Rapid *klee-klee-klee-klooee-klooee...* in display flight and from perch. [CD5:65]

Maps on page 108

1b
1a
1e
1f
1c
1d
2c
2f
2e
2d
2b
2a
3e
3f
3c
3d
3b
3a
4e
4f
4c
4d
4b
4a
5e
5f
5c
5d
5b
5a

PLATE 25: LARGE EAGLES

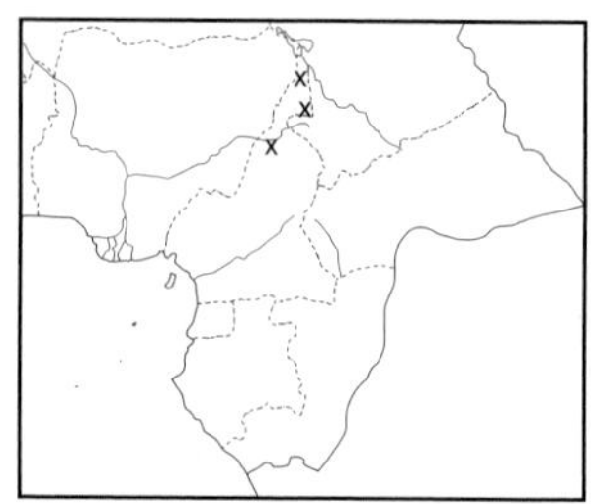

1. EASTERN IMPERIAL EAGLE

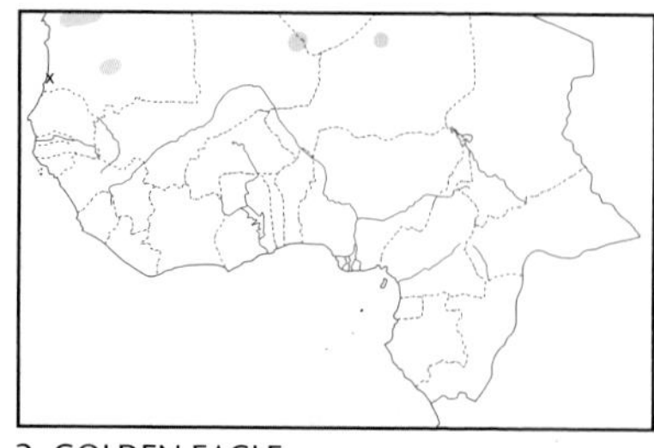

2. GOLDEN EAGLE

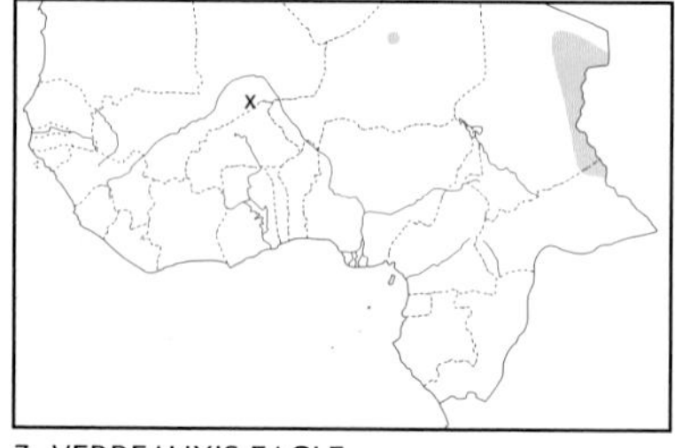

3. VERREAUX'S EAGLE

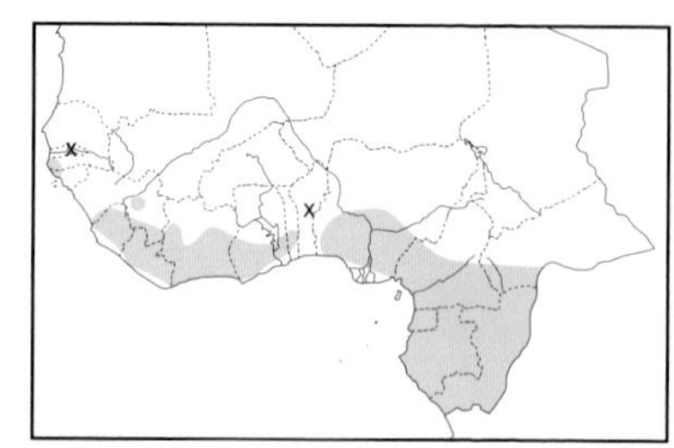

4. CROWNED EAGLE

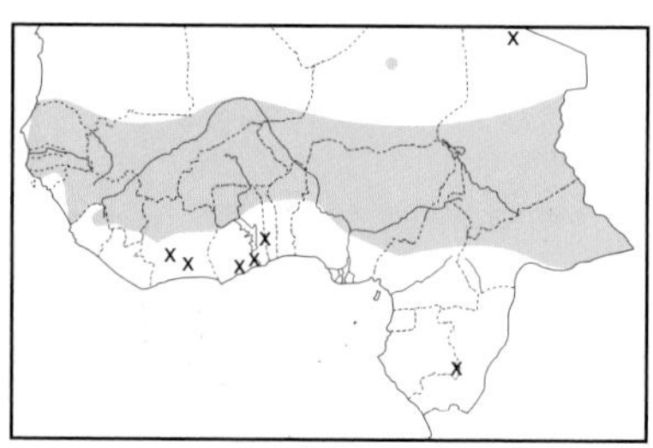

5. MARTIAL EAGLE

Plate on page 106

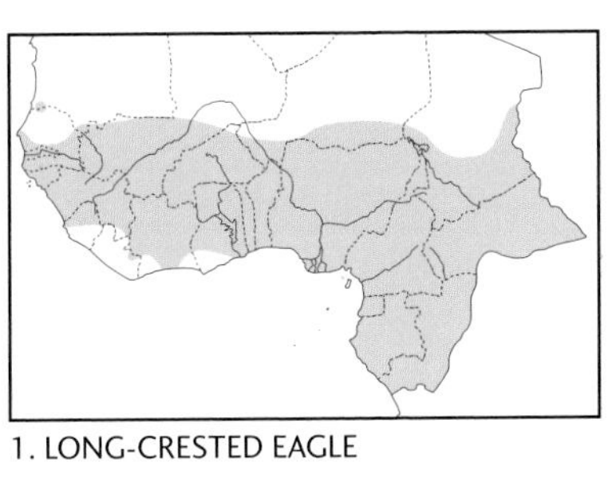
1. LONG-CRESTED EAGLE

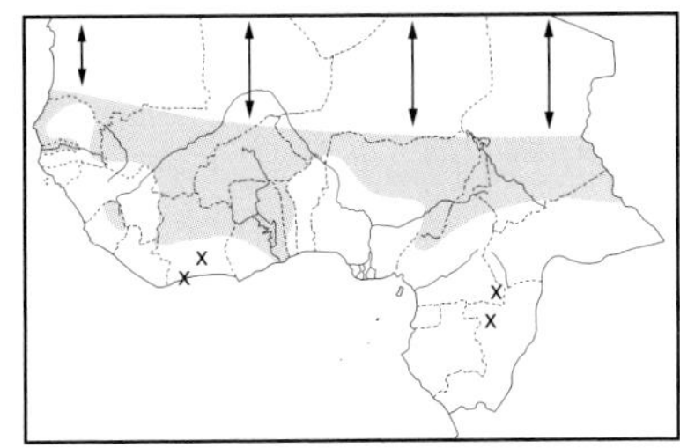
2. BOOTED EAGLE

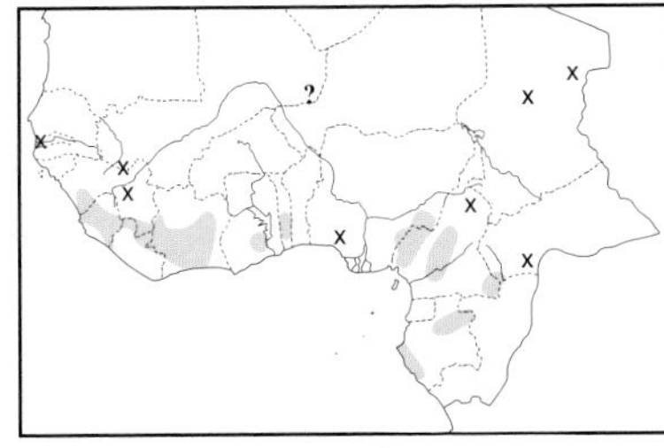
3. AYRES'S HAWK EAGLE

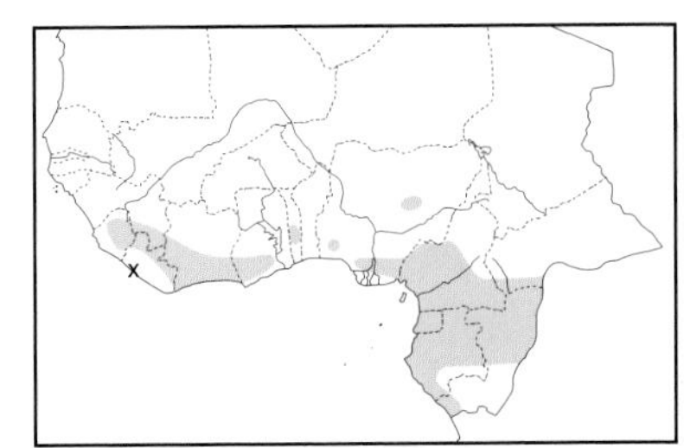
4. CASSIN'S HAWK EAGLE

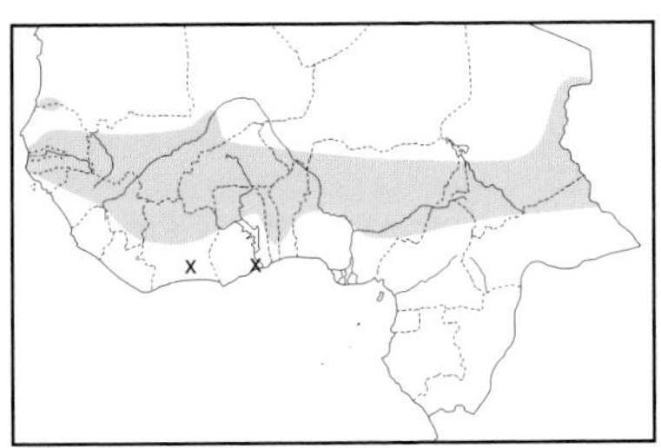
5. AFRICAN HAWK EAGLE

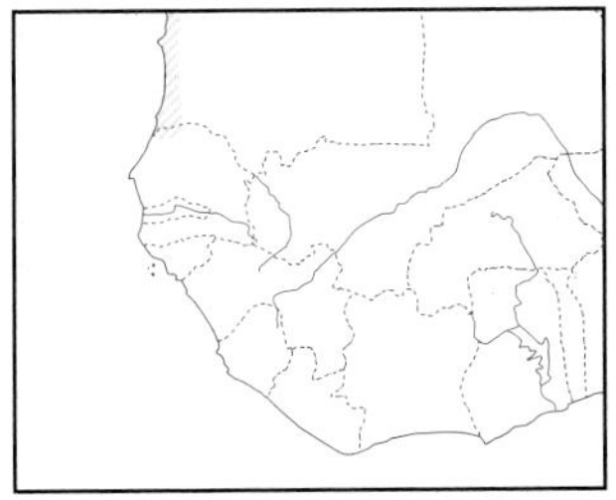
6. BONELLI'S EAGLE

Plate on page 110

PLATE 26: SMALL AND MEDIUM-SIZED EAGLES

1 LONG-CRESTED EAGLE *Lophaetus occipitalis* 53–58 cm; WS 115 cm **R u/f**
1a Ad Unmistakable. Entirely blackish; long, loose crest; yellow eye. Juvenile has shorter crest; eye grey. **1b–c Ad in flight** Large white 'windows' on primaries. ▲ Wooded savanna, cultivation, forest edge. ✤ Vocal. High-pitched, nasal *keeee-aah* and a series of sharp *kikikikikikeeah* with variations, from perch or in soaring display flight. Also short nasal *kwoh*. [CD5:62]

2 BOOTED EAGLE *Hieraaetus pennatus* 45–55 cm; WS 110–132 cm **P u**
2a Ad pale morph Brown above; creamy-white below. **2b Ad dark morph** Entirely dark brown. **2c Ad in flight** White 'landing lights' (small white spot at base of neck on leading edge of wing); whitish crescent on uppertail-coverts; broad pale band on upperwing-coverts. **2d Ad pale morph in flight** Blackish remiges; whitish underwing-coverts and body; unbarred, pale grey tail. **2e Ad dark morph in flight** Dark brown underwing-coverts and body; unbarred, pale grey tail. ▲ Sahel and savanna zones. ✤ Usually silent on migration. [CD1:96]

3 AYRES'S HAWK EAGLE *Hieraaetus ayresii* 45–60 cm; WS 124 cm **R u/s**
3a Ad Blackish-brown above; white below boldly streaked and blotched black. **3b Juv** Pale rufous to whitish head and underparts. **3c White-headed form** (age uncertain). **3d–e Ad in flight** White 'landing lights' (small white spot at base of neck on leading edge of wing). Below: heavily mottled black-and-white wing-coverts; heavily barred flight and tail feathers; black terminal tail band. **3f Juv in flight** Pale rufous underwing-coverts; blackish-barred remiges. ▲ Forest and its outliers. ✤ Usually silent. Melodious *whueeep-whip-whip-whip-whip-whueep* in aerial display; also *whip-whip-whip-...* [CD5:61]

4 CASSIN'S HAWK EAGLE *Spizaetus africanus* 50–61 cm; WS 120 cm **R s/lf**
4a Ad Blackish above; white below marked black on breast-sides, flanks and thighs. Compare Black Sparrowhawk (Plate 22:5). **4b Juv** Pale rufous head; mainly white underparts streaked blackish. **4c–d Ad in flight** Mainly dark above; blackish hood contrasts with white neck-sides; mainly black underwing-coverts; barred remiges. **4e Juv in flight** Rufous-streaked underwing-coverts. ▲ Forest. ✤ Vocal. Loud *ku-ku-wee* or *ku-wee ku-ku-wee* in soaring display flight; also drawn-out, high-pitched *weeeee-eh* from perch. [CD5:63]

5 AFRICAN HAWK EAGLE *Hieraaetus spilogaster* 60–74 cm; WS 142 cm **R u/s**
5a Ad Blackish above; white streaked black below; thighs unstreaked. **5b Juv** Rufous below. **5c–d Ad in flight** Pale greyish 'windows' on primaries above; black-mottled underwing-coverts; black trailing edge to wing; black terminal band to tail. **5e–f Juv in flight** Above, duller and browner version of adult. Below, rufous underwing-coverts bordered by black line; narrowly barred, grey tail. ▲ Savanna. ✤ Melodious *klu-klu-klu-kluee* or *kluee-kluee* rather reminiscent of Wahlberg's Eagle but shorter. [CD5:60]

6 BONELLI'S EAGLE *Hieraaetus fasciatus* 55–72 cm; WS 145–180 cm **P s/r**
6a Ad Dark brown above; white, lightly streaked dark below. **6b Juv** Rusty-brown below. **6c–d Ad in flight** Small white patch on mantle; no pale 'windows' on primaries above; white body contrasts with dark underwings; longish tail with black terminal band. **6e–f Juv in flight** Above, resembles 5e, but lacks obvious pale areas on primaries. Below, resembles 5f, but has dusky secondaries. ▲ Soars less than most other raptors, with wings held flat or slightly arched and tail often closed. Glides on slightly arched wings with carpal joints pressed forward and almost straight rear edge. ✤ Usually silent. [CD1:95]

Maps on page 109

1a
2b
2a
1b
1c
2d
2c
2e
3c
3b
4b
3f
3d
3a
3e
4a
4e
4c
5b
5a
4d
5c
5f
5e
5d
6b
6c
6f
6d
6a
6e

Small to medium-sized diurnal birds of prey with long, pointed wings and narrow, longish tail.

1 COMMON KESTREL *Falco tinnunculus* 30–38 cm; WS 65–80 cm **R/P/V* f/c**
1a Ad male *rufescens* (R) Grey head; rufous-brown upperparts heavily marked black; grey, barred tail. **1b Ad female *rufescens*** Brownish head streaked black; tail brown barred black. **1c Ad male *tinnunculus*** (P) Paler and less marked than 1a; tail unbarred. **1d–e Ad male *rufescens* in flight** Dark flight feathers; chestnut upperwing-coverts and upperparts; pale underwing densely barred and spotted dark. **1f–g Ad female *rufescens* in flight** Underwing more heavily barred than male. ▲ Open habitats. ✤ Sharp, piercing *kee-kee-kee-kee-*...; also shrill *krree-e-e-e-*... . [CD2:1b] Cape Verde forms ***alexandri*** and ***neglectus***: see Plate 145:6.

2 LESSER KESTREL *Falco naumanni* 25–33 cm; WS 58–74 cm **P u/f VU**
2a Ad male Plain grey head; plain rufous upperparts; grey wing panel; lightly marked underparts. **2b Ad female** As 1b but paler, with less-marked head and underparts. **2c–d Ad male in flight** Tricoloured upperwing; very pale underwing tipped dusky; central rectrices often slightly longer. **2e–f Ad female in flight** Underwing whiter than 1f–g; dusky tip more distinct. ▲ Open habitats. ✤ Usually silent away from breeding grounds. [CD2:1a]

3 FOX KESTREL *Falco alopex* 35–42 cm; WS *c.* 90 cm **R u/f**
3a Ad Very rufous, finely streaked black; tail narrowly barred black. Juvenile has streaking more prominent; barring on tail broader. **3b–c In flight** Long wings and tail; underwing-coverts pale rufous; underwing tipped black. ▲ Inselbergs in savanna. ✤ Sharp *kee-kee-kee-kee-*... near breeding sites. [CD5:67]

4 MERLIN *Falco columbarius* 24–33 cm; WS 50–69 cm **V**
4a Ad male Very small, compact; blue-grey upperparts. **4b Ad female** Brown upperparts fringed rufous or buff; tail barred. **4c Ad male in flight** Blue-grey above; tail with broad black subterminal band. **4d–e Ad female in flight** Dark brown above; coarsely barred underwing; barred tail. ▲ Open habitats. Palearctic vagrant. ✤ Usually silent. [CD2:3a]

5 BARBARY FALCON *Falco pelegrinoides* 32–45 cm; WS 76–100 cm **P/V r**
5a Ad As 6a but rufous on nape; moustachial stripe narrower; white cheek patch larger; underparts pale pinkish-buff narrowly and indistinctly barred brown. **5b Juv** As 6c but head with buff or rufous; underparts sandy-buff with finer streaking. **5c–d Ad in flight** Similar to 6d–e, but see 5a. **5e Juv in flight** Note uniformly marked, dark-tipped underwing with sandy-buff ground colour. ▲ Open habitats. ✤ Sharp *kek-kek-kek-*... [CD2:5b]

6 PEREGRINE FALCON *Falco peregrinus* 33–50 cm; WS 80–115 cm **R/P u/r**
6a Ad *minor* (R) Blackish top of head; broad moustachial stripe; densely spotted and barred black below. **6b Ad *peregrinus*** (P) Paler than 6a; underparts less heavily marked. **6c Juv** Brown above; heavily streaked below. **6d–e Ad in flight** Stocky; broad-based, pointed wings; relatively short tail. **6f Juv in flight** Streaked underparts; narrowly and uniformly barred underwing. ▲ Cliffs, open habitats. ✤ Loud, shrill *khyeh-khyeh-khyeh-*... Usually silent except near breeding site. [CD2: 5a]. Cape Verde form ***madens***: see Plate 145:5.

7 LANNER FALCON *Falco biarmicus* 38–52 cm; WS 90–115 cm **R/P u/f**
7a Ad *abyssinicus* Rufous crown and nape; narrow blackish moustachial stripe. **7b Juv** Browner; crown and nape paler; boldly streaked below. **7c–d Ad in flight** Long, rather blunt wings; pale underparts and underwing-coverts. **7e Juv in flight** Dark underwing-coverts contrast with paler flight feathers. ▲ Open habitats. ✤ Raucous *kreh-kreh-kreh-*... and shrill *kirrree-kirrree-*. Usually silent except near breeding site. [CD2:4a]

8 SAKER FALCON *Falco cherrug* 43–60 cm; WS 102–135 cm **P/V r EN**
8a Ad Pale head; streaked below, esp. on flanks; 'trousers' usually dark. **8b Juv** Head pattern usually more pronounced; underparts more boldly streaked. **8c–d Ad in flight** Underwing more contrasting than 7e. ▲ Open habitats. ✤ Usually silent. [CD2:4b]

 Maps on page 114

1c
1b
1a
1e
2d
3c
2f
1g
1d
3b
2b
2a
2e
2c
1f
5d
4e
3a
5e
5c
4d
4c
6f
4b
4a
5a
6b
6e
6d
5b
7e
8d
8c
6c
7d
6a
7c
7b
7a
8b
8a

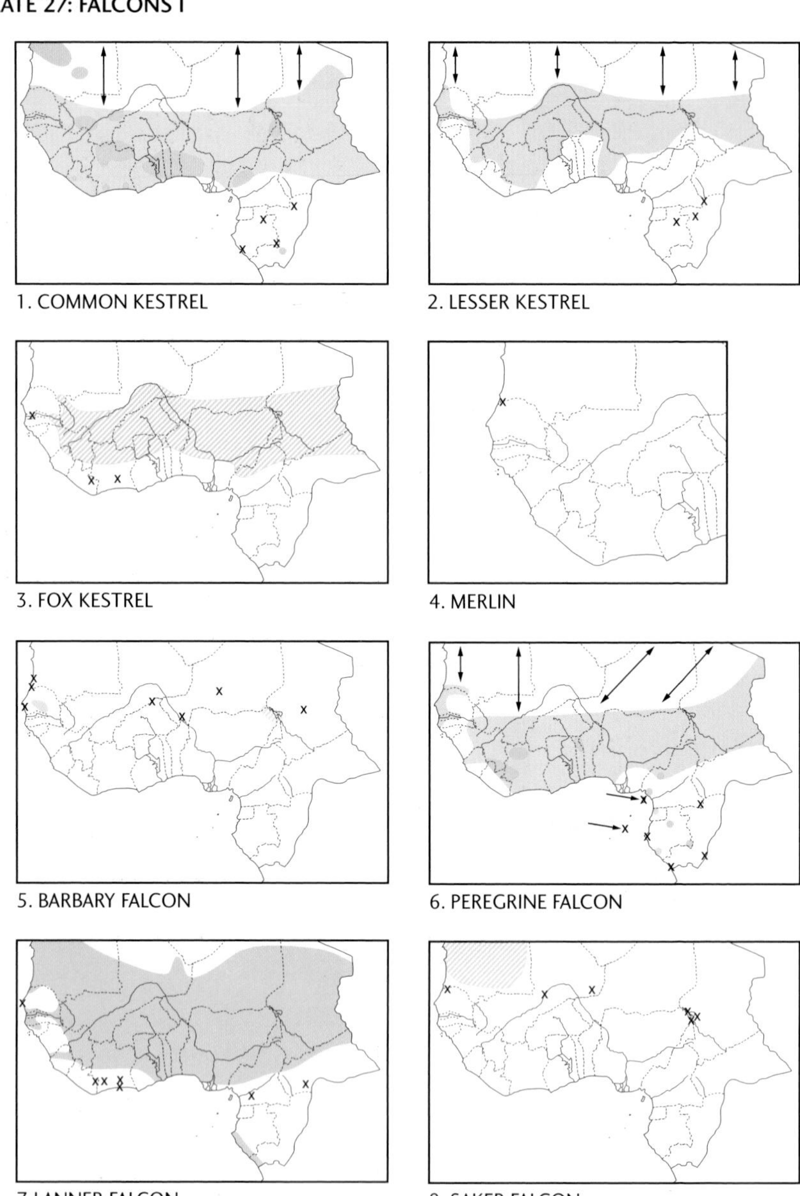

Plate on page 112

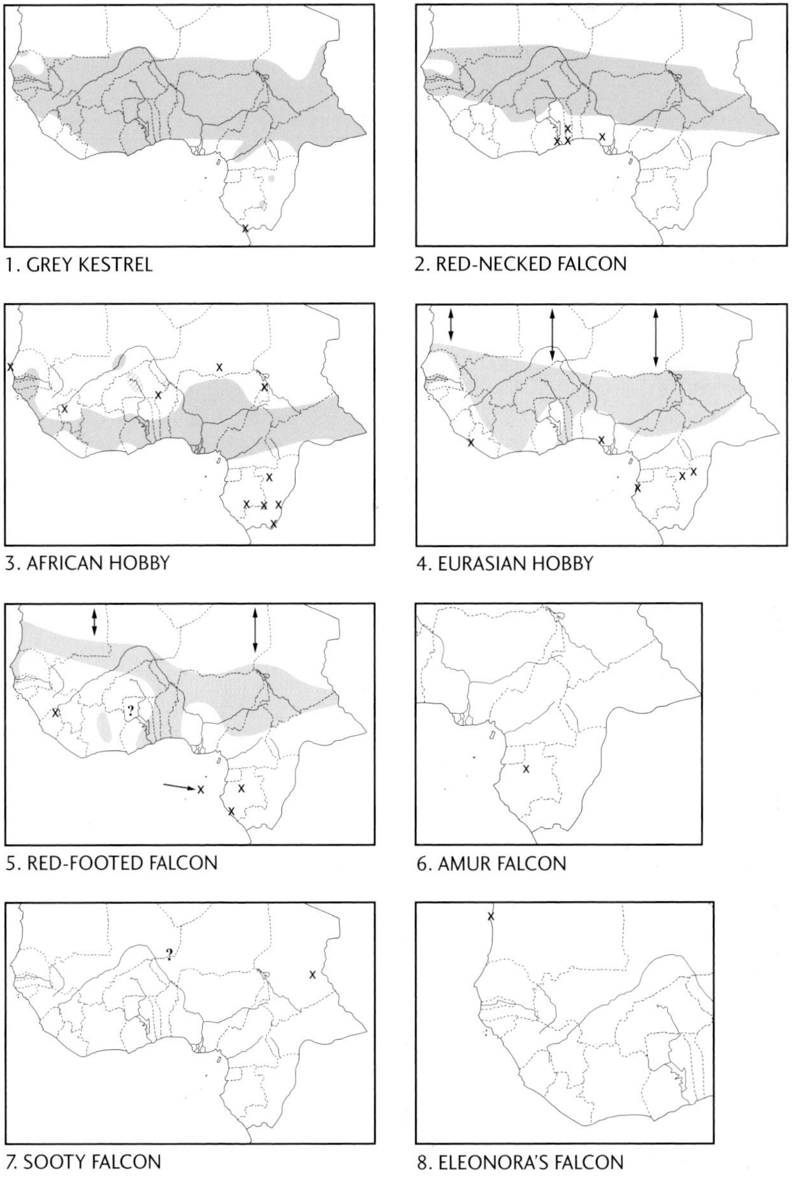

1. GREY KESTREL

2. RED-NECKED FALCON

3. AFRICAN HOBBY

4. EURASIAN HOBBY

5. RED-FOOTED FALCON

6. AMUR FALCON

7. SOOTY FALCON

8. ELEONORA'S FALCON

Plate on page 116

1 GREY KESTREL *Falco ardosiaceus* 30–39 cm; WS *c.* 70 cm R u/r
1a Ad Wholly slate-grey; cere, orbital ring and legs bright yellow. Juvenile very slightly tinged brown. **1b–c In flight** Slate-grey above and below; primaries blackish. ▲ Open habitats. ✤ Shrill, rasping, vibrant twitter. Usually silent. [CD5:68]

2 RED-NECKED FALCON *Falco chicquera* 29–36 cm; WS 65–80 cm R u
2a Ad Bright chestnut top of head; white cheeks; densely barred above and below. **2b Juv** Duller. **2c–d Ad in flight** Dark above, primaries blackish; barring below appears grey at distance. **2e Juv in flight** Pale rufous-buff underparts and underwing-coverts. ▲ Savanna, typically with palm trees. ✤ Shrill, harsh, scolding *kheep-kheep-kheep-*... and soft *k-krrree-up*. [CD5:70]

3 AFRICAN HOBBY *Falco cuvierii* 28–31 cm; WS *c.* 70 cm R s/lf
3a Ad Orange-rufous below; blackish-slate above; short moustachial stripe. **3b Juv** Duller; underparts heavily streaked black. **3c–d Ad in flight** Swift and slender; mainly orange-rufous below. **3e Juv in flight** Duller, streaky version of adult. ▲ Savanna, extensive clearings in forest zone. ✤ Shrill *kee-kee-kee-*... and short, sharp *kik*. [CD5:71]

4 EURASIAN HOBBY *Falco subbuteo* 29–36 cm; WS 70–92 cm P s/u
4a Ad Narrow black moustachial stripe contrasting with white cheeks and throat; boldly streaked underparts; rufous vent and thighs. **4b Juv** Browner; pale forehead; no rufous on vent and thighs. **4c–d Ad in flight** Long, pointed wings; medium-long tail. ▲ Open habitats. ✤ Usually silent away from breeding grounds. [CD2:3b]

5 RED-FOOTED FALCON *Falco vespertinus* 28–31 cm; WS 60–78 cm P r/u NT
5a Ad male Dark blue-grey; rufous vent and thighs; reddish cere, orbital ring and legs. **5b Ad female** Orangey crown, nape and underparts; small black mask. **5c Juv** Pale forehead; short moustachial streak; pale neck collar. Compare 4b. **5d–e Ad male in flight** Silvery-grey flight feathers contrast with rest of plumage. **5f–g Ad female in flight** Plain underwing-coverts; barred flight feathers. **5h Juv in flight** Less boldly streaked below than 4b; paler head pattern. ▲ Open habitats. ✤ Usually silent away from breeding grounds. [CD2:2a]

6 AMUR FALCON *Falco amurensis* 26–32 cm; WS 58–75 cm V
6a Ad male As 5a, but somewhat paler below. **6b Ad female** Slate-grey crown; short moustachial streak; whitish below barred and blotched dark. **6c Juv** Browner above than female; more streaked below. **6d–e Ad male in flight** As 5d–e, but has white underwing-coverts. **6f–g Ad female in flight** White underwing strongly barred black. **6h Juv in flight** Similar to adult female. ▲ Open habitats. Palearctic vagrant. Passage migrant to E Africa, wintering in S Africa. ✤ Usually silent.

7 SOOTY FALCON *Falco concolor* 32–38 cm; WS 73–90 cm V
7a Ad All blue-grey; much more slender, with longer wings and tail, than 1. **7b Juv** Head pattern as 4b, but cheeks and throat buffy; underparts brownish-buff. **7c Ad in flight** Slender; tail wedge shaped; underwings and tail blue-grey tipped dark. **7d Juv in flight** Note broad subterminal tail band and slender jizz. ▲ Open habitats. Palearctic vagrant. ✤ Usually silent away from breeding grounds.

8 ELEONORA'S FALCON *Falco eleonorae* 36–42 cm; WS 84–105 cm V+
8a Ad pale morph Well-defined black moustachial stripe; white cheeks and throat; orangey underparts narrowly streaked blackish. **8b Ad dark morph** Wholly brownish-black. **8c Juv** Browner than 8a; underparts buffish. **8d Ad pale morph in flight** Slender; long wings and tail; dark underwing-coverts. **8e Ad dark morph in flight** Underwing-coverts darker than flight feathers. **8f Juv in flight** Dark-looking underwing-coverts contrast with paler flight feathers. ▲ Coastal. Palearctic vagrant. ✤ Usually silent away from breeding grounds. [CD2:2b]

Maps on page 115

1a
1c
1b
2e
2d
2c
3e
3d
3c
2b
2a
3b
3a
5h
5g
5e
5f
5d
4d
4c
4b
4a
7d
7c
6h
6g
6e
6f
6d
5c
5b
5a
8f
6c
6b
6a
8e
7b
7a
8c
8b
8a
8d

Afrotropical family of medium-large terrestrial birds with small, bare head, relatively long neck and thickset body. Bare skin of head and neck often brightly coloured. Wings and tail short; feet strong. Sexes similar; males usually larger. Occur from forest to subdesert. Sedentary. Reluctantly take flight; prefer to run. Gregarious when not breeding. Often vocal. Roost in trees.

1 BLACK GUINEAFOWL *Agelastes niger* 40–43 cm R ls
1a Ad Entirely black with bare reddish head and throat. **1b Juv** Dull black with white belly. ▲ Lowland forest. ✤ Various quiet, twittering contact and feeding calls, constantly uttered, incl. hard *pit pit pit...* Song a fast, rising series of rather melodious, whistling notes, *huw hee-huwhee-huwhee-huwhee-huwhee-huwhee-wheet-wheet-wheet...* [CD5:73]

2 WHITE-BREASTED GUINEAFOWL *Agelastes meleagrides* 40–45 cm R r/lf VU
2a Ad Pinkish-red head and upper neck contrast with broad white collar. **2b Juv** Dull black; white belly; dark brownish head with two tawny crown-stripes. ▲ Lowland forest. ✤ Various quiet, twittering contact and feeding calls, constantly uttered, incl. *pit pit pit...* and short trill *prrirrr.* [CD5:72]

3 PLUMED GUINEAFOWL *Guttera plumifera* 45–51 cm R lf/s
3a Ad Bare grey-blue head; stiff feathery crest; plumage finely speckled bluish. **3b Juv** Dull upperparts; lightly scaled underparts. ▲ Lowland forest. ✤ Vocal. Loud, trumpeting *ku-ku-khep ku-ku-khep ku-ku-khep...* Flock produces discordant chorus of harsh *khep-khep-...* or *kha-kha-kha-...* calls when alarmed. Contact calls low and harsher than 1. [CD5:74]

4 CRESTED GUINEAFOWL *Guttera pucherani* 46–56 cm R lf
4a Ad *verreauxi* (Guinea-Bissau–SW Nigeria; SC Congo; CAR) Similar to 3 but has bare blue head, curly crest and black collar. **4b Ad *sclateri*** (SE Nigeria–SW Cameroon) As 4a but crest shorter in front. **4c Juv** Heavily scaled plumage with pale blue on mantle and tertials. ▲ Lowland and gallery forest. ✤ Contact call a soft clucking *chuk*. Sharp, nasal *kak!* interspersed with hard rattle when alarmed, *kak kak-kak kak-uk kak-uk kurr-r-r-r-r-k*, resembling 5 but lower pitched. [CD5:75]

5 HELMETED GUINEAFOWL *Numida meleagris* 53–63 cm R+ lc/s
5a Ad *galeatus* (S Mauritania–Cameroon; SE Gabon–Congo) Bare bluish head; small casque; red gape wattles. **5b Ad *meleagris*** (E Cameroon–CAR) Larger casque; blue gape wattles. **5c Juv** Tawny-buff head with two blackish lateral crown-stripes; no casque; dull grey-brown plumage speckled buffish-white and irregularly barred rusty-buff. ▲ Savanna; various open habitats. Introduced, Cape Verde, São Tomé and Annobón. ✤ Commonest call a series of hard, raucous notes interspersed with grating rattle *chek-chek-chek krrrrrr chek-chek-chek-...* [CD2:6a]

 Maps on page 120

1b
1a
2a
2b
3a
3b
4b
4a
5b
5a
4c
5c

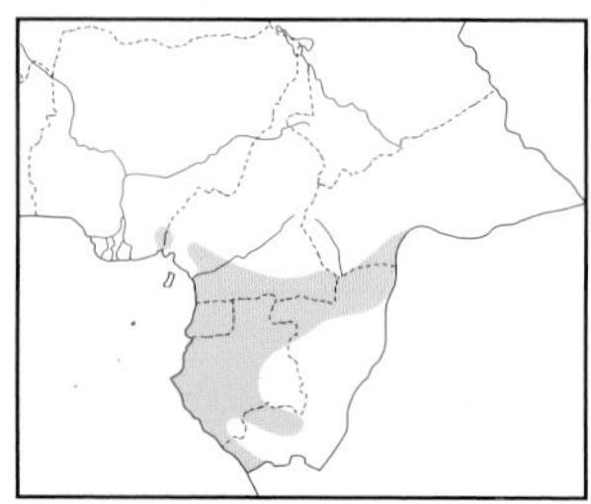
1. BLACK GUINEAFOWL

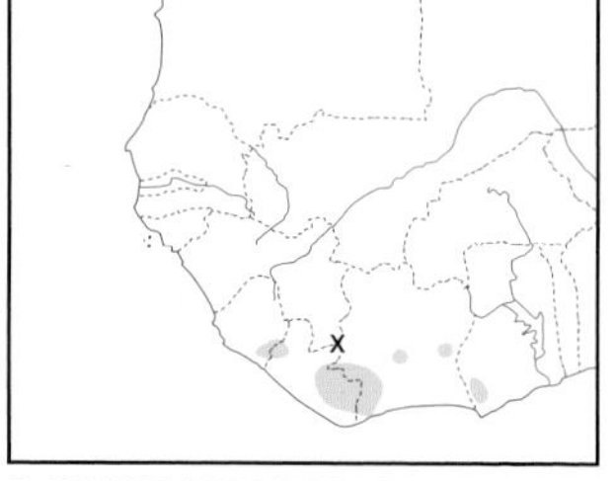
2. WHITE-BREASTED GUINEAFOWL

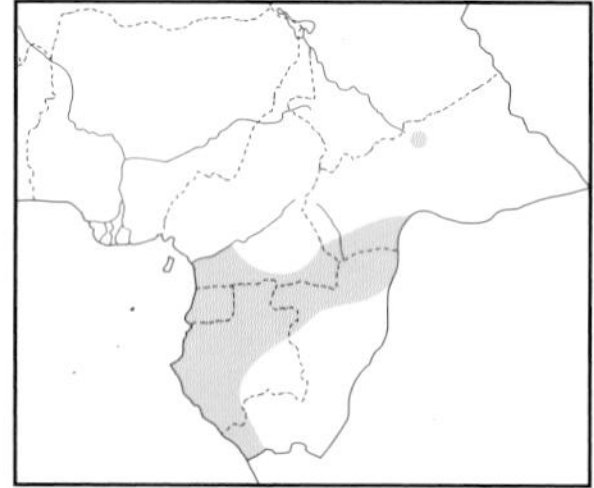
3. PLUMED GUINEAFOWL

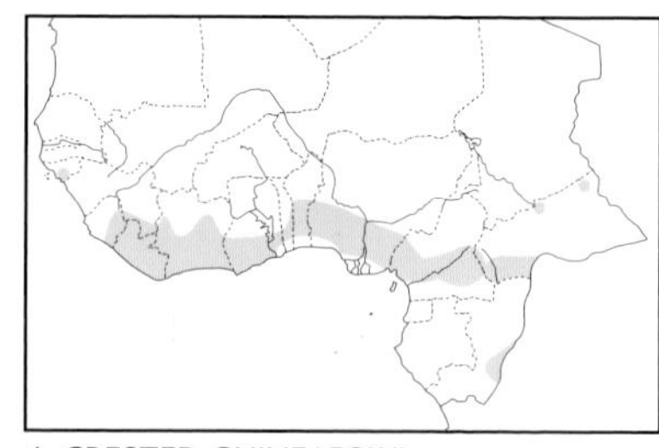
4. CRESTED GUINEAFOWL

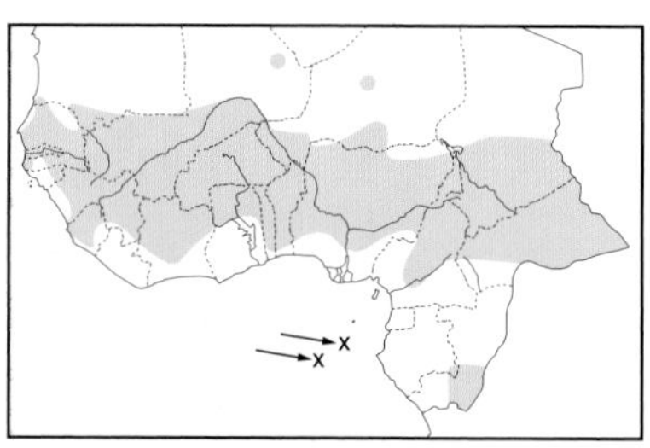
5. HELMETED GUINEAFOWL

Barbary Partridge

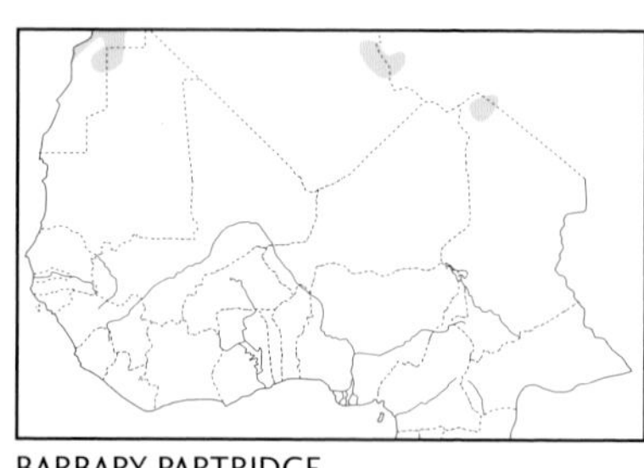
BARBARY PARTRIDGE

BARBARY PARTRIDGE *Alectoris barbara* 32–35 cm **R Ir**
Rufous-brown collar with white spots; boldly barred flanks; bright red bill and legs. ▲ Rocky slopes. N Mauritania. ✤ A rapid series of hoarse, impure *kruk* or *ktchuk* notes interspersed with occasional *kuk-kow* (or *chukor*), which breaks rhythm. [CD2:7a]

Plate on page 118

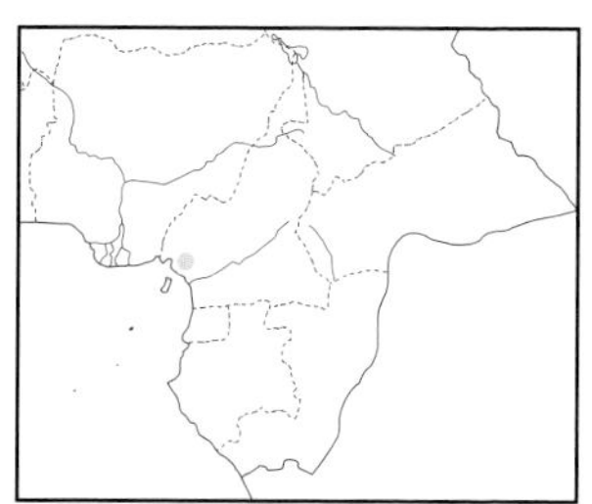

1. MOUNT CAMEROON FRANCOLIN

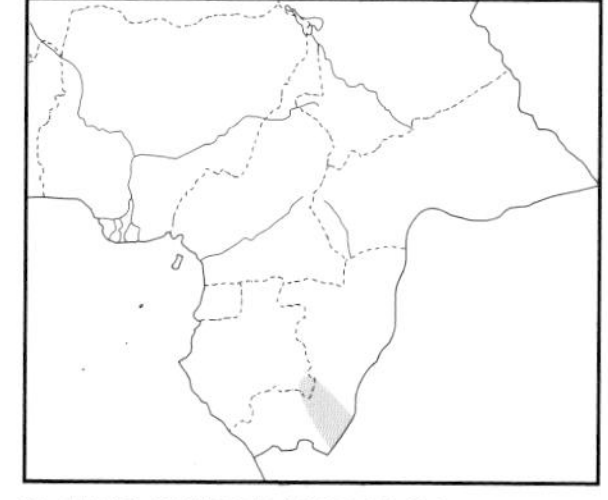

2. FINSCH'S FRANCOLIN

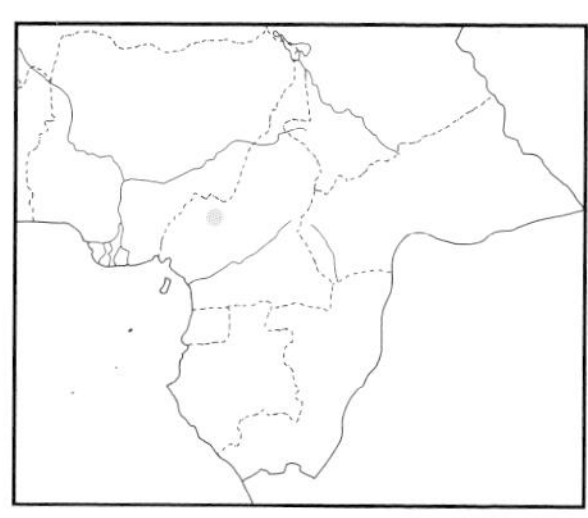

3. RING-NECKED FRANCOLIN

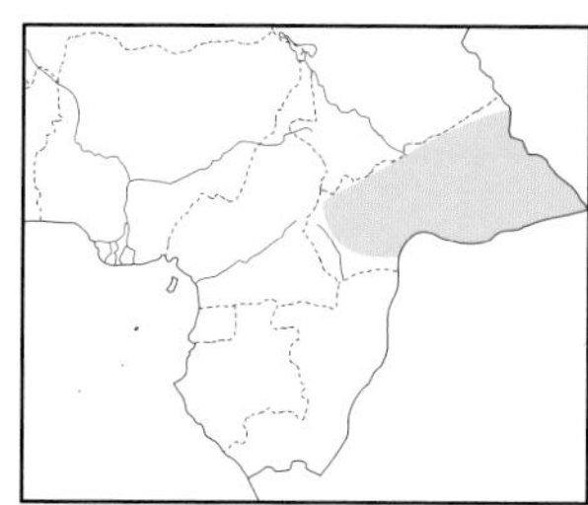

4. HEUGLIN'S FRANCOLIN

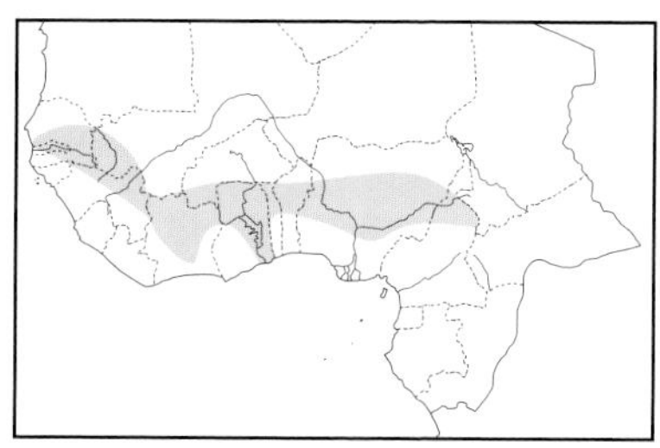

5. WHITE-THROATED FRANCOLIN

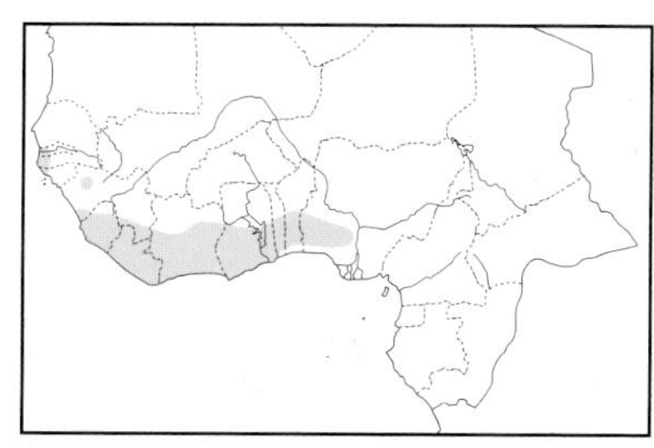

6. AHANTA FRANCOLIN

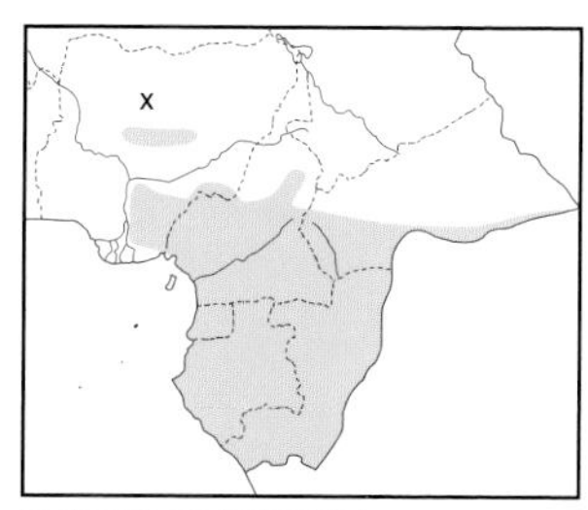

7. SCALY FRANCOLIN

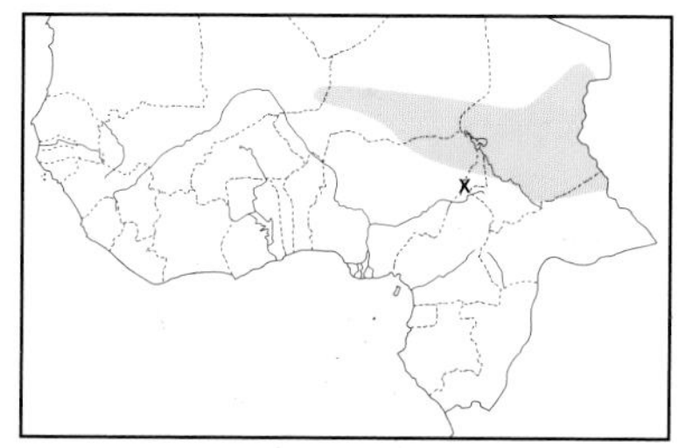

8. CLAPPERTON'S FRANCOLIN

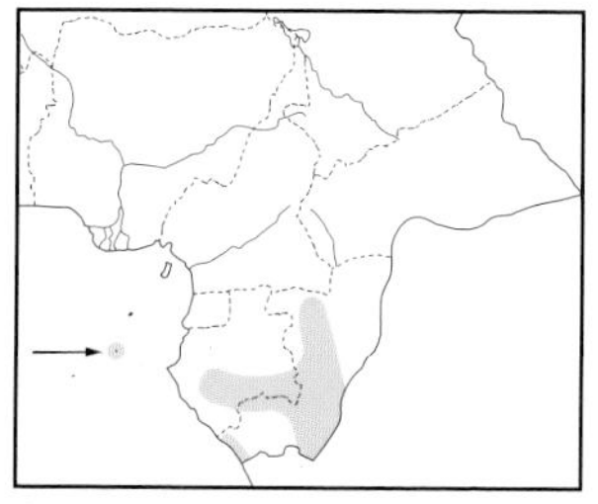

9. RED-NECKED FRANCOLIN

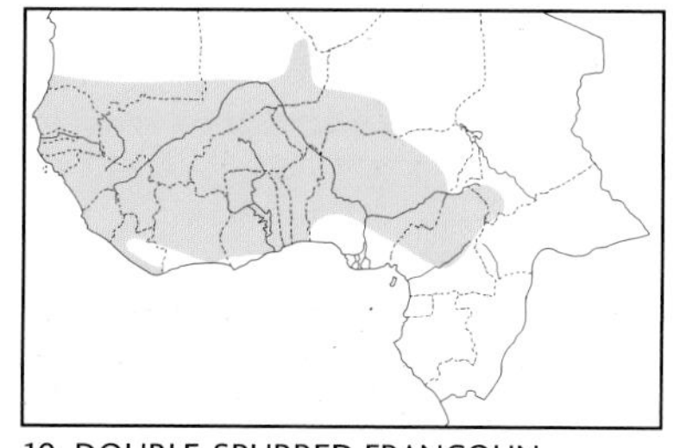

10. DOUBLE-SPURRED FRANCOLIN

PLATE 30: FRANCOLINS

Thickset terrestrial birds with small head, short, robust bill, short wings and tail, and strong feet. Usually in pairs or small groups. Reluctantly take flight; prefer to run. Identification features include bare part coloration, vocalisations, habitat and range. Calls often loud and raucous.

1 **MOUNT CAMEROON FRANCOLIN** *Francolinus camerunensis* *c.* 33 cm **R u EN**
1a Ad male Very dark; bright red bill, eye patch and legs. **1b Ad female** Streaked, whitish throat; whitish marked underparts. ▲ Montane forest. Endemic to Mt Cameroon (850–2100 m). ✤ Short series of loud, trumpeting (not grating) whistles, *KILU KILU KILU*, slower *KEE-ku KEE-ku KEE-ku* or combinations of both. Alarm an abrupt *khik*. [CD5:95]

2 **FINSCH'S FRANCOLIN** *Francolinus finschi* *c.* 35 cm **R lf/u**
2a Mainly rufous head. **2b In flight** Rufous in wings. ▲ Open grassland. ✤ Loud, repeated, rather high-pitched *kwit-e-kwee*, usually given at dusk. [CD5:86]

3 **RING-NECKED FRANCOLIN** *Francolinus streptophorus* 30–33 cm **R ls**
3a Ad male Black-and-white barring around neck and on upper breast. **3b Ad female** Finely barred rusty-buff above. ▲ Rocky, grassy slopes. W Cameroon (Foumban area). ✤ Distinctive, short series of melodious, fluty notes *thuuu*, *tee whiut tew-tewew*. [CD5:84]

4 **HEUGLIN'S FRANCOLIN** *Francolinus icterorhynchus* *c.* 32 cm **R c**
The only francolin with yellowish bill, eye patch and legs. ▲ Open and lightly wooded grassland, cultivation. ✤ Harsh, slow *kerak kerak kek* or faster *kerak-kerak-kerak-kerak-kerrr*. [CD5:93]

5 **WHITE-THROATED FRANCOLIN** *Francolinus albogularis* *c.* 23 cm **R u/r, l**
Small. **5a Ad male *albigularis*** (Senegambia–Ivory Coast) Whitish supercilium; white throat. Chestnut wings conspicuous in flight. **5b Ad female *albigularis*** Variably and finely barred dark on breast and flanks. **5c Ad male *buckleyi*** (E Ivory Coast–N Cameroon) Head buff to rusty-buff. **5d Ad female *buckleyi*** Wavy barring below extending onto belly. ▲ Wooded grassland. ✤ Very fast, high-pitched *kulikulikulikulikuli*... and shrill, rasping *kREEK-krik kREEk-krikikikikew*, accelerating at end and fading away. [CD5:82]

6 **AHANTA FRANCOLIN** *Francolinus ahantensis* *c.* 33 cm **R f/ls**
Dark brown streaked white on hindneck and mantle; orange-red bill and feet. ▲ Forest edge, second growth and thickets. Endemic. ✤ Repeated, loud, grating, rather high-pitched *kee-kee-keRRREE kee-kee-keRRREE*... usually in duet. [CD5:90]

7 **SCALY FRANCOLIN** *Francolinus squamatus* 30–33 cm **R c/f**
Dark grey-brown overall; scarlet legs. ▲ Forest edge, second growth, thickets. ✤ Series of loud, grating syllables increasing in volume *kerrrAAK kerrrAAK kerrrAAK*... with variations; also ascending *khiup-khiup-khiup-khiup khiupkiurrr khiupkiurrr*.... [CD5:89]

8 **CLAPPERTON'S FRANCOLIN** *Francolinus clappertoni* 30–33 cm **R lc**
Large; red eye patch; white supercilium; black bill; dusky-red legs. ▲ Arid bush and grassland. ✤ Short series of loud, grating *kerrrAK* or slower, more drawn-out *kerrroAH* calls, with variations, somewhat resembling 10 and 4. [CD5:94]

9 **RED-NECKED FRANCOLIN** *Francolinus afer* 35–41 cm **R lf/c**
9a Ad *cranchii* (S Gabon–Congo) Bare red throat. **9b Ad *afer*** (introduced São Tomé) Pale brown breast streaked black; rest of underparts blackish, feathers fringed white. ▲ Grassland. ✤ Loud, grating, rather low-pitched *kAARkukukuw kAARkukukuw*... and higher pitched squealing series increasing in volume and ending in drawn-out grating *koAARRK koAARRK koAARRK-kek koAARRK-kek KeRRRrrr*, with variations. [CD5:97]

10 **DOUBLE-SPURRED FRANCOLIN** *Francolinus bicalcaratus* 30–35 cm **R c**
Large; mainly brownish; white supercilium; greenish legs. ▲ Grassland, farmbush, scrub. ✤ Series of loud, grating syllables, *rrrraak kerRRAK kerRRAK*... [CD2:7b]

Maps on page 121

2b
2a
3b
1b
1a
3a
4
5a
5b
5d
5c
7
6
8
9b
9a
10

PLATE 31: BUTTONQUAILS, QUAILS AND SMALL FRANCOLINS

Buttonquails. Very small, quail-like terrestrial birds with short, rounded body and wings. Females larger and brighter than males. Secretive and difficult to flush; usually do not fly far.

1 LITTLE BUTTONQUAIL *Turnix sylvaticus* 14–16 cm **R/M s/lc**
1a Very small; orangey breast spotted black on sides; pale yellow eye. **1b In flight** Brown remiges contrast with pale coverts. ▲ Grassland, cultivation, other open habitats. ✤ Strange, low, far-carrying *hooooooo...* like distant lowing cow or foghorn. [CD2:8a]

2 BLACK-RUMPED BUTTONQUAIL *Turnix hottentottus* 14–16 cm **R/M/V lr**
2a Resembles 1a, but darker; orange-rufous breast-sides barred black and white; eye dark. **2b In flight** Black rump. ▲ Grassland. ✤ Low, resonant hoots similar to 1 but shorter and delivered in rather fast series *hooo, hooo, hooo, hooo,...* [CD6:1a]

3 QUAIL-PLOVER *Ortyxelos meiffrenii* 11–13 cm **R/M lu**
3a Pale sandy-rufous above; mainly whitish below. **3b In flight** Blackish remiges; diagonal band on greater coverts. Flight fluttering. ▲ Arid grassland. ✤ Mostly silent.

Quails. Very small and rounded. Sexually dimorphic. Secretive. Usually seen when flushed.

4 HARLEQUIN QUAIL *Coturnix delegorguei* 15–16 cm **M/V/R u**
4a Ad male Black-and-white head pattern; black-and-chestnut underparts. **4b Ad female** Darker than 6b; pale tawny underparts. ▲ Grassland, cultivation. Nomadic. ✤ Reminiscent of 6, but slightly slower, marginally higher pitched and more monotonous. [CD5:78]

5 BLUE QUAIL *Coturnix (chinensis) adansonii* 13–14 cm **R/M lu/r**
5a Ad male Dark slate-blue; chestnut on wings, flanks and uppertail-coverts. **5b Ad female** Heavily barred breast and flanks. ▲ Grassland, edges of cultivation. ✤ Male: series of piping whistles *kee-keew kee-keew kee-kee-kuh*, first note loudest. [CD5:77]

6 COMMON QUAIL *Coturnix coturnix* 16–18 cm **P/R* u/r**
6a Ad male Strong head pattern with blackish anchor on throat; breast-sides and flanks streaked buff, black and chestnut. **6b Ad female** Duller; no anchor. **6c In flight** Plain wings; streaked upperparts. ▲ Grassy habitats. ✤ Far-carrying, rhythmic *KWIk-ik-wik*. [CD2:6b]

Francolins (continued).

7 LATHAM'S FOREST FRANCOLIN *Francolinus lathami* 20–25 cm **R u/lf**
7a Ad male Bold head pattern; white-spotted underparts; yellow legs. **7b Ad female** Browner than male; belly white. **7c Juv** Brownish head; white throat. ▲ Forest interior. Secretive. ✤ Distinctive nocturnal calls include series of rather melodious notes, accelerating at end *krook-rookrookrookrookroo*; loud *krok! krokrorrrr...* Also loud, high-pitched *kirr-kikikiki...* in duet. [CD5:80]

8 SCHLEGEL'S FRANCOLIN *Francolinus schlegelii* 21–24 cm **R lu/r**
8a Ad male Rusty-orange head; dusky eye-stripe; vinous-chestnut upperparts. **8b Ad female** Duller; narrow barring on underparts more irregular. ▲ Open wooded grassland. ✤ Loud initial note, followed by accelerating series of more grating ones, fading at end, *KWEEK! kre-kre-krekrekrekew*; lower pitched and less shrill than those of 9. [CD5:83]

9 COQUI FRANCOLIN *Francolinus coqui* 20–25 cm **R r/lc**
9a Ad male *spinetorum* (S Mauritania–N Nigeria; r/u) Rufous-orange head; barred breast and flanks. **9b Ad female *spinetorum*** Note facial pattern. **9c Ad male *coqui*** (SW Gabon–Congo; lc) Bars extending onto belly. **9d Ad female *coqui*** Belly heavily barred. ▲ Grassland. ✤ Oft-repeated, rhythmic, shrill *kEE-kwi, kEE-kwi...*, and shrill, rasping *kREEK-krik kREEk-kREEk krikikikikew*, accelerating at end and fading away. [CD5:81]

10 STONE PARTRIDGE *Ptilopachus petrosus* c. 25 cm **R lc/u**
Bantam-like; dark brown; cocked tail. ▲ Various savanna habitats, often with rocks. ✤ An abruptly started, rising series of clear, fluty notes with cheerful, pleasant quality, *weet-weet-weet-weet-...*, often in duet or chorus and mostly at dawn and dusk. [CD5:79]

Maps on page 126

1b
2b
3b
1a
2a
3a
4a
4b
5a
5b
6c
6a
6b
7c
7b
7a
8a
8b
9c
9d
9a
9b
10

PLATE 31: BUTTONQUAILS, QUAILS AND SMALL FRANCOLINS

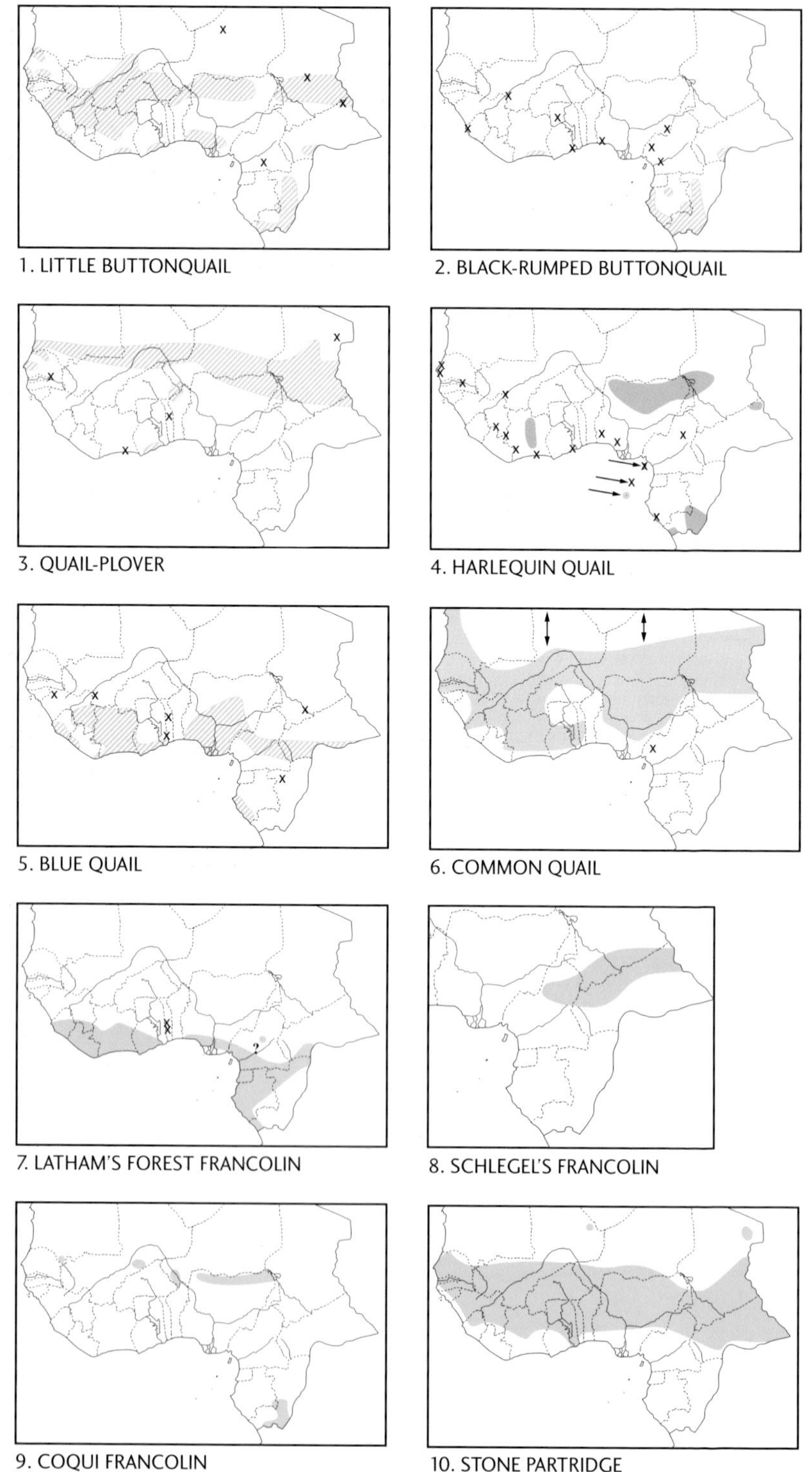

1. LITTLE BUTTONQUAIL

2. BLACK-RUMPED BUTTONQUAIL

3. QUAIL-PLOVER

4. HARLEQUIN QUAIL

5. BLUE QUAIL

6. COMMON QUAIL

7. LATHAM'S FOREST FRANCOLIN

8. SCHLEGEL'S FRANCOLIN

9. COQUI FRANCOLIN

10. STONE PARTRIDGE

Plate on page 124

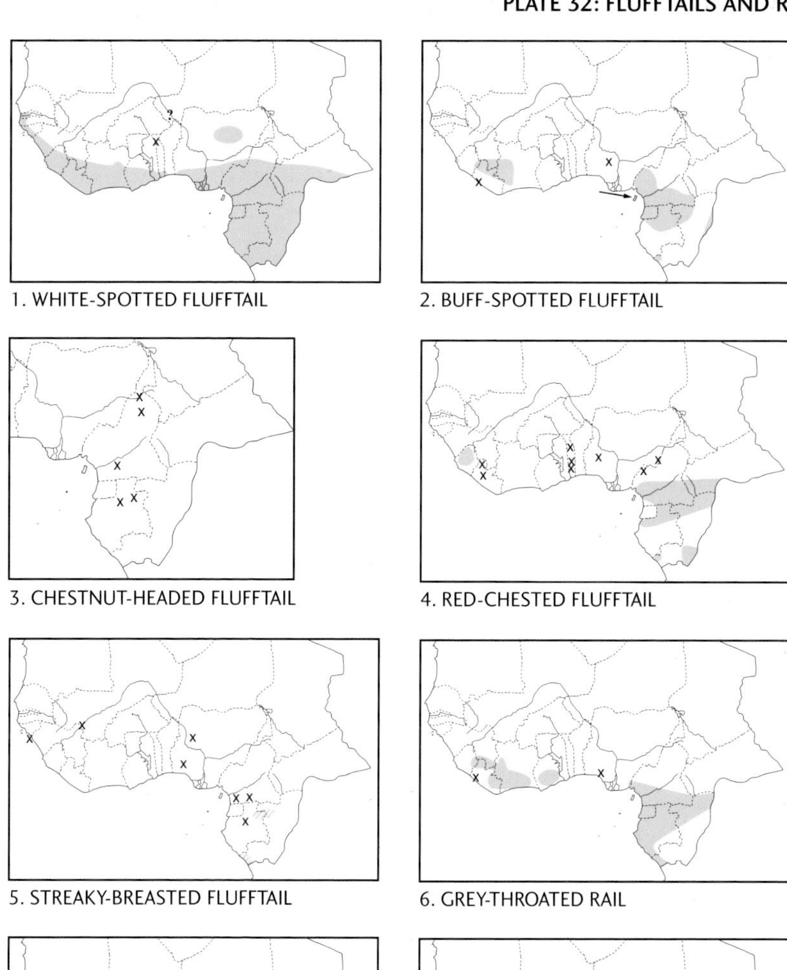

1. WHITE-SPOTTED FLUFFTAIL

2. BUFF-SPOTTED FLUFFTAIL

3. CHESTNUT-HEADED FLUFFTAIL

4. RED-CHESTED FLUFFTAIL

5. STREAKY-BREASTED FLUFFTAIL

6. GREY-THROATED RAIL

7. NKULENGU RAIL

8. AFRICAN WATER RAIL

Plate on page 128

Flufftails. Small, secretive rails. Vocalisations highly characteristic. Note, in males, amount of chestnut on head, neck and breast, markings on body (spots or streaks) and colour of tail.

1 WHITE-SPOTTED FLUFFTAIL *Sarothrura pulchra* 15 cm R c/s
1a Ad male Head, breast and tail chestnut; rest of plumage black densely spotted white. **1b Ad female** Black areas barred rufous-buff; tail chestnut barred black. **1c Juv** Unique among flufftails in being as respective adult but duller. ▲ Lowland forest, cultivation. ✤ Vocal. Song a rapid series of short, resonant notes *too-too-too-too-too-...* Second bird may join with similar song or low *krw krw krw...* When excited, a very fast, hard and shrill *kwipipipipipip...* [CD6:2b]

2 BUFF-SPOTTED FLUFFTAIL *Sarothrura elegans* 15 cm R/M r/lf
2a Ad male Upperparts spotted buff; tail rufous barred black. **2b Ad female** Brown above finely spotted buff; coarsely barred tawny-and-brown belly. **2c Juv** Sooty-brown; belly paler. ▲ Lowland forest, overgrown cultivation. ✤ Song (mainly at night) a low and hollow, foghorn-like *whoooooooooooooo...* Second bird may join with very fast, low *tugutugutugutugu...* [CD6:3a]

3 CHESTNUT-HEADED FLUFFTAIL *Sarothrura lugens* 15 cm R lr/u
3a Ad male Chestnut confined to head; rest of plumage black finely streaked white. **3b Ad female** Head streaked buff-brown; underparts spotted and streaked. **3c Juv** Mainly sooty-black. ▲ Swampy and grassy habitats. ✤ Song a series of moaning *hoo* notes, delivered at same or gradually increasing speed and intensity, then dying away and ending abruptly. [CD6:4a]

4 RED-CHESTED FLUFFTAIL *Sarothrura rufa* 15 cm R lc/r
4a Ad male Chestnut head and breast; rest of plumage black finely streaked white. **4b Ad female** Dark brown to buffish, densely barred. **4c Juv** Mainly sooty-black. ▲ Swampy and grassy habitats. ✤ Song a more or less regular series of *whoah* notes. Second bird may join with higher pitched notes. Also *hoo-du hoo-du hoo-du...* and fast *pulipulipulipulip...* [CD6:3b]

5 STREAKY-BREASTED FLUFFTAIL *Sarothrura boehmi* 15 cm R/V lc/r
5a Ad male Rufous paler and more extensive than in 3a, underparts whiter with coarser streaking. **5b Ad female** Brownish-black with scalloped and barred appearance. **5c Juv** Mainly sooty-black. ▲ Swampy and grassy habitats. ✤ Song a hollow *whoo* note, repeated in long series (up to 25x). Also a rapid, high-pitched, crescendo *kyeh-kyeh-kyeh-kyeh-...* [CD6:4b]

Rails. Medium-sized to large terrestrial and aquatic birds. Secretive. Sexes similar.

6 GREY-THROATED RAIL *Canirallus oculeus* 30 cm R s/lf
6a Ad White spots on edge of dark brown wing; grey face and throat; rufous-chestnut neck to upper belly and tail. **6b Juv** Dark brown head and underparts. ▲ Lowland forest. ✤ Ventriloquial song, a series of hollow *pl* notes, interspersed with soft *doo* or *dooah* notes after *c.* 20 seconds. Calls include loud, explosive booming of 3–6 notes on a descending scale of half tones *ooe-ooe-ooe* and short bursts of *dook-dook-dook* or muffled *thook-thook-....* [CD6:2a]

7 NKULENGU RAIL *Himantornis haematopus* *c.* 43 cm R u/lc
Large. Dark brownish scaled paler; long red legs. ▲ Lowland forest. ✤ Distinctive antiphonal song a loud, far-carrying, sonorous duet *koKAWkoKAW-hoHO*, repeated for several minutes and uttered mostly after sunset and before sunrise; also at night. Solitary birds give two loud, honking notes *HO-HO*, repeated at regular intervals. [CD6:1b]

8 AFRICAN WATER RAIL *Rallus caerulescens* 28 cm R lr
8a Ad Long red bill; red legs; head-sides and underparts grey. **8b Juv** Browner. ▲ Aquatic habitats. ✤ Vocal. A fast series of shrill notes, starting with rapid trill, speeding up, then gradually slowing and descending in intensity *trrrrri-kew-kew-kew-kew-kew-...* Others may join with low clucking and grunting notes. [CD6:5b]

 Maps on page 127

1b
1a
2a
2b
1c
3a
3b
2c
4c
3c
5c
4b
4a
5a
5b
6a
6b
7
8b
8a

Small to medium-sized terrestrial and aquatic birds. Flick short tail while walking. Legs and toes long; wings rounded. Flight over short distances fluttering and seemingly weak, but many are long-distance migrants. Most species skulking and difficult to observe.

1 BLACK CRAKE *Amaurornis flavirostra* 19–23 cm **R c**
1a Ad All black; bright yellow bill; red legs. **1b Juv** Duller; bill and legs dusky. ▲ Freshwater wetlands. Less shy than other crakes; often in the open. ✤ Vocal. A variety of clucking, purring and growling sounds, uttered in stereotyped duet. [CD6:7]

2 SPOTTED CRAKE *Porzana porzana* 22–24 cm **P r**
2a Ad male Yellow bill with orange-red base; plain buff undertail-coverts. **2b Ad female** Usually browner and more speckled than male. ▲ Marshes and similar habitats. Usually skulking but not particularly shy; also forages in the open. Flies reluctantly. Mainly active at dusk. ✤ Rhythmic series of far-carrying, explosive *hwit!* notes ('whiplash'), also heard in winter quarters, mainly at dusk and at night. Contact calls between wintering birds a hard *whee-up*. [CD2:10b]

3 LITTLE CRAKE *Porzana parva* 18–20 cm **P r**
3a Ad male Lime-green bill with red base; long primary projection (almost as long as exposed tertials, with at least 5 primary tips visible); blue-grey face and underparts; faint barring on flanks. **3b Ad female** Whitish throat; buffish underparts. ▲ Marshes and similar habitats. Usually secretive but not shy; sometimes in the open and allowing close approach. Only reluctantly flies. ✤ Probably silent in winter quarters. Song of male *ik ikik ik*, a rhythmic series *kik-kik-kik-kik-kik-*..., or an accelerating series of notes descending in pitch *kwek, kwek, kwek kwek kwek-kwek-kwek*..., mainly at night. [CD2:9b]

4 BAILLON'S CRAKE *Porzana pusilla* 17–19 cm **P/V f**
4a Ad male Green bill; short primary projection (primaries barely projecting beyond tertials, with rarely more than 3 primary tips visible); densely speckled upperparts; obvious barring on flanks. **4b Ad female** Throat and underparts often paler, ear-coverts brownish. ▲ Marshes and similar habitats. Very skulking and rarely in the open, but not shy and often tame. Creeps through dense vegetation flicking tail, only reluctantly flying short distances when flushed. Most active at dawn and dusk. ✤ Probably silent in winter quarters. Song of male a dry rattle *tk tk tk tk rrrrrkkkkkkk*, mainly at night. [CD2:10a]

5 STRIPED CRAKE *Aenigmatolimnas marginalis* 18–21 cm **M s/r**
5a Ad male Greenish bill; dark brown upperparts finely streaked white; russet rear flanks and undertail-coverts. **5b Ad female** Head and breast to flanks grey. **5c Juv** As male but duller and browner; no white streaking. ▲ Wet grasslands and similar habitats. Highly secretive. Very hard to flush, prefers creeping unobtrusively through vegetation. ✤ Usually silent. In breeding season a long dry rattle (like engine) *rrtktktktktktktktktk*..., mainly at night. Call a fast *krw-krw-krw-*... [CD6:6]

6 AFRICAN CRAKE *Crex egregia* 20–23 cm **M/R c/r**
6a Ad Face to breast grey; lower underparts barred black and white. **6b Juv** Browner below; barring less distinct. **6c In flight** Upperparts dark brown mottled black; flight feathers blackish. ▲ Grassy habitats. Active all day, but mostly at dawn and dusk. Skulking, but not shy. When flushed, flies reluctantly for short distance with dangling legs. ✤ Single hard *kluk, kruk, krw* or *kip*, occasionally in fast series. [CD6:5a]

7 CORN CRAKE *Crex crex* 27–30 cm **P r NT**
7a Mainly brownish-buff; broad brownish-grey supercilium. **7b In flight** Conspicuous chestnut wings (diagnostic). ▲ Grassy habitats, usually avoiding wet places. Skulking and rarely seen. When flushed, flies reluctantly over short distance with dangling legs. Most active at dawn and dusk. ✤ Silent in winter quarters. [CD2:8b]

 Maps on page 132

1a
1b
2a
3a
4a
2b
3b
4b
5b
5a
5c
6b
6c
6a
7b
7a

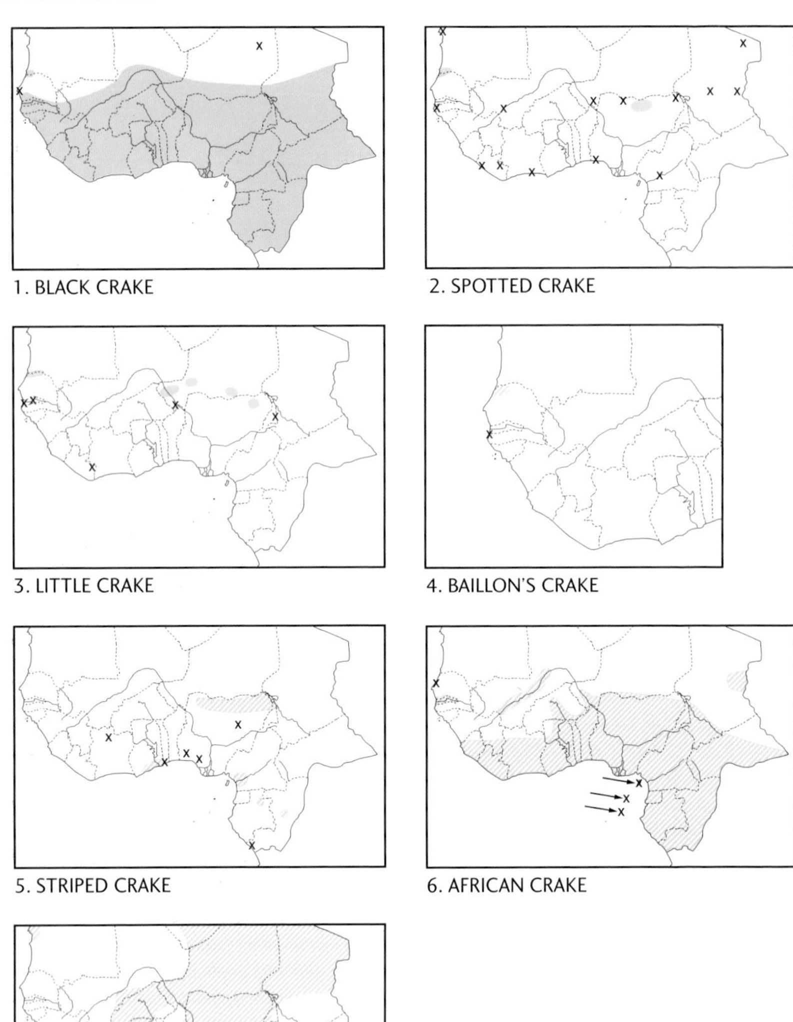

Plate on page 130

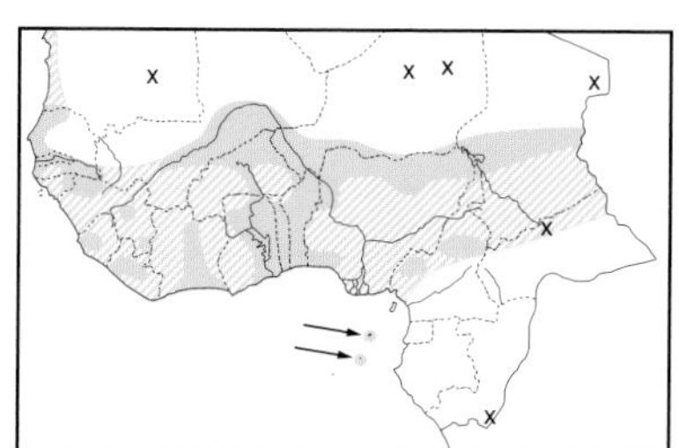
1. COMMON MOORHEN

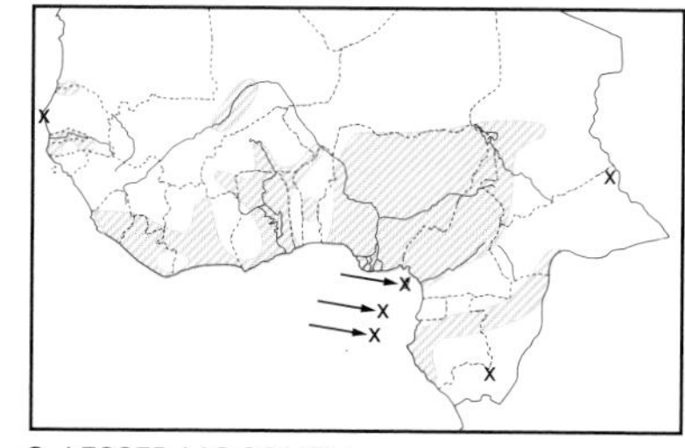
2. LESSER MOORHEN

3. ALLEN'S GALLINULE

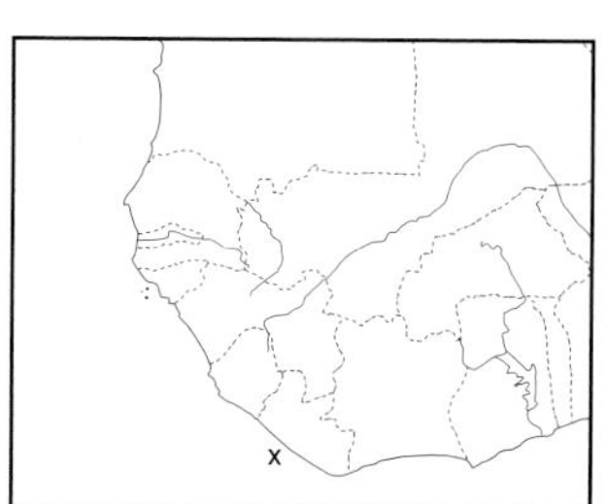
4. AMERICAN PURPLE GALLINULE

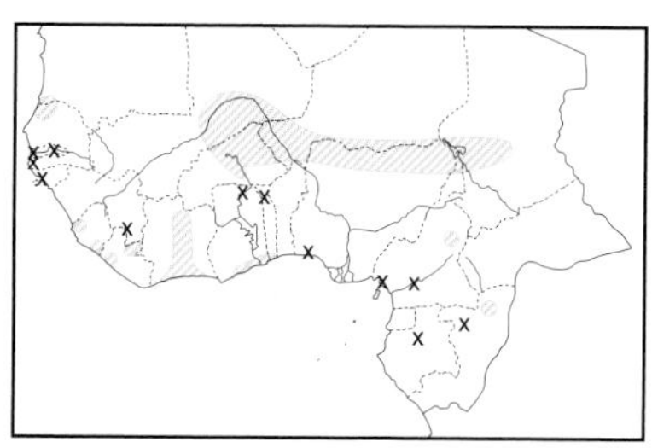
5. PURPLE SWAMPHEN

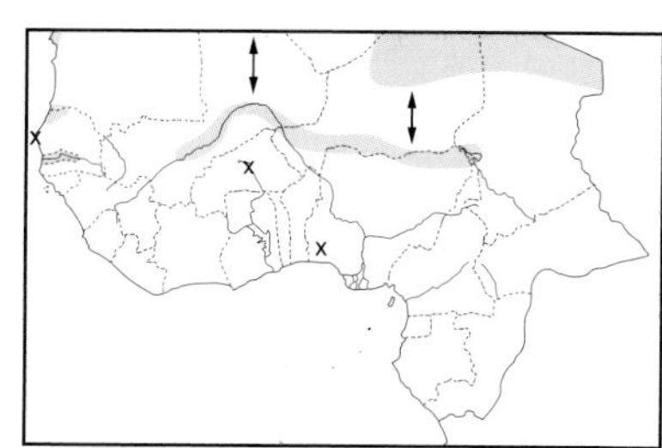
6. EURASIAN COOT

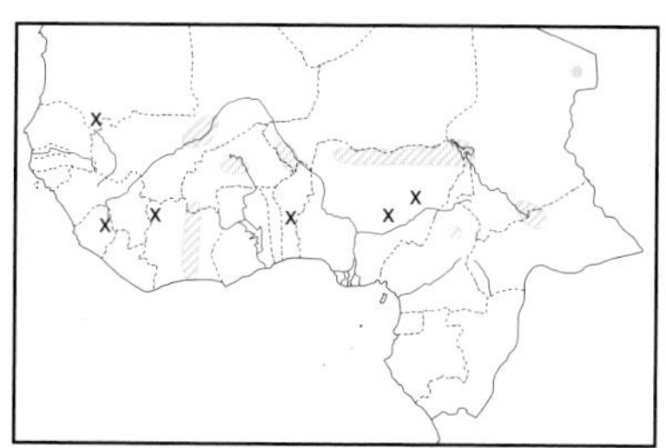
7. LESSER JACANA

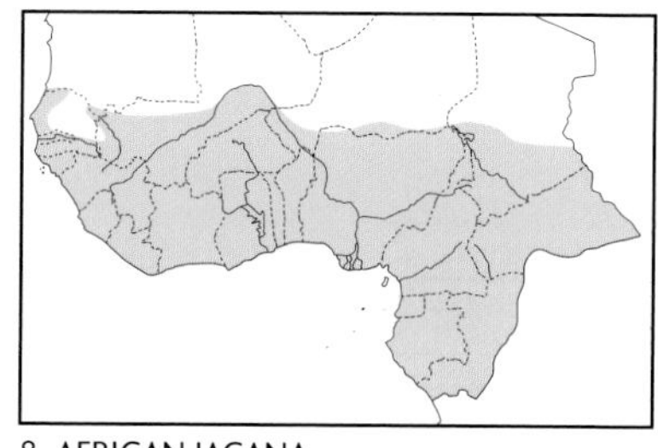
8. AFRICAN JACANA

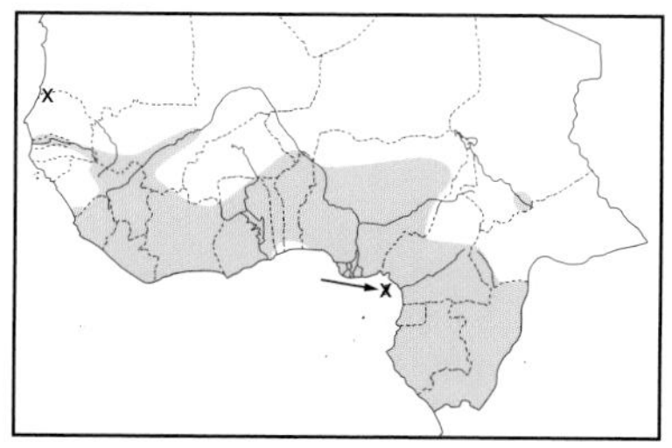
9. AFRICAN FINFOOT

PLATE 34: GALLINULES, JACANAS AND FINFOOT

Gallinules, moorhens and coots. Less secretive than rails and crakes. Frequent open water or clamber about in dense waterside vegetation.

1 COMMON MOORHEN *Gallinula chloropus* 30–36 cm R/P/V+ lf/r
1a Ad White line on flanks; red bill with yellow tip; greenish legs. **1b Juv** Mainly greyish-brown; throat and belly whitish; bill greenish-brown. ▲ Various freshwater wetlands. ✤ Sudden, single *yerrrrp* and short hard *kik*, sometimes in rapid series *kikikikikik-kik-kik-kik*... [CD2:11b]

2 LESSER MOORHEN *Gallinula angulata* 23 cm R/M lc/u
2a Ad As 1a, but smaller and greyer; bill yellow with red culmen. **2b Juv** Dark brown above; head-sides, neck and breast brownish-buff; bill dusky-yellow with darkish culmen. ▲ Various freshwater wetlands. ✤ Short, sharp *khup* and *prp*, also in series. [CD6:9]

3 ALLEN'S GALLINULE *Porphyrio alleni* 25 cm M/R u/lc
3a Ad Glossy purplish and dark green; red bill and legs; bluish frontal shield. **3b Ad female at start of breeding** Apple-green frontal shield (turquoise-blue in male). **3c Juv** Mainly buffish-brown; scaly upperparts. ▲ Various freshwater wetlands. ✤ Various hard, dry calls, singly or in series, e.g. *kuk kuk kuk kk* and *kip-kip-kip-kip-kirrrr*. In flight, high-pitched *kli-kli-kli-*... [CD6:8]

4 AMERICAN PURPLE GALLINULE *Porphyrio martinica* 33 cm V
Yellow-tipped red bill; yellow legs. ▲ American vagrant recorded off Liberian coast.

5 PURPLE SWAMPHEN *Porphyrio porphyrio* 45–50 cm R lc/r
P. p. madagascariensis (African Swamphen) **5a Ad** Large; glossy purple and blue-green; red bill, frontal shield and legs. **5b Juv** Paler and greyer; bare parts duskier. ▲ Various freshwater wetlands. ✤ Various nasal, groaning, clucking, hooting and wailing calls. [CD2:11a]

6 EURASIAN COOT *Fulica atra* 36–42 cm P r/lc
All black; white bill and frontal shield; lobed toes. ▲ Open waters. ✤ Loud, sharp *kut* or *khuk*. [CD2:12a]

Jacanas. Long-legged waterbirds with extremely long toes and claws, enabling them to walk on floating vegetation.

7 LESSER JACANA *Microparra capensis* 16 cm R/M lc/r
7a Ad Small size; chestnut crown; white supercilium; white underparts. Juvenile similar; rump blackish. Compare 8b. **7b In flight** Black flight feathers; white trailing edge; pale midwing panel; trailing feet. ▲ Various freshwater wetlands. ✤ Soft, rapid *whoop-whoop-whoop-whoop-whoop*, chattering *kikikikikikikiki*, softer *ku-ku-ku*, plaintive *shreee shree shree shree* or *see sree srrr*, and *soft chrr-chrr-chrr*. [CD6:18]

8 AFRICAN JACANA *Actophilornis africana* 30 cm R c
8a Ad Bright chestnut; white foreneck; blue bill and frontal shield; extremely long toes. **8b Juv** Underparts mainly white; white supercilium. ▲ Various freshwater wetlands. ✤ Various nasal, strident and grating sounds, incl. husky drawn-out whining *kyowrrr*, shorter, repeated *kreep-kreep-kreep*... and high-pitched *weep-weep-weep*... [CD6:17]

Finfoot. Unobtrusive grebe-like waterbird with long neck and small head. Inhabits forested rivers, lakes and lagoons with overhanging vegetation.

9 AFRICAN FINFOOT *Podica senegalensis* 50 cm R u/lc
9a Ad male *senegalensis* breeding (Senegambia–CAR and Congo) Red bill and feet; white line on neck-sides; finely spotted upperparts. **9b Ad male *camerunensis*** (S Cameroon, Gabon, Bioko) Much darker than 9a; few or no spots above; no white line on neck-sides. **9c Ad female** Browner; white throat and foreneck. **9d Juv** As female but paler, less spotted; lower neck to flanks tawny-buff. ▲ Throughout forest zone; also in wooded savanna. ✤ Calls include dry cackles and guttural notes. Usually silent. [CD6:13]

Maps on page 133

1a
1b
2a
2b
3a
3c
3b
4
5b
6
5a
7a
8b
8a
7b
9b
9a
9c
9d

1 **DEMOISELLE CRANE** *Anthropoides virgo* 90–100 cm; WS 155–180 cm **P r/s**
1a Ad White head plumes; black foreneck; long, narrow feathers over tail. **1b Imm** Duller. **1c In flight** Much as 2c; black on neck reaching breast; greyish tinge to inner primaries. ▲ Grasslands. ✤ Clear trumpeting, higher pitched than 2. [CD2:13b]

2 **COMMON CRANE** *Grus grus* 100–120 cm; WS 180–220 cm **V**
2a Ad Large; grey; white stripe on neck-sides; bushy feathers over tail. **2b Imm** Browner than adult. **2c In flight** Grey with black flight feathers; outstretched neck. Compare 1c. ▲ Open habitats. ✤ Far-carrying, nasal trumpeting, usually in flight. [CD2:13a]

3 **WATTLED CRANE** *Bugeranus carunculatus* *c.* 175 cm; WS 230–260 cm **V? VU**
White head and neck; dark grey cap; white feathered wattles below chin; red facial skin. ▲ Marshes, grasslands. Single old record, Guinea-Bissau (vagrant?). ✤ [CD6:10]

4 **BLACK CROWNED CRANE** *Balearica pavonina* 100 cm; WS 180–200 cm **R/M r/lf NT**
4a Ad *pavonina* (Senegambia–Chad) Dark; large white wing panel; straw-coloured crest. Bare area on head-sides half white, half pink. **4b Ad *ceciliae*** (E & C Chad) Bare cheek patch more extensively red. **4c Imm** Washed rusty; shorter crest; no bare area on head-sides. **4d In flight** Contrasting white wing-coverts. ▲ Open habitats. ✤ Loud trumpeting *honk* or *ka-wonk*. [CD6:11]

5 **SAVILE'S BUSTARD** *Lophotis savilei* 41 cm **R lf/r**
5a Ad male Small; short neck and legs; grey foreneck; black underparts. Compare 6. **5b Ad female** White throat; deep buff neck; broad white breast-band. **5c In flight** Blackish remiges separated from tawny upperwing-coverts by white line. ▲ Arid and semi-arid habitats. ✤ Clear, whistled *tuit-thit*. Male advertisement *tuit! tututututututu*, or *thut thut-thut-thut-thutututututut*; also series of frog-like notes in same rhythm. [CD6:14]

6 **BLACK-BELLIED BUSTARD** *Lissotis melanogaster* 60 cm **R/M f/r**
6a Ad male Long neck and legs; black throat and underparts. **6b Ad female** Mainly brownish-buff; slender. **6c Ad male in flight** Upperwing with largest area of white of any bustard. **6d Ad female in flight** Much less white in wing; flight feathers mainly blackish. ▲ Grassland, derived savanna. ✤ Generally silent. Male advertisement *vwok, rrorr-WIK!*. [CD6:16]

7 **WHITE-BELLIED BUSTARD** *Eupodotis senegalensis* 50–60 cm **R lf/r**
7a Ad male *senegalensis* (Sahel belt) Black throat patch; greyish-blue neck; tawny breast; white belly. **7b Ad male *mackenziei*** (Barrow's Bustard; SE Gabon–Congo) Tawny-buff hindneck. **7c Ad female** Throat white; neck mainly tawny-buff. **7d In flight** Tawny-buff above with mainly black remiges. ▲ Grassland, thorn scrub. ✤ Vocal. Loud *kuk-kwarrak*. [CD6:15]

8 **DENHAM'S BUSTARD** *Neotis denhami* 80–100 cm; WS 170–250 cm **R/M s/lf NT**
8a Ad male Large; rufous hindneck; grey foreneck; bold white wing markings. **8b Ad female** Smaller; head-sides, throat and foreneck tinged buffish. **8c In flight** Mainly dark above, with conspicuous white markings on wings. ▲ Wooded savanna. ✤ Mostly silent.

9 **ARABIAN BUSTARD** *Ardeotis arabs* 80–100 cm; WS 205–250 cm **R/M u/r**
9a Ad male Large; grey, thick-looking neck. Female smaller. **9b In flight** Mainly dark above; outer primaries and coverts mainly black. Compare 8c. ▲ Arid grassy plains, *Acacia* woodland. ✤ Generally silent. [CD2:15b]

10 **NUBIAN BUSTARD** *Neotis nuba* 50–70 cm; WS 140–180 cm **R/M u/r NT**
10a Ad male Fairly large and pale; rufous, black-edged crown; black throat. **10b Ad female** Head markings duller. **10c In flight** Black remiges; large white area on primaries; mainly white underwing. ▲ Dry thorn scrub. ✤ Described as a shrill *maqur*.

11 **HOUBARA BUSTARD** *Chlamydotis undulata* 55–65 cm; WS 135–170 cm **R u VU**
11a Black frill on neck-sides. **11b In flight** Black flight feathers; white patch on inner primaries. ▲ Semi-desert. Mauritania. ✤ Generally silent. [CD2:14b]

 Maps on page 138

1c
2c
4d
1b
1a
2b
2a
3
4c
4a
4b
8c
9b
10c
5c
6c
7d
6b
7b
7a
7c
5a
5b
6a
11b
8b
8a
6d
10b
10a
11a
9a

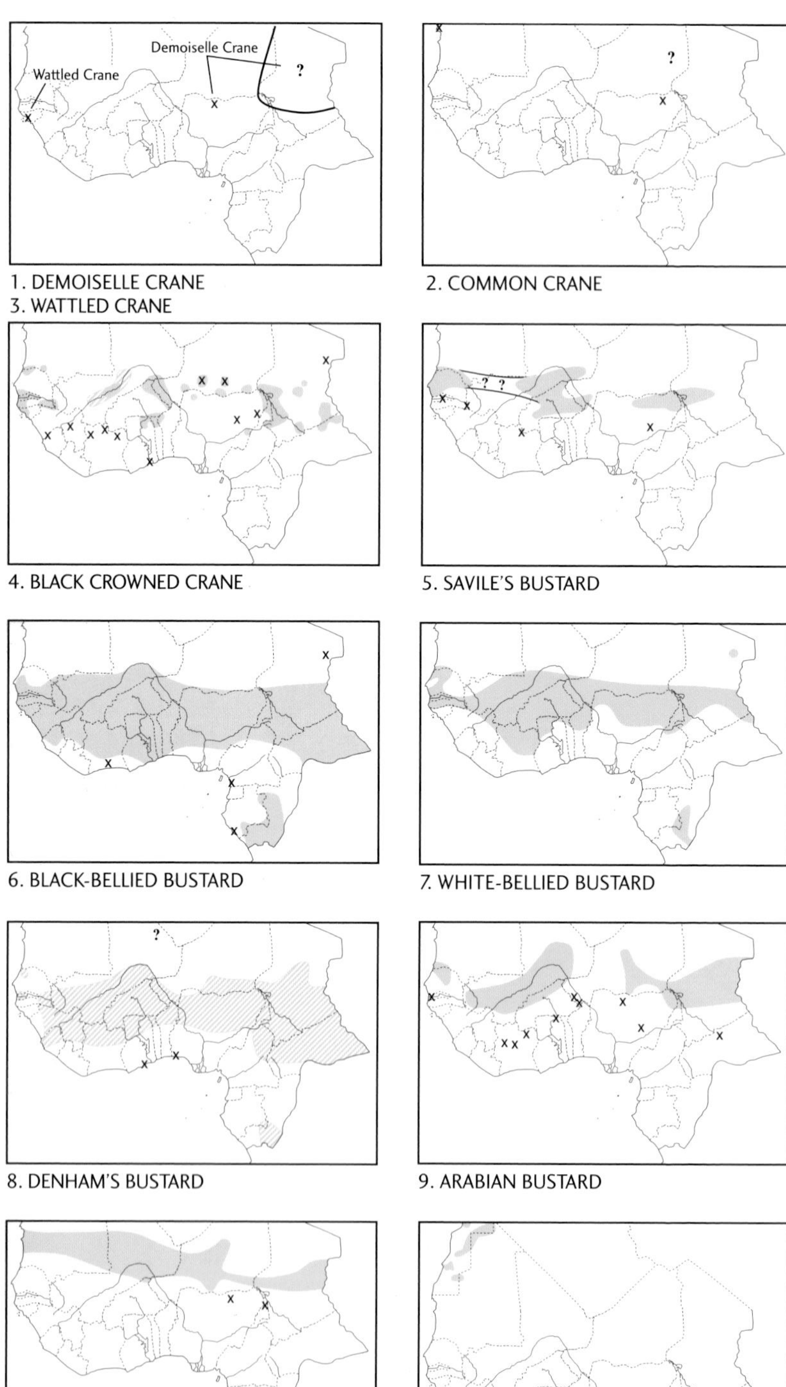

1. DEMOISELLE CRANE
3. WATTLED CRANE

2. COMMON CRANE

4. BLACK CROWNED CRANE

5. SAVILE'S BUSTARD

6. BLACK-BELLIED BUSTARD

7. WHITE-BELLIED BUSTARD

8. DENHAM'S BUSTARD

9. ARABIAN BUSTARD

10. NUBIAN BUSTARD

11. HOUBARA BUSTARD

Plate on page 136

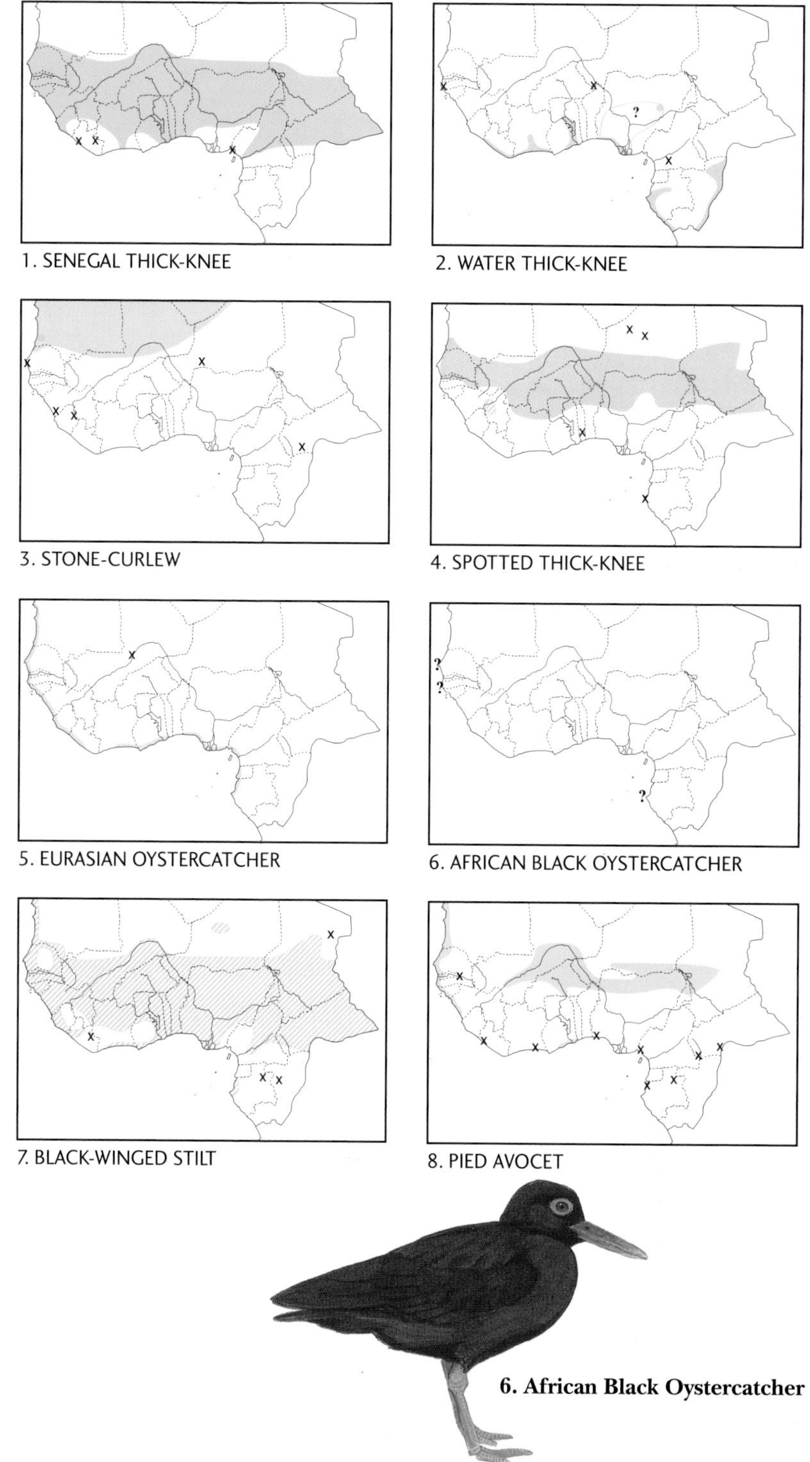

6. African Black Oystercatcher

Plate on page 140

PLATE 36: THICK-KNEES, OYSTERCATCHERS, STILT AND AVOCET

Thick-knees. Cryptically patterned, with large head and eyes. Crepuscular and nocturnal, often encountered on roads at night. Calls melodious and far carrying. Note wing pattern.

1 SENEGAL THICK-KNEE *Burhinus senegalensis* 32–39 cm; WS 75–80 cm **R/M c/f**
1a Broad, pale greyish wing panel bordered above by narrow black bar. Upperpart feathers with dark shafts. **1b In flight** Prominent white patches in black primaries; pale wing panel. ▲ River banks, lake shores, etc. ✤ A series of clear, piping notes, accelerating and increasing in volume, with last few notes fading away *pi pi pi-pi-pi-pi-pi-PII-PII-PII-pii-pii-pii-pii-piu*. [CD6:20]

2 WATER THICK-KNEE *Burhinus vermiculatus* 38–41 cm **R/M lc/r**
2a Greyish wing panel bordered above by narrow white bar highlighted by black above. Upperpart feathers with dark shafts and finely vermiculated (only visible at close range). **2b In flight** White primary patches; pale wing panel; feet project slightly beyond tail. ▲ River banks, lake shores, lagoons, etc. ✤ A series of plaintive, piping, whistled notes, first accelerating and rising in pitch and volume, then dying away with slower, drawn-out, plaintive notes, *pi-pi-pi-pi-pee-pee-PEE-PEE-PEE-PEE-PEE-PEE-PEE-peeu-peeeu-peeeu-peeeu-peeeu*. Also very rapid *pipipipipipipipipipi*. [CD6:21]

3 STONE-CURLEW *Burhinus oedicnemus* 38–45 cm; WS 76–88 cm **P/R u/r**
3a White wingbar bordered above and below by black. Bill yellow tipped black. **3b In flight** White primary patches; pale wing panel bordered by black-and-white bar. ▲ Arid, stony areas. Palearctic visitor (u/r). Resident, N Mauritania (r). ✤ Mostly silent in W Africa. [CD2:18a]

4 SPOTTED THICK-KNEE *Burhinus capensis* *c.* 43 cm **R u**
4a Ad Upperparts densely spotted; no bars or panel on wing. **4b Juv** Slightly duller; upperparts streaked. **4c In flight** Small white primary patches. ▲ Dry, open woodland, arid areas. ✤ A series of loud piping notes, rising in pitch and volume, then dying away, *pi-pi-pi-pi-plee-plee-PLEEW-PLEEW-WHEEW-WHEEW-wheew-wheew-...* and an accelerating, slightly rising series of similar notes *piu-piu-piu-piupiupiupiu...*; lower pitched than calls of 1 and 2. Also harsh notes and rapid *pi-pi-pi-...* in alarm. [CD6:22]

Oystercatchers. Stocky, pied shorebirds with stout bills.

5 EURASIAN OYSTERCATCHER *Haematopus ostralegus* 40–45 cm **P/V+ lc/r**
5a Ad non-breeding Bulky; black and white; red bill; white collar. **5b Ad breeding** As 5a but lacks white collar. **5c In flight** Broad white wingbar. ▲ Coast. ✤ Loud and vigorous, high-pitched *KLEEP!* or *K-PEEP!* Also sharp *kip kip kip...* [CD2:16b]

6 AFRICAN BLACK OYSTERCATCHER *Haematopus moquini* 42–45 cm **V? NT**
As 5 but plumage entirely black. S African species; occurrence in W Africa uncertain. See illustration on page 139.

Stilts and Avocets. Elegant, long-legged waders with pied plumage and slender bills.

7 BLACK-WINGED STILT *Himantopus himantopus* 35–40 cm; WS 67-83 cm **R+/P lc/u**
7a Ad Extraordinary long, pinkish legs; needle-like bill; head white or with variable amount of black. **7b Ad** Example of a dark-headed bird. **7c Juv** Duller; upperparts brownish with narrow buff fringes. **7d Ad in flight** Black upperwing; legs extend far beyond tail. ▲ Various aquatic habitats. ✤ Vocal in breeding season, otherwise rather silent. Calls varied, incl. sharp, fast *kyik kyik kyi...* grating *kreet kreet...*, more drawn-out *krrrrrt* and high-pitched *kip-kip-kip-...* [CD2:17a]

8 PIED AVOCET *Recurvirostra avosetta* 42–46 cm; WS 67–77 cm **P/V+ lf/r**
8a Slender, upturned black bill (diagnostic); grey-blue legs. **8b In flight** Distinctive black-and-white pattern. ▲ Mudflats, estuaries, lake margins. Typically feeds by sweeping bill from side to side through water or soft mud; also pecks items from mud. Frequently swims and up-ends like duck. ✤ Loud, clear *klup klup...* [CD2:17b]

1a
2a
3a
3b
4c
4b
4a
1b
2b
5b
5a
5c
7b
7a
7c
8a
7d
8b

PLATE 37: PHALAROPES, SNIPES, PRATINCOLES AND COURSERS

1 RED-NECKED PHALAROPE *Phalaropus lobatus* 17–19 cm; WS 30–34 cm **V+**
1a Ad non-breeding More dainty than 3; needle-like bill. **1b Juv** Rusty-yellow V on dark brown upperparts. ▲ Pelagic; various wetlands. ✤ Usually silent. [CD2:61]

2 WILSON'S PHALAROPE *Phalaropus tricolor* 22–24 cm **V**
2a Ad non-breeding Needle-like bill; very pale grey above; yellowish legs. **2b In flight** Plain upperparts and wings; white rump and uppertail-coverts. ▲ Various wetlands. N American vagrant, Ivory Coast. ✤ Usually silent. [CD2:60]

3 RED PHALAROPE *Phalaropus fulicarius* 20–22 cm; WS 36–41 cm **P+ r/u**
3a Ad non-breeding Bill thicker than 1; black eye patch; plain, pale upperparts. **3b In flight** White wingbar. ▲ Pelagic; various wetlands. ✤ Sharp, high-pitched *kip*. [CD2:62]

4 GREAT SNIPE *Gallinago media* 26–30 cm; WS 43–50 cm **P u/r NT**
4a Bulkier than 6; bill relatively shorter (5.5–7 cm); white wingbars; barred underparts. **4b In flight** White lines on upper wing; broad white tail corners. ▲ Various wetlands. Low, direct flight. ✤ Hoarse, gruff *shrt* or *krrrt*, occasionally given in flight. [CD2:43]

5 JACK SNIPE *Lymnocryptes minimus* 17–19 cm; WS 33–36 cm **P/V+ r/lu**
5a Small; shortish bill; two yellowish stripes on upperparts. **5b In flight** Note size and bill length. ▲ Various wetlands. Skulking. Flushes silently from underfoot. ✤ [CD2:41]

6 COMMON SNIPE *Gallinago gallinago* 23–28 cm; WS 39–45 cm **P+ u/c**
6a Cryptically patterned; boldly striped head; very long, straight bill (6–7.5 cm). **6b In flight** Narrow white trailing edge to wing; white edges to tail. Fast zigzagging flight. ▲ Various wetlands. ✤ Hoarse *rretch* when flushed and in flight. [CD2:42]

7 EGYPTIAN PLOVER *Pluvianus aegyptius* 20 cm **R/M lf/r**
7a Ad Distinctive pattern of black, white, grey and creamy-buff. **7b Juv** Some rusty on crown, mantle and wing-coverts. **7c In flight** Striking wing pattern with diagonal black band on white. ▲ Sandbars in large rivers. ✤ Harsh *chreek-chreek-chreek*. [CD6:23]

8 GREATER PAINTED-SNIPE *Rostratula benghalensis* 23–26 cm **R/M u/lc**
8a Ad male Rather rail-like; golden-buff 'spectacles' and V on mantle. **8b Ad female** Chestnut head and neck; white 'spectacles'; dark green upperparts. **8c Ad male in flight** Broad, rounded wings; dangling legs; slow wingbeats. **8d Ad female in flight**. ▲ Various wetlands. ✤ Usually silent. Breeding female utters low, sonorous *hooOOoo*. [CD6:19]

9 COLLARED PRATINCOLE *Glareola pratincola* 24–28 cm; WS 60–70 cm **R/M/P/V* lu**
9a Ad breeding Creamy-yellow throat bordered by black line. **9b Ad non-breeding** Duller; throat outline indistinct; breast mottled. **9c Juv** Scaly upperparts; throat pattern indistinct; breast blotched. **9d In flight** Long, pointed wings; chestnut underwing-coverts; narrow white trailing edge to inner wing; forked tail. ▲ Various dry and wet, open habitats. ✤ Sharp *kik* and *kirrik*; also piercing, rattling *krrrrrrret*. [CD2:19a]

10 BLACK-WINGED PRATINCOLE *Glareola nordmanni* 24–28 cm **P/V r/lc DD**
10a–b Ad breeding/non-breeding As 9a–b, but usually has shorter tail. **10c Juv** As 9c; differences as adult. **10d In flight** Darker above than 9d; underwing all dark; no white trailing edge to wing. ▲ Various dry and wet, open habitats. ✤ As 9 but drier. [CD2:19b]

11 GREY PRATINCOLE *Glareola cinerea* 18–20 cm **M lc/u**
11a Ad Pale grey above; pale chestnut nuchal collar. **11b Juv** Duller; head plainer. **11c In flight** Striking pattern; white patch on black outer wing. ▲ Sandbars in large rivers. ✤ Hoarse *zri* or *kree*, frequently in series.

Continued on page 144

2b
2a
3a
1b
3b
6a
1a
4a
5b
6b
5a
4b
8b
8d
7a
7b
8a
7c
8c
10d
9d
10b
9c
10c
9a
9b
11b
12d
10a
12a
12b
12c
11a
11c
14b
13a
14a
15a
15b
13b
14c
15c

Continued from page 142

12 ROCK PRATINCOLE *Glareola nuchalis* 18–20 cm **R/M c/**
12a Ad *liberiae* (from W Cameroon west) Chestnut nuchal collar; red, black-tipped bill; re
legs. **12b Ad *nuchalis*** (from W Cameroon east) White nuchal collar. **12c Juv** Duller; scal
upperparts; no nuchal collar. **12d In flight** White uppertail-coverts; shallowly forked, mainl
black tail. ▲ Rocky rivers. ✤ Sharp *kip*, clear *kweee*, shrill *krrreep krrrree*, etc. [CD6:26-27]

13 BRONZE-WINGED COURSER *Rhinoptilus chalcopterus* 25–29 cm **R/M? u/s**
13a Distinctive head pattern; black breast-band; long reddish legs. **13b In flight** Black fligh
feathers; white uppertail-coverts; black tail band. ▲ Wooded savanna. Nocturnal. ✤ Series o
3–4 ringing *hu thu-WHEH-hep*, at night. [CD6:25]

14 TEMMINCK'S COURSER *Cursorius temminckii* 19–21 cm **R u/l**
14a Ad Rufous crown; blackish patch on centre of chestnut belly. **14b Juv** Duller; hea
plainer; upperparts faintly scaled; belly patch more diffuse. **14c In flight** Black outer wing an
secondaries; black underwing. ▲ Savanna. ✤ Mostly silent. [CD6:24]

15 CREAM-COLOURED COURSER *Cursorius cursor* 21–24 cm; WS 51–57 cm **P/R+ u/**
15a Ad Sandy-cream; grey hindcrown; black V from eye to nape. **15b Juv** Duller; hea
plainer; upperparts faintly scaled. **15c In flight** Black outer wing; black underwing. ▲ Desert
semi-desert. ✤ Mostly silent. [CD2:18b]

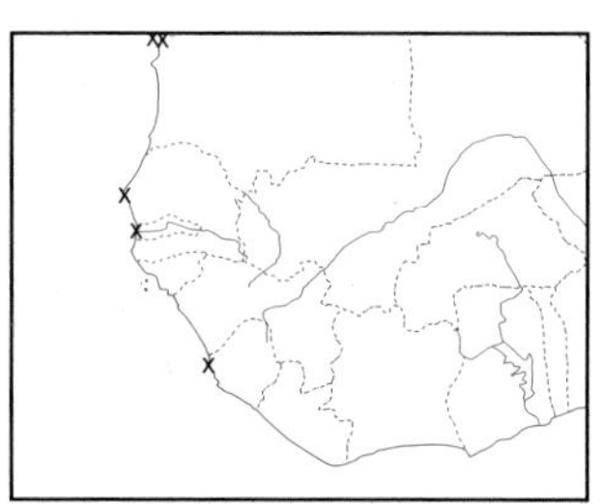

1. RED-NECKED PHALAROPE

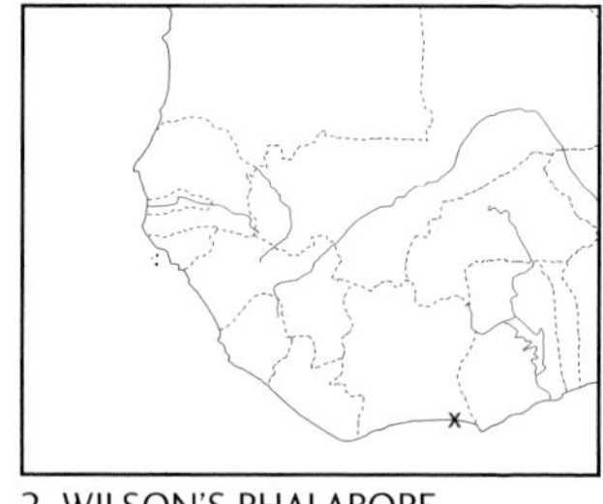

2. WILSON'S PHALAROPE

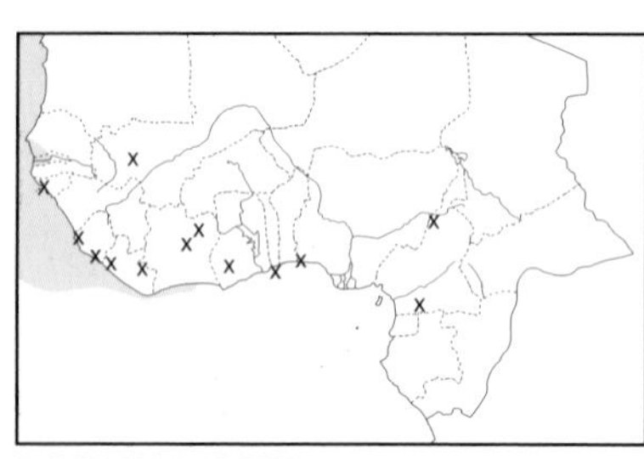

3. RED PHALAROPE

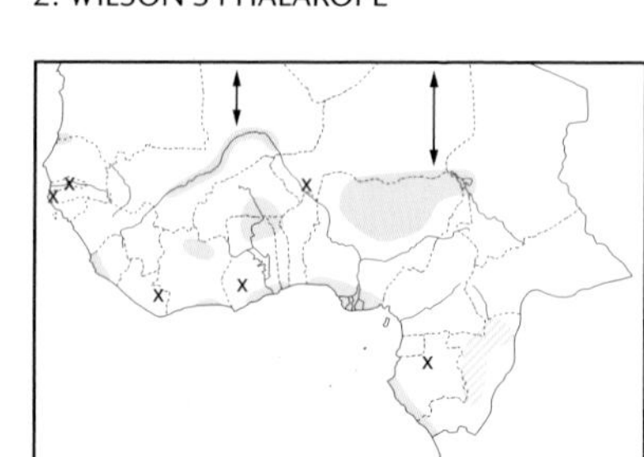

4. GREAT SNIPE

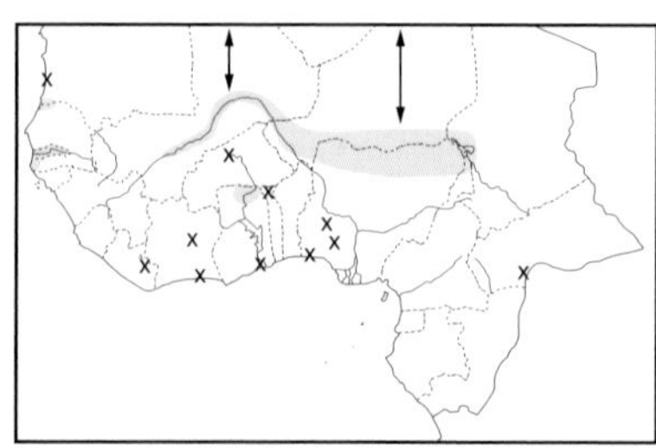

5. JACK SNIPE

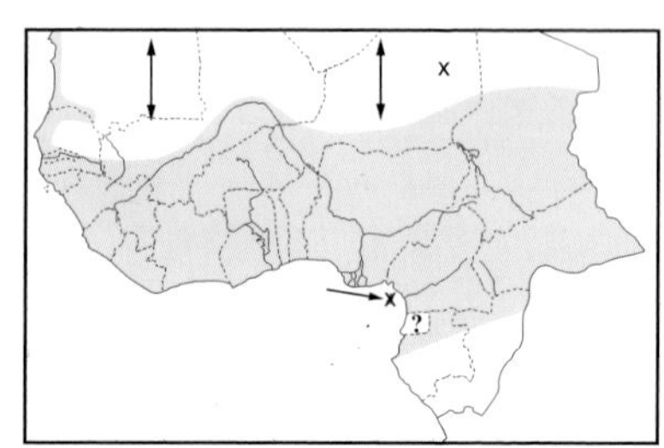

6. COMMON SNIPE

Plate on page 142

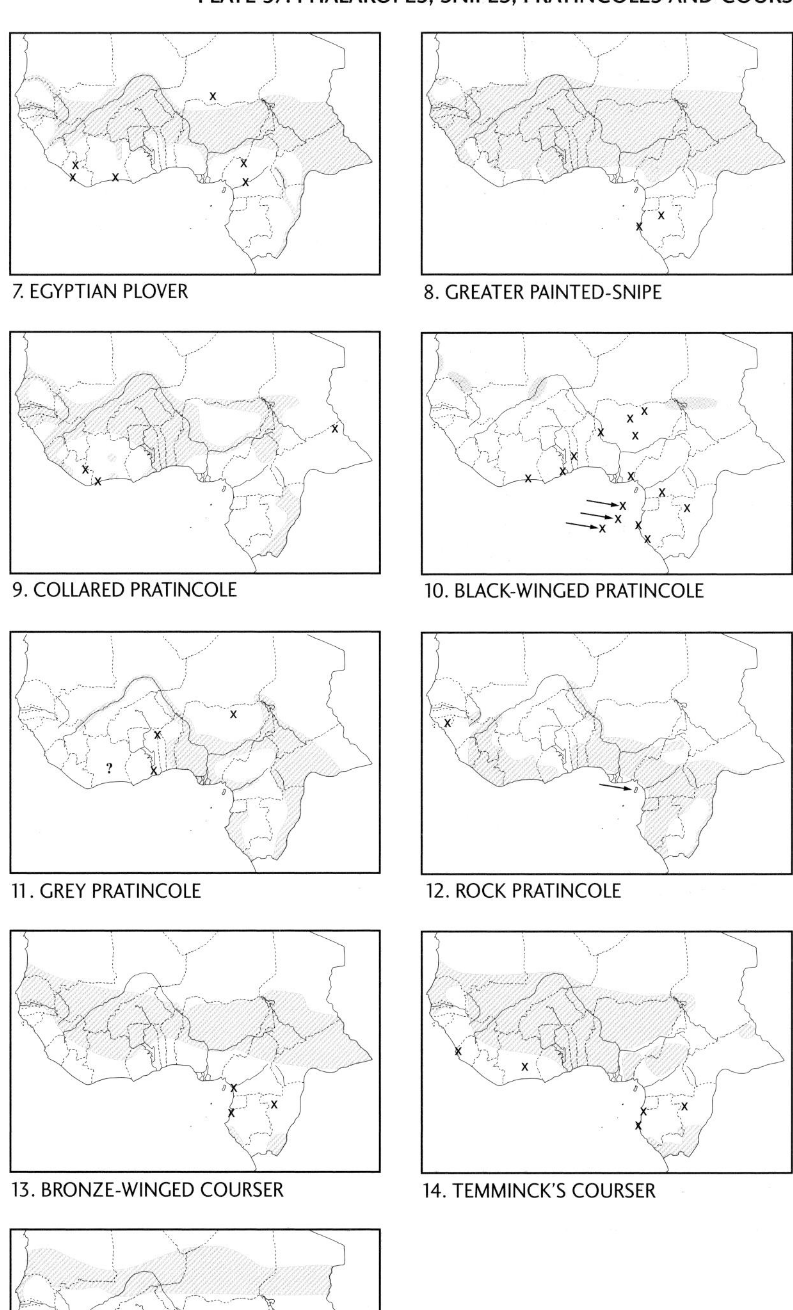

7. EGYPTIAN PLOVER

8. GREATER PAINTED-SNIPE

9. COLLARED PRATINCOLE

10. BLACK-WINGED PRATINCOLE

11. GREY PRATINCOLE

12. ROCK PRATINCOLE

13. BRONZE-WINGED COURSER

14. TEMMINCK'S COURSER

15. CREAM-COLOURED COURSER

Mostly well marked and easily identified. Several have wattles at base of bill or carpal spurs. Distinctive black-and-white wing pattern in flight. Flight rather slow and heavy.

1 LESSER BLACK-WINGED LAPWING *Vanellus lugubris* 22–26 cm **M u/f**
1a Ad Rather plain looking; white forehead. **1b Juv** Head pattern duller; wing feathers and scapulars fringed buff. **1c In flight** Broad white trailing edge to inner wing. ▲ Coastal savannas, lightly wooded grassland. ✤ Melodious *thi-HUwit*, shorter *thu-WIT* or *thi-whoo*, and longer *tihi-hooee*. [CD6:39]

2 BROWN-CHESTED LAPWING *Vanellus superciliosus* *c.* 23 cm **M lu**
2a Ad Dark chestnut breast-band; yellow wattle in front of eye. **2b Juv** Duller; upperparts scaled rusty. **2c In flight** Diagonal white band on wing. ▲ Short grassland, bare and recently burned ground near rivers and lakes, and other open areas in or near lightly wooded grassland. Breeds Nigeria–Cameroon, probably CAR. ✤ Harsh and shrill calls. [CD6:38]

3 WHITE-TAILED LAPWING *Vanellus leucurus* 26–29 cm **V**
3a Rather plain; slender; tail entirely white. **3b In flight** Broad diagonal white band on wing; white tail. ▲ Damp grassland, marshes, lakes. Frequently forages in shallow or deep water. Palearctic vagrant. ✤ Usually silent. [CD2:28]

4 BLACK-HEADED LAPWING *Vanellus tectus* *c.* 25 cm **R c/r**
4a Wispy black crest; black-tipped red bill; reddish legs. **4b In flight** Large white area from forewing across inner wing. ▲ Dry grassland with scattered bushes and patches of bare ground. Mostly active in early morning, in evening and at night, spending most of day resting in shade. ✤ Short, piercing *kir* and shrill *kwairr*. [CD6:35]

5 NORTHERN LAPWING *Vanellus vanellus* 28–31 cm **P/V+ s/r**
5a Wispy crest; dark glossy green above; black breast-band. **5b In flight** Strikingly broad, rounded wings; mainly dark above. ▲ Various open habitats, incl. beaches, mudflats and rice-fields. ✤ Hoarse, shrill *cheew-ep* or *pee-witt*. [CD2:29]

6 SPUR-WINGED LAPWING *Vanellus spinosus* 25–28 cm **R/V+ c/lu**
6a Top of head black; head-sides and neck white; underparts mainly black. **6b In flight** Diagonal white band from carpal joint to inner secondaries. ▲ Various wetlands, incl. marshes, rivers, lakes, reservoirs, estuaries, coastal saltpans and irrigated farmland. ✤ Sharp, metallic *kit*, often rapidly repeated. Territorial call a harsh *ti-ti-terr-it*. [CD6:37]

7 LONG-TOED LAPWING *Vanellus crassirostris* *c.* 31 cm **R lf**
7a White face and foreneck; black hindcrown to breast; red legs. **7b In flight** Mainly white wing-coverts. ▲ Lakes, ponds, river edges, swampy areas, short grass near water. Relatively long toes enable foraging on floating vegetation. ✤ Metallic *kik-k-k-k* and plaintive *wheet*. [CD6:41]

8 WHITE-HEADED LAPWING *Vanellus albiceps* 28–32 cm **R/M c/f**
8a White band on wing; large yellow wattles; greenish-yellow legs. **8b In flight** Wings mainly white with black on coverts and outer primaries. ▲ Sand banks, mud or rocks in rivers; also forages in grassy areas near water. ✤ Sharp, high-pitched *kip, kip, kip,...* and *kwip, kwip, kwip,...* [CD6:34]

9 AFRICAN WATTLED LAPWING *Vanellus senegallus* *c.* 34 cm **R/M c/u**
9a Ad *senegallus* Mainly pale brown; yellow wattles; long yellow legs. *V. s. lateralis* (Congo, vagrant) has belly darkening to blackish between legs; bill tipped black. **9b In flight** Broad white diagonal band on inner wing; black flight feathers. ▲ Damp grassland, grassy edges of lakes and rivers, marshes, open and lightly wooded grassland, usually (but not always) near water; also mangrove fringes. ✤ Shrill *kwip-kwip-kwip-...* and more nasal *ke-weep ke-weep...* [CD6:33]

Maps on page 148

1b
1a
2b
2a
1c
2c
3a
4a
5a
3b
4b
7a
6a
5b
7b
9a
6b
8a
9b
8b

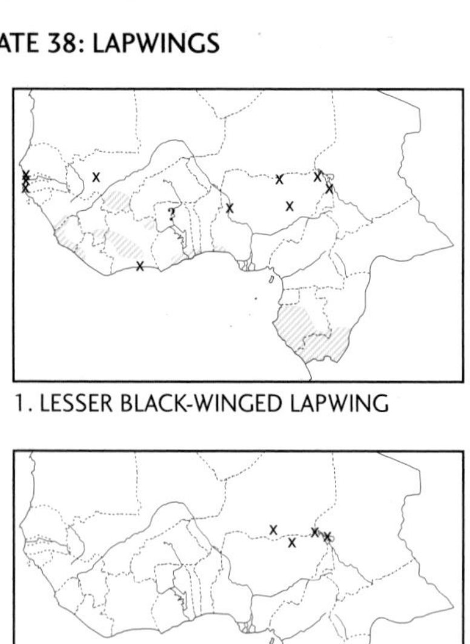

1. LESSER BLACK-WINGED LAPWING

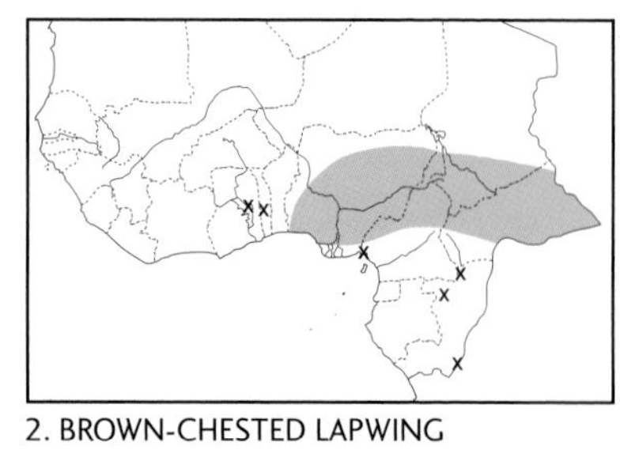

2. BROWN-CHESTED LAPWING

3. WHITE-TAILED LAPWING

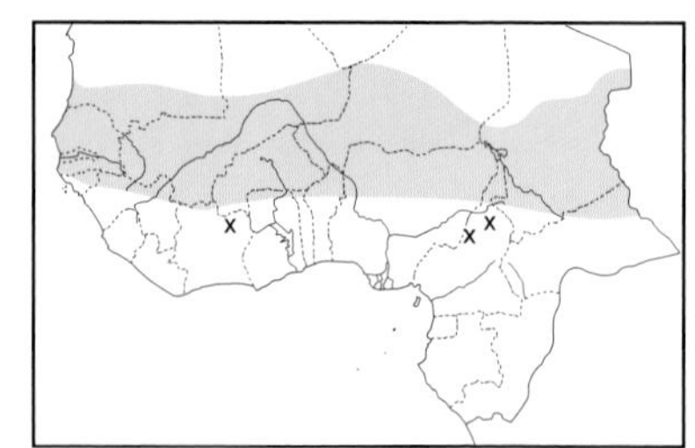

4. BLACK-HEADED LAPWING

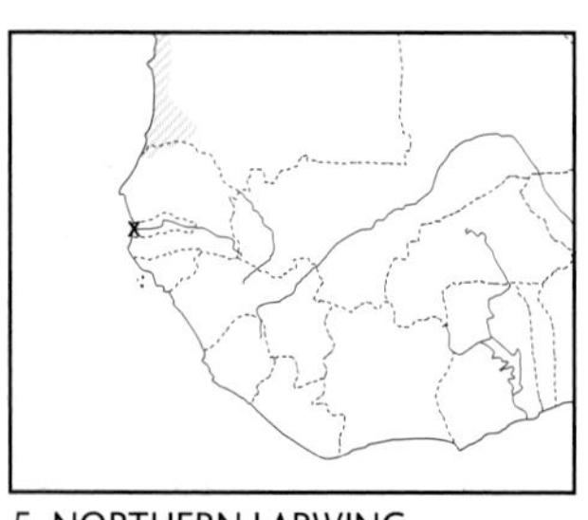

5. NORTHERN LAPWING

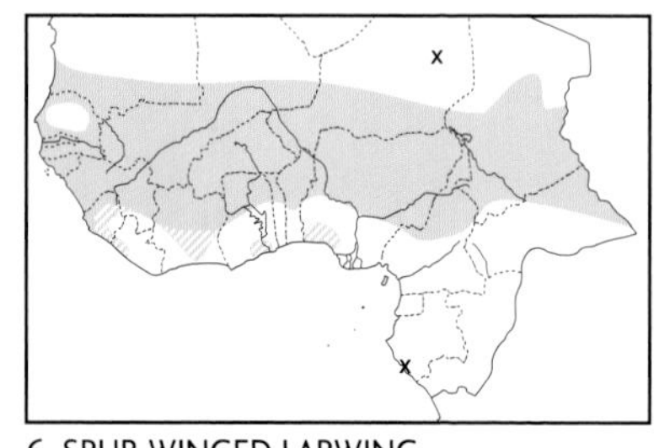

6. SPUR-WINGED LAPWING

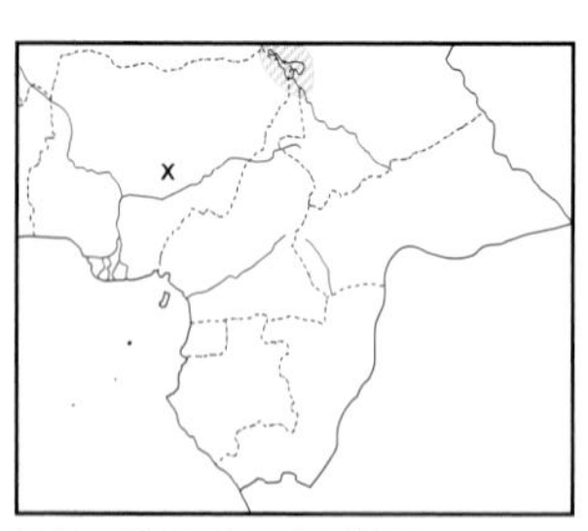

7. LONG-TOED LAPWING

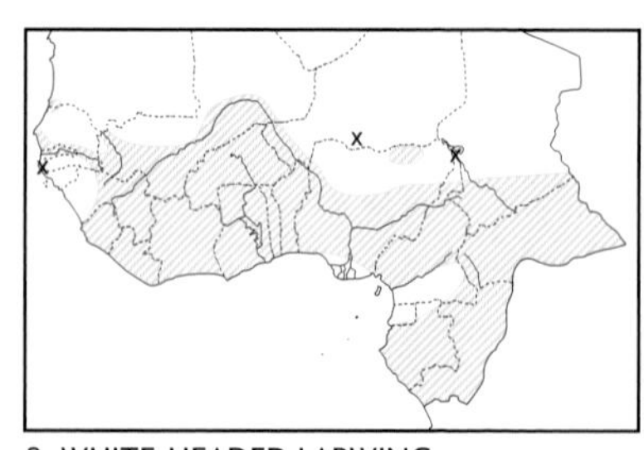

8. WHITE-HEADED LAPWING

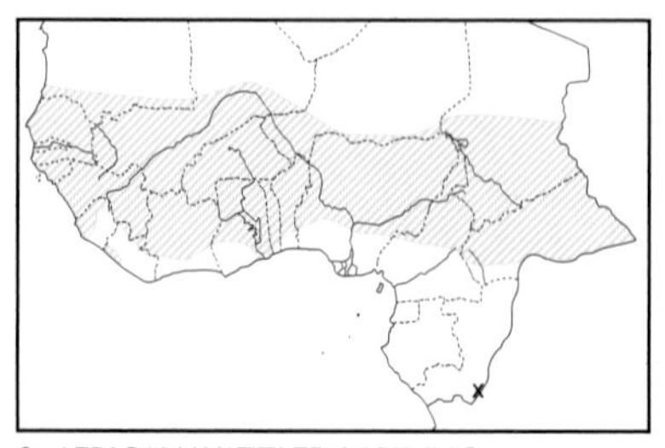

9. AFRICAN WATTLED LAPWING

Plate on page 146

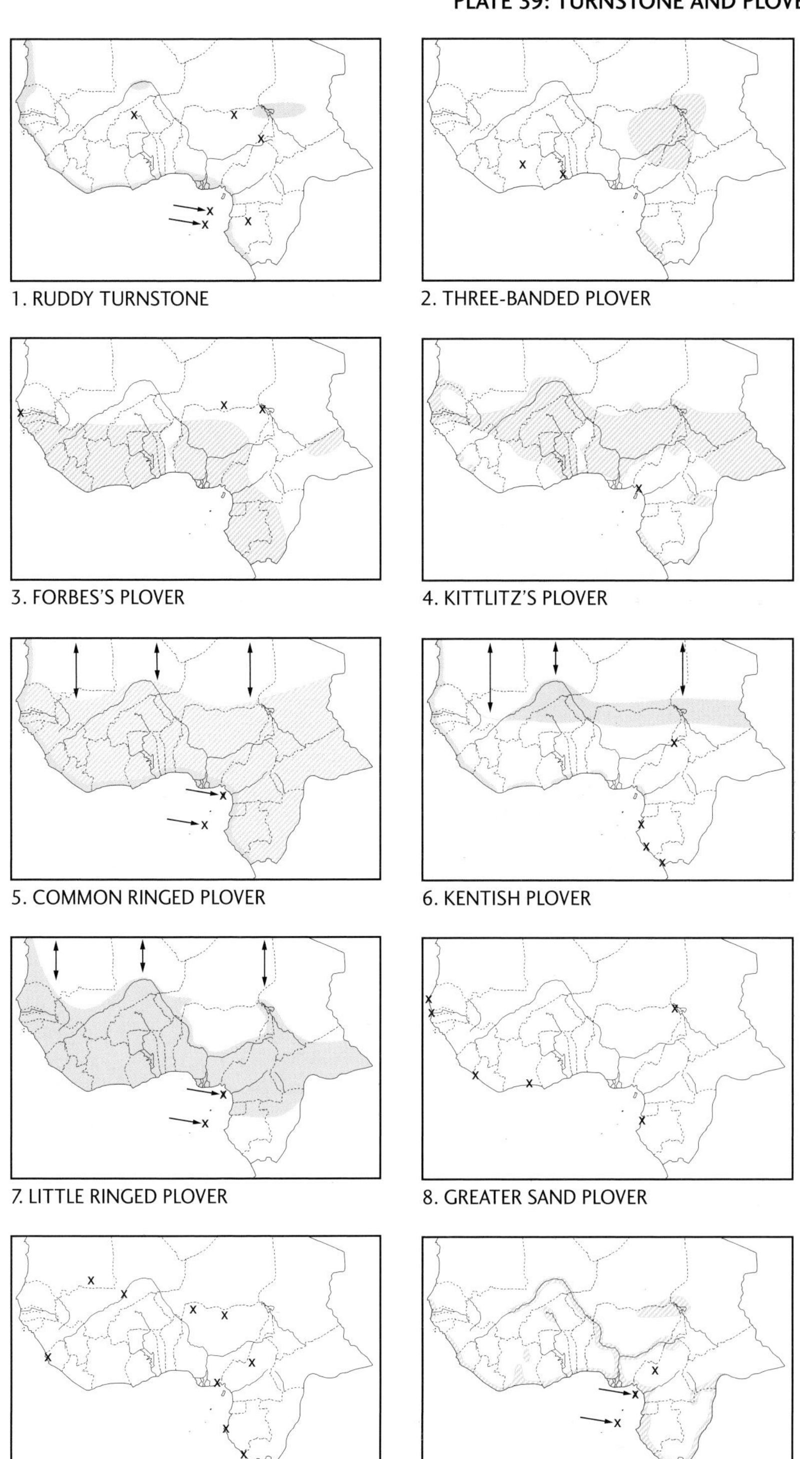

1. RUDDY TURNSTONE

2. THREE-BANDED PLOVER

3. FORBES'S PLOVER

4. KITTLITZ'S PLOVER

5. COMMON RINGED PLOVER

6. KENTISH PLOVER

7. LITTLE RINGED PLOVER

8. GREATER SAND PLOVER

9. CASPIAN PLOVER

10. WHITE-FRONTED PLOVER

1 RUDDY TURNSTONE *Arenaria interpres* 21–25 cm P/V+ c/u
1a Ad breeding Stocky; rufous-chestnut and black above; short, orange legs. **1b Ad non-breeding** Mottled blackish-brown. **1c Juv** Scaly upperparts. **1d In flight** Striking pied pattern. ▲ Coast. Vagrant inland. ✤ Low, nasal *tuk* and *tukatukatuk*. [CD2:59]

2 THREE-BANDED PLOVER *Charadrius tricollaris* 17–18 cm R lu/r
2a As 3 but forehead white; legs shorter. **2b In flight** Narrow white line on wings. ▲ Near water. ✤ Piercing *peeweet*, more rasping *kreep* and shrill *wik-wik*. [CD6:29]

3 FORBES'S PLOVER *Charadrius forbesi* *c.* 20 cm R/M lf/r
3a Ad Two blackish breast-bands; red orbital ring; white band from eye across nape. **3b Juv** Duller; band across nape buffish; upperparts scaled buff. **3c In flight** Appears dark; no white wingbar. ▲ Dry and wet habitats. ✤ Plaintive *pee-oo*, sometimes repeated, and sharp *pee-pee-pee-...* [CD6:30]

4 KITTLITZ'S PLOVER *Charadrius pecuarius* 14–16 cm R/M u/lc
4a Ad breeding Distinctive head pattern; warm-buff breast. **4b Ad non-breeding** Duller; supercilium orange-buff. Juvenile similar but upperparts scaly. **4c In flight** White patch on outer wing extends as narrow line on inner wing. ▲ Various open habitats. ✤ Mostly silent. Calls include *tuweet* or *pipeep*, hard *trip* and dry rattling *trrrrr*. [CD6:28]

5 COMMON RINGED PLOVER *Charadrius hiaticula* 18–20 cm P+ c/f
5a Ad breeding More compact than 7; stubby, black-tipped, orange bill; orange legs. **5b Ad non-breeding** White forehead and supercilium. **5c In flight** Prominent white wingbar. ▲ Mudflats, various open habitats. ✤ Mellow *too-EE* rising at end. [CD2:21]

6 KENTISH PLOVER *Charadrius alexandrinus* 15–17 cm R+/P c/r
6a Ad male breeding Rufous crown; narrow black patches on breast-sides. Compare 10. **6b Ad female/ad non-breeding** Duller; little or no rufous on crown. **6c In flight** White wingbar. ▲ Coastal and inland shores. ✤ Short *kip* or *kitip*. Alarm a hard *pirrrrr*. [CD2:22]

7 LITTLE RINGED PLOVER *Charadrius dubius* 16–18 cm P+ f/lc
7a Ad breeding Slimmer than 5; thinner, black bill; distinct yellow orbital ring; dull-coloured legs. **7b Ad non-breeding** Duller than 7a; plainer head than 5b. **7c In flight** No wingbar. Compare 5c. ▲ Various open habitats. ✤ Piping, rather plaintive *PEE-uu*. [CD2:20b]

8 GREATER SAND PLOVER *Charadrius leschenaultii* 20–23 cm V
8a Ad non-breeding Long greenish-grey legs; relatively long and heavy bill; no white nuchal collar. Compare Lesser Sand Plover (Plate 43:5). **8b In flight** White wingbar; white tail-sides; feet project beyond tail. ▲ Mudflats, grassland. Palearctic vagrant. ✤ Trilling *trrr* and *tirrirrilip*, often frequently repeated. [CD2:23]

9 CASPIAN PLOVER *Charadrius asiaticus* 19–21 cm V
9a Ad non-breeding Long, greyish to yellowish legs; broad grey-brown breast-band. **9b In flight** Faint wingbar; no white tail-sides; feet project beyond tail. ▲ Grassland. Palearctic vagrant. ✤ Short, sharp *chip* or *kwit*; also soft, repeated *tik*. [CD6:32]

10 WHITE-FRONTED PLOVER *Charadrius marginatus* *c.* 18 cm R lf/r
10a Ad male breeding Large white forehead and supercilium; orange-tinged breast. Female has less black on forecrown. **10b Ad non-breeding** Duller. **10c In flight** White wingbar; broad white tail-sides. ▲ Sandy coasts and inland shores. ✤ Low *wit* or *twirit*. Alarm a dry *trrrr*. [CD6:31]

 Maps on page 149

1c
1a
1b
1d
3c
3b
3a
2b
2a
4b
7c
4c
7a
5c
4a
5a
7b
5b
6a
6b
8b
6c
10b
9b
10a
8a
9a
10c

1 **EUROPEAN GOLDEN PLOVER** *Pluvialis apricaria* 26–29 cm **P/V r**
1a Ad breeding Golden-yellow tones above; white band bordering black below. **1b Ad non-breeding** Yellowish tones above; yellow-buff below. Compare 2 and 3. **1c In flight from above** Indistinct wingbar. **1d In flight from below** White underwing. Compare 2 and 3. ▲ Tidal mudflats. ✤ Soft, melodious *tluuee*. [CD2:26]

2 **PACIFIC GOLDEN PLOVER** *Pluvialis fulva* 23–26 cm **P/V r**
2a Ad breeding White 'shawl' from forehead to flanks. Slimmer than 1. **2b Ad non-breeding** Warm-buff plumage tones; primaries shorter and tertials longer than 3b; wings project slightly beyond tail. **2c In flight** Grey underwing; indistinct bar on upperwing. ▲ Various coastal wetlands, farmland. ✤ Clear, rapid *chu-it* or *kuweet*, reminiscent of Spotted Redshank, and more drawn-out and plaintive *klee-wee* and *klu-wee-up*.

3 **AMERICAN GOLDEN PLOVER** *Pluvialis dominica* 24–28 cm **V+**
3a Ad breeding White 'shawl' from forehead to breast-sides. **3b Ad non-breeding** Cold grey plumage tones; 4–5 primary tips project beyond relatively short tertials; wings clearly project beyond tail. In flight wing pattern as 2c. ▲ Various wetlands, farmland. N American vagrant. ✤ Rather mellow *tlu-ee*, with stress on first syllable, and *tlu-ee-uh*. [CD2:25]

4 **GREY PLOVER** *Pluvialis squatarola* 27–30 cm **P+ f/c**
4a Ad breeding Spangled black and white above; head-sides to belly black bordered white. **4b Ad non-breeding** Mainly dull greyish; diffuse supercilium. **4c Imm** As 4b but neatly chequered above; finely streaked below. 4d In flight from above Bold white wingbar; square white rump. **4e In flight from below** Black axillaries ('armpits'). Compare 1–3. ▲ Tidal mudflats. ✤ Far-carrying, clear, rather plaintive, drawn-out *tlueeee*. [CD2:27]

5 **EURASIAN DOTTEREL** *Charadrius morinellus* 20–23 cm **V**
5a Ad non-breeding Broad supercilia form V on nape; narrow pale breast-band; yellowish legs. **5b In flight** Plain upperwing. ▲ Open habitats. Palearctic vagrant. ✤ Mostly silent. [CD2:24]

6 **BLACK-TAILED GODWIT** *Limosa limosa* 37–44 cm **P/V+ c/r**
6a Ad breeding Long straight bill; long dark legs; rufous-orange head and neck. **6b Ad non-breeding** Plain grey above. Longer necked and legged than 7. **6c Juv** Rusty-orange fringes to feathers above. **6d In flight** Broad white wingbar; square white uppertail; black terminal tail band. ▲ Various wetlands. ✤ Short *kip* or *kip-kip-kip* and sharp *weeka-weeka*. [CD2:45]

7 **BAR-TAILED GODWIT** *Limosa lapponica* 37–41 cm **P+ c/s**
7a Ad breeding Slightly upturned bill; legs shorter than 6; underparts dark rufous. **7b Ad non-breeding** Pale brownish-grey above, appearing streaked. **7c Juv** Upperpart feathers notched buff. **7d In flight** Plain upperwing; white wedge on back; barred tail. Compare 8b. ▲ Tidal mudflats, sandy shores; occasionally inland. ✤ Generally silent. Flight call a low, nasal *kirruk* or *kvip*, and sharper, also nasal, *keweep*. [CD2:46]

8 **WHIMBREL** *Numenius phaeopus* 40–46 cm **P/V+ c**
8a Long decurved bill; boldly streaked head. **8b *N. p. phaeopus* in flight** (P) White wedge on back; wings more uniform than 7d. **8c *N. p. hudsonicus* in flight** (N American vagrant, Sierra Leone, Cape Verde) Upperparts all brownish (no white wedge); underwings dark (not white). ▲ Mainly coastal; occasionally inland. ✤ Clear, loud, rapid *bi-bi-bi-bi-bi-bi-bi*. Occasionally *kurrlee* (like 9) and hard, rasping *krrreep*. [CD2:47]

9 **EURASIAN CURLEW** *Numenius arquata* 50–60 cm **P+ c/s**
9a Larger and paler than 8; bill longer; no bold streaks on head. **9b In flight** Dark outer wing contrasts with paler inner wing. ▲ Various wetlands. ✤ Clear, liquid, bubbling *kur-lee* and *kwurrrr-lee*; also shorter *kwee-kwee*. [CD2:49]

Maps on page 154

1c
1a
1d
2a
2c
3a
1b
2b
3b
4d
4c
4e
5b
4b
4a
5a
6a
6d
7a
6c
6b
7c
7b
7d
9a
8a
8b
9b
8c

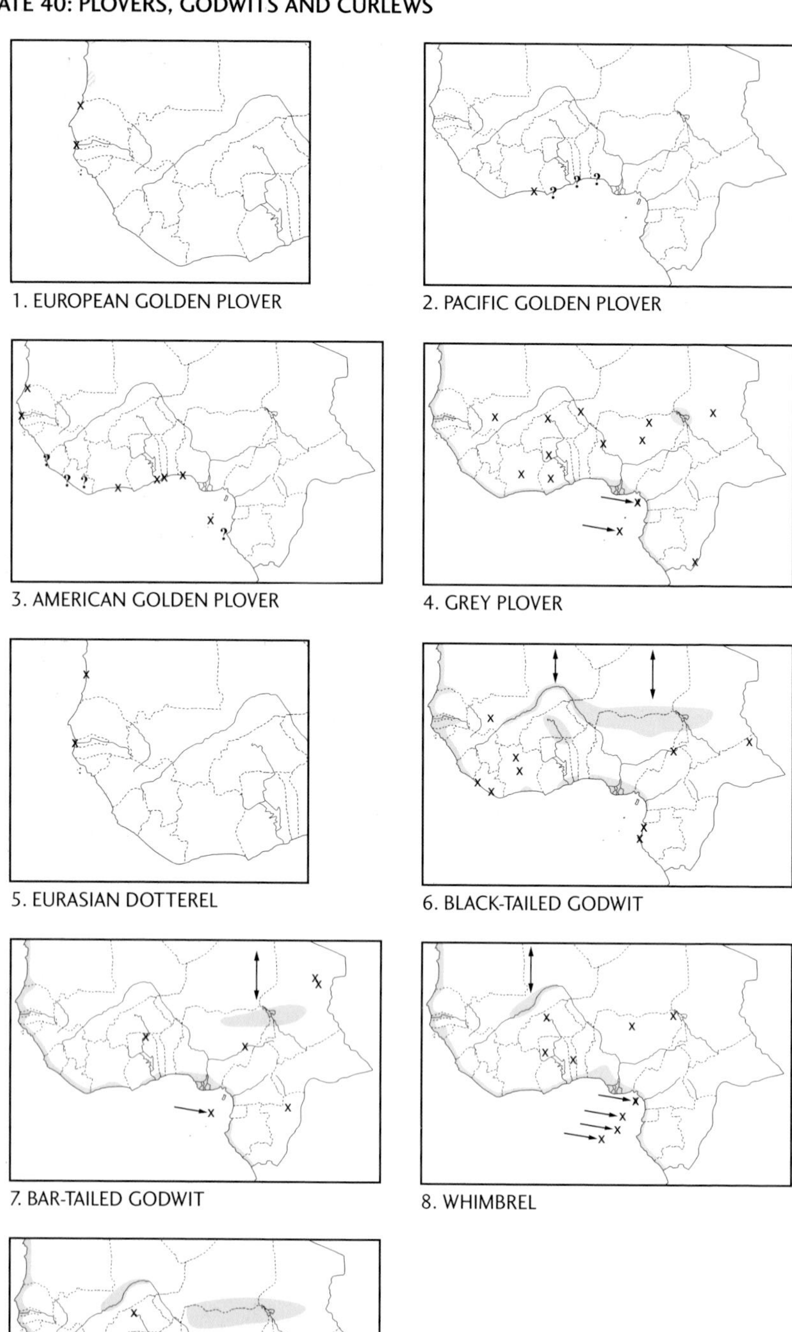

1. EUROPEAN GOLDEN PLOVER

2. PACIFIC GOLDEN PLOVER

3. AMERICAN GOLDEN PLOVER

4. GREY PLOVER

5. EURASIAN DOTTEREL

6. BLACK-TAILED GODWIT

7. BAR-TAILED GODWIT

8. WHIMBREL

9. EURASIAN CURLEW

Plate on page 152

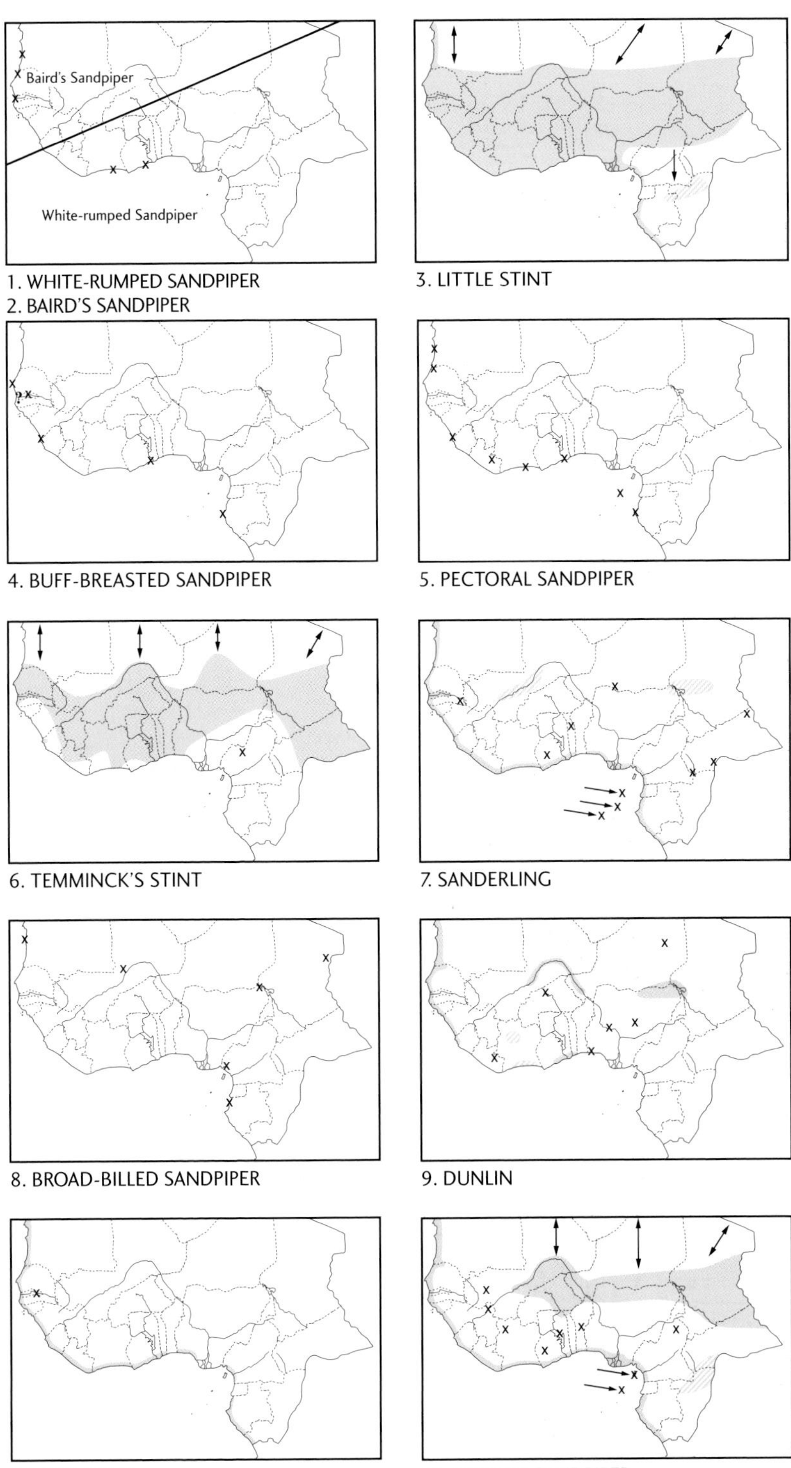

1. WHITE-RUMPED SANDPIPER
2. BAIRD'S SANDPIPER

3. LITTLE STINT

4. BUFF-BREASTED SANDPIPER

5. PECTORAL SANDPIPER

6. TEMMINCK'S STINT

7. SANDERLING

8. BROAD-BILLED SANDPIPER

9. DUNLIN

10. RED KNOT

11. CURLEW SANDPIPER

Plate on page 156

1 **WHITE-RUMPED SANDPIPER** *Calidris fuscicollis* 15–18 cm V
1a Ad non-breeding Long primary projection; wingtips clearly extend beyond tail. **1b In flight** White uppertail-coverts; dark tail. ▲ Various wetlands. N American vagrant. ✤ High-pitched, insect-like *tzreet*, often rapidly repeated. [CD2:34]

2 **BAIRD'S SANDPIPER** *Calidris bairdii* 14–17 cm V
2a Ad non-breeding Less prominent supercilium than 1a; slightly shorter, all-black bill. **2b In flight** Resembles 1b, but has dark-centred rump and uppertail-coverts. ▲ Various wetlands. N American vagrant. ✤ Soft, rolling *prrreet* or *kirrrp*, lower and softer than 11.

3 **LITTLE STINT** *Calidris minuta* 12–14 cm P+ c/f
3a Ad breeding Short, straight bill; black legs; rusty on head, upperparts and breast. **3b Ad non-breeding** Grey above; grey breast-sides. **3c Juv** Whitish Vs on mantle and scapulars. **3d In flight** Narrow white wingbar; grey outer tail. ▲ Various wetlands. ✤ Sharp *chit*. [CD2:32]

4 **BUFF-BREASTED SANDPIPER** *Tryngites subruficollis* 18–20 cm V NT
4a Ad non-breeding Sandy-buff below; small head; short bill; yellowish legs. **4b In flight** White underwing with dusky half crescent on greater primary-coverts. ▲ Wetlands, grassy areas. N American vagrant. ✤ Usually silent.

5 **PECTORAL SANDPIPER** *Calidris melanotos* 19–23 cm V+
5a Ad non-breeding Sharply demarcated streaked neck and upper breast; pale legs. **5b In flight** Indistinct narrow wingbar; white sides to rump and uppertail. ▲ Wetlands. N American/E Palearctic vagrant. ✤ Rolling, hoarse *krrrt*. [CD2:35]

6 **TEMMINCK'S STINT** *Calidris temminckii* 13–15 cm P/V* r/lf
6a Ad breeding Brownish to olive-yellow legs; more elongated than 3. **6b Ad non-breeding** Complete grey breast-band. Recalls miniature Common Sandpiper. **6c Juv** Buffish fringes above. **6d In flight** As 3d but outer tail white. When flushed, flies high and erratically. ▲ Various wetlands. ✤ High-pitched, dry trill *tirrr*. [CD2:33]

7 **SANDERLING** *Calidris alba* 18–21 cm P+ c/f
7a Ad breeding Black bill and legs; head and breast rufous marked black. **7b Ad non-breeding** Very pale grey above. **7c Juv** Chequered black and white above. **7d In flight** Bold white wingbar. ▲ Beaches. Runs very fast at edge of surf. ✤ Sharp *krit*. [CD2:31]

8 **BROAD-BILLED SANDPIPER** *Limicola falcinellus* 16–17 cm V/P r
8a Ad breeding 'Split' supercilium; boldly streaked breast. **8b Ad non-breeding** Stockier than 9; faint head pattern; dusky streaking above. **8c In flight** Narrow white wingbar; white sides to rump and uppertail. ▲ Mudflats, muddy edges. ✤ Dry *trrreet*. [CD2:39]

9 **DUNLIN** *Calidris alpina* 16–22 cm P+ c/r
9a Ad breeding Black belly patch. **9b Ad non-breeding** Less elegant than 11b; less evenly decurved bill; legs shorter. **9c Juv** Scaly upperparts; streaked breast; variably spotted belly-sides. **9d In flight** Narrow white wingbar; white sides to rump and uppertail. ▲ Various wetlands. ✤ Shrill, rasping *krrreet* or *treerrp*. [CD2:38]

10 **RED KNOT** *Calidris canutus* 23–25 cm P/V* c/s
10a Ad breeding Stocky; stout, straight bill; deep rufous face and underparts. **10b Ad non-breeding** Plain, pale grey above. **10c Juv** Scaly upperparts. **10d In flight** Narrow white wingbar; whitish-grey rump; grey tail. ▲ Mudflats. ✤ Low, hoarse *nut, nut, ...* [CD2:30]

11 **CURLEW SANDPIPER** *Calidris ferruginea* 18–23 cm P+ c/f
11a Ad breeding Evenly decurved bill; longish legs; rusty-red head and underparts. **11b Ad non-breeding** Grey above; prominent supercilium. Compare 9. **11c Juv** Scaly upperparts. **11d In flight** Broad white band on uppertail-coverts. ▲ Various wetlands. Small numbers inland. ✤ Rather pleasing, rippling *chirrup*, less grating than 9. [CD2:36]

Maps on page 155

1a
1b
2a
2b
3a
3b
3c
3d
4a
4b
5a
5b
6a
6b
6c
6d
7a
7b
7c
7d
8a
8b
8c
9a
9b
9c
9d
10a
10b
10c
10d
11a
11b
11c
11d

1 GREEN SANDPIPER *Tringa ochropus* 20–24 cm P+ c/f
1a Ad breeding Darker above than 2; distinct white eye-ring; greyish-green legs. **1b Ad non-breeding** Plainer. **1c In flight** Blackish upper- and underwings; pure white rump and belly. ▲ Various, mainly freshwater, wetlands. ✤ Clear, ringing *tlooeet weet-weet!* [CD2:55]

2 WOOD SANDPIPER *Tringa glareola* 19–21 cm P+ c
2a Ad breeding Slimmer than 1; densely speckled upperparts; distinct supercilium. **2b Ad non-breeding** As 2a, but less distinctly marked. **2c In flight** Pale grey underwing; plain upperwing; white rump. Compare 1c. ▲ Various wetlands. ✤ Loud, ringing *chiff-iff-iff* or *chipipip*. [CD2:56]

3 COMMON SANDPIPER *Actitis hypoleucos* 18–21 cm P+ c
3a Ad breeding As 3b but with dark markings above. **3b Ad non-breeding** Plain brown above; white wedge on sides of brown breast. **3c In flight** White wingbar. Low, fluttering flight on stiff, bowed wings. ▲ Various wetlands. Constantly bobs rear body. ✤ Rapid, clear, very high-pitched, piercing *tsee-wee-wee* in flight. Alarm *sweeeee-eet.* [CD2:58]

4 RUFF *Philomachus pugnax* male 26–32 cm, female 20–25 cm P+ c/r
4a Ad male Plumage variable (moulting male illustrated); head relatively small; legs orange-red to yellowish; mantle feathers appear loose and are often fluffed, giving hunch-backed appearance. **4b Ad male non-breeding** Grey-brown and scaly above. Example of white-headed bird. **4c Ad female non-breeding** Considerably smaller than male. **4d Juv** Neatly patterned upperparts. **4e In flight** Narrow white wingbar; white oval areas on tail-sides. ▲ Various wetlands. ✤ Usually silent. [CD2:40]

5 TEREK SANDPIPER *Xenus cinereus* 22–25 cm V
5a Ad non-breeding Gently upturned bill; shortish, pinkish to orange-yellow legs. **5b In flight** Dark leading edge and whitish trailing edge to wing; grey rump and tail. ▲ Various wetlands. Palearctic vagrant. ✤ Loud, melodious, ringing *eeb-eeb-eeb.* [CD2:57]

6 COMMON REDSHANK *Tringa totanus* 26–29 cm P+ c/s
6a Ad breeding Orange-red legs; red-based bill; densely streaked brown below. **6b Ad non-breeding** Plain grey-brown above; white eye-ring; greyish breast. **6c Juv** Brown above with buff spots and notches. **6d In flight** Broad white trailing edge to wing (distinctive); white wedge on back. ▲ Various wetlands. ✤ Mournful, drawn-out *tiuuu* and clear, ringing *tiu-lu tiu-lu-lu*, recalling 8 but higher pitched. [CD2:52]

7 SPOTTED REDSHANK *Tringa erythropus* 29–33 cm P+ f/r
7a Ad breeding Black; white spotting on upperparts. **7b Ad non-breeding** Pale grey above; grey wash to breast; legs and bill longer than 6. **7c Juv** Closely streaked and barred grey below. **7d Ad non-breeding in flight** Plain wings; white wedge on back; barred uppertail. ▲ Various wetlands. ✤ Loud, abrupt *chu-wit!* [CD2:51]

8 COMMON GREENSHANK *Tringa nebularia* 30–34 cm P+ c
8a Ad breeding Stout, slightly upturned bill; dull greenish legs. **8b Ad non-breeding** Plainer, greyer than 8a above; less streaked head and breast. **8c Juv** Darker than adult; streaky head and breast; white-fringed upperparts. **8d In flight** As 7d but wings darker, more contrasting; rump and tail whiter. ▲ Various wetlands. ✤ Loud, ringing *tiu-tiu-tiu*. Occasionally a rasping *kruip-kruip-kruip* when flushed, and rapid *chip-chip-chip*. [CD2:54]

9 LESSER YELLOWLEGS *Tringa flavipes* 23–25 cm V+
9a Ad non-breeding Slimmer than 6; bill finer and black; legs longer and yellow. **9b In flight** Square white rump. ▲ Various wetlands. N American vagrant. ✤ Clear *tew* or *tew-tew*, reminiscent of 8 but more subdued.

Continued on page 161

Maps on page 160

1a
1c
1b
2a
2c
4d
3a
2b
3b
4a
3c
4c
5b
4b
5a
6a
4e
6d
6c
7a
6b
7c
7b
7d
8a
10a
8c
10c
9a
8b
10b
8d
9b
10d

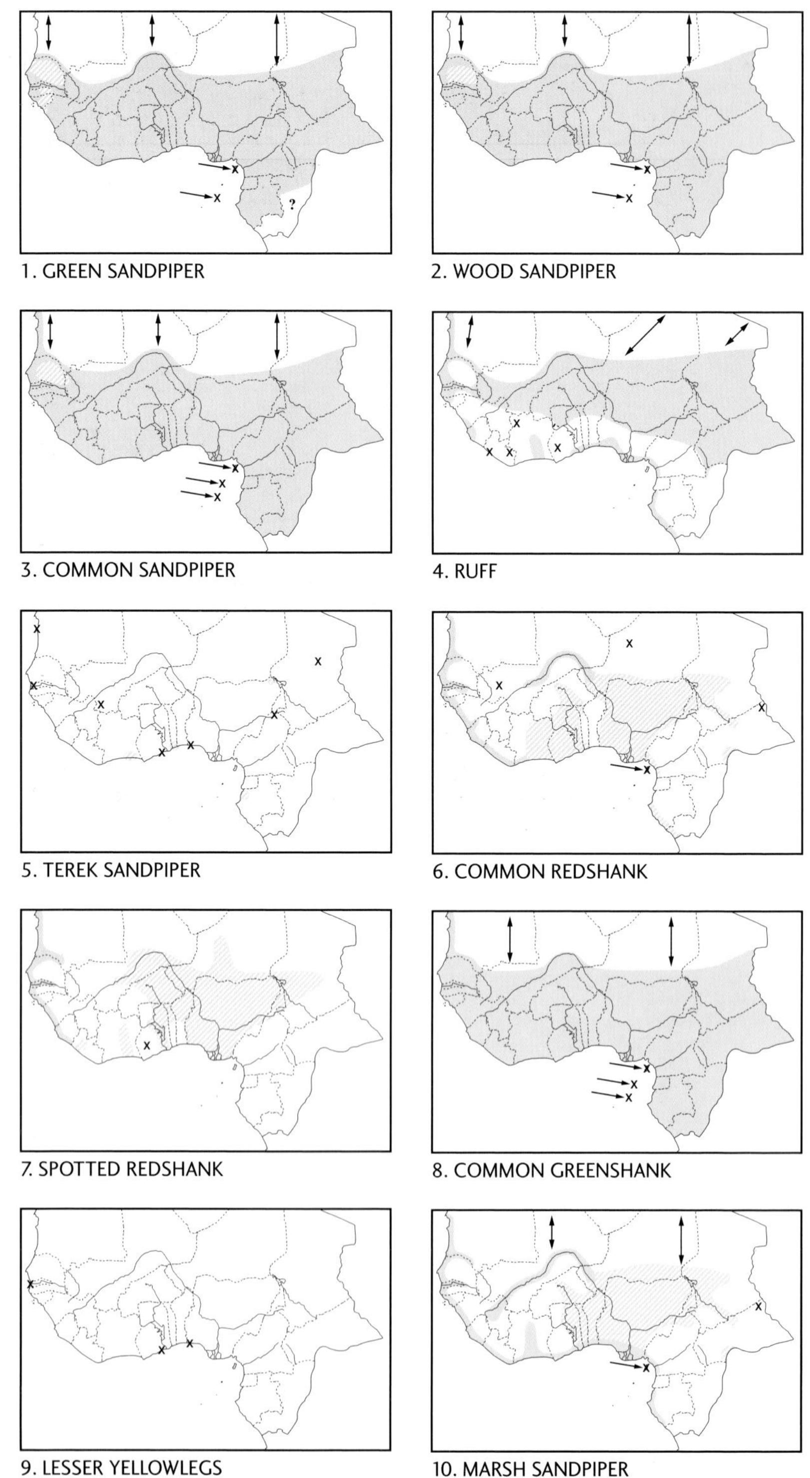

1. GREEN SANDPIPER

2. WOOD SANDPIPER

3. COMMON SANDPIPER

4. RUFF

5. TEREK SANDPIPER

6. COMMON REDSHANK

7. SPOTTED REDSHANK

8. COMMON GREENSHANK

9. LESSER YELLOWLEGS

10. MARSH SANDPIPER

Plate on page 158

Continued from page 158

10 MARSH SANDPIPER *Tringa stagnatilis* 22–25 cm **P/V* u/lf**
10a Ad breeding Straight, needle-like bill; long greenish to yellowish legs; spotted breast. **10b Ad non-breeding** White forehead and supercilium; plain grey above; white below. **10c Juv** Finely streaked crown and hindneck; white-fringed upperparts. **10d In flight** Long white wedge on back. ▲ Wetlands. ✤ Clear *keeuw*, singly or repeated up to 3–4 times, reminiscent of 8, but less ringing. Alarm a sharp, rapidly repeated *chip*. [CD2:53]

Continued from page 162

11 SLENDER-BILLED CURLEW *Numenius tenuirostris* 36–41 cm **V? CR**
11a All-black bill shorter, straighter and thinner than Eurasian Curlew (Plate 40:9) and Whimbrel (Plate 40:8); rounded spots on breast and flanks. **11b In flight** ▲ Extremely rare Palearctic species. Uncertain record, Lake Chad.

12 GREATER YELLOWLEGS *Tringa melanoleuca* 29–33 cm **V***
12a Ad non-breeding Resembles Common Greenshank (Plate 42:8) but legs bright yellow; plumage generally browner. **12b In flight** Square white rump. ▲ N American vagrant.

13 SOCIABLE LAPWING *Vanellus gregarius* 27–30 cm **CR**
13a Ad non-breeding Bold white supercilium; blackish legs. **13b In flight from above** Three-coloured wing pattern. **13c Ad breeding in flight from below** Black belly with deep chestnut rear part. ▲ Palearctic vagrant.

1 **LEAST SANDPIPER** *Calidris minutilla* 13–14 cm V*
1a Ad non-breeding Usually complete streaked breast-band; greenish to yellowish legs. **1b Juv** Recalls Little Stint (Plate 41:3c) but whitish V on mantle narrower and less distinct, primary projection shorter; legs greenish. ▲ N American vagrant.

2 **LONG-TOED STINT** *Calidris subminuta* 14–15 cm V
2a Ad non-breeding As 1a but slightly larger, with longer neck and longer legs and toes. **2b Juv** As 1b but brown-grey coverts fringed whitish (nor rufous). ▲ E Palearctic vagrant.

3 **SEMIPALMATED SANDPIPER** *Calidris pusilla* 13–15 cm V+
3a Ad non-breeding As Little Stint (Plate 41:3b) but has half-webbed toes and different call. Also slightly paler; bill rather heavier, thick-tipped. **3b Juv** Plumage more uniform than Little Stint (Plate 41:3c); no distinct white Vs on mantle. ▲ N American vagrant.

4 **PURPLE SANDPIPER** *Calidris maritima* 19–22 cm V
Ad non-breeding Stocky; darkish; short, yellowish legs. ▲ Palearctic vagrant.

5 **LESSER SAND PLOVER** *Charadrius mongolus* 17–19 cm V
Ad non-breeding *pamirensis* Very similar to Greater Sand Plover (Plate 39:8) but slightly smaller; bill shorter, blunter-tipped; head more rounded (with steeper forehead); legs blackish, shorter (esp. tibiae). Feet do not project beyond tail in flight. ▲ Mudflats. Palearctic vagrant. ✤ Clear, hard *krip* or *tirrip*.

6 **SEMIPALMATED PLOVER** *Charadrius semipalmatus* 16–17 cm V*
6a Ad breeding As Common Ringed Plover (Plate 39:5) but has small webs between all front toes; breast-band usually narrower; bill stubbier; supercilium behind eye indistinct or lacking. **6b Ad non-breeding** Duller; white post-ocular streak sometimes joins white forehead. ▲ N American vagrant.

7 **SOLITARY SANDPIPER** *Tringa solitaria* 18–21 cm V+
10a Ad non-breeding As Green Sandpiper (Plate 42:1) but tail dark centred. **10b In flight** Dark band on centre of rump and tail. ▲ N American vagrant.

8 **SPOTTED SANDPIPER** *Actitis macularius* 18–21 cm V+
8a Ad breeding Underparts boldly spotted black; black-tipped pinkish bill; pinkish legs. **8b Ad non-breeding** As Common Sandpiper (Plate 42:3) but tail shorter, legs paler. **8c In flight** Wingbar shorter than Common Sandpiper, not extending onto inner secondaries; usually less white on tail-sides. ▲ N American vagrant.

9 **SHORT-BILLED DOWITCHER** *Limnodromus griseus* 25–29 cm V?
9a Ad non-breeding Long, straight, blunt bill; prominent supercilium; greenish legs. **9b Juv** Breast and upperparts tinged rusty-buff; upperpart feathers neatly fringed. **9c In flight** Whitish trailing edge to wing, white oval on back and barred tail; feet barely project beyond tail. **9d Tail** Black bars narrower than white bars. **9e Juv tertial** Centre irregularly striped and barred buff. ▲ N American vagrant, claimed from Ghana.

For comparison:

LONG-BILLED DOWITCHER *Limnodromus scolopaceus* 27–30 cm
Extremely similar to 9. Differences include: **9f Tail** Black bars broader than white bars. **9g Juv tertial** Centre plain dark. ▲ N American species; potential vagrant.

10 **UPLAND SANDPIPER** *Bartramia longicauda* 28–32 cm V
Distinctive shape; small head; long, thin neck; shortish, straight bill; long tail. ▲ N American vagrant.

Continued on page 161

 Maps on page 166

1b
1a
2b
2a
3b
3a
4
5
6a
6b
7b
7a
8a
8b
8c
9d
9e
9g
9f
9b
9a
9c
10
11a
11b
12a
12b
13a
13b
13c

Gull-like seabirds with powerful, steady and fast flight. Predatory and piratical. Feed mainly by harassing other seabirds, especially gulls and terns, to force them to disgorge or drop their catch. Plumages variable and identification often problematic; important features include structure, proportions and flight action; also note presence or absence of tail projection and of prominent pale areas in plumage.

1 LONG-TAILED SKUA *Stercorarius longicaudus* 35–41 cm; WS 105–115 cm P/V+ s/r
1a Ad breeding Very long tail streamers (12–14 cm), creamy breast and face, grey belly. **1b Juv pale morph** Cold tones to plumage; very pale fringes to upperparts. **1c Ad breeding in flight** Black flight feathers (with two white shafts); grey upperparts; buoyant, almost tern-like flight. **1d Ad non-breeding in flight** Head duskier; vent barred; tail streamers shorter. **1e Juv pale morph in flight** Blunt tips of central tail feathers clearly project. **1f Juv dark morph in flight** Uppertail, underwing and vent barred. ▲ Winters mainly off Namibia and W South Africa. Rare/scarce or vagrant, Mauritania–Nigeria. Also Cape Verde and Gulf of Guinea seas. ✤ [CD2:65]

2 ARCTIC SKUA *Stercorarius parasiticus* 37–44 cm; WS 108–118 cm P+/V f/r
2a Ad breeding pale morph Pointed tail streamers (projecting 5–8.5 cm); grey breast-band. **2b Juv pale morph** Warm rufous-brown; often paler nuchal band (diagnostic). **2c Ad breeding pale morph in flight** Wings more slender than 3 (narrower at base); flight fast, falcon-like. **2d Ad breeding dark morph in flight** Blackish-brown; pale primary patches. **2e Ad non-breeding in flight** Head pattern less distinct; uppertail and vent barred. **2f Juv pale morph in flight** Points of central tail feathers project only slightly. **2g Juv dark morph in flight**. ▲ Winters mainly off Namibia and W South Africa. Frequent, Mauritania–Ghana. Rare/vagrant elsewhere. Also Cape Verde and Gulf of Guinea seas. ✤ [CD2:64]

3 POMARINE SKUA *Stercorarius pomarinus* 42–50 cm; WS 115–125 cm P+/V f/r
3a Ad breeding pale morph Spoon-shaped central tail feathers (projecting 5.5–11 cm); mottled breast-band. **3b Juv** Variable; lacks warm tones of typical juvenile Arctic; head unstreaked. **3c Ad breeding pale morph in flight** Noticeably heavier than 2; wings broad based; breast rounded. **3d Ad breeding dark morph in flight** Blackish-brown; white primary patches. Rare. **3e Ad non-breeding in flight** Uppertail and vent barred; tail streamers often lacking. **3f Juv pale morph in flight** Pale crescent at base of primaries creates double wing patch; barred areas on underwings, uppertail-coverts and vent contrastingly pale. **3g Juv dark morph in flight**. ▲ Winters mainly at sea between 20° and 8°N. Frequent, Mauritania–Ghana. Rare/vagrant elsewhere. Also Cape Verde and Gulf of Guinea seas. ✤ [CD2:63]

4 GREAT SKUA *Stercorarius skua* 50–58 cm; WS 125–140 cm P+/V r
4a Ad Dark; coarsely streaked; heavy. Juvenile similar but less streaky. **4b Ad in flight** Conspicuous white patches at base of primaries above and below. ▲ Main wintering area in Atlantic south to W Africa. Uncommon, Mauritania. Rare/vagrant Senegal–Nigeria, also Cape Verde seas and Annobón. ✤ [CD2:66]

5 SOUTH POLAR SKUA *Stercorarius maccormicki* 50–54 cm; WS 125–135 cm V
5a Ad dark morph in flight As 4, but slightly slimmer; cold greyish-brown; unstreaked; nape often pale. **5b Ad pale morph in flight** (drawn to smaller scale) Head and underparts pale buff-brown. ▲ Breeds Antarctica. Recorded off Senegal. Presumably regular passage migrant.

 Maps on page 167

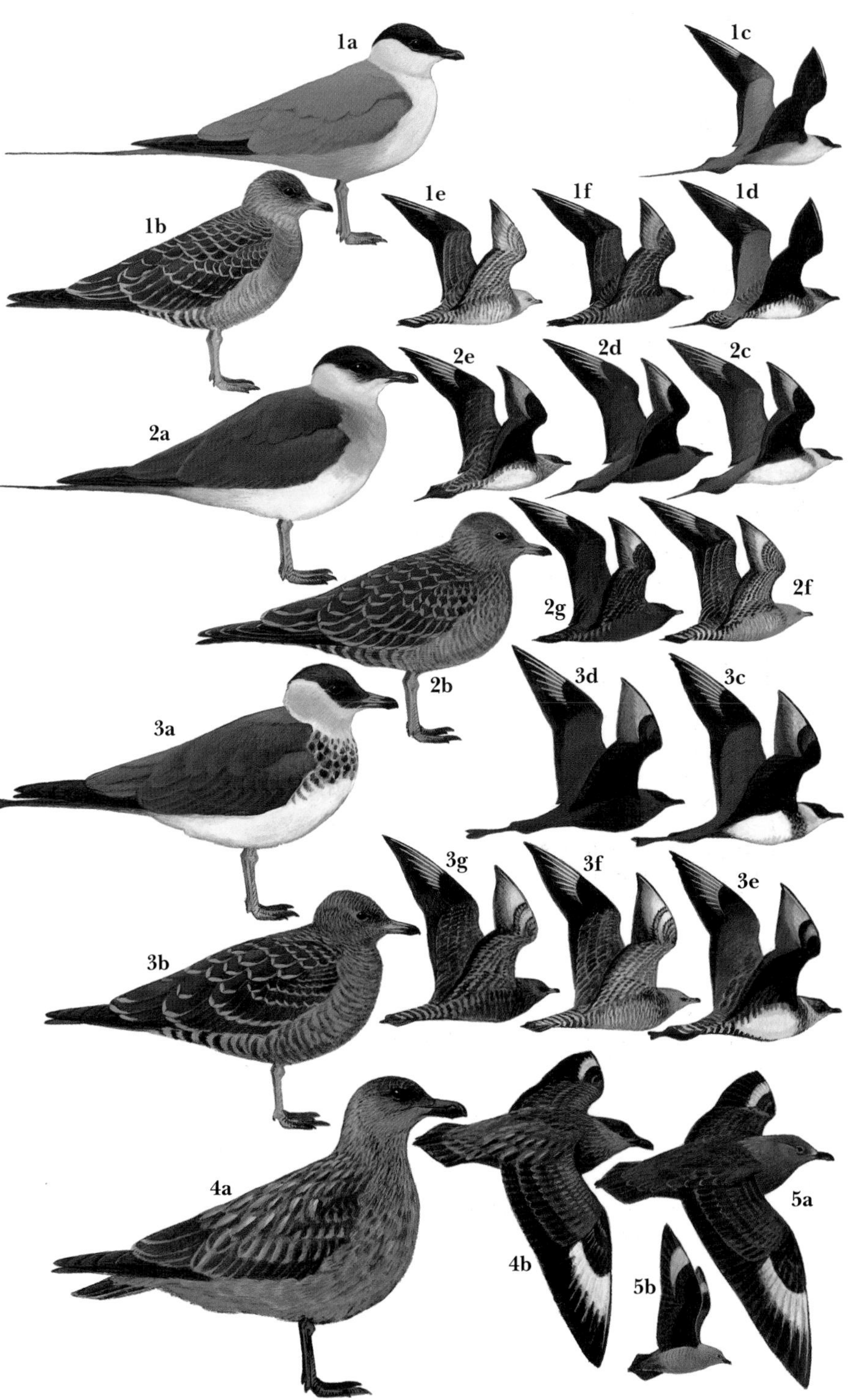
1a
1c
1e
1f
1d
1b
2e
2d
2c
2a
2g
2f
2b
3d
3c
3a
3g
3f
3e
3b
4a
5a
4b
5b

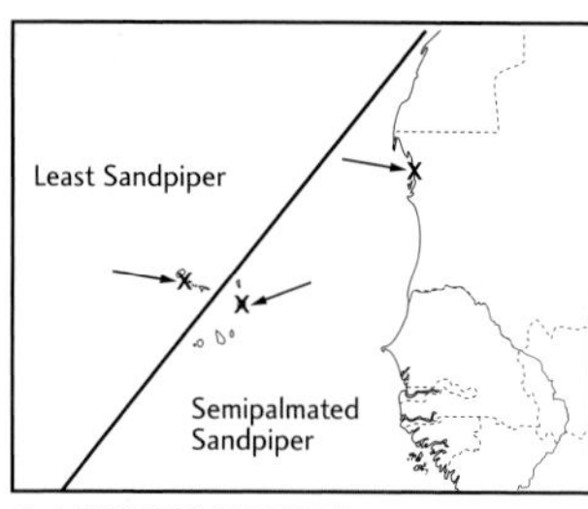

1. LEAST SANDPIPER
3. SEMIPALMATED SANDPIPER

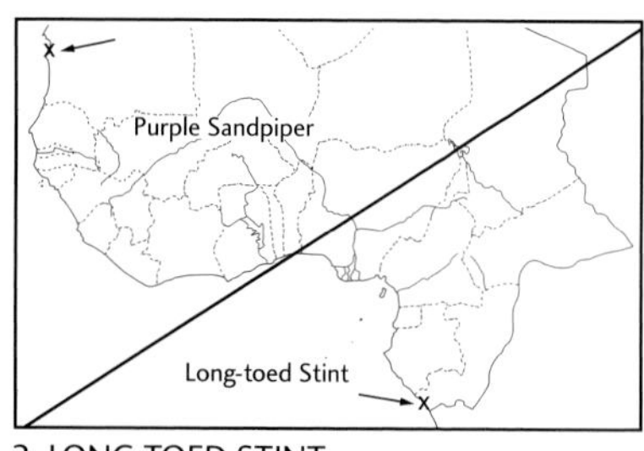

2. LONG-TOED STINT
4. PURPLE SANDPIPER

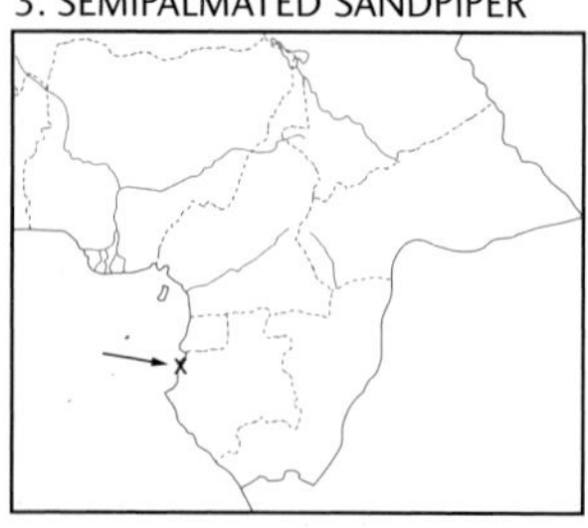

5. LESSER SAND PLOVER

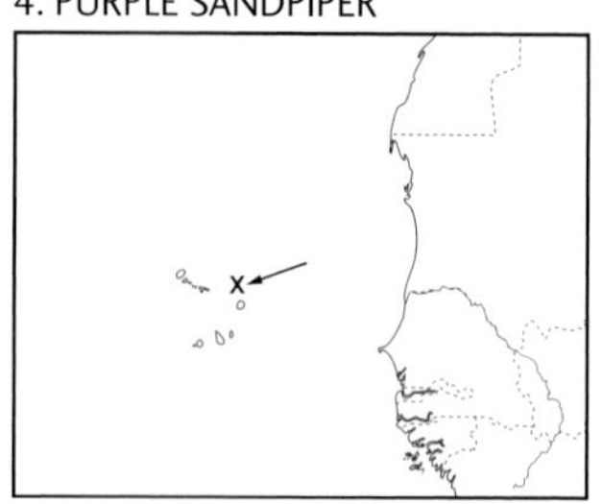

6. SEMIPALMATED PLOVER

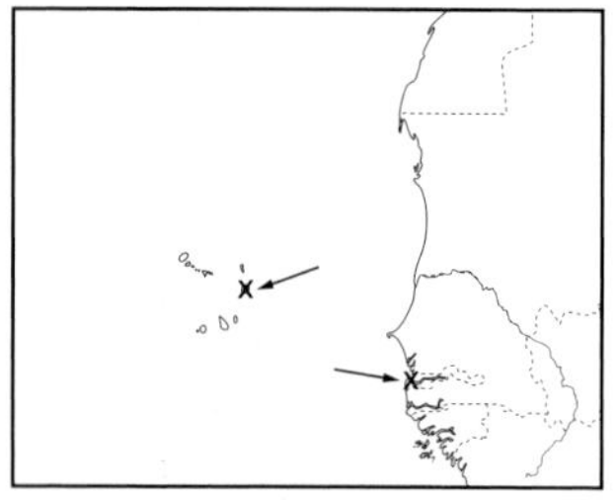

7. SOLITARY SANDPIPER

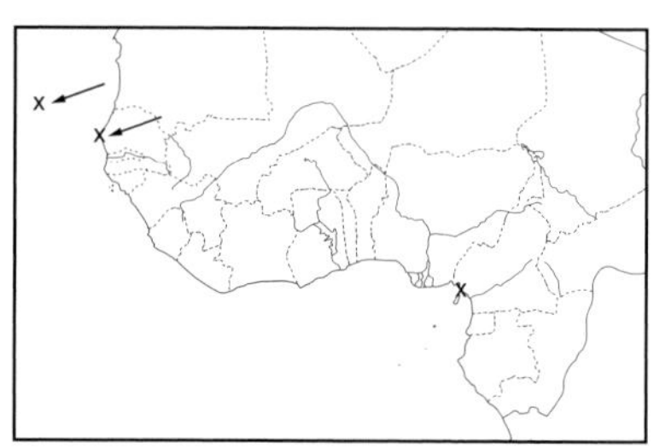

8. SPOTTED SANDPIPER

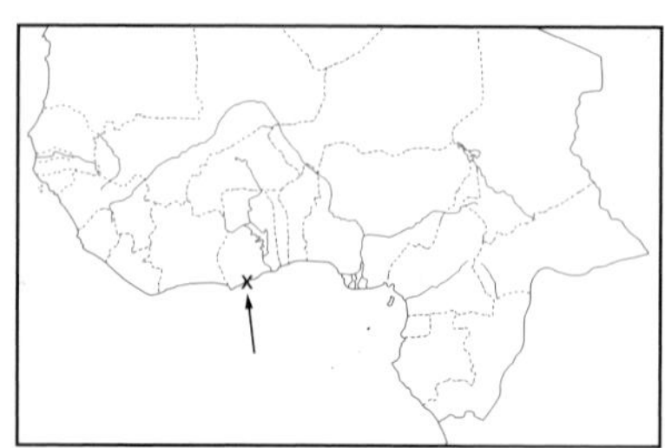

9. SHORT-BILLED DOWITCHER

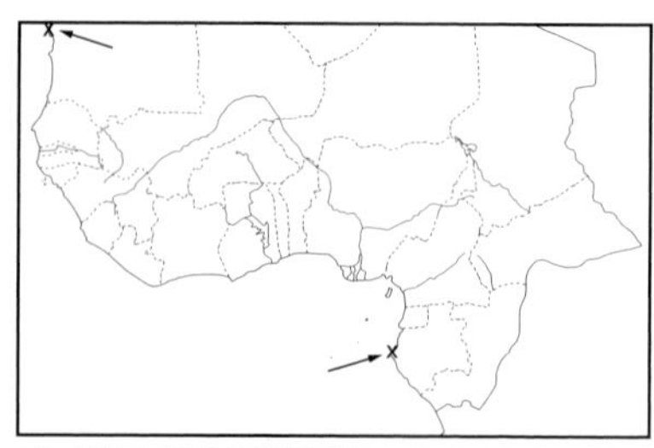

10. UPLAND SANDPIPER

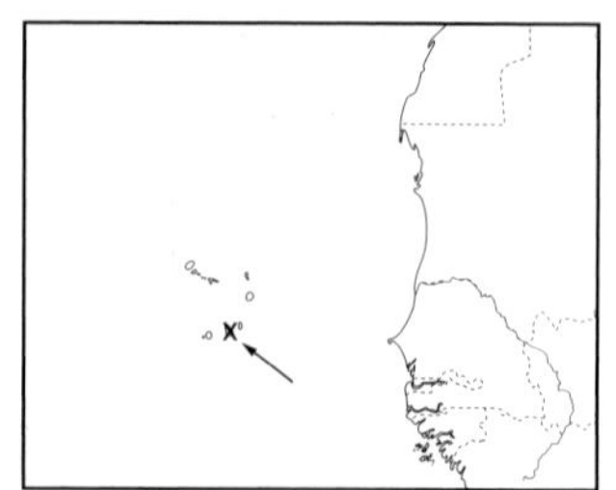

12. GREATER YELLOWLEGS

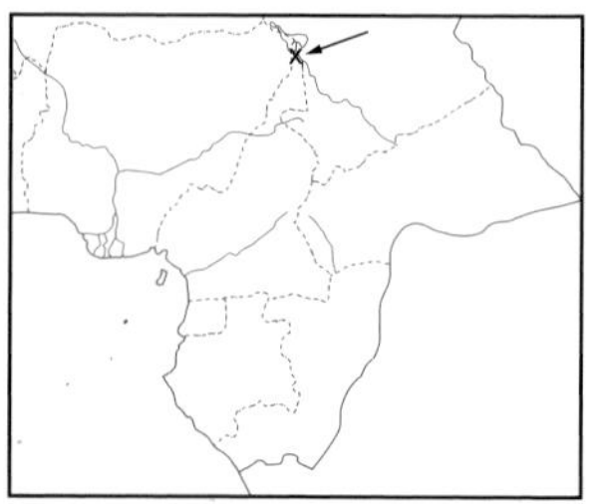

13. SOCIABLE LAPWING

Plate on page 162

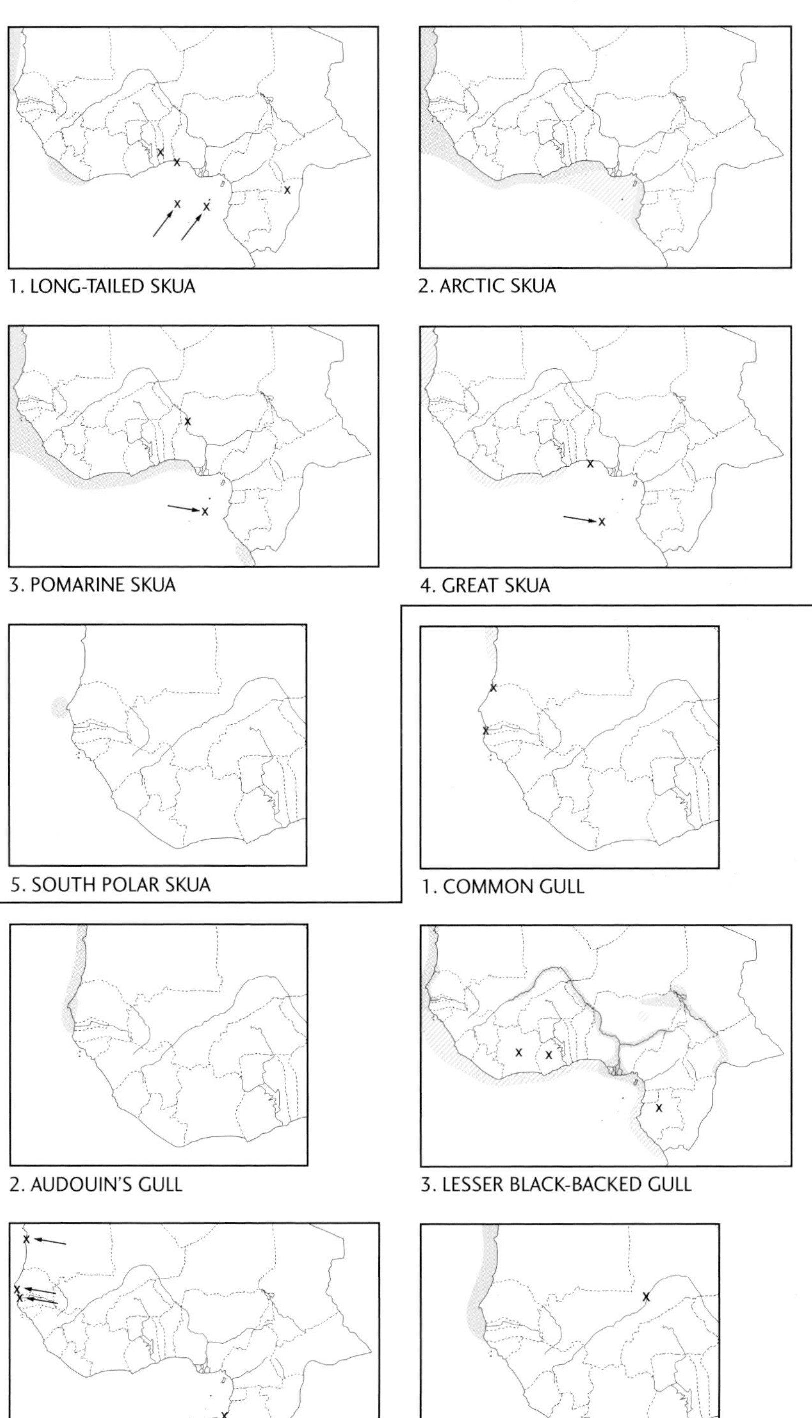

1. LONG-TAILED SKUA

2. ARCTIC SKUA

3. POMARINE SKUA

4. GREAT SKUA

5. SOUTH POLAR SKUA

1. COMMON GULL

2. AUDOUIN'S GULL

3. LESSER BLACK-BACKED GULL

4. KELP GULL

5. YELLOW-LEGGED GULL

Gulls. Robust, medium-sized to large birds of coast, oceans and inland waters with predominantly white plumage and grey or black wings and mantle. Distinct breeding, non-breeding and immature plumages. Adult plumage reached in 2nd–4th year; period of immaturity generally related to size. Tail white in adults; dark tail band indicative of immature. Sexes similar, male slightly larger. Flight strong and buoyant with slow wingbeats and much gliding and soaring. Often swim, rarely dive. Usually vocal and gregarious, breeding colonially. Most are omnivorous aquatic and littoral scavengers, fishers and predators. Several are long-distance migrants.

To identify gulls, note wing and head patterns, and bare-part coloration. Understanding moult greatly facilitates identification. First moult is post-juvenile: a partial moult, replacing head and body feathers and resulting in first-winter plumage. Subsequently, there are two moults per year: a partial one from winter to summer plumage, mainly replacing head and body feathers (flight feathers and all or most tail feathers thus still juvenile and much abraded in first-summer plumage), and a complete one after the breeding season, resulting in adult or immature winter plumages. The only exceptions to this rule are Sabine's Gull (partial moult in Nov–Dec and complete one in Feb–Apr) and Franklin's Gull (two complete moults per year).

1 COMMON GULL *Larus canus* 40–42 cm V

1a Ad breeding Rounded head; dark eye; dark blue-grey upperparts; yellow-green bill. **1b Ad non-breeding** Head has dusky markings; pale bill with dark band; greenish legs. **1c First-winter** Dusky head and breast; dusky-tipped pinkish bill; pink legs. ▲ Coastal. Palearctic vagrant, Mauritania–Senegambia. ✤ [CD2:73]

2 AUDOUIN'S GULL *Larus audouinii* 48–52 cm P r NT

2a Ad Pale grey upperparts; yellow-tipped dark red bill with dark subterminal band; olive-grey legs. **2b First-winter** Whitish head contrasts with grey-brown upperparts; dark grey legs. ▲ Coastal. ✤ [CD2:72]

3 LESSER BLACK-BACKED GULL *Larus fuscus* 52–67 cm P+ c/u

3a Ad *graellsi*/*fuscus* breeding Similar to non-breeding, but head and neck white. **3b Ad *graellsi* non-breeding** Slate-grey upperparts; grey-brown streaking on head and breast-sides; yellow legs. **3c Ad *fuscus* non-breeding** Blackish upperparts; much whiter head and breast-sides than 3b. **3d First-winter** Dark grey-brown; bill black; legs pink. ▲ Coast, major rivers, lakes. ✤ [CD2:74]

4 KELP GULL *Larus dominicanus* 58 cm M s

4a Ad More robust than 3; bill heavier; eye dark; legs olive. **4b First-winter** Mainly dark brown; bill black; pinkish-brown legs. ▲ Coastal. Migrant from southern hemisphere. ✤ [CD6:46]

5 YELLOW-LEGGED GULL *Larus cachinnans* 55–67 cm P+ u/r

5a Ad *michahellis* non-breeding (P) Ashy-grey upperparts; yellow bill with orange spot; yellow legs. **5b First-winter *michahellis*** Mainly brownish, with pale head and underparts; bill black; legs brownish-pink. **5c Ad *atlantis* non-breeding** (P) More heavily streaked on head and neck. **5d First-winter *atlantis*** All primaries dark, thus resembling 3d. ▲ Coastal.

✤ [CD2:76,77]

1c
1b
1a
2b
3b
2a
3a
3c
3d
4a
4b
5a
5b

1 COMMON TERN *Sterna hirundo* 31–35 cm; WS 77–98 cm R/P+/M c/r*
1a Ad breeding Black-tipped red bill; contrasting darker outer primaries; tail not projecting beyond wingtips. White line between cap and gape broader than in 2a. **1b Ad non-breeding** Dark bill with reddish base; white forehead; no tail streamers. **1c Juv** Gingery wash to forehead and scaly upperparts; blackish carpal bar. **1d Ad breeding in flight** Dusky wedge on outer wing. **1e First-winter in flight** Dusky carpal bar and secondary bar; dark outer primaries. ▲ Coastal. Breeds inland in Gabon. ✤ [CD2:88]

2 ARCTIC TERN *Sterna paradisaea* 33–35 cm; WS 75–85 cm P+ u/r
2a Ad breeding All-red bill; pale grey underparts; tail projecting beyond wingtips. **2b Ad non-breeding** Blackish bill (shorter than 1b); white forehead; no tail streamers. **2c Ad breeding in flight** Uniform upperwings; white underwings with tip bordered by neat black line. **2d First-winter in flight** Dusky carpal bar (less obvious than in 1e); very pale hindwing with broad white trailing edge. ▲ Coastal. ✤ [CD2:89]

3 ROSEATE TERN *Sterna dougallii* 33–38 cm; WS 72–80 cm P/V* u/r
3a Ad breeding Mainly black bill; very long tail projecting beyond wingtips. **3b Ad non-breeding** Much paler overall than 1b and 2b. **3c Ad breeding in flight** Very pale upperparts; dusky outer primaries. **3d First-winter in flight** Dusky carpal bar; whitish secondaries. ▲ Coastal/pelagic. ✤ [CD2:87]

4 BLACK-LEGGED KITTIWAKE *Rissa tridactyla* 38–40 cm P+/V r/s*
4a Ad breeding Head white; unmarked greenish-yellow bill. **4b Ad non-breeding** Back of head and neck smudged grey; blackish ear-spot. **4c First-winter** Blackish ear-spot and hindneck collar. ▲ Pelagic. Mauritania; probably regular offshore. ✤ [CD2:81]

5 FRANKLIN'S GULL *Larus pipixcan* 32–36 cm V
5a Ad breeding Differs from 6a in more rounded head, shorter bill, broader eye-crescents. **5b Ad non-breeding** Dark half-hood; broad eye-crescents. **5c First-winter** Differs from 6c in half-hood and white neck and underparts. ▲ Coastal. N American vagrant. ✤ [CD6:45]

6 LAUGHING GULL *Larus atricilla* 38–41 cm V
6a Ad breeding Black hood; white eye-crescents; dull red bill and legs. **6b Ad non-breeding** Dusky-marked head; dark grey upperparts; longish, blackish bill. **6c First-winter** Rear of head to breast and flanks dusky brownish-grey. ▲ Coastal. American vagrant. ✤ [CD6:44]

7 SABINE'S GULL *Larus sabini* 27–32 cm P/V*
7a Ad breeding Dark grey hood; yellow-tipped black bill. **7b Ad non-breeding** White head with blackish nape patch. **7c First-winter** Brownish wing-coverts; all-black bill. ▲ Pelagic.

8 LITTLE GULL *Larus minutus* 25–27 cm P/V s
8a Ad breeding Small size; black hood; no white eye-crescents. **8b Ad non-breeding** Dusky cap and ear-spot; pale grey upperparts. **8c First-winter** Dusky cap and ear-spot; black primaries and wingbar. ▲ Coastal. ✤ [CD2:68]

9 MEDITERRANEAN GULL *Larus melanocephalus* 36–38 cm P s/r
9a Ad breeding Slightly larger than 12a; bill stouter; hood extends on nape; legs longer. **9b Ad non-breeding** White wingtips; dark wedge behind eye. **9c First-winter** Dusky patch on ear-coverts; blunt bill. ▲ Coastal. ✤ [CD2:67]

10 SLENDER-BILLED GULL *Larus genei* 42–44 cm P/M/V+ lf
10a Ad breeding White head; pink flush to underparts; distinctive, elongated jizz. **10b Ad non-breeding** Faint dusky ear-spot. **10c First-winter** Brownish wing-coverts; dark tail band; pale bill and legs. ▲ Coastal. ✤ [CD2:71]

Continued on page 172

Maps on pages 172 and 173

1c
1b
1e
1d
1a
2b
2a
2d
2c
3b
3a
3d
3c
4c
5c
6c
4b
6b
5b
4a
7c
5a
6a
9c
8c
7b
9b
7a
8a
8b
9a
10c
11c
12c
10b
11b
12a
10a
11a
12b

Continued from page 170

11 GREY-HEADED GULL *Larus cirrocephalus* 39–42 cm **R/V c/s**
11a Ad breeding Pale grey hood; red bill and legs. **11b Ad non-breeding** White head, often with some greyish; faint dusky ear-spot. **11c First-winter** White head with dusky markings; brownish wing-coverts. ▲ Coast, major rivers. ✤ [CD2:69]

12 BLACK-HEADED GULL *Larus ridibundus* 34–37 cm **P+/V u/r**
12a Ad breeding Dark-brown hood; dark red bill, legs and feet. **12b Ad non-breeding** White head with dusky ear-spot. **12c First-winter** Brownish wing-coverts; dusky head markings. ▲ Coast, inland aquatic habitats. ✤ [CD2:70]

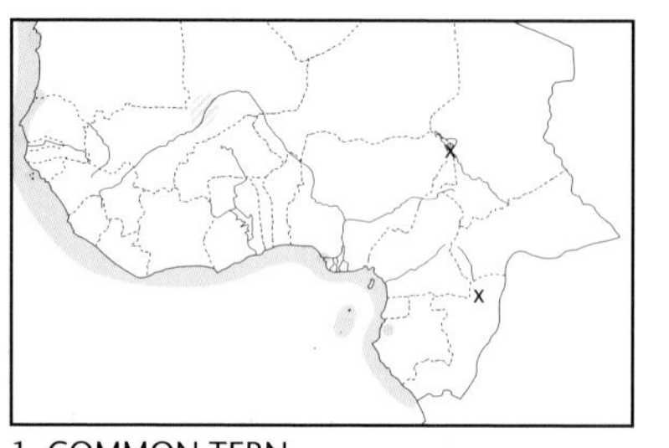

1. COMMON TERN

2. ARCTIC TERN

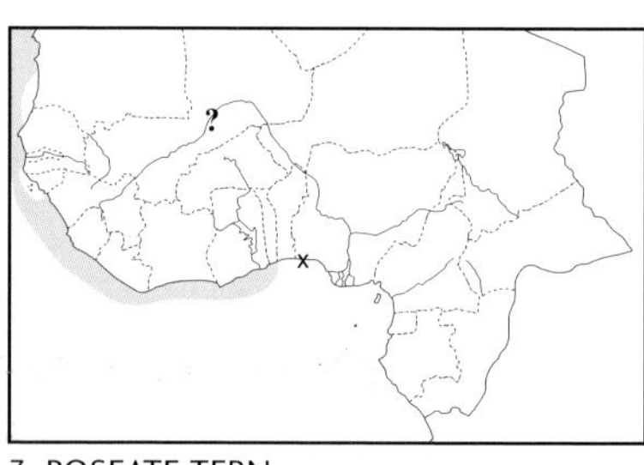

3. ROSEATE TERN

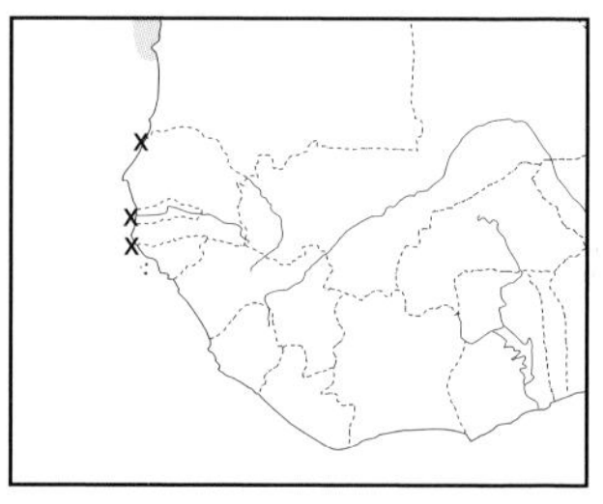

4. BLACK-LEGGED KITTIWAKE

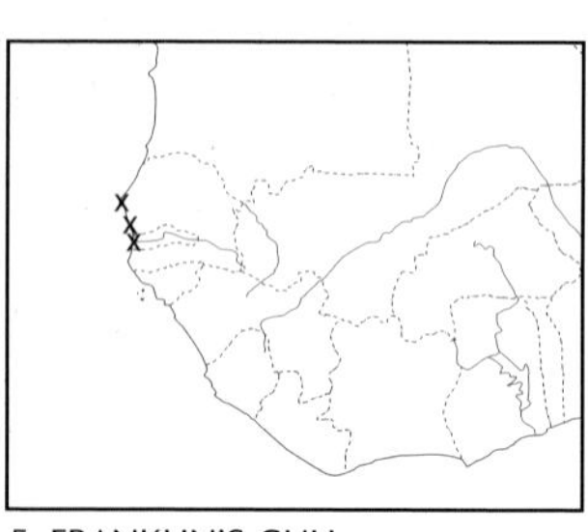

5. FRANKLIN'S GULL

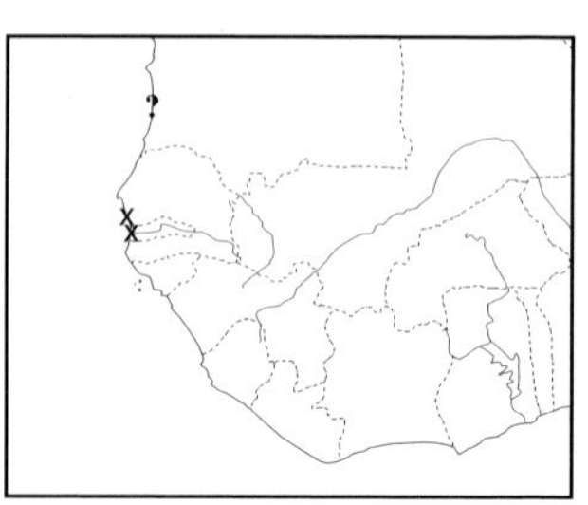

6. LAUGHING GULL

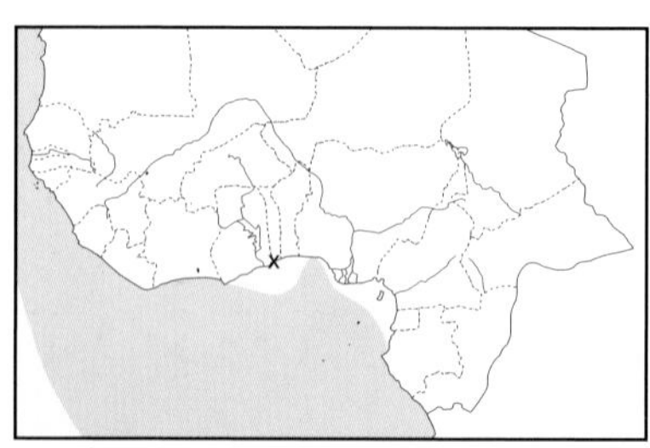

7. SABINE'S GULL

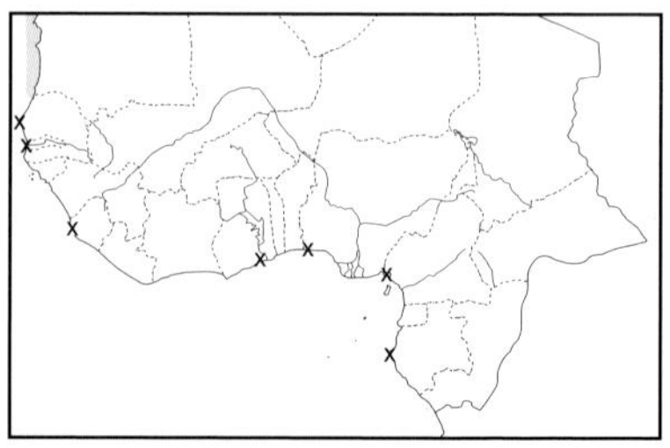

8. LITTLE GULL

Plate on page 170

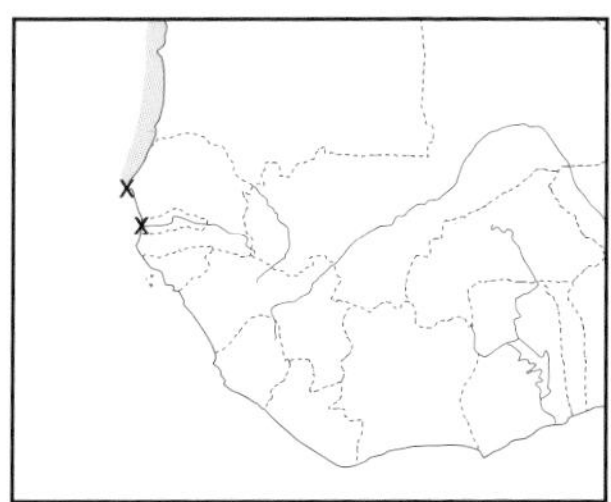
9. MEDITERRANEAN GULL

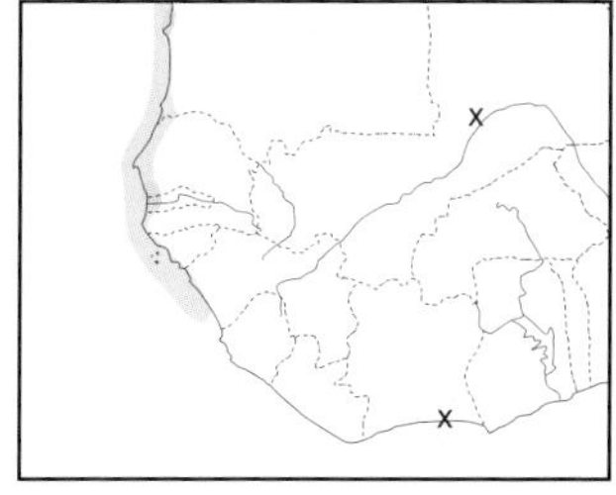
10. SLENDER-BILLED GULL

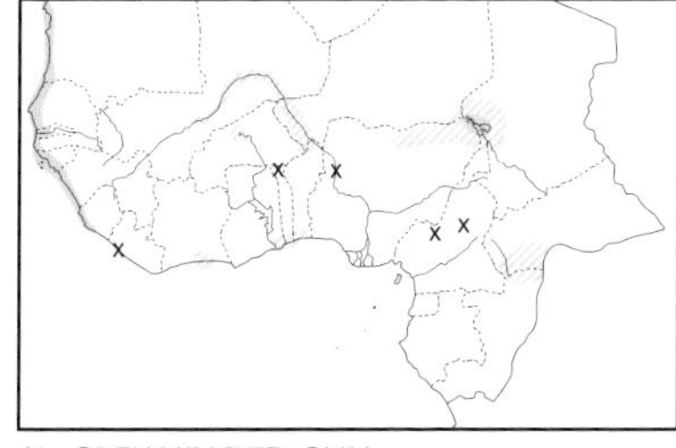
11. GREY-HEADED GULL

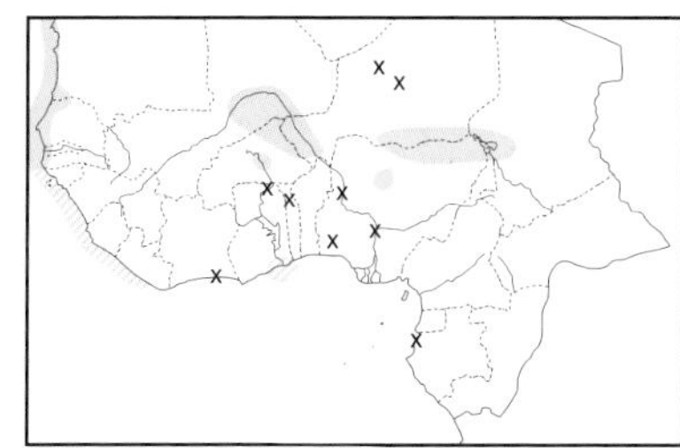
12. BLACK-HEADED GULL

PLATE 47: GULLS IN FLIGHT

Continued from page 174

13 **SLENDER-BILLED GULL** *Larus genei* WS 100–110 cm **P/M/V+ lf**
13a Ad non-breeding White leading edge to outer wing; elongated head and neck. Compare 11a. **13b First-winter** Pale brown carpal bar; blackish secondary bar and tail band.

14 **AUDOUIN'S GULL** *Larus audouinii* WS 115–140 cm **P r NT**
14a Ad Black wingtips sharply contrasting with rest of plumage; dark eye and bill. **14b First-winter** Blackish primaries and secondary bar; dark greater coverts; black tail; U-shaped band across rump.

PLATE 47: GULLS IN FLIGHT

See also Plates 45–46.

1 LITTLE GULL *Larus minutus* WS 75–80 cm **P/V s**
1a Ad non-breeding from above No black on wings; dusky cap and ear-spot. **1b Ad non-breeding from below** Blackish underwing with white trailing edge. **1c First-winter** Small size; blackish W on upperwings; black tail band. Compare 4b.

2 SABINE'S GULL *Larus sabini* WS 90–100 cm **P/V***
2a Ad non-breeding Triangular-patterned, tricoloured wings; slightly forked tail. **2b Juv** Same pattern, but grey areas browner; black tail band.

3 COMMON GULL *Larus canus* WS 110–130 cm **V**
3a Ad non-breeding Black wingtips with rather large white 'mirrors'. **3b First-winter** Blackish leading edge to outer wing, secondary bar and tail band. **3c Second-winter** Black leading edge to outer wing; small white 'mirrors'.

4 BLACK-LEGGED KITTIWAKE *Rissa tridactyla* WS 95–120 cm **P+/V r/s***
4a Ad non-breeding Clear-cut black wingtips. **4b First-winter** Neat black W on upperwings; blackish collar on nape; black tail band. Compare 1c.

5 LAUGHING GULL *Larus atricilla* WS 100–125 cm **V**
5a Ad non-breeding Slate-grey upperparts; black wingtips without 'mirrors'. **5b First-winter** Blackish primaries, secondary bar and tail band; dusky rear of head, breast and flanks.

6 FRANKLIN'S GULL *Larus pipixcan* WS 85–95 cm **V**
6a Ad non-breeding Black on wingtips surrounded by white; dark half-hood. **6b First-winter** Differs from 5b in dark half-hood, paler inner primaries, and narrower tail band, which does not extend to outermost tail feathers.

7 YELLOW-LEGGED GULL *Larus cachinnans* WS 138–155 cm **P+ u/r**
7a Ad non-breeding Ashy-grey upperparts; black wingtips with small white 'mirrors'. **7b First-winter** Brownish upperparts; outer wing, secondaries and broad tail band blackish. **7c Second-winter** Acquires ashy-grey mantle and white head and body. **7d Third-winter** As adult, but traces of immaturity on wings and tail.

8 LESSER BLACK-BACKED GULL *Larus fuscus* WS 135–155 cm **P+ c/u**
8a Ad breeding Blackish upperparts. **8b First-winter** Outer wing and secondaries blackish; greater coverts dark. **8c Second-winter** Acquires adult mantle colour and white head and body.

9 GREY-HEADED GULL *Larus cirrocephalus* WS 100–115 cm **R/V c/s**
9a Ad breeding Black wingtip bordered by white diagonal bar; grey hood. **9b First-winter** Black outer primaries, secondary bar and tail band; brown carpal bar. **9c Second-winter** Black on wingtip more extensive; dusky trailing edge.

10 KELP GULL *Larus dominicanus* WS 128–142 cm **M s**
10a Ad Black upperparts; dark eye; powerful bill. **10b First-winter** Very dark looking, with blackish outer wing and secondaries. **10c Second-winter** Largely white head; dark brown saddle; white tail base.

11 BLACK-HEADED GULL *Larus ridibundus* WS 100–110 cm **P+/V u/r**
11a Ad non-breeding Rather pointed wings; white leading edge to outer wing. **11b First-winter** Brown carpal bar; blackish secondary bar and tail band.

12 MEDITERRANEAN GULL *Larus melanocephalus* WS 92–100 cm **P s/r**
12a Ad non-breeding Unmarked wings; dusky patch on ear-coverts. **12b First-winter** Pale saddle; contrasting upperwing pattern; dark patch behind eye.

Continued on page 173

b
2b
3b
7b
2a
3c
1c
1a
3a
7c
4b
5a
7d
5b
4a
7a
8b
6a
6b
10b
9a
8c
9c
9b
8a
11a
10c
11b
10a
14a
12a
13b
12b
13a
14b

PLATE 48: LARGE AND MEDIUM-SIZED TERNS

Terns. Slender, small to medium-sized marine and freshwater birds with long, pointed wings, usually long, forked tail and slender, pointed bill. Distinct breeding and non-breeding plumages. Flight graceful, more buoyant and agile than gulls, on faster and deeper wingbeats and without any gliding or soaring. Feed by plunge-diving or plucking food from water; some (esp. marsh terns and Gull-billed Tern) also catch insects in the air. Most immatures of migratory Palearctic species spend first, sometimes also second, summer in winter quarters.

Main identification features include size, wing pattern, bill shape and colour, and call. Post-breeding moult of head, body and wing-coverts complete; pre-breeding partial.

1 GULL-BILLED TERN *Gelochelidon nilotica* 35–43 cm; WS 86–103 cm R/P/V* c/r
1a Ad breeding Black cap without crest; stout black bill. Compare 2. **1b Ad non-breeding** White head with black streak behind eye. **1c Juv** Faint brownish tinge to crown and upperparts; indistinctly marked above. **1d Ad breeding in flight** Upperparts, rump and tail pale grey; short tail. **1e First-winter in flight** Primaries worn and darker than in adult. ▲ Inland and coastal waters. ✤ Harsh, raucous *ger-vik* and *khaak*. Juvenile utters high-pitched *pe-eep* or fast *pe-pe-eep*. [CD2:82]

2 SANDWICH TERN *Sterna sandvicensis* 36–41 cm; WS 95–105 cm P+ c/r*
2a Ad breeding Black cap with shaggy crest; yellow-tipped black bill. **2b Ad non-breeding** Head white with black 'shawl'. **2c Ad non-breeding in flight** Very white looking; dark outer primaries; short tail. **2d First-winter in flight** Darkish carpal and secondary bars; darkish outer tail feathers. ▲ Coastal. ✤ Loud, rasping *kerrik* or *keerr-wit*. [CD2:86]

3 LESSER CRESTED TERN *Sterna bengalensis* 35–37 cm; WS 92–105 cm P s/lf
3a Ad breeding Black cap with shaggy crest, orange-red bill. **3b Ad non-breeding** Head white with black 'shawl'; orange bill. **3c Ad non-breeding in flight** Underwing with narrow dark trailing edge to outer primaries. **3d First-winter in flight** Dark outer primaries, dusky secondary bar. ▲ Coastal. ✤ Rasping *errik* and *krrr-eep*, resembling 2 but higher pitched. [CD2:85]

4 GREATER CRESTED TERN *Sterna bergii* 46–49 cm; WS 125–130 cm V?
4a Ad breeding Black cap with shaggy crest; white forehead; lemon-yellow bill. **4b Ad non-breeding** Black cap receding and speckled white; crest reduced. **4c Ad non-breeding in flight** Upperparts darker grey than in other large terns. **4d First-winter in flight** Upperwing with dark leading edge and dark secondary bar. ▲ Not recorded with certainty in W Africa. One doubtful claim from Nigeria.

5 ROYAL TERN *Sterna maxima* 45–50 cm; WS 125–135 cm R/M/V* c
5a Ad breeding Black cap with short crest; deep orange bill. Second largest tern. **5b Ad non-breeding** Black 'shawl' on hindcrown; pale orange bill. **5c Juv** Prominent carpal and secondary bars; dull yellowish legs. **5d Ad non-breeding in flight** Dark trailing edge to outer primaries. **5e First-winter in flight** Dark outer wing; three dark bars on innerwing; dark-tipped tail. ▲ Coastal. ✤ High-pitched, shrill *kee-err* and a loud, sharp *krryuk*, resembling 2 but deeper. [CD2:84]

6 CASPIAN TERN *Sterna caspia* 47–54 cm; WS 130–145 cm R/P/V* c/u
6a Ad breeding Large black cap with short rough crest; massive red bill. Largest tern. **6b Ad non-breeding** Dark cap densely streaked white. **6c Juv** Scaly pattern on upperparts; bill dull reddish tipped dusky. **6d Ad non-breeding in flight** Dark wedge on tip of underwing. **6e First-winter in flight** Dusky leading edge and secondary bar. ▲ Coastal; locally inland. ✤ Deep, hoarse *kraah-ap* or *rrha-ak*. [CD2:83]

Maps on page 178

1c
1e
1d
1b
1a
2d
2c
2a
2b
3d
3c
3a
3b
4a
4b
4d
5a
5b
5c
4c
5e
5d
6e
6d
6a
6b
6c

PLATE 48: LARGE AND MEDIUM-SIZED TERNS

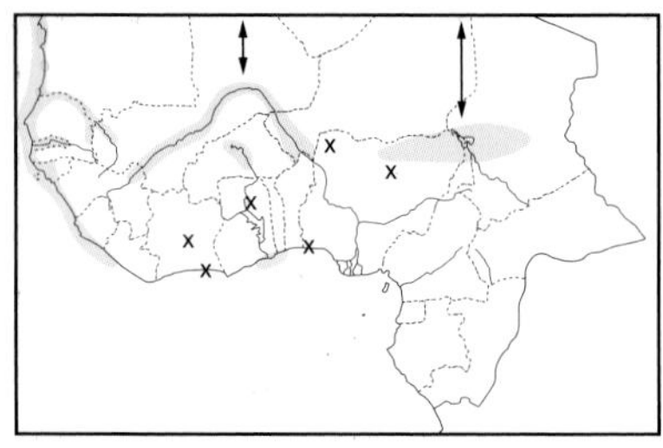

1. GULL-BILLED TERN

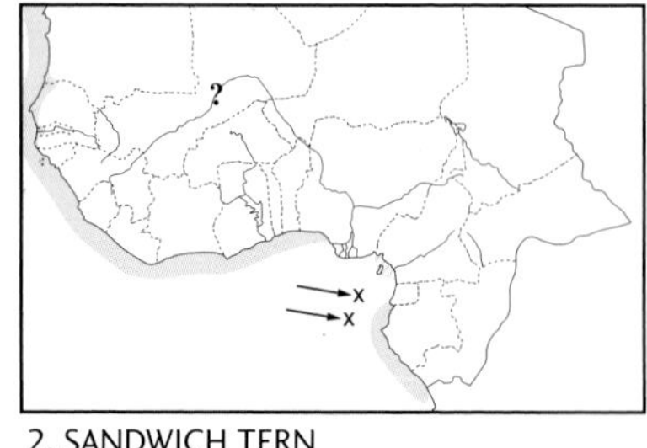

2. SANDWICH TERN

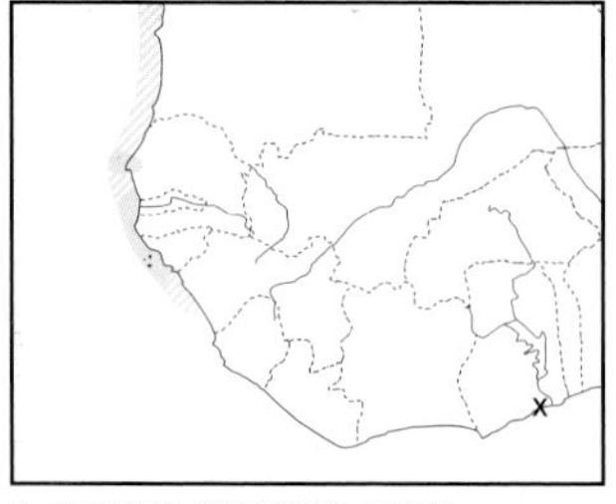

3. LESSER CRESTED TERN

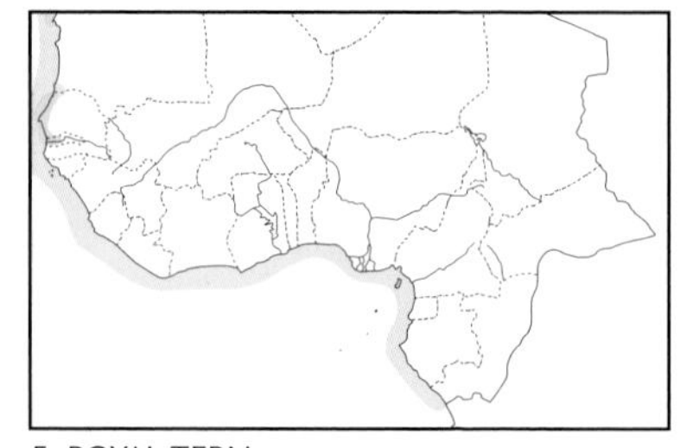

5. ROYAL TERN

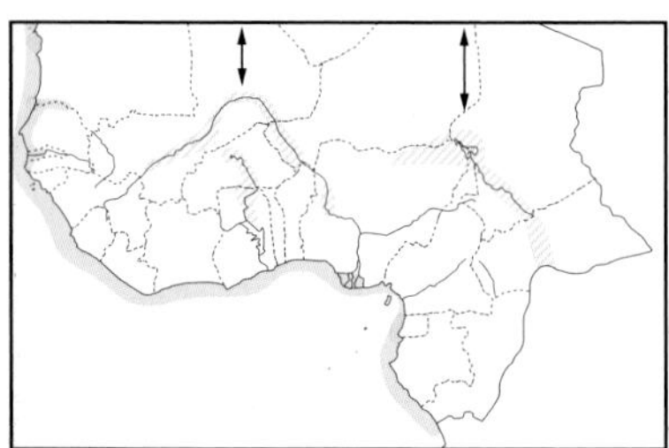

6. CASPIAN TERN

Plate on page 176

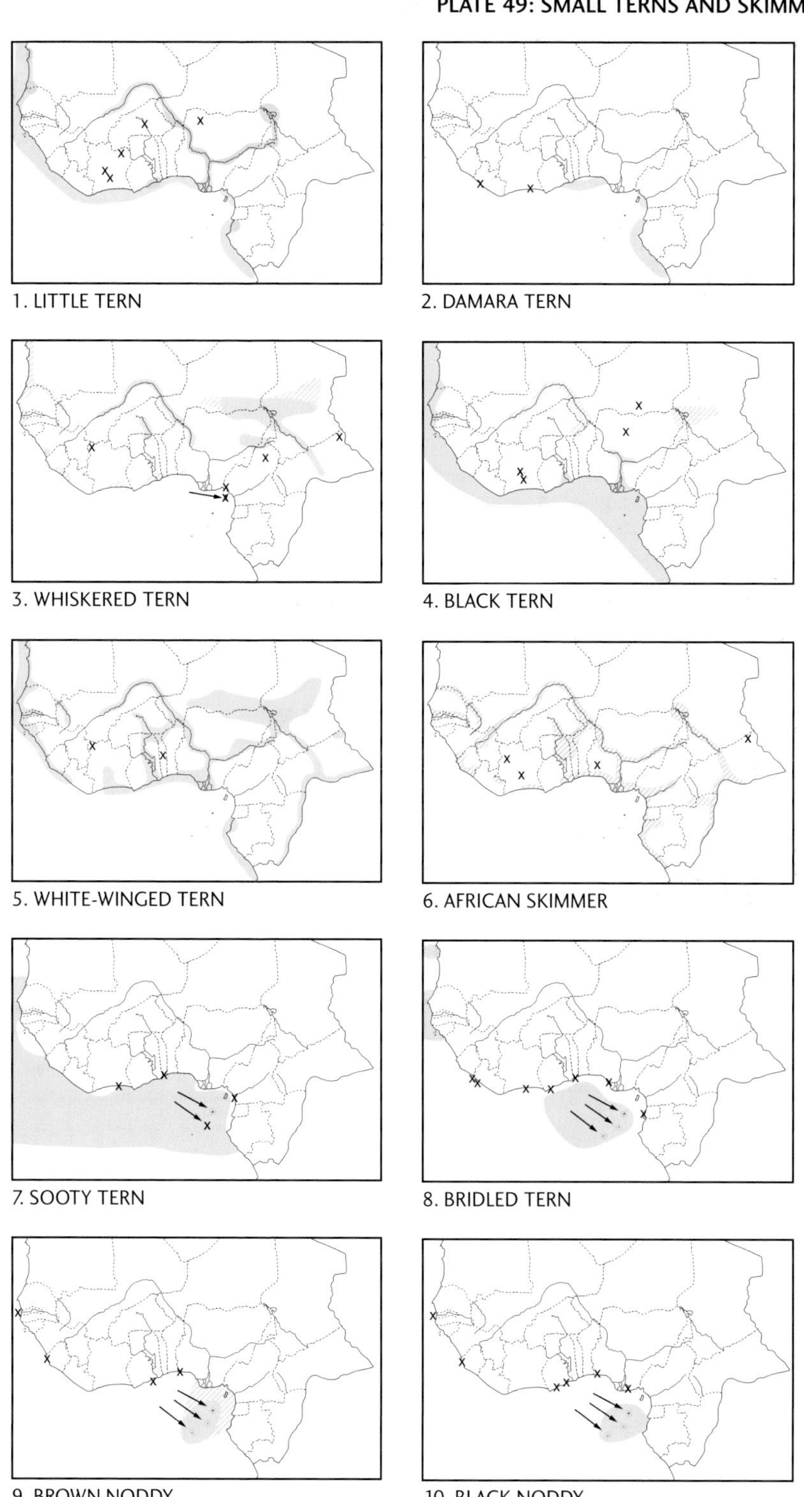

1. LITTLE TERN

2. DAMARA TERN

3. WHISKERED TERN

4. BLACK TERN

5. WHITE-WINGED TERN

6. AFRICAN SKIMMER

7. SOOTY TERN

8. BRIDLED TERN

9. BROWN NODDY

10. BLACK NODDY

Plate on page 180

PLATE 49: SMALL TERNS AND SKIMMER

1 LITTLE TERN *Sterna albifrons* 22–24 cm; WS 48–55 cm **R/M/P c/u**
1a Ad *guineae* breeding (R/M) Small size; yellow bill and legs; white forehead; white rump. **1b Ad *albifrons* breeding** (P) Yellow bill with clear black tip (in 1a little or no black); grey rump. **1c Ad non-breeding** Black bill; black 'shawl'; streaked crown; brownish legs. **1d Juv** Upperparts with pale fringes. **1e Ad *guineae* breeding in flight** Narrow dusky wedge on upperwing. Flight fluttering. **1f First-winter *albifrons* in flight** Dark carpal bar and leading edge. ▲ Coastal; inland along major rivers. ✤ [CD2:92]

2 DAMARA TERN *Sterna balaenarum* 23 cm; WS 51 cm **M r/s NT**
2a Ad breeding Small size; black cap; slightly decurved, black bill. **2b Ad non-breeding** White forehead. **2c Juv** Upperparts with pale fringes. **2d Ad breeding in flight** Dumpy; dusky leading edge to outer wing. **2e Ad non-breeding in flight** Dusky carpal bar and leading edge to outer wing. ▲ Coastal. S African migrant. ✤ [CD6:47]

3 WHISKERED TERN *Chlidonias hybrida* 23–25 cm; WS 74–78 cm **P/R c/s**
3a Ad breeding Black cap; grey underparts; contrasting white streak on head-sides. **3b Ad non-breeding** White underparts; black band on nape; streaked crown. **3c Ad breeding in flight** *Sterna* jizz; sooty-black belly. **3d Ad non-breeding in flight** Uniformly pale grey upperparts, rump and tail. Compare 4d. **3e Juv in flight** Dark brown saddle; dark bill and legs. **3f First-winter in flight** As 3e, but upperparts plain grey. ▲ Wetlands. ✤ [CD2:93]

4 BLACK TERN *Chlidonias niger* 22–24 cm; WS 64–68 cm **P c**
4a Ad breeding Black head and underparts; slate-grey above; black bill; darkish red legs. **4b Ad non-breeding** White head and underparts; black cap with large 'headphones'. **4c Ad breeding in flight** Contrasting pale grey wings and white undertail-coverts. **4d Ad non-breeding in flight** Dark patch on breast-sides; grey rump and tail. **4e Juv in flight** Darkish saddle only faintly contrasting; dark carpal bar. **4f First-winter in flight** As 4e, but upperparts mostly plain grey. ▲ Coastal. Some inland records. ✤ [CD2:94]

5 WHITE-WINGED TERN *Chlidonias leucopterus* 20–23 cm; WS 63–67 cm **P c/u**
5a Ad breeding Black; contrasting pale grey wings and white rump, vent and tail. **5b Ad non-breeding** White head and underparts; small black 'headphones' and streaking on rear crown. **5c Ad breeding in flight** Contrasting black underwing-coverts and pale grey remiges. **5d Ad non-breeding in flight** No breast patches. Compare 4d. **5e Juv in flight** Dark brown saddle; pale grey wings; white rump. **5f First-winter in flight** As 5e, but upperparts pale grey. ▲ Wetlands; also far from water. ✤ [CD2:95]

6 AFRICAN SKIMMER *Rynchops flavirostris* 36–42 cm; WS 125–135 cm **P/M lf/r NT**
6a Ad breeding Black above, white below; characteristic, long, orange-red bill. **6b Ad non-breeding** Browner above; whitish collar on lower hindneck. **6c Juv** Upperparts browner, fringed buffish; bill has dusky tip. **6d Ad breeding in flight** Long wings; short, forked tail. ▲ Large rivers, lakes, lagoons, coastal mudflats. Fishes by flying close to surface with tip of much longer lower mandible slicing through water. ✤ [CD6:50]

7 SOOTY TERN *Sterna fuscata* 33–36 cm; WS 82–94 cm **R/M lc**
7a Ad Broad white forehead extending to just above eye. Compare 8a. **7b Juv** Blackish; whitish lower underparts; white fringes to upperparts. **7c Ad in flight** All-black upperparts without neck collar. Compare 8c. **7d Imm in flight** Whitish underwing-coverts and lower underparts. ▲ Pelagic. Breeds mainly Príncipe. Disperses widely. ✤ [CD2:91]

8 BRIDLED TERN *Sterna anaethetus* 30–32 cm; WS 77–81 cm **R/M u**
8a Ad Narrow white forehead extending over and beyond eye. Compare 7a. **8b Juv** Paler; scalloped upperparts; dusky forehead. **8c Ad in flight** Grey collar usually separates black crown from brown upperparts. **8d Imm in flight** Mostly as adult, with white underwing-coverts and dark remiges. ▲ Pelagic. Breeds Mauritania, Senegal, São Tomé and Annobón. Disperses widely. ✤ [CD2:90]

Continued on page 182

1a
1b
1c
1d
1e
1f
2a
2b
2c
2d
2e
3a
3b
3c
3d
3f
3e
4a
4b
4c
4d
4f
4e
5a
5b
5c
5d
5f
5e
6a
6b
6c
6d
7a
7b
7c
7d
8a
8b
8c
8d
9a
9b
9c
9d
10a
10b
10c

Auks. Small to medium-sized seabirds with elongated bodies and strong, highly variable, bills. Plumages typically dark above and pale below. Wings short and narrow, used for both flight and swimming underwater; tail short. Sexes similar. Feed by diving. Flight direct and low with rapid, whirring wingbeats.

1 COMMON GUILLEMOT *Uria aalge* 38–43 cm V
Ad non-breeding Slender, pointed, black bill; dark line across ear-coverts; flanks variably streaked. ▲ Palearctic vagrant, NW Mauritania. ✤ [CD2:96]

2 RAZORBILL *Alca torda* 37–39 cm V
Ad non-breeding Heavy, blunt bill; pointed, often raised, tail. ▲ Palearctic vagrant, NW Mauritania. ✤ [CD2:97]

Gulls (continued).

3 BONAPARTE'S GULL *Larus philadelphia* 28–30 cm; WS 90–100 cm V
3a Ad non-breeding As Black-headed Gull (Plate 46:12b) but smaller; ear-spot more distinct; bill black. **3b First-winter** Bill all dark; legs pinkish. **3c Ad non-breeding in flight** Compare Black-headed Gull (Plate 47:11a). Underwing white; neat black edge to primaries. **3d First-winter in flight** Compare Black-headed Gull (Plate 47:11b). Primary coverts differently patterned; black trailing edge more distinct. ▲ N American vagrant.

4 RING-BILLED GULL *Larus delawarensis* 43–47 cm; WS 120–155 cm V+
4a Ad non-breeding Yellow bill with black subterminal band. Compare Common Gull (Plate 45:1). **4b First-winter** Typically heavily spotted on head, neck and breast-sides; pinkish, dark-tipped bill. **4c Ad non-breeding in flight** White mirrors smaller than in Common Gull (Plate 47:3a). **4d First-winter in flight** Grey on upperparts paler than in Common Gull (Plate 47:3b); tail band not clear-cut. ▲ N American vagrant.

5 GREAT BLACK-BACKED GULL *Larus marinus* 64–78 cm; WS 150–165 cm V
5a Ad non-breeding Very large and bulky; blackish upperparts; massive bill; pink legs. **5b First-winter** Whitsh head and breast; deep, heavy black bill. **5c Ad non-breeding in flight** Large mirrors on outer two primaries. **5d First-winter in flight** Dark flight feathers contrast with paler coverts. ▲ Palearctic vagrant.

Continued from page 180

9 BROWN NODDY *Anous stolidus* 38–40 cm; WS 77–85 cm R/V lc
9a Ad Dark brown with contrasting greyish-white cap. Compare 10a. **9b Juv** Almost entirely dark; buffish fringes to upperparts. **9c Imm** Grey restricted to forehead. **9d Ad in flight** Two-toned wing (esp. underwing). Compare 10c. ▲ Pelagic. Breeds Gulf of Guinea. ✤ [CD6:49]

10 BLACK NODDY *Anous minutus* 35–39 cm; WS 66–72 cm R/V lc
10a Ad Blacker than 9a, bill longer and more slender, cap whiter and more extensive. **10b Juv** White restricted to forehead and sharply demarcated. **10c Ad in flight** Uniformly blackish-brown. More slender, wingbeats faster than 9. ▲ Pelagic. Breeds Gulf of Guinea. ✤ [CD6:48]

 Maps on page 184

1
2
3d
3c
3b
3a
5c
4d
4c
5d
4b
4a
5b
5a

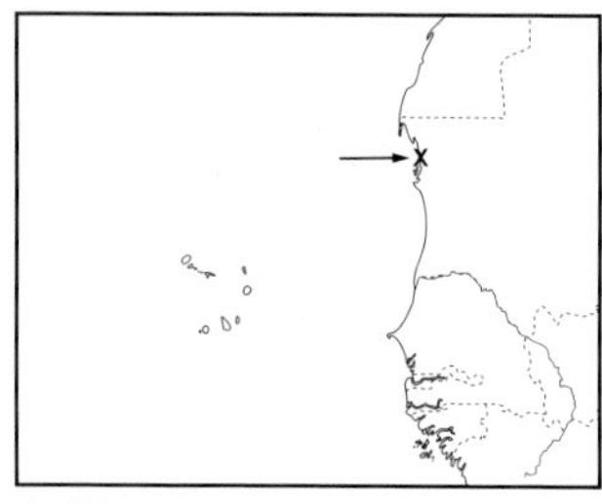

1. COMMON GUILLEMOT

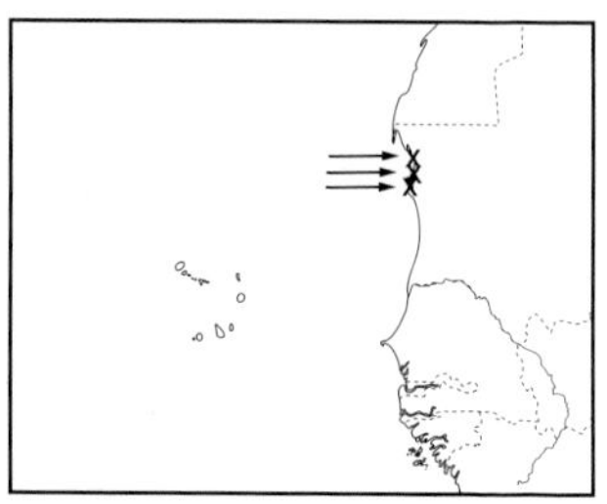

2. RAZORBILL

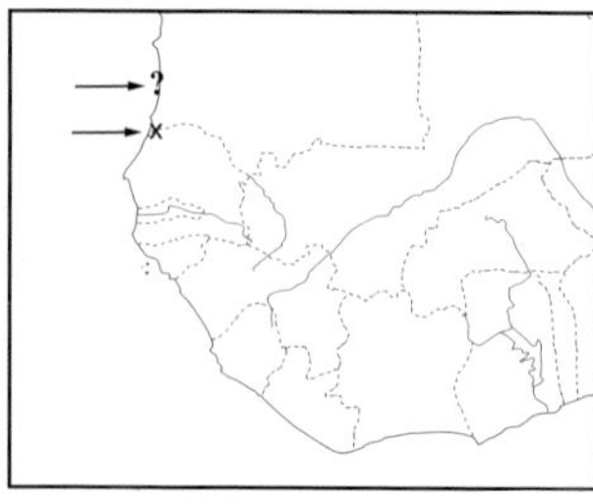

3. BONAPARTE'S GULL

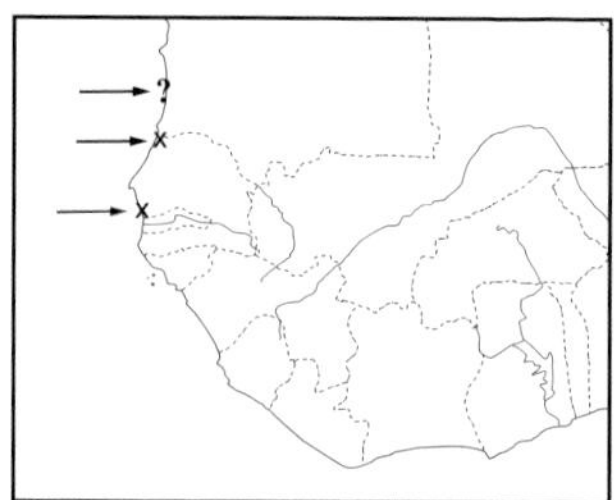

4. RING-BILLED GULL

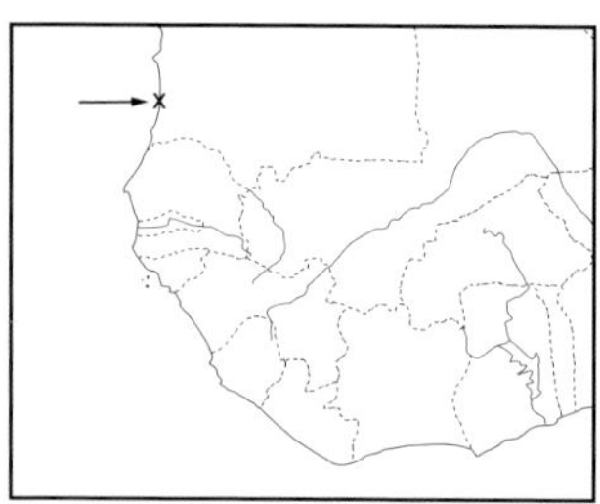

5. GREAT BLACK-BACKED GULL

Plate on page 182

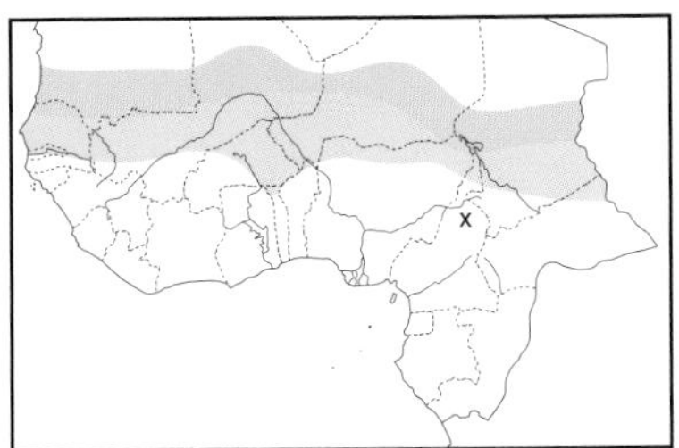

1. CHESTNUT-BELLIED SANDGROUSE

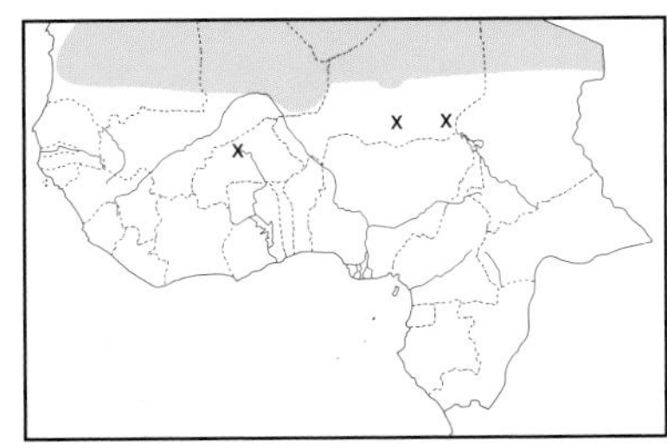

2. SPOTTED SANDGROUSE

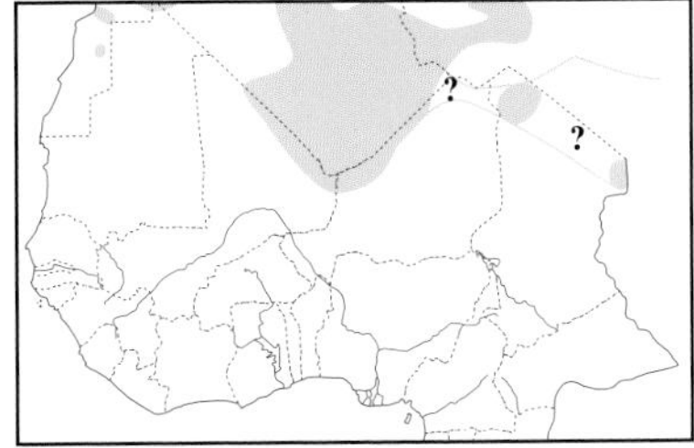

3. CROWNED SANDGROUSE

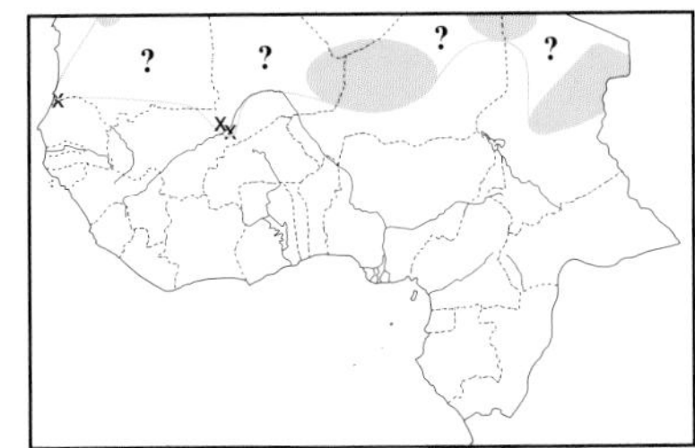

4. LICHTENSTEIN'S SANDGROUSE

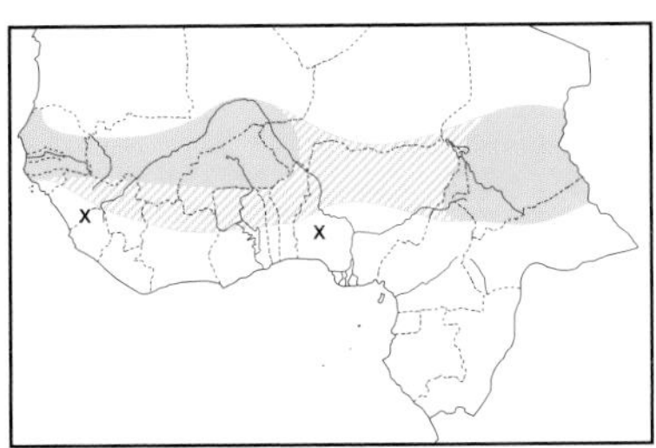

5. FOUR-BANDED SANDGROUSE

Medium-sized terrestrial species of arid habitats. Plumage cryptic, making them difficult to see on ground, wings and tail pointed, bill and legs short. Flight swift and direct, reminiscent of pigeons. Can cover considerable distances daily to visit favourite water holes, where they may gather in large flocks, usually at dawn or dusk. Calls are a good identification clue.

1 CHESTNUT-BELLIED SANDGROUSE *Pterocles exustus* 29–33 cm **R/M c/f**
1a Ad male Long, finely pointed tail; plain head and throat; narrow black breast-band; dark chestnut belly. **1b Ad female** Densely barred dark brown upperparts. **1c Ad male in flight** Entirely dark underwing; dark belly. **1d Ad female in flight.** ▲ Bare semi-desert, sandy arid scrub. Visits waterholes in early morning and before sunset in small or large flocks. Southward movements in dry season. ✤ In flight, a repeated guttural *kwit-gurut, kwit kwit-gurut...* or *kvitt-kerr-kerr...* [CD6:51]

2 SPOTTED SANDGROUSE *Pterocles senegallus* 29–35 cm **R u**
2a Ad male Long, finely pointed tail; dark streak on centre of pale belly; yellow-orange face. **2b Ad female** Spotted black upperparts and breast. **2c Ad male in flight** Dark flight feathers contrast with pale underwing-coverts. **2d Ad female in flight** Mainly pale upperwing with dusky trailing edge. ▲ Open, flat, patchily vegetated stony desert. Visits waterholes in early morning, where 100s may congregate. ✤ In flight, a frequently repeated, distinctive, staccato *wikow wik wikow...*, in chorus *wikowikowikowikowikow...* [CD3:2a]

3 CROWNED SANDGROUSE *Pterocles coronatus* 25–29 cm **R u**
3a Ad male Black facial mark; well-defined sandy spots on upperparts. **3b Ad female** Densely spotted and barred above and below. **3c Ad male in flight** Blackish flight feathers; pale upper- and underwing-coverts; short tail. **3d Ad female in flight.** ▲ Desert, esp. in stony areas. Gathers at waterholes in early morning. ✤ In flight, a fast guttural chatter *klak-klagarrarra klak-klak-klak-klagarra...* (or *chaga-chagarra...*). [CD3:2b]

4 LICHTENSTEIN'S SANDGROUSE *Pterocles lichtensteinii* 22–26 cm **R/M u**
4a Ad male Densely barred; black and white forehead; two narrow black bands on plain breast; orange bill. **4b Ad female** Finely and densely barred. **4c Ad male in flight** Blackish flight feathers; darkish underwings; short tail. **4d Ad female in flight.** ▲ Rocky and scrubby desert with scattered bushes. Gathers at freshwater source at dusk. ✤ Clear, liquid, sharp *k-kwio k-kwio....* or *kliuw kliuw...* (also transcribed as *KWEtal* or *kuitl*). Alarm a dry croaking *krre-krre-krre-krre-...* [CD3:3a]

5 FOUR-BANDED SANDGROUSE *Pterocles quadricinctus* 25–28 cm **R/M lc/f**
5a Ad male Black-and-white forehead; chestnut, white and black breast-bands. **5b Ad female** Plain face, throat and upper breast; upperparts barred chestnut and black. **5c Ad male in flight** Blackish flight feathers; grey underwing-coverts; short tail. **5d Ad female in flight** Blackish flight feathers; yellowish-buff upperwing-coverts. ▲ Dry wooded grassland, cultivation, open sandy scrub. Flocks gather at dusk and fly to waterholes, where sometimes congregate in numbers. Apparently largely nocturnal. ✤ In flight, a far-carrying whistled twittering *kik-krrr-reee* (also transcribed *wur-wulli* or *pirrou-ee*). Also an aerial song *whup whip twhu turuptu-wrree*, uttered at night. [CD6:52]

Maps on page 185

1d
1c
1b
1a
2d
2c
2b
2a
3d
3c
3b
3a
4d
4c
4b
4a
5d
5c
5b
5a

PLATE 52: PIGEONS AND DOVES I

Arboreal and terrestrial species with compact body, small head and strong, fast flight. Often produce loud and characteristic clapping sound on taking wing or in display. Feed on fruit or seeds. Calls distinctive and an excellent identification clue.

1 WESTERN BRONZE-NAPED PIGEON *Columba iriditorques* 28 cm R u/s
1a Ad male Dark blue-grey head; dark vinous breast and belly. Broad pale terminal tail band evident on landing. **1b Ad female** Cinnamon-rufous crown; greyish-chestnut underparts. ▲ Forest. ✤ One or two soft, barely audible, hesitating notes abruptly followed, after short pause, by 5–6 clear coos, last 2 descending *ehuu ehuu KOOOW KOOOW KOOW koo-koo*. [CD6:61]

2 BRUCE'S GREEN PIGEON *Treron waalia* 28–30 cm R f
Pale grey head and breast; yellow belly; whitish bill with bluish cere. ▲ Dry savanna, thorn scrub. ✤ A rapid series of hard, cracking notes (resembling slow opening of creaking door), followed by fluty whistles, suddenly rising, then descending and ending in some abrupt yapping grunts. [CD6:55]

3 AFRICAN GREEN PIGEON *Treron calvus* 25–28 cm R c
Green head, breast and belly; pale bill with large red base. ▲ Forest, wooded savanna. ✤ Rather soft, fluty trill, suddenly rising, then descending, and ending in throaty chuckling, creaking, barking and growling notes. [CD6:53]

4 CAMEROON OLIVE PIGEON *Columba sjostedti* 36–40 cm R c
4a Ad Very dark; speckled mantle and underparts; yellow bill and eye. **4b Juv** Much duller; scaly plumage. ▲ Montane forest and thickets. Endemic. ✤ Deep, drawn-out growl, followed by a quavering series of low, muffled coos. [CD6:64]

5 WHITE-NAPED PIGEON *Columba albinucha* 36 cm R lr NT
5a Ad Resembles 4a, but has large white or whitish nuchal patch; broad grey terminal tail band; red feet. **5b Juv** Much duller; white-fringed feathers on lower breast. ▲ Montane and transitional forest. ✤ Resembles 4.

6 AFEP PIGEON *Columba unicincta* 36–40 cm R f/s
Large; pale grey; red eye. In flight, contrasting pale grey band on middle of tail. ▲ Forest. ✤ Deep, drawn-out *hoooo*, uttered in series of up to 20, preceded by soft and barely audible vibrant guttural *oooorrr*. [CD6:67]

7 TAMBOURINE DOVE *Turtur tympanistria* 20–22 cm R c
7a Ad male White face and underparts; dark brown upperparts. In flight, rufous remiges. **7b Ad female** Greyish face and breast. **7c Juv** Duller, browner; scaly upperparts. ▲ Forest. ✤ A series of *hoo*s, starting rather slowly, then accelerating; similar to 9, but 'flatter' (not modulated), less 'bouncing'. [CD6:57]

8 LEMON DOVE *Aplopelia larvata* 24–25 cm R r/lu
8a Ad male *inornata* Whitish face; green-and-violet glossed hindcrown and mantle. **8b Ad female *inornata*** Cinnamon underparts. ▲ Forest. ✤ Monotonous series of 10–50 similar, low *hoot*s; speed variable. [CD6:63]
Gulf of Guinea forms ***simplex*** and ***principalis***: see Plate 146:4.

9 BLUE-HEADED WOOD DOVE *Turtur brehmeri* 25 cm R f
9a Ad Rufous-chestnut with blue-grey head. **9b Juv** Duller; indistinctly barred upperparts; no metallic wing spots. ▲ Forest. ✤ A series of plaintive *hoo*s, starting rather hesitantly, accelerating, then fading; very similar to 7, but more 'bouncing', esp. at end. [CD6:56]

10 SPECKLED PIGEON *Columba guinea* 35–40 cm R c
10a Ad Chestnut upperparts; spotted wing-coverts; pale grey rump; bare red eye patch. **10b Juv** Duller. ▲ Variety of open country, also towns. ✤ A fast, rising series of *oo* notes, increasing in pace and volume. [CD6:68]

Continued on page 190

Maps on pages 190 and 191

1b
1a
2
3
4b
4a
5a
5b
7c
6
7a
7b
9b
8b
9a
8a
10a
10b
11

Continued from page 188

11 ROCK DOVE *Columba livia* 31–34 cm **R+ lc/f**
Ad *gymnocycla* (widespread) Dark slate-grey; white rump; two black wingbars. *C. l. targia* (C Sahara) much paler, with grey rump. ▲ Rocky cliffs and mountains. ✤ Low, drawn-out, emphatic *rrooh* repeated several times. [CD3:6]

12 WOOD PIGEON *Columba palumbus* 40–45 cm **V**
Very large; white patch on neck-sides; broad white band on wings in flight; broad black terminal tail band. ▲ Palearctic vagrant, Mauritania. ✤ [CD3:4a]

12. Wood Pigeon

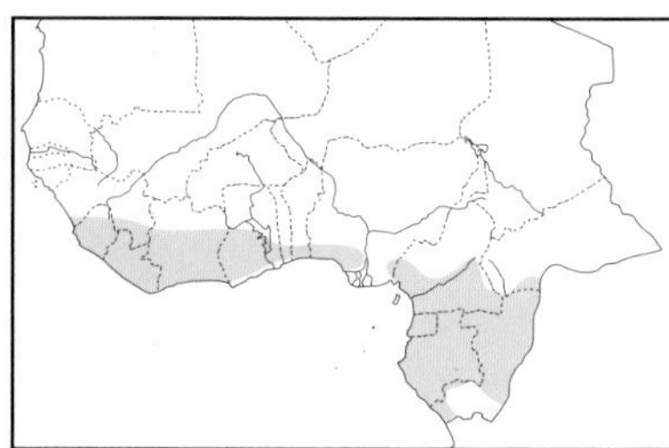
1. WESTERN BRONZE-NAPED PIGEON

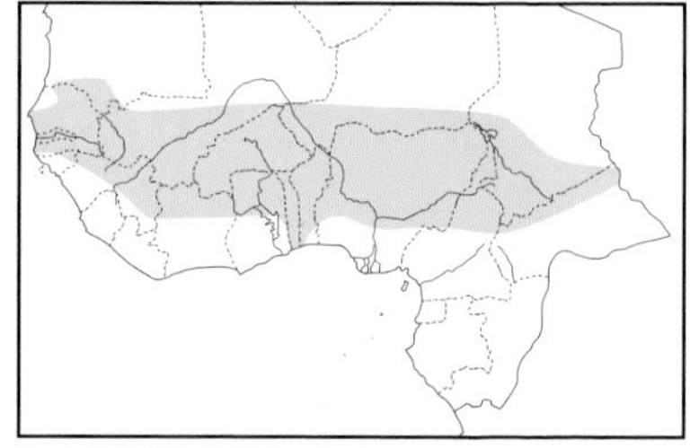
2. BRUCE'S GREEN PIGEON

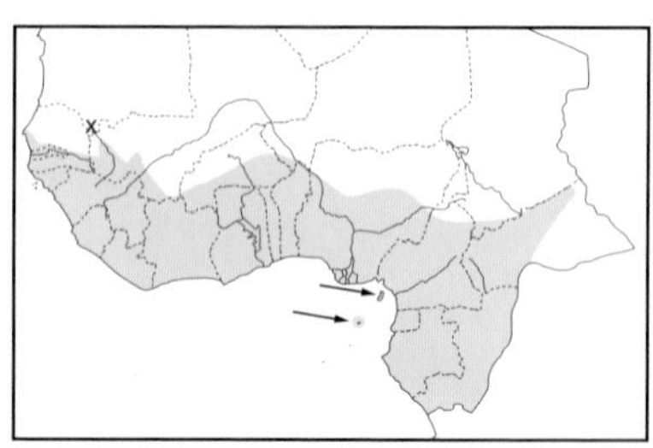
3. AFRICAN GREEN PIGEON

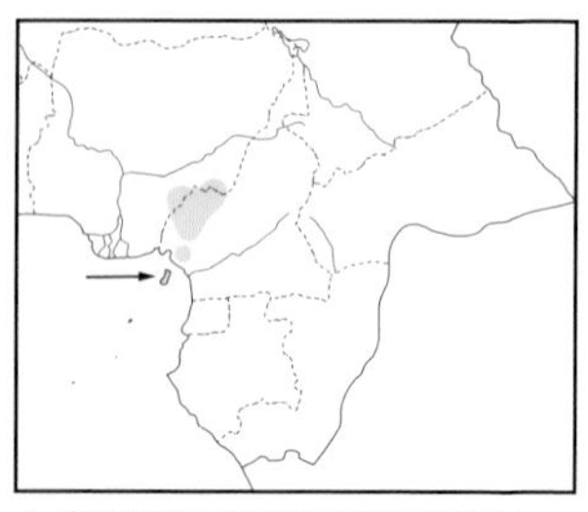
4. CAMEROON OLIVE PIGEON

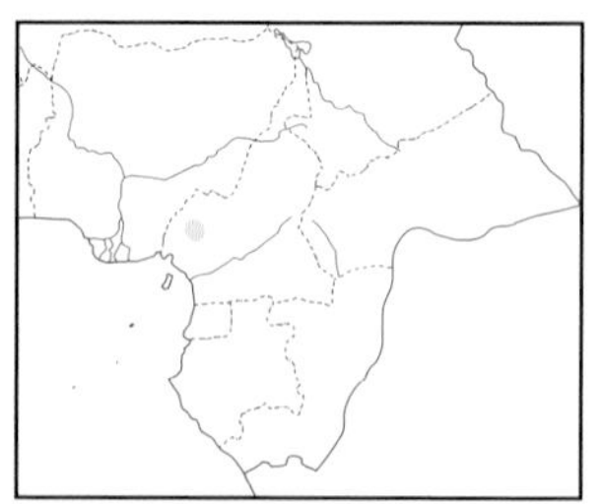
5. WHITE-NAPED PIGEON

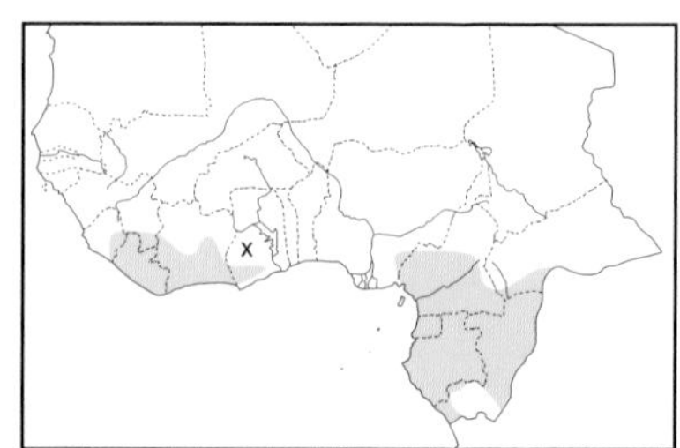
6. AFEP PIGEON

 Plate on page 188

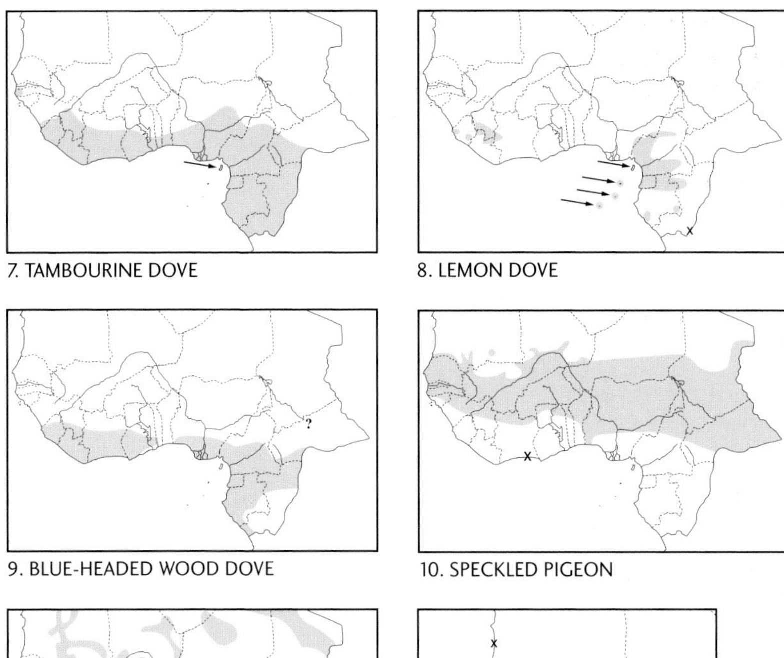

7. TAMBOURINE DOVE

8. LEMON DOVE

9. BLUE-HEADED WOOD DOVE

10. SPECKLED PIGEON

11. ROCK DOVE

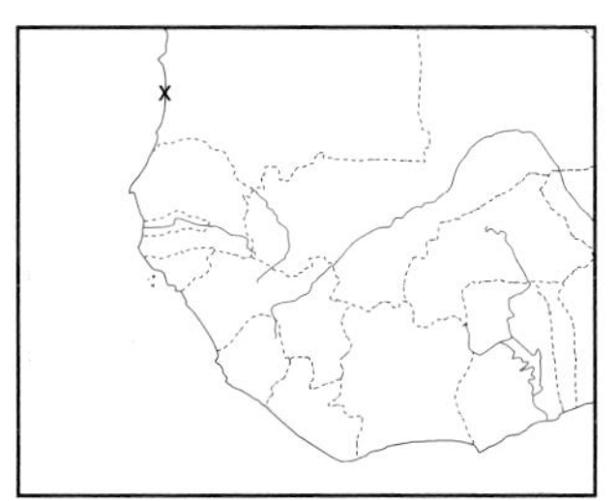

12. WOOD PIGEON

PLATE 53: PIGEONS AND DOVES II

1 EUROPEAN TURTLE DOVE *Streptopelia turtur* 26–28 cm **P+/R c/r**
1a Black-and-white striped neck patch; chequered upperparts. **1b In flight** Blue-grey wing panel; rufous shoulders; white terminal tail band. ▲ Dry savannas, farmland. ✤ Deep purring *rrrurrr rrrurrr..* Not recorded on migration. [CD3:8]

2 ADAMAWA TURTLE DOVE *Streptopelia hypopyrrha* 30–31 cm **R lc/u**
2a Blue-grey head to upper breast; broad black neck patch. **2b In flight** Dark overall; pale grey terminal tail band. ▲ Open woodland in rocky, hilly areas. ✤ Low, hard purring *rrrurrr rr-rrurr*, resembling 1, but sharper and deeper; sometimes with 4th syllable. [CD6:74]

3 AFRICAN COLLARED DOVE *Streptopelia roseogrisea* 27–29 cm **R/M c**
3a Pale grey-pink head and underparts; dark red eye. **3b–c In flight** Dark primaries contrast with pale upperwing-coverts; white underwing. ▲ Thornbush, arid farmland. ✤ Emphatic note followed by rolling purr *whooh rrrwhrrrooh* or *ooh krrruuuuu...* [CD6:73]

4 AFRICAN MOURNING DOVE *Streptopelia decipiens* 28–30 cm **R/M c**
4a Grey face and crown; yellowish eye; narrow red orbital ring. **4b In flight** White tail corners. ▲ Arid woodland. ✤ Emphatic note followed by quavering coo, *whoh! kho-o-o-o-o-o*; rapidly repeated *whoh whoh-hoo*. Quavering gargle *arh-r-r-r-r-rw* on landing. [CD6:70]

5 RED-EYED DOVE *Streptopelia semitorquata* 30–34 cm **R c**
5a Ad Grey crown; dark grey-brown upperparts; dark vinous-pink underparts. **5b Juv** Rufous-buff fringes to upperpart feathers; indistinct half-collar. **5c In flight** Black band on tail. ▲ Various habitats in forest and savanna zones. ✤ Rapid series of 6 short, rather nasal coos, with stress on 5th, *ho-hu ho-hu HOO-ho*. [CD6:69]

6 RING-NECKED DOVE *Streptopelia capicola* 25 cm **R ls/f**
6a Ad Blue-grey head washed pink; pale pinkish-grey underparts. **6b-c In flight** White tail corners; grey underwing. ▲ Open wooded grassland. ✤ Rhythmic, monotonous and far-carrying *ku-KOORRRR-ku*. [CD6:72]

7 VINACEOUS DOVE *Streptopelia vinacea* 25 cm **R/M c**
7a Pale vinous-pink head and underparts; dark eye. **7b–c In flight** White tail corners; pale grey underwing. ▲ Savanna. ✤ Fast, 3-syllable *wheh heh-ho wheh heh-ho...* [CD6:71]

8 LAUGHING DOVE *Streptopelia senegalensis* 23–25 cm **R c**
8a Rufous-brown above; blue-grey outer wing-coverts; black-speckled necklace. **8b In flight** Blue-grey wing panel; blue-grey back to uppertail-coverts. ▲ Open woodland, farmland, villages, towns. ✤ A single, emphatic note followed by a series of 5–7 hurried notes, first rising, then falling *hoo koHUHUhu-hoo*, likened to gentle laugh. [CD3:9]

9 NAMAQUA DOVE *Oena capensis* 22–26 cm **R/M/V+ c/r**
9a Ad male Black mask; long, pointed tail. **9b Ad female** Greyish face; dusky bill. **9c Juv** Scalloped upperparts. **9d Ad male in flight** Rufous primaries; long tail. ▲ Thornbush, dry grassland, cultivation. ✤ Short, soft note, followed by plaintive, drawn-out, emphatic and slightly rising syllable: *oh-whoooooah...* [CD3:3b]

10 EMERALD-SPOTTED WOOD DOVE *Turtur chalcospilos* 20 cm **R u**
10a Ad As 11a, but with dusky bill and metallic green wing spots. **10b Juv** As 11b, but with whitish tips to greater and median coverts. ▲ Wooded habitats. ✤ A series of soft, plaintive *hoo*s, very similar to 11, but slightly longer and higher pitched. [CD6:59]

11 BLUE-SPOTTED WOOD DOVE *Turtur afer* 20 cm **R c**
11a Ad Red bill with yellow tip; metallic blue wing spots. **11b Juv** Duller, browner; scaly upperparts; dusky bill. **11c In flight** Rufous primaries; two blackish bands across lower back. ▲ Various wooded habitats. ✤ Two muffled, plaintive syllables followed by a series of *hoo*s, initially rather hesitant and irregular, then accelerating, finally fading. [CD6:58]

Continued on page 194

 Maps on pages 194 and 195

1a
2b
2a
1b
3a
4b
4a
3c
3b
5a
5b
5c
6b
6a
7b
7a
7c
6c
8a
9c
9a
8b
9b
9d
10b
11c
11b
12b
12c
10a
11a
12a

Continued from page 192

12 BLACK-BILLED WOOD DOVE *Turtur abyssinicus* 20 cm **R c**
12a Ad As 11a, but slightly paler and greyer with black bill. **12b Juv** As 11b, but slightly paler and greyer. **12c In flight** More contrasting above than 11c. ▲ Dry savanna, thickets. ✤ A series of soft, plaintive *hoo*s, similar to 11, but initial notes more hesitant and drawn out. [CD6:60]

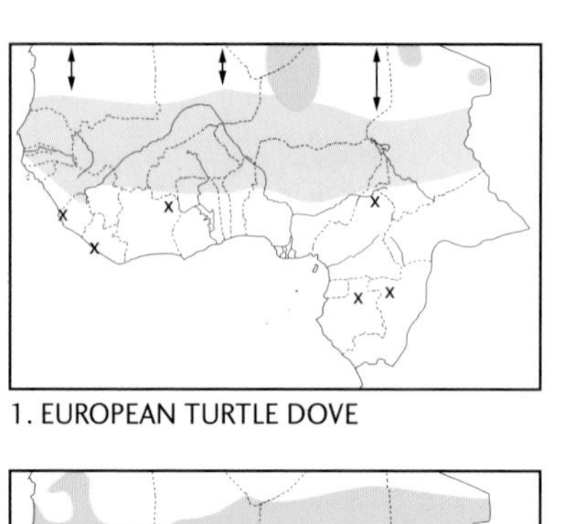

1. EUROPEAN TURTLE DOVE

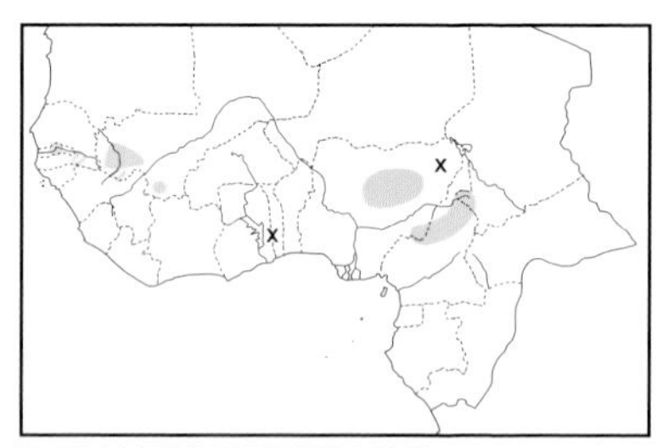

2. ADAMAWA TURTLE DOVE

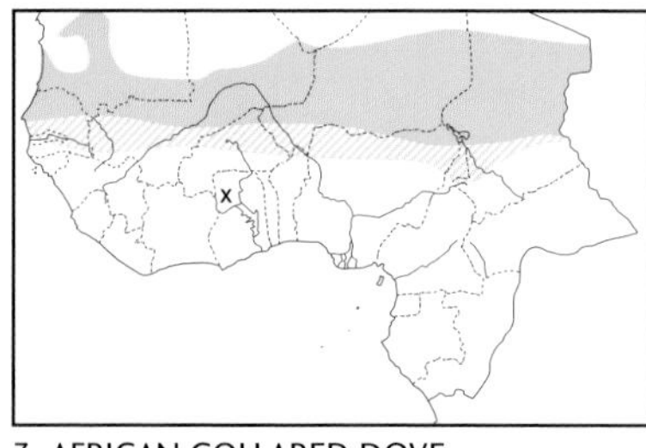

3. AFRICAN COLLARED DOVE

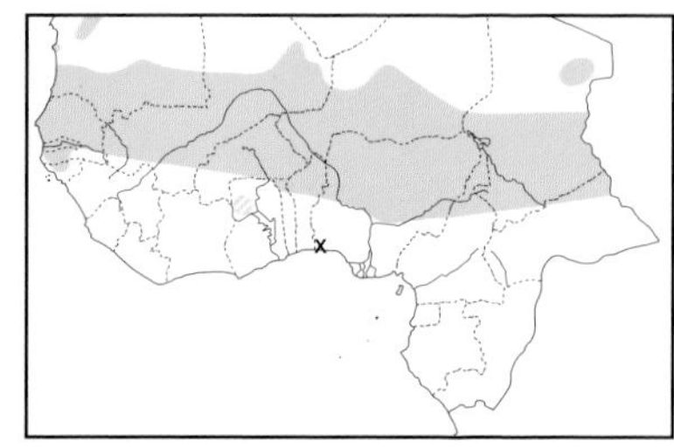

4. AFRICAN MOURNING DOVE

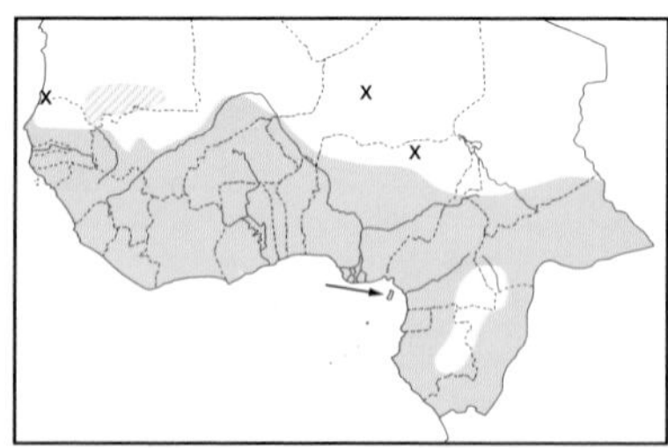

5. RED-EYED DOVE

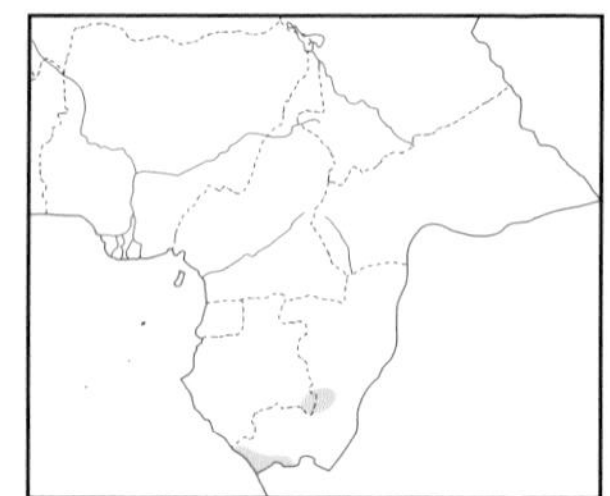

6. RING-NECKED DOVE

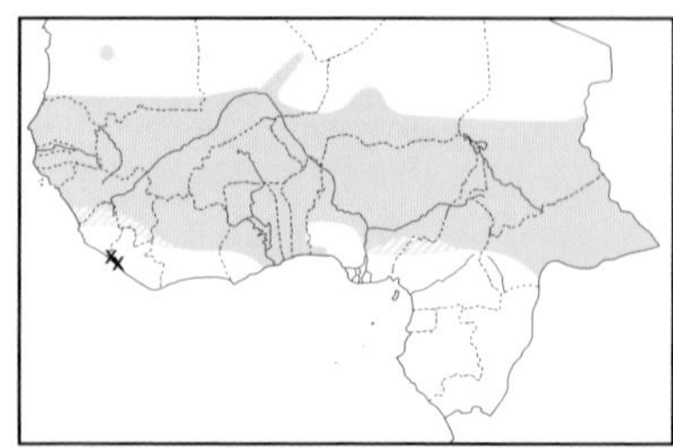

7. VINACEOUS DOVE

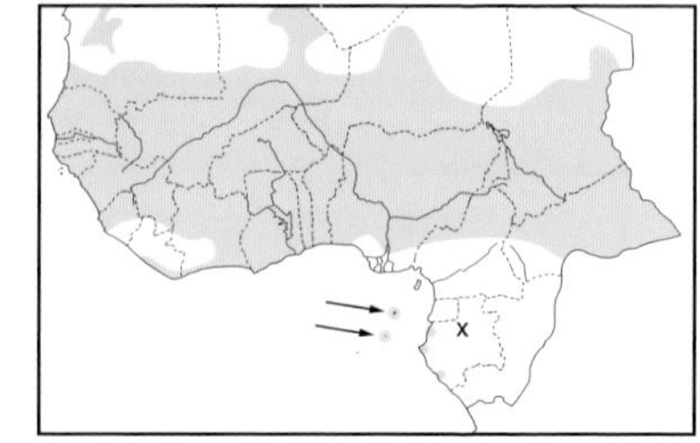

8. LAUGHING DOVE

Plate on page 192

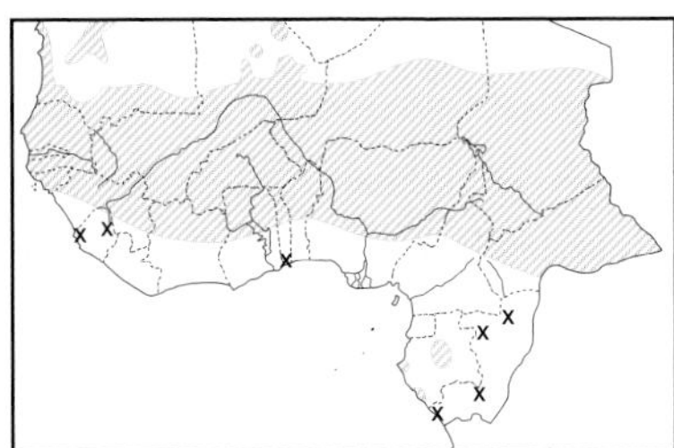

9. NAMAQUA DOVE

10. EMERALD-SPOTTED WOOD DOVE

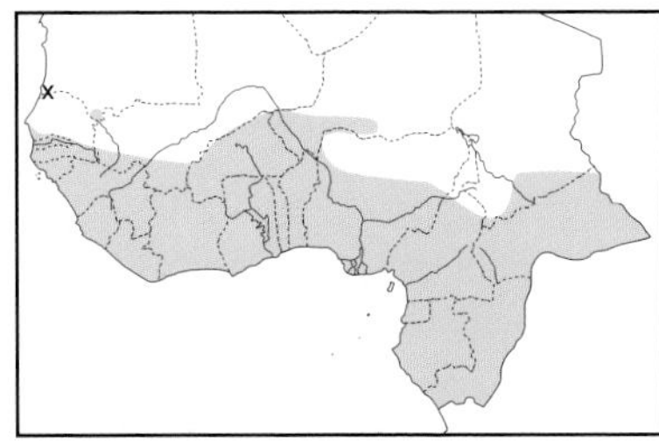

11. BLUE-SPOTTED WOOD DOVE

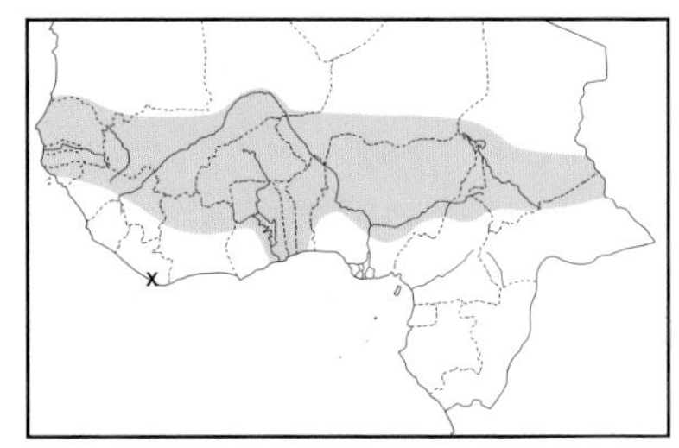

12. BLACK-BILLED WOOD DOVE

PLATE 54: PARROTS AND LOVEBIRDS

Mostly arboreal species with compact body and stout, strongly hooked and powerful bills, also used in climbing. Legs short, feet strong and zygodactyl, used like hands in feeding. Flight strong, fast and direct. Noisy, especially in flight. Calls include high-pitched screeches (parrots) and shrill twitterings or chirrupings (lovebirds). Rather gregarious. Feed mostly on fruits and seeds. Nest in tree holes. Several W African species sold as pets.

1 BLACK-COLLARED LOVEBIRD *Agapornis swindernianus* 13 cm **R r/lf**
1a Ad *swindernianus* (Liberia–Ghana) Small; narrow black collar on nape; yellowish band on mantle; black bill. **1b Ad *zenkeri*** (from S Cameroon east and south) Band on mantle orange-red. **1c Juv** All-green head; horn-coloured bill. ▲ Lowland forest canopy. May visit favoured fruiting trees daily at dawn. ✤ Distinctive shrill chirruping *srleeee* or *tchirrrlu*, in flight. [CD6:83]

2 RED-HEADED LOVEBIRD *Agapornis pullarius* 15 cm **R lc/u**
2a Ad male Red face and bill. **2b Ad female** Yellowish-orange face. **2c In flight** Triangular wings; short tail. ▲ Wooded grassland. Feeds mainly on grass seeds. ✤ Clear, loud *kl-eee*. [CD6:82]

3 ROSE-RINGED PARAKEET *Psittacula krameri* 38–42 cm **R+ f/c**
3a Ad male Long, pointed tail; red bill; head encircled by narrow ring. **3b Ad female** All-green head; tail shorter. ▲ Savanna, farmland, mangrove. Introduced, Cape Verde. ✤ Vocal. Loud, screeching calls, incl. shrill *kee-ak*. [CD6:84]

4 SENEGAL PARROT *Poicephalus senegalus* 23 cm **R c/u**
4a Ad *senegalus* (Senegambia–Guinea to NW Nigeria) Underparts deep yellow becoming orange in centre. **4b Ad *versteri*** (W Ivory Coast–SW Nigeria) Deep orange to scarlet underparts. *P. s. mesotypus* (NE Nigeria–SW Chad) intermediate. **4c Juv** Underparts mostly greenish. ▲ Savanna woodland, farmland with scattered trees. Endemic. ✤ Various harsh shrieks. [CD6:80]

5 MEYER'S PARROT *Poicephalus meyeri* 21–23 cm **R c**
5a Ad Ash-brown above; yellow patch on crown and on edge of wing. **5b Juv** No yellow on crown or wing. ▲ Savanna woodland. ✤ Harsh, high-pitched shrieks. [CD6:79]

6 NIAM-NIAM PARROT *Poicephalus crassus* 25 cm **R lu?**
Mainly grass-green with greyish-brown head and breast. ▲ Savanna woodland, forest–savanna mosaic; often near water. ✤ Short, sharp screech. [CD6:81]

7 RED-FRONTED PARROT *Poicephalus gulielmi* 26–30 cm **R r/u**
7a Ad Green head; orange or orange-red on forehead; yellowish-green rump. **7b Juv** Dusky head; no orange in plumage. ▲ Lowland forest. ✤ High-pitched screeches. [CD6:78]

8 BROWN-NECKED PARROT *Poicephalus robustus* 28–33 cm **R r/lc**
8a Ad male Large; greyish head; large bill. **8b Ad female** As 8a but with red forehead. Compare 7a. ▲ Savanna woodland; also mangroves. Undertakes regular daily flights from roosts to feeding areas. ✤ Strident, harsh *zzkeek*. [CD6:77]

9 GREY PARROT *Psittacus erithacus* 28–39 cm **R lc/s**
9a Ad *timneh* (west of Comoé R., Ivory Coast) Slate-grey; contrasting paler rump; dark maroon tail. **9b Ad *erithacus*** (east of Comoé R.) Very large; grey; bright scarlet tail. ▲ Lowland and gallery forest, wooded savanna, mangroves. Congregates in numbers at regular roosts. Partial to oil-palm fruits; may travel long distances in search of fruiting trees or minerals in saltpans. ✤ Vocal. The only parrot in the region to utter loud clear whistles, besides variety of high-pitched screeches, harsh and grating calls, and imitations. [CD6:76]

Maps on page 198

1b
1c
1a
2b
2c
2a
3b
3a
4c
4a
4b
5b
5a
6
7b
7a
8b
8a
9a
9b

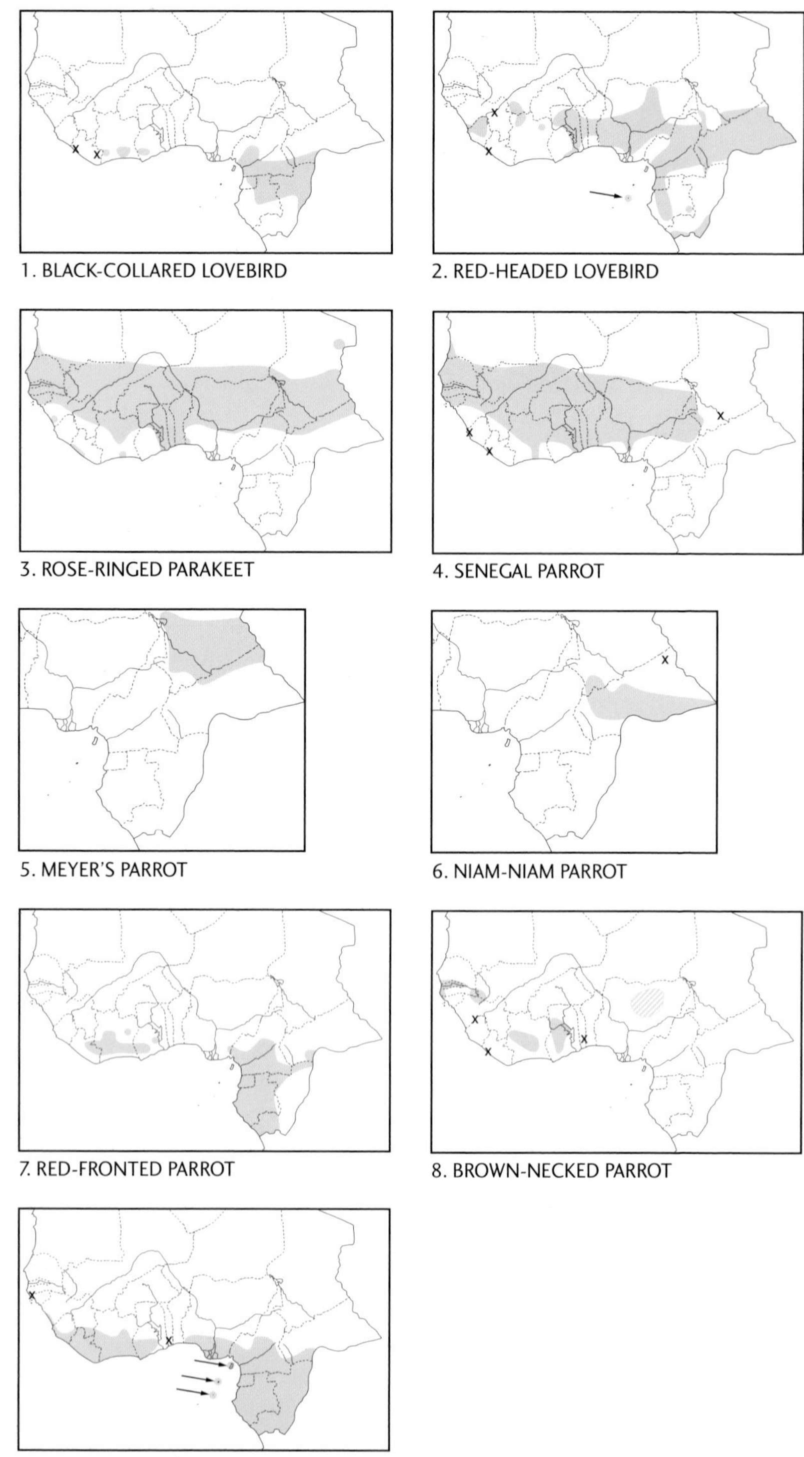

1. BLACK-COLLARED LOVEBIRD

2. RED-HEADED LOVEBIRD

3. ROSE-RINGED PARAKEET

4. SENEGAL PARROT

5. MEYER'S PARROT

6. NIAM-NIAM PARROT

7. RED-FRONTED PARROT

8. BROWN-NECKED PARROT

9. GREY PARROT

Plate on page 196

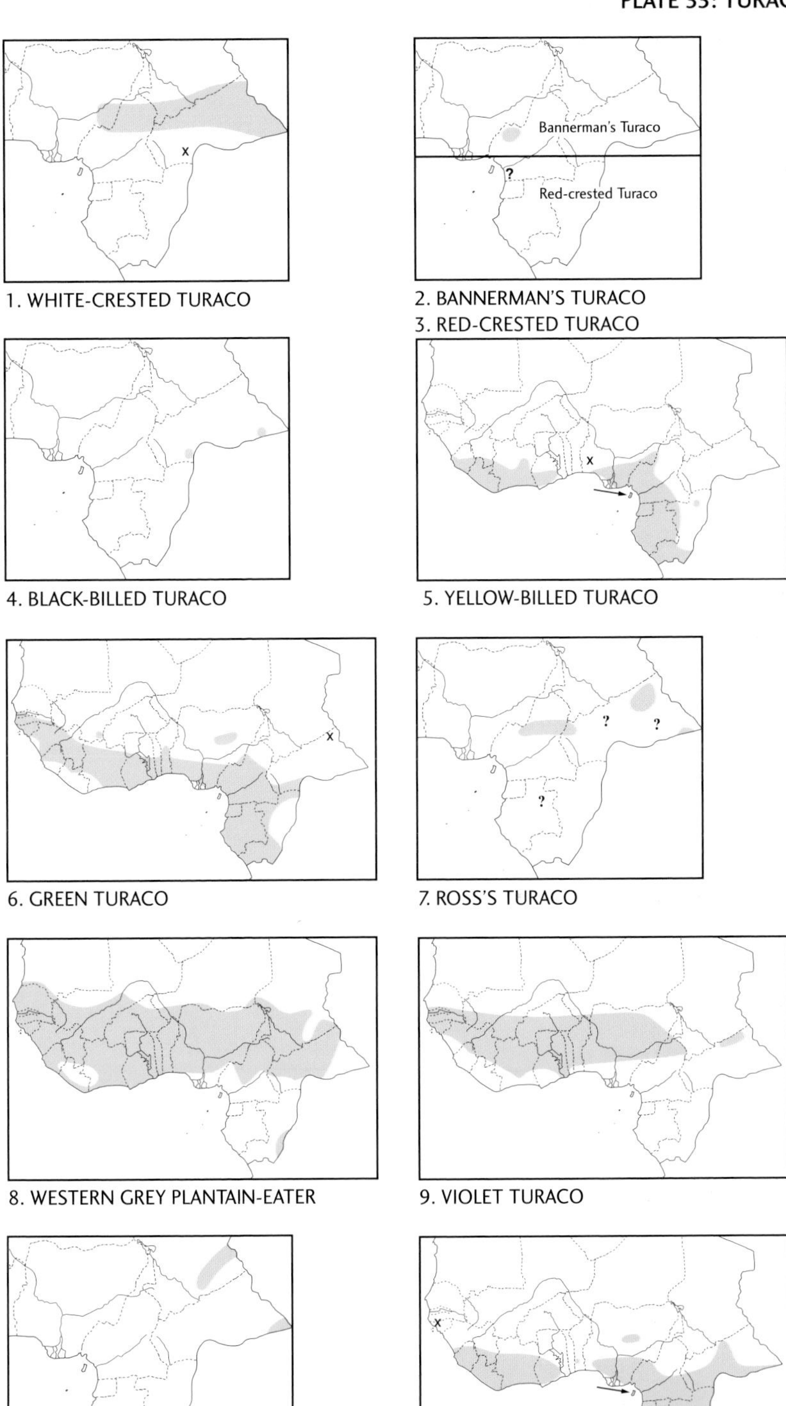

1. WHITE-CRESTED TURACO

2. BANNERMAN'S TURACO
3. RED-CRESTED TURACO

4. BLACK-BILLED TURACO

5. YELLOW-BILLED TURACO

6. GREEN TURACO

7. ROSS'S TURACO

8. WESTERN GREY PLANTAIN-EATER

9. VIOLET TURACO

10. EASTERN GREY PLANTAIN-EATER

11. GREAT BLUE TURACO

Plate on page 200

Arboreal species with crested heads and long tails. Flight weak, over short distances only and consisting of several flaps followed by a glide. Agile in trees, characteristically running and bounding along branches (except *Crinifer*). Feed principally on fruit. Endemic to Africa. Vocal. *Tauraco* have loud, raucous calls that are overall similar but may, with practice, be distinguished in most cases. The two *Musophaga* utter pleasant, rolling sounds. *Crinifer* have a variety of cackling and yapping calls. *Corythaeola* has a series of unmistakable, impressive calls.

1 **WHITE-CRESTED TURACO** *Tauraco leucolophus* — *c.* 40 cm **R lc**
Conspicuous white head, unique among turacos. ▲ Wooded savanna, gallery forest. ✤ Drawn-out *whooap* followed by a regular series of raucous *khaw* notes. [CD6:90]

2 **BANNERMAN'S TURACO** *Tauraco bannermani* — *c.* 43 cm **R lc EN**
Red crest; greyish face. ▲ Montane forest. Endemic. ✤ Single *whoop* followed by a regular series of raucous *khaw* notes. Softer and higher pitched than 6. [CD6:91]

3 **RED-CRESTED TURACO** *Tauraco erythrolophus* — *c.* 43 cm
Red crest; whitish face. ▲ Evergreen and gallery forest, Angola. No certain records. ✤ Single *whoop* followed by a regular series of raucous *khaw* notes. [CD6:92]

4 **BLACK-BILLED TURACO** *Tauraco schuetti* — *c.* 40 cm **R c**
White-tipped crest; black bill. ▲ Lowland forest, savanna woodland. ✤ As 6, but slightly slower. [CD6:88]

5 **YELLOW-BILLED TURACO** *Tauraco macrorhynchus* — 40–43 cm **R u/lc**
5a Ad *macrorhynchus* (Sierra Leone–Ghana) Predominantly yellow bill; black-and-white-tipped crest. **5b Ad *verreauxii*** (Nigeria–Congo) Red-tipped crest. **5c Juv, both races** All-green crest; dull-coloured bill. ▲ Lowland forest. ✤ Different from 6: starts abruptly with single loud, harsh, barking note, followed by a series of raucous *khaw* notes, starting rather fast, then progressively slowing down. [CD6:89]

6 **GREEN TURACO** *Tauraco persa* — 40–43 cm **R c**
6a Ad *persa* (from Ivory Coast east) All-green crest; dusky-red bill, white spot in front of eye. **6b Ad *buffoni*** (from Ivory Coast west) No white line below eye. **6c In flight** Mainly bright crimson flight feathers. ▲ Lowland and gallery forest, lower montane forest. ✤ Two or three rising notes *woop-woop* followed by a regular series of loud, raucous *khaw* notes. Also a repeated *rroooh* followed by *keh* in duet: *rroooh-keh rroooh-keh...* . [CD6:86]

7 **ROSS'S TURACO** *Musophaga rossae* — 51–54 cm **R lf**
Glossy blackish-violet; yellow bill and orbital skin; crimson crest. ▲ Gallery forest, forest edge. ✤ Similar to 9. [CD6:95]

8 **WESTERN GREY PLANTAIN-EATER** *Crinifer piscator* — *c.* 50 cm **R c**
Mainly grey and white; white wing patch conspicuous in flight; pale yellow bill. ▲ Gallery forest, savanna woodland, cultivated areas. ✤ A series of loud, high-pitched and rolling *kow-kow-kow-...* ending in chatter; cackling *kak-kak-kak-kalak-kalak...* [CD6:98]

9 **VIOLET TURACO** *Musophaga violacea* — 45–50 cm **R lf**
Glossy purple and black; yellow bill; scarlet orbital skin; crimson crown. ▲ Gallery forest, forest edge. Endemic. ✤ A melodious rolling series of far-carrying *koorroo* notes; often joined by second bird in asynchronous duet and eliciting response from others. [CD6:94]

10 **EASTERN GREY PLANTAIN-EATER** *Crinifer zonurus* — *c.* 50 cm **R c**
Outer tail feathers partially white. Compare 8. ▲ Open wooded savanna. ✤ A variety of cackling and yapping notes, similar to 8 but slower and lower pitched. [CD6:99]

11 **GREAT BLUE TURACO** *Corythaeola cristata* — 70–75 cm **R c/u**
Very large; long tail with broad black bar; blackish crest; yellow bill tipped reddish. ▲ Forest. In parties in canopy. ✤ Fast series of explosive *kok* notes and deep guttural *krraou*. [CD6:85]

Maps on page 199

1
2
3
4
5b
5a
5c
6c
6a
6b
7
8
9
10
11

Typical cuckoos. Arboreal with long, pointed wings and long tails, reminiscent of a small raptor in flight. Feed on insects and hairy caterpillars. Brood parasites. Many rather inconspicuous and some quite difficult to observe, but all have distinctive calls.

1 KLAAS'S CUCKOO *Chrysococcyx klaas* 18 cm M/R c
1a Ad male Small white streak behind eye; green patch on breast-sides; outer tail feathers mainly white. **1b Ad female** Variable; upperparts more bronzy; flanks finely barred; post-ocular streak buffy. **1c Juv** Barred green and russet above; fine olive-brown barring below; slight post-ocular patch. Compare 3b–c. ▲ Various wooded habitats. Parasitises warblers, sunbirds, flycatchers and white-eyes. ✤ Plaintive *huee-ti huee-ti* or *whee-ee chew whee-ee chew*. [CD7:15]

2 DIDRIC CUCKOO *Chrysococcyx caprius* 19 cm M/R c
2a Ad male White eye-stripe; white spots on wings and outer tail feathers; short malar stripe; red orbital ring. **2b Ad female** Variable; upperparts more bronzy than male; red orbital ring. **2c Juv** Highly variable (rufous morph illustrated); coral-red bill. ▲ Various open and wooded habitats (not closed forest). Conspicuous (unusual for a cuckoo). Parasitises mainly weavers. ✤ Plaintive *deea deea deedrik*. Also rapid *di-di-di-di-di*. Female a *deea-deea-deea*. [CD7:16]

3 AFRICAN EMERALD CUCKOO *Chrysococcyx cupreus* 23 cm R/M c/s
3a Ad male Upperparts bright glossy green; belly golden-yellow. **3b Ad female** Barred russet and green above; boldly barred bronzy-green below. **3c Juv** As female, but crown and nape feathers green fringed white. ▲ Lowland and gallery forest, dense woodland. Known hosts include bulbuls, illadopsises, flycatchers, sunbirds and weavers. ✤ Far-carrying, clear and melodious *ptiu*, *tiu-ut*; a characteristic sound of African rainforest. Also an explosive, stuttering *tiuw tu-tu-tu*... Female utters clear *tiuw* and *huu tu-tu*. [CD7:13]

4 YELLOW-THROATED CUCKOO *Chrysococcyx flavigularis* 18 cm R r/u
4a Ad male Darkish, with conspicuous golden-yellow stripe on throat; outer tail feathers white with subterminal dark bar. **4b Ad female** Face and underparts buffish narrowly barred darkish. **4c Juv** Russet fringes to feathers of upperparts; tawnier head and underparts. ▲ Forest. Hosts unknown. ✤ A rapid series of *c.* 10 pure whistled notes on same pitch *tiu tee-tee-tee-tee-tee-*... Also a distinctive, loud, clear, double whistle *hee-huu*. [CD7:14]

5 GREAT SPOTTED CUCKOO *Clamator glandarius* 35–40 cm M/R/P/V+ s/lc
5a Ad Spotted upperparts; long white-tipped tail; grey crest; creamy underparts. **5b Juv** Darker; rudimentary crest; rufous primaries. ▲ Semi-arid open woodland. Moves south in dry season. Parasitises mostly Pied Crow. ✤ Vocal in breeding season. Loud *kweeow kweeow kweeow*, excited *kiu-ku-ku-ker* and variety of harsh and chattering calls. [CD3:10]

6 JACOBIN CUCKOO *Oxylophus jacobinus* 33 cm M lc
6a Ad Black above, white below; crest; white wing patch; long tail. **6b Juv** Duskier; smaller crest. ▲ Dry savanna woodland. Moves south in dry season, crossing equator. Parasitises mainly Common Bulbul. ✤ A series of clear *keew* notes (as 7 but slightly higher pitched) ending in staccato chatter. [CD7:1]

7 LEVAILLANT'S CUCKOO *Oxylophus levaillantii* 38–40 cm M c/u
7a Ad As 6a, but larger and with heavily streaked throat and breast. **7b Juv** Duskier; streaks finer; crest smaller. ▲ Dense savanna woodland, forest edge, adjacent cultivated areas. Moves south in dry season. Parasitises babblers. ✤ A series of clear, loud *KEEow* notes, ending in staccato chatter (like machine gun burst). [CD7:2]

Maps on page 204

1a
1b
1c
5a
2a
5b
2b
2c
6a
6b
3c
3a
7a
3b
4c
7b
4b
4a

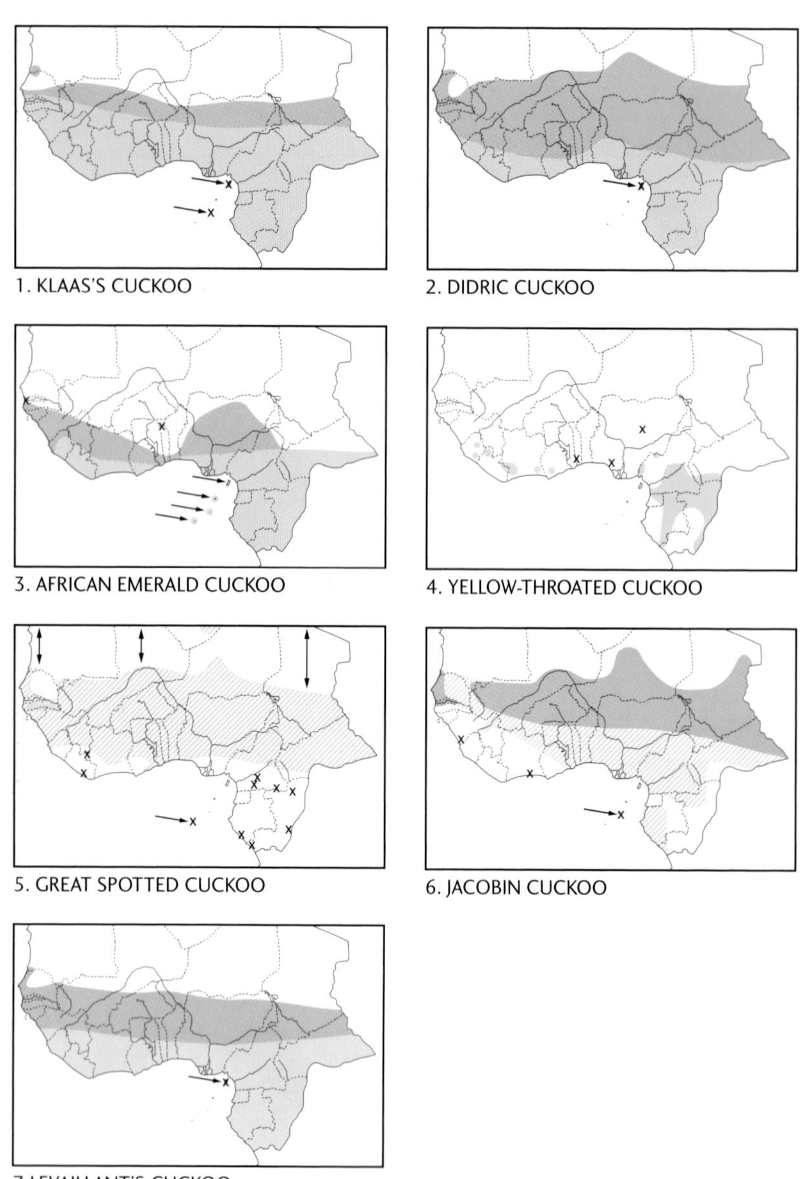

1. KLAAS'S CUCKOO

2. DIDRIC CUCKOO

3. AFRICAN EMERALD CUCKOO

4. YELLOW-THROATED CUCKOO

5. GREAT SPOTTED CUCKOO

6. JACOBIN CUCKOO

7. LEVAILLANT'S CUCKOO

Plate on page 202

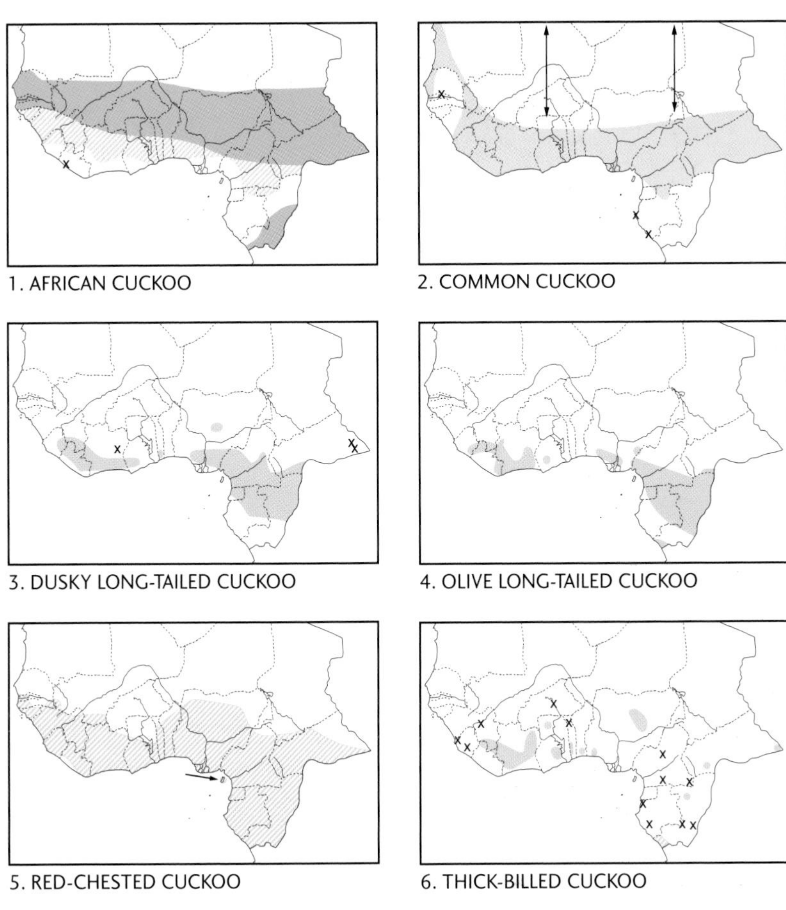

1. AFRICAN CUCKOO

2. COMMON CUCKOO

3. DUSKY LONG-TAILED CUCKOO

4. OLIVE LONG-TAILED CUCKOO

5. RED-CHESTED CUCKOO

6. THICK-BILLED CUCKOO

7. BLACK CUCKOO

1 AFRICAN CUCKOO *Cuculus gularis* 31–33 cm **M u/c**
1a Ad male Grey upperparts and breast; barred underparts; bill yellow tipped black. **1b Ad female** May have more buffish underparts, esp. throat and breast. **1c Juv** Grey upperparts finely barred white; dirty buff throat and breast. **1d In flight** Long pointed wings; long tail. May recall small hawk. ▲ Wooded savanna. Moves north with rains. Parasitises Fork-tailed Drongo. ✤ Frequently repeated, far-carrying *oo-OO*, with stress on second, slightly higher pitched syllable (as flat, inverted song of 2). [CD7:6]

2 COMMON CUCKOO *Cuculus canorus* 32–36 cm **P/V* s**
2a Ad male As 1a, but bill mainly dusky with yellow base. Female similar but with rusty-brown tinge. **2b Ad female hepatic form** Rufous upperparts; barred above and below. Rare. ▲ Wooded and derived savanna, open areas in forest. ✤ Silent in Africa. [CD3:11]

3 DUSKY LONG-TAILED CUCKOO *Cercococcyx mechowi* 33 cm **R r/lc**
3a Ad Slender, small bodied, with very long tail. Blackish-grey above; boldly barred below; rich buff undertail-coverts. **3b Juv** Dark brownish-grey upperparts with dark rufous fringes to feathers; underparts boldly barred. ▲ Forest. Hosts unknown. ✤ Two song types. In Upper Guinea (east to W Cameroon), first type consists of 3 rising notes, *hu hee wheeu*; second, a less frequently uttered whinnying series of rather plaintive notes, first accelerating, then slowing and descending *tiutiutiutiutiutiui-tiu-tiu-tiu-...*, reminiscent of *Halcyon* kingfisher. In Lower Guinea, first type is faster, with 3 similar, less melodious notes *wheet-wheet-wheet*; second, a fast, descending *wheewheewheewheewhee...*, almost twice as fast as equivalent in Upper Guinea. [CD7:9–10]

4 OLIVE LONG-TAILED CUCKOO *Cercococcyx olivinus* 33 cm **R r/lc**
4a Ad Jizz as 3. Dark greyish-brown above; boldly barred below; pale buff undertail-coverts. **4b Juv** Underparts initially streaked, progressively replaced by barring. ▲ Forest. Hosts unknown. ✤ A series of 3 notes *whit tiuw-tiuw* (first note inaudible at distance). Also a long, rising series *teeru-teeru-teeru-teeru-...*, appearing ever more impatient, ceasing abruptly. [CD7:11]

5 RED-CHESTED CUCKOO *Cuculus solitarius* 28–31 cm **M f/u**
5a Ad Slate-grey upperparts and throat; chestnut-red breast; barred belly and undertail-coverts. **5b Juv** Very dark; narrow white crescents; white nape patch; belly and undertail-coverts boldly barred on buffish-white. ▲ Forest, wooded savanna. Parasitises mainly Turdidae. ✤ Loud and descending 3-note *WHIT-whit-teew*, frequently uttered. [CD7:4]

6 THICK-BILLED CUCKOO *Pachycoccyx audeberti* 36 cm **R r**
6a Ad Slate-grey or slate-brown above, white below. **6b Juv** Boldly spotted above, white below. ▲ Wooded savanna, gallery forest, forest edge. Restless, often flying from treetop to treetop. Undertakes noisy, buoyant display flight over long distances. Parasitises helmet-shrikes. ✤ A series of clear, far-carrying *whuee-di* or *hwee-wik*. Also rather soft *weedidi weedidi weedidi* in flight. [CD7:3]

7 BLACK CUCKOO *Cuculus clamosus* 28–31 cm **M/R u/lc**
7a Ad *gabonensis* (Forest, E Sierra Leone–SE Guinea to CAR–Congo) Variable, but always very dark; dark chestnut throat and breast; boldly barred below. **7b Ad *gabonensis*** A less strongly coloured bird. **7c Ad *gabonensis*** A bird with unbarred throat. **7d Ad *clamosus*** (Woodland, S Senegambia to Chad–CAR and Congo) All black; white-tipped tail. Underparts may be barred. **7e Juv, both races** Wholly dull black. ▲ In S Africa recorded to parasitise mainly shrikes. ✤ A series of 3 hesitant notes, each with more emphasis and higher pitched than the preceding, final note sometimes repeated after short pause: *who whuu whee, wheee*. Also a wild, bubbling trill, rising and falling and gradually dying away *lululululululuWHIRlulululu-WHIRluluWHIRlu...* [CD7:5]

Maps on page 205

1b
1d
1a
2b
1c
2a
3a
3b
5b
4a
4b
5a
7c
7b
7a
6b
6a
7d
7e

PLATE 58: YELLOWBILL AND COUCALS

Non-parasitic cuckoos. Yellowbill is arboreal and mainly insectivorous. **Coucals** are semi-terrestrial, stoutly built birds with rounded wings and long, broad tails. Flight clumsy and not sustained. Mainly insectivorous, but also take small vertebrates. Utter rapid series of deep, hollow hoots and liquid bubbling notes (like water poured from a bottle).

1 YELLOWBILL *Ceuthmochares aereus* 33 cm **R c**
Wholly slate-grey with yellow bill. **1a Ad *aereus*** (east of lower Niger R.) Bluish gloss to upperparts and tail. **1b Ad *flavirostris*** (west of lower Niger R.) Purplish gloss. **1c Juv** Upper mandible brownish; plumage duller. ▲ Forest (even where heavily degraded). At all levels, mostly mid-stratum and lower canopy, unobtrusively creeping through dense leafy vegetation with creepers. Joins mixed-species flocks. ✤ A series of loud, explosive *kuk* notes, first uttered slowly, but rapidly gathering speed *kuk, kuk, kuk kuk kukkukkukkukukukkkkkkkrrrrrr*. Single *kuk* notes resemble those of Western Nicator. Also soft, plaintive *mweeeew*. [CD7:17]

2 BLACK COUCAL *Centropus grillii* 30–35 cm **M/R u/lc**
2a Ad breeding Black, with rufous wings. **2b Juv** Pale; head and mantle streaky; rest of upperparts barred. Non-breeding adult similar. ▲ Moist grassland, marshy areas, shrub. ✤ In breeding season, a long series of variable speed *wok-wok, wok-wok, wok-wok,* ... or *po-op po-op po-op*... and faster *popopopopopopop*..., from exposed perch. Also 'water-bottle' song, similar to 3 but higher pitched, faster and not rising at end. [CD7:21]

3 SENEGAL COUCAL *Centropus senegalensis* 36–40 cm; WS *c.* 50 cm **R c**
3a Ad Head and tail glossy black; upperparts rufous-brown; underparts whitish. **3b Juv** Strongly barred upperparts; head dark and streaky; underparts tawny. **3c Ad '*epomidis*' morph** Head and throat black; breast and belly chestnut. ▲ Roadsides, farmland, gardens and variety of open habitats with tall grass and bushes. Clambers in low vegetation, often in thick cover, runs and hops on ground, or flies clumsily for short distance, crash-landing into bush. ✤ A characteristic, accelerating series of hollow *hoots*, first rising in pitch, then dying away. Often in duet, both songs delivered in similar rhythm but different pitch. Speed and pitch variable, slowest and deepest resembling song of 6. Calls include nasal *gook* (alarm) and fast, dry *k-t-k-t-k-t-k-t-k-t-k-*... [CD7:22–23]

4 BLUE-HEADED COUCAL *Centropus monachus* 45 cm **R u/c**
4a Ad As 3a but with strong blue gloss to head to mantle; also slightly larger and heavier looking. **4b Juv** Similar to 3b, but with black tail and faint buff barring on uppertail-coverts. ▲ Swampy places, forest edge, savanna woodland. ✤ Very similar to 3 but two initial notes clearly distinct from rest of phrase. Songs of duetting pair have different pitch but same rhythm. [CD7:24]

5 GABON COUCAL *Centropus anselli* 48–58 cm **R c**
5a Ad Dark; underparts tawny. **5b Juv** Barred throat. ▲ Forest. ✤ A series of deep, resonant *hoots*, similar to those of 6. Songs of duetting pair have different pitch and rhythm. Vocalisations often higher pitched and more varied than other coucals, including more drawn-out syllables and series of rapid notes with 'yelping' quality. [CD7:19]

6 BLACK-THROATED COUCAL *Centropus leucogaster* 48–58 cm **R f**
6a Ad Head to breast black; rest of underparts white. **6b Juv** Duskier, with barred upperparts and tail. ▲ Forest interior and edges. ✤ A series of 10–20 deep, resonant *hoots*; similar to 3 in structure, but lower and slower, never accelerating at end. Songs of duetting pair have similar pitch but different rhythm. [CD7:18]

7 WHITE-BROWED COUCAL *Centropus superciliosus* 40 cm **R? r**
The only coucal with a pale supercilium. ▲ Tall grass, shrubbery. ✤ An accelerating series of hollow *hoots*, similar to, but normally faster than 3. [CD7:25]. See illustration on page 210.

1c
1a
1b
2b
2a
3b
3a
3c
4b
4a
5b
5a
6b
6a

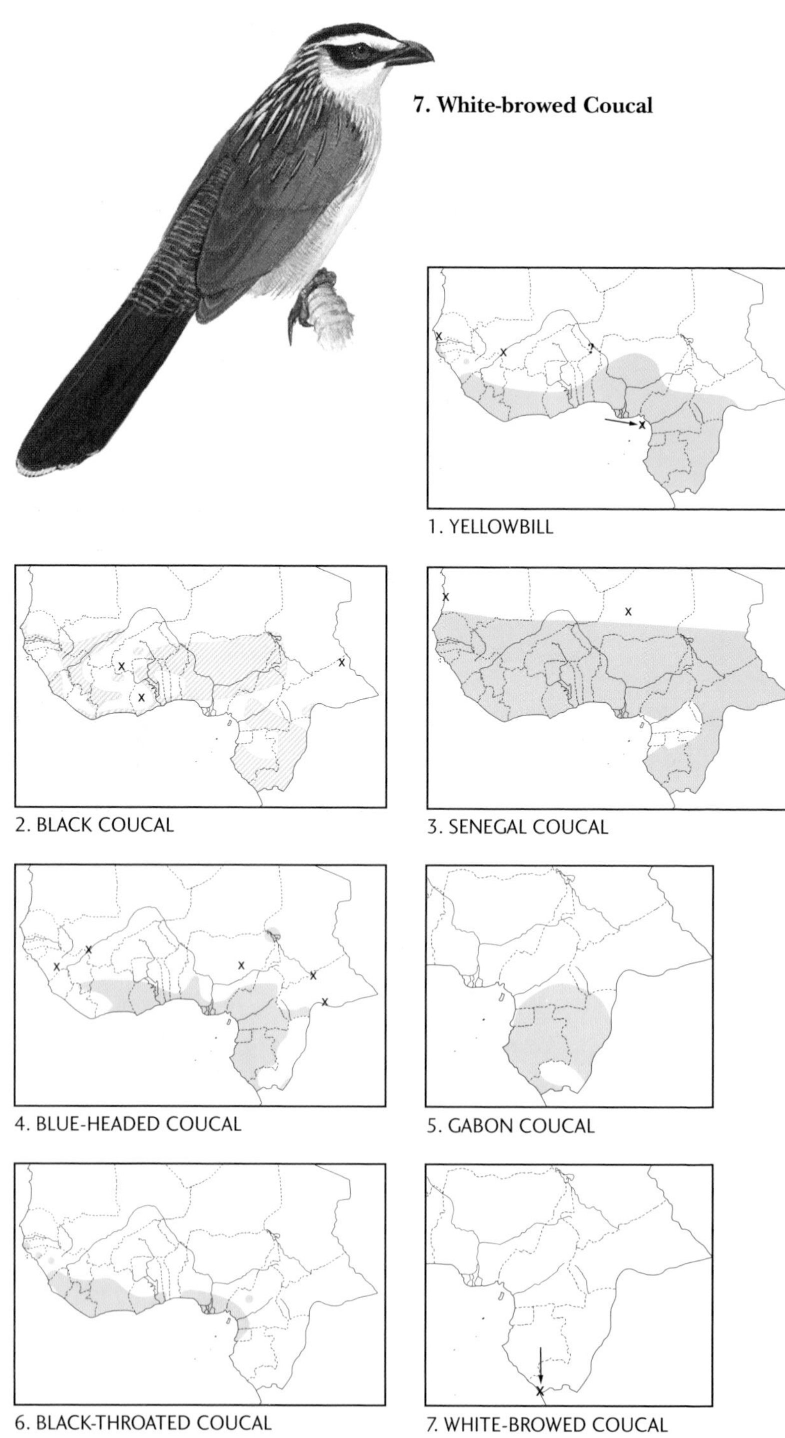

1. YELLOWBILL

2. BLACK COUCAL

3. SENEGAL COUCAL

4. BLUE-HEADED COUCAL

5. GABON COUCAL

6. BLACK-THROATED COUCAL

7. WHITE-BROWED COUCAL

Plate on page 208

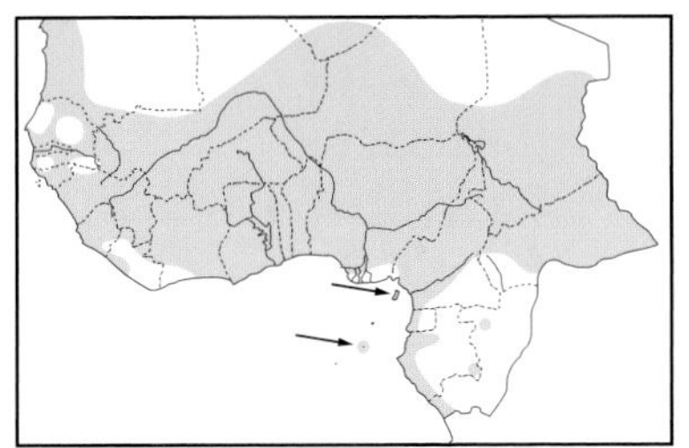

1. BARN OWL

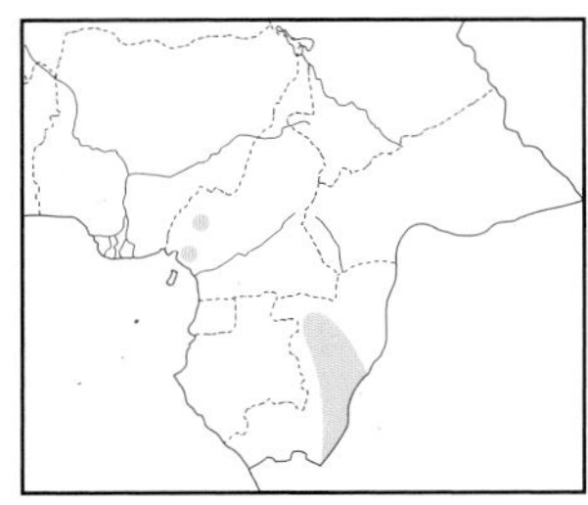

2. AFRICAN GRASS OWL

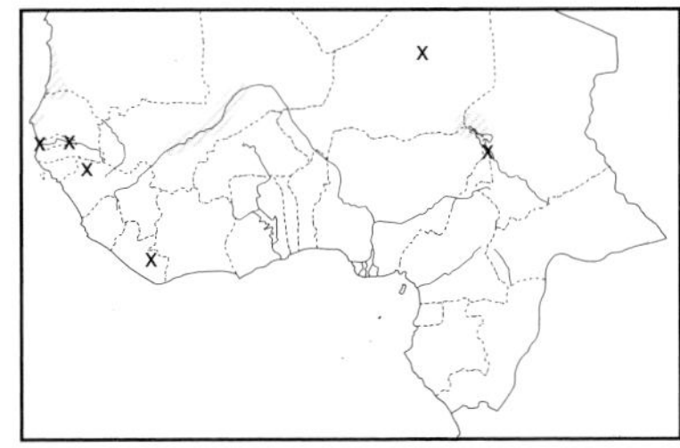

3. SHORT-EARED OWL

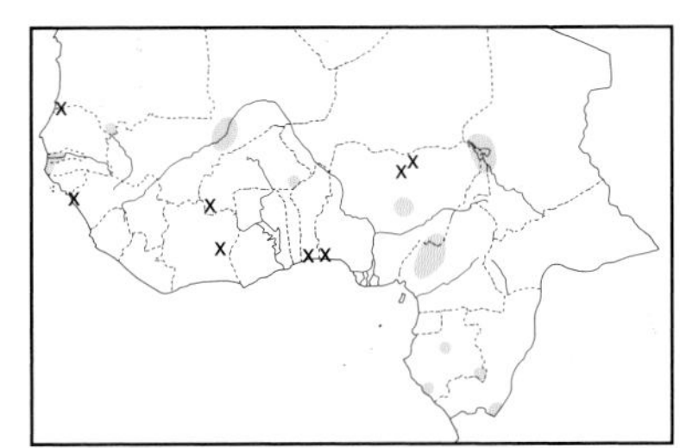

4. MARSH OWL

5. RUFOUS FISHING OWL

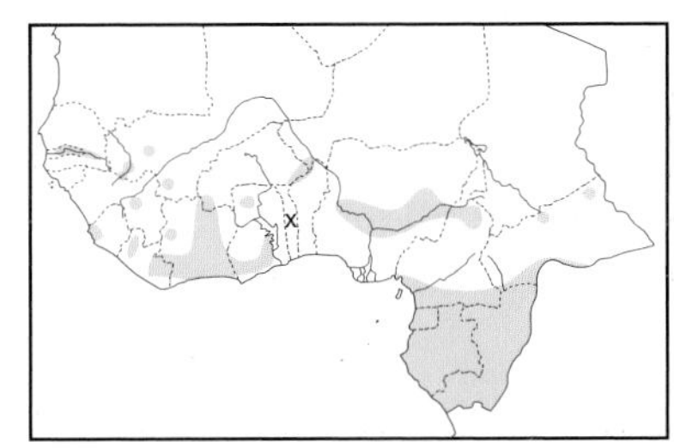

6. PEL'S FISHING OWL

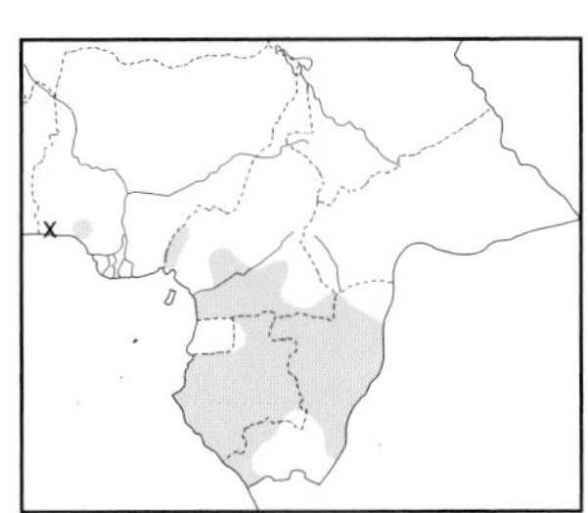

7. VERMICULATED FISHING OWL

Barn owls. Medium-sized nocturnal birds of prey with a distinctive heart-shaped, pale facial disc. Flight buoyant and silent. Perch upright.

1 BARN OWL *Tyto alba* 33–36 cm; WS 80–95 cm R+ u/f
1a Ad *affinis* (Mainland and Bioko) Pale grey mixed with golden-buff above; white below, variably washed buff. Nominate *alba* (Aïr, Niger) paler overall. **1b Ad *thomensis*** (São Tomé) Darker; face pale brownish; underparts golden-brown. *T. a. detorta* (Cape Verde) similar; face cinnamon. **1c Ad *affinis* in flight** Very pale overall. ▲ Various habitats; not in closed forest; often associated with man. Nocturnal and crepuscular. ✤ Hoarse, drawn-out screech *shree-eeeeeee*, in flight. [CD3:12]

2 AFRICAN GRASS OWL *Tyto capensis* 36–40 cm R r/lf
2a Ad Dark brown above; white face; whitish to pale rufous below; speckled breast. **2b Juv** Dusky-russet face; rufous-buff below. **2c In flight** Pale patch at base of primaries. ▲ Dambos, montane grassland. Mainly nocturnal. When flushed, flies short distance with dangling legs before dropping into cover of high grass. ✤ Usually silent. Hoarse, drawn-out screech, recalling 1 but softer. [CD7:26]

True owls. Small to large, mainly nocturnal birds of prey with characteristic large, rounded heads and large, forward-facing eyes within flat facial disc. Plumage mostly cryptic, many species with ear-tufts. Calls distinctive; usually most vocal just after dark and before dawn.

3 SHORT-EARED OWL *Asio flammeus* 35–40 cm; WS 95–110 cm P/V+ r/u
3a Sandy-buff marked dark brown and tawny above; whitish face; yellow eyes. **3b In flight** Long wings; dark carpal patch; buff patch at base of primaries; black wingtips. ▲ Various open habitats. Crepuscular, nocturnal and often diurnal. ✤ Usually silent in Africa. [CD3:18]

4 MARSH OWL *Asio capensis* 30–35 cm; WS 80–95 cm R/M r/lf
4a Plain dark brown; rounded head; large dark eyes. **4b In flight** Large pale patch at base of primaries. ▲ Open grassland. Mainly crepuscular and nocturnal, occasionally diurnal. ✤ Various harsh croaks and rasping calls, uttered in flight and on ground. [CD3:19]

5 RUFOUS FISHING OWL *Scotopelia ussheri* 43–51 cm R r/s EN
Pale rufous below with fine streaks; darker above; upperparts plain. ▲ Riverine forest, mangrove. Nocturnal. ✤ Single, deep, drawn-out wailing hoot, repeated at 1-minute intervals. [CD7:37]

6 PEL'S FISHING OWL *Scotopelia peli* 55–63 cm R r/lf
Very large; orange-rufous; barred above; plain face; large dark eyes; no ear-tufts. ▲ Large forested rivers, swamps. Largely nocturnal. Roosts by day in shady spot in large tree near water. Flushes with noisy wingbeats. Catches fish from low perch. ✤ Far-carrying, deep sonorous hoot, sometimes preceded and often followed by low grunt *hooommm-hut*. Pair may duet, male starting with low grunting *uh-uh-uhu...* reaching higher *hoommm*, answered by deeper hoot of female. [CD7:36]

7 VERMICULATED FISHING OWL *Scotopelia bouvieri* 43–51 cm R r/lf
Rufous above streaked and vermiculated dusky-brown; whitish below boldly streaked dark. ▲ Forested rivers, swamp forest. Nocturnal. ✤ A series of short, abrupt notes *kroh! woh!-woh! woh!-woh! woh!-woh!* and a deep, drawn-out, croaking wail *krooOOoah*. [CD7:38]

 Maps on page 211

1b
1a
2a
2b
3a
3b
1c
2c
4b
4a
5
6
7

1 **SJÖSTEDT'S (CHESTNUT-BACKED) OWLET** *Glaucidium sjostedti* 24–28 cm **R r/c**
Large rounded head; finely barred head and mantle; finely barred cinnamon-rufous underparts. ▲ Forest interior. Nocturnal and partially diurnal. ✤ A descending and accelerating series of vibrant notes *krroow krroow krroow-rroo-rroo-rroo* and a short, rapid series of similar hoots *ho-ho-ho-ho*. [CD7:43]

2 **AFRICAN BARRED OWLET** *Glaucidium capense* 20–23 cm **R r/lf**
Dumpy; dark brown upperparts barred tawny; white underparts barred and spotted dark brown. ▲ Forest. Nocturnal and partially diurnal. Note: W African forms (*etchecopari* and, probably, *castaneum*) sometimes considered specifically distinct as Chestnut Owlet *G. castaneum*. ✤ A fairly rapid, rhythmic series of 6–18 plain whistled hoots *hoot-hoot-hoot-hoot-hoot-*.... Also similar, but slightly accelerating series with more vibrant notes. [CD7:41]

3 **RED-CHESTED OWLET** *Glaucidium tephronotum* 20–24 cm **R r/u**
3a Ad *tephronotum* (Upper Guinea forest) Rather long tail with three white spots on centre; white underparts spotted brown; rufous-chestnut breast-sides and flanks. **3b Ad *pycrafti*** (Lower Guinea forest) Darker above; less rufous below; spots blackish. ▲ Forest. Nocturnal and partially diurnal. ✤ A rhythmic, rather slow series of whistles *huut huut huut huut* ... (repeated at intervals of *c.* 1 second); length and speed of series variable. [CD7:40]

4 **PEARL-SPOTTED OWLET** *Glaucidium perlatum* 19–21 cm **R u/lc**
4a Brown spotted white above; white heavily streaked rufous-brown below; tail rather long. **4b Back of head** Two dark 'eye-spots' forming false face. ▲ Wooded savanna. Partially diurnal; regularly calls by day. ✤ A distinctive, rhythmically rising series of short notes followed by loud whistling climax *hu-hu-hu-hu-HU-HU-HU TEEEUW TEEEUW TEEEUW*... [CD7:39]

5 **EUROPEAN SCOPS OWL** *Otus scops* 18–20 cm; WS 47–54 cm **P r/f**
Very similar to 6. In hand: outermost (10th) primary longer than 5th. ▲ Various bushy and wooded habitats. ✤ Pure, whistled *pyuu*. Normally silent in Africa. [CD3:13]

6 **AFRICAN SCOPS OWL** *Otus senegalensis* 16–18 cm **R u/lc**
Cryptic plumage; prominent ear-tufts; short tail. ▲ Wooded grassland. Nocturnal and crepuscular. ✤ Single, short, vibrant *prr-u-u-p*, often repeated, with pauses of 5–10 seconds. Generally starts calling at dusk, but may do so up to one hour before. [CD7:28]

7 **SANDY SCOPS OWL** *Otus icterorhynchus* 18–22 cm **R u/r**
7a Ad *icterorhynchus* (Upper Guinea forest) Cinnamon-brown speckled white; short ear-tufts. **7b Ad *holerythrus*** (Lower Guinea forest) More rufous; no white spots on wing-coverts. ▲ Forest. Nocturnal. ✤ Single, drawn-out, descending whistle *wheeoo*, repeated with pauses of a few seconds. [CD7:27]

8 **NORTHERN WHITE-FACED OWL** *Ptilopsis leucotis* 23–28 cm **R f**
White face boldly rimmed with black; large orange eyes; prominent ear-tufts. ▲ Various wooded habitats, mainly wooded grassland. Nocturnal and crepuscular. ✤ Two rather low, melodious, fluting syllables *whoh whoow*, repeated at intervals of 4–8 seconds. Also a lower and slightly faster *whoh thohoow*. [CD7:30]

9 **SOUTHERN WHITE-FACED OWL** *Ptilopsis (leucotis) granti* 23–28 cm **R f**
As 8 but darker; greyer; black markings more pronounced. ▲ Various wooded habitats, mainly wooded grassland. Nocturnal and crepuscular. ✤ Rapid, muffled bubbling followed by a clear, melodious hoot *kdkdkdkdkdkd-whOOw*.

10 **LITTLE OWL** *Athene noctua* 21–25 cm **R lu**
Sandy-brown above dappled white; flat head; 'frowning' face. ▲ Stony desert, esp. in montane areas. Partially diurnal. ✤ A sharp *keeow* and *wooow*. Male advertising call a rather low, mellow *ko-ooep*. Call of female higher pitched and more nasal. Alarm, a loud, sharp *kyip kyip*. [CD3:15]

Maps on page 216

1
3b
3a
2
4a
4b
6
5
7a
7b
8
10
9

PLATE 60: OWLS II

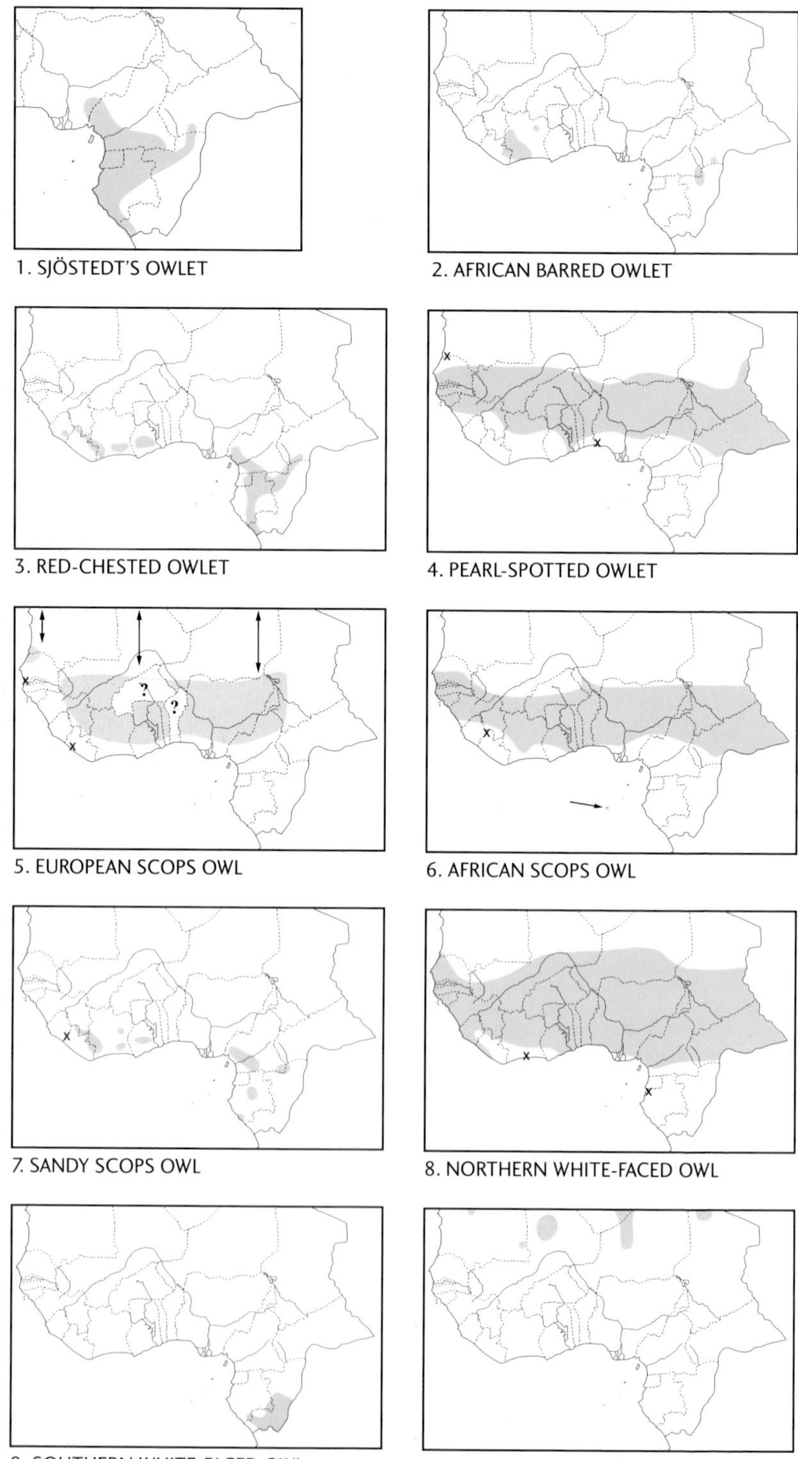

1. SJÖSTEDT'S OWLET

2. AFRICAN BARRED OWLET

3. RED-CHESTED OWLET

4. PEARL-SPOTTED OWLET

5. EUROPEAN SCOPS OWL

6. AFRICAN SCOPS OWL

7. SANDY SCOPS OWL

8. NORTHERN WHITE-FACED OWL

9. SOUTHERN WHITE-FACED OWL

10. LITTLE OWL

Plate on page 214

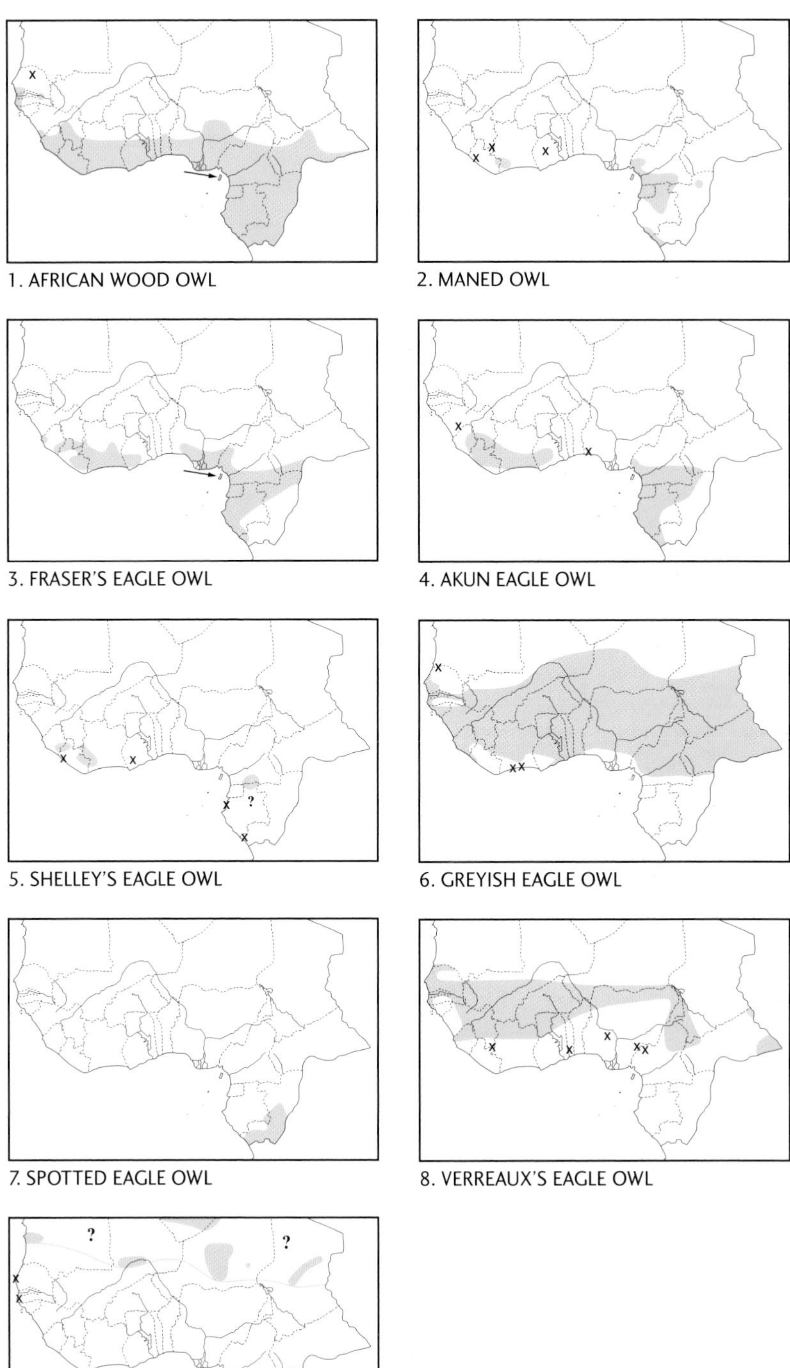

1. AFRICAN WOOD OWL

2. MANED OWL

3. FRASER'S EAGLE OWL

4. AKUN EAGLE OWL

5. SHELLEY'S EAGLE OWL

6. GREYISH EAGLE OWL

7. SPOTTED EAGLE OWL

8. VERREAUX'S EAGLE OWL

9. DESERT EAGLE OWL

Plate on page 218

1 **AFRICAN WOOD OWL** *Strix woodfordii* 33–35 cm **R c**
Round head; barred underparts; large dark eyes. ▲ Forest, dense woodland, plantations. Often in pairs. Nocturnal. ✤ Characteristic, single, drawn-out *whoOOow* or *oowhEEoo* and rhythmic *hu-hoo, hu-hoo hoo-hu-hoo*. Duets; female higher pitched. Male may answer female *whoOOow* by low *hooo*. [CD7:44]

2 **MANED OWL** *Jubula lettii* 34–37 cm **R r DD**
Rufous-brown; bushy ear-tufts; rufous face rimmed with black; deep yellow to crimson eyes. ▲ Forest, esp. sites with dense lianas. ✤ Song unknown. Probably consists of single hoot followed by a series.

3 **FRASER'S EAGLE OWL** *Bubo poensis* 38–45 cm **R u/f**
Dark rufous; narrowly barred above and below; ear-tufts; dark eyes. ▲ Forest. Nocturnal. ✤ Plaintive, drawn-out *whuah* or *whooaah* and an often long, low purring or grunting trill, rising and falling in tempo and volume, ceasing abruptly, recalling sound of small engine *kudakudakudakudakdkdkdkdkdkdrdrdrdrd...* [CD7:33]

4 **AKUN EAGLE OWL** *Bubo leucostictus* 40–46 cm **R r/lf**
Dark brown; underparts barred and blotched dark; two-toned ear-tufts; yellow eyes. ▲ Forest. Nocturnal. ✤ Reminiscent of 3, with similar low grunting trill, but even more plaintive wail *wheeaah!*; also utters kind of barking and series of low *ro, ro, ro,...* Calls mainly just after dusk and before dawn. [CD7:32]

5 **SHELLEY'S EAGLE OWL** *Bubo shelleyi* 54–61 cm **R r NT**
Very large; dark brown above; whitish below heavily barred dark brown; prominent ear-tufts. ▲ Forest. Nocturnal. ✤ Single, loud, high-pitched, plaintive scream *KEEEEOOOOUW!* Reported to call mainly at and just after dusk and before dawn. [CD7:34]

6 **GREYISH (VERMICULATED) EAGLE OWL** *Bubo (africanus) cinerascens* 43–48 cm **R u/f**
Large; grey-brown; two-toned ear-tufts; finely barred underparts; dark blotches on breast; dark eyes. ▲ From rocky desert outcrops to wooded grassland and forest edge. Often on dirt roads at night. ✤ Male utters short, abrupt hoot, followed after short pause by second, lower and drawn-out hoot *HO!, hoooo*. Female has trisyllabic *ho, hohooo*, middle note higher. Often duets. Calls mainly at dawn and dusk. [CD7:31]

7 **SPOTTED EAGLE OWL** *Bubo africanus* 43–48 cm **R u/f**
As 6 but greyer overall; yellow eyes. ▲ Wooded grassland, forest edge. Often on dirt roads at night. ✤ As 6.

8 **VERREAUX'S EAGLE OWL** *Bubo lacteus* 58–65 cm **R u/r**
Very large; greyish; pink eyelids; dark eyes; ear-tufts seldom raised. ▲ Wooded grassland, riverine woodland. ✤ A short, irregular series of low, discontented grunts. Calls mainly at dusk; also not infrequently during day. [CD7:35]

9 **DESERT EAGLE OWL** *Bubo (bubo) ascalaphus* 46–50 cm **R u**
Tawny-rufous streaked and blotched dark and pale; prominent ear-tufts; orange eyes. ▲ Desert and sub-desert, esp. with rocky outcrops and in oases. Mainly nocturnal. ✤ Far-carrying, deep, sonorous hooting *WHOO-oo*, second syllable lower pitched; repeated at intervals of 10–15 seconds. During courtship, trisyllabic *WHOO-hoo-hoo*. Duets; female higher pitched. Calls at dusk. [CD3:14]

 Maps on page 217

1
2
3
4
5
6
7
8
9

Crepuscular and nocturnal with very cryptic plumage, long wings, moderately long tail, tiny bill and huge gape for catching insects on the wing. Normally not seen during the day, except when flushed from underfoot. Identification of many species problematic; important features include overall coloration, markings on scapulars and wing-coverts, presence of nuchal collar, and amount of white (male) or buff (female) in wings and tail, but best identified by voice. Typically sing at or just after dusk and just before dawn; often also throughout moonlit nights. Often on dirt roads at night; orange reflection of eyes in headlights of car visible at considerable distance.

1 BROWN NIGHTJAR *Veles binotatus* 21–23 cm R r
1a Small; very dark. **1b In flight** No white in wings and tail. ▲ Forest interior and edge. ✤ Song a long series of *kyup* or *kliou* notes. [CD7:46]

2 BATES'S NIGHTJAR *Caprimulgus batesi* 29–31 cm R u/lf
2a Large; very dark; relatively long tail. **2b Ad male in flight** Small white wing patch; white tail corners. **2c Ad female in flight** No wing patch; buff tail corners. ▲ Forest and forest edge. ✤ Song an initial clear, yelping note, followed after a short pause by a rapid series of 2–12 similar notes, *whow! whow-whow-whow-whow whow! whow-whow...* resembling Freckled Nightjar but more monotonous. Flight call a low, guttural *ugh-ugh-ugh*. [CD7:45]

3 BLACK-SHOULDERED NIGHTJAR *Caprimulgus nigriscapularis* 23–25 cm R f/r
3a Rufous-brown; blackish shoulder; rusty-buff line on scapulars. **3b Ad male in flight** White wing patch; white tail corners. **3c Ad female in flight** Wing patch and tail corners slightly smaller. ▲ Woodland, gallery forest, forest edge. ✤ Song a clear, melodious drawn-out note followed by tremulous whistling *kiiiuu-iup kiiurrrr*, sometimes preceded by fast *whoap-whoap-whoap-whoap-...* Flight call a low *chuk* often repeated rapidly 2–3 times. [CD7:52]

4 FIERY-NECKED NIGHTJAR *Caprimulgus pectoralis* 23–25 cm R lc
4a As 3a but paler, greyer brown; wing-coverts more uniform. **4b Ad male in flight** White wing patch (slightly larger than 3b); white tail corners. **4c Ad female in flight** Buffish wing patch and tail corners. ▲ Woodland, gallery forest, forest edge. ✤ Song a clear, melodious drawn-out note followed by tremulous whistling *kyoo-yiup kyuiurrrr*, very similar to 3 but first note more modulated, often preceded by fast *whoa-whoa-whoa-whoa-...* (slightly slower than 3). [CD7:51]

5 RUFOUS-CHEEKED NIGHTJAR *Caprimulgus rufigena* 23–24 cm M f/r
5a Grey-brown; pale line on scapulars; buff-tipped wing-coverts. **5b Ad male in flight** White wing patch; white tail corners. **5c Ad female in flight** Wing patch smaller, usually buffy; no white in tail. ▲ Wooded grasslands. ✤ Probably silent in W Africa. [CD7:57]

6 EUROPEAN NIGHTJAR *Caprimulgus europaeus* 24–28 cm; WS 52–60 cm P u/r
6a Dark greyish; pale wingbar; pale line on scapulars; whitish moustachial stripe. **6b Ad male in flight** White wing patch; white tail corners. **6c Ad female in flight** No white in wings and tail. ▲ Various open and lightly wooded habitats. ✤ Normally silent in W Africa. [CD3:22]

7 RED-NECKED NIGHTJAR *Caprimulgus ruficollis* 30–34 cm; WS 60–65 cm P r/u
7a Large; rufous nuchal collar; whitish moustachial stripe; large white throat patch. **7b Ad male in flight** White wing patch; white tail corners. **7c Ad female in flight** Buffish wing patch and tail corners. ▲ Various open and wooded habitats. ✤ Song a series of rapid, mechanical-sounding *kotok-kotok-kotok-kotok-kotok-kotok-...* accelerating in long *kotokotokotokotokoto-koto...* then slowing down again. [CD3:21]

8 LONG-TAILED NIGHTJAR *Caprimulgus climacurus* 28–43 cm R/M c
8a Ad male Small; long, graduated tail; white wingbar. **8b Ad male in flight** White wing patches; white trailing edge to wing. **8c Ad female in flight** Tail shorter than in male; buffish wing patch. ▲ Semi-desert to forest zone. ✤ Song a monotonously sustained, hard reeling with single, fast rhythm *rrerrrrrrrrrrrrrrrrrrrrrr...* Female has similar, slightly lower pitched *rrorrrrrrr...* In flight, a repeated *chiong-chiong..* [CD7:48]

Maps on page 222

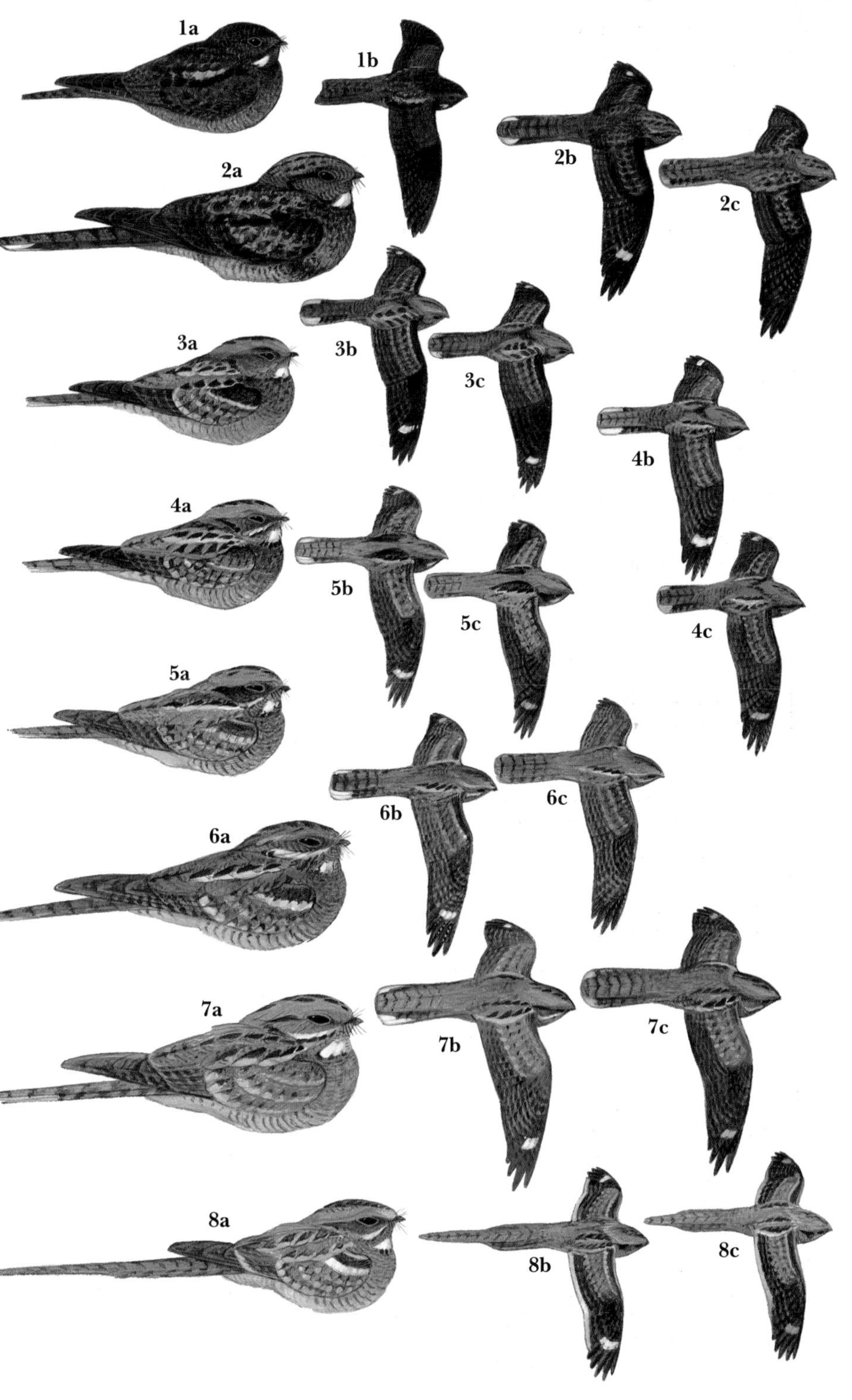
1a
1b
2a
2b
2c
3a
3b
3c
4a
4b
4c
5a
5b
5c
6a
6b
6c
7a
7b
7c
8a
8b
8c

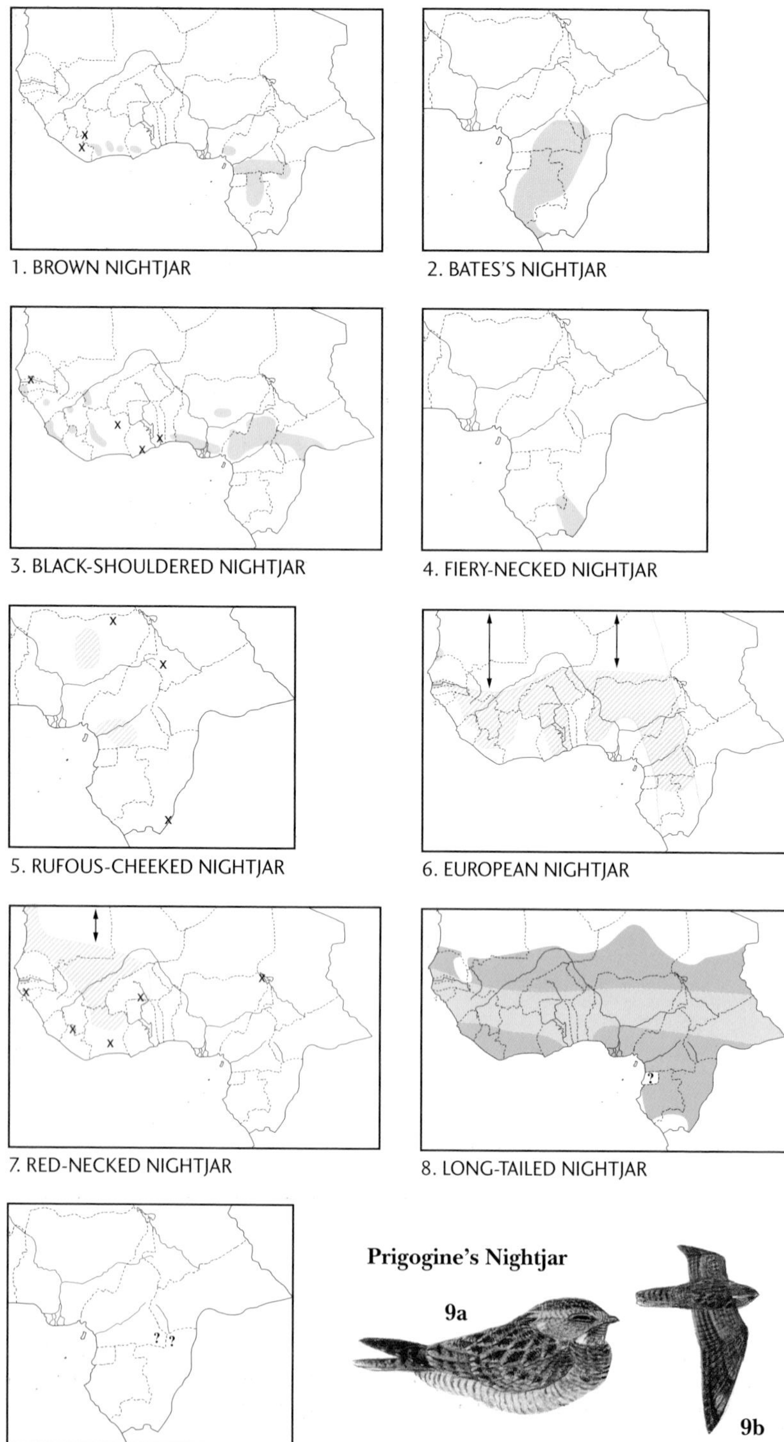

1. BROWN NIGHTJAR

2. BATES'S NIGHTJAR

3. BLACK-SHOULDERED NIGHTJAR

4. FIERY-NECKED NIGHTJAR

5. RUFOUS-CHEEKED NIGHTJAR

6. EUROPEAN NIGHTJAR

7. RED-NECKED NIGHTJAR

8. LONG-TAILED NIGHTJAR

9. PRIGOGINE'S NIGHTJAR

Plate on page 220

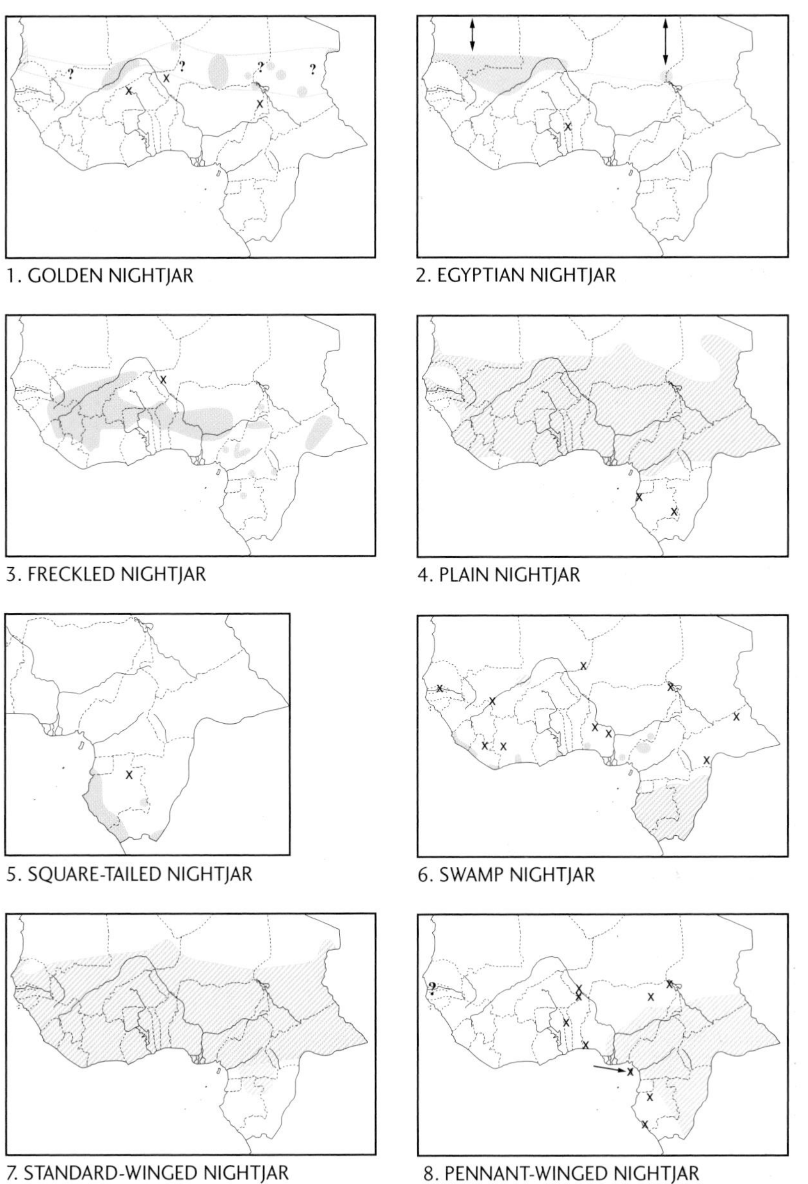

1. GOLDEN NIGHTJAR

2. EGYPTIAN NIGHTJAR

3. FRECKLED NIGHTJAR

4. PLAIN NIGHTJAR

5. SQUARE-TAILED NIGHTJAR

6. SWAMP NIGHTJAR

7. STANDARD-WINGED NIGHTJAR

8. PENNANT-WINGED NIGHTJAR

---

9 PRIGOGINE'S NIGHTJAR *Caprimulgus prigoginei* 19 cm **R? EN**
9a Ad female Small; darkish; blotched and speckled blackish, rufous, tawny and buff; relatively short tail. **9b Ad female in flight** No white patch in wings; narrow white tail corners. ▲ Forest, SE Cameroon–N Congo?

Plate on page 224

1 GOLDEN NIGHTJAR *Caprimulgus eximius* 23–25 cm **R u/r**
1a Golden-buff with irregular white spots, each bordered and speckled black. **1b Ad male in flight** Large white wing patch; white corners to tail. **1c Ad female in flight** Smaller and buffish wing patch and tail corners. ▲ Sahel. ✤ Song a monotonously sustained, hard reeling.

2 EGYPTIAN NIGHTJAR *Caprimulgus aegyptius* 24–27 cm; WS 53–58 cm **P u/lf**
2a Rather uniform, pale greyish plumage. **2b In flight** No white in wings and tail; dark brownish outer wing. ▲ Desert, semi-desert. ✤ Mostly silent in W Africa. Song, given on ground, a long, very rapid, regular series of *krow* notes, *krowkrowkrowkrowkrowkrowkrowkrow*..., resembling sound of running engine. [CD3:20]

3 FRECKLED NIGHTJAR *Caprimulgus tristigma* 26–28 cm **R u/lf**
3a Dark greyish or blackish-brown finely speckled white or buffish. **3b Ad male in flight** Small white wing patch; white tail corners. **3c Ad female in flight** Wing patch smaller; no white in tail. ▲ Rocky outcrops. ✤ Song a series of yelping notes *whaow! whaow! whaow!*... and *aow-whaow! aow-whaow! aow-whaow!*..., also a more barking *wah!-wah!-wah!-*... Flight call *wok*. [CD7:56]

4 PLAIN NIGHTJAR *Caprimulgus inornatus* 22–23 cm **M/R c/f**
4a Ad Variably coloured but fairly uniform plumage; row of small black spots on scapulars. **4b Ad** Rufous-coloured bird. **4c Ad male in flight** White wing patch and tail corners. Cinnamon-coloured individual illustrated. **4d Ad female in flight** Buff wing patch and tail corners. Grey-coloured individual. ▲ Various open and wooded habitats. ✤ Song a monotonous, sustained, hard reeling *rrorrrrrrrrrrrrrrr*... *rrerrrrrrrrrrrrrrrrrrrrr*... similar to but noticeably lower pitched and slower than that of Long-tailed Nightjar. In flight, chuckling *kwakow*. On ground, low *chuk*. [CD7:55]

5 SQUARE-TAILED NIGHTJAR *Caprimulgus fossii* 23–24 cm **R c**
5a White wingbar; buffish line on scapulars. **5b Ad male in flight** White wing patch; white trailing edge to wing; white-edged tail. **5c Ad female in flight** Wingbar, wing patch, trailing edge and tail edge buffish. ▲ Grasslands, open woodland, farmland. ✤ Song a far-carrying, sustained reel alternating with a more sputtering rattle *rrerrrrrrrr-rreheheheheh-rrerrrrrrrr-rrehehehe-heh-*... Flight call a yelping *whaoop!* or *wowaw!* [CD7:49]

6 SWAMP NIGHTJAR *Caprimulgus natalensis* 20–24 cm **R lr/f**
6a Long pale supercilium and moustachial stripe; dark head-sides; large blackish markings on scapulars and wing-coverts; round buff spots on breast. **6b Ad male in flight** White wing patch; white-edged tail. **6c Ad female in flight** Buffish wing patch; buff-edged tail. ▲ Damp and dry grasslands. ✤ Song a long, regular series of rapid notes *chuk-chuk-chuk-chuk-chuk-chuk-*... or *tuk-tuk-tuk-tuk-tuk-*..., resembling knocking on dry wood, and faster *chukukukukukukukuk*. Also a sharp note, often repeated a few times, *kip kip-kip-kip*; a highly distinctive, variable series of melodious, tremulous notes *whoa-whulululu whoa-whuwhu whoa-whulululu whoa!* in flight, and fast *rrrrukukukukukukuk-whoalulululu*. [CD7:47]

7 STANDARD-WINGED NIGHTJAR *Macrodipteryx longipennis* 21–22 cm **M f/c**
7a Ad male non-breeding/ad female Rufous nuchal collar; buffish line on scapulars; buffish throat. **7b Ad male in flight** Extremely elongated primary with bare shaft and blackish vane at tip. ▲ Various open, wooded habitats. ✤ Song a fast, high-pitched, shrill, insect-like *tsikitsikitsikitsikitsiki*... and slower *tseepetseepetseep etseepetseepe*... Flight call a low *kuk*. [CD7:58]

8 PENNANT-WINGED NIGHTJAR *Macrodipteryx vexillarius* 24–28 cm **M lc/s**
8a Ad female As 7a, but larger, slightly darker above, white throat patch, paler and more barred lower underparts. **8b Ad male in flight** Broad white band on black remiges; extremely elongated whitish primary. **8c Ad female in flight** No white in wings and tail. ▲ Wooded grassland. ✤ Probably silent in W Africa. Song a fast, high-pitched, shrill, insect-like *tsitsitsits-itsitsitsitsitsitsi*... recalling 7. Flight calls *wheeeo* and *chup*. [CD7:59]

 Maps on page 223

1a
1b
1c
2a
2b
3a
3b
3c
4a
4b
4c
4d
5a
5b
5c
6a
6b
6c
7a
7b
8a
8b
8c

Highly aerial, anchor-shaped birds with mostly dark plumage. Flight swift and effortless on long, stiff wings. Catch insects on the wing; never perch. Often gregarious and vocal. Some difficult to identify; useful features include size, silhouette, rump pattern, tail shape, habitat and locality.

1 MOTTLED SPINETAIL *Telacanthura ussheri* 13–14 cm **R s/lc**
1a Above Broad white rump. Compare 8 and 9. **1b Below** Pale, mottled throat; narrow whitish area on vent. ▲ Wooded savanna, open areas in forest zone. ✤ [CD7:62]

2 SABINE'S SPINETAIL *Rhaphidura sabini* 10.5–11.5 cm **R lc/u**
2a Above White rump and uppertail-coverts; small. **2b Below** White belly and undertail-coverts; tail almost completely hidden. ▲ Rainforest. ✤ [CD7:60]

3 CASSIN'S SPINETAIL *Neafrapus cassini* 12–13 cm **R lf**
3a Above Narrow white rump band; very short tail. **3b Below** White belly and undertail-coverts; distinctly notched wings. Fluttering, bat-like flight. ▲ Rainforest. ✤ [CD7:63]

4 BÖHM'S SPINETAIL *Neafrapus boehmi* 9–10 cm **M ls**
4a Above As 3 but much smaller, rump band broader; tail extremely short. **4b Below** Throat paler than 3. ▲ Rainforest. NE Gabon. ✤ [CD7:64]

5 BLACK SPINETAIL *Telacanthura melanopygia* 15–17 cm **R r/lc**
Wholly blackish; throat mottled dusky. ▲ Rainforest. ✤ [CD7:61]

6 PALLID SWIFT *Apus pallidus* 16–18 cm; WS 39–44 cm **P/R/V* lc/r**
Paler than 7, with larger throat patch; inner wing contrasting with darker outer wing. ▲ Over various habitats. ✤ As 7 but often slightly deeper and disyllabic. [CD3:23]

7 COMMON SWIFT *Apus apus* 17–18.5 cm; WS 40–44 cm **P+ c/u***
Uniformly blackish-brown; forked tail; ill-defined throat patch. Compare 6 and 10. ▲ Over all habitats. ✤ High-pitched, shrill, screaming *srreeee*. [CD3:24]

8 LITTLE SWIFT *Apus affinis* 12–13.5 cm; WS 32–34 cm **R c**
8a Above Broad white rump patch; square tail. Compare 1 and 9. **8b Below** White throat. ▲ Mainly in towns. ✤ Rapid, high-pitched twittering, frequently uttered. [CD3:26]

9 HORUS SWIFT *Apus horus* 15 cm **M/R? r/lu**
9a Above Broad white rump patch; slightly forked tail. Compare 8 and 13. **9b Below** White throat. **9c Dark morph** ('*toulsoni*'; Gabon–Congo) Dark brown rump; greyer throat. ▲ Wooded savanna. ✤ Remarkable trill *krrweepeeo*... or *przweeew*... [CD7:68]

10 AFRICAN BLACK SWIFT *Apus barbatus* 16 cm **R lu/r**
10a Above As 7 but darker; contrasting paler inner wing panel. **10b Below** As 7 but shallower fork to tail. **10c Head** Compare 11. ▲ Various habitats. ✤ As 7 but more toneless (due to higher frequency), sounding like harsh, buzzing trill *szrirrrzr*. [CD7:67]

11 FERNANDO PO SWIFT *Apus (barbatus) sladeniae* 16 cm **R lr DD**
Blacker than 10; little or no greyish-white on throat. ▲ Mainly highlands.

12 PLAIN SWIFT *Apus unicolor* 14–15.5 cm; WS 35–39 cm **V+**
Smaller and slimmer than 7 and 6; throat darker; flight feathers semi-translucent from below; flight more erratic. ▲ Palearctic vagrant. ✤ Similar to 7.

13 WHITE-RUMPED SWIFT *Apus caffer* 14–15.5 cm; WS 33–37 cm **R/M r/lu**
13a Above Long, deeply forked tail (often held closed); U-shaped rump patch. **13b Below** White throat. ▲ Savanna, farmland. ✤ [CD3:25]

14 AFRICAN PALM SWIFT *Cypsiurus parvus* 16 cm, incl. tail of up to 9 cm **R c**
Wholly greyish; long wings and tail. ▲ Near palm trees, frequently in towns. ✤ Rapid, high-pitched, sibilant twittering *srrit-ititititit*, frequently uttered. [CD7:66]

Continued on page 228

Maps on pages 228 and 229

1b
2b
3b
4b
5
6
2a
4a
1a
3a
7
8b
9c
10a
12
10c
8a
11
9b
9a
15
10b
17
13b
14
16
18
13a

15 BATES'S SWIFT *Apus batesi* 14 cm **R r/lc**
Wholly glossy black; slender tail. Compare 7, 10 and 18. ▲ Forest. ✤ Usually silent.

16 MOTTLED SWIFT *Tachymarptis aequatorialis* 23 cm **R/M r/lu**
Very large. All dark except whitish throat and greyish-looking belly. ▲ Crags, escarpments. Disperses widely. ✤ [CD7:69]

17 ALPINE SWIFT *Tachymarptis melba* 20–23 cm; WS 51–58 cm **P+/M f/r**
Very large. White underparts with dark brown breast-band. ▲ Variety of habitats.✤ [CD3:27]

18 SCARCE SWIFT *Schoutedenapus myoptilus* 16.5 cm **R lr/s**
Dark brown; deeply forked tail (often held closed); pale throat; rest of underparts slightly paler than upperparts. ▲ Montane forest; also lower. ✤ Diagnostic clicking call. [CD7:65]

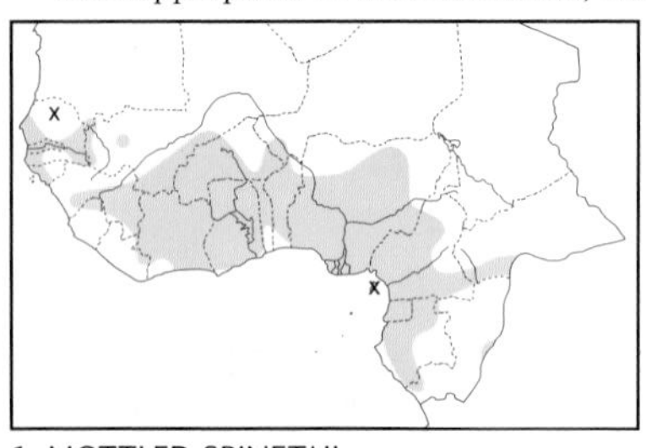
1. MOTTLED SPINETAIL

2. SABINE'S SPINETAIL

3. CASSIN'S SPINETAIL

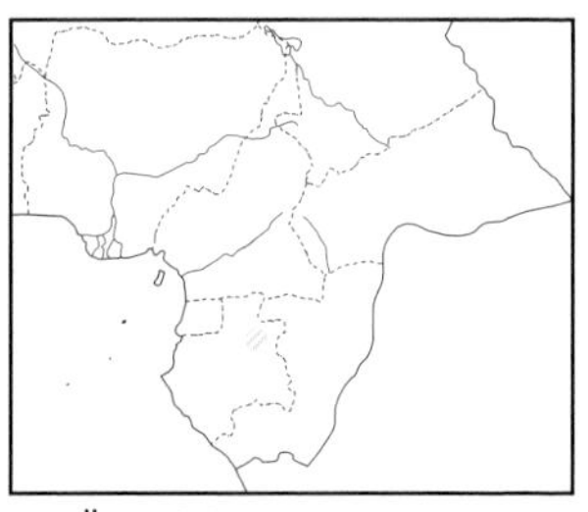
4. BÖHM'S SPINETAIL

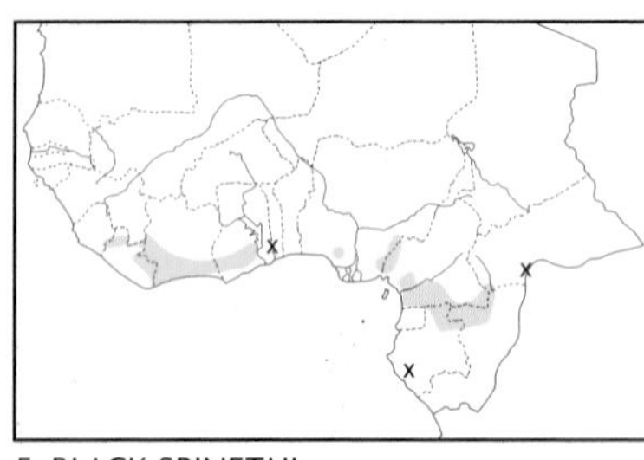
5. BLACK SPINETAIL

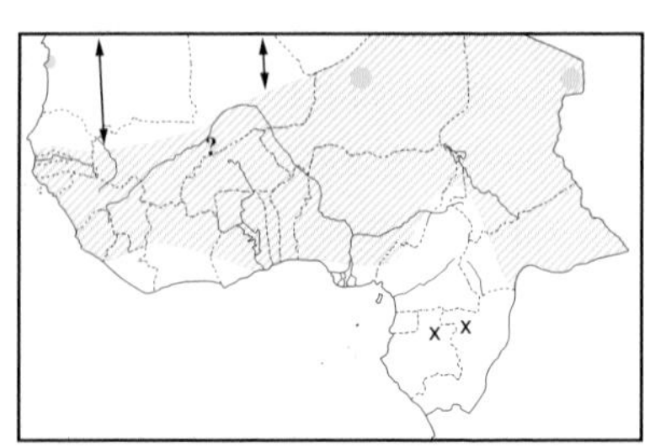
6. PALLID SWIFT

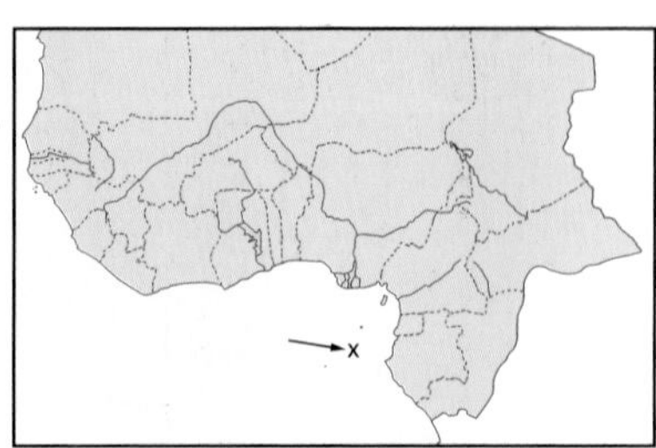
7. COMMON SWIFT

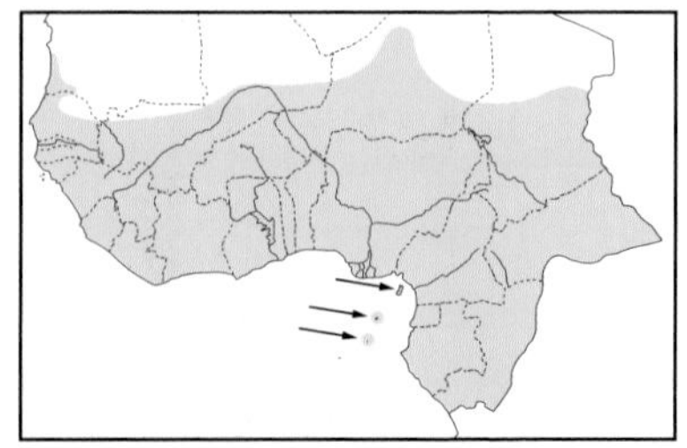
8. LITTLE SWIFT

 Plate on page 226

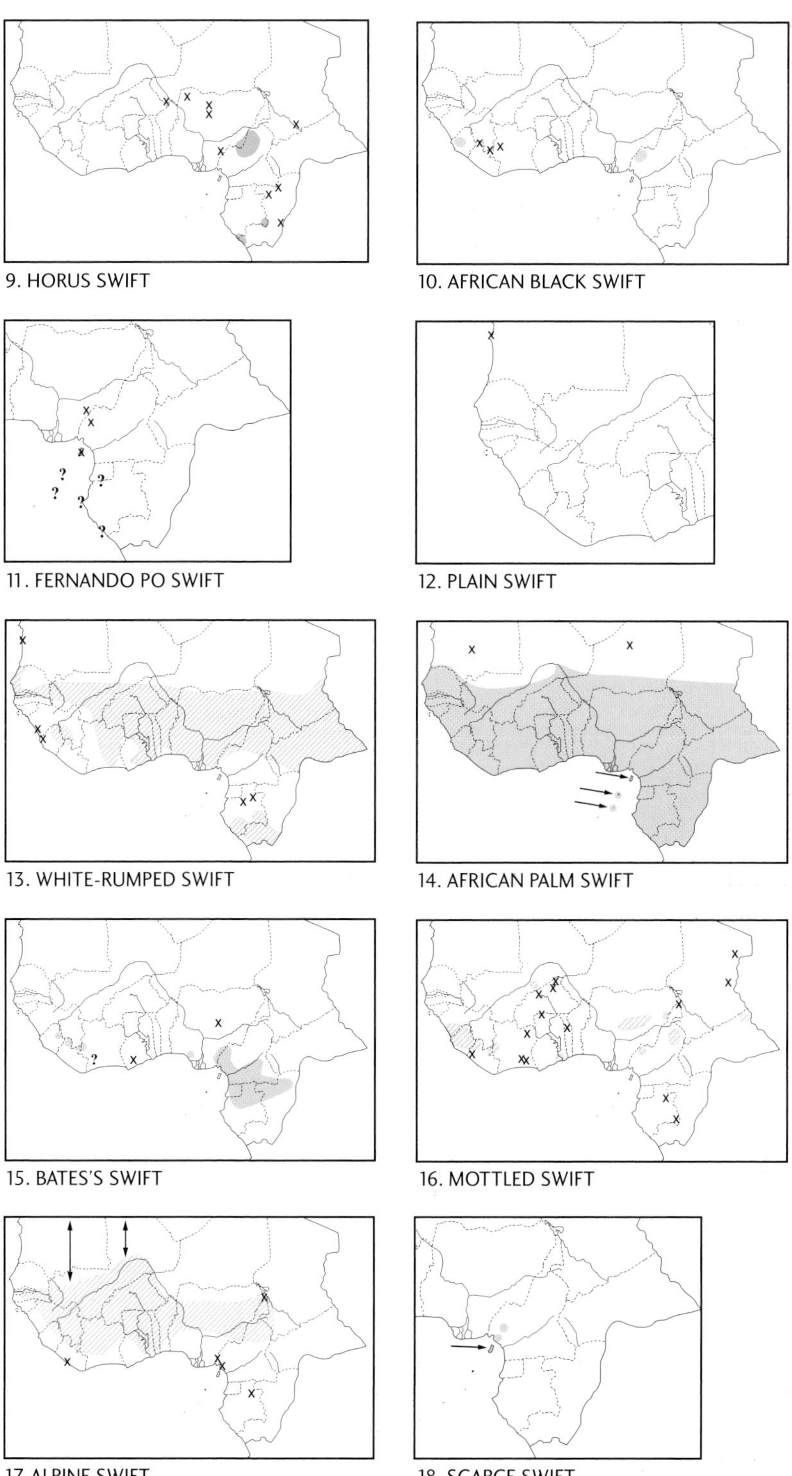

9. HORUS SWIFT

10. AFRICAN BLACK SWIFT

11. FERNANDO PO SWIFT

12. PLAIN SWIFT

13. WHITE-RUMPED SWIFT

14. AFRICAN PALM SWIFT

15. BATES'S SWIFT

16. MOTTLED SWIFT

17. ALPINE SWIFT

18. SCARCE SWIFT

Mousebirds. Buffish-brown or greyish with long, stiff, graduated tails and short, erectile crests. Always in small parties. Flight on short, whirring wings, fast and direct with long glides. Climb well. Feed mainly on fruit and leaves. Endemic to Africa.

1 BLUE-NAPED MOUSEBIRD *Urocolius macrourus* 33–38 cm R c/u
1a Ad Very long, slender tail; blue nape patch; red on bill and around eye. **1b Juv** No blue on nape; shorter crest; dull-coloured bare parts. ▲ Arid open country, thorn scrub. ✤ Far-carrying, long, clear whistle *pwheeeeeee*; also shorter notes including *kwee*, *pwee-u* and *kruw*. [CD7:71]

2 SPECKLED MOUSEBIRD *Colius striatus* 30–36 cm R c/lf
2a Ad Very long tail; bushy crest. **2b Juv** Shorter tail and crest; pale stripe on back. ▲ Various open and wooded habitats. In small, noisy parties, actively foraging by clambering mouse-like through bushes, often hanging vertically on branches. ✤ Sibilant twittering calls *tsiu*, *whseet-whut*, *tsi-ui* etc, constantly uttered at rest and in flight. Alarm a buzzing *tzik-tzik*. [CD7:72]

3 RED-BACKED MOUSEBIRD *Colius castanotus* c. 33 cm ?
Chestnut lower back and rump; no bare whitish spot behind eye. ▲ No certain records; claim from S Gabon not accepted. Locally common in coastal strip, Angola.

Trogons. Arboreal forest species with brightly coloured plumage and long, broad tails. Feed on insects, which they catch on vegetation during swift, swooping flight, reminiscent of drongos. Nest in tree-holes. Easily overlooked despite their brilliant coloration as they perch motionless for long periods at mid-level in forest shade.

4 NARINA'S TROGON *Apaloderma narina* 29–34 cm R s/lc
4a Ad male *brachyurum* (from Nigeria east) Bare patches on head-sides green; outer tail feathers white. **4b Ad female *brachyurum*** Forehead to breast greyish; rest of underparts pink. **4c Ad male *constantia*** (from Nigeria west) Bare patch at gape yellow. **4d Ad female *constantia*** As 4b but lacking fine vermiculations on breast. **4e Juv *constantia*** White and buff spots on wing. ▲ Forest. ✤ A series of paired, moaning, soft and dove-like hoots, with stress on second syllable and repeated up to 14 times, starting hesitantly, then slightly rising in volume and giving impression of being forced out *who-ot-WHO poe-WHO poe-WHO poe-WHO...* [CD7:74]

5 BAR-TAILED TROGON *Apaloderma vittatum* 28 cm R u/lf
5a Ad male Darker than 4 and 6; bare patches at gape and below eye orange-yellow; outer tail feathers closely barred black and white. **5b Ad female** Head and throat dark brownish. **5c Juv** White or buff spots on wing. ▲ Montane forest. ✤ A series of 7–14 high-pitched, yelping *kew* or *kiup* notes, starting softly and increasing in volume. Also a clear, drawn-out whistle, descending in scale *whueeeeuw* (easily imitated). [CD7:73]

6 BARE-CHEEKED TROGON *Apaloderma aequatoriale* 28 cm R u/f
6a Ad male Bare patches on head-sides yellow and more conspicuous than in 4. No overlap with 4c. **6b Ad female** Throat and breast brownish. **6c Juv** White or buff spots on wing. ▲ Lowland forest. ✤ A series of 6–8 plaintive *hoo* notes, with first note longer and stressed, the following being shorter and gradually dying away; reminiscent of Blue-spotted Wood Dove. [CD7:75]

 Maps on page 232

1b
1a
2a
2b
3
4b
4a
4d
5a
5b
5c
4e
4c
6a
6b
6c

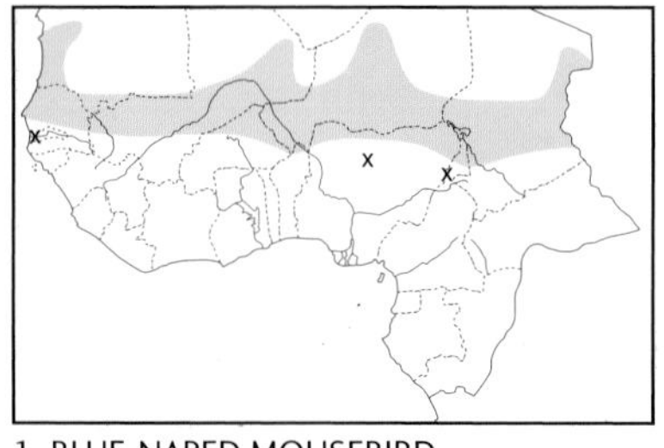

1. BLUE-NAPED MOUSEBIRD

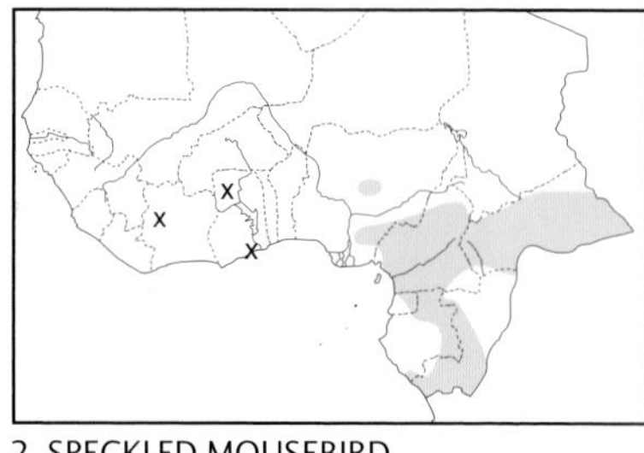

2. SPECKLED MOUSEBIRD

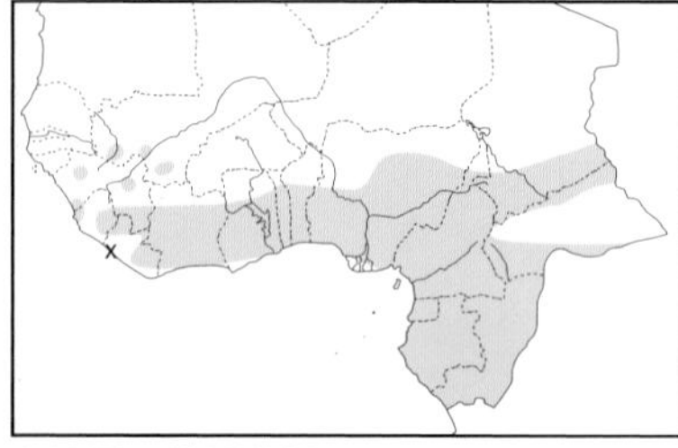

4. NARINA'S TROGON

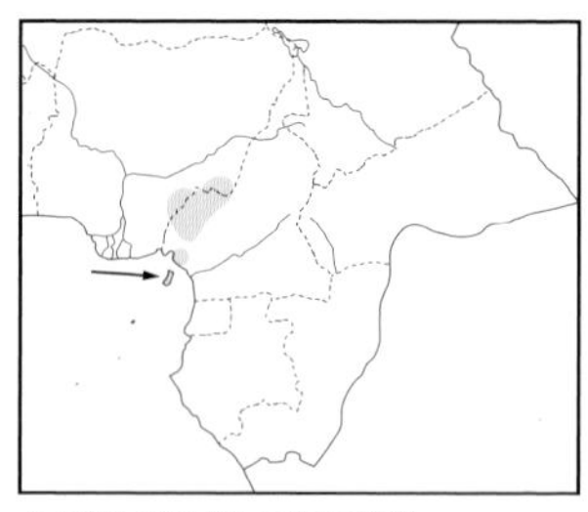

5. BAR-TAILED TROGON

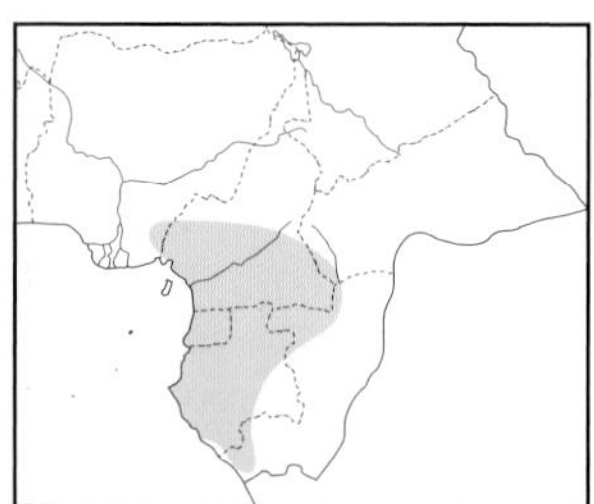

6. BARE-CHEEKED TROGON

Plate on page 230

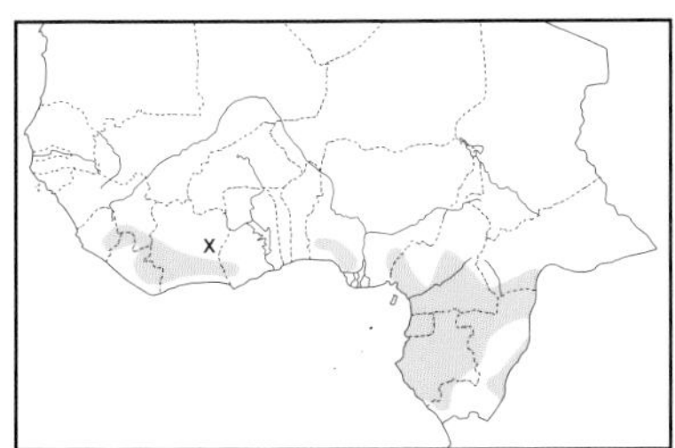

1. AFRICAN DWARF KINGFISHER

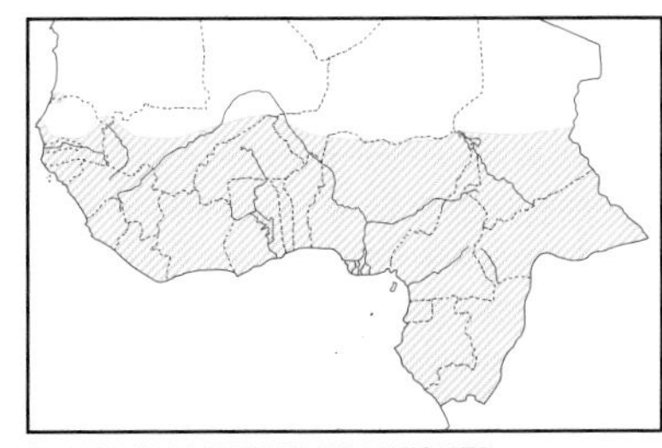

2. AFRICAN PYGMY KINGFISHER

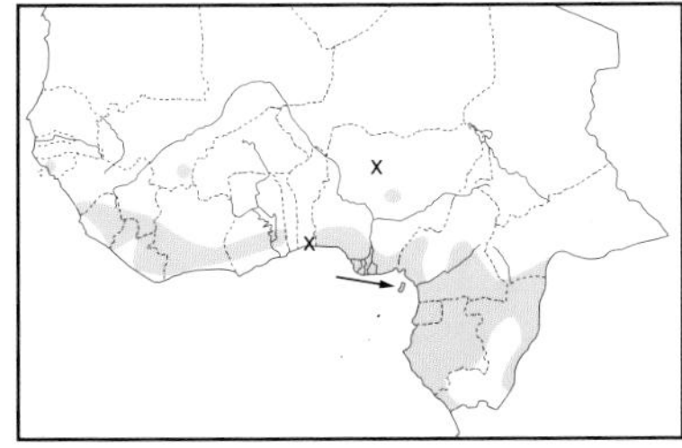

3. WHITE-BELLIED KINGFISHER

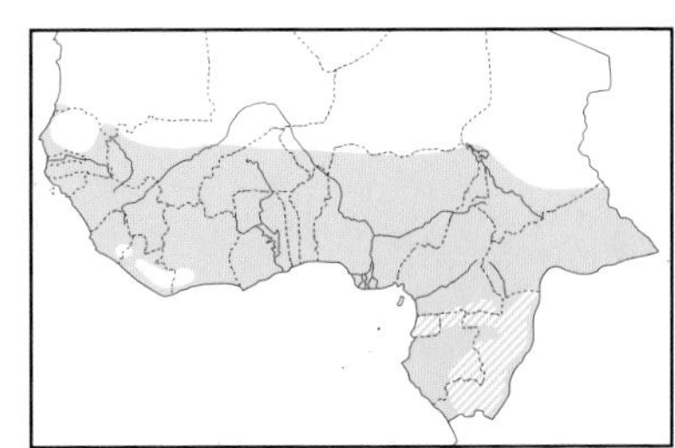

4. MALACHITE KINGFISHER

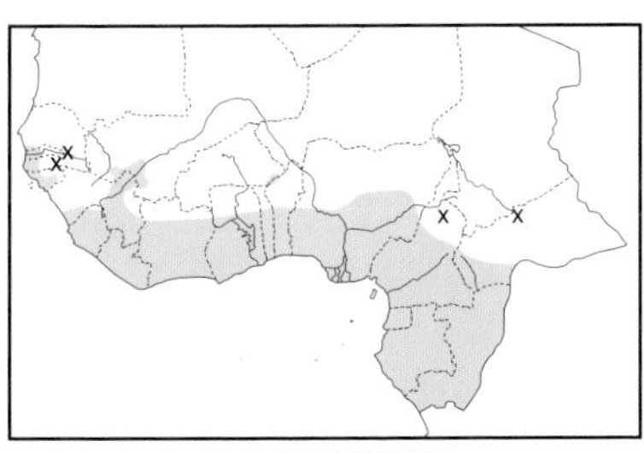

5. SHINING-BLUE KINGFISHER

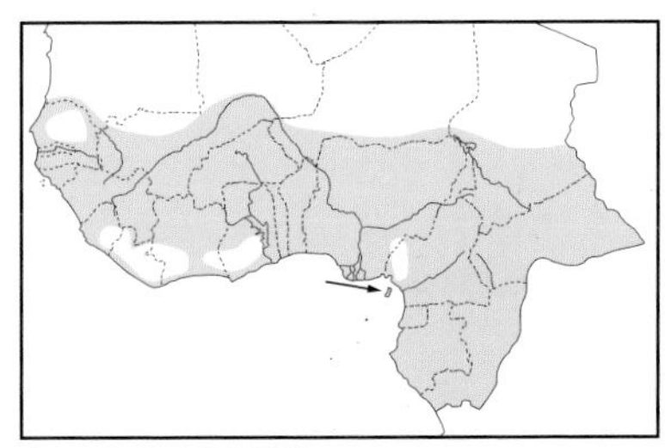

6. PIED KINGFISHER

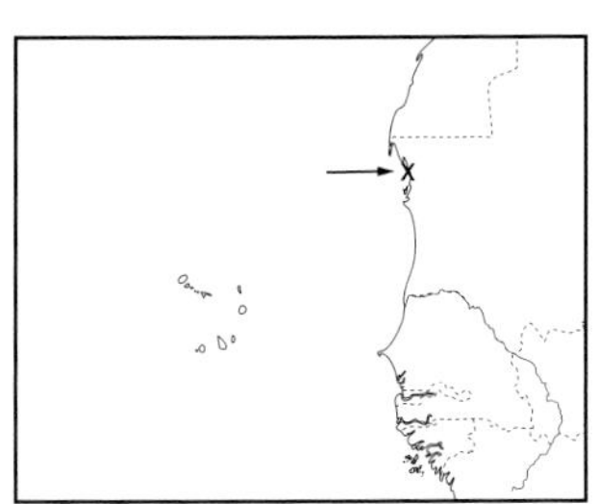

7. COMMON KINGFISHER

Characterised by compact silhouette with large head, long dagger-shaped bill, short body and very short legs. Most species brightly coloured. Occur in a variety of habitats. Aquatic species feed mainly on fish captured by plunge-diving; terrestrial species subsist on insects and reptiles on which they swoop from a perch. Flight fast and direct. Nest in tree or earth holes.

1 AFRICAN DWARF KINGFISHER *Ceyx lecontei* 10 cm **R r/u**
1a Ad Tiny. Rufous crown; black forehead. **1b Juv** Blackish crown and upperparts with blue-tipped feathers; bill blackish tipped white. ▲ Dense rainforest and its edge. Perches at all levels, mainly low. Unobtrusive. Occasionally in mixed-species flocks and attending ant columns. ✤ High-pitched *tseep*. [CD7:82]

2 AFRICAN PYGMY KINGFISHER *Ceyx pictus* 12 cm **R/M c/u**
2a Ad Very small. Blue-and-black crown; rufous supercilium; violet wash to ear-coverts. **2b Juv** Duller; mottled upperparts; bill black tipped white. ▲ Various habitats; not usually near water. ✤ Single sharp, high-pitched *tseet*, in flight. [CD7:83]

3 WHITE-BELLIED KINGFISHER *Alcedo leucogaster* 13 cm **R u/s**
3a Ad *leucogaster* (Nigeria to Gabon–CAR; Bioko) Underparts mainly white; head-sides, breast and flanks rufous-chestnut. *A. l. leopoldi* has blue superciliary area; intergrades with 3a in Congo R. basin. **3b Juv *leucogaster*** Duller; black, white-tipped bill. **3c Ad *bowdleri*** (Guinea-Bissau–Togo; Benin, this race?) Rufous on lores more extensive than in 3a. ▲ Small forest streams. Shy. Usually seen flying low and fast over water or perching low. ✤ High-pitched, vigorous *(t)seee*, in flight; reminiscent of 2 but louder, more piercing.

PRÍNCIPE KINGFISHER *Alcedo (leucogaster) nais* See Plate 147:2.

4 MALACHITE KINGFISHER *Alcedo cristata* 13 cm **R/M c**
4a Ad Brilliant violet-blue above; blue-and-black crest. **4b Juv** Duller; bill black. ▲ Lakes, pools, marshes, lagoons, slow-running rivers. Not shy. ✤ Shrill *pseek*, in flight. [CD7:84]

SÃO TOMÉ KINGFISHER *Alcedo (cristata) thomensis* See Plate 147:1.

5 SHINING-BLUE KINGFISHER *Alcedo quadribrachys* 16 cm **R s/c**
5a Ad male *quadribrachys* (Senegambia–WC Nigeria) Dark blue upperparts; all-black bill; mainly rufous underparts. **5b Ad female *quadribrachys*** Some dark red on base of lower mandible. **5c Juv *quadribrachys*** Bill tipped whitish; underparts pale and mottled. **5d Ad male *guentheri*** (SW Nigeria to CAR–Congo) Mantle to uppertail-coverts paler, more contrasting with rest of upperparts. ▲ Forested aquatic habitats. Easily overlooked; mostly seen in low, fast flight over water. ✤ Shrill, high-pitched *tseep*, in flight. [CD7:86]

6 PIED KINGFISHER *Ceryle rudis* 25–27 cm **R c**
6a Ad male Black and white; double breast-band. **6b Ad female** Single incomplete breast-band. **6c Ad male hovering**. ▲ Aquatic habitats. In pairs or small parties. Feeds almost exclusively on fish, caught by plunge-diving from low perch or hovering flight. ✤ Vocal. Calls include high-pitched, chattering *kwik-kwik* uttered in flight or on perch, *chikrr-chekrr...* (threat call) and *kittle-te-ker* (when flying from perch). [CD7:88]

7 COMMON KINGFISHER *Alcedo atthis* 15–16 cm **V?**
Brilliant blue above; broad green-blue malar stripe. ▲ Aquatic habitats. Palearctic vagrant claimed from N Mauritania.

1a
1b
2a
2b
3c
3b
3a
4b
4a
5c
5b
5a
5d
6c
6b
6a
7

1 **CHOCOLATE-BACKED KINGFISHER** *Halcyon badia* 21 cm **R u/lf**
1a Ad Dark brown upperparts; clean white underparts; azure rump and tail. **1b Juv** Black bill, tipped whitish. ▲ Rainforest. Perches quietly at mid-height. Not associated with water. ✤ Song starts with single, weak *wheet* followed after short pause by a series of 12–17 fluty *hu* notes, initially slightly increasing in volume and speed, then slowing down, becoming more plaintive and trailing off *wheet, huhuhuhuHuHuHUhuhu-hu-hu-hu...* [CD7:76]

2 **BROWN-HOODED KINGFISHER** *Halcyon albiventris* 22 cm **R f**
2a Ad male Streaked brownish head; lightly streaked, dirty buff underparts; heavy bill. **2b Ad female** Back browner. **2c Juv** Duller; bill dusky, tipped whitish. ▲ Woodland. Perches quietly for long periods. ✤ A series of 3–5 weak, descending notes *kweep eep eep eep*. Alarm a loud, harsh chatter *chrrrit-chrrit-chrrit-...* Displaying birds utter loud, excited *kik-kik-kik-kik-...* [CD7:77]

3 **WOODLAND KINGFISHER** *Halcyon senegalensis* 22 cm **M/R c**
3a Ad *senegalensis* (Senegambia–Chad and CAR; savanna zone, migratory) Bright azure-blue upperparts; black wing-coverts; grey head and mantle; greyish-white underparts. **3b Ad *fuscopileus*** (Sierra Leone–Congo and S CAR, also Bioko; forest zone, sedentary) Darker crown; greyer breast and mantle. **3c Juv** Duller; bill dusky. ▲ Woodlands, forest clearings, farmland, mangroves, gardens. Perches on high branch or other, often exposed, vantage point. Conspicuous, attracting attention by loud song. ✤ Distinctive and easily learnt song vigorous and explosive, consisting of single sharp initial note, followed after short pause by hard descending trill *PTIK TIRRRRrrrrrr*; uttered throughout the day. Alarm a fast *kee-kee-kee-kee-...* [CD7:80]

4 **GREY-HEADED KINGFISHER** *Halcyon leucocephala* 22 cm **M/R* f/c***
4a Ad Grey head; bright chestnut belly. **4b Juv** Dusky bill; buff or pale chestnut belly. ▲ Woodland, cultivation. Perches quietly for long periods. Displaying pairs circle above trees, calling. ✤ Call a loud, scolding and explosive *CHeK!* or *KHE!*, frequently uttered in short series. Song a weak descending trill *chichichichi-chi-chiu*, usually given in flight. [CD7:78]

5 **BLUE-BREASTED KINGFISHER** *Halcyon malimbica* 25 cm **R u/c**
5a Ad Black back, scapulars and wing-coverts; black streak through eye. Compare 3. **5b Juv** Head and underparts tinged buff. ▲ Forest, densely wooded savanna. Mainly keeps in deep shade below canopy. ✤ Song starts with single, abrupt *chiup!*, followed after short pause by a series of piping *pu* notes, initially accelerating and ascending slightly, then descending and slowing down: *chiup! pu-pupupuPUPUUpuu-puu puu puu*; uttered throughout the day, mostly from perch, but also in display flight. Alarm a raucous *chup, chup-chup-chup* or *KIAH, KIAH-KIAH.* [CD7:79]
Príncipe form ***dryas***: see Plate 146:9.

6 **STRIPED KINGFISHER** *Halcyon chelicuti* 17 cm **R f**
Rather drab coloured. Streaked crown and underparts; black streak through eye. ▲ Dry savanna woodland and bush. Perches unobtrusively on branch at mid-height. ✤ Song a series of loud, far-carrying *KEE-RRRUU KEE-RRRUU KEE-RRRUU...* [CD7:81]

7 **GIANT KINGFISHER** *Megaceryle maxima* 42–46 cm **R f**
Largest kingfisher. Massive, black bill; white-speckled, slate-grey upperparts. Unmistakable, but beware of confusing flying bird with Green-backed Heron (Plate 8:4), which has similar size, upperpart coloration and call. **7a Ad male** Chestnut breast. **7b Ad female** Chestnut belly. ▲ Wooded rivers, lagoons and lakes, mangroves; also rocky and sandy seashores. ✤ Loud *KEK!* uttered singly or in series. [CD7:87]

Maps on page 238

1b
1a
2c
2b
2a
3b
3c
3a
4b
4a
5b
5a
7b
6
7a

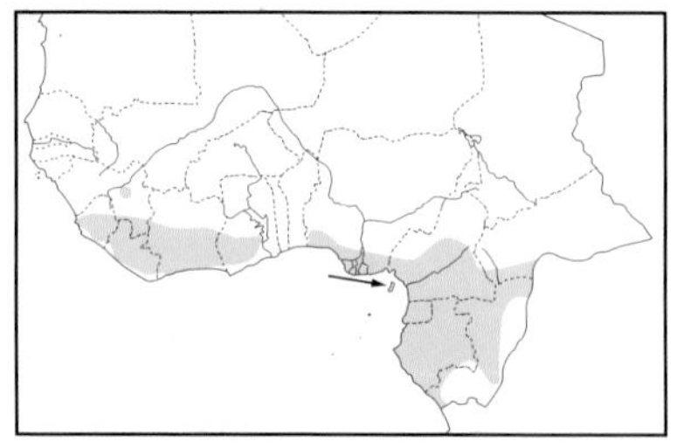

1. CHOCOLATE-BACKED KINGFISHER

2. BROWN-HOODED KINGFISHER

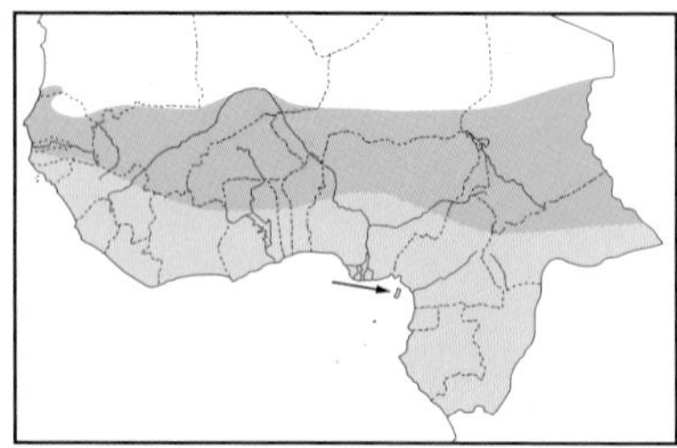

3. WOODLAND KINGFISHER

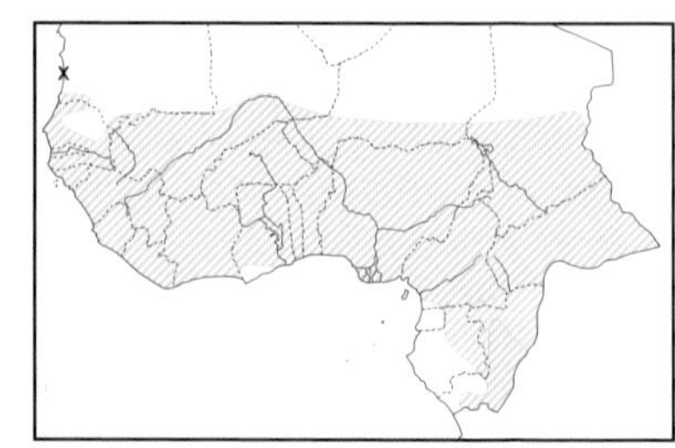

4. GREY-HEADED KINGFISHER

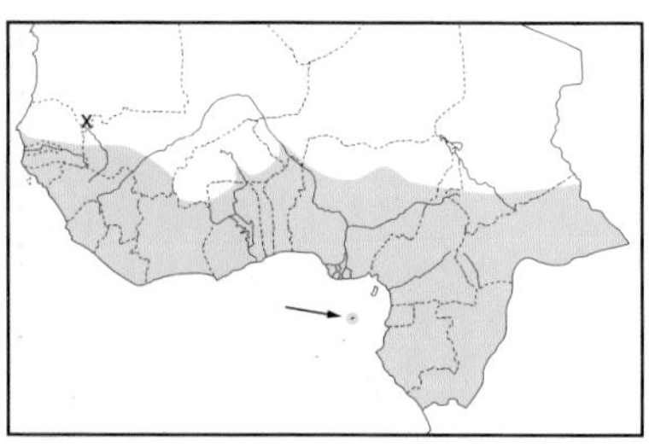

5. BLUE-BREASTED KINGFISHER

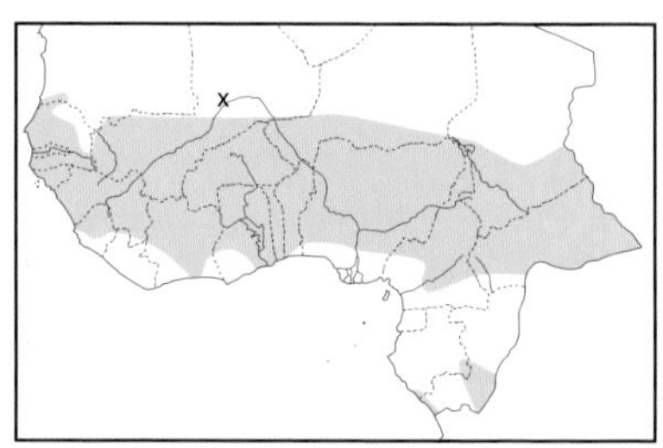

6. STRIPED KINGFISHER

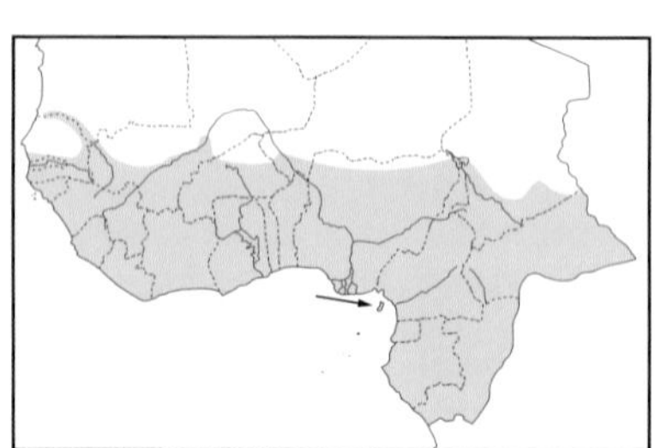

7. GIANT KINGFISHER

Plate on page 236

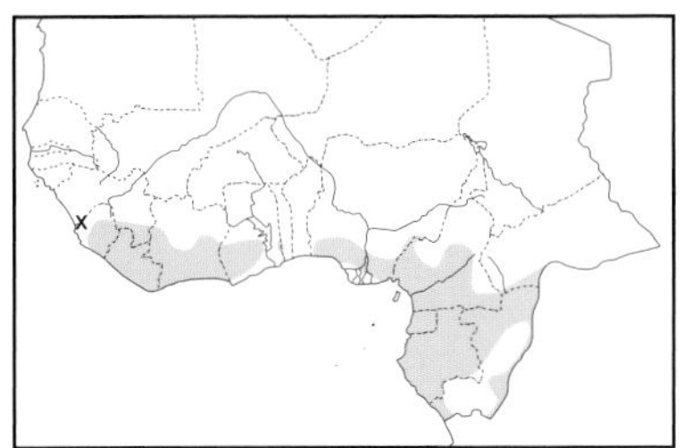
1. BLACK BEE-EATER

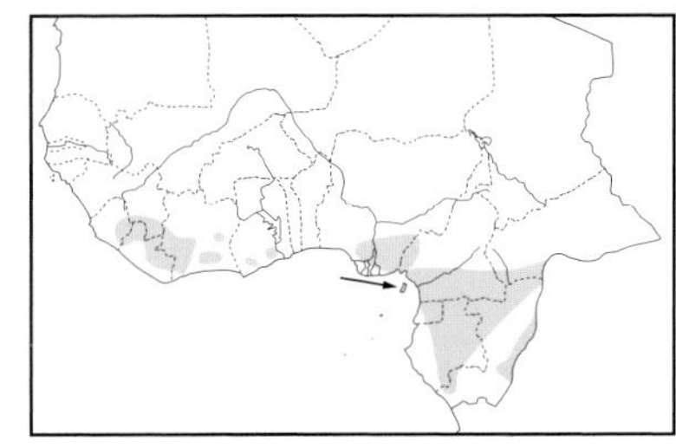
2. BLUE-HEADED BEE-EATER

3. BLACK-HEADED BEE-EATER

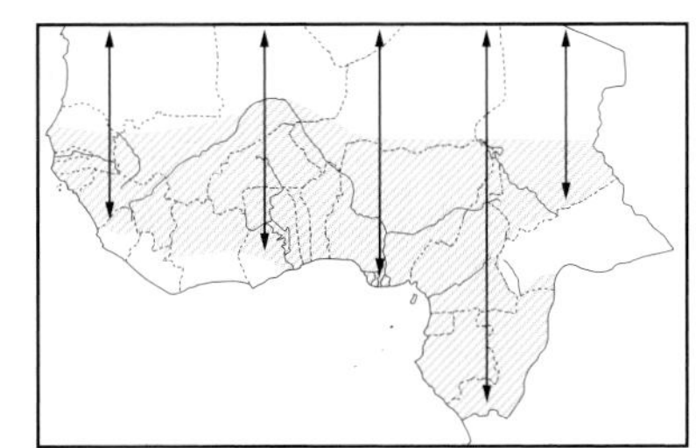
4. EUROPEAN BEE-EATER

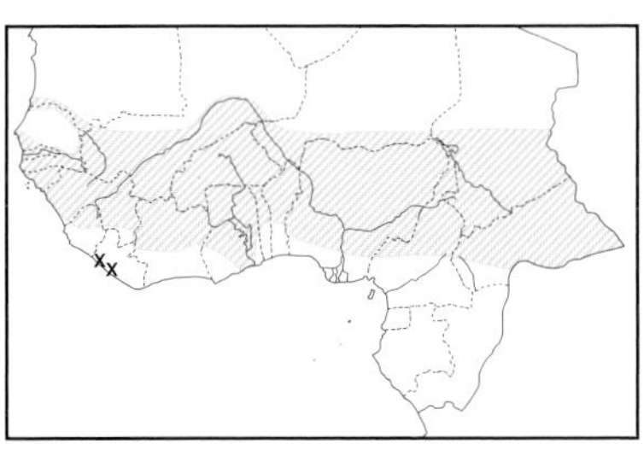
5. NORTHERN CARMINE BEE-EATER

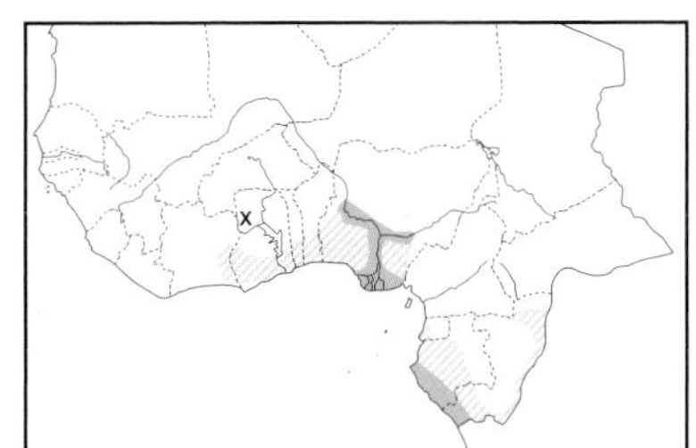
6. ROSY BEE-EATER

Slender, mostly brightly coloured species with long, pointed, slightly decurved bills and triangular wings. Feed mainly on wasps and bees, either caught in short sallies from a perch or in lengthy hawking flights. Some migratory and gregarious, others sedentary and occurring in pairs or family parties. Nest in burrows excavated in banks or level ground.

1 BLACK BEE-EATER *Merops gularis* 20 cm R s/lc
Black upperparts; azure-blue rump; scarlet throat; blue-streaked breast and belly. **1a Ad *gularis*** (Sierra Leone–Nigeria) Blue supercilium. **1b Juv *gularis*** Duskier than adult; no red throat. **1c Ad *australis*** (Cameroon to CAR–Congo) No supercilium. **1d In flight** Black with conspicuous azure-blue rump. ▲ Clearings, forest edges. Perches conspicuously and usually quite high on dead branches. ✤ Rather loud and distinctive *klip* or *wik*, not infrequently uttered. [CD7:91]

2 BLUE-HEADED BEE-EATER *Merops muelleri* 19 cm R s/r
Deep purple-blue; dark chestnut back and wings; small red throat patch. **2a Ad *mentalis*** (west of Douala, Cameroon) Short, blunt tail streamers. **2b Juv *mentalis*** Duskier than adult; no streamers. **2c Ad *muelleri*** (south and east of Douala, Cameroon) Whitish forehead; no streamers. **2d In flight** Appears mainly dark. ▲ Forest. Perches on lianas and thin bare branches, usually in mid-strata, but occasionally very low. Wags tail in short arc. ✤ Contact call a discreet *slip* or *sip*, frequently uttered. Infrequently a high-pitched *kee-klip* or *ptii-wit*. Also bill-snapping and soft, muffled, hoarse little sounds. [CD7:90]

3 BLACK-HEADED BEE-EATER *Merops breweri* 25–28 cm R s/r
3a Ad Large. Black head; green upperparts; cinnamon breast and belly; long central tail feathers. **3b Juv** Duskier than adult; no streamers. **3c In flight** Green, with black head; tail has some cinnamon; wingtips rather broad. ▲ Forest edge, open woodland. Rather inconspicuous. ✤ Mainly silent. [CD7:89]

4 EUROPEAN BEE-EATER *Merops apiaster* 23–25 cm; WS 36–40 cm P/V* lc/s
4a Ad male Chestnut crown and mantle; golden scapulars; yellow throat; bluish underparts; tail streamers. **4b Ad female** As male, but less chestnut on wing; scapulars suffused with green. **4c In flight** Chestnut mantle and coverts; golden scapulars. ▲ Various open habitats; avoids forest zone. Gregarious. Flight graceful, wheeling; rapid wingbeats alternating with gliding. ✤ Melodious, liquid, remarkably far-carrying *prruuip*, *pru-ik* or *kwirip* (calls of Blue-cheeked and Rosy Bee-eaters very similar and only separable with practice). Also *pik-pik*. [CD3:30]

5 NORTHERN CARMINE BEE-EATER *Merops nubicus* 24–27 cm M f/lc
5a Ad Carmine-pink; dark bluish head; very long tail streamers. **5b Juv** Duskier than adult; short tail streamers. **5c In flight** Blue head, rump and undertail-coverts contrast with striking carmine plumage. ▲ Dry and open savanna. Gregarious. Forages in 'sailing', twisting flight. Also feeds from perch. Partial to grasshoppers and locusts and therefore attracted to bush fires. ✤ Rather unmusical *klienk*, *klunk* or *terk*, often repeated and sometimes followed by *ki-ki-ki-ki-*... [CD8:3a]

6 ROSY BEE-EATER *Merops malimbicus* 22–25 cm M lc/r
6a Ad Grey above; reddish-pink below; white streak below black mask; tail streamers. **6b Juv** Duskier; no streamers. **6c In flight** Contrasting under- and upperparts; streamers. ▲ Savanna woodland; also over forest; often near water. Gregarious. Very local breeder Nigeria, S Gabon and S Congo, nesting on sandbars in large rivers, often in huge colonies; disperses as far west as E Ivory Coast. ✤ Rather hoarse *pru* or *krrp*, shorter, more abrupt, less melodious than call of 4. Alarm *wik*. [CD8:2b]

 Maps on page 239

2d
4c
5c
1d
3c
6c
1c
1b
1a
2c
2b
2a
4b
3b
3a
4a
5b
5a
6b
6a

PLATE 69: BEE-EATERS II

1 BLUE-BREASTED BEE-EATER *Merops variegatus* 17 cm R lu
1a Ad *variegatus* (S Cameroon to Congo–SW CAR) Yellow throat; purple-blue gorget; white spot on neck-sides. Compare 2. **1b Juv *variegatus*** No gorget; pale greenish breast. **1c Ad *loringi*** (highlands of E Nigeria–W Cameroon) Blue supercilium. **1d In flight** Cinnamon fight feathers tipped black; primaries tinged green. ▲ Various open habitats. Behaviour as 2. ✤ Quiet *prru* or *tup*, uttered singly or in series, *prru-tup prru-tup-tup ptup prru*... [CD7:93]

2 LITTLE BEE-EATER *Merops pusillus* 15–17 cm R c/s
2a Ad *pusillus* (S Mauritania–Liberia, east to Chad and CAR) Small. Yellow throat; black gorget; nearly square tail. Compare 1. **2b Juv *pusillus*** No black gorget; pale greenish breast. **2c Ad *meridionalis*** (S Gabon–Congo) Short, narrow blue supercilium. **2d In flight** Cinnamon flight feathers; black trailing edge to wing. ▲ Various open habitats. Perches on long grass stem or low shrub, occasionally higher. ✤ Quiet *slip* or *sip* and hard, clipped *tsip* or *klip*, uttered singly or in series. Also high-pitched, sibilant *siddle-iddle-ip*, *d'jee* in high-intensity greeting display. [CD7:92]

3 RED-THROATED BEE-EATER *Merops bulocki* 20–22 cm R lc
3a Ad Bright red throat; deep blue undertail-coverts. **3b Ad** Rare yellow-throated variant. **3c Juv** Duller. **3d In flight** Black trailing edge to wing; tail green with some buff, square. ▲ Savanna woodland. Gregarious. Breeds colonially in sand banks. ✤ Vocal. Calls include *wip*, querulous *kirrup*, *kwirrup* or *krrip*, delivered in trilling series in greeting, and nasal *kweep*, uttered in flight and from perch. [CD7:96]

4 WHITE-FRONTED BEE-EATER *Merops bullockoides* 22–24 cm R lc
4a Ad White forehead; white line below black mask; red throat. Juvenile duller. **4b In flight** Black trailing edge to wing; deep blue vent; square bluish-green tail. ▲ Wooded grassland. Coastal strip, Gabon–Congo. Breeds colonially. ✤ Vocal. Call a distinctive nasal *wèèh* or *wèèh-up*, very different from other bee-eaters. Alarm a sharp *waark*. [CD7:97]

5 SWALLOW-TAILED BEE-EATER *Merops hirundineus* 20–22 cm R/M s/lf
5a Ad Strongly forked, pale blue tail; yellow throat; blue gorget. **5b Juv** Throat greenish-white; no gorget. **5c In flight** Mainly green with forked, white-tipped tail. ▲ Savanna woodland. Rather silent and unobtrusive. ✤ Subdued *weeerp-weeerp*, *tip-tip*, or *diddle-diddle-ip*, typically bee-eater-like in quality. [CD7:95]

6 WHITE-THROATED BEE-EATER *Merops albicollis* 19–21 cm M c/f
6a Ad Very long tail streamers (up to 12 cm); distinctive black-and-white head pattern. **6b Juv** Duller; no streamers; throat tinged yellowish. **6c In flight** Long streamers conspicuous; wings green and pale rufous with black trailing edge. ▲ From Sahel (during rains) to forest zone (dry season). Seasonally common/frequent. Gregarious. ✤ Vocal. Pleasing, melodious and far-carrying *pruuee*, *trooee* or *pruik*, reminiscent of European Bee-eater but distinctly higher pitched and delivered in longer series. [CD7:98]

7 BLUE-CHEEKED BEE-EATER *Merops persicus* 24–28 cm; WS 35–39 cm M/P/V* lf/r
7a Ad Large. Bright green; very long tail streamers (up to 11 cm). **7b Juv** Duller; no streamers. **7c In flight** All green above; underwing rufous with narrow dusky trailing edge. ▲ Very local breeder at desert edge. Winters in bushy grassland and open farmland. Gregarious. ✤ Far-carrying *prri-ip*, *dirrip* or *pririk*, very similar to European Bee-eater but harder, less liquid and shorter. [CD3:29]

8 LITTLE GREEN BEE-EATER *Merops orientalis* 16–18 cm R c/u
8a Ad Small. All green; very long tail streamers; narrow black gorget. **8b Juv** Duller; no streamers; no gorget. **8c In flight** Note small size and long tail streamers. ▲ Subdesert and woodland. Perches low. ✤ Hard, rapid and slightly buzzing trill *trrri-trrri-trrri* or *kree-kree-kree-*... Alarm a staccato *ti-ti-ti* or *ti-ik*. [CD8:1a]

Maps on page 244

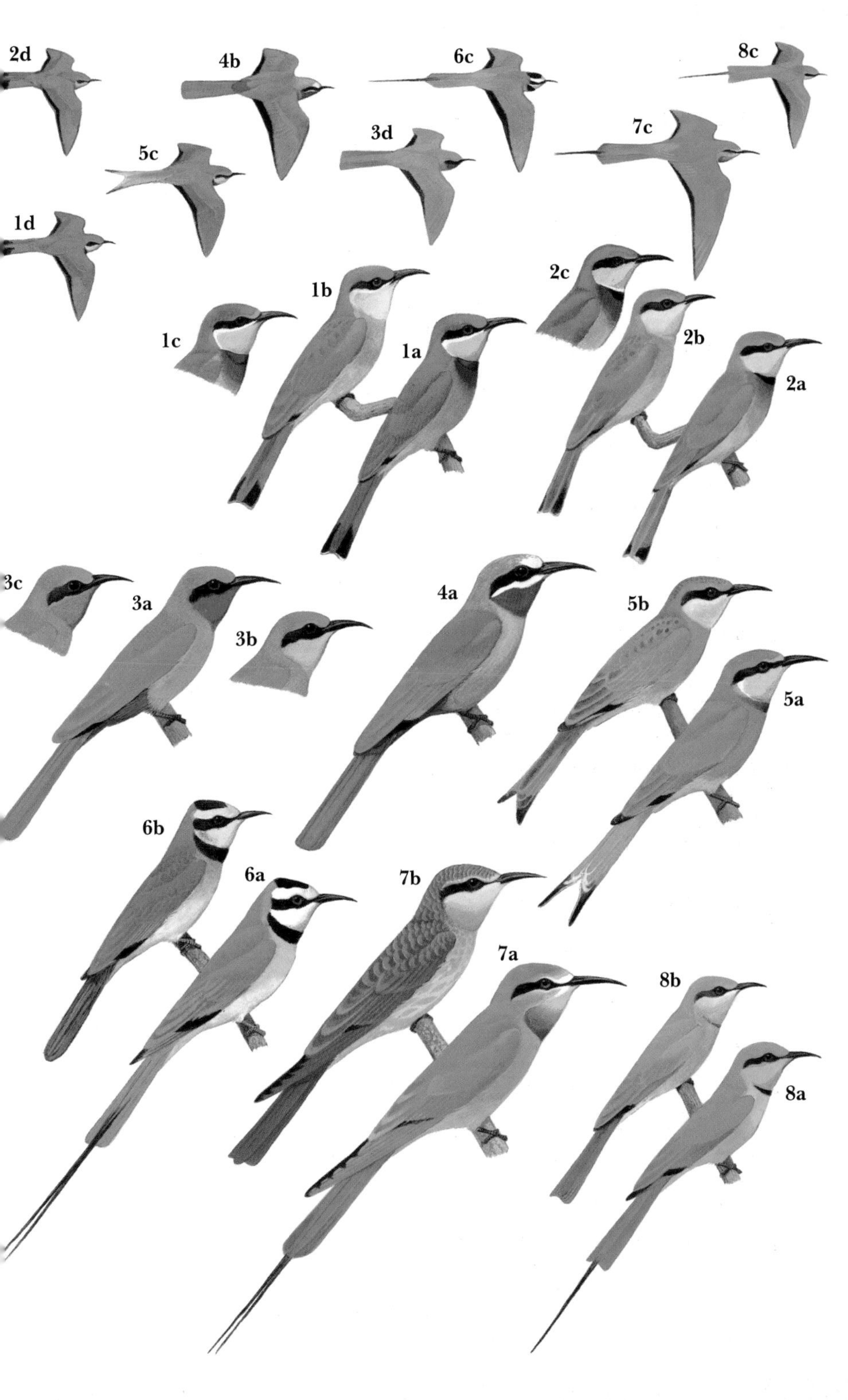

2d
4b
6c
8c
5c
3d
7c
1d
2c
1b
1c
1a
2b
2a
3c
3a
4a
5b
3b
5a
6b
6a
7b
7a
8b
8a

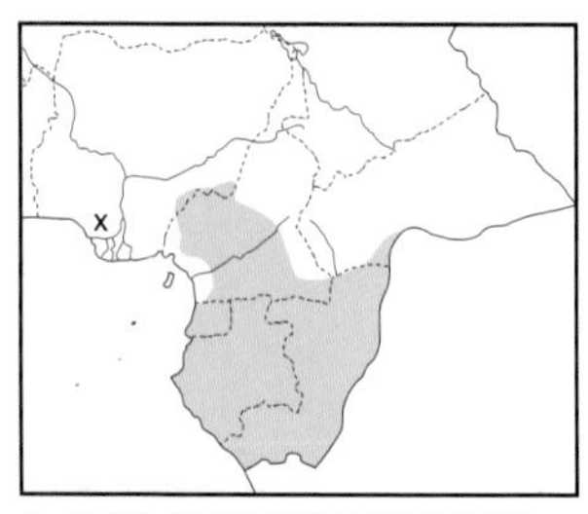

1. BLUE-BREASTED BEE-EATER

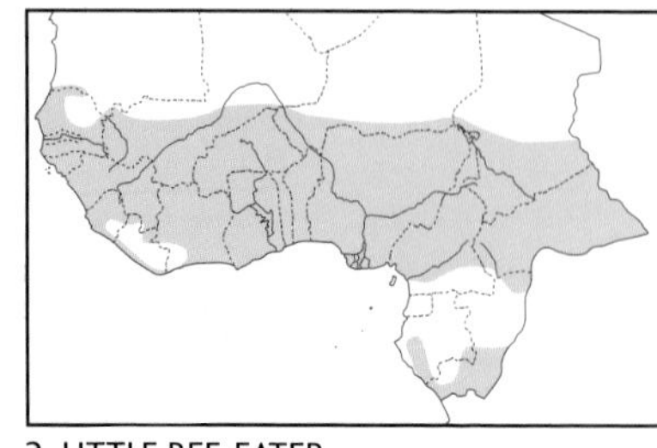

2. LITTLE BEE-EATER

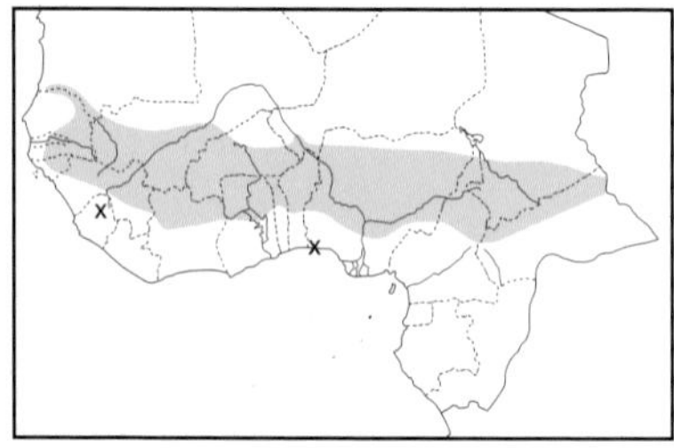

3. RED-THROATED BEE-EATER

4. WHITE-FRONTED BEE-EATER

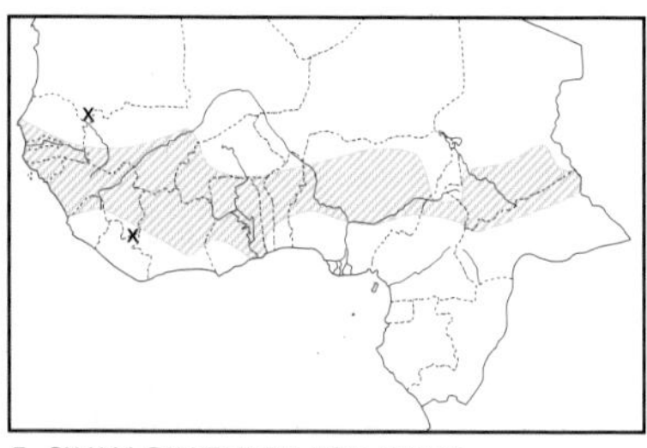

5. SWALLOW-TAILED BEE-EATER

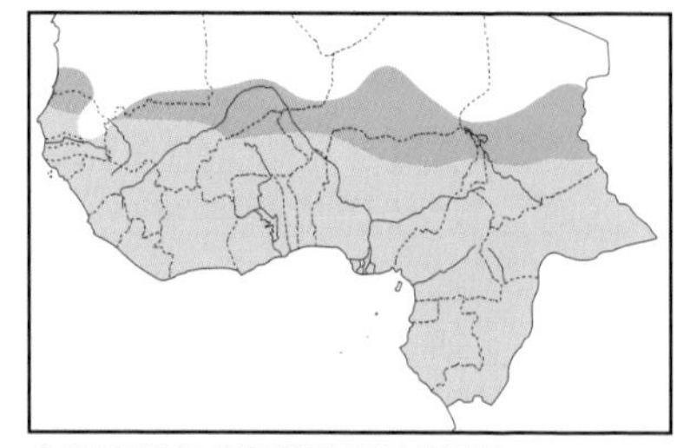

6. WHITE-THROATED BEE-EATER

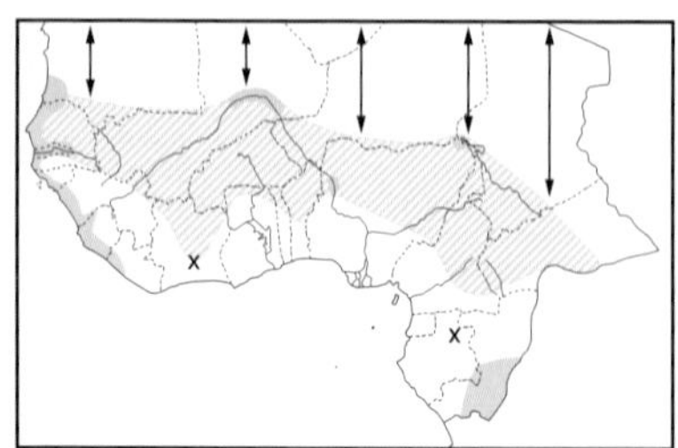

7. BLUE-CHEEKED BEE-EATER

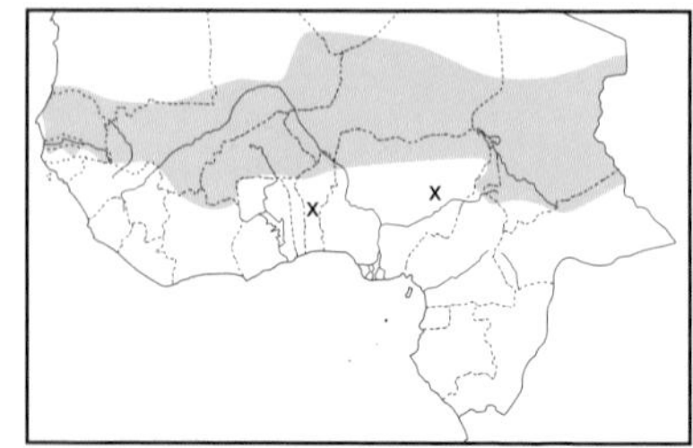

8. LITTLE GREEN BEE-EATER

Plate on page 242

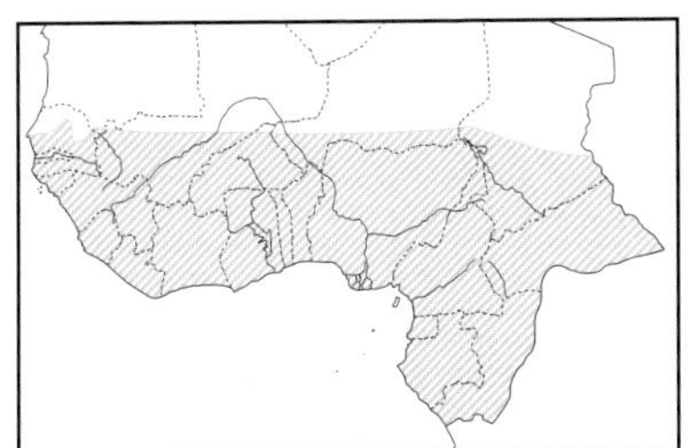

1. BROAD-BILLED ROLLER

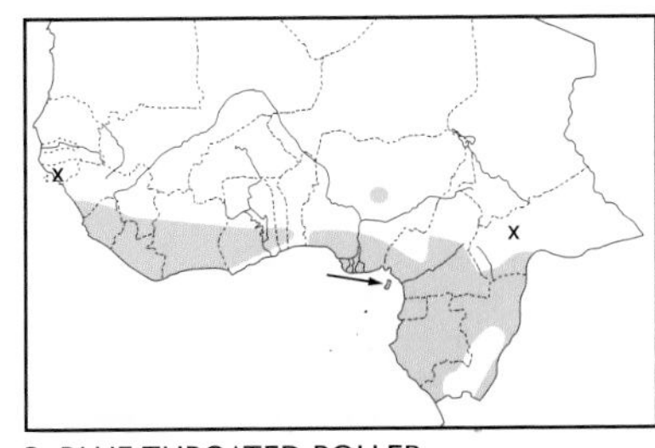

2. BLUE-THROATED ROLLER

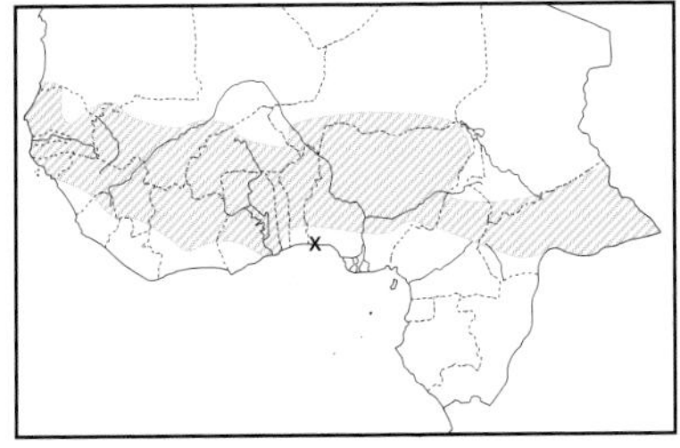

3. RUFOUS-CROWNED ROLLER

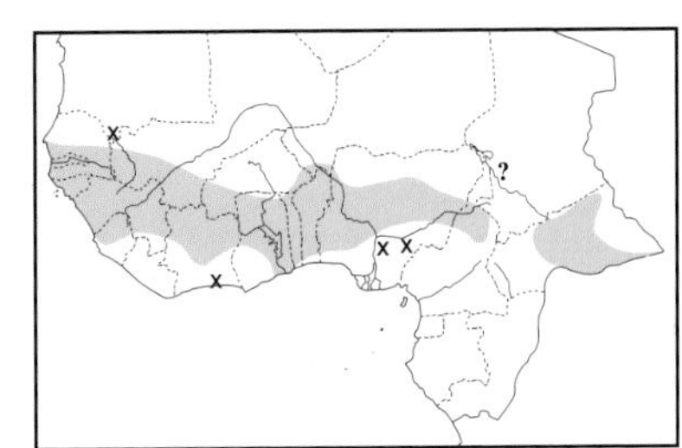

4. BLUE-BELLIED ROLLER

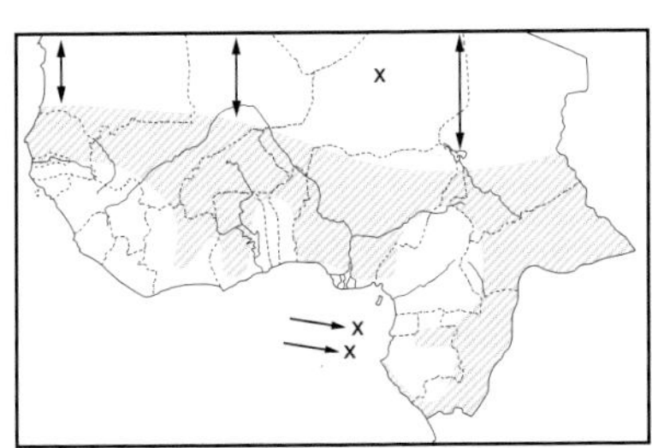

5. EUROPEAN ROLLER

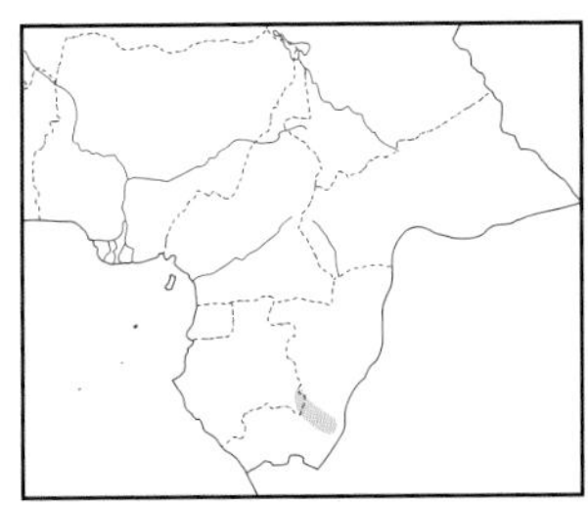

6. LILAC-BREASTED ROLLER

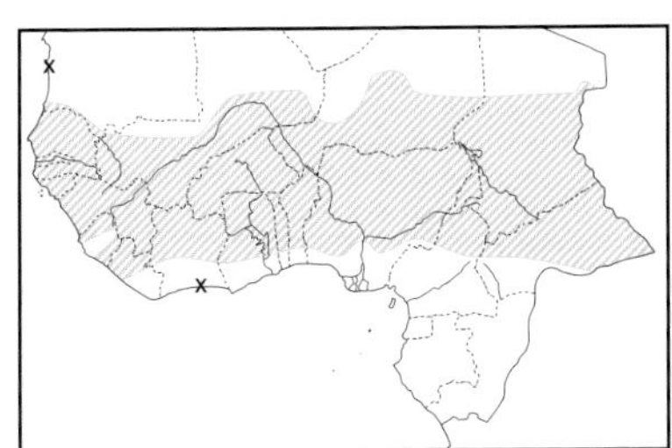

7. ABYSSINIAN ROLLER

Robust, conspicuous birds with colourful plumage, large heads and stout bills, slightly hooked at tip. Hunt from perch, swooping to ground to catch large insects or small vertebrates (*Coracias*), or pursuing insects in the air (*Eurystomus*). Vocal and pugnacious. Voice harsh.

1 BROAD-BILLED ROLLER *Eurystomus glaucurus* 29–30 cm M/V+ c

1a Ad Chestnut above; deep lilac below; short yellow bill; mainly azure-blue tail and undertail. Compare 2. **1b Juv** Duskier; underparts mottled brown and bluish; bill partly dusky. **1c Ad in flight** Pointed wings; dark blue greater coverts. ▲ Various wooded habitats, from forest to Sahel. Perches on tall treetops, occasionally hawking insects high in the air. Assembles in wheeling feeding parties in late afternoon and at dusk. Pointed wings combined with powerful and aerobatic flight may recall falcon. ✤ Vocal. Sharp, nasal *kek*, sometimes developing into harsh 'laughing' *kekekekekek-k-k-k-k-r-r-r-r*; also variety of other guttural, grating and growling notes, uttered in series. [CD8:7a]

2 BLUE-THROATED ROLLER *Eurystomus gularis* 25 cm R u/f

2a Ad Dark chestnut; blue throat patch; chestnut undertail. Compare 1. **2b Juv** Duskier; lower breast and belly greyish-blue mottled dusky. **2c Ad in flight** Silhouette as 1; dark chestnut greater coverts. ▲ Forest. Perches conspicuously on bare branch from canopy, occasionally hawking for insects. Gathers in feeding flocks in late afternoon, often with 1. ✤ Various strident notes, e.g. *khlee* and *grwree*, sometimes developing into a series, *khleep-khleep-khleep-*... or *kikikikikik*... [CD8:6b]

3 RUFOUS-CROWNED ROLLER *Coracias naevius* 35–40 cm R/M u/r

3a Ad Large; conspicuous white supercilium; brownish-pink underparts streaked white; square tail. **3b Juv** Duller, more olive. **3c Ad in flight** Purplish fight feathers; lilac upperwing-coverts. ▲ Open woodlands. Usually singly, perching on high vantage point. Less vocal than other rollers. ✤ Various cackling, muffled and nasal notes, uttered singly or in rapid series, less harsh than other rollers and sometimes reminiscent of Green Wood-hoopoe. [CD8:4a]

4 BLUE-BELLIED ROLLER *Coracias cyanogaster* 28–30 cm R/M lu/s

4a Ad Pale buffish head and breast; purplish-blue belly. Juvenile duller, lacking streamers. **4b In flight** Pale blue bar on purplish-blue wings; pale blue tail with long tail streamers. ▲ Wooded savanna. Perches conspicuously atop trees, electricity poles, etc. ✤ Vocal. Fast, sharp *keh-keh-keh-keh-k-r-r-r-r-r-r* and an accelerating, descending, scolding 'laugh' *HEheheheheh*... [CD8:4b]

5 EUROPEAN ROLLER *Coracias garrulus* 29–32 cm; WS 52–57 cm P/V* u/r NT

5a Ad Mainly bright pale blue and chestnut; square tail. Juvenile paler and duller until moult (Nov–Dec). **5b In flight** Blackish flight feathers. ▲ Open wooded habitats. ✤ Mostly silent in W Africa. [CD3:31]

6 LILAC-BREASTED ROLLER *Coracias caudatus* 28–30 cm R f

6a Ad Lilac throat and breast streaked white; long tail streamers (up to 8 cm). Juvenile duller, more buffish, lacking streamers. **6b In flight** Purplish flight feathers. ▲ Wooded grassland. ✤ Loud, harsh rattling notes *wrek, wrek*... or sharp *keh! keh!*..., in display flight developing into a series. [CD8:5b]

7 ABYSSINIAN ROLLER *Coracias abyssinicus* 28–30 cm M c

7a Ad Mainly turquoise and chestnut; very long tail streamers (up to 12 cm). **7b Juv** Duller; no streamers. Compare 5. **7c Ad in flight** Flight feathers, forewing, rump and uppertail-coverts deep violet-blue. ▲ Dry wooded habitats. Noisy and conspicuous, perching on exposed vantage point. ✤ Harsh, explosive, scolding screech *kwrèèèèh* or *kèèèèèhh* uttered from perch; sharp *kek*, mainly in flight. In display flight a rapid series of similar harsh notes. [CD8:5a]

Maps on page 245

1c
1b
1a
2c
2b
2a
3c
3b
3a
4b
4a
5b
5a
6a
6b
7a
7b
7c

Wood-hoopoes. Slender with mainly dark glossy plumage, slender, decurved bills and long graduated tails. Arboreal and agile, often hanging upside-down while probing bark for insects. Larger species conspicuous, noisy and usually gregarious; smaller ones quieter, mostly solitary or in pairs. Nest in tree holes. Endemic to Africa.

1 GREEN WOOD-HOOPOE *Phoeniculus purpureus* 35–40 cm **R u/lc**
1a Ad Large; long graduated tail; red or red-tipped bill; red feet. **1b Juv** Smaller and duller; bill and feet black. **1c Ad in flight** Double white wingbar. ▲ Open wooded habitats. Noisy and gregarious, typically in groups of 3–12 (or more). ✤ Single, loud, cracked *whak* or *kuk* and loud, high-pitched cackling, usually started by one and soon accelerating and developing into resonant 'laughter' when other group members join in. [CD8:8b]

2 WHITE-HEADED WOOD-HOOPOE *Phoeniculus bollei* 30–35 cm **R u**
2a Ad *bollei* (SE Guinea–S Ghana, and S Nigeria–CAR) Buffish-white head; red bill and feet; no white on wings and tail. **2b Juv *bollei*** Smaller; variable: a dusky-headed individual. **2c Juv *bollei*** A pale-headed individual. **2d Ad *okuensis*** (Mt Oku, W Cameroon) Less buffish-white on head. ▲ Lowland, montane and large gallery forest; occasionally woodland. Noisy, but less so than 1. ✤ Fast, rippling, chattering twitter, often by several birds together. Also a frequent, high-pitched *kuk*, uttered singly or in rapid series. [CD8:8a]

3 BLACK WOOD-HOOPOE *Rhinopomastus aterrimus* 23 cm **R u/lf**
3a Ad *aterrimus* (S Mauritania–Sierra Leone, east to Chad and CAR) Violet-black; bill and feet black. **3b Ad *aterrimus* in flight** Double white wingbar. **3c Ad *anchietae*** (S Gabon–SC Congo) No white on primary-coverts; subterminal white spots on outer two tail feathers. ▲ Open wooded habitats. In trees and bushes, occasionally low, even on ground. Joins mixed-species flocks. ✤ A series of plaintive, fairly loud or subdued notes *kwheep-kwheep-kwheep-...* [CD8:9b]

4 FOREST WOOD-HOOPOE *Phoeniculus castaneiceps* 28 cm **R s/u**
4a Ad *castaneiceps* (Liberia–SE Guinea to S Ghana; SW Nigeria) Chestnut head; pale bill; black feet. Juvenile duller. **4b Ad male *brunneiceps*** (S Cameroon to SW CAR–N Congo) Head glossy bottle-green (as illustrated), pale brownish or buffish-white with feathers narrowly fringed darker. **4c Ad female *brunneiceps*** Head whitish (as illustrated) or pale brownish. **4d Juv *brunneiceps*** Duller; buffish-white head; feathers narrowly fringed darker. ▲ Dense rainforest, semi-deciduous forest. Forages high up; occasionally in mixed-species flocks. ✤ A series of up to 20 plaintive notes *kweep-wheep-wheep-wheep-....-wheew* (similar to 3 and recalling raptor). [CD8:7b]

Hoopoes. Very distinctive, with slender decurved bills, broad rounded wings and erectile, fan-shaped crests. Feed on ground. Nest in holes.

5 HOOPOE *Upupa epops* 25–29 cm; WS 44–48 cm **R/M/P/V+ u/c**
Unmistakable. Sandy or cinnamon; black-and-white wings and tail. **5a Ad *epops*** (Palearctic passage migrant and winter visitor; Mauritania–Sierra Leone to Chad; also Gabon. Vagrant, Cape Verde) Palest race. **5b Ad *epops* in flight** Broad white band on primaries and four white bands on secondaries and coverts. *U. e. major* (NE Chad) slightly duller that 5a-b; less white in secondaries; tail band narrower; belly more streaked. **5c Ad *senegalensis*** (resident; Mauritania–Sierra Leone to Chad; also Liberia, where rare) Slightly darker than 5a. **5d Ad *senegalensis* in flight** More white in secondaries than 5b, forming patches. **5e Ad male *africana* in flight** (Gabon–Congo) No white on primaries; body rich cinnamon. *U. e. waibeli* (south of *senegalensis* in Cameroon, S Chad, CAR) intermediate between 5c–d and 5e; primaries as 5c–d. Females slightly duller and smaller; *africana* rather paler, with less white in secondaries (like nominate). Juveniles duller still, with shorter crest and bill. Note: African breeding races (or *africana* only) occasionally treated as separate species (African Hoopoe), but morphological, behavioural and vocal similarities do not suggest specific distinctness. ▲ Wooded savanna, dry scrub. ✤ Low and soft, but far-carrying *hoop-oop-oop* (sometimes 2 or 4 notes). [CD3:32]

Maps on page 250

1a
1c
2a
2d
2b
1b
2c
3c
3a
3b
4a
4c
4d
4b
5b
5c
5e
5d
5a

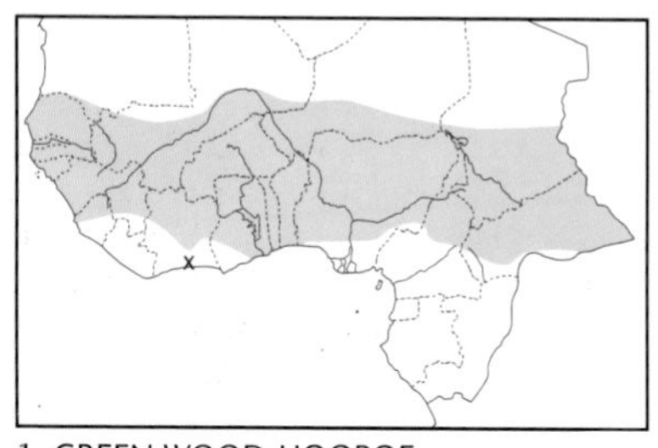

1. GREEN WOOD-HOOPOE

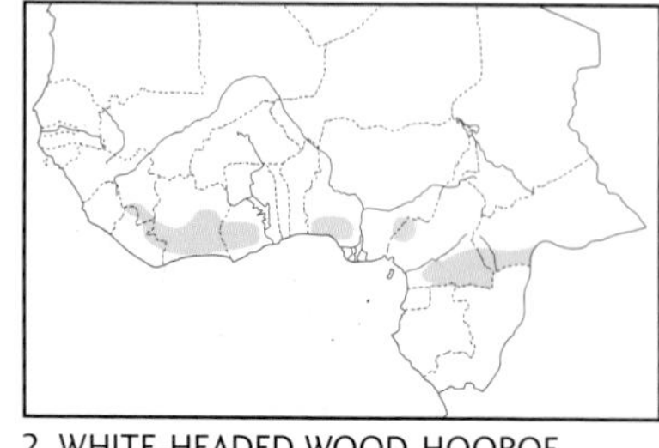

2. WHITE-HEADED WOOD-HOOPOE

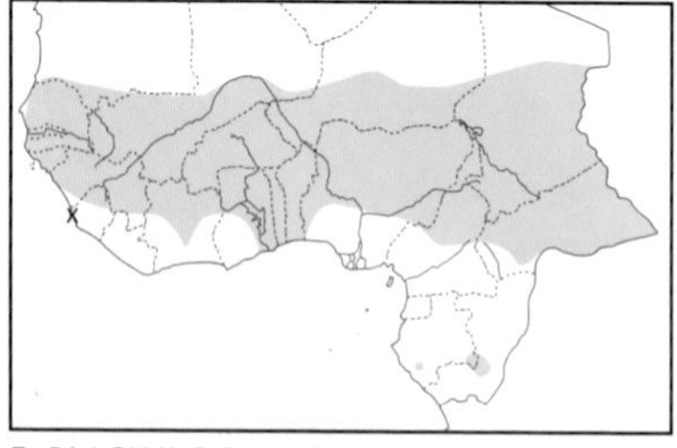

3. BLACK WOOD-HOOPOE

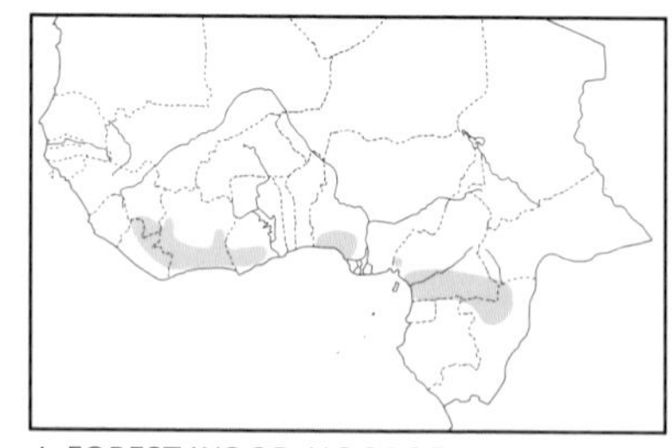

4. FOREST WOOD-HOOPOE

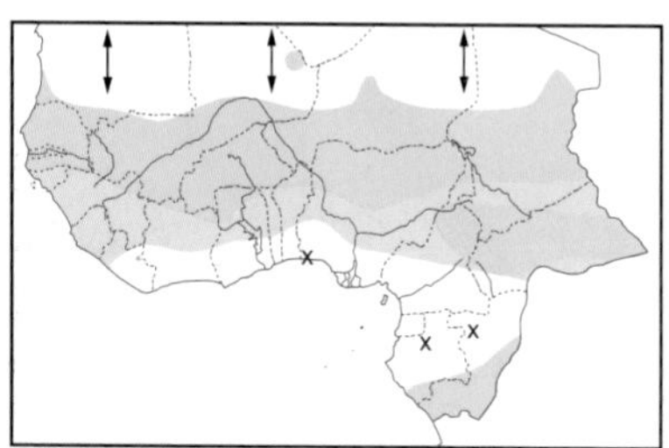

5. HOOPOE

Plate on page 248

Continued from page 252

7 RED-BILLED HORNBILL *Tockus erythrorhynchus* 40–48 cm **R c/f**

7a Ad male *erythrorhynchus* (Northern Red-billed Hornbill) Mainly white head; white spots on wing-coverts; slender red bill; yellowish to pink orbital ring. **7b Ad female *erythrorhynchus*** Bill smaller. **7c Juv *erythrorhynchus*** Bill smaller and duller. **7d Ad male *kempi*** (Western Red-billed Hornbill; from W Mali west) As 7a but orbital ring black. ▲ Savanna woodland, thorn scrub. ✤ A long series of clucking notes, increasing in tempo and volume (as if getting more excited) *uk uk uk uk-uk-uk-UK-UK-UK-UK-UK-uhWUK uhWUK UK-UK-uhWUK...* Several birds may join in excited clucking chorus. [CD8:12b]

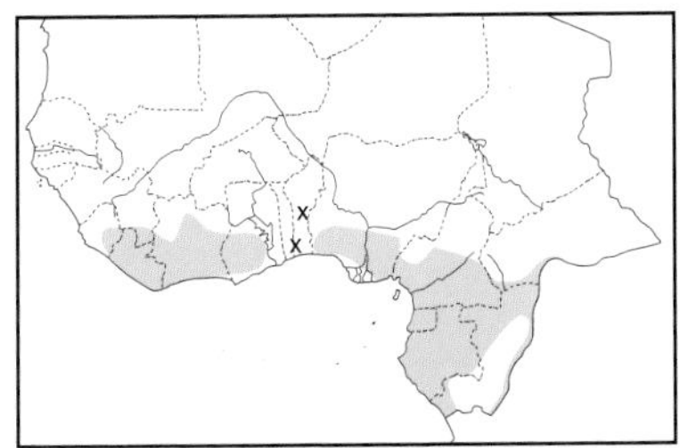

1. RED-BILLED DWARF HORNBILL

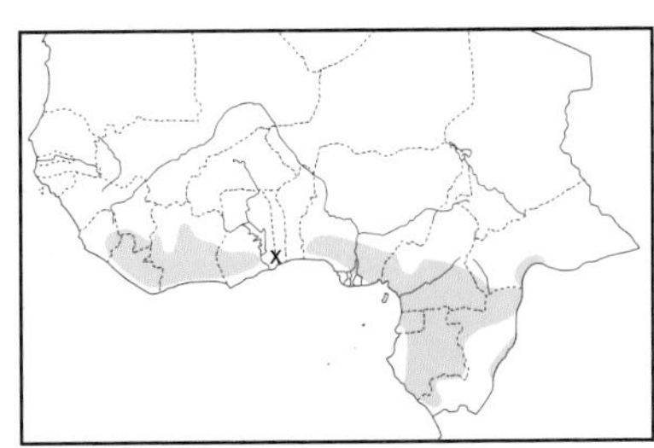

2. BLACK DWARF HORNBILL

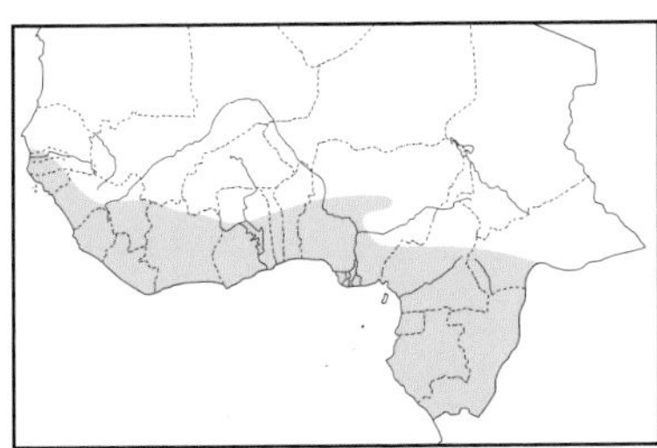

3. AFRICAN PIED HORNBILL

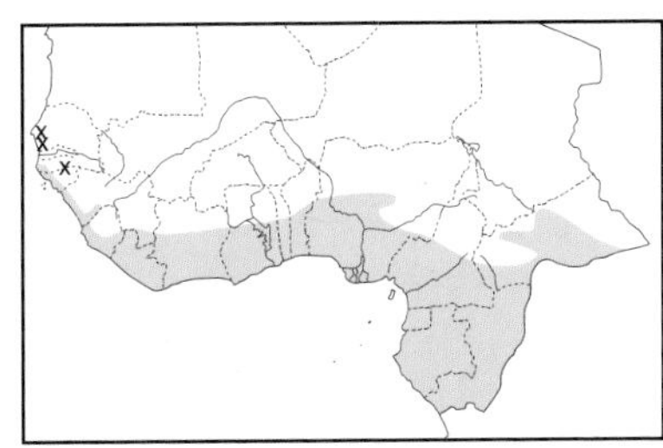

4. PIPING HORNBILL

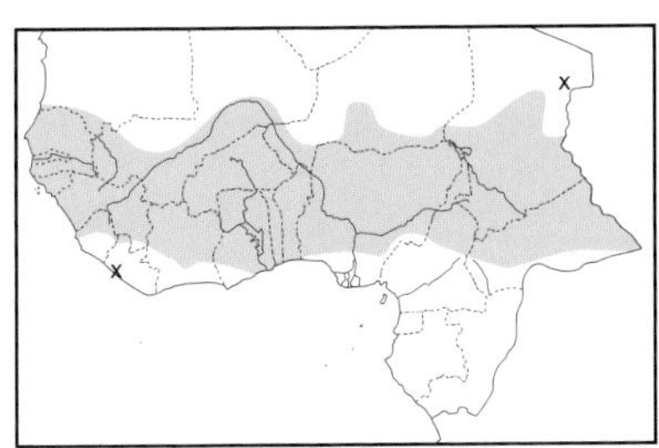

6. AFRICAN GREY HORNBILL

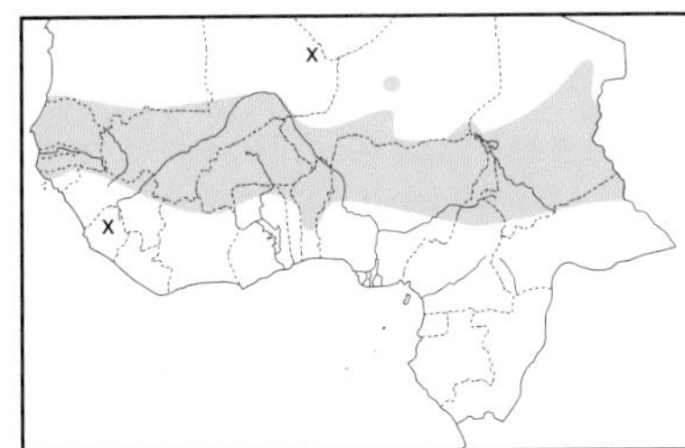

7. RED-BILLED HORNBILL

Plate on page 252

Medium-sized to very large, with large decurved bills surmounted by horny casque of variable size. Tail long. Sexes separable by shape and pattern of casque, coloration of bare parts and/or plumage. Most species arboreal, frugivorous and insectivorous. Many may make long daily and/or seasonal movements in response to fruit availability. Fly with neck outstretched. Nest in tree holes; entrance sealed from within by female (except in ground hornbills), leaving a narrow slit through which she is fed by her mate throughout incubation and until the young are partially grown. Most species vocal, with distinctive calls. Those on this plate are small to medium-sized with small casque, limited to low ridge in most.

1 RED-BILLED DWARF HORNBILL *Tockus camurus* 34–39 cm R f
1a Ad male Small; brown above; red bill. **1b Ad female** Red bill tipped black. ▲ Forest. Joins mixed-species flocks. ✤ Vocal. A distinctive, far-carrying series of slightly mournful, melancholy calls, first rising, then falling, *koo-kio-kio-kio-kio-kio-kio-kio*. [CD8:12a]

2 BLACK DWARF HORNBILL *Tockus hartlaubi* 35–39 cm R u
2a Ad male *hartlaubi* (S Sierra Leone–Togo; S Nigeria–Congo) Small; black above; white supercilium; black bill tipped dark red. **2b Ad female *hartlaubi*** Bill entirely black. **2c Ad male *granti*** (SW CAR–Congo) Wing-coverts and inner secondaries tipped white; more red on bill. ▲ Forest. Unobtrusive. ✤ Rather silent. Fairly loud *ee-ep ee-ep*, frequently repeated. Also a series of rather soft notes *kwu-wu-wu-wu-...* ending in rising *kwee-kukwee-kukWEE*. [CD8:11b]

3 AFRICAN PIED HORNBILL *Tockus fasciatus* 48–55 cm R c
3a Ad male *semifasciatus* (from Owerri area, Nigeria, west) Broad black wings; outer tail feathers 3–4 broadly tipped white; cream-yellow bill tipped black. Female smaller, with slightly smaller bill. **3b Ad male *fasciatus*** (from extreme E Nigeria east) Outer tail feathers 3–4 mostly white; bill tipped dark red below. **3c Ad female *fasciatus*** Bill tipped black. **3d Juv** Bill entirely creamy; in both races white in tail restricted to broad tips to outer tail feathers 3–4. ▲ Forest and adjacent wooded habitats. ✤ Vocal. A series of high-pitched, scolding whistles varying in pitch and tempo *pyi-pyi-pyi-pyi-PYI-PYI-PYI-...*; in display ending with drawn-out *pieeu*. [CD8:13a]

4 PIPING HORNBILL *Bycanistes fistulator* 50–60 cm R f/c
4a Ad male *fistulator* (from W Nigeria west) Wings black with broad white tips to secondaries; outer tail feathers broadly tipped white; grooved bill dusky with pale base and tip. **4b Ad female *fistulator*** Bill smaller. **4c Juv *fistulator*** Bill smaller and entirely dusky. **4d Ad male *sharpii*** (E Nigeria–Congo) White in wings extending to inner primaries; outer tail feathers entirely white. **4e Ad male *duboisi*** (SC Cameroon–CAR) Outer primaries also white; well-developed casque. ▲ Lowland and gallery forest. ✤ Vocal. Distinctive, loud, harsh, nasal laughing. Also shrill piping *peep-peep-peep*. [CD8:17]

5 CROWNED HORNBILL *Tockus alboterminatus* 48–55 cm
Ad male Dusky-brown above; red bill with yellow band at base. Female has smaller bill. ▲ Not recorded in region; reaches coastal forest patches in Congo-Kinshasa and Cabinda. ✤ A series of shrill whistles varying in pitch and tempo, recalling 3. [CD8:13b]

6 AFRICAN GREY HORNBILL *Tockus nasutus* 45–51 cm M/R c
6a Ad male Dull greyish; white supercilium; black bill with creamy patch at base. **6b Ad female** Smaller bill tipped dark red; rest of upper mandible pale yellow. **6c Juv** Smaller, entirely dusky-grey bill. ▲ Savanna woodland. ✤ Rather melancholy *pee-o*, often uttered in series, and a series of far-carrying, rhythmic, piping notes *pee-pee-pee-pee-PEE-PEE PEE-pyew PEE-pyew PEE-PEE-pyew-pee...* [CD8:15]

Continued on page 251

1a
1b
3c
3d
3b
2c
2a
2b
3a
4a
5
4b
4c
4d
6b
6c
6a
4e
7b
7a
7c
7d

PLATE 73: HORNBILLS IN FLIGHT

All figures are of adult males. See also Plates 72 and 74.

1 AFRICAN GREY HORNBILL *Tockus nasutus* **M/R c**
Dull greyish; outer tail feathers broadly tipped white.

2 RED-BILLED HORNBILL *Tockus erythrorhynchus* **R c/f**
T. e. erythrorhynchus Head-sides and neck white; upperparts blackish-brown with white streak on central back and white spots on wing-coverts; tail black with white outer feathers; others broadly tipped white, except two central pairs; bill red.

3 WHITE-CRESTED HORNBILL *Tropicranus albocristatus* **R f/lc**
T. a. albocristatus Crown and head-sides white; upper- and underparts black; very long, graduated, white-tipped tail.

4 RED-BILLED DWARF HORNBILL *Tockus camurus* **R f**
Small. Mainly rufous-brown; lower underparts white. Wing-coverts tipped white, forming two bars; flight feathers edged pale rufous-brown to whitish. Tail feathers tipped white except central pair. Bill red.

5 BLACK DWARF HORNBILL *Tockus hartlaubi* **R u**
Small, black. Tail graduated; all feathers tipped white except central pair.

6 AFRICAN PIED HORNBILL *Tockus fasciatus* **R c**
6a ***T. f. semifasciatus*** Broad black wings; outer tail feathers 3–4 broadly tipped white. **6b** ***T. f. fasciatus*** Outer tail feathers 3–4 mostly white.

7 PIPING HORNBILL *Bycanistes fistulator* **R f/c**
7a ***B. f. fistulator*** Wings black with broad white tips to secondaries; outer tail feathers broadly tipped white. **7b** ***B. f. sharpii*** White in wings extending to inner primaries; outer tail feathers entirely white. **7c** ***B. f. duboisi*** Outer primaries also white; well-developed casque.

8 BROWN-CHEEKED HORNBILL *Bycanistes cylindricus* **R r/lf NT**
Wings black with outer half of flight feathers white; broad black band on tail; upper part of thighs black; pale bill and casque.

9 WHITE-THIGHED HORNBILL *Bycanistes albotibialis* **R u/lc**
Black band on tail narrower than in 8; thighs entirely white; bill mainly dusky; casque pale and differently shaped to 8.

10 BLACK-AND-WHITE-CASQUED HORNBILL *Bycanistes subcylindricus* **R lc/r**
B. s. subcylindricus Wings black with black-based white inner primaries and secondaries; central tail feathers black; blackish bill with two-toned casque.

11 YELLOW-CASQUED HORNBILL *Ceratogymna elata* **R r/lu NT**
Black with white outer tail feathers; bill blackish; upper part of casque cream. Wingbeats heavier than those of 12 (Yellow-casqueds weigh distinctly more than Black-casqueds).

12 BLACK-CASQUED HORNBILL *Ceratogymna atrata* **R lc/r**
Black with broad white tips to outer tail feathers; bill and casque entirely black.

13 ABYSSINIAN GROUND HORNBILL *Bucorvus abyssinicus* **R r/lf**
Huge; black with white primaries.

1
2
3
4
5
6a
6b
7a
7b
7c
8
9
10
11
12
13

All species on this plate, except White-crested Hornbill, are very large, black-and-white birds, with large decurved bills surmounted by high horny casques. Their wingbeats make a distinctive loud swishing noise.

1 BROWN-CHEEKED HORNBILL *Bycanistes cylindricus* 65–77 cm R r/lf NT
1a Ad male Broad black band on tail; upper part of thighs black; pale bill and casque. **1b Ad female** Bill and casque smaller. **1c Juv** Bill smaller, without casque. ▲ Forest. Endemic. ✤ Calls include high-pitched *pfeet!* or *khlee!*, sounding like squeaky hinge or rusty water pump, and harsh notes, uttered singly or in series at variable speed. Distinctly higher pitched and less raucous than calls of 2. [CD8:20]

2 WHITE-THIGHED HORNBILL *Bycanistes albotibialis* 65–77 cm R u/lc
2a Ad male Black band on tail narrower than in 1a; thighs entirely white; bill mainly dusky; casque pale and differently shaped to 1a. **2b Ad female** Bill and casque smaller and dusky. ▲ Forest. ✤ Loud, harsh, barking *gak!* or *rrhoak!* uttered singly or in slow, descending series. Also an abrupt *kekh!* [CD8:19]

3 YELLOW-CASQUED HORNBILL *Ceratogymna elata* 70–90 cm R r/lu NT
3a Ad male Black with white outer tail feathers; bill blackish; upper part of casque cream. **3b Ad female** As 4b but with pale bill, paler rufous head and tail pattern as 3a. ▲ Forest. Endemic. ✤ Far-carrying, resonant, nasal trumpeting. Less raucous and plaintive, more fluting and slower than calls of 4. [CD8:22]

4 BLACK-CASQUED HORNBILL *Ceratogymna atrata* 70–90 cm R r/lc
4a Ad male Black with broad white tips to outer tail feathers; bill and casque entirely black. Compare 3a. **4b Ad female** Rufous-brown head and neck; smaller bill and casque. Compare 3b. ▲ Forest. Endemic. ✤ Far-carrying, resonant, strikingly plaintive trumpeting (sounding like wailing child), often given in flight. Female calls are much higher pitched than male's. More raucous and plaintive than calls of 3. [CD8:21]

5 BLACK-AND-WHITE-CASQUED HORNBILL *Bycanistes subcylindricus* 65–78 cm R lc/r
5a Ad male *subcylindricus* (west of Niger R.) Central tail feathers black; blackish bill with two-toned casque. **5b Ad female *subcylindricus*** Smaller bill and casque. **5c Juv *subcylindricus*** Smaller, entirely dusky bill without casque. **5d Ad male *subquadratus*** (east of Niger R.) Casque with larger pale area. ▲ Forest. ✤ Calls loud and nasal, including abrupt *keh!* or *heh!* and a slow series of sarcastic laughing *hah-hah-hah-hah-hah-...* Calls of subspecies may differ; more research required. [CD8:18]

6 WHITE-CRESTED HORNBILL *Tropicranus albocristatus* 70–80 cm R f/lc
6a Ad male *albocristatus* (from W Ivory Coast west) White crown and head-sides; very long, graduated, white-tipped tail. **6b Ad male *cassini*** (from Nigeria east) Head-sides black; flight feathers, greater coverts and scapulars tipped white. **6c Ad male *macrourus*** (E Ivory Coast–Benin) As 6a but white of head extending to throat and neck. **6d Juv** Smaller; bill entirely darkish without casque. ▲ Forest. Forages in dense foliage. Joins mixed-species flocks and associates with monkeys for the small animals they disturb. ✤ Unlike other hornbills. A series of rather soft, plaintive sounds culminating in a distinctive, drawn-out wail *oooooooooaah!*. [CD8:11a]

7 ABYSSINIAN GROUND HORNBILL *Bucorvus abyssinicus* 90–110 cm R r/lf
7a Ad male Huge; black with white primaries (conspicuous in flight); bare red skin on neck and throat. **7b Ad female** Smaller; bare skin entirely blue. **7c Juv** Dark sooty-brown; small bill; undeveloped casque. ▲ Savanna. Walks in search of prey; only occasionally flies. ✤ Deep, far-carrying, booming *uu-uh, uh-uh-uh*, from ground or perch, esp. at dawn. [CD8:10a]

Maps on page 258

1a
1b
1c
2a
2b
3a
3b
4a
4b
5a
5b
5c
5d
6a
6b
6c
6d
7a
7b
7c

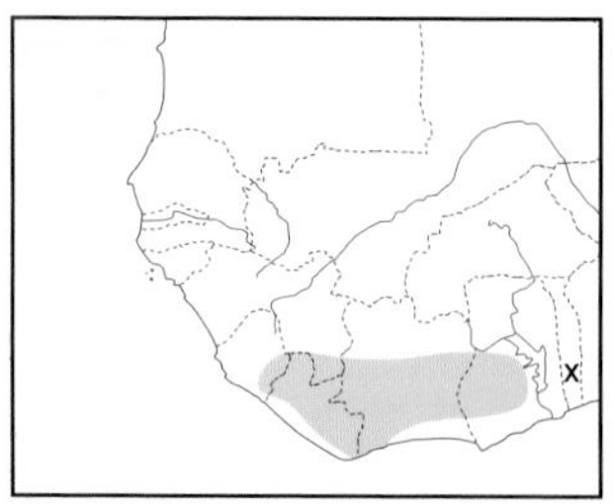

1. BROWN-CHEEKED HORNBILL

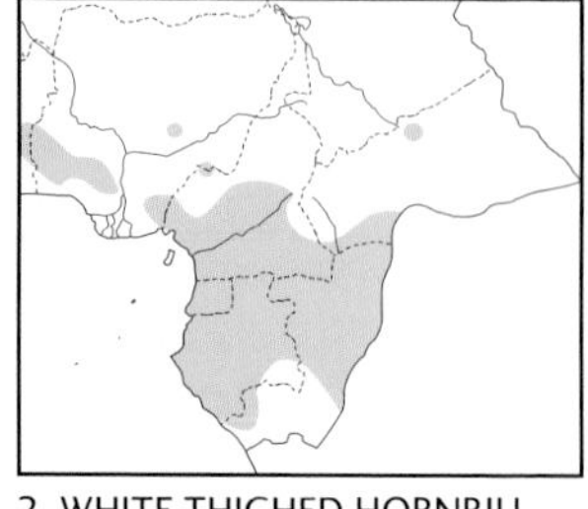

2. WHITE-THIGHED HORNBILL

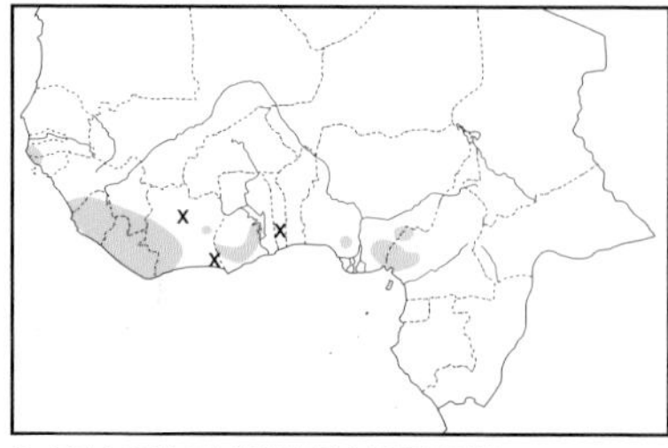

3. YELLOW-CASQUED HORNBILL

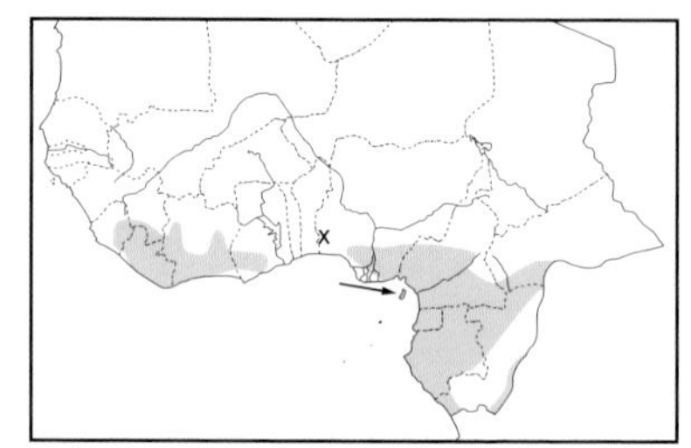

4. BLACK-CASQUED HORNBILL

5. BLACK-AND-WHITE-CASQUED HORNBILL

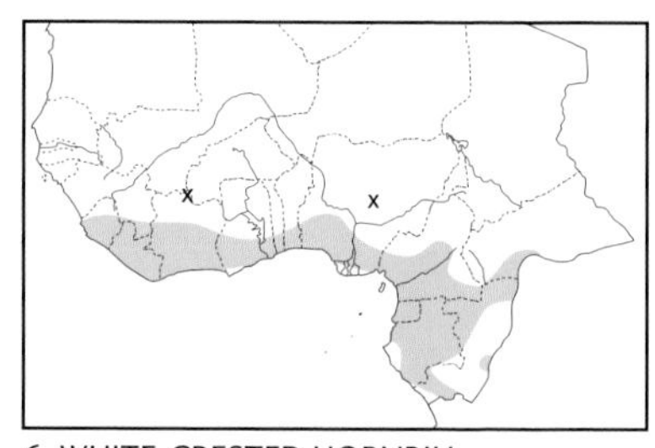

6. WHITE-CRESTED HORNBILL

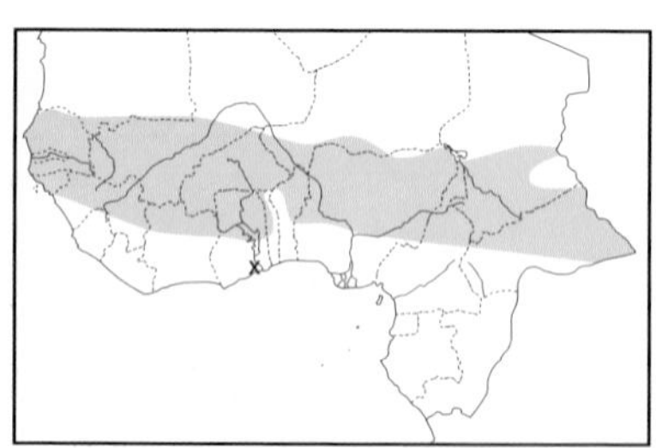

7. ABYSSINIAN GROUND HORNBILL

Plate on page 256

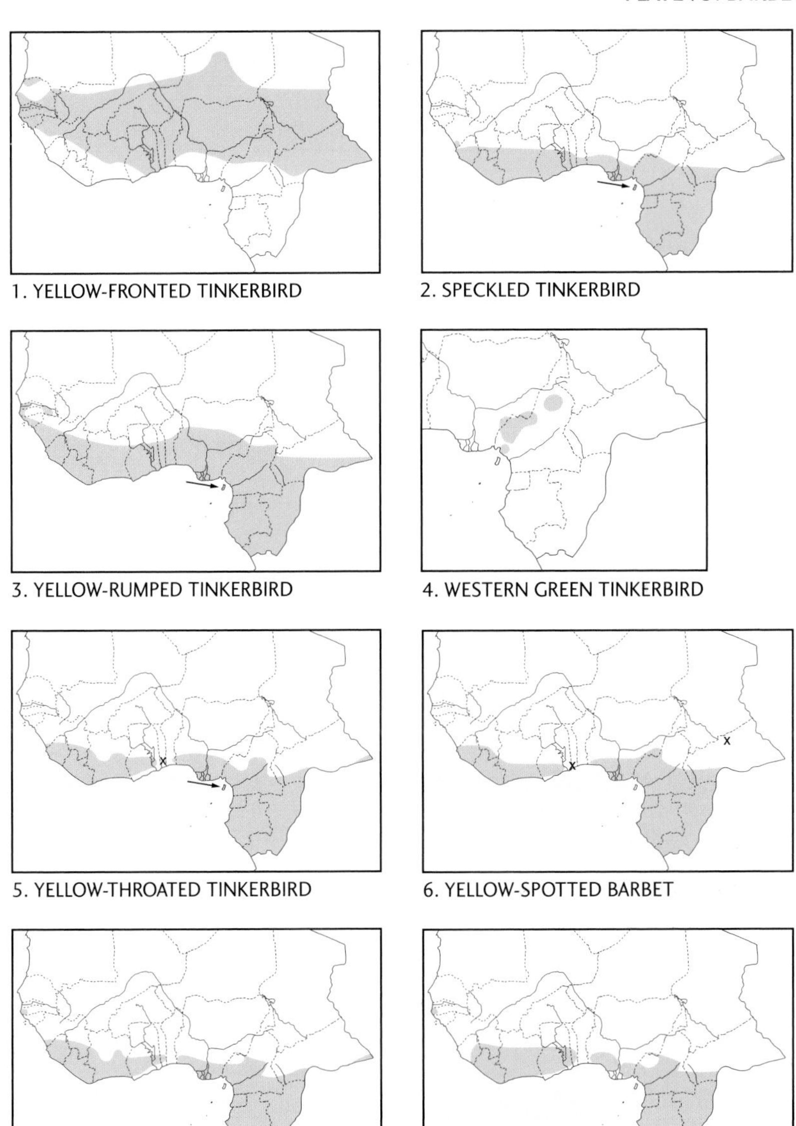

1. YELLOW-FRONTED TINKERBIRD

2. SPECKLED TINKERBIRD

3. YELLOW-RUMPED TINKERBIRD

4. WESTERN GREEN TINKERBIRD

5. YELLOW-THROATED TINKERBIRD

6. YELLOW-SPOTTED BARBET

7. RED-RUMPED TINKERBIRD

8. HAIRY-BREASTED BARBET

Plate on page 260

Stocky, with large heads and heavy bills. Mostly arboreal and frugivorous; nest in self-excavated holes in trees. Calls characteristic, typically consisting of a monotonously repeated single note.

1 YELLOW-FRONTED TINKERBIRD *Pogoniulus chrysoconus* 11.5 cm **R c**
Golden-yellow forecrown; boldly black-and-white streaked face and upperparts. ▲ Wooded savanna. ✤ Monotonous series of *poop*s repeated for long periods without pause. Also series of hoarse *kwèp* notes and *kwrrrr kwrrrr...* trills in variable series. [CD8:32]

2 SPECKLED TINKERBIRD *Pogoniulus scolopaceus* 13 cm **R c**
Dark brown; scaly upperparts; pale eye. ▲ Forest, forest edge, clearings, bush. Usually quite tame and foraging lower than most other barbets. ✤ A series of *kwip* notes, first a single note, repeated a few times, then a doubled *kwip-ip*, also repeated, then tripled, gradually 4–5 rapid notes. Also various other sounds, always delivered in rhythmic series: an insect- or toad-like *kwibibibbbbbt*; trills *kukkrrrrr...*, hoarse, high-pitched *hyep*, and *poop-poop-poop-...* resembling other tinkerbirds. [CD8:28]

3 YELLOW-RUMPED TINKERBIRD *Pogoniulus bilineatus* 10 cm **R c**
Black upperparts; bold white facial stripes; yellow rump; white throat. ▲ Forest, woodland. Canopy or mid-level. ✤ Monotonous, far-carrying, rhythmic series of 3–6 (sometimes more) *kok* or *poop* notes, uttered with short pause between each series and endlessly repeated, even during the heat of the day. Also series of *krrw* notes and rattling *kkkkkk*. [CD8:31]

4 WESTERN GREEN TINKERBIRD *Pogoniulus coryphaeus* 9 cm **R lc**
Black upperparts; yellow stripe from crown to rump. ▲ Montane forest. Usually forages in middle and higher strata of fruiting trees, hanging tit-like on branches. ✤ Rather metallic *kwip* or *pwip* in descending series at varying speed. Also a trill *kirrrrik*. [CD8:27]

5 YELLOW-THROATED TINKERBIRD *Pogoniulus subsulphureus* 10 cm **R c**
5a Ad *chrysopyga* (from Ghana west) As 3; distinguished by faster song. **5b Ad *flavimentum*** (S Nigeria (Togo?) to Congo–SE CAR) Yellow throat; yellow facial stripes. Nominate (Bioko) similar. ▲ Forest. ✤ Rhythmic series of *poop* notes similar to 3, but higher pitched, noticeably faster (5/sec versus 3/sec) and usually in longer series. Also an even series of *krrrw* notes and accelerating *kwip kwip-kwipkwipkwipipip*. [CD8:30]

6 YELLOW-SPOTTED BARBET *Buccanodon duchaillui* 15 cm **R c/u**
Red forehead to crown; yellow stripe on head-sides. ▲ Forest. Usually singly, in canopy. Several may gather in fruiting tree. Occasionally joins mixed-species flocks. ✤ In west of range, a series of 7–10 accelerating *oop* notes. In east of range, a characteristic purring *rrurrrrrr...* (lasting 1–2 seconds). [CD8:33]

7 RED-RUMPED TINKERBIRD *Pogoniulus atroflavus* 13 cm **R u/f**
Red rump; three yellow facial stripes. ▲ Forest. Forages at all levels, but usually high. ✤ Rhythmic series of *poop* calls, resembling 3, but slower, lower pitched (*c.* 1/sec) and in much longer series. Also series of trills *kirrr kirrr kirrr...* or *krukukkk...* [CD8:29]

8 HAIRY-BREASTED BARBET *Tricholaema hirsuta* 16.5 cm **R lc**
8a Ad *hirsuta* (from Ghana west) Head and throat black; two bold facial stripes. Intergrading with 8c in Togo and W Nigeria. **8b Ad *ansorgii*** (E Cameroon–CAR) Throat whitish streaked black; nape spotted yellow. **8c Ad *flavipunctata*** (S Nigeria to N & C Gabon) Head freckled without facial stripes; plumage browner. *T. h. angolensis* (S Gabon, W & C Congo) similar but browner, esp. on belly. ▲ Forest and forest edge. ✤ Deep *hoop*s, uttered in continuous, rhythmic series at variable speed (most often *c.* 1-2/sec, but also much faster). Also accelerating *hoop hoop-hoophoopoopoopoopop*. [CD8:34–35]

 Maps on page 259

1
2
3
4
5a
5b
6
7
8c
8b
8a

PLATE 76: BARBETS II

1 **NAKED-FACED BARBET** *Gymnobucco calvus* 17–18 cm **R lc**
1a Ad Dusky-brown; bare blackish head; sparse short bristles at gape and nostrils. **1b Juv** Head more feathered; darkish-tipped bill. ▲ Forest and forest edge. Gregarious. ✤ Sharp, explosive *KYEW!*, singly or in loose series. Rattling *kirrrrrrr*... and *kreepipipipppp*... [CD8:25]

2 **BRISTLE-NOSED BARBET** *Gymnobucco peli* 17–18 cm **R u/lc**
2a Ad As 1a, but with nasal tufts and paler bill. **2b Juv** Head feathered; nasal tufts small or absent. ▲ As 1. ✤ Hard *KYEW!*, shorter, higher pitched than 1. Noisy, rattling *kreepipipppppp*... [CD8:24]

3 **SLADEN'S BARBET** *Gymnobucco sladeni* 17–18 cm **R?**
Grey throat and neck-sides; buffish nasal tufts; black bill; reddish eye. ▲ Forest. 1 record.

4 **GREY-THROATED BARBET** *Gymnobucco bonapartei* 16.5–17.5 cm **R lc**
4a Ad Greyish head and throat; brownish nasal tufts; blackish bill; brown to red eye. **4b Juv** Bill yellowish with dusky tip; no nasal tufts. ▲ Forest and forest edge. Gregarious. ✤ Hoarse *whewp*, rattling *kripipipipppp*..., nasal *nyaaa* and buzzy *spszsz*. [CD8:23]

5 **WHITE-HEADED BARBET** *Lybius leucocephalus* 15.0–16.5 cm **R lc/s**
Ad *adamauae* All-white head and breast; white rump. ▲ Woodland, cultivation. ✤ High-pitched *pyup* or *kyip*, singly or in series. Duet may end in *ki-ki-ki-ki-ki-*... [CD8:39]

6 **BLACK-BACKED BARBET** *Lybius minor* 16.5 cm **R s/lc**
6a Ad *minor* (S Gabon–Congo) Red forehead and crown; pinkish belly. **6b Ad *macclounii*** (extralimital?) Head-sides white; white scapulars form V. Intermediates occur. ▲ Woodland, gallery forest. ✤ Calls *kyek*, singly or in series, *kik-ik-ik-ik*. Song a buzzy trill *krrrrriiiii*. [CD8:42]

7 **BLACK-BILLED BARBET** *Lybius guifsobalito* 15 cm **V**
Black with red on face to upper breast; yellow-edged flight feathers. ▲ Open savanna woodland. ✤ Duet a series of *hik-kup-oot*. Calls include *kek*, nasal *kaw* etc. [CD8:40]

8 **VIEILLOT'S BARBET** *Lybius vieilloti* 15 cm **R c/u**
Red head; pale yellow underparts; red spots from throat to central belly. ▲ Savanna woodland, thorn scrub. ✤ Duet a rhythmic series of far-carrying, melodious *poop* calls, preceded by chattering or purring notes *kekkekkekkkk-urrrrr poop poop poop-eh-poop epoop poop-eh*... [CD8:38]

9 **YELLOW-BREASTED BARBET** *Trachyphonus margaritatus* 20 cm **R lc**
Pink bill; black cap; white-spotted upperparts. **9a Ad male** Back breast-patch. **9b Ad female** No black on breast. ▲ Dry wooded grassland, thorn bush, desert edge. ✤ Vocal. Duet starts with clear *pwewp* notes and evolves in buoyant, rolling *pwewp-up kwewp-up tew-kwip-to*... [CD8:48]

10 **YELLOW-BILLED BARBET** *Trachylaemus purpuratus* 24 cm **R f**
Yellow bill; blackish upperparts; long tail. **10a Ad *purpuratus*** (from SE Nigeria east) Lower breast and belly blotched black; yellow orbital patch. **10b Ad *togoensis*** (SW Nigeria–E Ghana) Red on head brighter than 10c; throat streaked white; greyish-blue orbital patch. **10c Ad *goffinii*** (from Ghana west) No black on underparts. ▲ Forest and forest edge. ✤ Long series of low-pitched *hoops* (*c.* 1/sec); 10c slower than 10a. [CD8:46]

11 **BLACK-BREASTED BARBET** *Lybius rolleti* 26.5 cm **R u/lc**
Massive ivory bill; mainly black plumage; red belly; white flanks. ▲ Wooded grassland, cultivation. ✤ Low rasping and growling notes, and sharp *kak!*. [CD8:45]

12 **BEARDED BARBET** *Lybius dubius* 25 cm **R lf**
Massive yellowish bill; black breast-band; no wingbar. Compare 13. ▲ Woodland. Endemic. ✤ Harsh, low, grating notes. [CD8:44]

13 **DOUBLE-TOOTHED BARBET** *Lybius bidentatus* 23 cm **R f**
Heavy whitish bill; no breast-band; red (or pinkish to whitish) wingbar; grey legs and feet. ▲ Woodland. ✤ Loud *KEK!* or *KZEK!*. Song a long purring, frog-like *errrrrrrrr*. CD8:43]

 Maps on page 264

1b
1a
2b
2a
4b
4a
3
6a
5
7
8
6b
10a
10c
9a
9b
10b
11
13
12

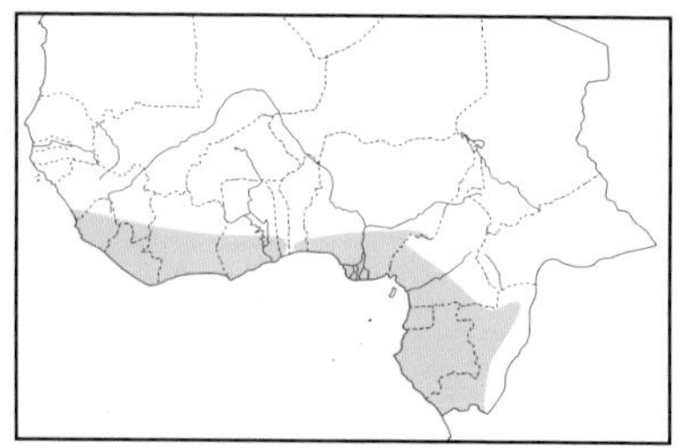
1. NAKED-FACED BARBET

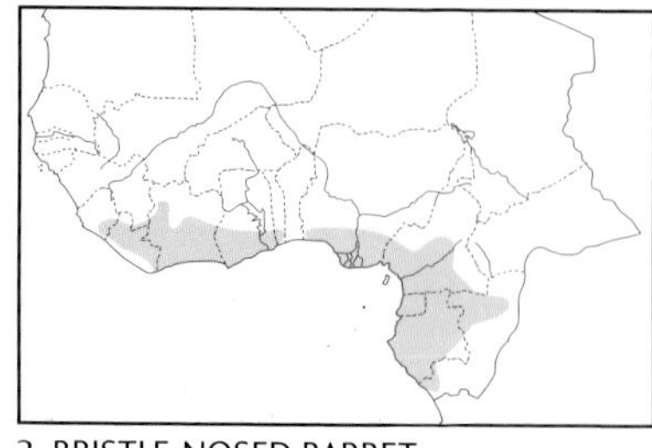
2. BRISTLE-NOSED BARBET

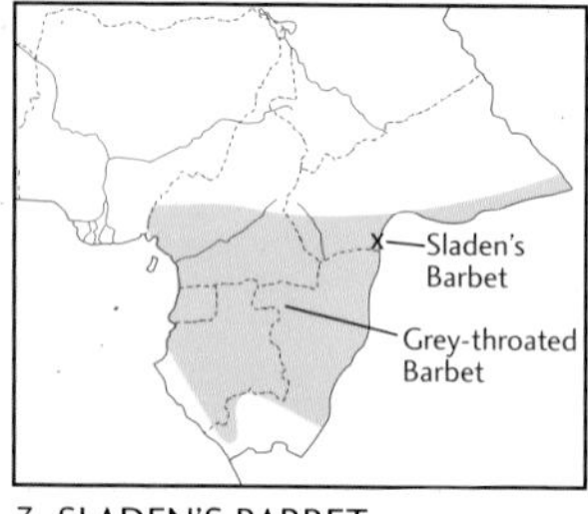

3. SLADEN'S BARBET
4. GREY-THROATED BARBET

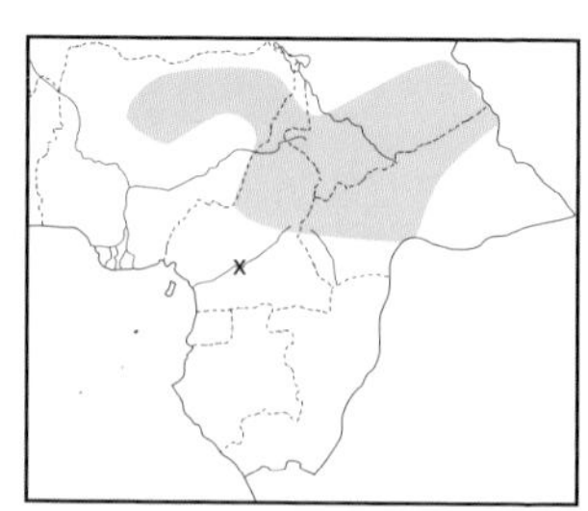
5. WHITE-HEADED BARBET

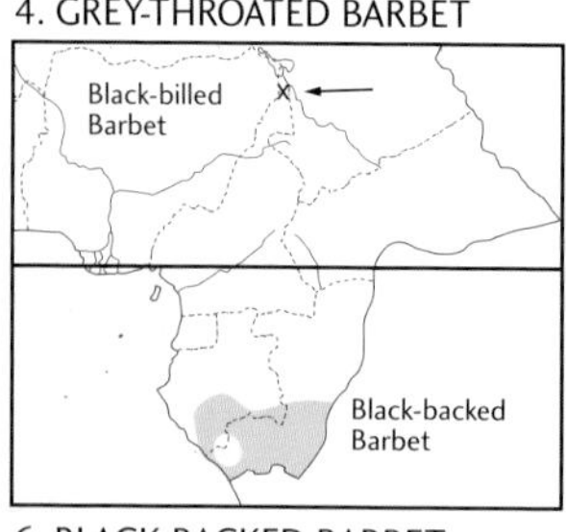

6. BLACK-BACKED BARBET
7. BLACK-BILLED BARBET

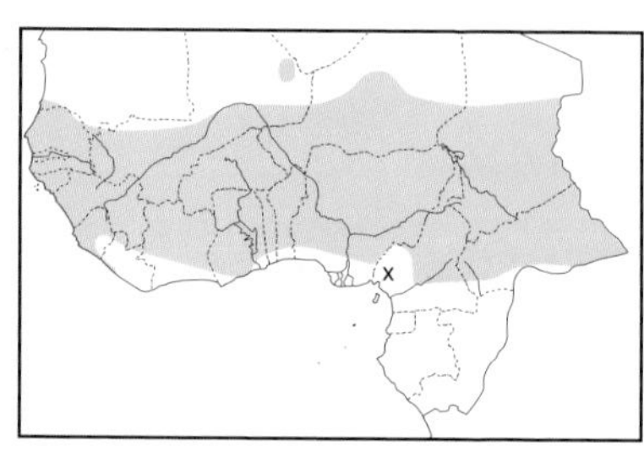
8. VIEILLOT'S BARBET

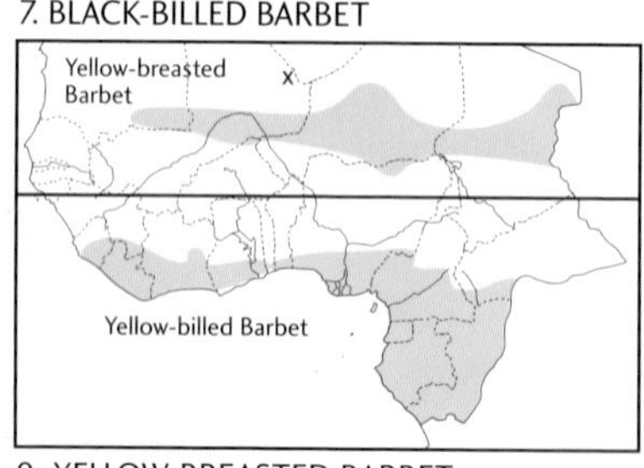

9. YELLOW-BREASTED BARBET
10. YELLOW-BILLED BARBET

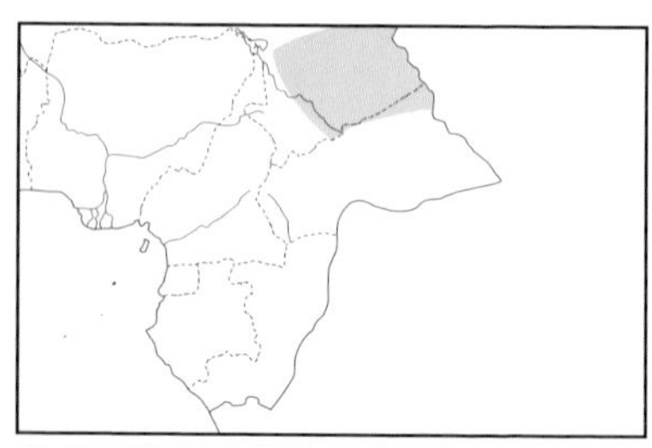
11. BLACK-BREASTED BARBET

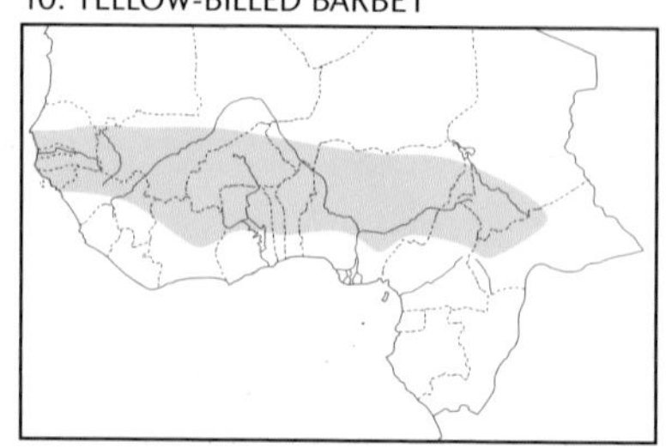
12. BEARDED BARBET

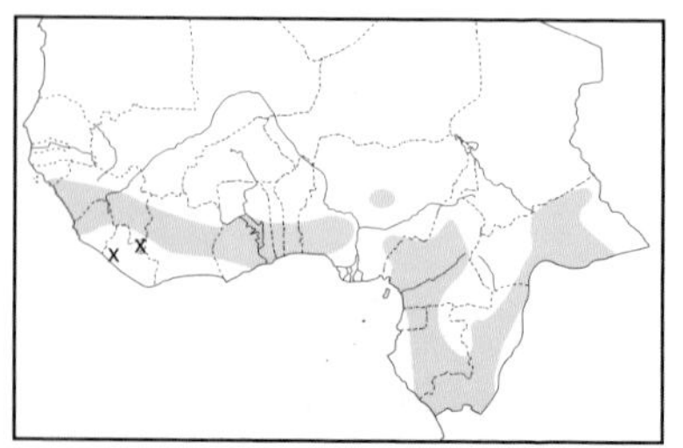
13. DOUBLE-TOOTHED BARBET

Plate on page 262

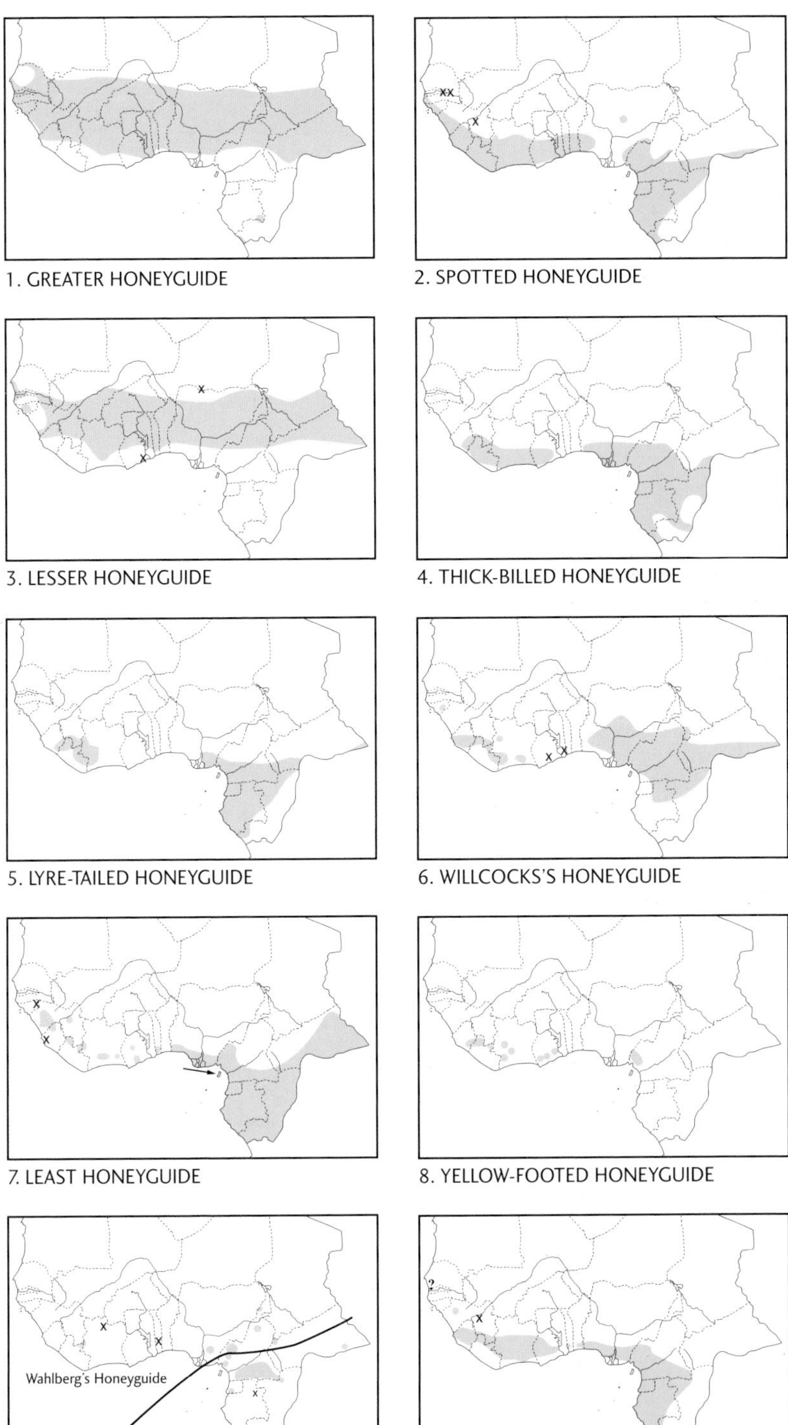

1. GREATER HONEYGUIDE

2. SPOTTED HONEYGUIDE

3. LESSER HONEYGUIDE

4. THICK-BILLED HONEYGUIDE

5. LYRE-TAILED HONEYGUIDE

6. WILLCOCKS'S HONEYGUIDE

7. LEAST HONEYGUIDE

8. YELLOW-FOOTED HONEYGUIDE

9. ZENKER'S HONEYGUIDE
10. WAHLBERG'S HONEYGUIDE

11. CASSIN'S HONEYBIRD

Drab coloured and generally inconspicuous. Outer tail feathers white and generally dark tipped, often conspicuous in flight. Feed on wax, insects and fruit and are brood parasites.

1 GREATER HONEYGUIDE *Indicator indicator* 18.0–19.5 cm R s/lc
1a Ad male Black throat; whitish ear patch; pink bill. **1b Ad female** Duller than 1a; no distinctive head pattern; bill dark. **1c Juv** Pale to deep yellow throat and breast; bill dark. ▲ Savanna woodland. ✤ Distinctive, far-carrying series of *WHIT-birr WHIT-birr..* ending with single *WHIT*. 'Guiding' call a loud, fast chattering. [CD8:55]

2 SPOTTED HONEYGUIDE *Indicator maculatus* 16.5–17.5 cm R r/lf
2a Ad *maculatus* (from SW Nigeria west) Olive underparts spotted creamy-green. *I. m. stictithorax* (from SE Nigeria east and south) has spots paler, more distinct; head-sides finely streaked. **2b Juv** Speckled forehead; spotting below more extensive and distinct. ▲ Evergreen and gallery forest. ✤ Purring *brrrrrrrr....* [CD8:53]

3 LESSER HONEYGUIDE *Indicator minor* 14–15 cm R u
Stubby bill; grey head; indistinctly streaked above. Compare 4, 6 and 7. ▲ Wooded savanna. ✤ Far-carrying, deliberate, rhythmic series of 10–30 identical notes *wrip wrip wrip wrip* ... introduced by faint *pee-yew* or *tyeew*; rhythm variable. [CD8:56]

4 THICK-BILLED HONEYGUIDE *Indicator (minor) conirostris* 14–15 cm R u
As 3; typically more distinctly streaked above, darker grey below. Juveniles (of 3 and 7 also) lack loral and submoustachial marks. ▲ Forest. ✤ As 3. [CD8:57]

5 LYRE-TAILED HONEYGUIDE *Melichneutes robustus* 16.5–17.5 cm R r/u
Distinctly shaped tail. Rarely seen, but sound diagnostic. ▲ Lowland, lower montane and gallery forest. ✤ In display flight an accelerating series of 10–30 nasal sounds, first rising in pitch, then fading *heyih heyih heyih-heyih-heiheiheihei...*; extremely difficult to locate. [CD8:52]

6 WILLCOCKS'S HONEYGUIDE *Indicator willcocksi* 11–13 cm R r/u
Olive wash on grey breast; no submoustachial stripe; dusky streaks on flanks. Compare 7, 3 and 4. ▲ Forest, gallery forest, dense woodland. ✤ Long rhythmic series of identical notes, resembling 3 and 7 in structure but interspersed with a distinctive snapping note, and lacking soft introductory syllable *p-wEEw-Pk p-wEEw-Pk p-wEEw-Pk...* or *huwEEw-TK huwEEw-TK huwEEw-TK* (*Pk* and *TK* = snapping note). [CD8:59]

7 LEAST HONEYGUIDE *Indicator exilis* 11–13 cm R u/r
Distinctly streaked above and on flanks; dark submoustachial stripe; narrow white line on lores. Compare 6 and 4. ▲ Forest. ✤ Far-carrying, deliberate, rhythmic series of 10–30 identical notes *wrEEu wrEEu wrEEu wrEEu* ... resembling 3 but less abrupt and less snapping. [CD8:58]

8 YELLOW-FOOTED HONEYGUIDE *Melignomon eisentrauti* 14–15 cm R r DD
Bill and legs yellowish; off-white undertail-coverts; tail mainly white from below. ▲ Forest. Endemic. ✤ A series of *c.* 13 clear, emphatic notes, slightly descending and slowing down at end, *tuu-i tuu-i tuu-i tuu-i ... tuu tuu tuu.*

9 ZENKER'S HONEYGUIDE *Melignomon zenkeri* 14–15 cm R r/lf
Darkish, bulbul-like. Wholly greyish-olive below; graduated tail mainly dark; bicoloured bill (dark above, yellow below); yellowish legs. Compare 8. ▲ Forest. ✤ Rhythmic series of identical, whistled notes *wEEu wEEu wEEu wEEu* ..., first rising, then gradually fading. [CD8:51]

10 WAHLBERG'S HONEYBIRD *Prodotiscus regulus* 12–13 cm R r
Resembles 11 but browner; outer tail feathers tipped dark (all white in juvenile). ▲ Woodland, bush. ✤ Buzzy *tsrrr-tsrrr-...*, uttered in flight. Song *tsrrrrrrrrrrrrrrrrrrrrr..* [CD8:50]

11 CASSIN'S HONEYBIRD *Prodotiscus insignis* 11–12 cm R s/u
Flycatcher- or warbler-like; outer tail feathers all white; fine bill. ▲ Evergreen and gallery forest. ✤ Distinctive buzzy *tsrrr-tsrrr-...*, uttered in flight.

Maps on page 265

1c
1b
1a
3
2b
2a
4
6
5
7
9
8
10
11

PLATE 78: WOODPECKERS I

Small to large birds with strong, straight, pointed bills. Characteristically cling to trees, using stiff tail as prop. Largely insectivorous. Often in pairs. Sexes differ in head pattern, males usually with (more) red on head. Nest in self-excavated holes in trees or termitaria.

1 GOLDEN-TAILED WOODPECKER *Campethera abingoni* 20 cm **R** r/u
1a Ad male Dull green upperparts; barred back; streaked underparts; red crown and moustachial stripe. **1b Ad female** Forecrown and moustachial stripe black speckled white. ▲ Gallery forest, woodland. ✤ Single, drawn-out *k-heeeeew*. [CD8:66]

2 FINE-SPOTTED WOODPECKER *Campethera punctuligera* 21 cm **R** lc/u
2a Ad male Pale yellow underparts; breast finely spotted blackish; red crown and moustachial stripe. **2b Ad female** Forecrown black speckled white; indistinct, speckled moustachial stripe. **2c Juv** Forecrown plain black; moustachial stripe blacker than in adult female. ▲ Woodland. ✤ A ringing series, often starting with a few *kip* calls: *kip-kip-kip-kieeh-kieeh-kieeh*, often uttered simultaneously by pair. [CD8:63]

3 NUBIAN WOODPECKER *Campethera nubica* 20 cm **R** u
3a Ad male Olive-brown upperparts boldly barred and spotted whitish; breast and flanks boldly spotted; red crown and moustachial stripe. **3b Ad female** Forecrown and moustachial stripe black speckled white. ▲ Open woodland. ✤ Similar to 2. An accelerating series of ringing, strident notes *kieeh, kieeh kieeh kieeh kieeh-kieeh-kieeh kieeh kiee*. [CD8:64]

4 TULLBERG'S WOODPECKER *Campethera tullbergi* 18–20 cm **R** u
4a Ad male Plain green above; diagnostic red carpal patch; finely spotted face and underparts; red forecrown. **4b Ad female** Forecrown black speckled white. ▲ Moist montane forest. ✤ Single *kreeer* or *kweeh*. [CD8:69]

5 GABON WOODPECKER *Dendropicos gabonensis* *c*. 15 cm **R** u/lc
5a Ad male *lugubris* (Melancholy Woodpecker; Sierra Leone–Togo) Small; plain upperparts; heavily streaked underparts; head-sides and malar stripe olive-brown; hindcrown and nape red. **5b Ad female *lugubris*** Forehead and crown dark olive. **5c Ad *reichenowi*** (S Nigeria–SW Cameroon) Intermediate between 5a and 5d. **5d Ad male *gabonensis*** (Gabon–Congo) Boldly spotted underparts; head-sides streaked; red crown. **5e Ad female *gabonensis*** Forehead and crown black. ▲ Forest edge, second growth. ✤ Diagnostic, rapid, high-pitched trill, *krititititititti* or *trrreeeeeee*. Also series of loud *kree* notes, resembling Cardinal Woodpecker's, *kwik-ik-ik*... and *kreek-rrek-rrek-rrek*... [CD8:74]

6 GREEN-BACKED WOODPECKER *Campethera cailliautii* *c*. 16 cm **R** u/c
6a Ad male Plain green upperparts; densely barred underparts; greenish tail; red forehead and crown. **6b Ad female** Forecrown speckled black. ▲ Forest, forest edge and adjacent woodland. ✤ Drawn-out, plaintive *huweeeeh* or *wheeeee*, often in series of 3–5. [CD8:68]

7 LITTLE GREEN WOODPECKER *Campethera maculosa* *c*. 16 cm **R** u/lc
7a Ad male Golden-olive upperparts; boldly barred underparts; black tail; red forehead and crown. **7b Ad female** Forehead and crown barred black and white. ▲ Forest and forest edge. Endemic. ✤ Drawn-out, plaintive *huweeeeh*, very similar to 6, and shorter, harsh *whee* and hard *kewik*. [CD8:67]

8 BROWN-EARED WOODPECKER *Campethera caroli* 18–19 cm **R** lc/u
8a Ad male Dark olive; distinctive brown ear patch; spotted buffish below; some red feathers on nape. **8b Ad female** Entire crown and nape olive. ▲ Forest. ✤ Distinctive, single *huu-wEEEEuu*, plaintive and remarkably prolonged. [CD8:71]

9 BUFF-SPOTTED WOODPECKER *Campethera nivosa* 14–16 cm **R** lc/r
9a Ad male Dark olive; densely spotted buffish below; red patch on nape. **9b Ad female** Nape olive. ▲ Forest. ✤ Mostly silent. Occasionally a soft *peeer*. [CD8:70]

Maps on page 270

1b
2b
3a
1a
2a
4a
2c
3b
4b
5e
5b
6b
6a
5d
5a
5c
7b
8b
9b
7a
8a
9a

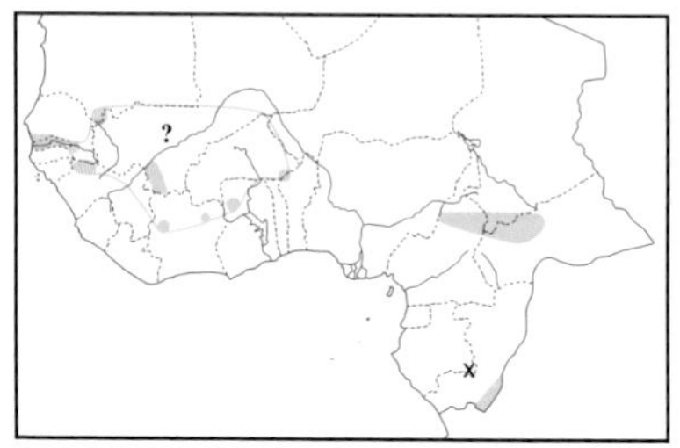

1. GOLDEN-TAILED WOODPECKER

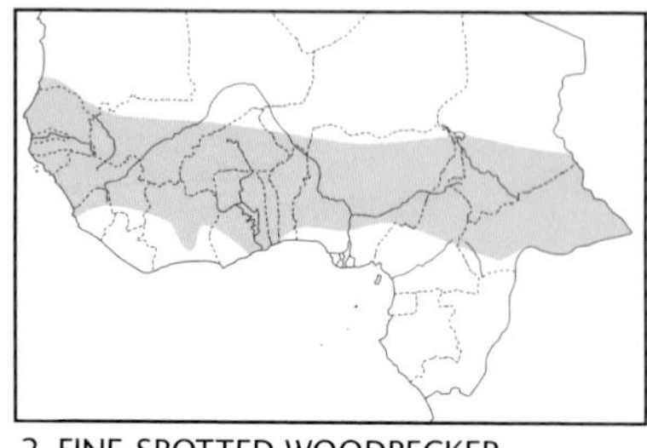

2. FINE-SPOTTED WOODPECKER

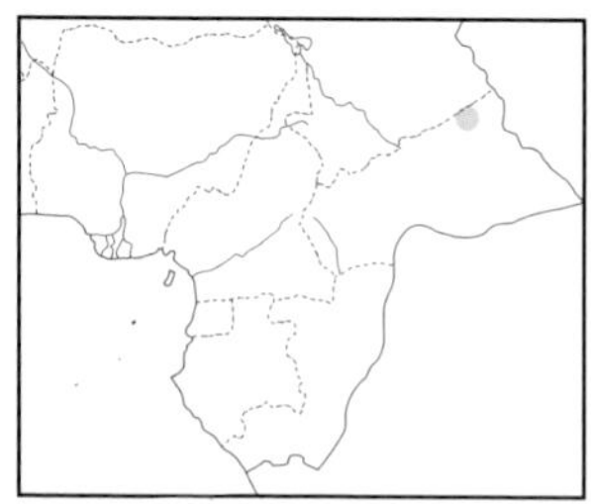

3. NUBIAN WOODPECKER

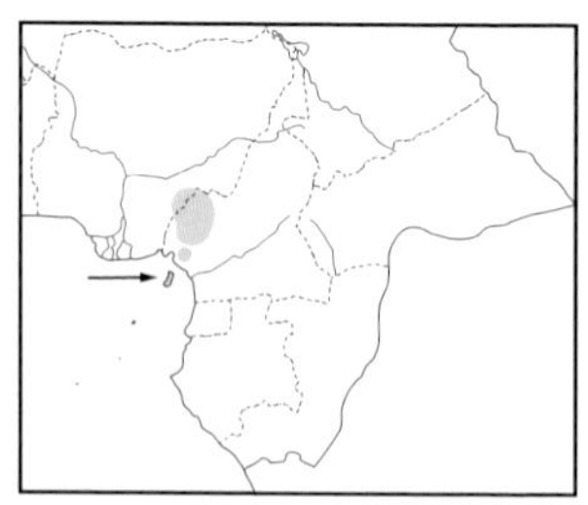

4. TULLBERG'S WOODPECKER

5. GABON WOODPECKER

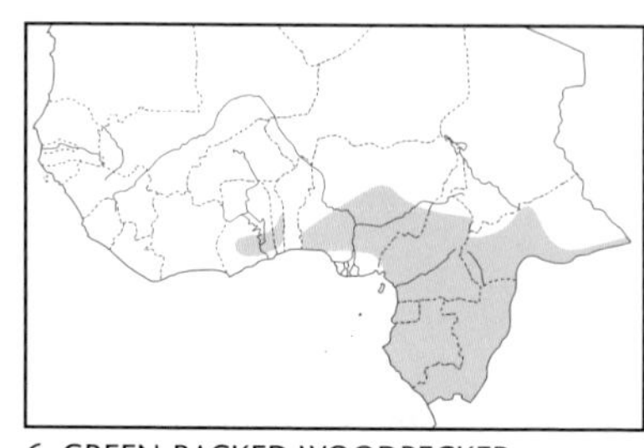

6. GREEN-BACKED WOODPECKER

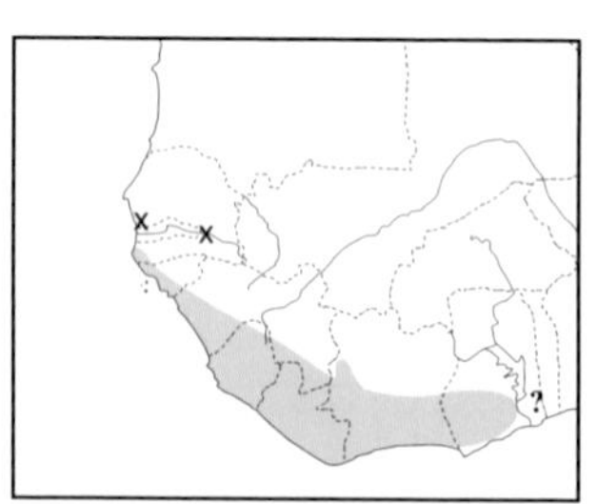

7. LITTLE GREEN WOODPECKER

8. BROWN-EARED WOODPECKER

9. BUFF-SPOTTED WOODPECKER

Plate on page 268

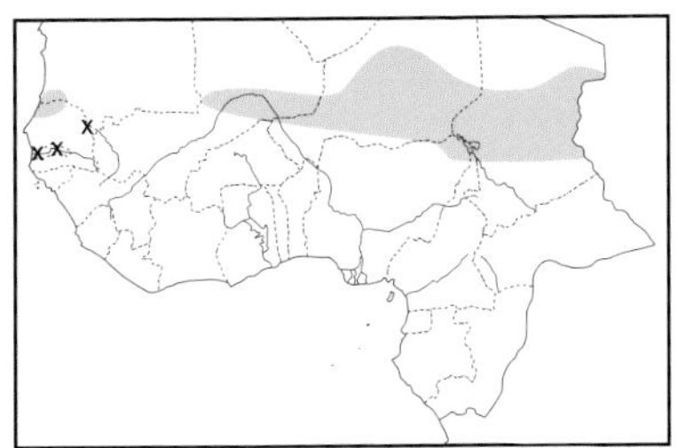

1. LITTLE GREY WOODPECKER

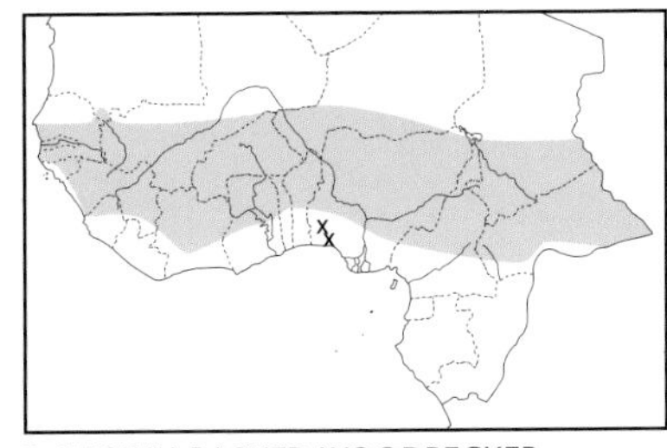

2. BROWN-BACKED WOODPECKER

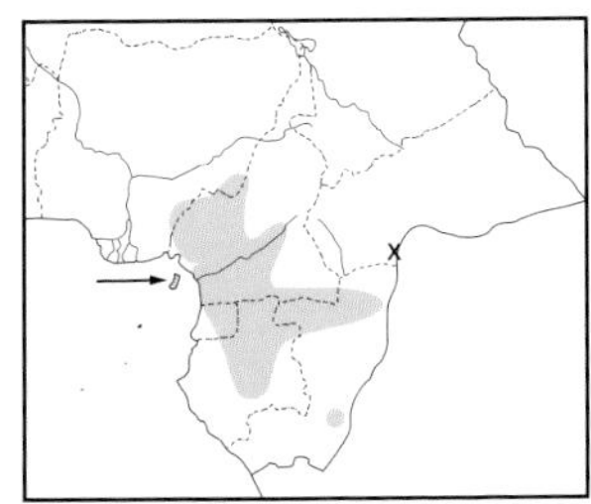

3. ELLIOT'S WOODPECKER

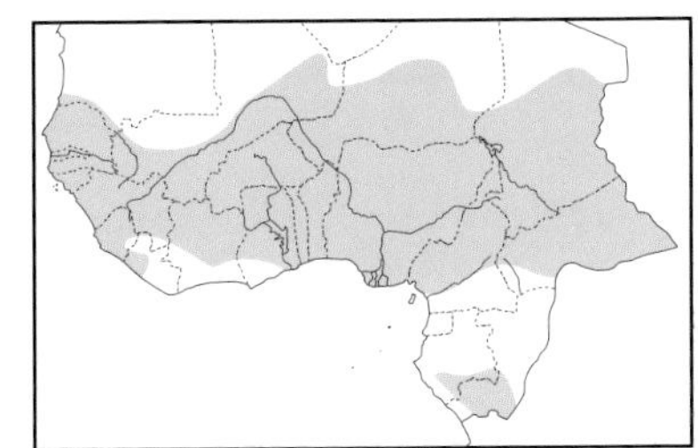

4. GREY WOODPECKER

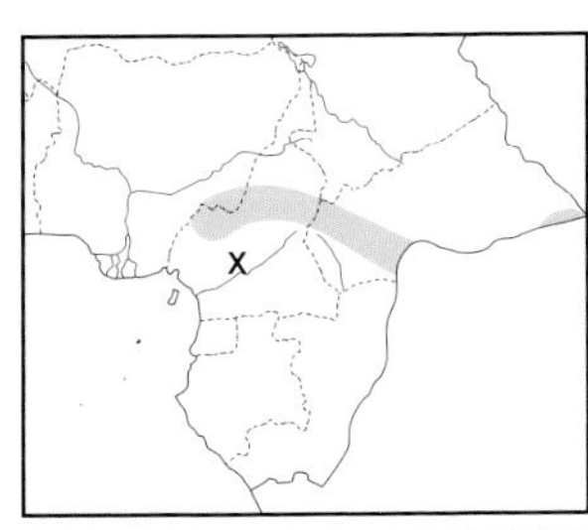

5. SPECKLE-BREASTED WOODPECKER

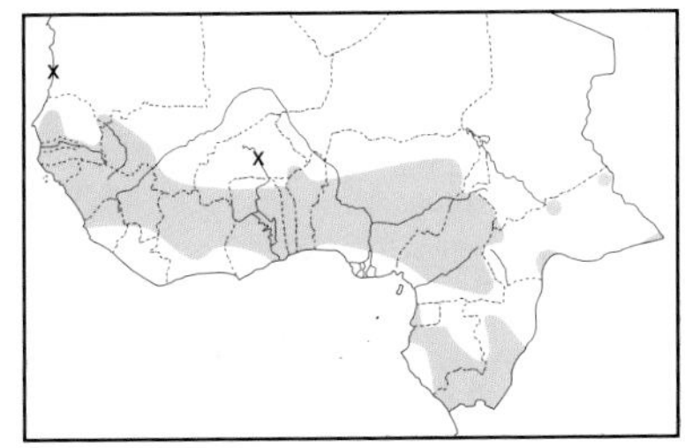

6. CARDINAL WOODPECKER

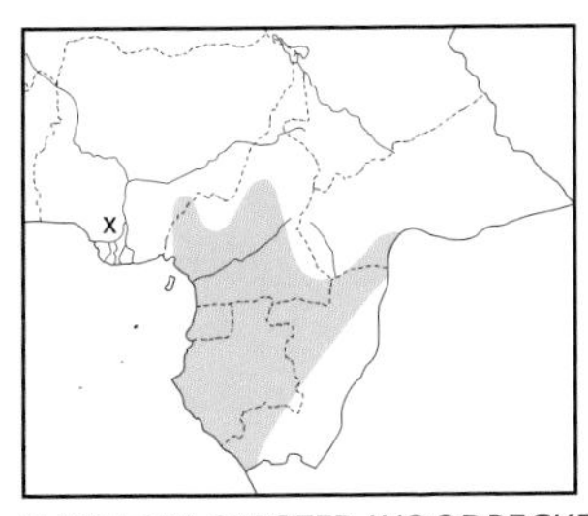

7. YELLOW-CRESTED WOODPECKER

8. BEARDED WOODPECKER

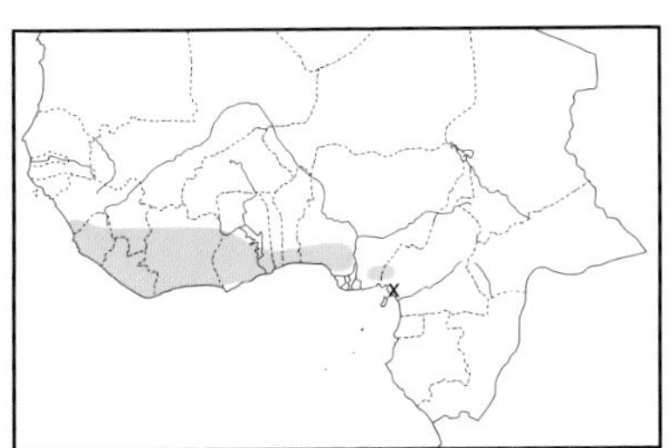

9. FIRE-BELLIED WOODPECKER

Plate on page 272

1 **LITTLE GREY WOODPECKER** *Dendropicos elachus* 11–13 cm **R u/r**
1a Ad male Very small and pale; barred back; red rump; red hindcrown and nape. **1b Ad female** Crown and nape dull brown. ▲ Arid country. ✤ Sharp, very rapid rattling *kree-kree-kree-kree-...* similar to 6, but harder and faster. [CD8:72]

2 **BROWN-BACKED WOODPECKER** *Dendropicos obsoletus* 13–14 cm **R u/f**
2a Ad male Small; greyish-brown above; plain back; wings and tail heavily spotted white; underparts whitish and only faintly streaked; red hindcrown. **2b Ad female** Entire crown dark brown. ▲ Woodland. ✤ Rapid *ki-ki-keeew-keeew-keeew* and *krreet-krreet-krreet*. [CD8:82]

3 **ELLIOT'S WOODPECKER** *Dendropicos elliotii* 19–21 cm **R u/lc**
3a Ad male *elliotii* (lowland Cameroon to Gabon–NW Congo) Plain green upperparts; black forehead and forecrown; red hindcrown and nape; plain pale greenish head-sides; heavily streaked underparts. **3b Ad male *johnstoni*** (montane areas SE Nigeria–W Cameroon, Bioko) Underparts mainly plain; red hindcrown and nape. **3c Ad female *johnstoni*** Crown and nape all black. ▲ Forest. ✤ Shrill *kree-kree-kree* and softer *kiwik-kiwik-kiwik*. Rather silent. Drumming similar to 7 but softer. [CD8:79]

4 **GREY WOODPECKER** *Dendropicos goertae* *c.* 20 cm **R f/c**
4a Ad male Grey head and underparts; golden-olive upperparts; red rump; orangey streak on belly; red crown and nape. **4b Ad female** Entire head grey. ▲ Woodland, forest edge. ✤ Shrill, loud *krreet-krreet-krreet*, descending in volume and sometimes preceded by *kik-kik-kik-*; fast and hard *kwik-wik-wik-wik* resembling 8 but higher pitched; *kee-krrirrt-krrirrt...* and variations. [CD8:80]

5 **SPECKLE-BREASTED WOODPECKER** *Dendropicos poecilolaemus* 14–15 cm **R r/lc**
5a Ad male Small; distinctive fine spotting on breast; red hindcrown. **5b Ad female** Entire crown black. ▲ Forest edges, light woodland. ✤ Hard, rapid series of *k-ret* notes. [CD8:73]

6 **CARDINAL WOODPECKER** *Dendropicos fuscescens* 14–15 cm **R s/lc**
6a Ad male *lafresnayi* (from Nigeria west) Small; faintly barred back; lightly streaked below; red hindcrown and nape. *D. f. sharpii* (from Cameroon east and south) has more heavily streaked underparts. **6b Ad female** Hindcrown and nape black. ▲ Woodland. ✤ Rapid series of harsh, rattling notes *kree-kree-kree-kree...* Other rapidly uttered notes include *kwik-ik-ik-ik...* [CD8:75]

7 **YELLOW-CRESTED WOODPECKER** *Dendropicos xantholophus* 20–23 cm **R r/lc**
7a Ad male Plain dark olive above; densely spotted and barred below; bold head markings; yellow hindcrown diagnostic (but often hard to see). **7b Ad female** Forehead and crown all black. ▲ Forest, dense woodland. ✤ Calls include *kreeerr, kreeerr, kreeerr* followed by short, descending series of *kweek* notes. Drumming, by both sexes, loud and rather slow. [CD8:78]

8 **BEARDED WOODPECKER** *Dendropicos namaquus* 23–25 cm **R f**
8a Ad male Large and dark; bold head markings; golden rump and tail tips; red hindcrown. **8b Ad female** Forehead speckled, hindcrown black. ▲ Woodland. ✤ Loud and rapid *wik-wik-wik-wik...* Drumming, by both sexes, loud and rather slow (single taps clearly noticeable). [CD8:76]

9 **FIRE-BELLIED WOODPECKER** *Dendropicos pyrrhogaster* *c.* 24 cm **R f/r**
9a Ad male Large; red rump and centre of underparts; bold head markings; red crown. **9b Ad female** Forehead and crown all black. ▲ Forest and forest edge. Endemic. ✤ Hard, short *kweep* or *kwip*, accelerating into rattling *kwipibibibibib*. Drumming loud, in rather fast and short bursts. [CD8:77]

Maps on page 271

1b
2b
3c
1a
2a
3b
3a
4b
4a
5b
6b
5a
6a
7b
8b
9b
7a
8a
9a

PLATE 80: WRYNECKS, PICULET, BROADBILLS AND PITTAS

Wrynecks have cryptic plumage and soft tails, which are not used as brace when clinging to trees. Feed mainly on ants. Nest in natural cavities or old woodpecker or barbet holes.

1 EURASIAN WRYNECK *Jynx torquilla* 16–18 cm **P u/r**
Cryptically patterned, bark-like plumage. ▲ Savanna. ✤ Usually silent in winter quarters; occasionally a series of loud *kièh-kièh-kièh-*..., recalling small falcon. [CD3:33]

2 RED-THROATED WRYNECK *Jynx ruficollis* 18–19 cm **R u/lc**
Brown mottled and barred above; distinctive chestnut breast patch. **2a Ad *pulchricollis*** (SE Nigeria–CAR) Chestnut patch extends onto lower throat. **2b Ad *ruficollis*** (SE Gabon–Congo) Chestnut patch larger, extending onto chin. ▲ Open woodland. ✤ A series of 2–12 hard *kweeh* notes, slower than 1. [CD8:61]

Piculets. Tiny woodpecker-like birds with short bills and very short tails, which are not used as brace when clinging to bark. Nest in self-excavated holes in trees and vines.

3 AFRICAN PICULET *Sasia africana* 8–9 cm **R r/lc**
Tiny, appearing tail-less. Sharp pointed bill; bare red skin around eye; red feet. **3a Ad male** Orange-red forehead. **3b Ad female** Olive-green forehead. ▲ Forest. ✤ Very fast, high-pitched, piercing little trill *tsiririririri*, reminiscent of bat or insect. [CD8:62]

Broadbills. Small, flycatcher-like forest birds with broad gapes, large heads and short tails. Possess unique short elliptical display flight from favoured branches, accompanied by distinctive rattling. Otherwise very inconspicuous.

4 AFRICAN BROADBILL *Smithornis capensis* 13 cm **R s/r**
4a Ad male *camarunensis* (from E Nigeria east) Streaked underparts; black crown. White patch on back conspicuous in display flight. *S. c. delacouri* (Sierra Leone–Togo) is slightly paler; nape greyer. **4b Ad female *camarunensis*** Crown greyish-brown. ▲ Secondary and gallery forest, thickets. ✤ Weak plaintive *hweee*. Fast, strident rattle, similar to 5 but lower pitched, produced during display. [CD8:85]

5 RUFOUS-SIDED BROADBILL *Smithornis rufolateralis* 11.5 cm **R u/lf**
5a Ad male Bright rufous-orange patch on breast-sides; black head; white spots on wing-coverts. White patch on back conspicuous in display flight. **5b Ad female** Breast patches duller; head brown. ▲ Forest interior. ✤ Thin, plaintive *theew* or *huiiii* often followed by display flight. Fast, strident rattling *tttt-trrrrrrrrree* in display. [CD8:84]

6 GREY-HEADED BROADBILL *Smithornis sharpei* 15 cm **R lf/r**
6a Ad male Bright rufous breast; dark grey head with rufous loral spot and malar stripe; undertail-coverts rufous. **6b Ad female** Washed-out rufous on breast; pale grey head. ▲ Lowland and montane forest. ✤ Thin, plaintive *theew* or *huiiii*. Fast, strident rattle in display flight resembles 4 and 5, but is rather shorter and lower pitched than 5. [CD8:83]

Pittas. Brilliantly coloured but very secretive terrestrial forest species with long, strong legs and short tails.

7 AFRICAN PITTA *Pitta angolensis* 17–22 cm **M/R r**
7a Ad Unmistakable. Robust and brightly coloured. Cinnamon-buff breast; azure-blue rump. **7b Juv** Duller and darker version of 7a; brownish underparts; no wing markings. **7c Ad in flight** White wing patch. ▲ Evergreen and semi-deciduous forest. ✤ Low frog-like croak. In display a loud, abrupt, rising and rather melodious whistle *prrruueep!* or *krrooit!* introduced by wing noise. Mechanical sounding wingbeats *prrt* audible at long range. [CD8:87]

8 GREEN-BREASTED PITTA *Pitta reichenowii* 19–21 cm **R r**
Similar to 7a, but with green breast. ▲ Old secondary forest, thickets. ✤ Call a pure, rather plaintive whistle *huu* ... *huu* ... *huu* ... uttered in series of variable speed and resembling call of Fire-crested Alethe. Short *prrrt* in flight. Song similar to 7. [CD8:86]

Maps on page 276

1
2b
2a
3b
3a
4b
4a
6b
6a
5b
5a
7c
8
7a
7b

PLATE 80: WRYNECKS, PICULET, BROADBILLS AND PITTAS

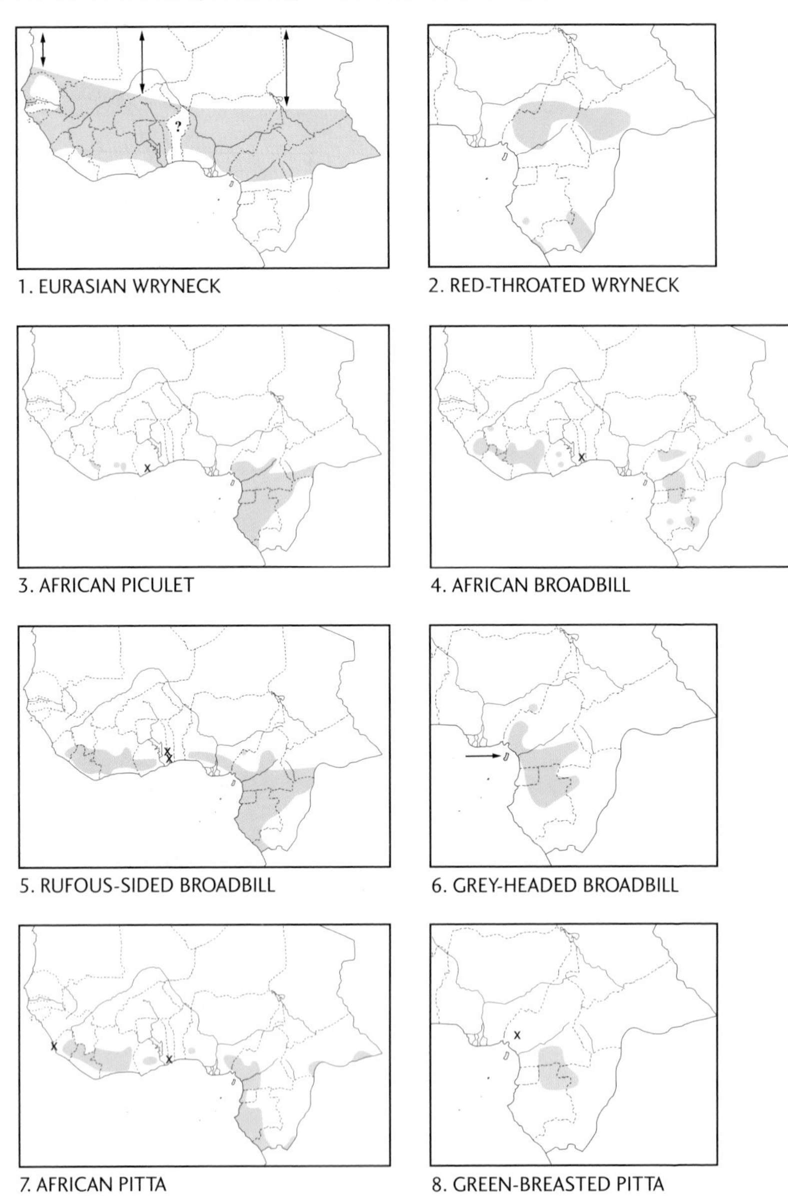

1. EURASIAN WRYNECK

2. RED-THROATED WRYNECK

3. AFRICAN PICULET

4. AFRICAN BROADBILL

5. RUFOUS-SIDED BROADBILL

6. GREY-HEADED BROADBILL

7. AFRICAN PITTA

8. GREEN-BREASTED PITTA

Continued from page 278

10 THEKLA LARK *Galerida theklae* 17 cm V

Very similar to 8. Differences include slightly shorter crest, better marked head pattern, bolder streaking on whiter breast, contrasting cinnamon-tinged rump. Often perches atop bushes. ▲ Dry open country. Palearctic vagrant (rare resident?). ✤ [CD3:47]

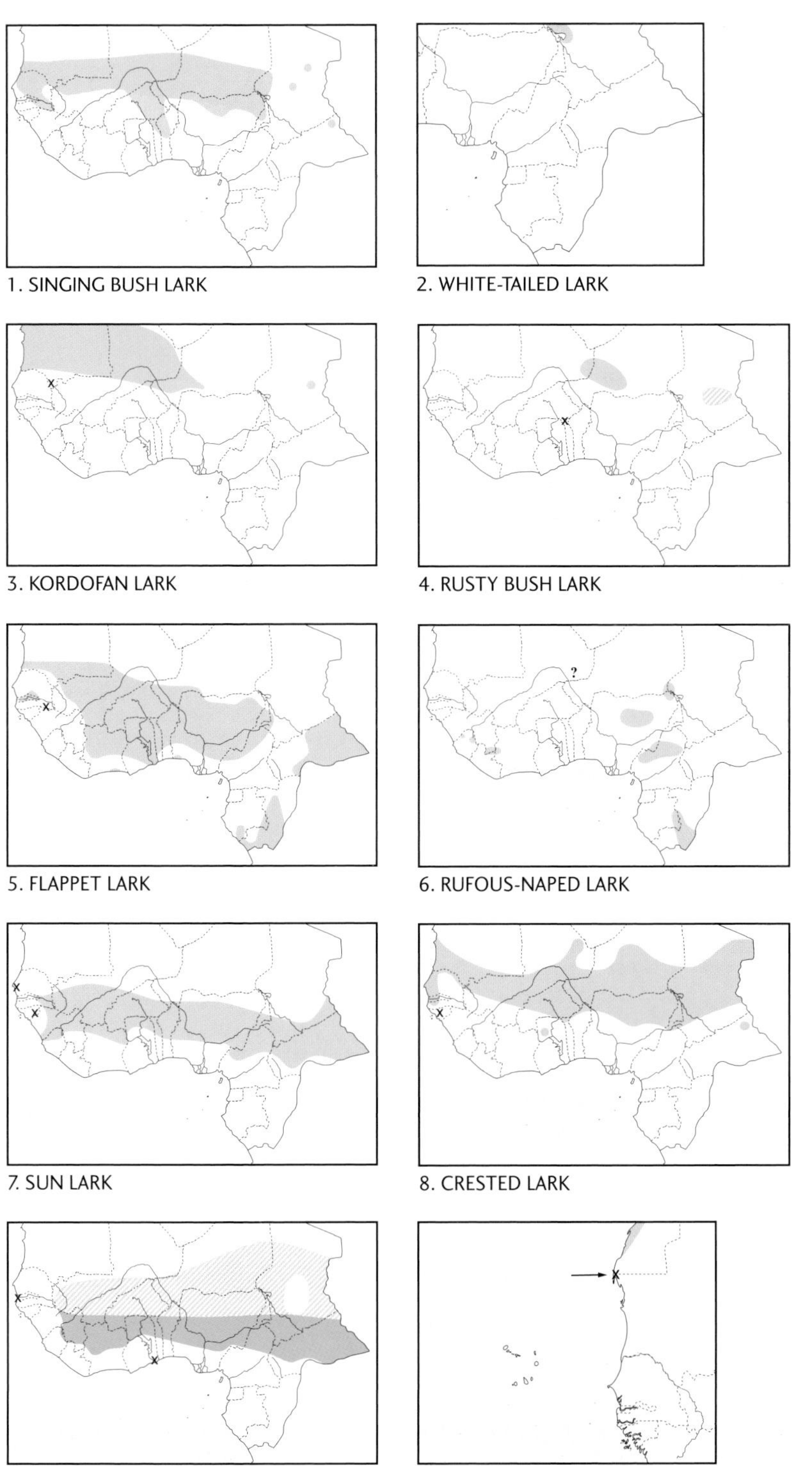

1. SINGING BUSH LARK

2. WHITE-TAILED LARK

3. KORDOFAN LARK

4. RUSTY BUSH LARK

5. FLAPPET LARK

6. RUFOUS-NAPED LARK

7. SUN LARK

8. CRESTED LARK

9. RUFOUS-RUMPED LARK

10. THEKLA LARK

Plate on page 278

Terrestrial birds with cryptic plumage pattern and coloration, often varying within a species to match dominant shade of soil. Fresh and worn plumage often quite different: with wear, pale feather tips and fringes abrade, leaving only dark centres, resulting in more uniform plumage.

1 SINGING BUSH LARK *Mirafra cantillans* 13 cm **R/M lf**
Sandy grey-brown above, streaked dusky; whitish below; pale bill. ▲ Dry, open grassland, semi-arid thornbush. ✤ Rapid medley of short phrases of varied notes; same phrase often repeated several times; from ground or low perch, or in fluttering display flight. [CD8:88a]

2 WHITE-TAILED LARK *Mirafra albicauda* 13 cm **R r**
Blackish, scaly-looking above; short tail with white outer feathers; rufous wing panel. ▲ Open grassland. ✤ Harsh notes and some whistles, lacking 1's trills, in high display flight. [CD8:89]

3 KORDOFAN LARK *Mirafra cordofanica* 14 cm **R/M u/r**
Pale sandy-rufous above; tricoloured tail (rufous, black and white); stout whitish bill. ▲ Sub-desert with red sandy soil. ✤ Series of short, varied phrases with trills, chirps, whistled notes and imitations; from ground, bush tops or in high display flight. [CD8:88b]

4 RUSTY BUSH LARK *Mirafra rufa* 13–14 cm **R lf**
Rufous-brown above, variably streaked; no white in comparatively long tail; bicoloured bill. ▲ Dry, open, rocky bush. ✤ Song, in cruising display flight, described as 'pleasing'.

5 FLAPPET LARK *Mirafra rufocinnamomea* 13–14 cm **R lu/c**
Plumage variable. In flight, rufous wing patch. **5a Ad *tigrina*** (E Cameroon–E CAR) Rufous above and below. *M. r. serlei* (SE Nigeria) slightly deeper rufous above. **5b Ad *buckleyi*** (from Nigeria west) Cinnamon-brown streaked blackish above. *M. r. schoutedeni* (Gabon–Congo, S CAR) pale brown, lightly marked above. ▲ Grassy habitats. ✤ Dry, rattling *prrrrrrp*, produced by rapid wing-flaps during undulating display flight (diagnostic). [CD8:93]

6 RUFOUS-NAPED LARK *Mirafra africana* 16–18 cm **R lu/c**
Short crest. In flight, rufous wing patch, relatively short tail. Plumage variable. **6a Ad *henrici*** (Sierra Leone; Mt Nimba) Upperpart feathers blackish fringed sandy-rufous; boldly streaked breast. *M. a. batesi* (E Mali, Nigeria, SE Niger) slightly paler. **6b Ad *stresemanni*** (N Cameroon) Rufous-brown lightly streaked black above; rich cinnamon below. *M. a. bamendae* (W Cameroon) similar below; more like *henrici* above. **6c Ad *malbranti*** (SE Gabon–Congo) Dull brown above; almost unstreaked, buffish, below. ▲ Various open habitats. ✤ From low perch, 2–3 clear whistles, endlessly repeated. Song flight with whistles, chirps and trills, also imitations. [CD8:90-91]

7 SUN LARK *Galerida modesta* 14 cm **R lf/c**
Prominent supercilium; streaked cap; heavily streaked upperparts and breast. Plumage variable (4 races). **7a Ad *modesta*** (almost throughout) Palest race; sandy-rufous above. **7b Ad *nigrita*** (Guinea–Sierra Leone) Darkest race. ▲ Open grassy habitats. ✤ Short, fast, grating warble; from ground or bush, or in flight. More varied and sustained song in hovering display flight. [CD8:97]

8 CRESTED LARK *Galerida cristata* 17 cm **R f/lc**
Stocky; spiky crest; relatively short tail; slightly decurved longish bill. Plumage variable (4–5 races). **8a Ad *senegallensis*** Pale grey-brown above; off-white below; streaked breast. **8b Ad *jordansi*** Most rufous race; indistinct small spots on breast. ▲ Open habitats. ✤ Liquid *doo-leeoo* and *diui*. Song varied and sustained, with clear whistles and throaty twitters. Also fast and clear *tee-titee-titiu*. From perch or ground and in flight. [CD3:46]

9 RUFOUS-RUMPED LARK *Pinarocorys erythropygia* 18 cm **M lf**
Large, slender. Rufous rump and tail edges; heavily streaked breast; well-marked head pattern. ▲ Open savanna woodland, farmland. ✤ Short, often repeated series of clear, far-carrying whistles with some variation, in high display flight with dangling legs. [CD8:94]

Continued on page 276

1
2
3
4
5a
5b
6c
6b
6a
7b
7a
8a
9
8b
10

1 **BAR-TAILED LARK** *Ammomanes cinctura* 14 cm **R+/M lu/c**
1a Plain, sandy-coloured above; pale bill. **1b In flight** Pale orange-rufous tail with black terminal bar; pale orange-rufous in wings. ▲ Desert. Erratic. ✤ Monotonously repeated, clear, high-pitched whistle *tsu-wheeh*, *tsu-wheeh*,... mostly in undulating display flight. [CD3:40]

2 **DESERT LARK** *Ammomanes deserti* 15–16 cm **R lc**
2a Greyer than 1; bill stouter and darker. Also compare 3. **2b In flight** Orange-rufous tail with broad, ill-defined blackish terminal bar. ▲ Arid country. ✤ Repeated, simple series of trilling syllables *trreewrrurrip*, *trreewrrurrip*, ... , etc. from perch or in short display flight. [CD3:41]

3 **DUNN'S LARK** *Eremalauda dunni* 14 cm **R u/r**
3a Resembles 1, but faintly streaked when fresh; bill stouter, stubbier. **3b In flight** Black outer tail contrasting with pale centre. Compare 1b. ▲ Desert, sub-desert. ✤ From ground or low perch, a fast *wit-wit-wtrrreedridridridrree*; more variable in display flight. [CD3:44]

4 **LESSER SHORT-TOED LARK** *Calandrella rufescens* 13 cm **P r**
Resembles 5, but with distinct primary projection; variably streaked breast without black patches; shorter, stubbier bill. ▲ Arid, open country. ✤ Rippling *prrrrt*... [CD3:43]

5 **GREATER SHORT-TOED LARK** *Calandrella brachydactyla* 13–14 cm **P/V+ u/lc**
Streaked above; broad pale supercilium; long tertials almost completely cloaking primaries; variable dark patch on breast-sides. ▲ Arid and semi-arid country. ✤ Hard, dry *prrt*, *prrt-trrt*. Song in wintering grounds a weak jumble of harsh and buzzy notes. [CD3:42]

6 **RED-CAPPED LARK** *Calandrella cinerea* 14–15 cm **R/M lu**
Distinctive: rufous crown; white supercilium; rufous patch on breast-sides. ▲ Grass- and farmland. ✤ Short, varied jumble of high-pitched notes and trills, including imitations of other birds, in high display flight. [CD8:96]

7 **CHESTNUT-BACKED SPARROW LARK** *Eremopterix leucotis* 12 cm **R/M c**
7a Ad male Black head; white patches on ear-coverts and nape; chestnut upperparts; black underparts and underwing. **7b Ad female** Dusky mottled head; pale buff below; breast mottled brownish. ▲ Semi-arid country. Moves north with rains. ✤ Rapid phrase with harsh, grating notes, from ground or low perch and in flight. [CD8:98]

8 **BLACK-CROWNED SPARROW LARK** *Eremopterix nigriceps* 11–12 cm **R+ u/lc**
8a Ad male White forehead; pale grey-brown upperparts; black underparts. **8b Ad female** Pale sandy-cinnamon above; whitish below; pale cinnamon breast-band. **8c Ad female in flight** Black underwing-coverts diagnostic. ▲ Arid and semi-arid country. ✤ Abrupt, buzzing *eezp*. Pleasant, simple, rapid phrase of high, clear whistles, usually in flight [CD8:99]

9 **EURASIAN SKYLARK** *Alauda arvensis* 16-18 cm **V**
Robust; streaked; short crest. In flight, white trailing edge to wing and white outer tail feathers. ▲ Palearctic vagrant. ✤ Rippling trruwee or chirrup in flight. [CD3:49]

10 **TEMMINCK'S HORNED LARK** *Eremophila bilopha* 14 cm **P/R? r**
Pale sandy; contrasting face pattern; thin horns on crown-sides. ▲ Desert, sub-desert. ✤ Calls *tsip*, *seeuu* and *sweeeup*. [CD3:52]

11 **GREATER HOOPOE LARK** *Alaemon alaudipes* 19–22 cm **R+/M c/u**
11a Large; sandy; long decurved bill; long legs. **11b In flight** Black-and-white wing pattern diagnostic. ▲ Sandy desert. ✤ Piping, accelerating and ascending series of pure drawn-out whistles, occasionally with short buzzy trill. [CD3:37]

12 **THICK-BILLED LARK** *Rhamphocoris clotbey* 17 cm **P r**
12a Large, stubby bill; heavily spotted underparts. **12b In flight** Large head; long wings with black-and-white pattern; relatively short tail. ▲ Stony sub-desert. ✤ Short jingle of clear, sweet and tinkling notes. [CD3:38]

 Maps on page 282

1b
1a
2b
2a
3b
3a
4
5
6
7b
7a
8a
8c
9
10
8b
11b
11a
12b
12a

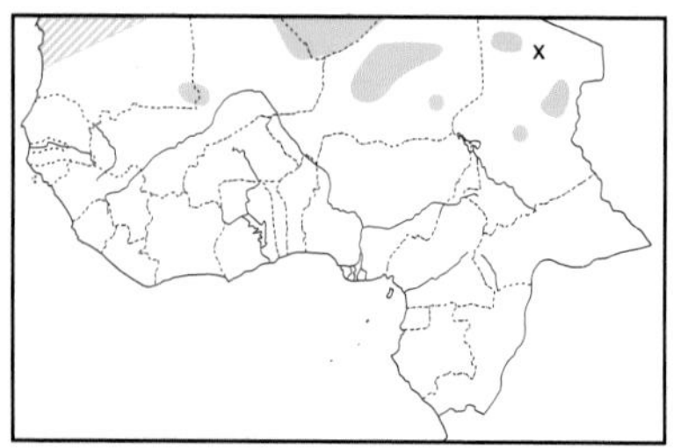
1. BAR-TAILED LARK

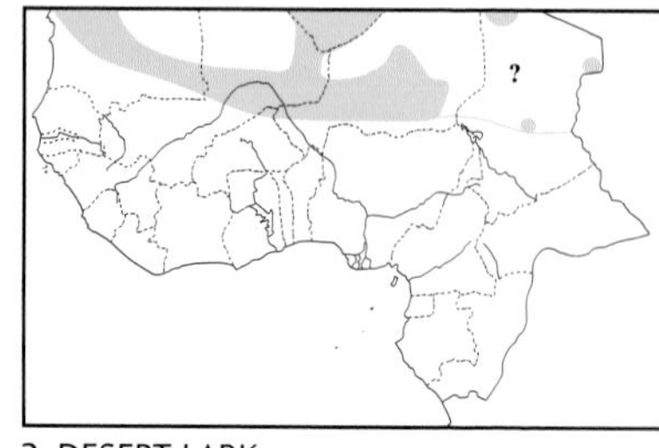
2. DESERT LARK

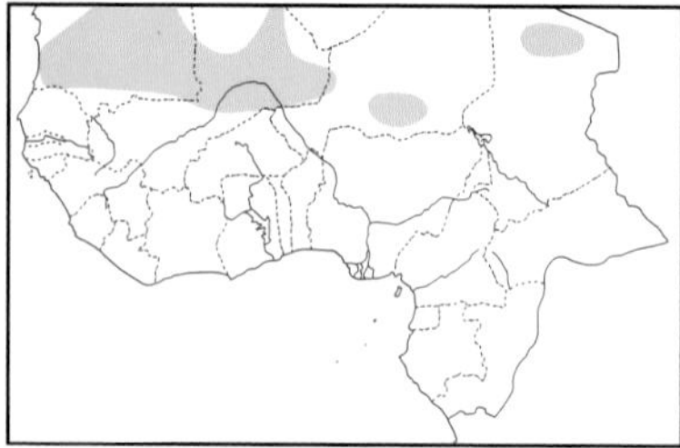
3. DUNN'S LARK

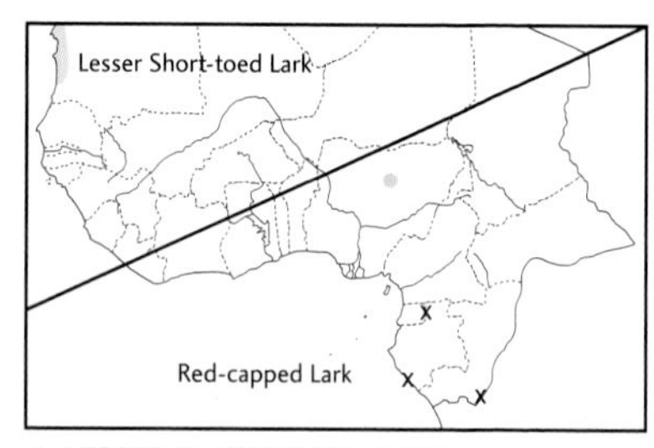

4. LESSER SHORT-TOED LARK
6. RED-CAPPED LARK

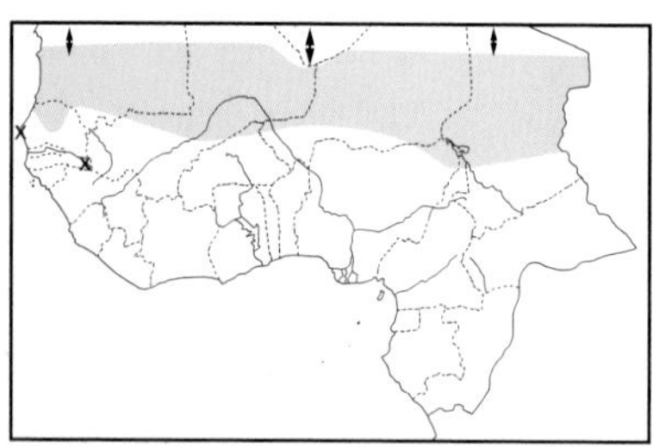
5. GREATER SHORT-TOED LARK

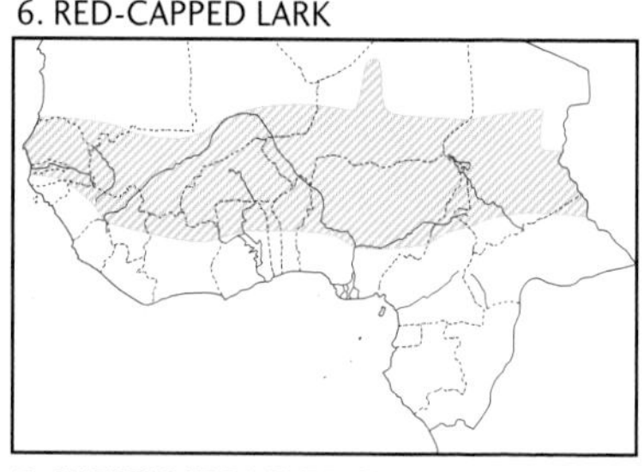
7. CHESTNUT-BACKED SPARROW LARK

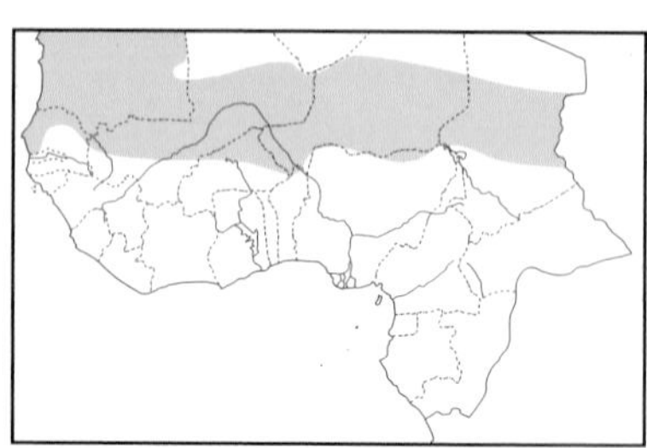
8. BLACK-CROWNED SPARROW LARK

9. EURASIAN SKYLARK
12. THICK-BILLED LARK

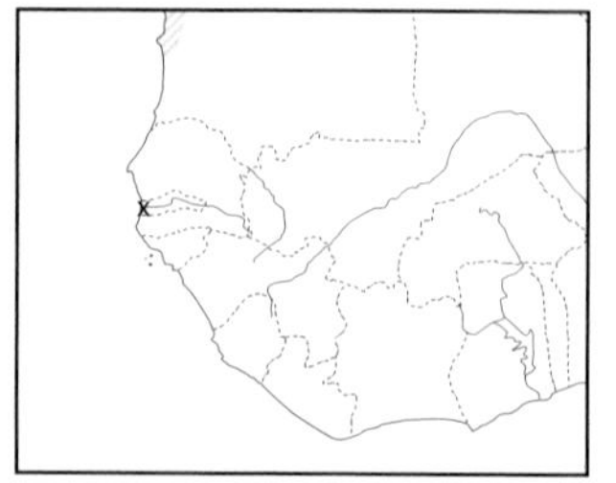
10. TEMMINCK'S HORNED LARK

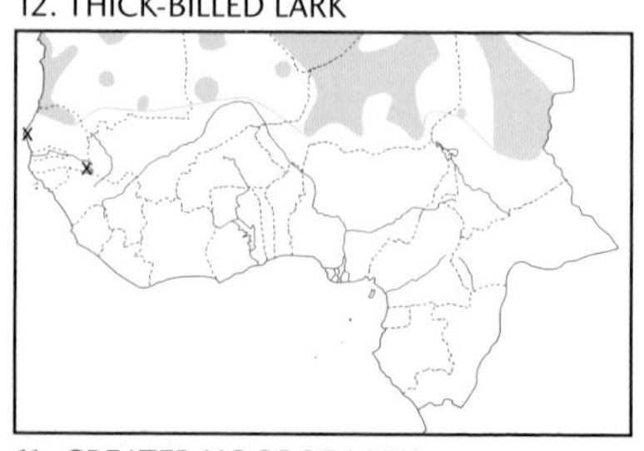
11. GREATER HOOPOE LARK

Plate on page 280

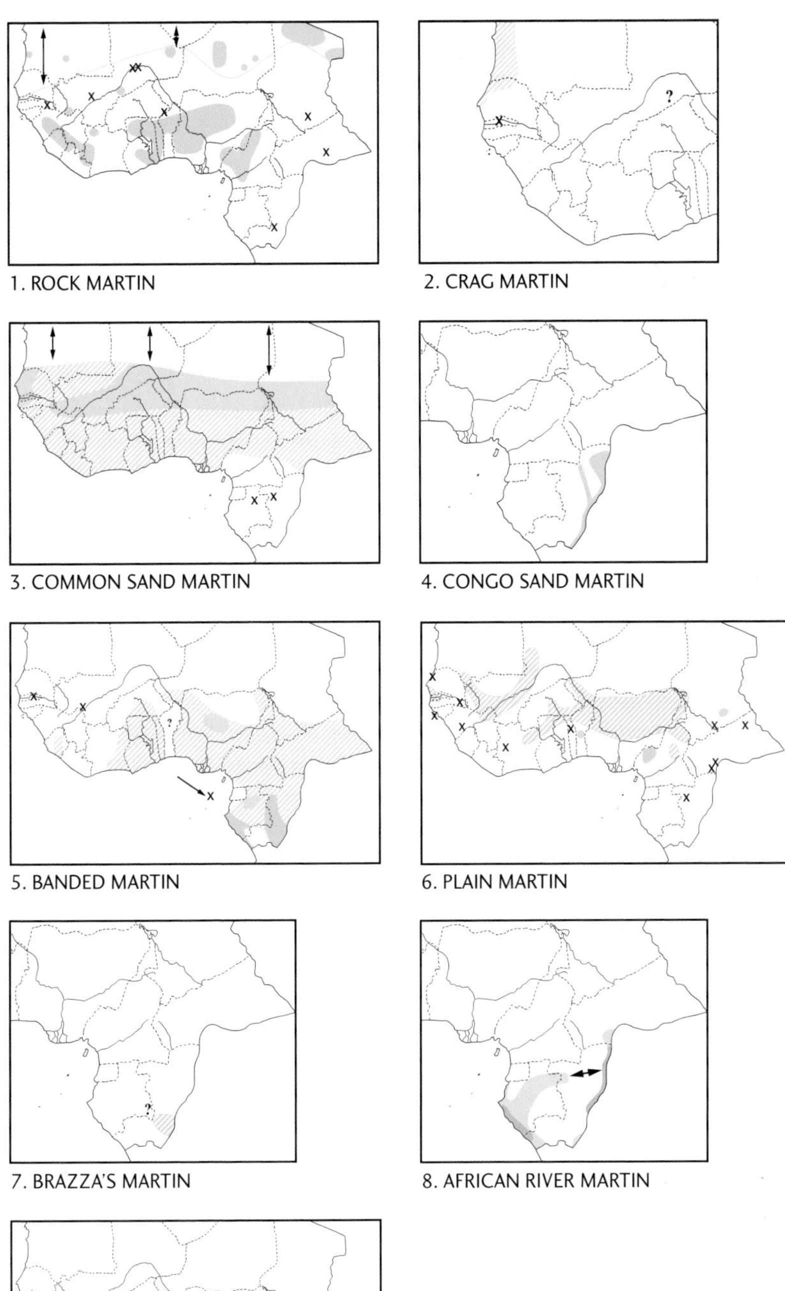

1. ROCK MARTIN

2. CRAG MARTIN

3. COMMON SAND MARTIN

4. CONGO SAND MARTIN

5. BANDED MARTIN

6. PLAIN MARTIN

7. BRAZZA'S MARTIN

8. AFRICAN RIVER MARTIN

9. WHITE-THROATED BLUE SWALLOW

Plate on page 284

PLATE 83: SWALLOWS I

Distinctive, highly specialised aerial insectivores with slender body, short neck and pointed wings. Bill very short, but gape broad; legs short. In some, tail deeply forked with long streamers, often with white 'windows' (white patches on inner webs of feathers). Sexes similar; in long-tailed species, females have shorter outer tail feathers. Juveniles similar but duller. Regularly perch on wires and bare branches. Most gregarious when not breeding (some also when breeding), often forming mixed groups with other hirundines. Many are migratory.

1 ROCK MARTIN *Hirundo fuligula* 12–13 cm **R/P lf/c**
1a Tropical W African races (almost throughout, except range of 1c) Dark brown above, paler below. **1b In flight** White patches in tail; no contrasting wing-coverts. **1c *Obsoleta* group** (Pale Crag Martin; N Mauritania, Mali, N Niger, NE Chad) Paler and greyer than 1a. **1d In flight** Resembles 2, but much less contrasting underwing-coverts. ▲ Rocky outcrops, cliffs, gorges, villages, towns. ✤ Mostly silent. [CD3:56]

2 CRAG MARTIN *Hirundo rupestris* 14–15 cm **P r**
2a Grey-brown above; pale buffish below, brownish on lower abdomen. **2b In flight** Dark underwing-coverts contrast strongly with paler flight feathers. ▲ Open habitats. ✤ Mostly silent. [CD3:57]

3 COMMON SAND MARTIN *Riparia riparia* 12–13 cm **P/V+ r/lc**
3a Brown above; pure white below; well-defined brown breast-band. **3b In flight** White underparts with contrasting breast-band; tail slightly forked. ▲ Open habitats. ✤ Dry, rasping *chrrp*. [CD3:54]

4 CONGO SAND MARTIN *Riparia congica* 11–12 cm **R lc**
4a Resembles 3, but brown breast-band much more diffuse. **4b In flight** White underparts with diffuse breast-band; tail almost square. ▲ Rivers. Congo (Congo R., lower Ubangi R., Sangha R.). ✤ No information.

5 BANDED MARTIN *Riparia cincta* 17 cm **M/R/V u/r**
5a Large; brown above, white below; brown breast-band; short white supercilium. **5b In flight** Large size conspicuous; slow and erratic, tern-like flight. ▲ Grassy savanna. ✤ Mostly silent. Call *chrip*. [CD9:6]

6 PLAIN MARTIN *Riparia paludicola* 11–12 cm **R/M/V+ lf/r**
6a Brown above; greyish-brown throat and breast. **6b In flight** Brownish throat and breast; rest of underparts white. ▲ Over or near rivers and lakes; in off-season also away from water. Usually in small flocks. ✤ Calls include rasping *chtrrr* and harsh *steeh*. Song a soft twitter. [CD3:53]

7 BRAZZA'S MARTIN *Phedina brazzae* 12 cm **R/M? r DD**
7a Dark brown above; white below, heavily streaked dark brown. **7b In flight** Streaked underparts; dark brown underwing; square tail. ▲ Forested rivers. ✤ No information.

8 AFRICAN RIVER MARTIN *Pseudochelidon eurystomina* 14 cm **R/M DD**
8a All black; reddish bill and eye; large head. **8b In flight** Compact, triangular silhouette produced by broad-based wings and shortish tail. ▲ Rivers, sandy grassland. Migrates over forest, sometimes in huge numbers. ✤ Hard, grating *dzreh dzreh*... or *chèrr chèrr*... [CD9:1]

9 WHITE-THROATED BLUE SWALLOW *Hirundo nigrita* 12 cm **R lc**
9a Dark glossy purple-blue; white throat. **9b In flight** White patches in tail. Flies low and fast over water. ▲ Forested rivers. ✤ Vigorous *weetch*, hard *vwhit* and soft *whit*. Song a soft, dry trill *prl-trrrrrr* mixed with disharmonic notes.

Maps on page 283

1a
1b
1c
1d
2a
2b
3a
3b
4a
4b
5a
5b
6a
6b
7a
7b
8a
8b
9a
9b

1 **BARN SWALLOW** *Hirundo rustica* 15–19 cm **P+ c/f**
1a Ad breeding Dark rufous forehead and throat; broad dark blue breast-band. Compare 4. **1b Ad non-breeding** Duller; rufous on throat may become very pale (even whitish); no tail streamers. **1c Ad breeding in flight** Long tail streamers; whitish underparts. ▲ Various habitats. ✤ Vigorous *whit-whit*; also a sharp *siflit*. [CD3:58]

2 **WHITE-THROATED SWALLOW** *Hirundo albigularis* 14–17 cm **V**
2a Rufous forehead; white throat; blue breast-band (sometimes broken). **2b In flight** Dark breast-band contrasting with white throat. ▲ Vagrant from southern Africa. ✤ [CD9:22]

3 **GREY-RUMPED SWALLOW** *Pseudhirundo griseopyga* 14 cm **R/M r/u**
3a Small; dark above; white below; long tail streamers. **3b–c In flight** Grey rump; no white in tail. ▲ Grassy and wooded savanna. ✤ Mostly silent. Soft, nasal *chwèèp*. [CD9:7]

4 **RED-CHESTED SWALLOW** *Hirundo lucida* 15 cm **R/M u/lc**
4a As 1a, but rufous reaches to upper breast; breast-band narrower. **4b In flight** As 1c, but tail streamers shorter; more white in tail; underparts whiter. ▲ Open habitats. ✤ Similar to 1. [CD9:24]

5 **ANGOLA SWALLOW** *Hirundo angolensis* 14–15 cm **R/M? u**
5a As 4a, but with ashy underparts. **5b In flight** Dull ashy below, incl. underwing-coverts. Flight rather slow. ▲ Open habitats. ✤ Song a soft twittering. [CD9:23]

6 **ETHIOPIAN SWALLOW** *Hirundo aethiopica* 13 cm **R/M u/lc**
6a Rufous forehead; buff-white throat; dark blue patch on breast-sides. Compare 7. **6b In flight** White below; no breast-band; long tail streamers. ▲ Open habitats. ✤ Soft *chit* or *cheep*. Song a melodious twittering. [CD9:21]

7 **WIRE-TAILED SWALLOW** *Hirundo smithii* 14 cm **R/M u**
7a Rufous cap; white underparts; no breast-band. **7b In flight** White below; dark blue patches on lower flanks; wire-like tail streamers. ▲ Usually near water. ✤ Mostly silent. [CD9:18]

8 **GREATER STRIPED SWALLOW** *Hirundo cucullata* 18–20 cm **V**
8a As 9 but streaking much finer; cap darker rufous; ear-coverts white. **8b In flight** Rump paler than in 9; streaking only visible at close range. ▲ S African vagrant, S Congo. ✤ [CD9:11]

9 **LESSER STRIPED SWALLOW** *Hirundo abyssinica* 15–19 cm **R/M f/lc**
9a Ad *puella* (Senegambia–N Cameroon) White underparts finely streaked dark. In *bannermani* (NE CAR) streaking much finer; rufous areas paler. **9b Ad *maxima*** (SE Nigeria–SW CAR) Streaks broader and blacker. In *unitatis* (Gabon–Congo) streaking intermediate between 9a and 9b. **9c–d In flight** Rufous head and rump; streaked underparts; long tail streamers. ▲ Open wooded habitats. ✤ Wheezy *cheeew*. Song varied and nasal. [CD9:10]

10 **RED-RUMPED SWALLOW** *Hirundo daurica* 16–17 cm **R/M/P u/lc**
10a Ad *domicella* (Senegambia–Chad) Dark blue cap; rufous collar; creamy-white underparts. **10b–c Ad *domicella* in flight** Rufous rump; long tail streamers, often curving inwards; clear-cut black undertail-coverts; no white in tail. **10d Ad *kumboensis*** (highlands Sierra Leone, W Ivory Coast, Cameroon) Pale rufous underparts. **10e Ad *rufula*** (Palearctic migrant in north) Creamy-buff underparts with faint streaks. **10f Ad *rufula* in flight** Pale forehead; two-toned rump. ▲ Savanna. ✤ Soft *djuit*. Song varied. [CD3:55]

11 **MOSQUE SWALLOW** *Hirundo senegalensis* 21–23 cm **R/M s/lc**
11a Dark blue cap; rufous collar; pale throat; orange-rufous underparts. **11b–c In flight** Large; orange-rufous rump; tail streamers shorter than 12; whitish underwing-coverts; rufous undertail-coverts; no white in tail. ▲ Open habitats. ✤ Nasal *nyaa*; a piping, reedy note, recalling sound of tiny trumpet; guttural notes. [CD9:9]

Continued on page 289

Maps on page 288

1b
1c
1a
2b
2a
3c
3a
3b
4a
5a
7a
5b
6b
6a
8a
9c
7b
9b
4b
9a
8b
10b
10e
10d
10a
9d
10f
10c
12b
11a
12c
11b
11c
12a

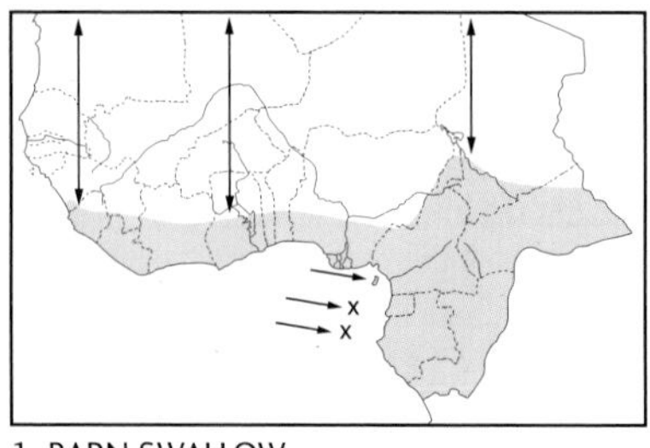

1. BARN SWALLOW

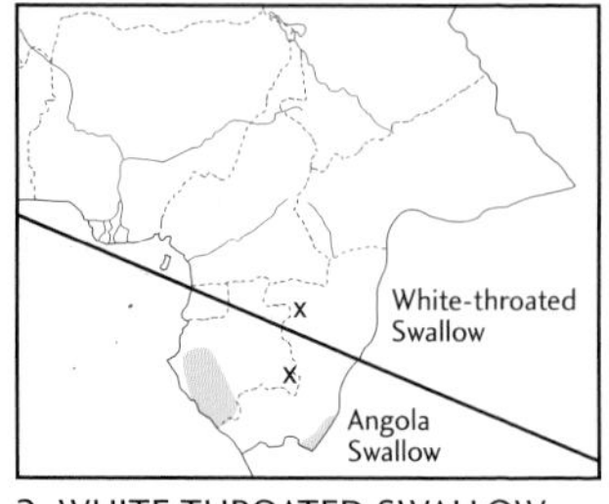

2. WHITE-THROATED SWALLOW
5. ANGOLA SWALLOW

3. GREY-RUMPED SWALLOW

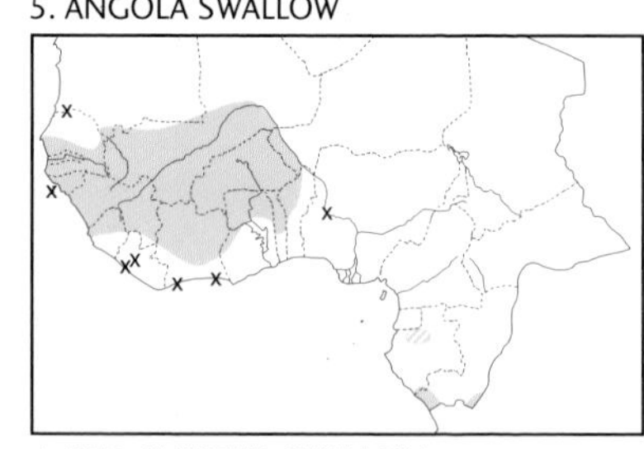

4. RED-CHESTED SWALLOW

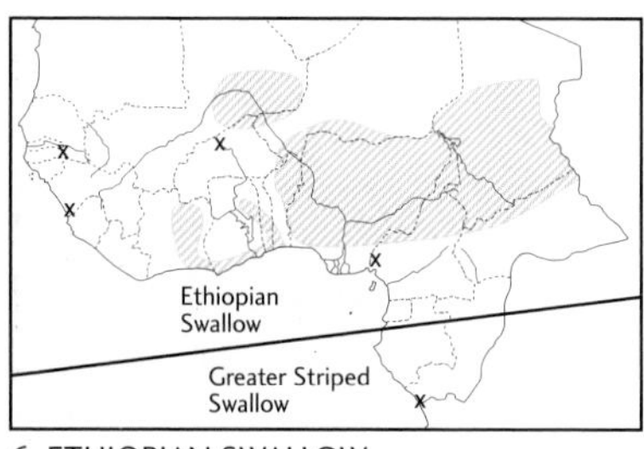

6. ETHIOPIAN SWALLOW
8. GREATER STRIPED SWALLOW

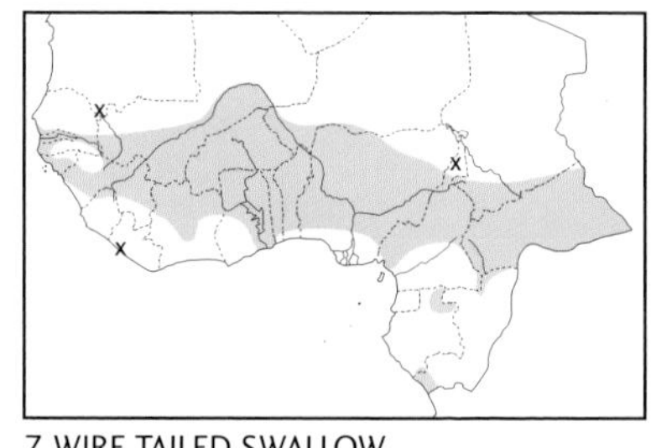

7. WIRE-TAILED SWALLOW

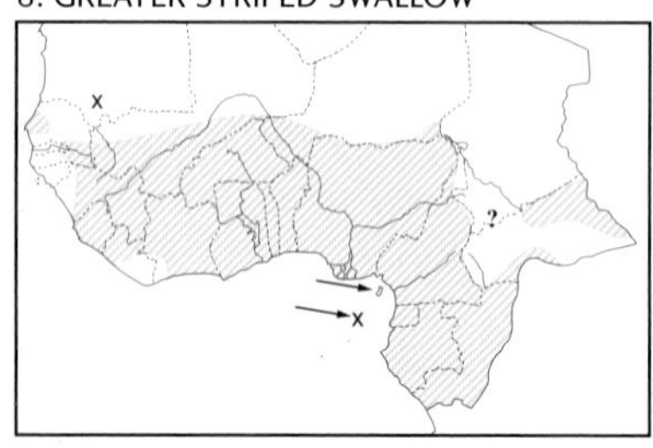

9. LESSER STRIPED SWALLOW

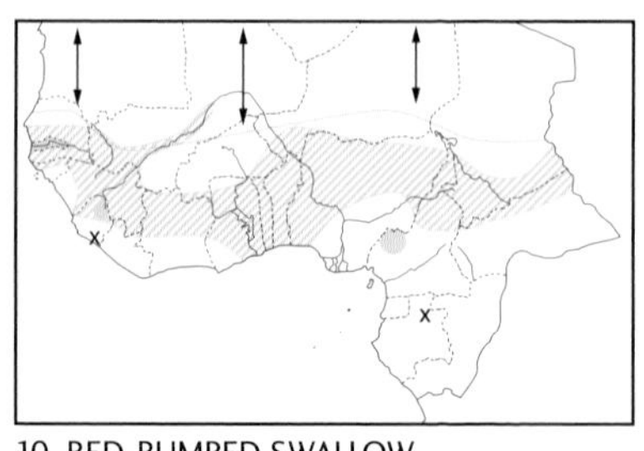

10. RED-RUMPED SWALLOW

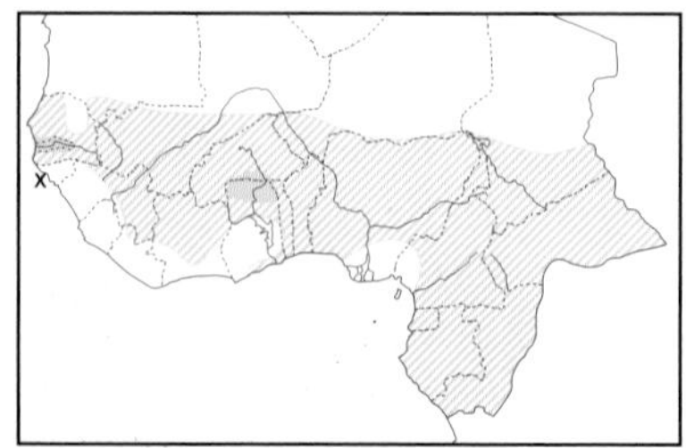

11. MOSQUE SWALLOW

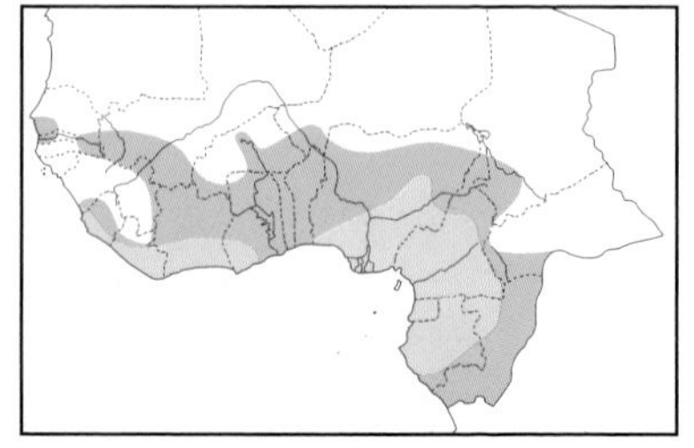

12. RUFOUS-CHESTED SWALLOW

Plate on page 286

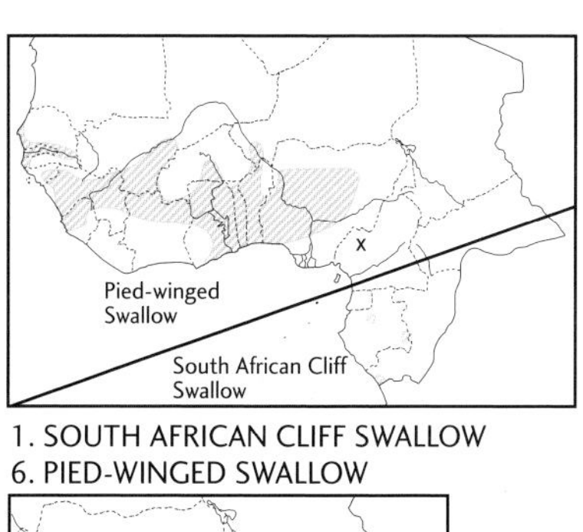

1. SOUTH AFRICAN CLIFF SWALLOW
6. PIED-WINGED SWALLOW

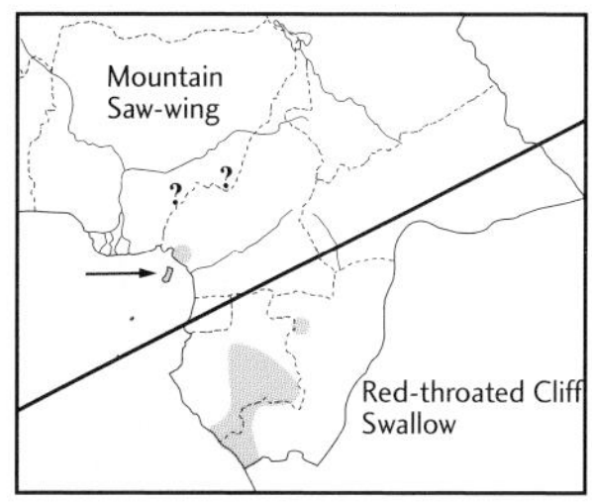

2. RED-THROATED CLIFF SWALLOW
8. MOUNTAIN SAW-WING

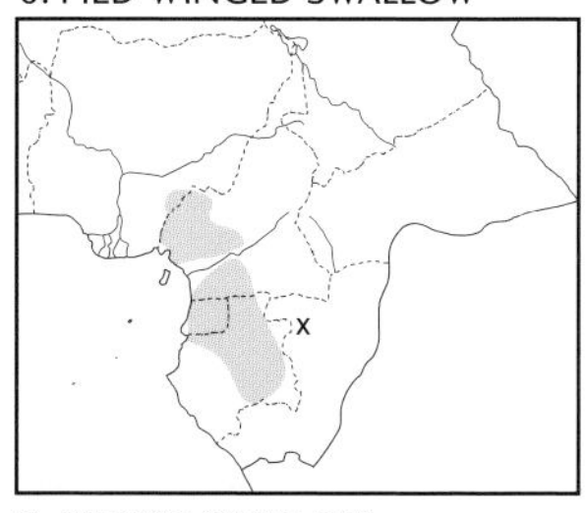

3. FOREST SWALLOW

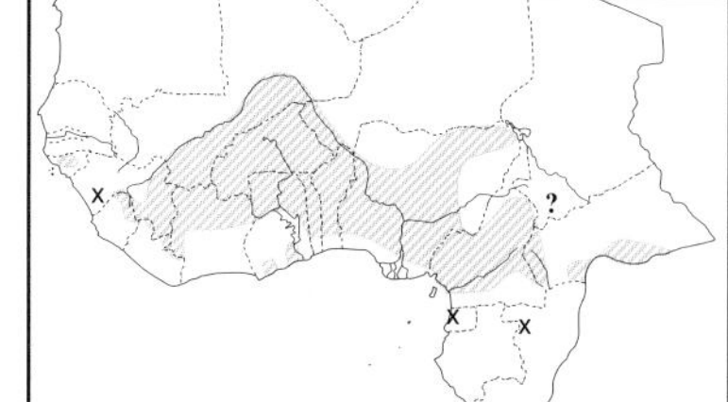

4. PREUSS'S CLIFF SWALLOW

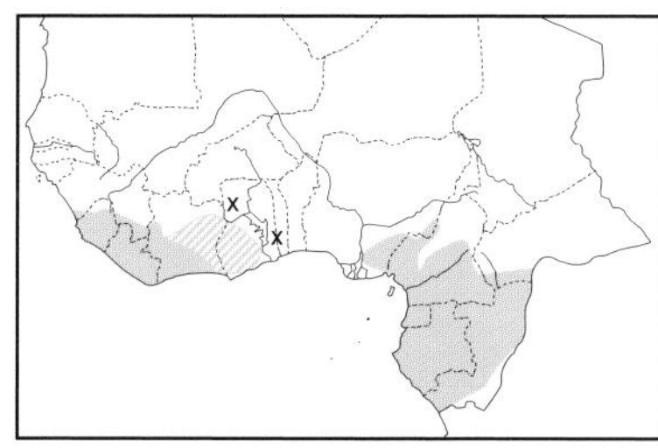

5. SQUARE-TAILED SAW-WING

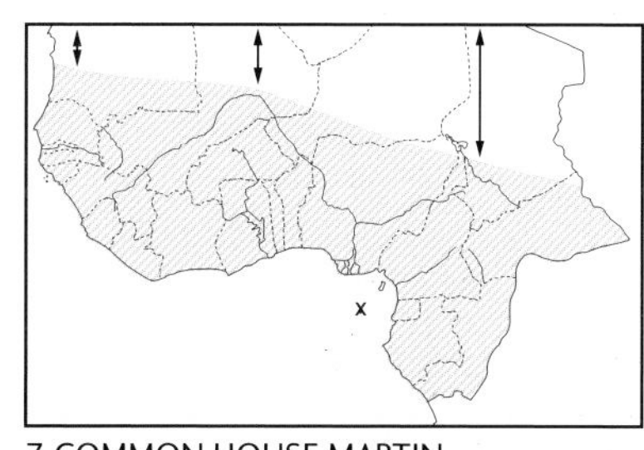

7. COMMON HOUSE MARTIN

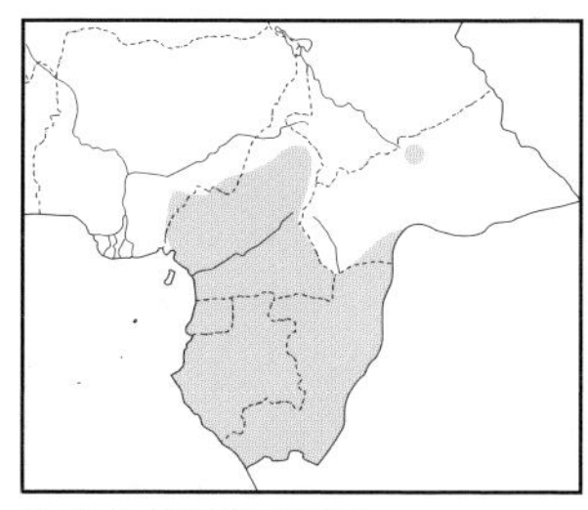

9. BLACK SAW-WING

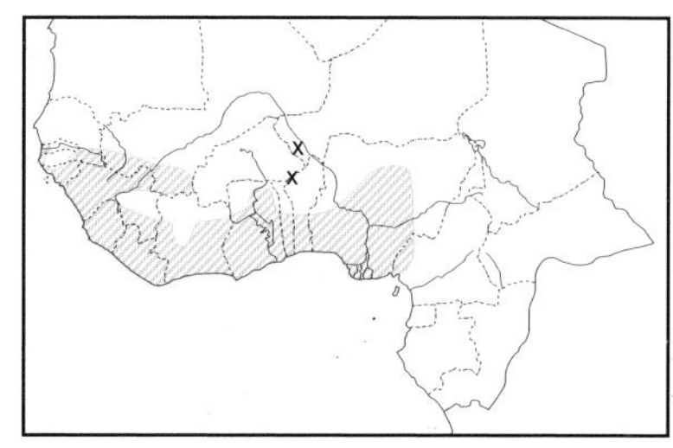

10. FANTI SAW-WING

Continued from page 286

12 **RUFOUS-CHESTED SWALLOW** *Hirundo semirufa* 18–21 cm **R/M u/lc**
12a Dark blue hood; no complete collar; orange-rufous underparts. **12b-c In flight** Orange-rufous rump; very long tail streamers; rufous underwing-coverts; orange-rufous undertail-coverts; white patches in tail. ▲ Open habitats. ✤ Gurgling *trlrrrrr* or *chip-chip-chleeeeurrrr*; also soft *dee-uuuh*, short *chip* and high-pitched *weet-weet*. [CD9:8]

1 SOUTH AFRICAN CLIFF SWALLOW *Hirundo spilodera* 14 cm **M s**
1a Very dark above; pale rufous/buffish-white below; speckled throat and breast. **1b–c In flight** Rufous rump; squarish tail; throat and breast often appear darkish; no white in tail. ▲ Open habitats. ✤ Call *prrp-prrp*, in flight. [CD9:15]

2 RED-THROATED CLIFF SWALLOW *Hirundo rufigula* 12 cm **R/M lf**
2a Ad Rufous throat; streaky ear-coverts. **2b Juv** Duller; browner above; rufous areas paler. **2c–d Ad in flight** Rufous rump; quarish tail with white patches below. **2e Juv in flight** Duller than adult. ▲ Various habitats, often near rocks and rivers. ✤ Rasping *prrp-prrp*, in flight. Song a rapid twittering. [CD9:14]

3 FOREST SWALLOW *Hirundo fuliginosa* 11 cm **R/M? lf**
3a Small; dull blackish-brown; rusty tinge on throat. **3b In flight** All dark; slightly notched tail. Flight fast with much gliding. Compare 5. ▲ Forest. Endemic. ✤ Mostly silent.

4 PREUSS'S CLIFF SWALLOW *Hirundo preussi* 12 cm **R/M r/lc**
4a Ad Blue-black above; pale buff below; small rufous patch behind eye. **4b Juv** Duller; dark brown above; throat and breast washed brownish. **4c–d Ad in flight** Pale buff rump. Resembles dull or dirty version of 7. Below: pale buff; small white patches in tail. **4e Juv in flight** Duller than adult. ▲ Savanna. ✤ Call *prrp-prrp*, in flight. [CD9:13]

5 SQUARE-TAILED SAW-WING *Psalidoprocne nitens* 11 cm **R c/r**
5a Wholly black. **5b In flight** Notched tail. Flight slow and fluttering. Compare 3. ▲ Forest. ✤ Soft, rather hoarse *pzzuit*, *pseeru*, *pruruit* and *psit*. [CD9:2]

6 PIED-WINGED SWALLOW *Hirundo leucosoma* 12 cm **R s/lf**
6a Glossy steel-blue above; white below; elongated white wing patches. **6b–c In flight** Long white patches on inner wing (diagnostic). Below white; forked tail with white patches. ▲ Wooded savanna, farmbush. ✤ Mostly silent. [CD9:19]

7 COMMON HOUSE MARTIN *Delichon urbicum* 13–15 cm **P u/lc**
7a Glossy blue-black above; white below. **7b–c In flight** Broad white rump (often mottled pale buff-brown in non-breeding). Below white; forked tail. ▲ Open habitats. ✤ Hard *prrt prrpt*. [CD3:59]

8 MOUNTAIN SAW-WING *Psalidoprocne fuliginosa* 12 cm **R lc**
8a Dull dark brown; tail fork *c.* 1.5–2.5 cm. **8b In flight** Grey-brown underwing-coverts. ▲ Forest clearings, montane grassland. Mt Cameroon, Bioko. Endemic. ✤ Soft *see-su*. [CD9:5, 12]

9 BLACK SAW-WING *Psalidoprocne pristoptera* 13 cm **R/M c/u**
9a Ad *petiti* (Petit's Saw-wing; E Nigeria–Congo) Wholly black with bronze-brown gloss; long tail streamers; tail fork *c.* 2.5–3.5 cm. **9b Ad *petiti* in flight** Whitish underwing-coverts. **9c Ad *chalybea*** (Shari Saw-wing; N CAR; possibly also N & C Cameroon) Greenish gloss; tail fork *c.* 3.8-5.4 cm. **9d Ad *chalybea* in flight** Grey underwing-coverts. ▲ Savanna, montane grassland, farmbush. ✤ Mainly silent. Soft, nasal *sheeu*. [CD9:4]

10 FANTI SAW-WING *Psalidoprocne obscura* 17 cm **R/M c**
10a Wholly glossy black; very long tail streamers; tail fork male 5.0–7.5 cm, female 3.0–6.5 cm. **10b In flight** Black underwing. ▲ Forest clearings, tracks and edges, farmbush, woodland, savanna. Endemic. ✤ Mainly silent. [CD9:3]

 Maps on page 289

1c
1a
1b
2d
2b
2a
2c
2e
4d
3a
4c
3b
4e
4b
4a
6c
7c
5b
6b
7b
5a
6a
7a
9d
9c
10b
8b
8a
9b
10a
9a

PLATE 86: PIPITS

Terrestrial insectivores with rather cryptic, variably streaked brown plumage. Slightly resemble larks, but slimmer. Large species conspicuously long legged. Non-breeding plumage usually similar to breeding; pale tips and fringes to wing feathers broader when fresh. For identification, note extent of streaking on upper- and underparts, colour of outer tail feathers, and calls.

1 TREE PIPIT *Anthus trivialis* 15 cm **P/V+ c/u**
Boldly streaked on breast, thinly on flanks; plain rump. ▲ Various open wooded habitats. ✤ Call a single *tzeep* in flight. [CD3:66]

2 MEADOW PIPIT *Anthus pratensis* 14.5 cm **P u/r**
Bold streaking on breast extending onto flanks; plain rump. ▲ Various open habitats. ✤ Call a sharp *eest-eest-eest*. [CD3:68]

3 RED-THROATED PIPIT *Anthus cervinus* 15 cm **P c/s**
3a Ad breeding Distinctive brick-red or orange-buff face and throat. **3b Ad non-breeding** Boldly streaked above and below; three lines of streaks extending onto flanks; streaked rump; moustachial stripe ending in blotch. ▲ Grassland, cultivation. ✤ Call *speeeh*, longer, less rasping and more piercing than 1. [CD3:67]

4 TAWNY PIPIT *Anthus campestris* 16–17 cm **P/V+ r/u**
Palest, sandy-coloured pipit; contrasting dark centres to median wing-coverts. ▲ Open habitats. Breeding suspected, SW Mauritania. ✤ Sparrow-like *chlip* and *cheeup*. [CD3:64]

5 SHORT-TAILED PIPIT *Anthus brachyurus* 12–13 cm **R lu**
Smallest pipit. Heavily streaked; short narrow tail. ▲ Open grassland. Skulking; difficult to flush or see on the ground. ✤ Nasal *tseep*. Song, nasal *chirrup* notes with buzzing wing-snaps in cruising display flight. [CD9:33]

6 PLAIN-BACKED PIPIT *Anthus leucophrys* 17 cm **R u/lc**
6a Ad *zenkeri* (S Mali–Ghana east) Plain dark brown above; warm cinnamon-buff below; buff outer tail. *A. l. bohndorffi* (SE Gabon, lower Congo R.) paler below. **6b Ad *gouldii*** (west of *zenkeri*) Darker above; much paler below; distinct streaks on breast. *A. l. ansorgei* (S Mauritania–Guinea-Bissau) greyer above. ▲ Various open habitats. ✤ Soft *chee-chee* on take-off. Song a rather slow and monotonous series of single or alternating *swree* and *chirrup* notes, given from ground or low perch. [CD9: 30]

7 LONG-BILLED PIPIT *Anthus similis* 17–19 cm **R u/lc**
7a Ad *bannermani* (highlands SW Mali–Sierra Leone to W Cameroon) Heavily streaked above and below; streaks extending onto flanks; buffish-white outer tail. **7b Ad *asbenaicus*** (C & E Mali, Niger) Paler, more sandy-buff; almost no streaks on underparts. ▲ Rocky slopes. ✤ Rather soft *chee* or *djeep*. Song a simple series of 2–3 detached, sparrow-like notes repeated at intervals *chriu shree chewee* or *tsreep shree shruw*; from ground, tree or large rock, or in slow, fluttering display flight. [CD9: 28]

8 WOODLAND PIPIT *Anthus nyassae* 17–18 cm **R lc**
Ad *schoutedeni* Streaking above more diffuse than 7a; below, only streaked on breast. ▲ Wooded grassland; grassy hills with exposed rocks. ✤ Song a simple, sustained series of detached chirruping notes, e.g. *tswee twuree shree srwee chruee...* [CD9:29]

9 GRASSLAND PIPIT *Anthus cinnamomeus* 16–18 cm **R/V lc**
Ad *lynesi* (highlands SE Nigeria–W Cameroon; vagrant N Nigeria, W Chad) Large; streaked upperparts and breast; bold face pattern; pale lores; buff-white outer tail. *A. r. camaroonensis* (Mt Cameroon, Mt Manenguba) paler below. Race of those in Liberia, Ghana, NE Nigeria and W Chad unknown. ▲ Various open habitats, incl. grassland and cultivation. ✤ Hard *chip* when flushed. Song a rapid series of 3–5 identical notes *shree-shree-shree-shree-...*, from low perch or in display flight. [CD3:63]

Continued on page 295

Maps on page 294

1
2
3a
3b
4
5
6b
6a
7b
7a
8
11
9
10

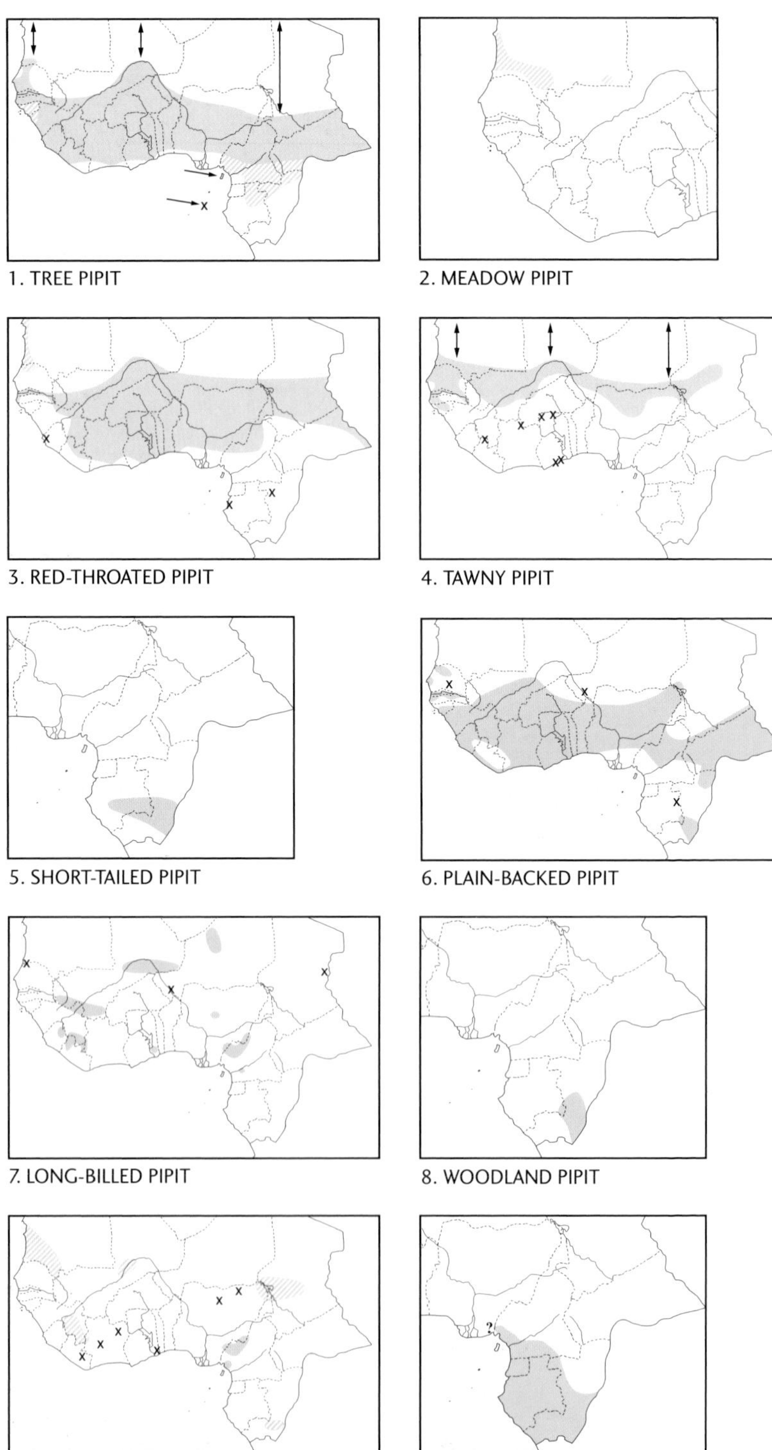

1. TREE PIPIT

2. MEADOW PIPIT

3. RED-THROATED PIPIT

4. TAWNY PIPIT

5. SHORT-TAILED PIPIT

6. PLAIN-BACKED PIPIT

7. LONG-BILLED PIPIT

8. WOODLAND PIPIT

9. GRASSLAND PIPIT

11. LONG-LEGGED PIPIT

Plate on page 292

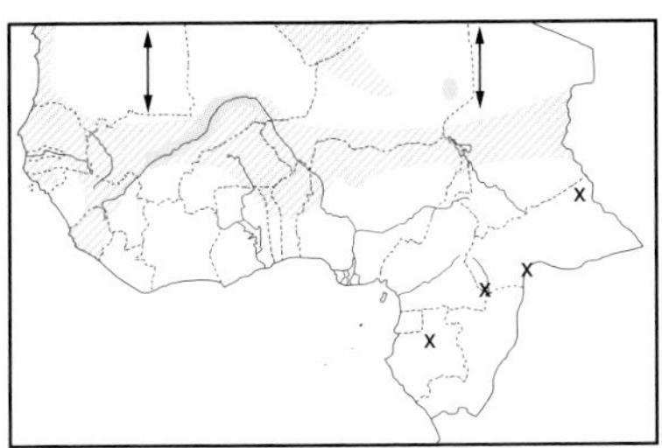

1. WHITE WAGTAIL

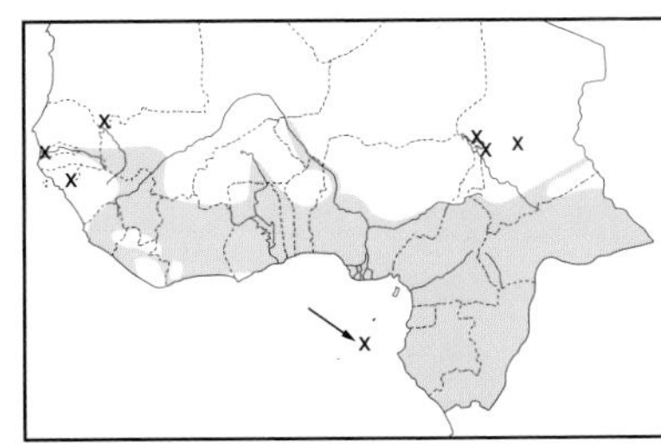

2. AFRICAN PIED WAGTAIL

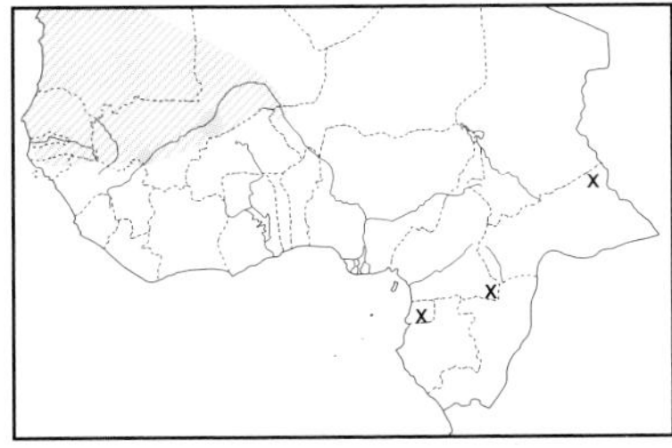

3. GREY WAGTAIL

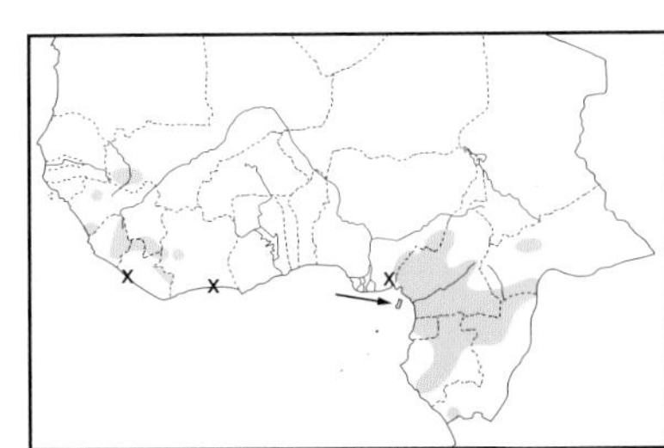

4. MOUNTAIN WAGTAIL

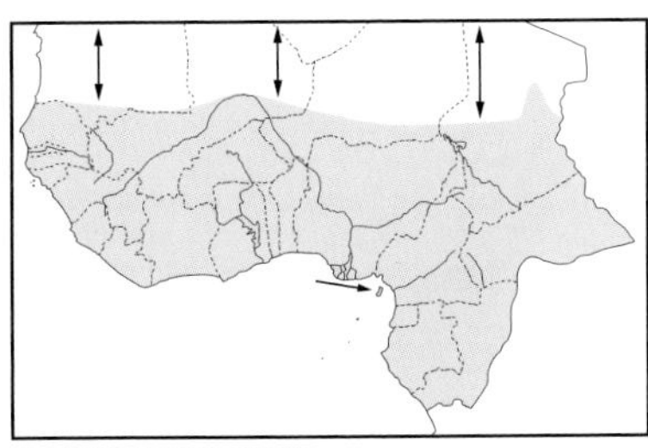

5. YELLOW WAGTAIL

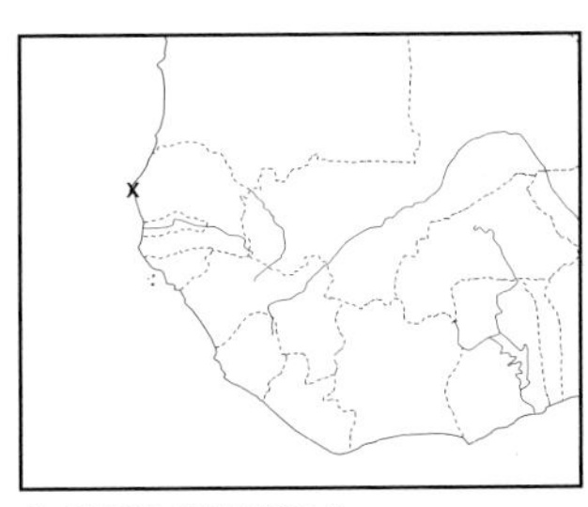

6. CITRINE WAGTAIL

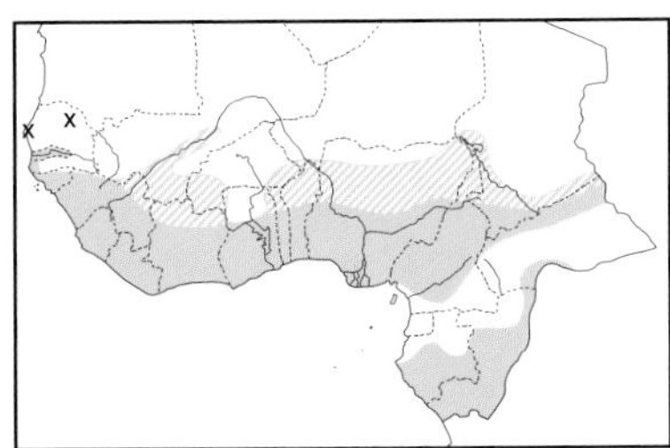

7. YELLOW-THROATED LONGCLAW

Continued from page 292

10 RICHARD'S PIPIT *Anthus richardi* 17–20 cm **P? r**
Larger than 9; wing edgings more sandy-buff; underparts whiter; outer tail feathers pure white. ▲ Mauritania, Mali? ✤ Rather loud, abrupt, grating *pshreep* or *pshriu.* [CD3:63]

11 LONG-LEGGED PIPIT *Anthus pallidiventris* *c.* 18 cm **R lc**
Large; long legs; relatively short tail; prominent supercilium; plain greyish-brown above; buff outer tail; indistinct streaks on breast. ▲ Various grassy habitats. ✤ Flight call *psee-ip* or *ch-seep.* Song, frequently repeated chirrupy notes. [CD9: 32]

Wagtails. Small, slender, terrestrial insectivores with long, frequently pumped tails.

1 WHITE WAGTAIL *Motacilla alba* 18 cm P+ c/u
1a Ad male breeding Black throat; grey upperparts. **1b Ad male non-breeding** White face and throat; black breast-band. ▲ Various open habitats. ✤ Sharp *chissik*; alarm *chik*. [CD3:62]

2 AFRICAN PIED WAGTAIL *Motacilla aguimp* 19 cm R c/u
2a Ad Black and white; large white wing patch. **2b Juv** Much duller; black replaced by dark grey-brown. ▲ Near water and habitation. ✤ Clear, whistled *tluwsee*. Song a varied series of quite melodious, short phrases of clear piping and whistling notes. [CD9:27]

3 GREY WAGTAIL *Motacilla cinerea* 18–19 cm P/V r
3a Ad male breeding Very long tail; black throat; wholly yellow below. **3b Ad male non-breeding** Grey above; yellow on rump and vent. ▲ Various habitats, normally near water. ✤ Disyllabic *chzizik*, similar to 1 but shorter, more metallic. [CD3:61]

4 MOUNTAIN WAGTAIL *Motacilla clara* 18–19 cm R s/u
Grey above; white below; narrow blackish breast-band. ▲ Along swift, rocky forest streams and lakes. ✤ Loud, hard, ringing trill *tsrrrrup* and drawn-out, high-pitched *tseeeet*. Song a sustained medley of vigorous, rather melodious phrases with high-pitched, sibilant notes and ringing trills, e.g. *tsu-tseeee-seeeuu, tsrrrup-tsrrrup-tsrrrup, tseeeee, tsrrrrreeeuw, tsee-tsu-tsee-uw, trrrip-trrrip, tsweu-tswee-tsee-uw*... [CD9:26]

5 YELLOW WAGTAIL *Motacilla flava* 16–17 cm P c
Illustrated forms are all in breeding plumage. Non-breeding similar to female *flava* but duller. **5a Ad male *flava*** Blue-grey head; white supercilium; wholly yellow underparts. **5b Ad female *flava*** Duller; less distinct head pattern; paler below. **5c Ad male *flavissima*** Yellow-green head. **5d Ad male *beema*** Pale blue-grey head (paler than 5a); throat sometimes white. **5e Ad male *iberiae*** Dark grey head; narrow supercilium; white throat. **5f Ad male *cinereocapilla*** Similar to 5e, but supercilium absent or vestigial. **5g Ad male *thunbergi*** Top of head slate-grey; head-sides blackish; supercilium indistinct, often absent. **5h Ad male *melanogrisea*** Black head bordered below by white line. **5i Ad male *feldegg*** (Black-headed Wagtail) As 5h, but without white line. ▲ Various, generally open, habitats. ✤ Rather loud *pseew*. [CD3:60]

6 CITRINE WAGTAIL *Motacilla citreola* 17–18 cm V
6a Ad male breeding Bright lemon-yellow head; black nuchal band; grey upperparts; two white wingbars. **6b First-winter** Dull ash-grey above; two white wingbars; broad whitish ear-covert surround; narrow, weak, dusky malar stripe. ▲ Palearctic vagrant, Senegal. ✤ Sharp, rasping *tsrreep*.

Longclaws. Robust ground-dwellers, with relatively short tail, strong legs and long claw on hind toe.

7 YELLOW-THROATED LONGCLAW *Macronyx croceus* 20–22 cm R lc/s
7a Ad Yellow below; black necklace. **7b Juv** Duller; necklace buff with dark markings. **7c In flight** White tail corners. ▲ Various grassy habitats. ✤ Far-carrying, melodious and repeated whistle *teeuwheee* or *twee-eu*, uttered by both sexes and by male in display flight, sometimes followed by *tiri-tiri-ti* or variation. [CD9:36]

Maps on page 295

1a
1b
2b
2a
3a
3b
3c
4
5b
5c
5e
5a
5d
5g
5f
5h
5i
6b
6a
7a
7c
7b

Unobtrusive arboreal birds. Mainly singly or in pairs, quietly gleaning foliage in search of insects, usually in canopy.

1 BLUE CUCKOO-SHRIKE *Coracina azurea* 21 cm R u/c
1a Ad male Deep glossy blue; black face; reddish eye. **1b Ad female** Duller; less black on face. ▲ Rainforest. Regularly in mixed bird parties. ✤ Rather vocal. Hoarse, nasal *chwee-ep* or *chuee*, short *chup*, and series of *tuk* notes, first accelerating, then slowing down. Song a variable series of loud and clear whistling notes, e.g. *pooeet-pooi-pooeet-peeoo*, a repeated *chup-peeeo* or *peeeoo*, often interspersed with call notes. [CD9:44]

2 GREY CUCKOO-SHRIKE *Coracina caesia* 23 cm R f
2a Ad male Slate-grey; black lores. **2b Ad female** No black on lores. **2c Juv** Barred dusky and whitish above and below. ▲ Montane forest. ✤ Thin, sharp, high-pitched *seeeeeu*. Song a jumble of high-pitched notes interspersed with call note. [CD9:42]

3 WHITE-BREASTED CUCKOO-SHRIKE *Coracina pectoralis* 25 cm R u/f
3a Ad male Pale grey above; lower breast to undertail-coverts white; black lores. **3b Ad female** Lores paler; throat whitish bordered paler grey. ▲ Savanna woodland. Regularly joins mixed-species flocks. ✤ Rather thin, high-pitched *seeeu*, easily passing unnoticed. Various other calls. [CD9:43]

4 RED-SHOULDERED CUCKOO-SHRIKE *Campephaga phoenicea* 20 cm R/M u/lf
4a Ad male Glossy blue-black; scarlet shoulder patch. **4b Ad male** Orange-shouldered variant. Golden-yellow shoulders also recorded. **4c Ad female** Greyish olive-brown above; white barred black below. ▲ Forest patches, gallery forest, woodland. Occasionally in mixed-species flocks. ✤ Single clear loud whistle *heeew*, double *huu-tseew* and short *tsuk*. Song a jumble of high-pitched and scratchy notes, somewhat reminiscent of sunbird. [CD9:39]

5 PURPLE-THROATED CUCKOO-SHRIKE *Campephaga quiscalina* 20 cm R u/r
5a Ad male Black glossed blue-green above; head-sides to breast glossed purple. **5b Ad female** Grey head; plain olive upperparts; bright yellow underparts. **5c Juv** Variably barred above and below. ▲ Evergreen and gallery forest, forest patches. Occasionally in mixed-species flocks. ✤ Vigorous, far-carrying *tsee-up*; female has shorter *tseeu*. Song a rather fast and vigorous series of melodious syllables *slueet-slueet-swit-wit slueet-slueet-swit-wit tluw-tluweew* ... [CD9:41]

6 PETIT'S CUCKOO-SHRIKE *Campephaga petiti* 20 cm R u
6a Ad male Entirely glossy blue-black; yellow or orange gape wattles usually conspicuous. **6b Ad female** Mainly yellow; heavily barred blackish above; usually some bars on breast. **6c Juv** More olive above than adult female; spotted below. ▲ Montane forest, gallery forest, forest patches. ✤ Short whistled *seep*. Song a high-pitched *psiuu*, *tsi-tsi*; also a high, scratchy warbling *sueet-sueet*, *siueet-seet-seet-sireet*.

7 EASTERN WATTLED CUCKOO-SHRIKE *Lobotos oriolinus* 19 cm R r DD
7a Ad male Glossy blue-black head; large gape wattles; yellow tinged orange below. **7b Ad female** Duller; no orange tinge below. ▲ Primary and tall secondary rainforest. Joins mixed-species flocks. ✤ Mainly silent.

8 WESTERN WATTLED CUCKOO-SHRIKE *Lobotos lobatus* 19 cm R r VU
8a Ad male As 7a but orange-chestnut suffused with yellow below. **8b Ad female** Yellow below. ▲ Evergreen and semi-deciduous forest. Joins mixed-species flocks. Endemic. ✤ Short *tsik*, in flight.

Maps on page 300

1b
1a
2a
3b
3a
2c
2b
4c
4a
4b
5c
5b
5a
6b
6c
6a
7b
7a
8b
8a

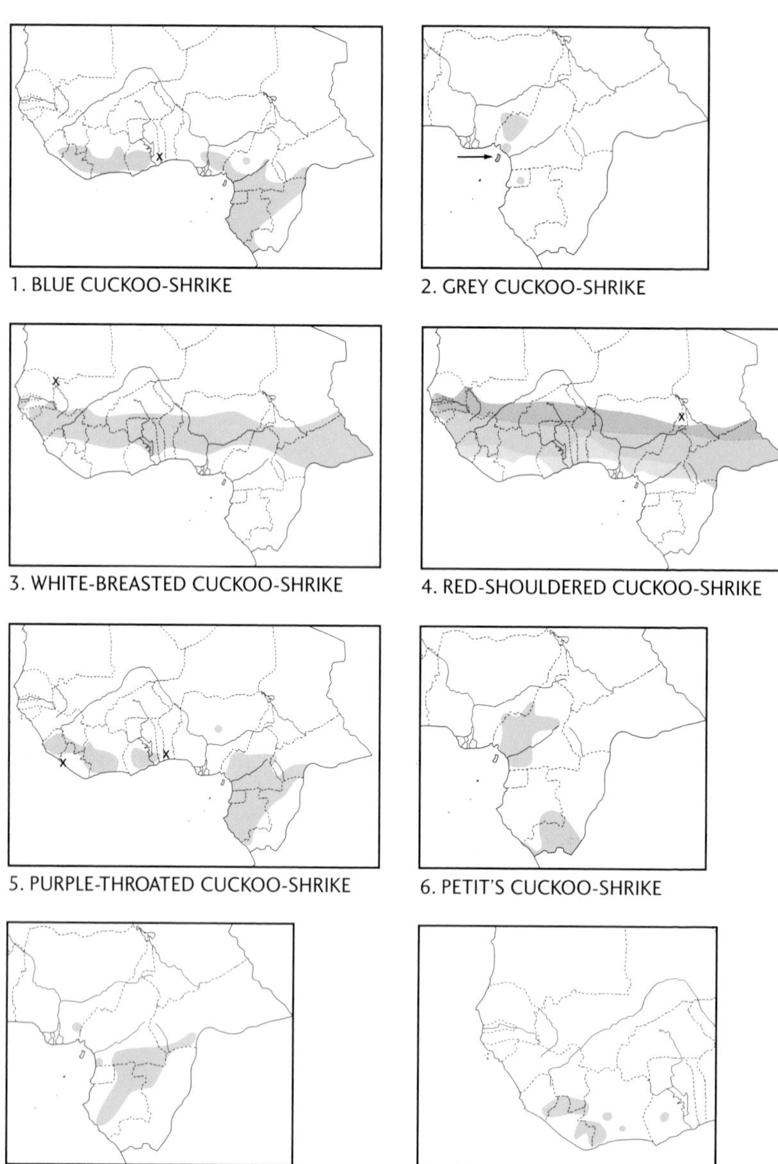

1. BLUE CUCKOO-SHRIKE

2. GREY CUCKOO-SHRIKE

3. WHITE-BREASTED CUCKOO-SHRIKE

4. RED-SHOULDERED CUCKOO-SHRIKE

5. PURPLE-THROATED CUCKOO-SHRIKE

6. PETIT'S CUCKOO-SHRIKE

7. EASTERN WATTLED CUCKOO-SHRIKE

8. WESTERN WATTLED CUCKOO-SHRIKE

Plate on page 298

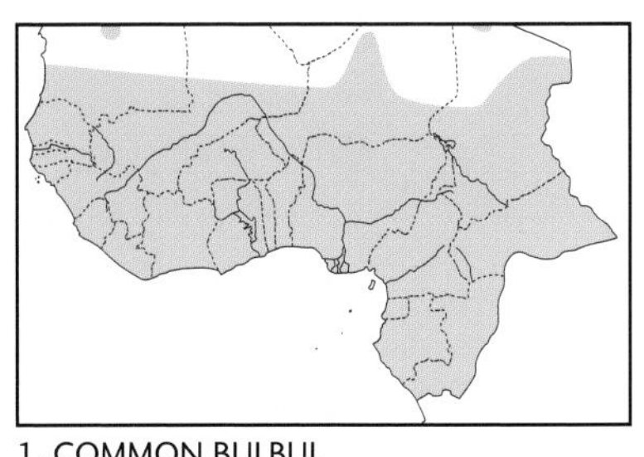

1. COMMON BULBUL

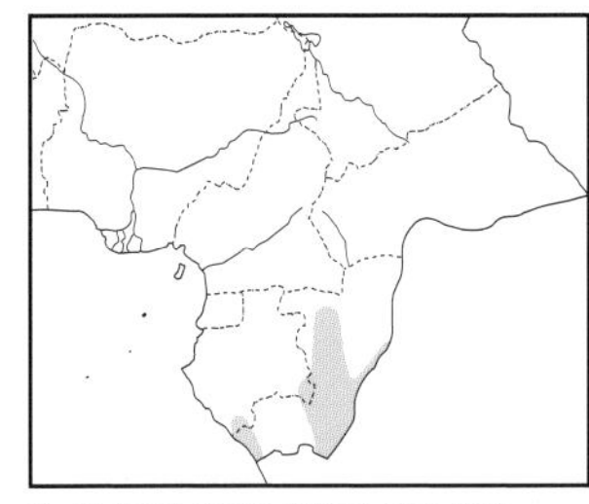

2. BLACK-COLLARED BULBUL

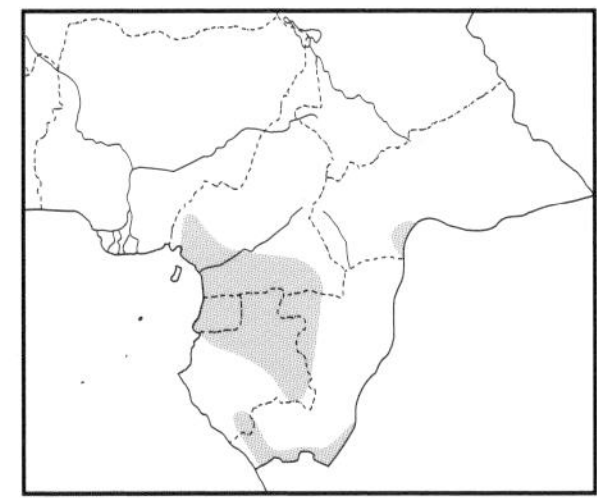

3. YELLOW-NECKED GREENBUL

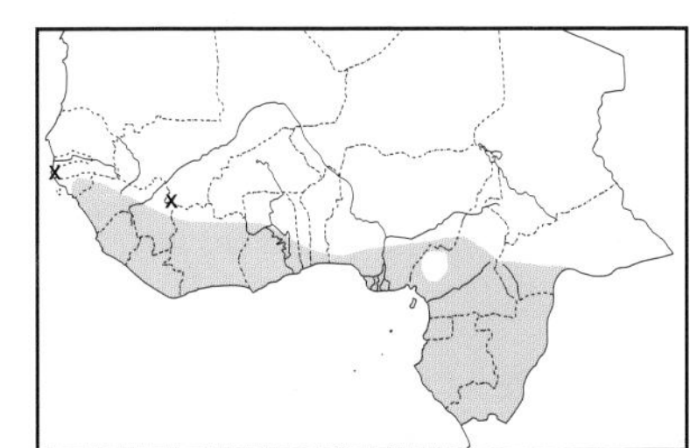

4. SIMPLE LEAFLOVE

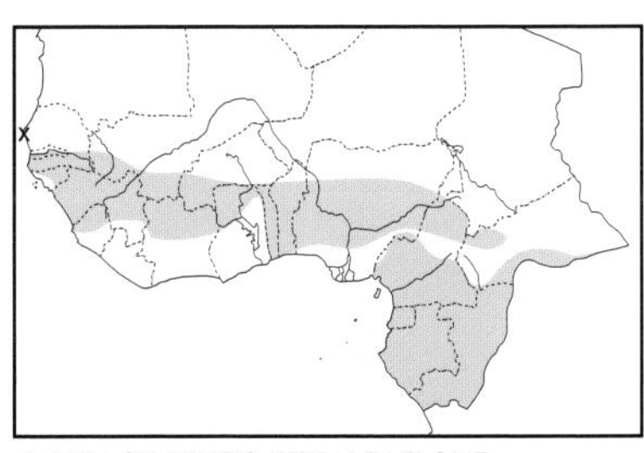

5. YELLOW-THROATED LEAFLOVE

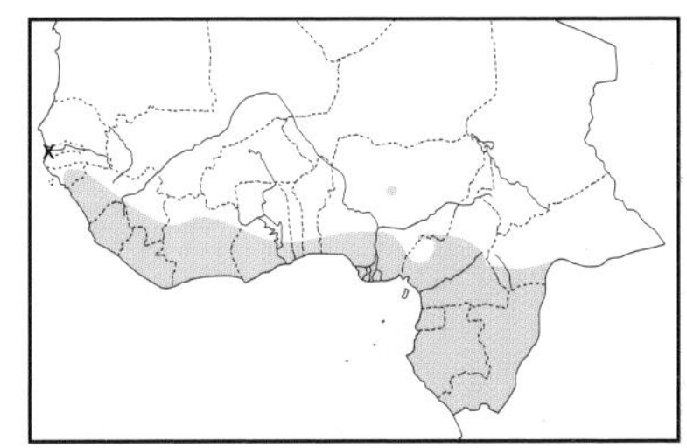

6. SWAMP PALM BULBUL

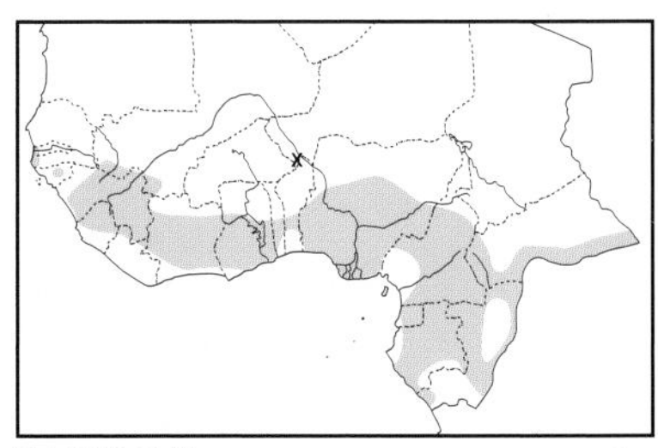

7. LEAFLOVE

Small to medium-sized (13–26 cm), mostly arboreal birds. Sexes similar, males usually slightly larger; juveniles/immatures similar to, but generally duller than, adults. Family is well represented in Africa; all species but one are endemic. Most are inconspicuously coloured in various shades and combinations of dull green, grey, brown and yellow, and are associated with forest habitat. Frugivorous and insectivorous; occasionally flycatch. Mainly sedentary. Those on this plate are among the easiest of the family to identify.

1 COMMON BULBUL *Pycnonotus barbatus* 18–20 cm R c
Africa's ubiquitous, most conspicuous and familiar bird. Mainly brown, darker on head, paler below. **1a *P. b. inornatus*** (almost entire region) Dark brown breast merges gradually with paler belly. **1b *P. b. arsinoe*** (E Chad) Breast well demarcated from pale belly. **1c *P. b. tricolor*** (E Cameroon, Congo) Undertail-coverts bright yellow. In *P. b. gabonensis* (C Nigeria–C Cameroon to Gabon–S Congo) undertail-coverts are white tinged with variable amount of yellow. ▲ Almost all habitats except closed forest and treeless desert. Pugnaceous and noisy. ✤ Song a cheerful phrase of 3–6 notes, e.g. *chuk chuk twirulup* or *chuk twee tu twuri* or *kwik kweek kwuyu*, etc.; most adequately rendered by popular *quick quick, doctor, quick!* Calls include a variety of chattering and ringing notes. [CD3:71]

2 BLACK-COLLARED BULBUL *Neolestes torquatus* 16 cm R lf
Grey top of head; olive-green upperparts and tail; whitish underparts with black breast-band. ▲ Lightly wooded savanna. Usually forages low. Sings from top of low trees and bushes. ✤ Rapid subdued babble, somewhat reminiscent of 1. Also a quavering *twee-dududu*. [CD9:87]

3 YELLOW-NECKED GREENBUL *Chlorocichla falkensteini* 18 cm R lc/r
Bright yellow throat contrasting with olive-green upperparts. ▲ Farmbush, forest patches and edges. ✤ Song a string of nasal notes *kik-kuk-ku-KWEE-uk-wik-e-wik-kup*, with slight variations. Song and chattering calls similar in quality to those of 4. [CD9:62]

4 SIMPLE LEAFLOVE *Chlorocichla simplex* 21 cm R lc
White throat; broken white eye-ring. ▲ Farmbush, forest regrowth, thickets. Skulking. ✤ Song a subdued nasal chattering. Call a scolding *wherr* and short, clipped *kwit*! [CD9:63]

5 YELLOW-THROATED LEAFLOVE *Chlorocichla flavicollis* 22.5 cm R f/s
Large. Mainly dark brown. **5a *C. f. flavicollis*** (Senegambia–Cameroon) Yellow throat. **5b *C. f. soror*** (Cameroon to CAR–Congo) Whitish throat. ▲ Woodland, thicket, swamp forest. Mainly at lower and mid-levels. ✤ Call a loud, nasal *chow* and *kyip*. Song a string of *chow*s interspersed with shorter notes; in quality reminiscent of 6. [CD9:64]

6 SWAMP PALM BULBUL *Thescelocichla leucopleura* 23 cm R f/lc
Large. Creamy belly; outer tail feathers boldly tipped white. ▲ Lowland and gallery forest. Partial to palm trees. In small, mobile and noisy parties. ✤ Raucous nasal cackling like tape-recorded conversation being played too fast. Also other loud, nasal, scolding calls. [CD9:65]

7 LEAFLOVE *Pyrrhurus scandens* 22 cm R lc/s
Pale grey head; pale chestnut tail. ▲ In small, noisy groups in forest fringing rivers and swampy places within forest; also forest patches in savanna. At all levels, though mainly in canopy and middle stratum. ✤ Song starts with some subdued, nasal notes, then bursts into loud, resonant, pleasant conversational cackling, including *kyop-kee-kyop-kyop-ke-kyop--...*, uttered for long periods by family groups. [CD9:66]

 Maps on page 301

1a
1b
1c
2
3
5b
4
5a
6
7

1 **GOLDEN GREENBUL** *Calyptocichla serina* 18 cm R r/lf
Bright olive-green above; mainly yellow below; pale pinkish bill. ▲ Forest canopy. ✤ Short *tsip* and, in flight, thin *see*. Song a clear, short *tiup-chiweew*; many variations. [CD9:56]

***Andropadus* bulbuls** are small to medium-sized forest inhabitants with mainly featureless olive-green plumage; underparts paler. Essentially frugivorous, hence rarely with mixed-species flocks. Vocalisations an important aid to identification.

2 **SLENDER-BILLED GREENBUL** *Andropadus gracilirostris* 18 cm R c
Olive-brown above; brownish-grey below. ▲ Canopy of forest and cultivation with large trees. ✤ Clear, drawn-out whistle *tseeeeu*. Also a seasonal song of 4–5 whistled notes *whee-ti-twheew-ti-twhee* or *whuut-hEET whuut-hEET*..., with slight variations. [CD9:54]

3 **CAMEROON MONTANE GREENBUL** *Andropadus montanus* 18 cm R s/lc NT
Wholly olive-green. ▲ Montane forest; forages usually low. Endemic. ✤ Song a rather subdued, nasal, husky babble ending with faster, rather cheerful, chuckling phrase *churp-churp-churp-chipurchipurcherr*. Call a low *kerr* and rapidly repeated, nasal *chup*. [CD9:45]

4 **LITTLE GREY GREENBUL** *Andropadus gracilis* 16 cm R f/c
Small. Head and throat olive-grey; underparts yellowish-olive; narrow white eye-ring. ▲ Upper mid-level of forest habitats. ✤ Song consists of 4–5 rapid, jaunty notes *wheet wu-wheet-wu-wheet*; also a quieter *tehu-tehee-tee*. Call a short *tyuk*. [CD9:50]

5 **ANSORGE'S GREENBUL** *Andropadus ansorgei* 16 cm R u/lc
As 4 but underparts lack any yellow; flanks and vent ginger. ▲ Forest; usually at upper mid-level. ✤ Three-note song *wheet-whuut-whit* or *tiu-wheet-tweet*, resembles 4, but lacks its sprightliness; last syllable may be dropped. Also a harsh, flat trill *rititititititit*. [CD9:51]

6 **WESTERN MOUNTAIN GREENBUL** *Andropadus tephrolaemus* 18 cm R c
Grey head and throat; bright olive-green upperparts; yellowish-olive underparts. ▲ At all levels of montane forest. Joins mixed-species flocks; clings to tree trunks. Endemic. ✤ Song a monotonous, steady series of notes, same form repeated continuously, e.g. *whup-wheep-whip-whupchipup* and *whup-wheep-whup-wheep-whup-wheep-*...; also faster series of nasal, chirruping notes. Call a nasal, scolding *whee-up* or *dzut-dzuwi*. [CD9:47]

7 **LITTLE GREENBUL** *Andropadus virens* 16.5 cm R c
7a Ad Wholly dirty olive-green without distinguishing markings. **7b Juv** Lower mandible and gape yellow; similar to 9b, but legs dark. ▲ Forest edge, rank scrub, abandoned cultivation, thicket, etc.; lower level. ✤ Song a few subdued chuckling notes followed by a rapidly ascending, pleasant bubbling warble, increasing in volume and abruptly ending on a clear, high, rising note. Call a dry *kuk-kuk-kuk*... [CD9:49]

8 **CAMEROON SOMBRE GREENBUL** *Andropadus curvirostris* 17 cm R f/c
Very similar to 7; voice best distinction. Usually with narrow, broken, white eye-ring. ▲ Lower mid-level of forest, esp. edges. ✤ Song of *A. c. leoninus* (from Ghana west) unarresting but distinctive *tiuwhee-tiu triiiiii* or *su-hi-oo triiii*, with stress on final harsh trill, often also given separately. Also a hard *wrrrrit* and longer *wrrrrrititit*. Nominate has quite different, 3-note *wheet-tiuwhee-tuu* and *wheet-tu-twhee*; also harsh trill *triiiii*. [CD9:52–53]

9 **YELLOW-WHISKERED GREENBUL** *Andropadus latirostris* 17 cm R c
9a Ad Wholly olive-green; bright yellow 'whiskers'; pinkish legs. **9b Imm** No or only small 'whiskers'; leg colour as adult. Compare 7. ▲ At all levels of various forest types and dense scrub. ✤ Song a monotonous series of *c.* 12 *chruk* notes, slightly increasing in volume. Call a repeated *chuk* and rapid rattling *ditditditdit*... (alarm). [CD9:55]

Continued on page 307

1
2
3
4
5
6
7a
7b
8
9b
9a
10b
10a
11

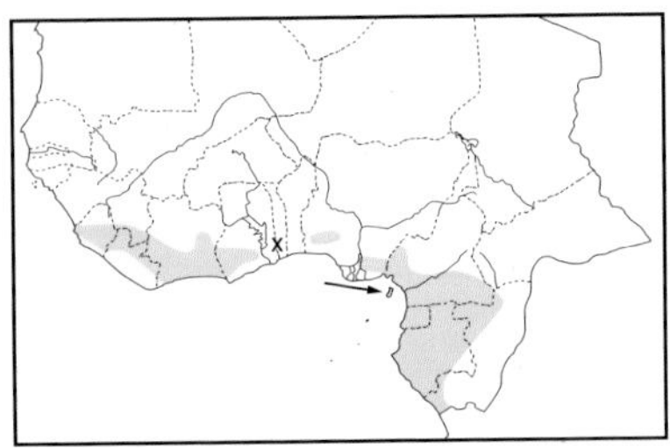
1. GOLDEN GREENBUL

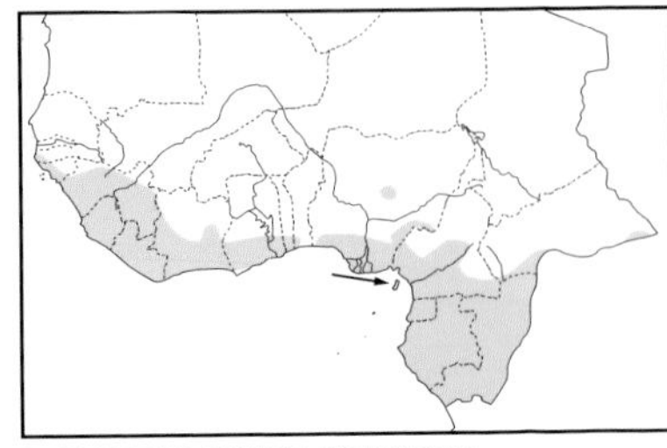
2. SLENDER-BILLED GREENBUL

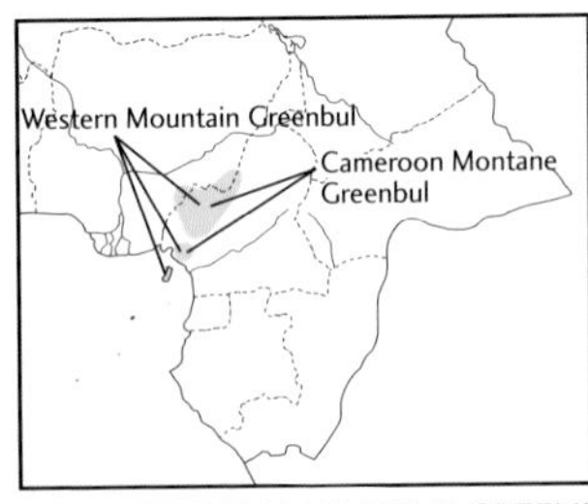

3. CAMEROON MONTANE GREENBUL
6. WESTERN MOUNTAIN GREENBUL

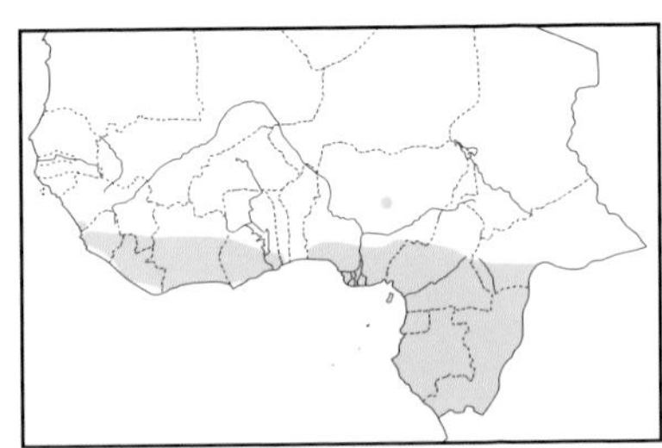
4. LITTLE GREY GREENBUL

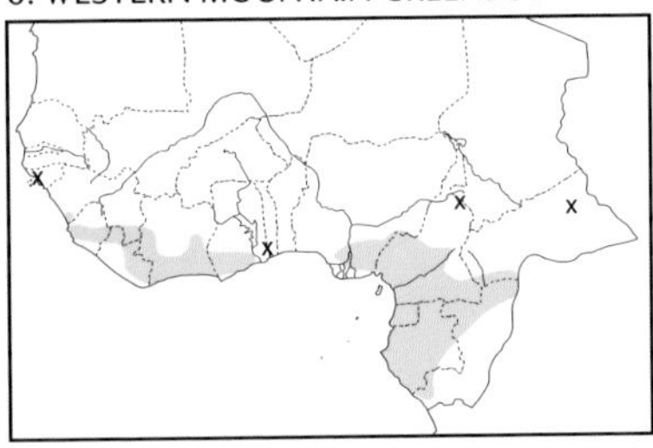
5. ANSORGE'S GREENBUL

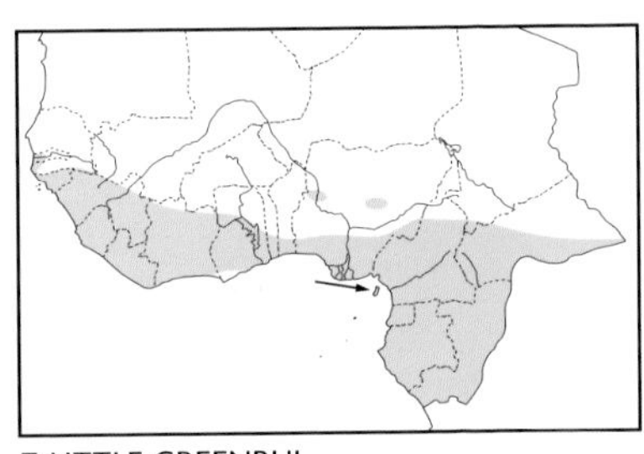
7. LITTLE GREENBUL

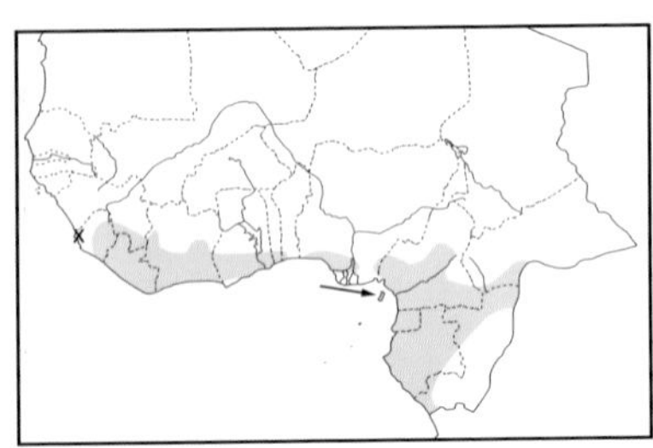
8. CAMEROON SOMBRE GREENBUL

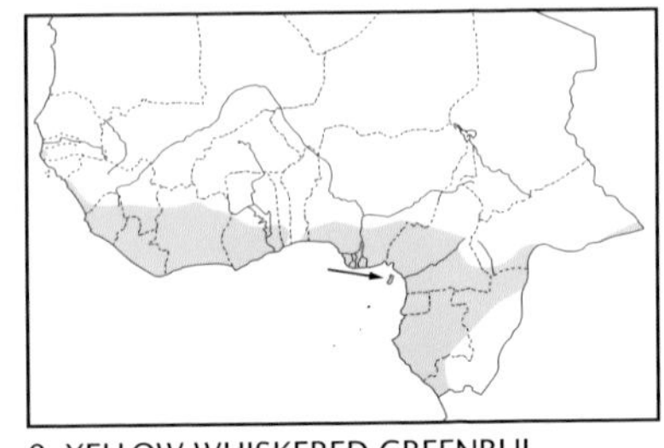
9. YELLOW-WHISKERED GREENBUL

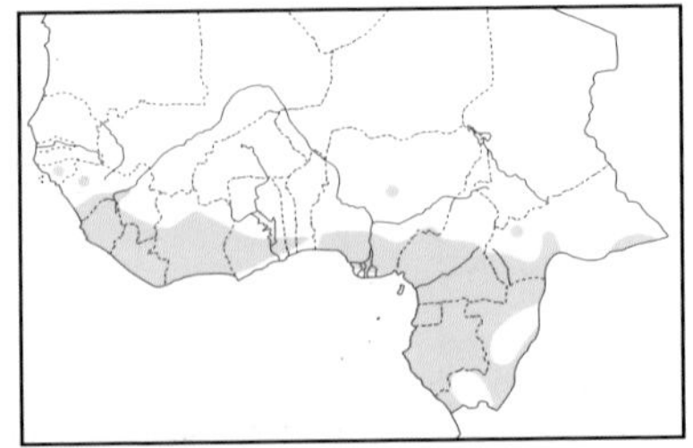
10. HONEYGUIDE GREENBUL

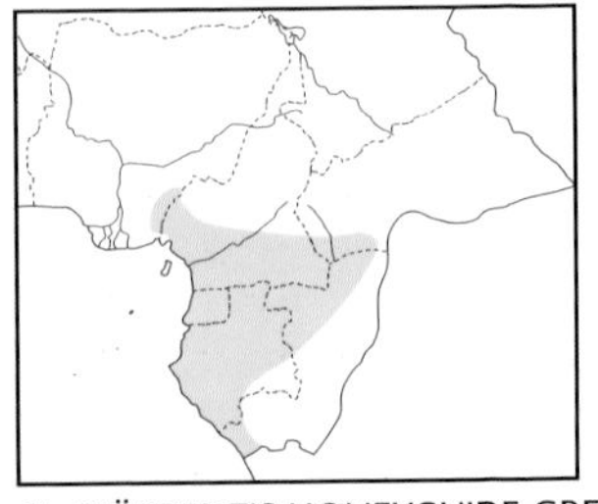
11. SJÖSTEDT'S HONEYGUIDE GREENBUL

Plate on page 304

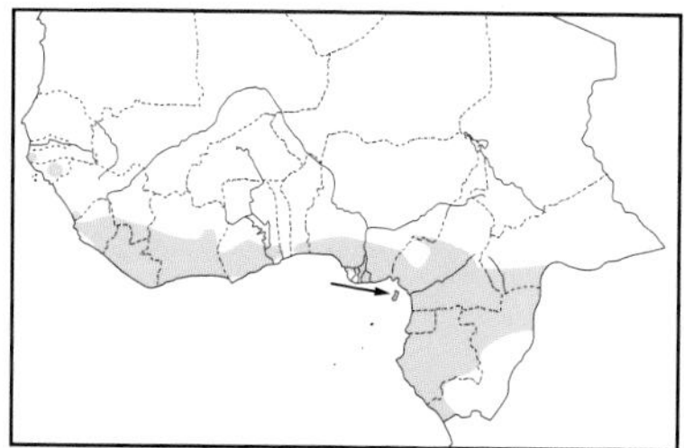

1. RED-TAILED GREENBUL

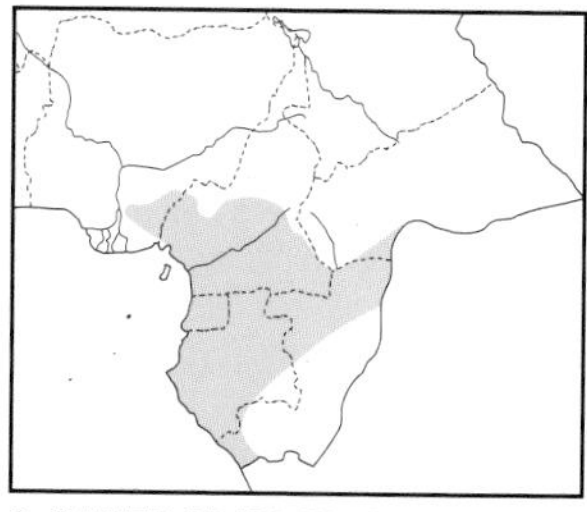

2. WHITE-BEARDED GREENBUL

3. YELLOW-BEARDED GREENBUL

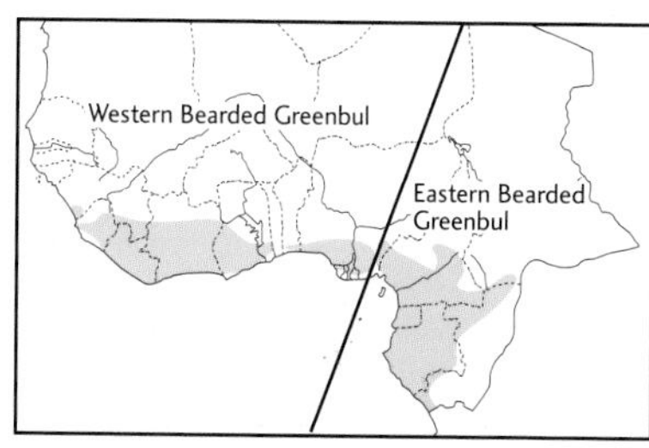

4. WESTERN BEARDED GREENBUL
5. EASTERN BEARDED GREENBUL

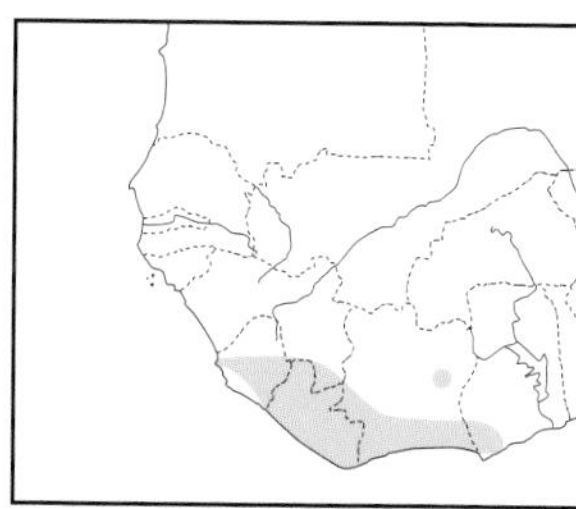

6. GREEN-TAILED BRISTLEBILL

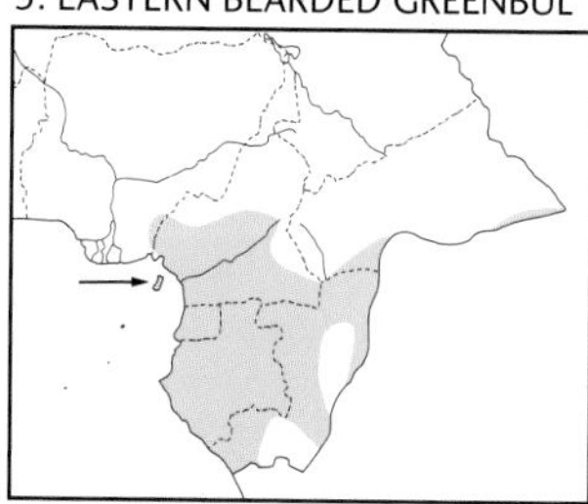

7. LESSER BRISTLEBILL

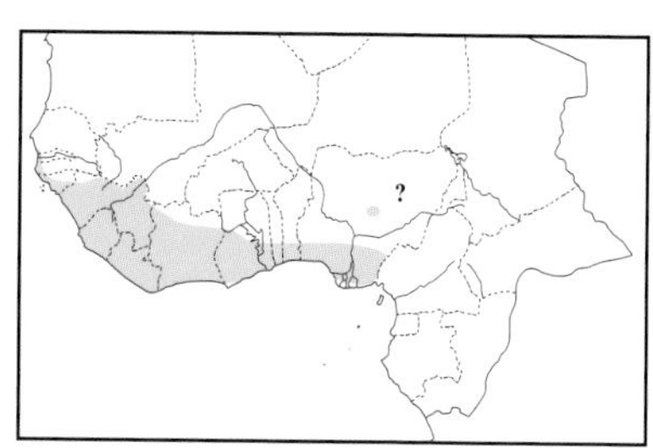

8. GREY-HEADED BRISTLEBILL

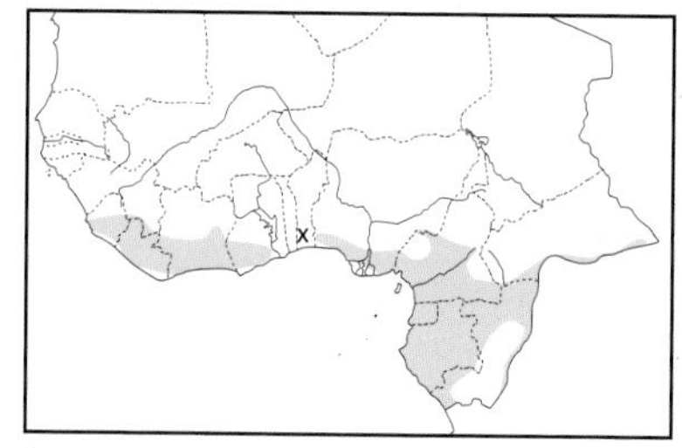

9. RED-TAILED BRISTLEBILL

Continued from page 304

10 **HONEYGUIDE GREENBUL** *Baeopogon indicator* 19 cm **R c/f**
Dark and stocky. Outer tail feathers white tipped blackish (in adult) or wholly white (in immature), recalling honeyguide (Plate 77). **10a Ad** Whitish eye. **10b Juv** Dark eye. ▲ Forest canopy and upper mid-level, esp. at edges. Vocal. ✤ Varied. Clear melodious whistles; often abbreviated as a hurried *vik-vik-view* or *tiu-liuuw*. Also a mewing note. [CD9:57]

11 **SJÖSTEDT'S HONEYGUIDE GREENBUL** *Baeopogon clamans* 19 cm **R lc**
As 10 but underparts paler and buffier; white outer tail feathers lack dark tips. ▲ Mid-levels and lower canopy of lowland forest. ✤ Loud, hard, nasal *whEw!*, often running into short, rapid, nasal babble *teeturuteetutwhee*. [CD9:58]

***Criniger* Bearded bulbuls.** Medium-sized to large; distinguished by long white or yellow throat feathers, which are frequently puffed out, forming a 'beard'. Inhabit lower and mid-strata of closed forest. Usually in pairs or small groups, frequently joining mixed-species flocks. Vocal.

1 RED-TAILED GREENBUL *Criniger calurus* 20 cm R f/c
1a *C. c. verreauxi* (SW Senegal–SW Nigeria) Grey head; white 'beard'; central breast and belly yellow; olive-green tail. **1b *C. c. calurus*** (S Nigeria to SW CAR–Congo; Bioko) Dull rufous tail. Compare 3 and 5. ▲ Forest. ✤ Song a cheerful rising *chup-chup-chwirulup*. Also a 3-note *tsik-tyu-tyip* and rapid series *tyu-tyutyutyu-tyip-tyip-tyip*... Alarm *tsik*. [CD9:84]

2 WHITE-BEARDED GREENBUL *Criniger ndussumensis* 18 cm R u/lc
Indistinguishable in the field from 1b, but voice as 3. ▲ Forest. ✤ Three short, harsh, relatively low syllables *whut-chruw-chruw* or *chuk-ker-chyer*, indistinguishable from 3; similar in structure to 1 but lower pitched and lacking its cheerfulness. Alarm *tsik* (as 1). [CD9:86]

3 YELLOW-BEARDED GREENBUL *Criniger olivaceus* 18 cm R r/lf VU
Olive-green head and upperparts; bright yellow throat; olive-green breast and flanks; yellowish belly. Compare 4a and Icterine Greenbul (Plate 92:1). ▲ Forest. Endemic. ✤ As 2. [CD9:85]

4 WESTERN BEARDED GREENBUL *Criniger barbatus* 22 cm R f/c
4a *C. b. barbatus* (Guinea–W Togo) Olive-brown head and upperparts; yellow 'beard'; mottled grey and olive breast; dirty olive-yellow belly; dull olive-chestnut tail. **4b *C. b. ansorgeanus*** (S Nigeria) White chin; very pale yellow throat; bright rufous tail. ▲ Forest. Endemic. ✤ Clear, slightly quavering whistle introduced by lower note *teruu twEEEur*. Aggressive song more rapid, 'impatient' *chiwee-chiWEE-WEEur*. Calls include hard *tsyuk* and loud *KYUW*. [CD9:82]

5 EASTERN BEARDED GREENBUL *Criniger chloronotus* 22 cm R c
Grey head and breast; white 'beard'; bright rufous tail; yellow on underparts pale, confined to belly. Compare 1b. ▲ Forest. ✤ Rather soft, mournful, quavering 2-note song, quite different from 4. Alarm a weak chatter. [CD9:83]

***Bleda* Bristlebills.** Large skulkers of forest undergrowth, with olive-green upperparts, yellow underparts, and stout bills and legs. Join mixed-species flocks; attend ant columns. Vocal.

6 GREEN-TAILED BRISTLEBILL *Bleda eximius* 21.5–23.0 cm R r/s VU
As 7 but without yellow loral spot; yellow tips to outer tail feathers narrower. ▲ Forest. Endemic. ✤ Song a series of pure, whistled notes *hee-huu-huu-hu-heeu* and *hu-hu-heeuu-heeuu*, often slightly vibrating and rather similar in tone to 9. Calls include abrupt, nasal *kyop*, sometimes followed by fast *kiuwkiuwkiuwkiuwkiuw*... [CD9:79]

7 LESSER BRISTLEBILL *Bleda notatus* 19.5–21.0 cm R c
7a *B. n. notatus* (SE Nigeria–SW CAR, Congo) Entirely olive-green above; yellow spot in front of eye; four outer tail feathers tipped yellow (as 8); bare blue skin above eye. **7b *B. n. ugandae*** (SE CAR) Loral spot duller; eye yellow. ▲ Forest. ✤ Song tremulous and descending, slightly resembling 9 but lacking its melancholic quality. Calls include hard, oft-repeated *chup* and *chiup*, sometimes followed by rattle *trrrrrttttt*, and *wheew*. [CD9:81]

8 GREY-HEADED BRISTLEBILL *Bleda canicapillus* 20.5–22.0 cm R c
Slate-grey head; olive-green tail feathers tipped yellow on four outer pairs. Compare 6 and 7. ▲ Forest and gallery extensions. Endemic. ✤ Varied. Song far carrying with cheerful, ringing quality, typically a loud initial note, a brief pause, then a series on descending scale. Other notes include loud *CHEEup* and *kyuw*, often in long series. [CD9:80]

9 RED-TAILED BRISTLEBILL *Bleda syndactylus* 21.5–23.0 cm R u/c
Olive-green above; bright rufous tail; sulphur-yellow below; bare blue skin above eye. ▲ Forest. ✤ Song ends with a series of pure, vibrant syllables on same pitch *turrruuu turrruuu turrruuu*. Calls include nasal *kyow* or *pyeeuw* and hard *chup*. [CD9:78]

Maps on page 307

1a
1b
2
6
3
7a
7b
4b
4a
8
5
9

***Phyllastrephus* bulbuls** are generally small to medium-sized, with long, slender bills and russet tails. In family parties in lower and mid-strata of forest. Insectivorous; frequent members of mixed-species flocks. No arresting vocalisations.

1 **ICTERINE GREENBUL** *Phyllastrephus icterinus* 15–16 cm **R f/c**
Small. Olive-green above; drab yellowish below; dull rufous tail. ▲ Forest interior. Gregarious; usually in mixed-species flocks, at lower and mid-levels. ✤ Rather fast nasal chatter, slowing at end, easily passing unnoticed. Alarm a nasal trill. [CD9:73]

2 **WHITE-THROATED GREENBUL** *Phyllastrephus albigularis* 17 cm **R u/lc**
White throat contrasts with grey head and olive-grey breast; pale yellow belly. ▲ Forest. Usually low, but also in middle and upper levels. ✤ Variable, rapid series of clear, scolding notes, usually rising, then dying away, often preceded by loud, rolled *turrr*. [CD9:75]

3 **XAVIER'S GREENBUL** *Phyllastrephus xavieri* 16–18 cm **R lc/u**
Extremely similar to 1 but voice diagnostic. Males slightly larger. ▲ Forest, at all levels. ✤ Short, nasal *kwah, kwah, kwah,... kwahkwah* and more drawn-out, squeaky *kwèèèh*, uttered in shorter or longer series of similar or combined notes (*kwah-kèèh*). [CD9:74]

4 **BAUMANN'S GREENBUL** *Phyllastrephus baumanni* 18 cm **R u/r DD**
Olive-brown above; pale olive-grey below; rusty tail. ▲ Lower levels of semi-deciduous mid-altitude forest, gallery forest, thicket. Skulking. Endemic. ✤ Calls include series of loud *week* and *wik* notes, and scolding *chèrrr*. Song consists of 2-4 slightly nasal, rising notes *whu whee wheew* followed by some scolding *wik* or *chewik* notes.

5 **LIBERIAN GREENBUL** *Phyllastrephus leucolepis* *c.* 16 cm **R lr CR**
As 1 but with whitish spots on upperparts. ▲ Rainforest. Endemic. ✤ Unknown.

6 **CAMEROON OLIVE GREENBUL** *Phyllastrephus poensis* 18 cm **R lc**
Nondescript. Brown above; longish rufous tail; pale below with whitish throat. ▲ Montane forest. In dense cover, mainly low down. Endemic. ✤ Low, grating *chrrr-chrrr-chrrr-*.... Song a series of unmusical notes; also loud, scolding notes. [CD9:72]

7 **GREY-HEADED GREENBUL** *Phyllastrephus poliocephalus* 20–23 cm **R lc NT**
Large. Grey head; olive-green upperparts; white throat; bright yellow underparts. ▲ Submontane forest. Usually in small, noisy groups at mid-elevations. Typically in mixed-species flocks. Endemic. ✤ Loud, rather harsh *churp*, often repeated, and fast *churp-p-p*. [CD9:77]

8 **PALE OLIVE GREENBUL** *Phyllastrephus fulviventris* 19 cm **R**
Olive-brown above; dull rufous tail; creamy-white throat does not contrast with rest of underparts. ▲ Gallery forest, dense bush; mostly in undergrowth. Skulking. Possibly coastal S Gabon and S Congo. ✤ Nasal, chattering note. [CD9:69]

9 **SPOTTED GREENBUL** *Ixonotus guttatus* 17 cm **R lc/s**
9a Small. Upperparts spotted white; underparts and outer tail feathers white. ▲ Forest, forest edge. In noisy, mobile groups in canopy. Frequently raises one wing alternately (**9b**). ✤ Dry chirping or 'ticking' call, like noise made by electrical spark, constantly uttered. [CD9:59]

10 **WESTERN NICATOR** *Nicator chloris* 21–24 cm **R c/s**
Large. Olive-green above with bold golden spots; pale grey below; ashy-white throat. Heavy bill. ▲ Forest and gallery extensions. ✤ Powerful, melodious and very varied song comprising an explosive crescendo of notes, clear whistles, guttural rattles, etc. Occasionally mimics other species. Calls a loud, abrupt *tok!*, often in long, accelerating series. [CD14:1a]

11 **YELLOW-THROATED NICATOR** *Nicator vireo* 17–19 cm **R f**
As 10 but smaller and with yellow throat; yellow supraloral streak; grey face. ▲ Middle and lower levels of forest. Skulking. ✤ Loud series of resonant, explosive notes e.g. *ko-kwee-ko-ko-ko-kwee-kuk-kuk*, with variations. Also angry *gwrrrrrrr*. [CD14:1b]

 Maps on page 312

1
2
3
4
5
6
7
8
9a
9b
10
11

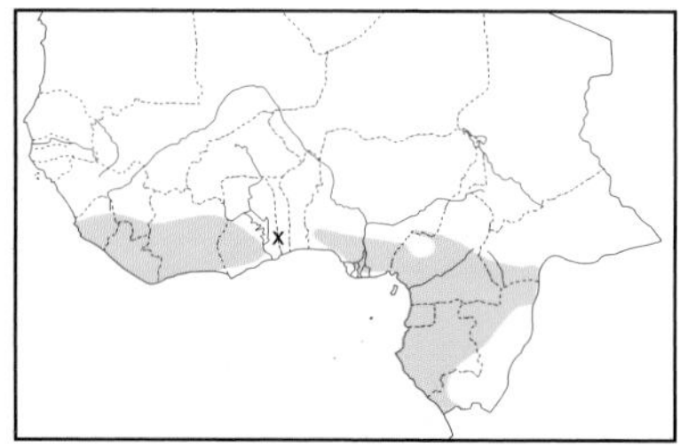

1. ICTERINE GREENBUL

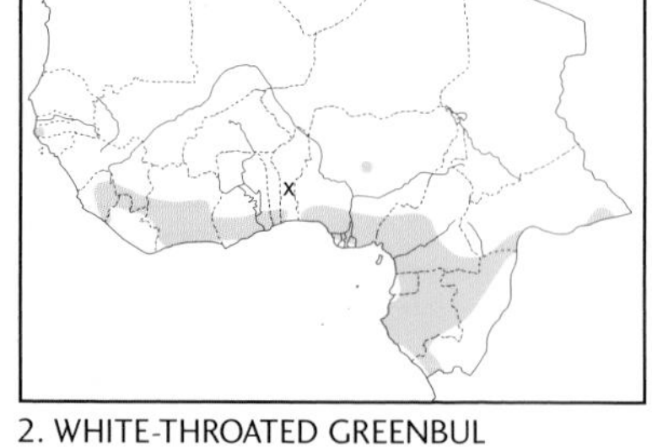

2. WHITE-THROATED GREENBUL

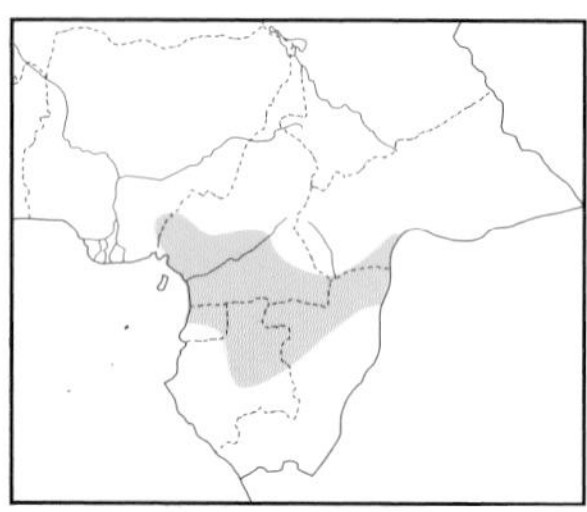

3. XAVIER'S GREENBUL

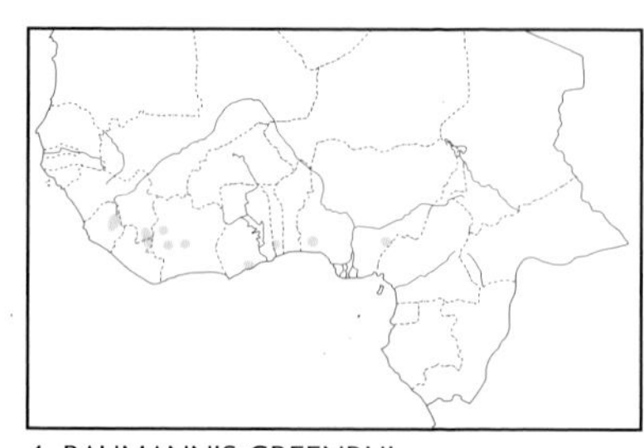

4. BAUMANN'S GREENBUL

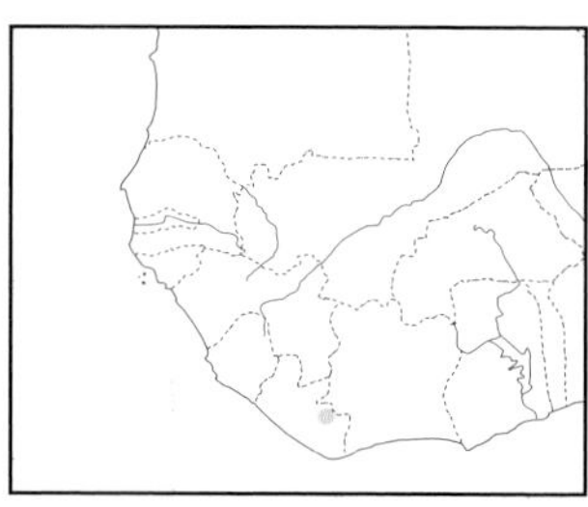

5. LIBERIAN GREENBUL

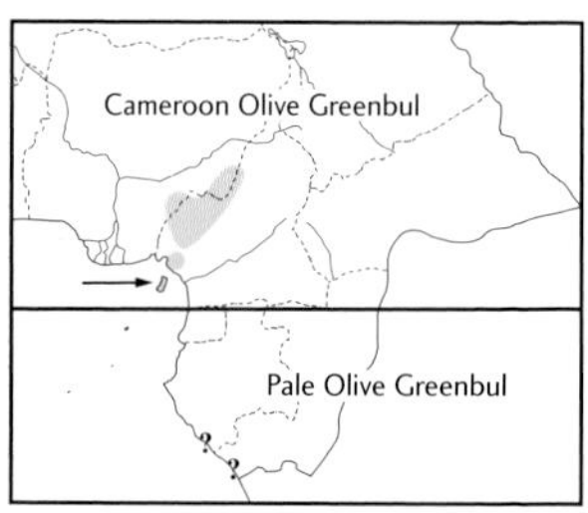

6. CAMEROON OLIVE GREENBUL
8. PALE OLIVE GREENBUL

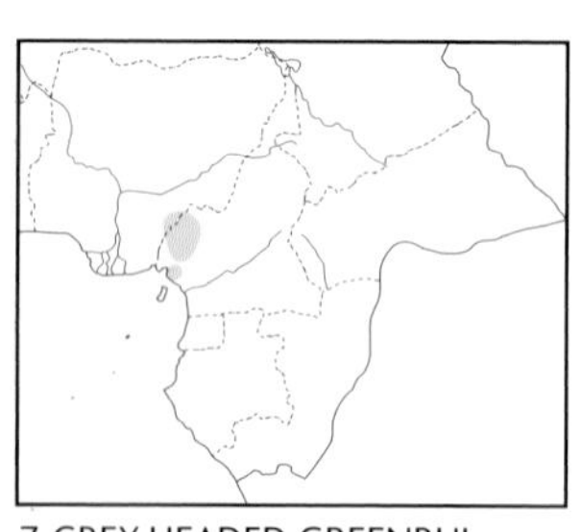

7. GREY-HEADED GREENBUL

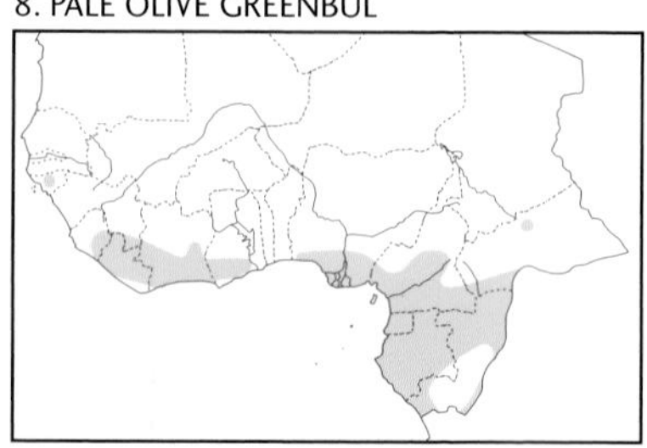

9. SPOTTED GREENBUL

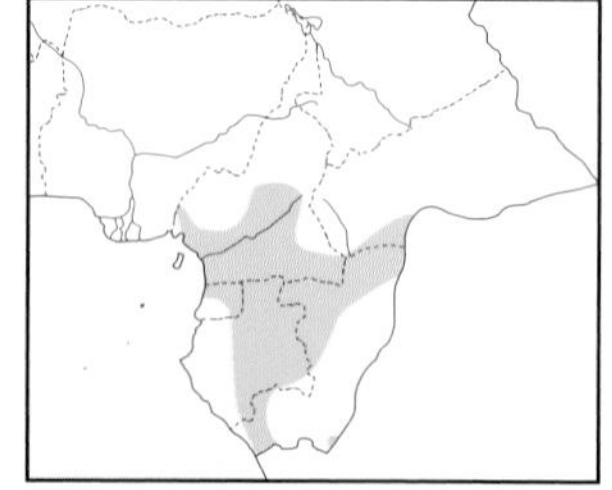

10. YELLOW-THROATED NICATOR

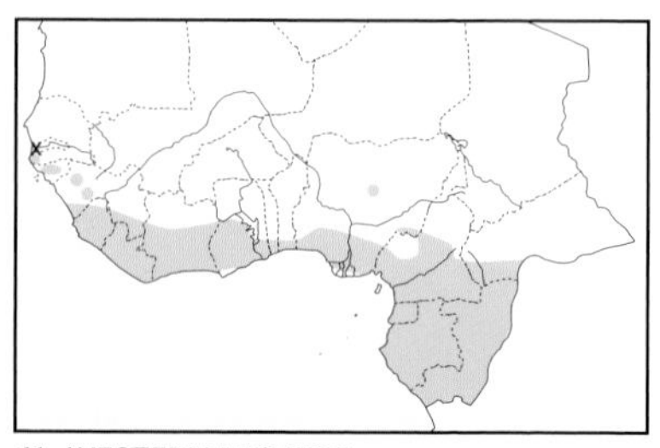

11. WESTERN NICATOR

Plate on page 310

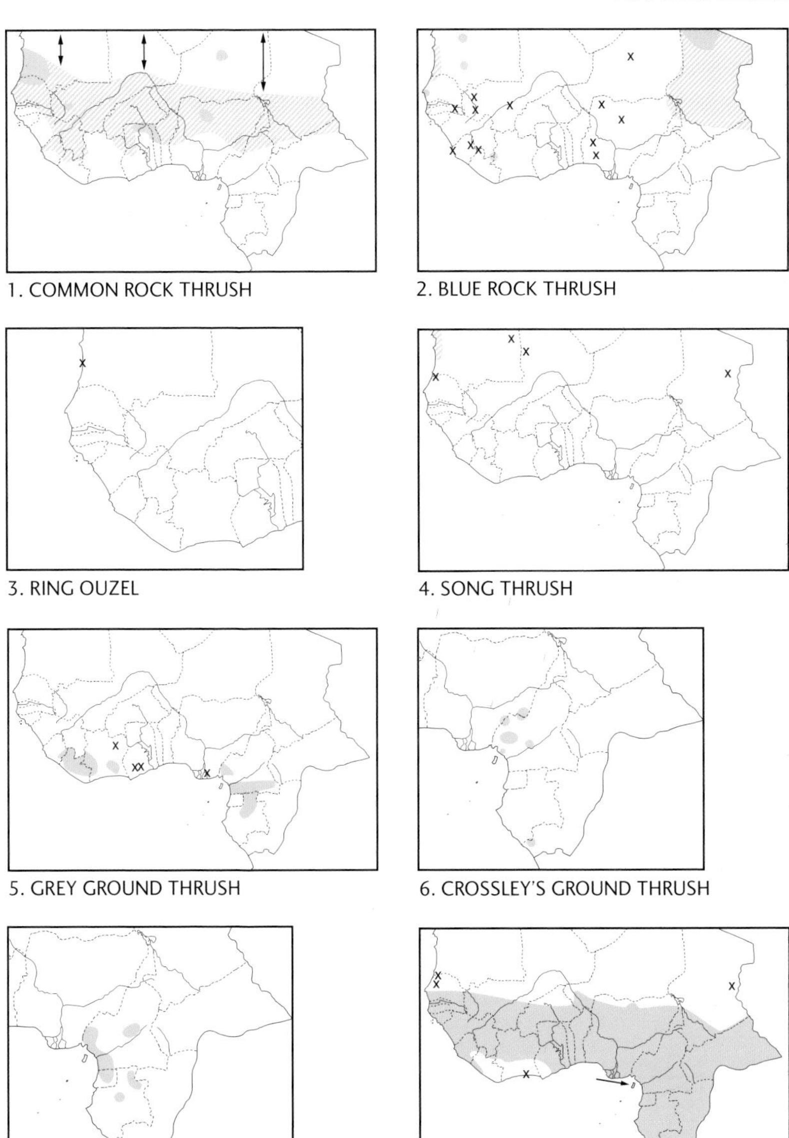

1. COMMON ROCK THRUSH

2. BLUE ROCK THRUSH

3. RING OUZEL

4. SONG THRUSH

5. GREY GROUND THRUSH

6. CROSSLEY'S GROUND THRUSH

7. BLACK-EARED GROUND THRUSH

8. AFRICAN THRUSH

Large and diverse family of small to medium-sized songbirds (Plates 93–98) with usually strong legs and feet and mostly square or rounded tails. Sexes usually similar, with variable dimorphism, the female being duller. Broad range of vocalisations; some are notable songsters.

1 COMMON ROCK THRUSH *Monticola saxatilis* 18.5 cm **P u/r**
Short, orange-rufous tail; dark wings. **1a Ad male breeding** Blue head; white patch on back; orange-rufous underparts. **1b Ad male non-breeding** Dark brown, mottled upperparts; crescent-marked underparts. **1c Ad female** As 1b but paler. Compare 2b. ▲ Various habitats. ✤ Short *chak*. Mostly silent in winter quarters. [CD3:96]

2 BLUE ROCK THRUSH *Monticola solitarius* 20 cm **P/R s/r**
2a Ad male Wholly dark blue. **2b Ad female** Dark brown upperparts and tail; barred underparts. Compare 1c. ▲ Mainly rocky habitats. ✤ Mainly silent. Calls include *tak-tak* (contact), *chuk-chuk* (alarm) and high-pitched *tsee*. [CD3:97]

3 RING OUZEL *Turdus torquatus* 23–24 cm **V**
3a Ad male Black; white crescent-shaped breast-band; yellow bill. **3b Ad female** Browner than male; breast-band less distinct. ▲ Palearctic vagrant. ✤ Loud *chak* or *chakchakchak*. [CD4:2a]

4 SONG THRUSH *Turdus philomelos* 23 cm **P/V+ r**
Brown above; boldly spotted below. ▲ Flight fast and direct. ✤ Short, dry *tsik* in flight. [CD3:99]

5 GREY GROUND THRUSH *Zoothera princei* 21 cm **R r**
Two black patches on head-sides; two white wingbars. ▲ Lowland rainforest. Joins mixed-species flocks; sometimes attends ant swarms. Extremely shy. ✤ Sharp, high-pitched *ssrrii* or *sssirrr* and thin *seeep* (alarm). Song unknown. [CD10:34]

6 CROSSLEY'S GROUND THRUSH *Zoothera crossleyi* 21.5 cm **R r/lf NT**
Resembles 7, but with blackish mask; also distinctly larger. ▲ Montane and submontane forest, often in ravines. Shy. ✤ Alarm a thin, high-pitched *seeep*. Song, from ground or high, hidden perch, varied and far carrying, consisting of short, melodious phrases, repeated several times at regular intervals, until a new motif is introduced. [CD10:31]

7 BLACK-EARED GROUND THRUSH *Zoothera camaronensis* *c.* 18 cm **R r**
Pattern as 5, but bright orange-rufous on face and underparts. ▲ Lowland rainforest. Very shy. ✤ Thin, high-pitched *ssreee* (probably alarm) and *tssrrr*. Song unknown. [CD10:33]

8 AFRICAN THRUSH *Turdus pelios* 23 cm **R c/f**
Typical thrush with yellow bill. Races differ more or less in depth of plumage coloration. **8a Ad *chiguancoides*** (Senegal–W Ghana) Palest race; upperparts dull grey-brown; underparts pale brown and white; flanks washed pale orange-buff. *T. p. saturatus* (W Ghana–Congo), nominate *pelios* (E Cameroon, Chad–N CAR) and *centralis* (S CAR–Congo) have underparts somewhat deeper coloured, with more distinct streaks on throat. **8b Ad *nigrilorum*** (Mt Cameroon) Darkest race; browner; no orange on flanks. *T. p. poensis* (Bioko) similar, but slightly paler. **8c Juv *chiguancoides*** Rufous wash over face and breast; pale tips to coverts; crescentic markings on breast and flanks. ▲ Various wooded habitats; not in rainforest. Forages on ground. ✤ Song far carrying, melodious and variable, with series of short repetitions of simple motifs, e.g. *toolee toolee toolee weetyuuwee weetyuuwee teewit teewit teewit swrreep swrreep leepoo leepoo leepoo churp churp churp*... etc. Call a hard *chuk*, often repeated in dry series *chukukukukuk*... [CD10:40]

 Maps on page 313

1c
1a
2b
1b
3b
2a
3a
4
5
6
8b
7
8c
8a

1 FOREST ROBIN *Stiphrornis erythrothorax* 13 cm R f

1a Ad *erythrothorax* (Sierra Leone–Nigeria) Dark olive-brown above; white spot in front of eye; bright orange breast. **1b Juv *erythrothorax*** Spotted rufous above; pale buffish throat; blackish breast mottled rufous. **1c Imm *erythrothorax*** Gradually as adult but duller; throat whitish. **1d Ad *xanthogaster*** (Cameroon to CAR–N Congo) Sooty-grey above; breast paler than 1a. **1e Ad *gabonensis*** (SW Cameroon–SW Congo; Bioko) Above as 1d, but below as 1a. **1f Ad *sanghensis*** (Sangha Forest Robin; SW CAR; DD) Breast yellow-orange; lower underparts pale yellow. ▲ Lowland forest. ✤ Variable, fast and sweet song with high-pitched whistled motifs. Calls: low *karrrr*, whistled *whi-whiuuu*. [CD9:89]

2 BOCAGE'S AKALAT *Sheppardia bocagei* 13 cm R f/r

2a Ad As 3a but head-sides orange-rufous. **2b Juv** As 3b; younger bird illustrated. ▲ Montane and submontane forest. Secretive. ✤ Song a quiet, mournful series of 7–10 sweet whistles, *hu-hee-hu-hluwee-hu-hlu-whee-hu-hee...* Calls a soft twittering and sibilant ratchet note. [CD9:91]

3 LOWLAND AKALAT *Sheppardia cyornithopsis* 13 cm R r/lf

3a Ad Olive-brown above; orange-rufous below. **3b Juv** Dark brown spotted rufous above and below; belly whitish with some dusky scalloping. ▲ Forest. ✤ Song, two soft whistles in alternation *whee, whiu, whee, whiu,* or series of *whiu*. Call *tiee*; alarm a low *krrr*. [CD9:92]

4 WHITE-TAILED ALETHE *Alethe diademata* 18 cm R f

4a Ad Dark chestnut above; rusty-orange crown-stripe; white tail corners. **4b Juv** (older bird than 5b illustrated) As 5b but with whitish tail corners. ▲ Forest, gallery forest, forest patches in savanna. ✤ Three sweet ascending whistles, *huu hee hueee*; third note sometimes absent. Calls: whistled *huu*, often in long series; harsh notes. [CD10:3b]

5 FIRE-CRESTED ALETHE *Alethe (diademata) castanea* 18 cm R f

5a Ad As 4a but brighter above; no white in tail. **5b Juv** Spotted orange-rufous above; underparts orange-rufous scalloped blackish on throat and breast; lower underparts whiten with age. ▲ As 4. From Nigeria east. ✤ As 4. [CD10:3b]

6 BROWN-CHESTED ALETHE *Alethe poliocephala* 16 cm R f

6a Ad *poliocephala* (W Guinea–Ghana) Grey-black crown; white supercilium; brown head-sides; chestnut upperparts. **6b Juv *poliocephala*** Upperparts with large orange-rufous spots; dirty white underparts with orange-rufous breast scalloped blackish. **6c Ad *compsonota*** (SW Nigeria to SW CAR–Congo; Bioko) Head-sides dark grey. *A. p. carruthersi* (SE CAR) has crown browner; head-sides brownish. ▲ Forest, gallery forest, forest patches in savanna. Secretive. ✤ Soft *karr-karr*. Generally silent. [CD10:4b]

7 WHITE-TAILED ANT THRUSH *Neocossyphus poensis* 20 cm R u/lc

Dark. White tail corners. Thrush-like jizz. Pumps tail like chat. ▲ Lowland forest. Terrestrial; secretive. ✤ Clear *huweeeeet*, sibilant *tseeeuw* and dry *prrt prrt*. [CD10:6]

8 FINSCH'S FLYCATCHER THRUSH *Stizorhina finschi* 18 cm R f/r

White tail corners. Resembles large flycatcher in jizz and foraging behaviour. Flicks outer tail feathers sideways. ▲ Lowland forest. Endemic. ✤ Harsh croaking *truwee-trueet*. Song a rather slow series of 4 melodious, whistled notes *hooee, hooee hooee-huEE*. [CD10:8]

9 RED-TAILED ANT THRUSH *Neocossyphus rufus* 22 cm R f/r

Large; rufous. Rufous outer tail feathers. Thrush-like jizz. ▲ Lowland forest. Terrestrial; secretive. ✤ Sibilant whistle *pseeeuuw*. Dry *prrt prrt* around ant swarms and on take-off. Song, two clear whistles followed by trill *tseee wheh tsisisisisisisrrru*. [CD10:5]

10 RUFOUS FLYCATCHER THRUSH *Stizorhina fraseri* 18 cm R f/c

Resembles 8 in jizz and actions, but outer tail feathers rufous. ▲ Lowland forest. ✤ Whistling *tsweetweetweetweet*, rapid *trrwit-rrwit-rrwit*, and various hoarse notes. Song a rather slow, slightly rising series of 4 whistled notes, *trwee tu-trwee-twee*. [CD10:7]

 Maps on page 318

1b
1c
1d
1a
2b
2a
3b
3a
1e
1f
6b
6a
6c
4b
5b
5a
4a
8
7
9
10

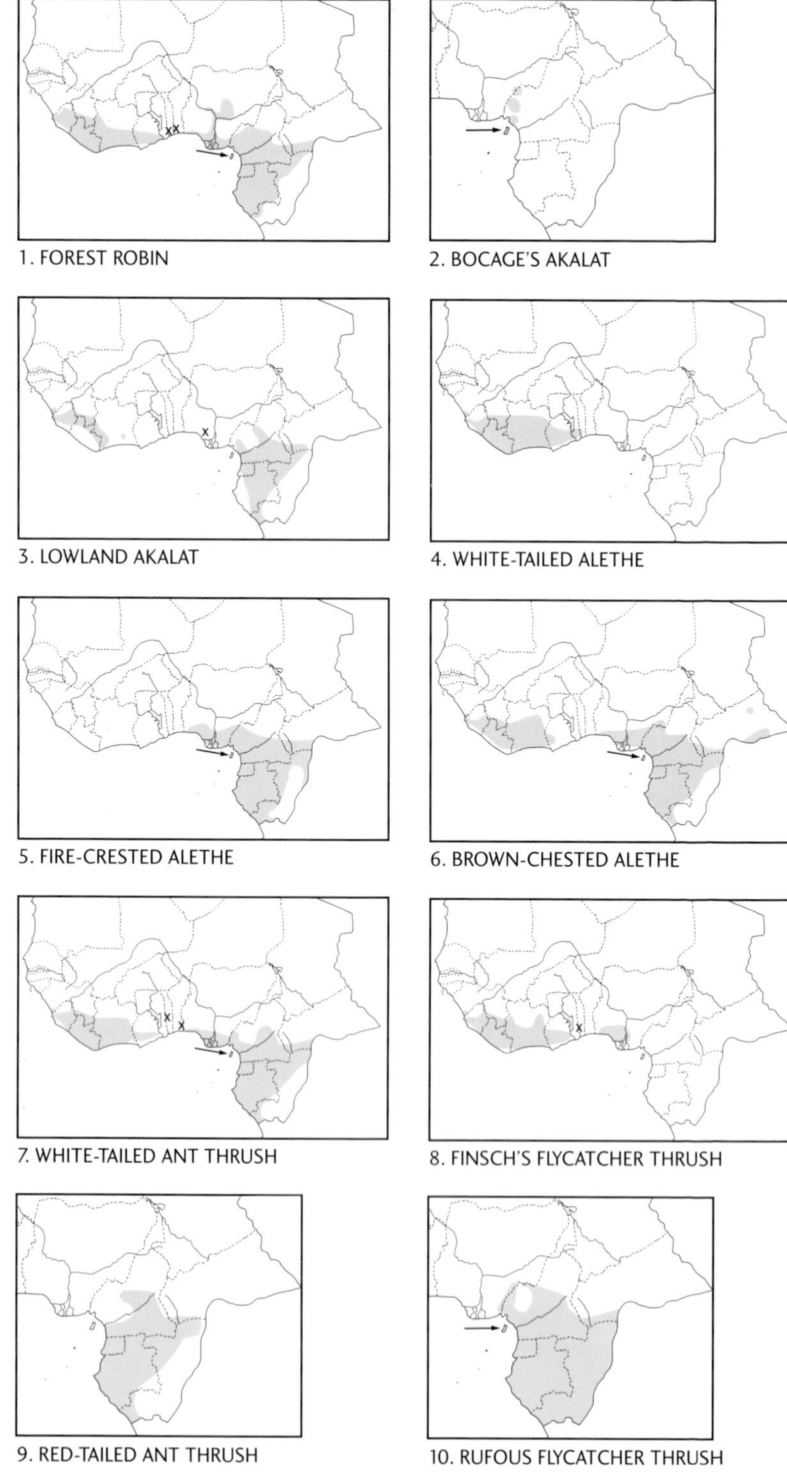

Plate on page 316

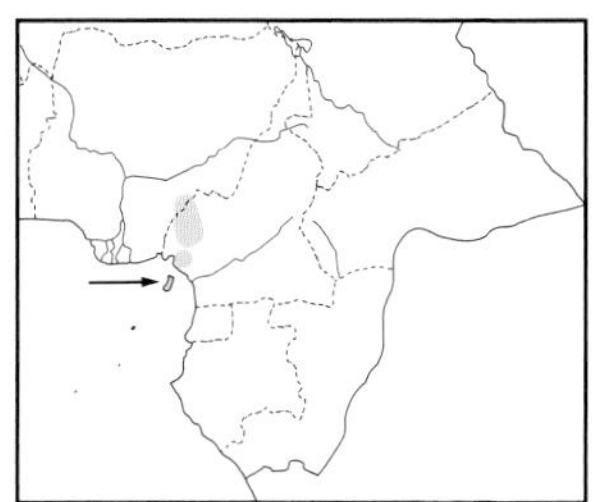

1. WHITE-BELLIED ROBIN CHAT

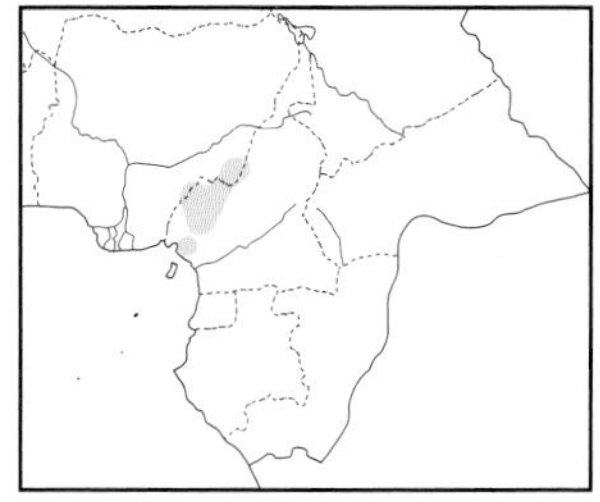

2. MOUNTAIN ROBIN CHAT

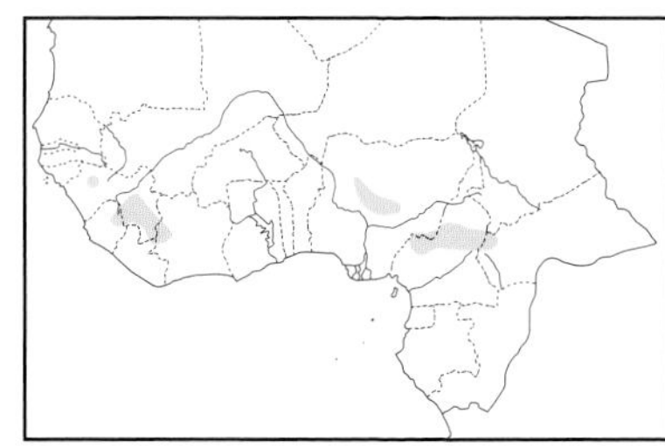

3. GREY-WINGED ROBIN CHAT

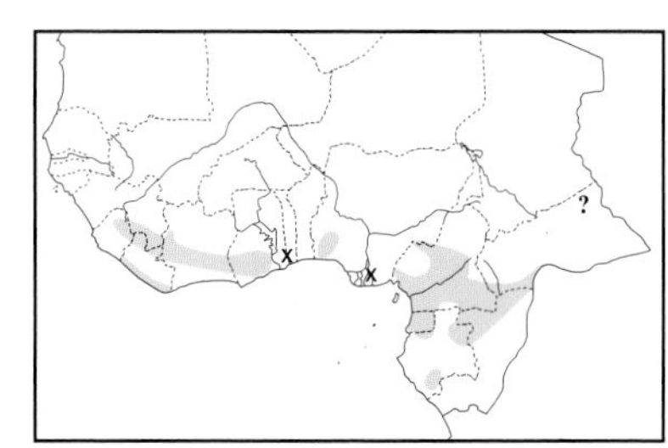

4. BLUE-SHOULDERED ROBIN CHAT

5. RED-CAPPED ROBIN CHAT

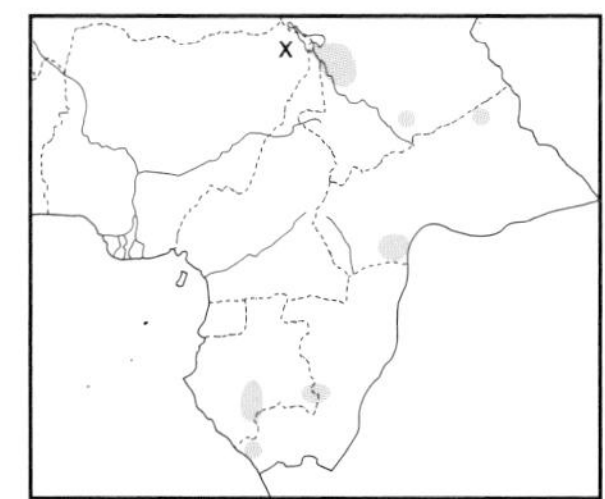

6. WHITE-BROWED ROBIN CHAT

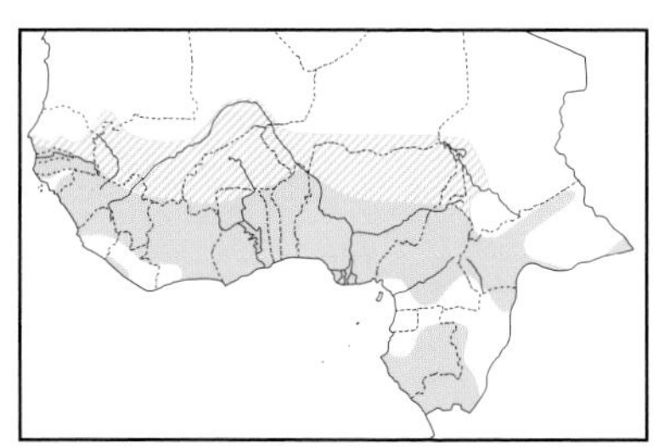

7. SNOWY-CROWNED ROBIN CHAT

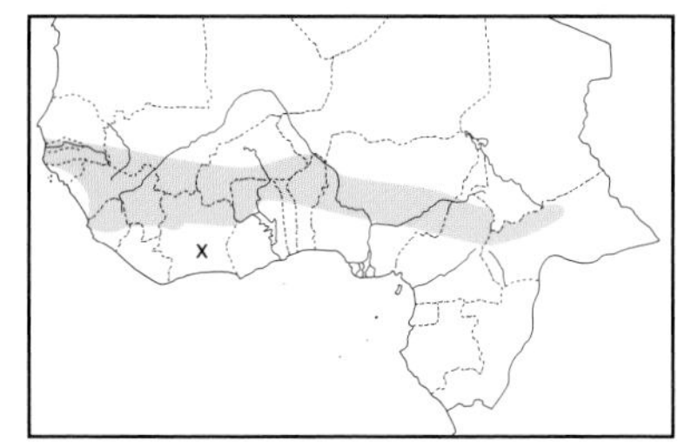

8. WHITE-CROWNED ROBIN CHAT

Colourful thrushes with orange-rufous underparts and orange-rufous tails with, typically, black central feathers. Most have loud, very melodious and varied songs with much mimicry.

1 WHITE-BELLIED ROBIN CHAT *Cossyphicula roberti* 13 cm R u/lc
1a Ad Resembles akalats (Plate 94:2–3) but tail black and red. **1b Juv** Dark brown spotted rufous above; rufous-buff scalloped black below. ▲ Montane and submontane forest. Quiet and unobtrusive. ✤ Song a fast series of 6 rather high-pitched whistled notes *tsu-ti-tu-ti-tu-tu*, frequently repeated without pause. Alarm a fast *ti-ti-ti-ti-ti-*... [CD9:95]

2 MOUNTAIN ROBIN CHAT *Cossypha isabellae* 15 cm R f
2a Ad *isabellae* (Mt Cameroon) Narrow white supercilium; darker above than 2b; contrasting orange-rufous rump. **2b Ad *batesi*** (SE Nigeria–W Cameroon) Olive-brown above becoming rufous-brown on rump; lower belly off-white. **2c Juv** Mainly dark brown spotted rufous. ▲ Montane and submontane forest. Endemic. ✤ Song a fairly loud, rather tuneless two-note trill *tsri-tsrrrrr* repeated in rapid succession. Call a guttural *grrr*. [CD9:96]

3 GREY-WINGED ROBIN CHAT *Cossypha polioptera* 15 cm R r/lc
3a Ad Long white supercilium underlined by black eye-stripe; uniformly rufous tail. **3b Juv** Duller; no supercilium; rufous spots on head and wing-coverts. ▲ Mid-elevation and gallery forest. ✤ Song varied and melodious, including imitations. Higher pitched than song of other robin chats that include imitations. [CD10:1a]

4 BLUE-SHOULDERED ROBIN CHAT *Cossypha cyanocampter* 16.5 cm R u
4a Ad Long white supercilium; blue shoulder patch. **4b Juv** Duller; no supercilium; rufous spots on head and wing-coverts. ▲ Forest, gallery forest. Skulking. ✤ Song very varied, melodious and sustained, including perfect imitations of other birds and human whistles. Lower pitched and slower than songs of 3 and 7. Call a loud *trurr*. [CD10:1b]

5 RED-CAPPED ROBIN CHAT *Cossypha natalensis* 17 cm R/M r/lu
5a Ad Orange-rufous head; beady eye. **5b Juv** Dark brown spotted rufous above; rufous-brown scalloped dark brown below. ▲ Wooded habitats. Skulking. ✤ Song rich and melodious, consisting of whistled phrases and often including many imitations. Calls include a rather plaintive *whuh ti-eh* (second syllable nasal) and slightly trilled *prree prrup*, monotonously repeated; alarm a guttural *grrr*. [CD10:2a]

6 WHITE-BROWED ROBIN CHAT *Cossypha heuglini* 20 cm R u/lf
6a Ad Distinctive, long white supercilium. **6b Juv** Dark brown spotted rufous above; rufous-buff scalloped dark brown below. ▲ Riverine forest, thickets in savanna. ✤ Song loud, varied and melodious, with simple, whistled phrases characteristically starting quietly, then increasing in volume. Does not usually include imitations. Members of pair duet or sing antiphonally, female uttering high-pitched *tseeeee*. Alarm a harsh, rattling *tsrek-tsrek*. [CD9:99]

7 SNOWY-CROWNED ROBIN CHAT *Cossypha niveicapilla* 20.5–22.0 cm R/M lc/u
7a Ad Large; white crown; rufous hind-collar. Compare 8. **7b Juv** Blackish spotted rufous above; rusty-buff scalloped blackish below. ▲ Thickets in savanna, forest edges and clearings, overgrown farmland, gallery forest, gardens. ✤ Vocal. Song rich, melodious and sustained, with characteristic fast delivery and including many imitations; some motifs repeated a few times, often with slight variations and thus somewhat reminiscent of African Thrush. Calls include *heeee* (contact) and guttural *krrr* (alarm). [CD10:2b]

8 WHITE-CROWNED ROBIN CHAT *Cossypha albicapilla* 26 cm R u/lc
8a Ad *albicapilla* (Senegambia–N Ivory Coast) Very large; white crown; no rufous hind-collar; long tail. **8b Ad *giffardi*** (S Burkina Faso–NW CAR) Crown mainly black with white crescents. ▲ Thickets in savanna, riverine scrub, large overgrown gardens. ✤ Song varied, sustained and very fast, including scratchy notes, trills and occasional imitations, but not as varied as that of 7. Call a penetrating *sweeuee*. [CD10:3a]

 Maps on page 319

1b
1a
2c
2b
2a
3b
3a
4b
4a
5b
5a
6b
6a
7b
7a
8b
8a

PLATE 96: PALEARCTIC CHATS, PALM THRUSH AND SCRUB ROBINS

1 BLUETHROAT *Luscinia svecica* 14 cm **P/V u**
In all plumages: diagnostic rufous sides to basal half of outer tail feathers (conspicuous in flight). **1a Ad male *svecica*** (P) Blue throat and breast; rufous spot in centre. **1b Ad female *svecica*** Buff-white throat bordered by dappled black breast-band. **1c Ad male *cyanecula*** (P) As 1a, but with white spot in centre of blue. ▲ Scrub. ✤ Call a hard *tak* and *hueet*. [CD3:79]

2 COMMON REDSTART *Phoenicurus phoenicurus* 14 cm **P/V* u/lf**
In all plumages: almost constantly shivering, bright rufous tail. **2a Ad male breeding** Blue-grey above; black face; orange-rufous underparts. **2b Ad male non-breeding** Pale feather tips partially obscure breeding plumage. **2c Ad female** Pale grey-brown above; mainly buffish below. Compare 4b. ▲ Wooded habitats. ✤ Call *huweet* (*tuk-tuk*). [CD3:82]

3 EUROPEAN ROBIN *Erithacus rubecula* 14 cm **V**
Orange-red face and breast. ▲ Palearctic vagrant. ✤ Dry *tik*, often in fast series. [CD3:77]

4 BLACK REDSTART *Phoenicurus ochruros* 14 cm **P/V+ r**
4a Ad male Dark grey above; white wing panel; bright rufous tail; black face and breast. **4b Ad female** Entirely slate-grey (greyer than 2c); tail as male. ▲ Arid and semi-arid habitats. ✤ Call *tsip*, often followed by *tak-tak-tak*. [CD3:81]

5 MOUSSIER'S REDSTART *Phoenicurus moussieri* 12–13 cm **V**
Ad male non-breeding Long white supercilium; white wing patch; rusty-red underparts and tail. ▲ Palearctic vagrant. ✤ [CD3:83]

6 RUFOUS-TAILED PALM THRUSH *Cichladusa ruficauda* 18 cm **R u/lc**
Rufous-brown above; rufous tail; pale greyish supercilium and head-sides. ▲ Palm savanna, gardens. ✤ Loud melodious whistled song. Alarm a harsh *chrrr*. [CD10:10]

7 THRUSH NIGHTINGALE *Luscinia luscinia* 16.5 cm **V**
As 8 but less brightly coloured; upperparts and tail darker; indistinctly mottled darkish below. ▲ Palearctic vagrant. ✤ Low *krrrr*, high-pitched *heet*, short *tak*. [CD9:94]

8 COMMON NIGHTINGALE *Luscinia megarhynchos* 16.5 cm **P/V* lc/u**
Open face; rufous tail. ▲ Scrub, farmbush, etc. Usually in dense cover. ✤ High *hueet*, hard *tak*, low *karrr*. Song loud, rich and melodious; often uttered in winter quarters. [CD3:78]

Scrub robins. Characteristic, fairly long, broad, graduated and white-tipped tails, typically held cocked above back. Songs melodious.

9 FOREST SCRUB ROBIN *Cercotrichas leucosticta* 15.0–16.5 cm **R s**
Note head pattern, white spots on bend of wing. ▲ Lowland forest. ✤ Song sweet, with variable, high-pitched, whistled phrases. Calls *chuk* and *chit-chit-chit*. [CD10:12]

10 WHITE-BROWED SCRUB ROBIN *Cercotrichas leucophrys* 15 cm **R lc**
Warm brown above becoming rufous on rump; two white wingbars; streaked breast. ▲ Wooded savanna, thicket edges. ✤ Song loud and clear, with short whistled phrases constantly repeated. Alarm a hard *chrrrr*. [CD10:15]

11 BROWN-BACKED SCRUB ROBIN *Cercotrichas hartlaubi* 15 cm **R s/f**
Cold dark grey-brown above; bright rufous rump; clear-cut blackish terminal band to rufous tail; breast indistinctly greyish. ▲ Wooded savanna. ✤ Song loud and clear, with repeated short whistled phrases, from perch and in short flight. [CD10:14]

Continued on page 325

1c
1b
1a
2a
2c
2b
3
4b
4a
6
9
5
10
7
8
11
12b
13
12a
12c

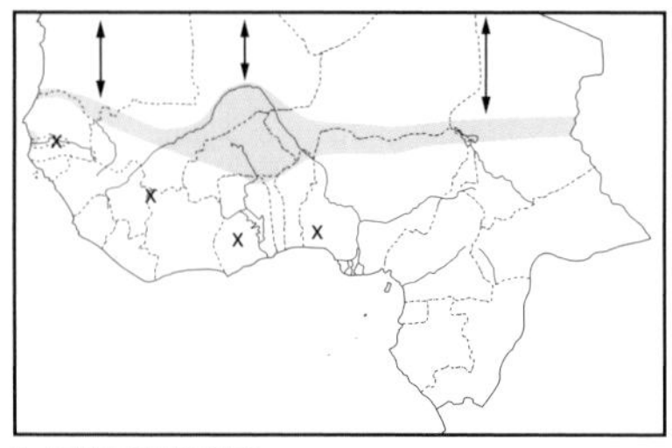

1. BLUETHROAT

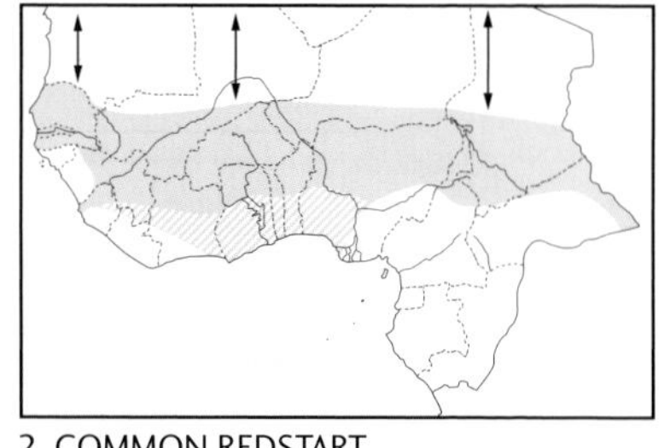

2. COMMON REDSTART

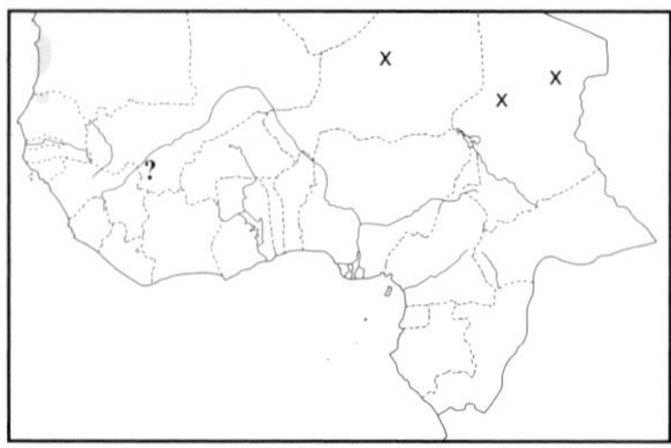

4. BLACK REDSTART

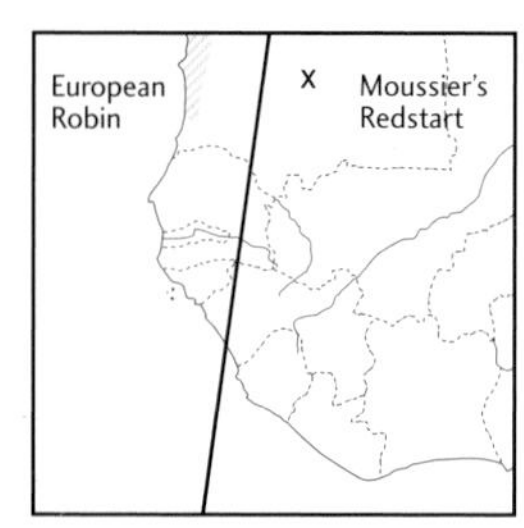

3. EUROPEAN ROBIN
5. MOUSSIER'S REDSTART

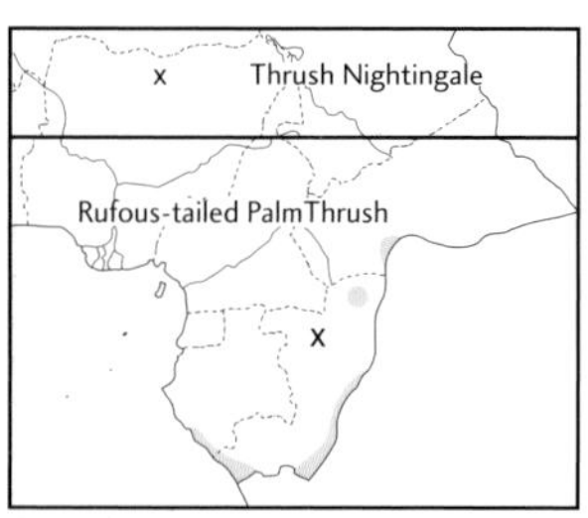

6. RUFOUS-TAILED PALM THRUSH
7. THRUSH NIGHTINGALE

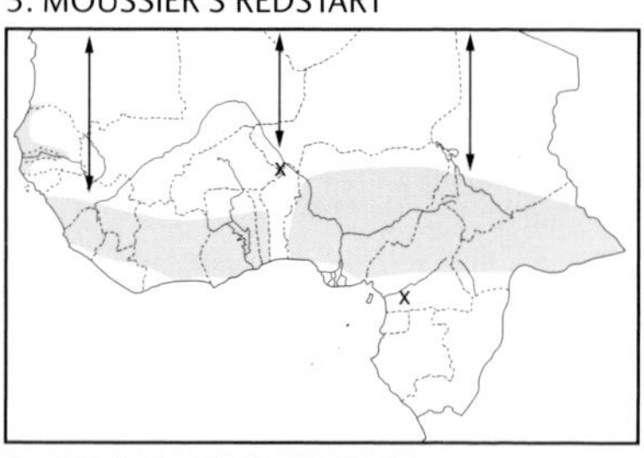

8. COMMON NIGHTINGALE

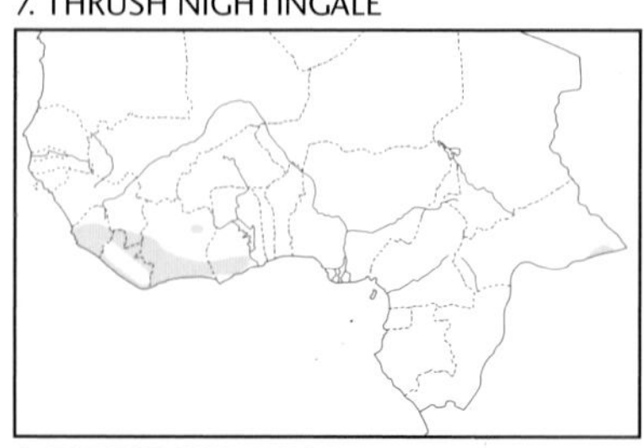

9. FOREST SCRUB ROBIN

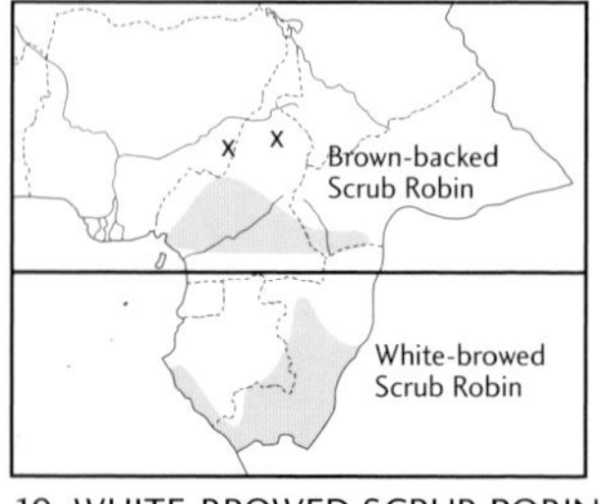

10. WHITE-BROWED SCRUB ROBIN
11. BROWN-BACKED SCRUB ROBIN

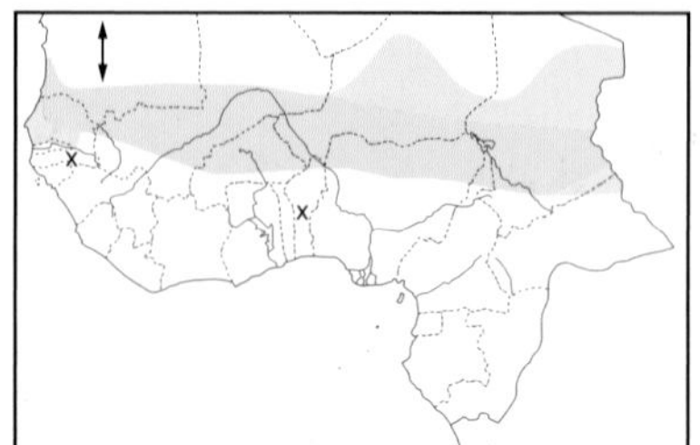

12. RUFOUS SCRUB ROBIN

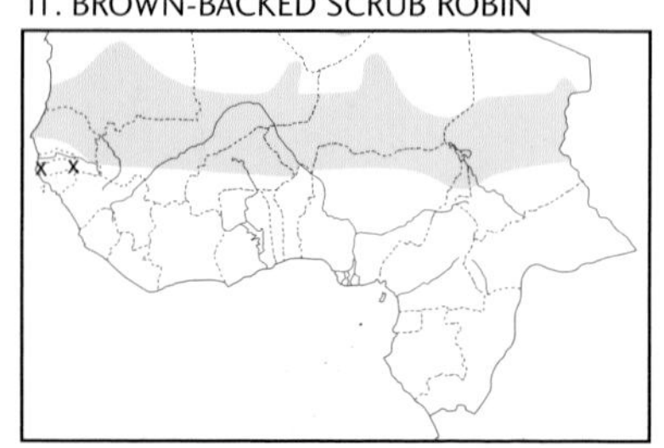

13. BLACK SCRUB ROBIN

Plate on page 322

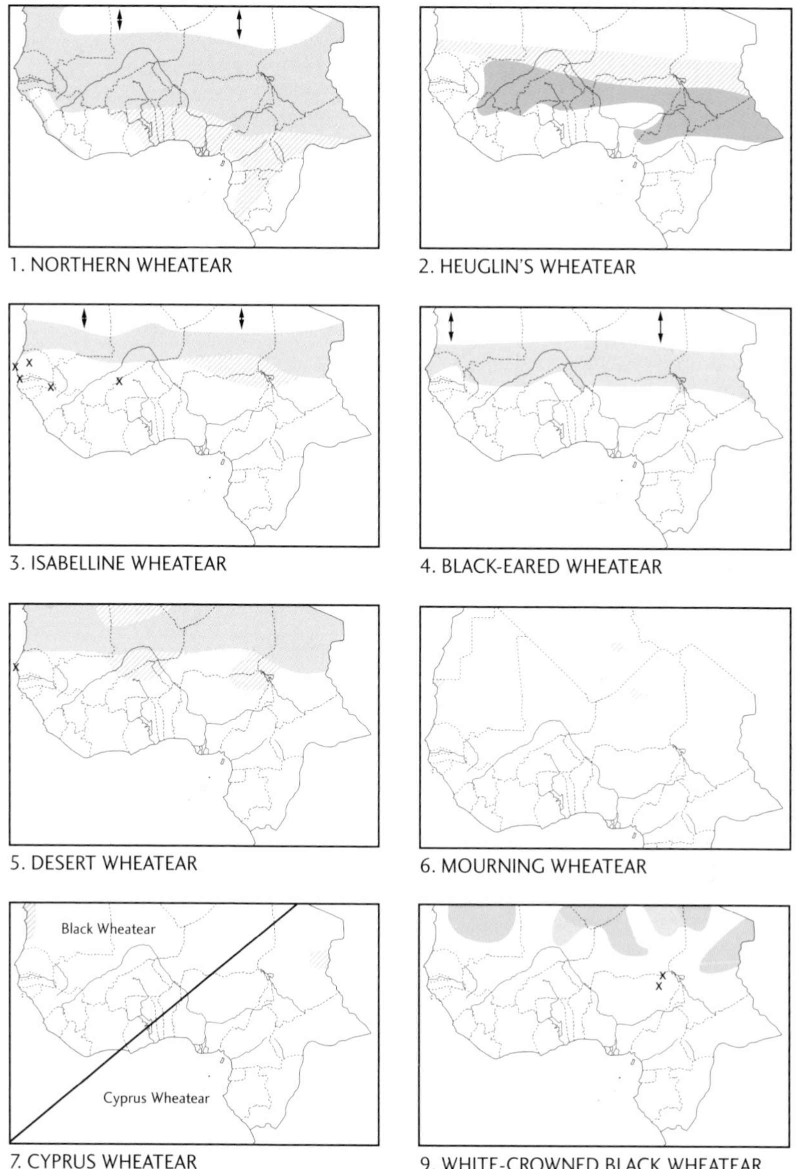

1. NORTHERN WHEATEAR
2. HEUGLIN'S WHEATEAR
3. ISABELLINE WHEATEAR
4. BLACK-EARED WHEATEAR
5. DESERT WHEATEAR
6. MOURNING WHEATEAR
7. CYPRUS WHEATEAR
8. BLACK WHEATEAR
9. WHITE-CROWNED BLACK WHEATEAR

Continued from page 322

12 RUFOUS SCRUB ROBIN *Cercotrichas galactotes* 15 cm **R/P lc**

12a Plain sandy rufous-brown above; black-and-white tipped tail. **12b Tail *C. g. minor*** (R). **12c Tail *C. g. galactotes*** (P) Subterminal black band broader than 12b. ▲ Semi-arid habitats. ✤ Song loud and sustained, with short whistled phrases, regularly repeated; from perch or in display flight. Calls a hard *tek tek*, low *chrrr*. [CD3:80]

13 BLACK SCRUB ROBIN *Cercotrichas podobe* 18 cm **R f**

Long, fan-shaped tail boldly tipped white. Rufous in wing visible in flight. ▲ Arid country, oases. ✤ Far carrying song with fairly short, varied phrases, resembling 12. [CD10:16]

PLATE 97: WHEATEARS

Small, ground-dwelling insectivores. Black-and-white tail pattern an important identification mark. Adult males usually distinctive, but females and immatures often difficult to identify. Usually silent; most common call a hard *chak*.

1 NORTHERN WHEATEAR *Oenanthe oenanthe* 15–16 cm **P+ c/r**
1a Ad male non-breeding Dull grey-brown above; variable dark mask; whitish supercilium. **1b Ad female non-breeding** Browner than 1a; no mask. **1c Ad male *seebohmi* breeding** (winters SW Mauritania, N Senegal, Mali) Black face and throat. **1d Tail** T-patterned; terminal band of even width on distal third. ▲ Various open habitats. Common in north, uncommon/rare further south. ✤ [CD3:89-90]

2 HEUGLIN'S WHEATEAR *Oenanthe (bottae) heuglini* 14–15 cm **M/R? s/lf**
2a Dark brown above; rufous-buff below. **2b Tail** T-patterned; broad terminal band covering *c.* half of tail. ▲ Degraded savanna, dry farmland, rocky hillsides, inselbergs. ✤ Song consists of varied phrases, including imitations. [CD10:19]

3 ISABELLINE WHEATEAR *Oenanthe isabellina* 16–17 cm **P c/u**
3a Plain, sandy coloured; blackish alula; robust; long legged. Compare 1b. **3b Tail** T-patterned; broad terminal band of even width covering *c.* half of tail. ▲ Dry open country. ✤ [CD3:95]

4 BLACK-EARED WHEATEAR *Oenanthe hispanica* 14.5 cm **P u/r**
4a Ad male *hispanica* non-breeding (from Mali west) Rich sandy-buff; black mask not connected with black wings. **4b Ad male *hispanica* non-breeding** Black-throated form. Similar to 4a. **4c Ad female *hispanica* non-breeding** Sandy-brown; indistinct mask; dark brown wings broadly fringed buffish. **4d Ad male *melanoleuca* non-breeding** (from Mali east) As 4e, but has white throat. **4e Ad male *melanoleuca* non-breeding** Black-throated form. Upperparts washed grey. **4f Tail** T-patterned; terminal band uneven, sometimes broken. ▲ Semi-desert, dry savanna. ✤ [CD3:91]

5 DESERT WHEATEAR *Oenanthe deserti* 14–15 cm **P c/s**
5a Ad male non-breeding Sandy; black face and throat connected to black wings. **5b Ad female non-breeding** Deep sandy-buff; no black on face. **5c Tail** Black, with very little white at base. ▲ Desert, semi-desert. ✤ [CD3:94]

6 MOURNING WHEATEAR *Oenanthe lugens* 14.5 cm **V**
6a Ad male non-breeding Mainly black and white; orange-buff undertail-coverts. **6b Ad female non-breeding** Dull grey version of 6a. **6c Tail** T-patterned; terminal band of even width on distal third. ▲ Desert. N African vagrant, Mauritania, Niger (Aïr). ✤ [CD3:92]

7 CYPRUS WHEATEAR *Oenanthe (pleschanka) cypriaca* 13.5 cm **P u**
7a Ad male non-breeding Dull brownish crown (paler with wear); black face, throat and upperparts. **7b Ad female non-breeding** Dull brownish crown (not paler with wear). **7c Tail** T-patterned; terminal band uneven, black increasing towards outer feathers. ▲ Dry, open country. EC Chad. ✤ [CD10:17]

8 BLACK WHEATEAR *Oenanthe leucura* 18 cm **R/P? r**
8a Ad male All black except tail base and tail. **8b Ad female** As male, but browner. Juvenile similar. **8c Tail** T-patterned. ▲ Desert. ✤ [CD3:88]

9 WHITE-CROWNED BLACK WHEATEAR *Oenanthe leucopyga* 17 cm **R lc**
9a Ad Black, with white crown, white tail base and vent. **9b Juv** Dark brown; no white on crown. **9c Ad tail** White with black on distal half of central feather pair. **9d Juv/ad variant tail** As 9c, but with dark smudges near tip. ▲ Desert. Local migrant. ✤ [CD3:87]

Maps on page 325

1c
1b
1a
2b
2a
3a
1d
3b
4c
5c
4b
4a
5b
5a
4f
4d
4e
9d
9c
6b
6c
6a
7b
7a
9b
8b
8a
9a
8c
7c

1 WHINCHAT *Saxicola rubetra* 12.5 cm **P/V+ u/c**
1a Ad male breeding Broad white supercilium; head-sides blackish; dark tail with white patch at sides of base (conspicuous in flight). Female similar but paler. **1b Ad non-breeding** Very buffish overall; head-sides brownish. ▲ Open habitats. ✤ Call (*whu*) *tek-tek*. Usually silent. [CD3:86]

2 AFRICAN STONECHAT *Saxicola torquatus* 12.5 cm **R u/lc**
2a Ad male *salax* (SE Nigeria east and south; Bioko) Black head and upperparts; white neck-sides and underparts; chestnut breast patch. In flight, white wing patch and rump conspicuous. *S. t. moptanus* (Senegal and inner Niger deltas) intermediate between 2a and 2b. **2b Ad male *nebularum*** (Mt Nimba) Chestnut breast patch more extensive than 2a. **2c Ad female** Streaked dark brown above; no supercilium. Compare 1. ▲ Open habitats. ✤ Call (*wheet*) *trek-trek*. Song: short clear, twittering and scratchy phrases. [CD3:84]

3 COMMON STONECHAT *Saxicola (torquatus) rubicola* 12.5 cm **P/V**
Ad male non-breeding Browner above than 2; rump rusty-brown mottled white. ▲ Open habitats. Palearctic migrant, coastal Mauritania. Vagrant, Mali, Niger, Chad. ✤ [CD3:84]

4 BLACKSTART *Cercomela melanura* 14 cm **R u/lc**
Plain sandy-brown above; black tail. ▲ Rocky areas in arid country. ✤ Song a short, rather pleasant warble. Call a harsh note; alarm a high-pitched *hiih*. [CD10:22]

5 BROWN-TAILED ROCK CHAT *Cercomela scotocerca* 13 cm **R lf**
Small, nondescript; dark brown tail. ▲ Dry rocky country. ✤ Song a rapid, chirruping phrase, frequently repeated with slight variations. [CD10:21]

6 WHITE-FRONTED BLACK CHAT *Myrmecocichla albifrons* 15 cm **R u/lc**
6a Ad male *frontalis* (west of range) Entirely black; white patch on forehead. **6b Ad male *limbata*** (from E Cameroon east) Some white on shoulder. **6c Ad female** As male but without any white. ▲ Open savanna woodland. ✤ Song: short phrases with sharp and rolling notes, including frequently repeated *uwheetirr*; sometimes mimics. Call a penetrating *heet*. [CD10:26]

7 FAMILIAR CHAT *Cercomela familiaris* 14 cm **R u/lc**
Plain brown; conspicuous rufous rump and tail. ▲ Rocky areas in savanna. ✤ Song a soft series of whistled and chattering notes. Call *whee* (*chak-chak*); also *cher-cher*. [CD10:20]

8 CLIFF CHAT *Myrmecocichla cinnamomeiventris* 20 cm **R s/lf**
8a Ad male *bambarae* (west of range) Black and rufous with small white shoulder patch. **8b Ad female *bambarae*** As male but duller; no white. **8c Ad male *cavernicola*** (west of range) Distinct shoulder patch; narrow pale line below black breast. **8d Ad male *coronata*** (east of range) White crown; pale area below breast. **8e Ad female *coronata*** Head rufous-grey, paler than 8b; entirely rufous below. ▲ Rocky areas in savanna. ✤ Song melodious, with varied whistles or rapid imitations. Call a penetrating *seeeo* and *seeu-seeu*. [CD10:28-29]

9 NORTHERN ANTEATER CHAT *Myrmecocichla aethiops* 19 cm **R lc/f**
9a Entirely sooty-brown. **9b In flight** Large white patch on primaries. ▲ Open grassland, farmland with scattered bushes. ✤ Song, from perch, a varied mixture of clear whistles and short, hard trills. Call a penetrating *tseeu* and *heeh*. [CD10:24]

10 SOOTY CHAT *Myrmecocichla nigra* 16 cm **R lu/c**
10a Ad male Entirely glossy black; white shoulder patches. **10b Ad female** Entirely blackish-brown. **10c Ad male in flight** White shoulder patches conspicuous. ▲ Grassland. ✤ Song: clear whistles, occasionally with short trill or imitations. Call: various whistles. [CD10:25]

11 CONGO MOOR CHAT *Myrmecocichla tholloni* 18–19 cm **R lf**
11a Robust; head-sides and throat mainly dirty white; underparts brown-grey. **11b In flight** White wing patch; white rump. ▲ Grassland. ✤ Song consists of short, clear, melodious whistles interspersed with rolled *chiurrr*. Alarm a sharp *peep*. [CD10:23]

 Maps on page 330

1b
1a
2b
2d
2a
3
4
5
6a
6c
6b
7
8c
8b
8a
8e
8d
9a
9b
10c
10b
10a
11b
11a

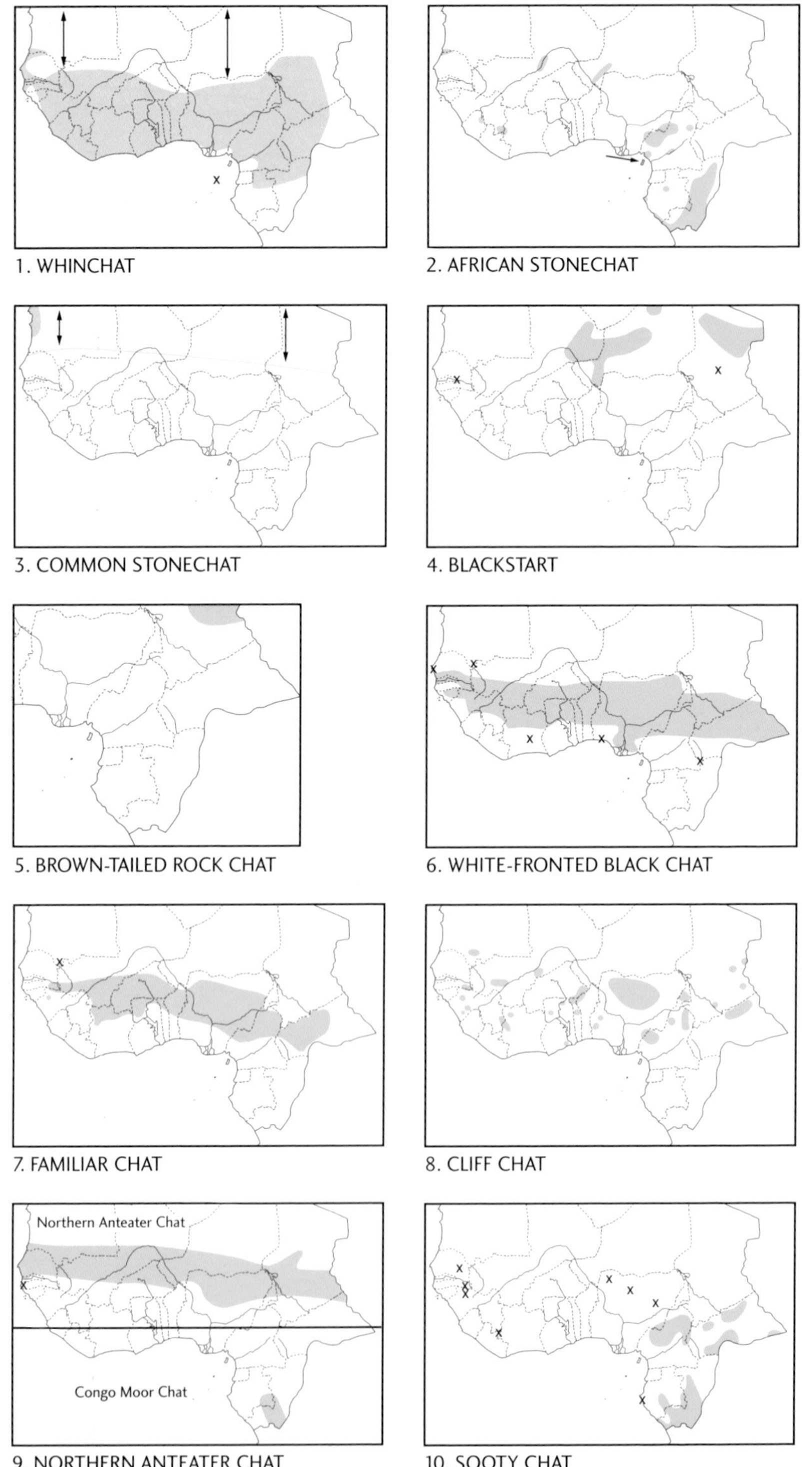

1. WHINCHAT

2. AFRICAN STONECHAT

3. COMMON STONECHAT

4. BLACKSTART

5. BROWN-TAILED ROCK CHAT

6. WHITE-FRONTED BLACK CHAT

7. FAMILIAR CHAT

8. CLIFF CHAT

9. NORTHERN ANTEATER CHAT
11. CONGO MOOR CHAT

10. SOOTY CHAT

Plate on page 328

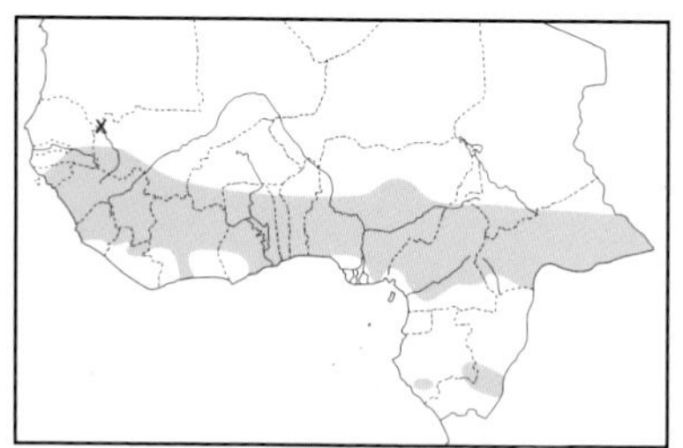
1. AFRICAN MOUSTACHED WARBLER

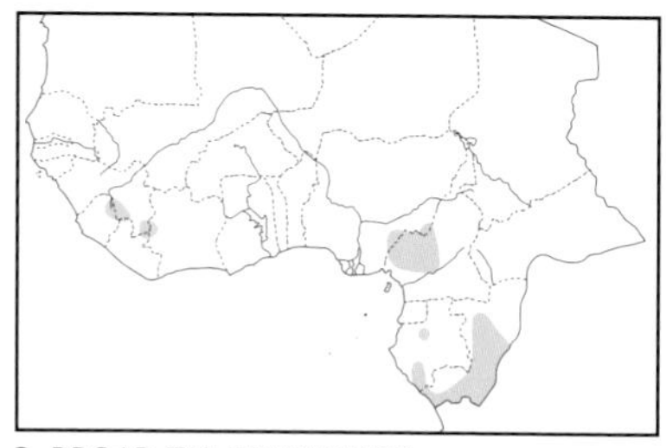
2. BROAD-TAILED WARBLER

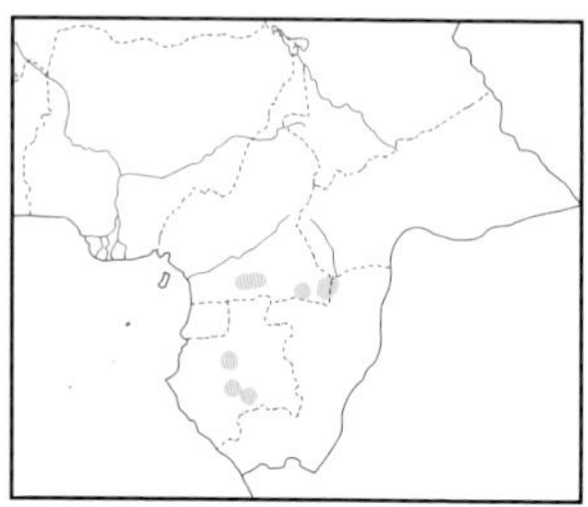
3. DJA RIVER WARBLER

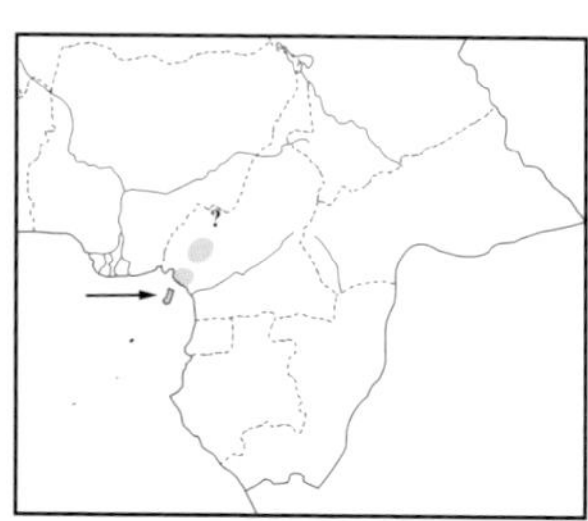
4. EVERGREEN FOREST WARBLER

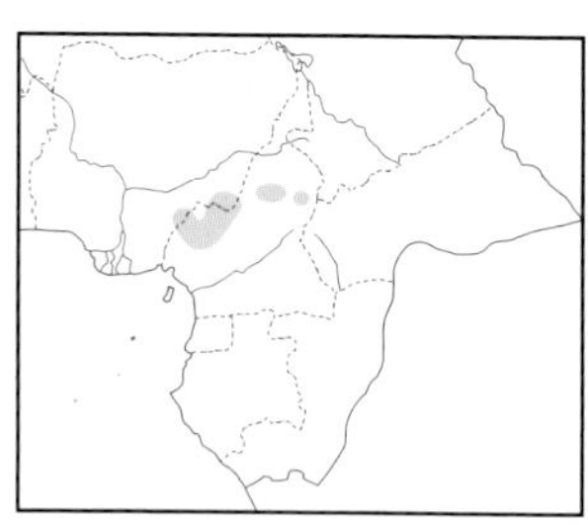
5. BANGWA FOREST WARBLER

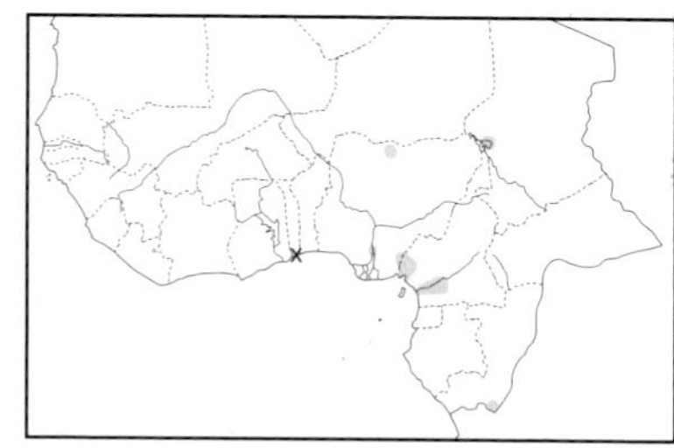
6. LITTLE RUSH WARBLER

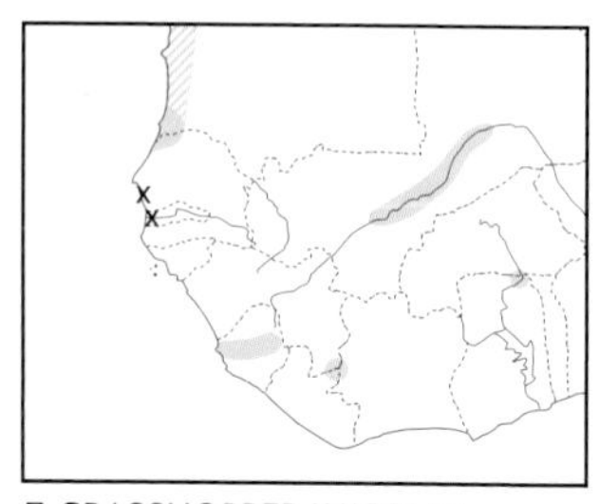
7. GRASSHOPPER WARBLER

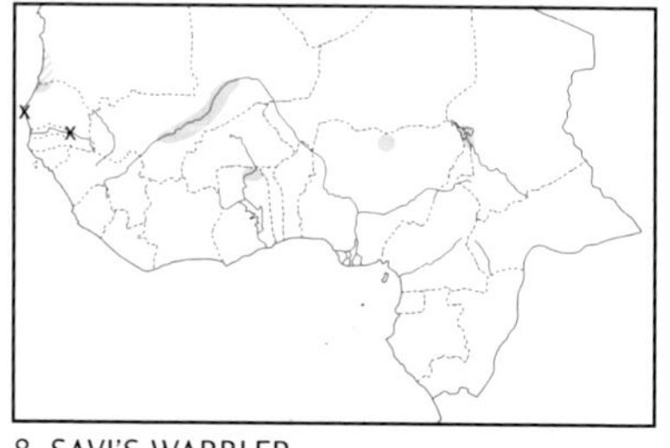
8. SAVI'S WARBLER

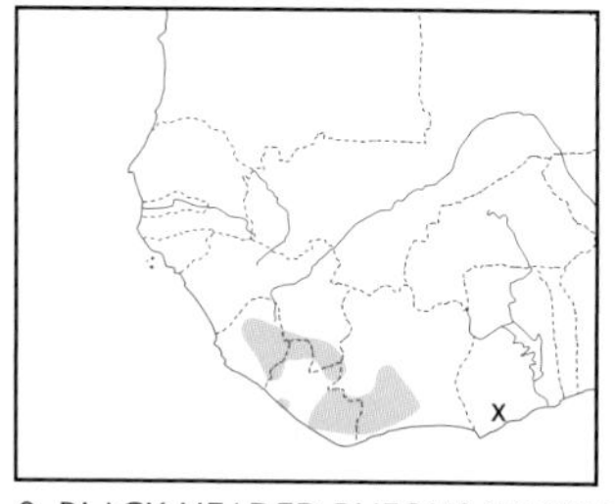
9. BLACK-HEADED RUFOUS WARBLER

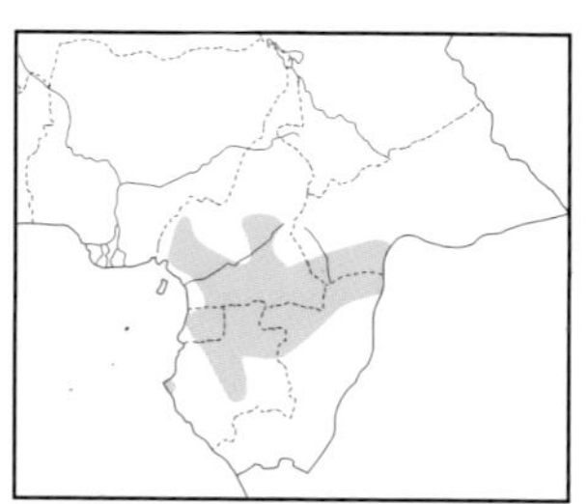
10. BLACK-FACED RUFOUS WARBLER

Plate on page 332

PLATE 99: WARBLERS I

Large, varied family of mostly small songbirds (Plates 99–109). Sexes similar, except in *Sylvia* and *Apalis*; juveniles unspotted. Mainly insectivorous. Broad range of vocalisations. Occur in all habitats.

1 AFRICAN MOUSTACHED WARBLER *Melocichla mentalis* 19–20 cm R f
Large and bulky; black malar stripe; broad tail. ▲ Rank herbage in savanna. ✤ Song a distinctive, cheerful and vigorous *tup-tup-twiddle-diddle-dee*. [CD11:38]

2 BROAD-TAILED WARBLER *Schoenicola platyurus* 17 cm R lu
Long, broad, blackish tail. ▲ Grassy habitats. ✤ Slow, deliberate series of similar, high-pitched whistles *heeet, heeet, heeet,* ... In song flight a dry *tzit...tzit...tzit...* [CD11:57]

Genus *Bradypterus*. Drab-coloured skulkers of reedbeds, rank herbage, scrub and forest undergrowth. Songs distinctive.

3 DJA RIVER WARBLER *Bradypterus grandis* 18.5 cm R r NT
Resembles 6, but much larger; throat streaked. ▲ Forest edge, tall grassland, marshland. Near or on ground, creeping mouse-like through dense rank herbage. Very difficult to see. Endemic. ✤ Song consists of 4 introductory notes, increasing in speed, followed by drawn-out trill *sweep-sweep-sweepswip-rrrrrurrrr*. [CD10:48]

4 EVERGREEN FOREST WARBLER *Bradypterus lopezi* 13.0–14.5 cm R lf
Darkish. Tail usually heavily worn. ▲ Undergrowth of montane forest. ✤ Song a vigorous, rapid, rhythmic series of 3–9 identical notes. Speed and notes variable. Call a hard *chrrk*, vibrant *pirr* and loud *klik* (alarm). [CD10:49–50]

5 BANGWA FOREST WARBLER *Bradypterus (lopezi) bangwaensis* 14–15 cm R lf NT
Mainly rufous; throat white. ▲ Dense vegetation in montane areas. Endemic. ✤ Song a rhythmic series of identical notes, increasing in volume; speed variable. Call a rattling *krrr*. [CD10:51]

6 LITTLE RUSH WARBLER *Bradypterus baboecala* 15 cm R lr
Dark brown upperparts. ▲ Reedbeds, dense aquatic vegetation, rank herbage. ✤ Song a distinctive, loud series of identical, dry, single notes, accelerating at end and ceasing abruptly *kruk, kruk kruk kruk-krukukukukukuk*. Alarm a nasal *meew*. [CD10:43]

Genus *Locustella*. Dull-coloured with round, almost graduated tails and long undertail-coverts. Skulk low in dense vegetation.

7 GRASSHOPPER WARBLER *Locustella naevia* 12.5 cm P/V+ s/r
Streaked upperparts; graduated tail; faint supercilium. ▲ Dense vegetation. Secretive. ✤ Call a short *twit* or *pit*. [CD4:3b]

8 SAVI'S WARBLER *Locustella luscinioides* 14 cm P/V+ r/s
Plain brown upperparts; broad, rounded tail. ▲ Dense scrub, herbage near or in water. ✤ Call a sharp *tswik*. [CD4:4]

Genus *Bathmocercus*. Secretive species of dense forest undergrowth. Songs distinctive.

9 BLACK-HEADED RUFOUS WARBLER *Bathmocercus cerviniventris* 13 cm R r/f NT
9a Ad male Black head, throat and centre of breast; upperparts rufous-brown. **9b Ad female** Paler; white chin and 'whiskers'. ▲ Lowland and gallery forest. In dense vegetation along paths and glades, often near water. Endemic. ✤ Song simple but variable, consisting of endlessly repeated series of 2–3 clear, vigorous whistles. [CD11:29]

10 BLACK-FACED RUFOUS WARBLER *Bathmocercus rufus* 13 cm R lf
10a Ad male Black mask extending to centre of breast; upperparts rufous-chestnut. **10b Ad female** Upperparts dark grey. ▲ Lowland and montane forest. ✤ Song: monotonous series of penetrating, high-pitched whistles (hard to locate), e.g. *HEEEEET, HEEEEET, ... ; HEET-HEET-... ; HEE-HEE hu-EE....* Call *chip*. [CD11:30]

Maps on page 331

1
2
3
4
5
6
7
8
9b
10b
10a
9a

Genus *Acrocephalus*. Upperparts mainly brown; underparts paler; tail rounded; bill prominent.

1 SEDGE WARBLER *Acrocephalus schoenobaenus* 13 cm P lc
Bold creamy supercilium; streaked upperparts; unstreaked, rusty rump. ▲ Reedbeds, moist rank vegetation. ✤ Song a fast series of extremely varied rasping and musical notes, including clear trills and imitations. Call *tuk*. Alarm *trrr*. [CD4:6]

2 AQUATIC WARBLER *Acrocephalus paludicola* 13 cm P r VU
Creamy-buff crown-stripe; heavily streaked, straw-coloured upperparts; spiky tail. ▲ Reedbeds, moist herbage. ✤ Call a sharp *krrr krrr*. [CD4:5]

3 GREAT REED WARBLER *Acrocephalus arundinaceus* 19–20 cm P s/f
Large. Prominent supercilium; stout bill. ▲ Scrub, dense herbage, swamps, etc. ✤ Song a rather slow series of far-carrying, low, raucous, grating notes interspersed with high squeaks, repeated 2–3 times. Call a hard *krek*. [CD4:9]

4 LESSER SWAMP WARBLER *Acrocephalus gracilirostris* 14–15 cm R r
Dark brown above; distinct supercilium. ▲ Reedbeds and similar moist vegetation. ✤ Song: a few notes followed by a fast trilling or bubbling series. Call *chuk!* and *cheruk*. [CD10:54]

5 GREATER SWAMP WARBLER *Acrocephalus rufescens* 18 cm R ls/f
Dark brown above; greyish-white below; no supercilium. ▲ Reedbeds, marshes, moist herbage. ✤ Song: loud croaking, churring and gurgling notes, very similar to 4, but slightly lower pitched and more guttural. Call a harsh *chrr*, low *kreeok* and *chok!* [CD10:55]

6 AFRICAN REED WARBLER *Acrocephalus (scirpaceus) baeticatus* 12.0–12.5 cm R/M lf
As 7, but wings shorter, primaries not projecting beyond rump. ▲ Reedbeds; also drier habitats. Status inadequately known through similarity with 7. ✤ As 7. [CD10:52]

7 EUROPEAN REED WARBLER *Acrocephalus scirpaceus* 12.5–13.0 cm P s/c
Plain brown upperparts; rusty-tinged rump; indistinct supercilium. Note primary projection. ▲ Scrub, dense herbage, etc. ✤ Song a series of slow, grating and nasal syllables repeated 2–5 times. Call *krrrr* and *chrrrt*. Alarm a sharp *tek*, *tek* and *krrrt*. [CD4:7]

8 MARSH WARBLER *Acrocephalus palustris* 12.5–13.0 cm V
Very similar to 7; typically more uniform olive grey-brown above; less rusty rump; slightly longer wings; 7-8 pale-tipped dark primaries visible; legs usually paler. ▲ Palearctic vagrant ✤ Calls: sharp *tek*, low *tuk*, grating *krrr* (alarm). [CD10:53]

Continued on page 336

1
2
3
4
5
6
7
8
9
10
11
12
13
14

Continued from page 334

Genus *Chloropeta*. Upperparts olive; underparts yellow. Bill broad, flat, flycatcher-like.

9 AFRICAN YELLOW WARBLER *Chloropeta natalensis* 14 cm R lu/f
Bright yellow underparts. ▲ Rank herbage, reeds, dense forest edge. ✤ Song: a few dry notes followed by a fast series of throaty and varied notes. Alarm *tsk!* or *chrr!* [CD10:57]

Genus *Hippolais*. Tail square; bill prominent. Arboreal.

10 WESTERN OLIVACEOUS WARBLER *Hippolais (pallida) opaca* 12–14 cm P/V+ f/r
Pale geyish-brown above. Elongated jizz. Holds tail still. ▲ Scrub, various wooded habitats. ✤ Song a rapid, pleasant warble. Call *chek!* and *churr-churr*. [CD4:11]

11 EASTERN OLIVACEOUS WARBLER *Hippolais pallida* 12.0–13.5 cm P/R/M f/r
As 10 but slightly paler and smaller; bill narrower. Frequently pumps tail. ▲ Arid scrub, wooded habitats. ✤ Song a scratchy phrase, going up and down in pitch. Call as 10.

12 MELODIOUS WARBLER *Hippolais polyglotta* 13 cm P c/u
Olive-green upperparts; yellow underparts; orangey bill. Compare 14. ▲ Various wooded and scrubby habitats. ✤ Song a rapid, sustained and varied babble, sometimes including imitations. Call a chattering, sparrow-like *tchèèèrr* or *tret-tret-*... [CD4:12]

13 OLIVE-TREE WARBLER *Hippolais olivetorum* 15.5 cm V
Large; greyish. ▲ Dry bush country. ✤ Call a hard *tsek* or *tuk*. [CD10:59]

14 ICTERINE WARBLER *Hippolais icterina* 13.5 cm P f/r
Resembles 12, but has longer primary projection and pale wing panel. Crown peaked. ▲ Open woodland, scrub. ✤ Song a very varied and sustained jumble of harsh, melodious and discordant, often repeated notes. Call *tek* and *deedeweet*. [CD10:60]

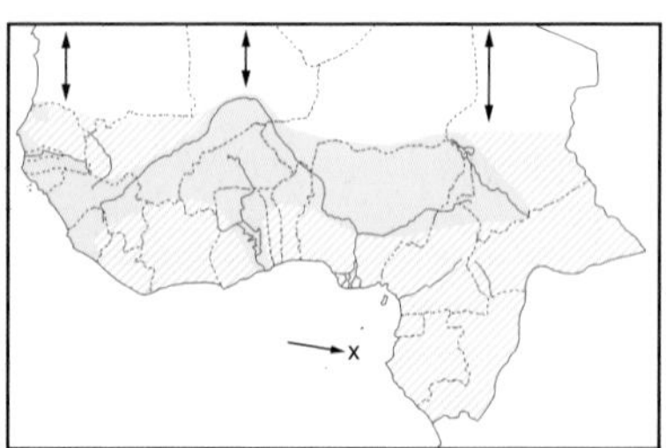

1. SEDGE WARBLER

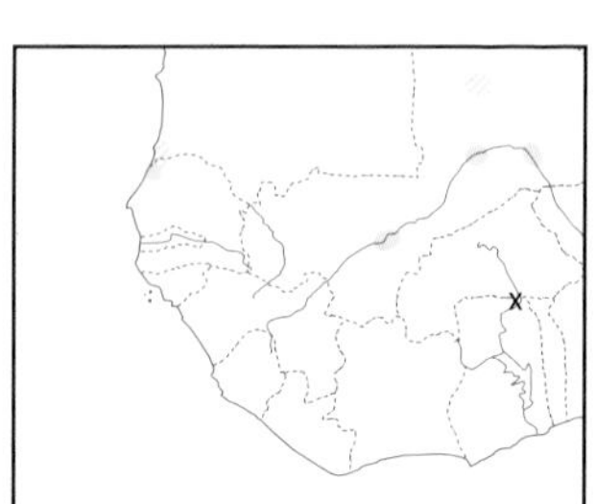

2. AQUATIC WARBLER

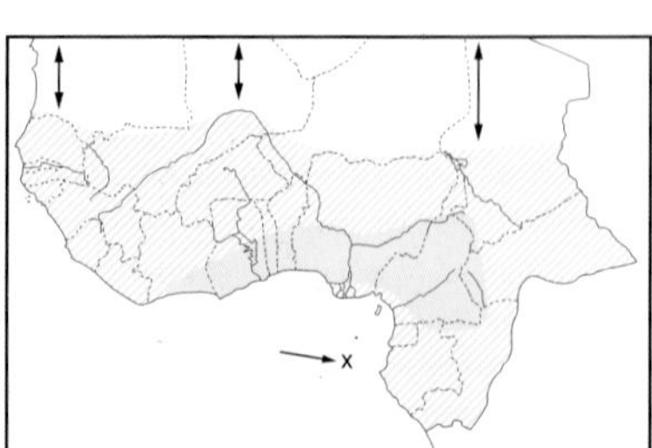

3. GREAT REED WARBLER

4. LESSER SWAMP WARBLER

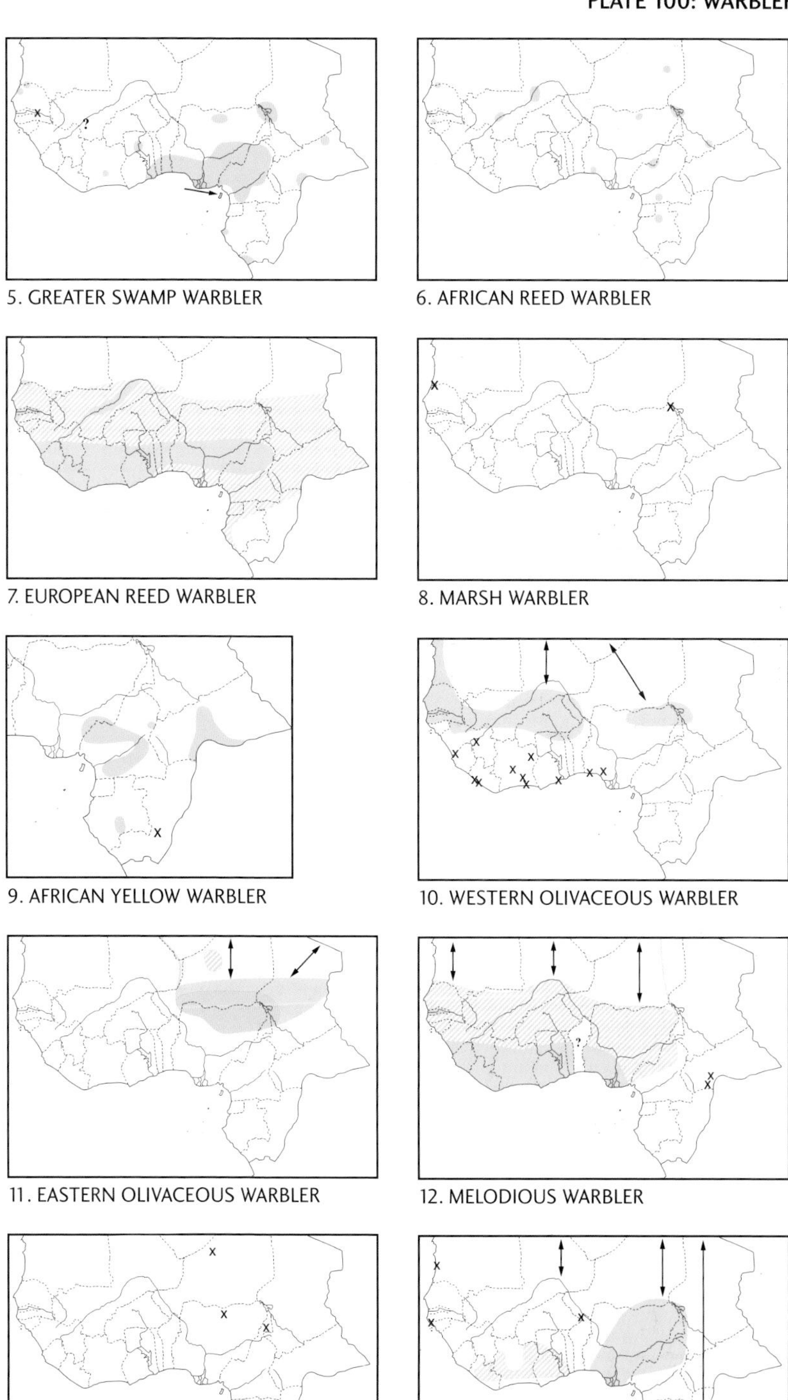

5. GREATER SWAMP WARBLER

6. AFRICAN REED WARBLER

7. EUROPEAN REED WARBLER

8. MARSH WARBLER

9. AFRICAN YELLOW WARBLER

10. WESTERN OLIVACEOUS WARBLER

11. EASTERN OLIVACEOUS WARBLER

12. MELODIOUS WARBLER

13. OLIVE-TREE WARBLER

14. ICTERINE WARBLER

PLATE 101: WARBLERS III

Genus *Sylvia*. Palearctic scrub and woodland warblers. Sexes usually different. Females/immatures of some difficult to separate (see Plate 102). Characteristic vocalisations a sharp *tak* and a harsh churr. Song a sweet pleasant warble.

1 GARDEN WARBLER *Sylvia borin* — 13.0–14.5 cm **P c/f**
Featureless, plain grey-brown and buff; gentle, open face; rather short, stubby bill, rounded crown; greyish legs. Lack of any distinctive features is in itself diagnostic. ▲ Various wooded and bushy habitats. ✤ Song a sustained, melodious, even warble of largely mellow notes; in winter quarters often subdued. [CD4:14]

2 BARRED WARBLER *Sylvia nisoria* — 15.5–17.0 cm **V**
2a Ad Robust; crescentic barring on breast-sides and flanks. **2b First-winter** Pale fringes to wing feathers and uppertail-coverts; dark crescents on undertail-coverts. ▲ Dry bush and woodland. Skulking, shy and rather slow. Frequently raises crown and flicks tail. Flight strong and usually heavy. ✤ [CD10:61]

3 LESSER WHITETHROAT *Sylvia curruca* — 12.5–13.5 cm **P r/lc**
Greyish upperparts; whitish underparts; dark ear-coverts (forming usually distinct mask, barely noticeable in some). ▲ Dry scrub, *Acacia* woodland. Skulking; often low. ✤ Song a brief muffled warble followed by fast, far-carrying rattle *djedjedjedjedjedje*. [CD4:17]

4 COMMON WHITETHROAT *Sylvia communis* — 13–15 cm **P c/r**
4a Ad male Rusty wing panel; greyish head; white outer tail feathers. **4b Ad female** Brownish head. ▲ Dry scrub, wooded savanna, gardens. ✤ Calls include sharp *tak*, grating *charrr* and nasal *wed wed wed*. Song a brief, hurried, scratchy warble. [CD4:16]

5 BLACKCAP *Sylvia atricapilla* — 13.5–15.0 cm **P/R* s/lc**
5a Ad male Plain grey; black cap. **5b Ad female** Plain grey; chestnut cap. ▲ Various wooded habitats. ✤ Song a rich warble starting with a hurried subdued jumble followed by clear whistles and ending in a loud flourish. In winter quarters mostly a subdued, less varied, but more sustained version. [CD4:15]

6 WESTERN ORPHEAN WARBLER *Sylvia hortensis* — 15–16 cm **P u/r**
Large, stout. Black or dusky face; white throat. **6a Ad male** Blackish hood; pale eyes. **6b Ad female** Grey-brown crown. First-winter similar, but paler; crown brown as upperparts; outer tail feathers tinged buffish. ▲ Dry scrub, wooded savanna. Not shy, though often remains within cover, quietly gleaning insects in tops of trees and bushes. In flight may recall chat. ✤ [CD4:13]

Maps on page 340

1
2a
2b
3
5b
4b
5a
4a
6b
6a

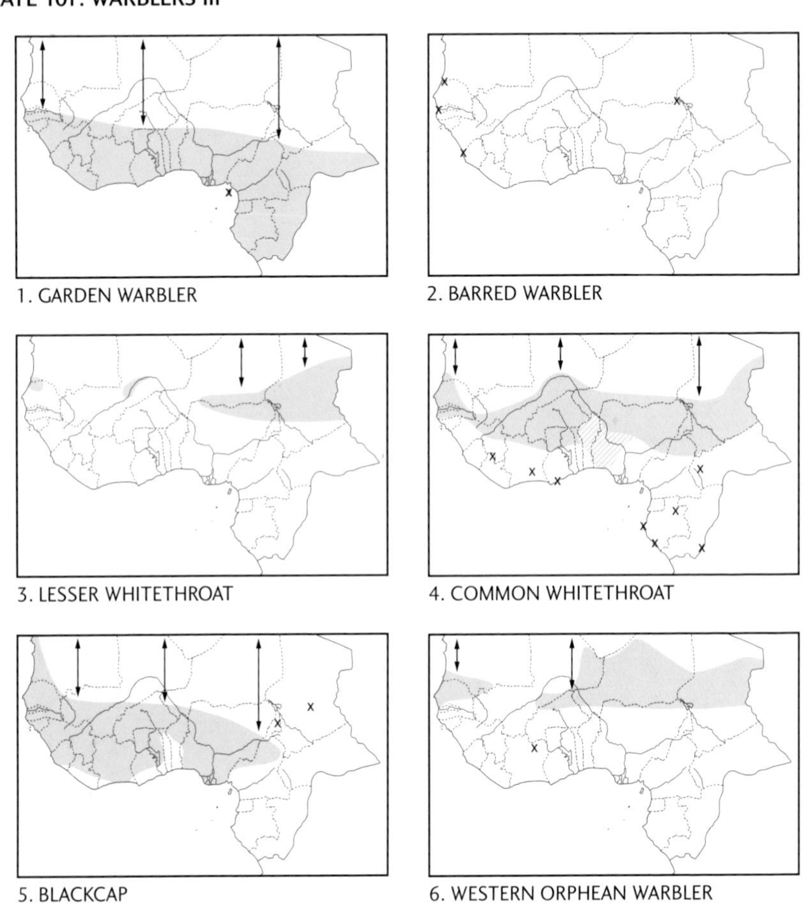

Plate on page 338

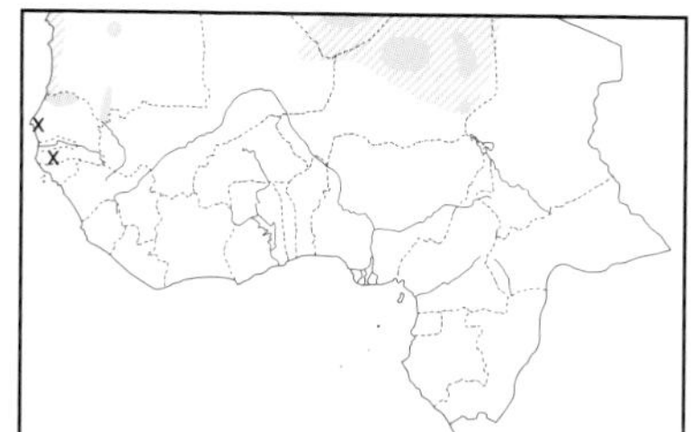

1. SARDINIAN WARBLER

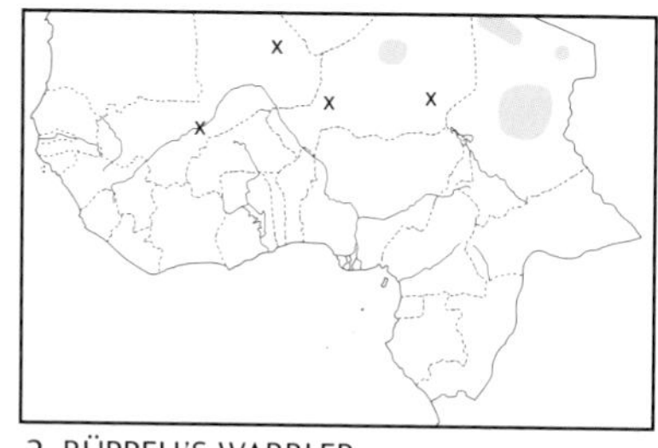

2. RÜPPELL'S WARBLER

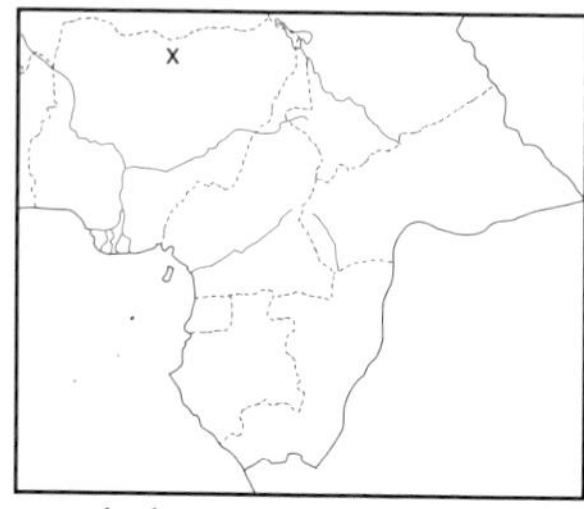

3. MÉNÉTRIES'S WARBLER

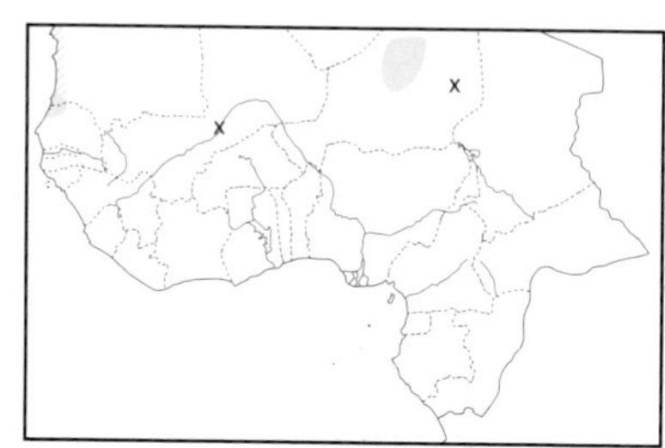

4. SPECTACLED WARBLER

5. AFRICAN DESERT WARBLER

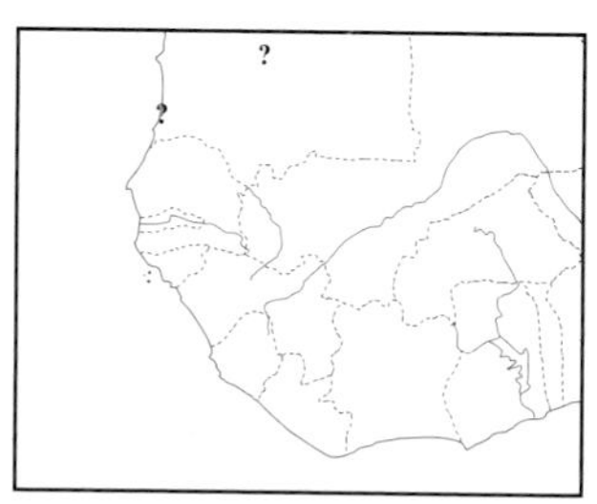

6. TRISTRAM'S WARBLER

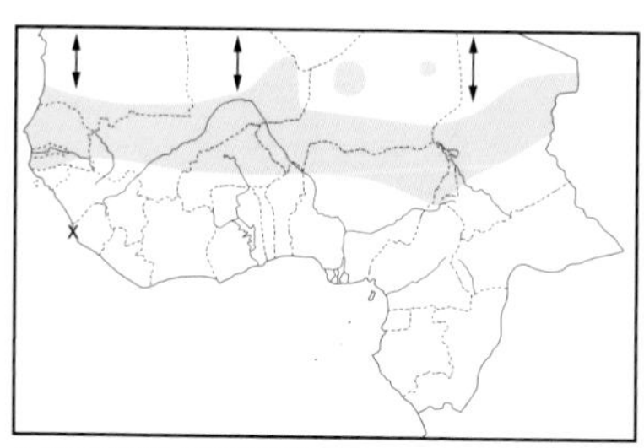

7. SUBALPINE WARBLER

Plate on page 342

1 **SARDINIAN WARBLER** *Sylvia melanocephala* 13.5 cm **P/V+ u**

1a Ad male Black hood; white throat; conspicuous, red orbital ring. **1b Ad female** Dusky-grey hood; brown upperparts. ▲ Bush, wooded savanna, gardens. Skulking but not particularly shy; inquisitive. Frequently cocks tail. ✤ Call a harsh rattle *chretetikitikitik*..., fast, harsh *treet-treet* or *chreet* notes, and hard *tsek*. Song an even, pleasing warble, incorporating harsh calls. [CD4:19]

2 **RÜPPELL'S WARBLER** *Sylvia rueppelli* 14 cm **P/V f/c**

2a Ad male Black throat; white submoustachial stripe. **2b Ad female** Greyish head; pale fringes to wing feathers. Variable. ▲ Low bushes in very arid country. Skulking but not shy. ✤ Call a sparrow-like chatter; also sharp *tak*. Song a series of short, fast, dry, chattering phrases. [CD10:62]

3 **MÉNÉTRIES'S WARBLER** *Sylvia mystacea* 13.5 cm **V**

3a Ad male Dull black crown to ear-coverts; pinkish-tinged underparts. **3b Ad female** Sandy grey-brown upperparts; pale fringes to wing feathers. ▲ Dry scrub. Skulking and restless, constantly flicking tail in all directions. ✤ Call a sharp *tak*, harsh *chrrrr* and sparrow-like chattering similar to 2.

4 **SPECTACLED WARBLER** *Sylvia conspicillata* 12–13 cm **P/R* lc/r**

Resembles miniature Common Whitethroat (Plate 101:4). Orange-rufous wing panel. **4a Ad male** Grey head; black area from bill to eye; white eye-ring; white throat. **4b Ad female** Browner; blackish tail contrasts with rest of upperparts. ▲ Dry scrub. Moves restlessly through and between low bushes, cocking tail and moving it sideways. Frequently hops on ground. ✤ Call a distinctive harsh rattle *trrrrrrr*; also *tek*. Song a typical, short, fast, high-pitched *Sylvia* chatter. [CD4:21]

5 **AFRICAN DESERT WARBLER** *Sylvia (nana) deserti* 11.5–12.5 cm **R/V* r/lu**

Small and pale. Rufous tail; yellowish eyes, bill and legs. ▲ Desert scrub. Usually seen scurrying across sand before disappearing into low scrub. Not particularly shy. Tail held horizontally or slightly cocked. ✤ Song consists of short phrases usually starting with *krrrr* call and followed by a jaunty warble ending with a rising whistle. [CD4:18]

6 **TRISTRAM'S WARBLER** *Sylvia deserticola* 12 cm **P?**

6a Ad male Slate-grey above; brick-red below; rusty wing panel. **6b Imm** Browner above; paler below. ▲ Desert scrub. Restless and skulking, moving through scrub with cocked tail. Possibly rare migrant to Mauritania. ✤ Call a sharp *trk* or *trk-it*, frequently uttered. [CD4:22]

7 **SUBALPINE WARBLER** *Sylvia cantillans* 12–13 cm **P/V+ c/s**

7a Ad male *cantillans* (throughout) Blue-grey upperparts; rusty-orange throat, breast and flanks; white moustache; red orbital ring. *S. c. inornata* (Senegal–W Niger) is more orange-brown below; *albistriata* (from E Mali eastwards) more chestnut-brown, less orange and often flecked white, contrasting with pale rufous flanks and more extensive white on underparts; *moltonii* (breeds Mediterranean islands; winter range unknown) pale brownish-pink below, without or with very little orange or chestnut tone. **7b Ad female** Pale grey-brown upperparts; pinkish-buff to buffish-white underparts; paler orbital ring. **7c Imm** Buffish-brown above; buff below; pale fringes to wing feathers; pale orbital ring. ▲ Dry scrub; also gardens, mangroves. Frequently cocks tail. ✤ Call a hard but quiet *tek* (nominate and *inornata*), *trek-trek* (*albistriata*) or rattling *trrrt* (*moltonii*). Song (occasionally uttered in winter quarters) consists of regularly spaced phrases of a fast, clear and scratchy warble. [CD4:20]

Maps on page 341

1b
2b
1a
2a
3b
4b
3a
4a
5
7c
6b
7b
6a
7a

1 **WOOD WARBLER** *Phylloscopus sibilatrix* 11.0–12.5 cm **P c/f**
Throat and upper breast yellow; rest of underparts white. Long primary projection. ▲ Forest and adjacent savanna. ✤ Song an accelerating series of *sip* notes ending in shivering trill, often preceded by clear, piping *piu-piu-piu-piu*. Call *piu*. [CD4:27]

2 **WILLOW WARBLER** *Phylloscopus trochilus* 11.0–12.5 cm **P+ c**
Olive-green upperparts; yellowish underparts and supercilium; pale brownish legs. ▲ Various wooded habitats. ✤ Soft, plaintive *hooeet*. Song starts with a few faint notes, grows louder, then fades and ends in short flourish. [CD4:31]

3 **WESTERN BONELLI'S WARBLER** *Phylloscopus bonelli* 10.5–12.0 cm **P/V+ c/f**
Pale; bland face; yellowish-green wings and rump contrast with greyish upperparts. ▲ Dry scrub, bushy savanna. ✤ Call *hoo-eet*. Song a short, shivering trill, reminiscent of finale of 1, but slower, lower pitched and lacking acceleration. [CD4:26]

4 **COMMON CHIFFCHAFF** *Phylloscopus collybita* 10–12 cm **P c/r**
More olive-brown and with shorter primary projection than 2; dark legs. ▲ Open, dry and wooded habitats. ✤ Call *hweet*. Song a rhythmic *chiff-chaff-chiff-chiff-chaff-...* [CD4:28]

5 **IBERIAN CHIFFCHAFF** *Phylloscopus (collybita) ibericus* 10–12 cm **P**
Very similar to 4 and not safely separable unless voice heard. Upperparts greener with faint or no brownish tinge; head-sides washed yellow and usually lacking buff-brown tinge; supercilium usually more distinct and more lemon-yellow; underparts on average cleaner lemon-yellow and white with, typically, vivid lemon-yellow (occasionally pale yellow) undertail-coverts. ▲ As 4. ✤ Slightly nasal, downslurred *fiuu*. Song rhythmic and ending in short trill *tup-tup-tup-weet-weet-tsu-tchutututu*. [CD4:30]

6 **GREEN HYLIA** *Hylia prasina* 11.5 cm **R c**
Dark olive-green above; conspicuous dirty yellowish-white supercilium. ▲ Forest. ✤ Loud, clear, double whistle *hee-hee* and short dry rattle. [CD11:61]

7 **BLACK-CAPPED WOODLAND WARBLER** *Phylloscopus herberti* *c.* 9 cm **R c/f**
Black top of head; black eye-stripe; golden-green upperparts. ▲ Montane and submontane forest. Endemic. ✤ Song a short, rapid, melodious whistle or warble. Much faster, shorter and lower pitched than song of 9. [CD10:68]

8 **UGANDA WOODLAND WARBLER** *Phylloscopus budongoensis* 10 cm **R lc**
Well-defined whitish supercilium; plain olive-green above; greyish-white below. ▲ Forest. Mainly canopy. ✤ Song a distinctive, clear phrase of 4–8 sweet, high-pitched notes. [CD10:65]

9 **YELLOW-BROWED WARBLER** *Phylloscopus inornatus* 10 cm **V**
Long, pale yellow supercilium; double yellowish wingbar. ▲ Palearctic vagrant. ✤ Sharp *sweest*.

10 **RUFOUS-CROWNED EREMOMELA** *Eremomela badiceps* 11 cm **R c**
8a Ad Bright rufous-chestnut cap; grey upperparts; black gorget. **8b Juv** Olive above; no cap nor gorget (or just a hint of these); yellowish underparts. ▲ Forest. Mainly canopy. ✤ Soft little contact calls. Also short dry trills *trr trr trr trr* .... [CD11:45]

11 **GREEN-CAPPED EREMOMELA** *Eremomela scotops* 11 cm **R u/lc**
Grey-green above; lemon-yellow below; dark lores; pale eye. ▲ Wooded grassland. ✤ A twittering trill. Dawn song a rhythmic, monotonous *twurp-twurp-twurp-...* [CD11:41]

12 **BUFF-BELLIED WARBLER** *Phyllolais pulchella* 11.5 cm **R u/lc**
Nondescript. Longish, graduated tail. ▲ Dry *Acacia* woodland. ✤ Song a dry, ascending, insect-like rattling *zrrrrrt zrrrrrt*. Calls *cht-cht-cht-cht* and *tzr tzr chit*. [CD11:26]

Continued on page 346

1
2
5
3
4
6
7
8
9
10b
11
10a
12
13
14
15
16

Continued from page 344

13 **SENEGAL EREMOMELA** *Eremomela pusilla* 10 cm **R c**
Greyish head; bright lemon-yellow lower breast and belly. ▲ Savanna woodland. ✤ Cheerful little trill. Dawn song a rhythmic, monotonous *whirp-whirp-whirp-*... [CD11:44]

14 **GREEN-BACKED EREMOMELA** *Eremomela canescens* 11 cm **R c**
Clear white supercilium; dark eye-stripe; head more contrasting with upperparts than 13. ▲ Wooded grassland. ✤ Unmusical but cheerful chittering; a harsh, double note. [CD11:43]

15 **YELLOW-BELLIED EREMOMELA** *Eremomela icteropygialis* 10 cm **R c/u**
Pale greyish above; lemon-yellow belly. ▲ Arid *Acacia* scrub. ✤ Short, clear song, similar to Northern Crombec's. [CD11:39]

16 **SALVADORI'S EREMOMELA** *Eremomela (icteropygialis) salvadorii* 11 cm **R lc**
Greenish mantle; belly brighter yellow than 15. ▲ Wooded grassland. ✤ As 15. [CD11:40]

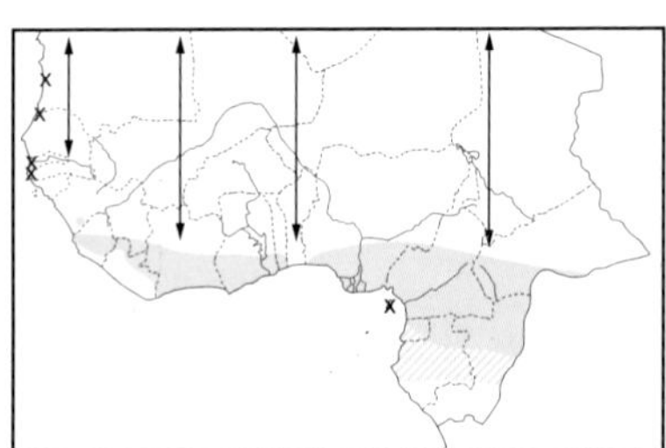
1. WOOD WARBLER

2. WILLOW WARBLER

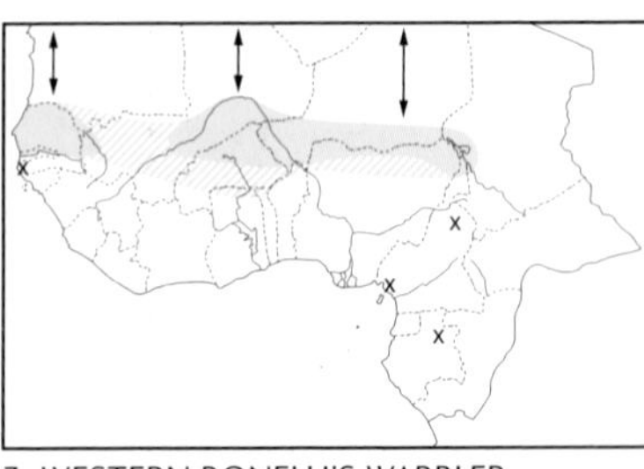
3. WESTERN BONELLI'S WARBLER

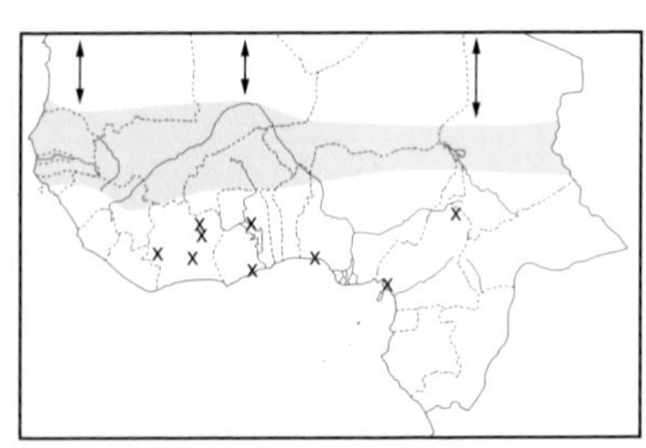
4. COMMON CHIFFCHAFF

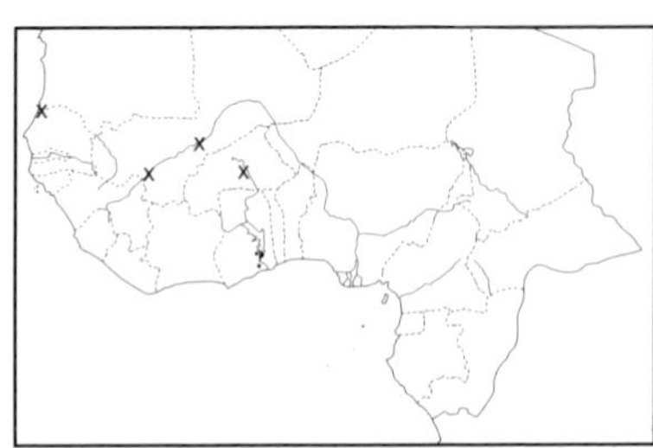
5. IBERIAN CHIFFCHAFF

6. GREEN HYLIA

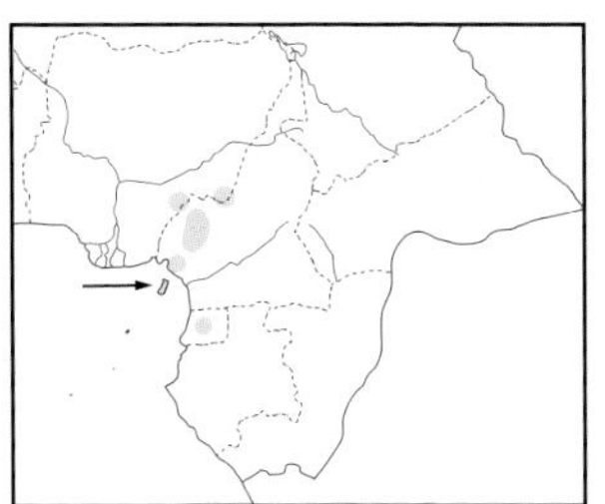
7. BLACK-CAPPED WOODLAND WARBLER

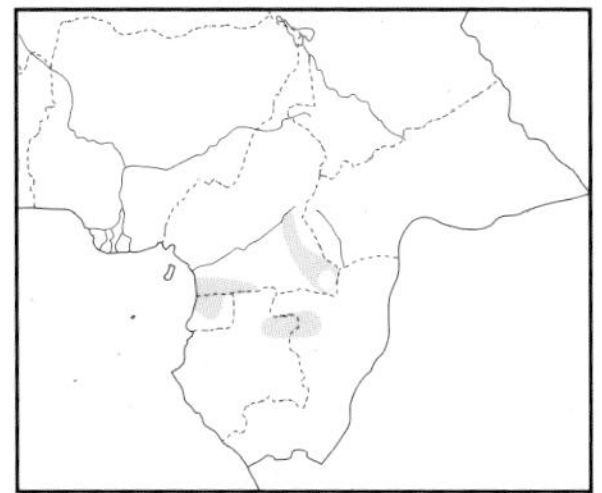
8. UGANDA WOODLAND WARBLER

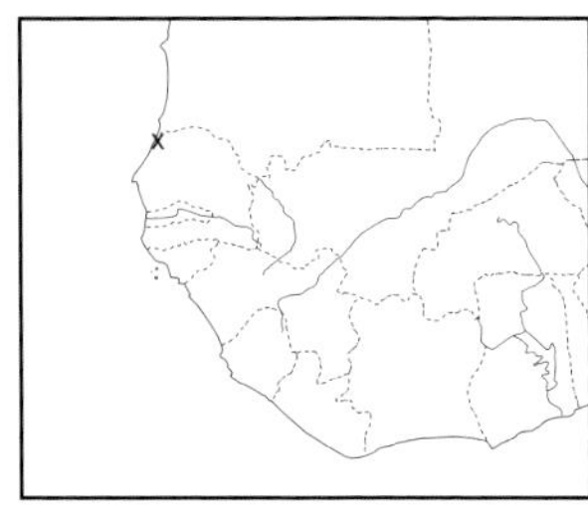
9. YELLOW-BROWED WARBLER

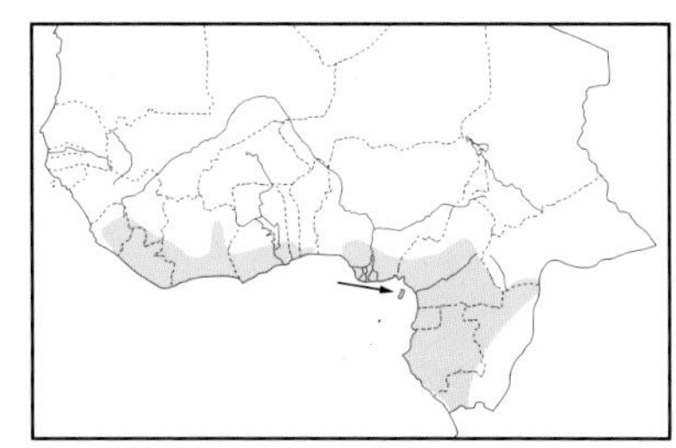
10. RUFOUS-CROWNED EREMOMELA

11. GREEN-CAPPED EREMOMELA

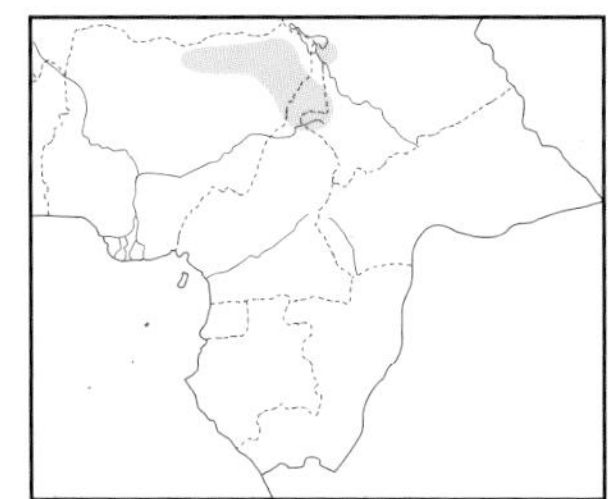
12. BUFF-BELLIED WARBLER

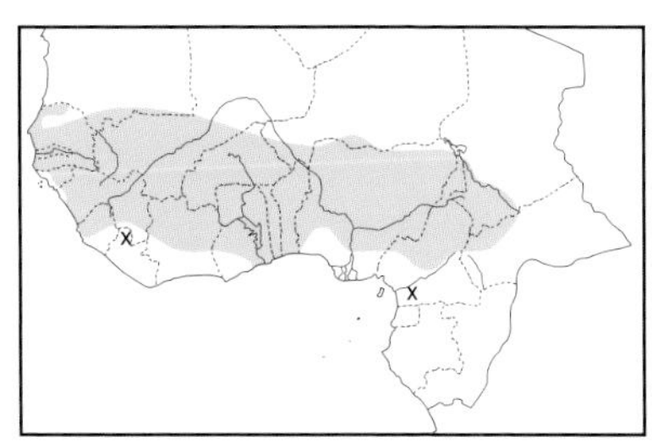
13. SENEGAL EREMOMELA

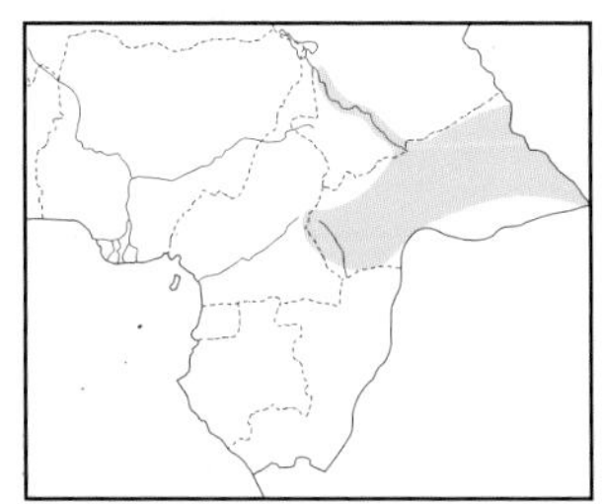
14. GREEN-BACKED EREMOMELA

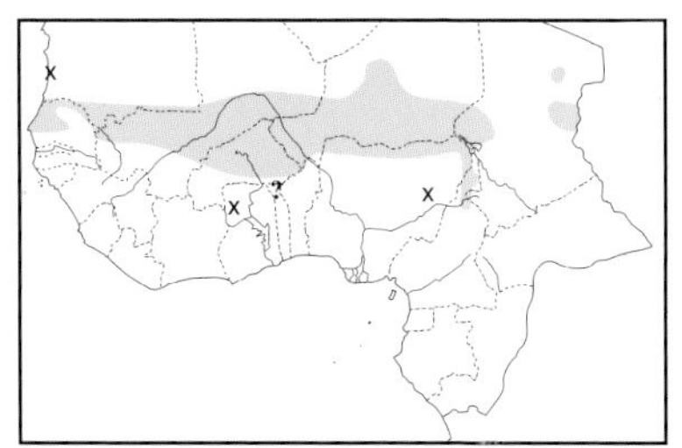
15. YELLOW-BELLIED EREMOMELA

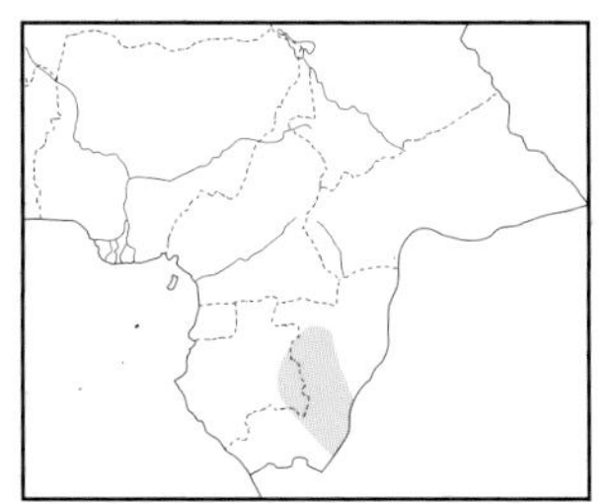
16. SALVADORI'S EREMOMELA

Genus *Sylvietta* (crombecs). Very small size and extremely short tail create distinctive jizz. Often in mixed-species flocks.

1 GREEN CROMBEC *Sylvietta virens* 9 cm R c
1a Ad *virens* (east of lower Niger R.) Head brown; upperparts greenish; throat and breast pale brown. **1b Ad *flaviventris*** (west of lower Niger R.) Lower breast and upper belly yellow. **1c Juv *flaviventris*** Entire underparts yellow. ▲ Second growth, dense scrub, tangles along tracks and clearings. Mostly to mid-levels. ✤ Song a short, sweet, rather high-pitched, clear whistle, descending in scale. Call a dry *prrt, prrt*. [CD11:48]

2 NORTHERN CROMBEC *Sylvietta brachyura* 9 cm R c/u
Grey above; tawny below; buffish supercilium; dusky eye-stripe. ▲ Dry and wooded savanna. ✤ Song a short, sweet, clear warble. Call a sharp, double note. [CD11:50]

3 RED-CAPPED CROMBEC *Sylvietta ruficapilla* 10–12 cm R u
Pale greyish above; rufous on head and breast. ▲ Wooded grassland. ✤ Song a clear little warble ending with two clear notes, with emphasis on last. Call *chik*. [CD11:46]

4 LEMON-BELLIED CROMBEC *Sylvietta denti* 8 cm R f
4a Ad *denti* (Nigeria–Congo) Olive-green above; yellow below; breast tinged olive. **4b Ad *hardyi*** (Guinea–Ghana) Brighter yellow below. ▲ Forest. Mostly in canopy. ✤ Song a rapid series of identical, high-pitched notes (usually up to 7), *tswee-tswee-tswee-tswee-*... easily passing unnoticed, or of two notes on different pitch *tsee-tsu-tsee-tsu-tsee-tsu-*... [CD11:49]

Genus *Hypergerus*. Large, unmistakable warbler with long, graduated tail.

5 ORIOLE WARBLER *Hypergerus atriceps* 20 cm R lc/r
Black, scaly head; green upperparts; yellow underparts. ▲ Gallery forest, thicket, mangroves. Endemic. ✤ Song a short, rapid and frequently repeated series of loud, melodious whistles *hu-hee-tee-teehu* and longer *hu-hee-tee-tu-hwu-hwuu* or *hu-heehee-heehu*, sometimes ending with descending *whee-whuu-whuu-whuu*; with variations. Female utters a *rikitikitikitik thuwthuwthuw* in duet. [CD11:36]

Genus *Hyliota*. Quite distinct, aberrant warblers with relatively short tails. In canopy of forest or savanna woodland; often in mixed-species flocks. Formerly regarded as flycatchers.

6 VIOLET-BACKED HYLIOTA *Hyliota violacea* 12.5 cm R r/lf
6a Ad male *violacea* (Nigeria to CAR–Congo) Deep violet-blue above; mainly white below; white patch of variable size on inner greater coverts (lacking in western *nehrkorni*). **6b Ad female *violacea*** Throat and breast orange. ▲ Forest. ✤ Series of dry *tik-tik-tik-tik-*... on take-off. Song a series of 4–5 sweet whistles *see-su-su-wit see-su-su-wit-wu* ... [CD11:60]

7 SOUTHERN HYLIOTA *Hyliota australis* 12.5 cm R r
7a Ad male As 8a, but upperparts sooty-black; no white edges to tertials and secondaries. **7b Ad female** As 8b, but note tertial pattern; upperparts dark brown. ▲ Extremely rare: puzzling record in secondary forest in Rumpi Hills, Cameroon. [Note: some birds observed in S Cameroon (Campo, Mar 1999) had large white wing patch, similar in size to that of this species. Precise identity of these unclear.] ✤ Twittering contact calls. Squeaky whistles, followed by a trilling warble. [CD11:59]

8 YELLOW-BELLIED HYLIOTA *Hyliota flavigaster* 12.5 cm R lf/s
8a Ad male Glossy blue-black above; white wing patch; pale yellowish-buff below. **8b Ad female** Duller and paler; upperparts dark grey-brown. ▲ Savanna woodland. ✤ Usually silent. Call *twep twep* ... Song a repeated *tiu-wheep tiu-wheep* ... or higher-pitched *tseep hweet tseep-tseep hweet*... Also dry trills. [CD11:58]

Maps on page 350

1c
2
1a
1b
3
4b
4a
5
6b
6a
7b
8b
7a
8a

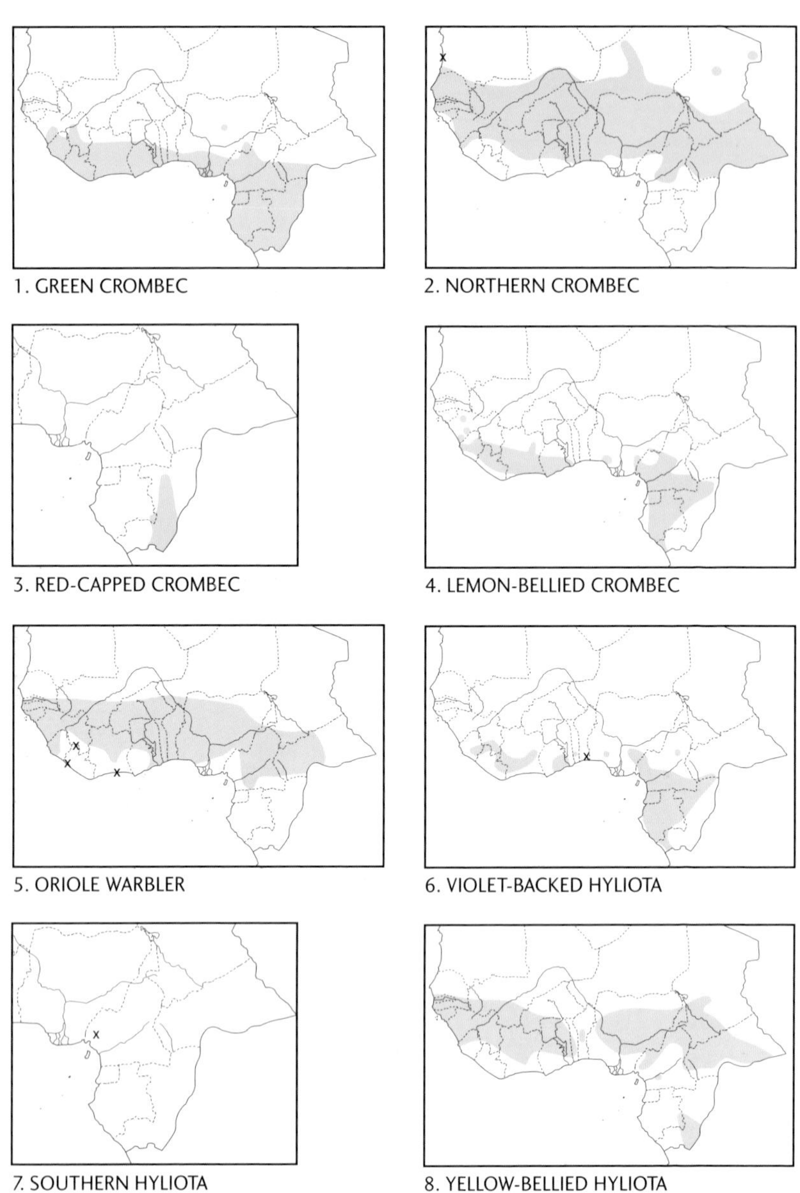

1. GREEN CROMBEC

2. NORTHERN CROMBEC

3. RED-CAPPED CROMBEC

4. LEMON-BELLIED CROMBEC

5. ORIOLE WARBLER

6. VIOLET-BACKED HYLIOTA

7. SOUTHERN HYLIOTA

8. YELLOW-BELLIED HYLIOTA

Plate on page 348

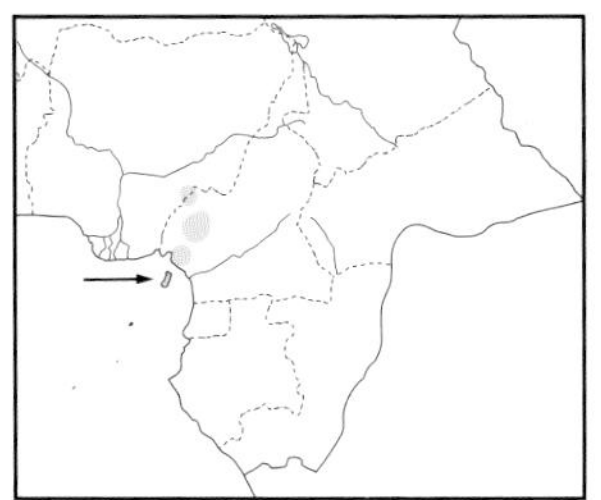
1. WHITE-TAILED WARBLER

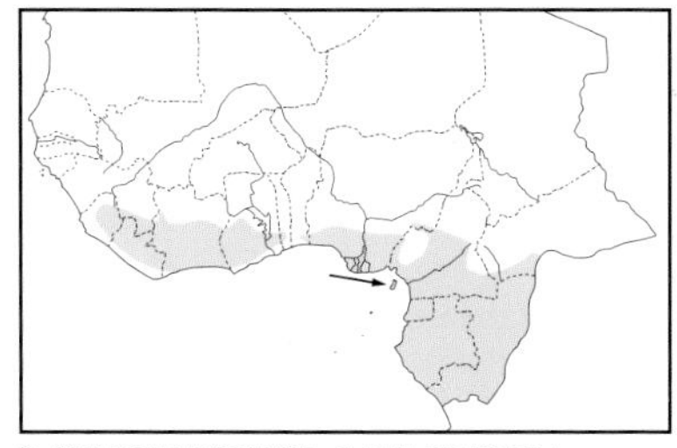
2. YELLOW-BROWED CAMAROPTERA

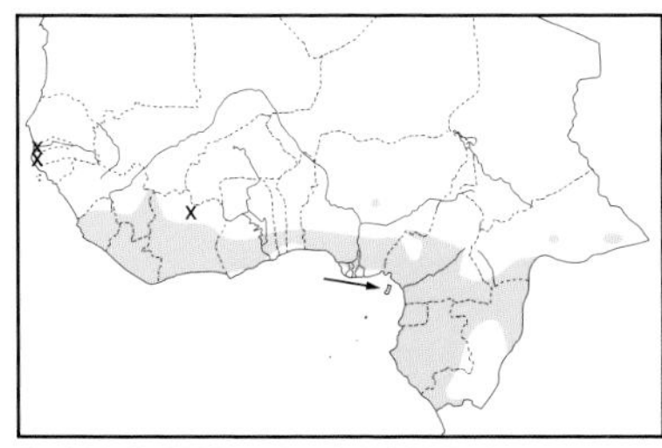
3. OLIVE-GREEN CAMAROPTERA

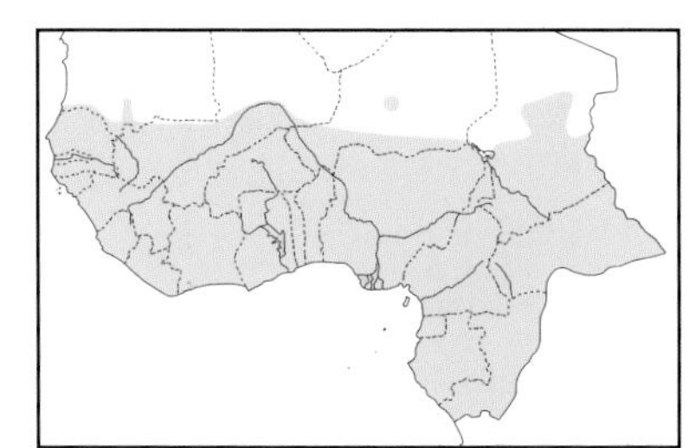
4. GREY-BACKED CAMAROPTERA

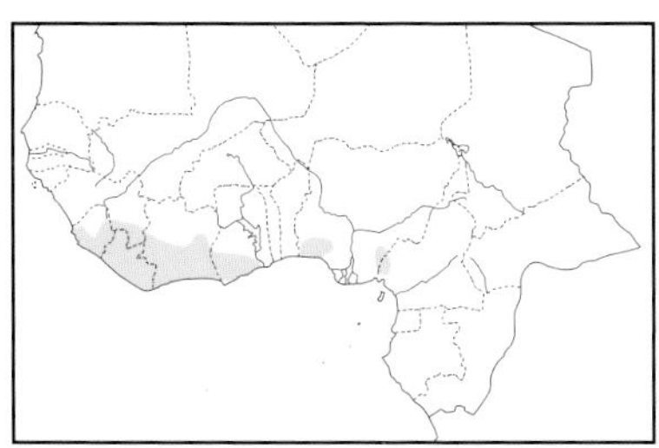
6. KEMP'S LONGBILL

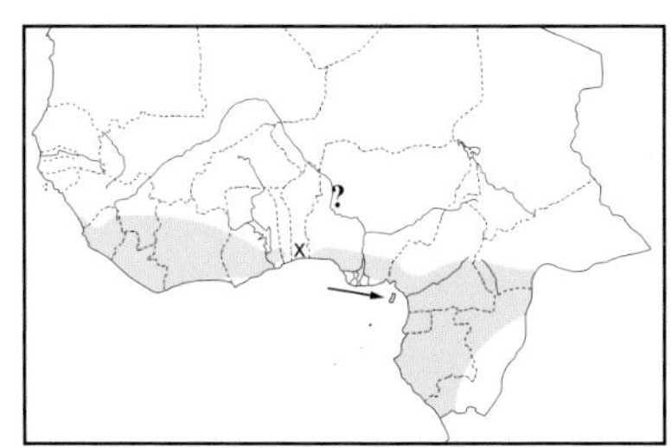

7. GREY LONGBILL

8. YELLOW LONGBILL

Plate on page 352

1 WHITE-TAILED WARBLER *Poliolais lopezi* 10 cm R lf NT
Tail short, dark brown above, white below. **1a Ad male *alexanderi*** (Mt Cameroon) Dark grey with olive wash. **1b Ad female *alexanderi*** Forehead and head-sides rufous. **1c Ad male *manengubae/lopezi*** (SE Nigeria–W Cameroon, Bioko) Wholly sooty-grey. ▲ Montane forest undergrowth. Endemic. ✤ Song a regular series of clear, high-pitched, single notes, each repeated several times. Call a clear *hee-huuw hee-huuw...* Alarm a plaintive, high-pitched *peep*. [CD11:28]

Genus *Camaroptera*. Small, with rather long and straight bills; colour of thighs contrasting with belly. Skulk in dense vegetation, mostly low down. Have distinctive mewing calls.

2 YELLOW-BROWED CAMAROPTERA *Camaroptera superciliaris* 11 cm R u/c
2a Ad Bright green above; supercilium, ear-coverts, thighs and undertail-coverts bright yellow. When singing, throat of male swells and reveals bare blue patch at sides. **2b Juv** Throat and breast yellow. ▲ Forest edge. ✤ Song a distinctive double nasal note, sounding like *koa-koa* or *kweh-kweh*. [CD11:34]

3 OLIVE-GREEN CAMAROPTERA *Camaroptera chloronota* 11 cm R lf
Skulking, but song loud and distinctive. **3a Ad *chloronota*** (Togo to CAR–Congo) Dark olive-green above; greyish-white below; ear-coverts greyish. **3b Ad *kelsalli*** (Guinea–Ghana) Ear-coverts tinged rufous. ▲ Lowland forest and savanna outliers. ✤ Song a remarkable, loud and sustained series of a single note uttered at same pitch and without interruption for up to several minutes before ceasing abruptly. [CD11:35]

4 GREY-BACKED CAMAROPTERA *Camaroptera brachyura* 11.5 cm R c
4a Ad *tincta* (forest zone) Grey, with yellowish-green wings. **4b Ad *brevicaudata* non-breeding** (savanna zone) Ashy-brown above; paler below. Breeding plumage as 4a but slightly paler. **4c Juv** Olive-green above; lemon below. ▲ Dense shrubbery in various habitats. One of the most widespread and common warblers in the region. ✤ Call a distinctive mewing note. Song variable, but always a rhythmic repetition (variable in speed) of a single note, e.g. *churrup-churrup-churrup-...*, *churp-churp-churp-...*, *chrup chrup chrup ...*, *kechup-kechup-...*, etc. [CD11:32]

5 MIOMBO WREN WARBLER *Calamonastes undosus* 13 cm
Wholly dark grey; rather long tail constantly cocked. ▲ Savanna woodland; Cabinda and lower Congo R. Not (yet?) recorded in our region. ✤ [CD11:33]

Genus *Macrosphenus* (longbills). Aberrant warblers with long straight bill, sharply hooked at tip, short wings and tail, and long, loose feathers on rump and flanks. Occur in dense undergrowth, tangles and vines of forest and forest edge. Song distinctive.

6 KEMP'S LONGBILL *Macrosphenus kempi* 13 cm R u/s
6a Ad Dark brownish above; orange-rufous flanks and undertail-coverts; yellow eye. **6b Juv** Olive-green above; yellower below; lemon-yellow throat. ▲ Forest. Endemic. ✤ Song a distinctive, melodious *tee tuwe-tuwe-tuwe-tuwe tee*, with variations. Also a fast series of rather tuneless, scratchy, husky and nasal notes. [CD11:56]

7 GREY LONGBILL *Macrosphenus concolor* 11.5 cm R c
Wholly olive-green; long, straight bill. ▲ Forest. ✤ Song a lively, sustained, extremely rapid warble; also slower versions of short, repeated phrases. Less frequent song consists of very rapid, repetitive imitations of other forest species. [CD11:54]

8 YELLOW LONGBILL *Macrosphenus flavicans* 13 cm R u/lf
Mainly olive-green above; whitish throat; rest of underparts bright olive-yellow; eye yellow. ▲ Forest. ✤ Song a series of single, clear whistles, descending in scale and slightly accelerating at end, *hee hu hu hu hu hu hu-hu-hu*. [CD11:55]

Maps on page 351

1b
1c
2b
1a
2a
4c
3b
4b
4a
3a
6b
6a
5
7
8

Rather slender, active species with long, graduated tails.

1 RIVER PRINIA *Prinia fluviatilis* *c.* 12 cm **R**
As 2, but somewhat greyer above and whiter below. ▲ Waterside vegetation. Local; distribution inadequately known. ✤ Song a rapid, rhythmic series of a single, high-pitched note. Not reminiscent of 2; notes more drawn-out, rhythm much less variable. [CD11:3]

2 TAWNY-FLANKED PRINIA *Prinia subflava* 11–12 cm **R c**
2a Ad breeding Brownish above; buff supercilium; pale tawny flanks. **2b Ad non-breeding** (savanna only) Paler, rusty-brown above; tail longer; bill horn coloured. ▲ Various grassy and bushy habitats. ✤ Song a monotonous, rhythmic series of a single note; speed and tone variable, e.g. sharp *tzreep tzreep tzreep* ... and fast *plip-plip-plip-plip-*... Alarm a harsh *zbeee*. [CD11:2]

3 WHITE-CHINNED PRINIA *Schistolais leucopogon* 14 cm **R u/f**
Entirely grey except black mask and contrasting creamy-white throat. ▲ Forest edge, farmbush. ✤ Rather harsh *chi-chik*, constantly uttered but easily passing unnoticed. Also *djuweet*. Song, always in duet or group, a jumble of *chi-chik* calls. [CD11:6]

4 SIERRA LEONE PRINIA *Schistolais leontica* 13 cm **R lu/r VU**
Dark ash-grey; whitish eye; buff flanks and undertail-coverts. ▲ Forest edge in hilly areas. Endemic. ✤ Call *psit* or *pit*. Song, given in unsynchronised duet, a rapid, high-pitched *sipsipsip-sipsip*... with second bird uttering lower, nasal *bur-bur-bur-bur-*... [CD11:7]

5 BANDED PRINIA *Prinia bairdii* 11.5 cm **R u/lc**
5a Ad Blackish-brown above with white spots on wings; boldly barred below. **5b Juv** Duller; underparts mainly greyish-brown, without barring. ▲ Forest edges. ✤ Song a loud, shrill, fast series of single notes *plee-plee-plee-plee-*... and more rapid *pliplipliplipliplipli*..., uttered from cover; also *wheet-wheet-trlrrrr*. [CD11:5]

6 RED-WINGED WARBLER *Heliolais erythropterus* 12 cm **R f/u**
6a Ad breeding Grey head and upperparts; bright rufous wings. **6b Ad non-breeding** Pale vinous-rufous upperparts; horn-coloured bill. ▲ Wooded grassland. ✤ Song a rapid, monotonous series of high *pseep* or *thu-weet* notes; in duet second bird utters dry chatter. Calls include thin *tseek* notes and little dry churring trill. [CD11:25]

7 GREEN LONGTAIL *Urolais epichlorus* 15 cm **R lc**
Bright green above; tail long and usually heavily worn. ▲ Montane and submontane forest. Endemic. ✤ Song an ascending *peeeep-peep-peep-peep-pip-pip-pip*; also a long rhythmic series of a single sharp note *tsip-tsip-tsip-*.... [CD11:24]

8 STREAKED SCRUB WARBLER *Scotocerca inquieta* 10 cm **R ls**
Cocked, white-tipped tail; finely streaked crown; broad pale supercilium. ▲ Desert scrub. ✤ Song a dry *dzit dzit* followed by melodious, liquid *deedle doleedle doleed*. [CD4:25]

9 RED-WINGED GREY WARBLER *Drymocichla incana* 14 cm **R lu/r**
Pale grey; conspicuous rufous wing panel. ▲ Wooded grassland. ✤ Song includes series of loud, clear *kweeup kweeup kweeup*... and repeated *kwup* ending in chatter. Duets. [CD11:27]

10 RED-FRONTED WARBLER *Urorhipis rufifrons* 11 cm **R u**
Mouse-brown above; rufous forehead. ▲ Desert scrub, arid bush. ✤ Song a simple series of chirping notes *tik tik tik tik* ... and rhythmic *tsyep-tsyep-tsyep-tsyep-*... Alarm a sharp *seep-seep* or *tzii* and trill *spispihehehe*. [CD11:23]

11 CRICKET WARBLER *Spiloptila clamans* *c.* 11.5 cm **R u/lf**
Pale; long tail; black-and-white pattern on wing-coverts. **11a Ad male** Hindcrown and nape pale grey. **11b Ad female** Hindcrown and nape cinnamon. ▲ Thorn scrub. ✤ Song: rhythmic, insect-like trills and series of single, high-pitched notes. Duets. [CD11:22]

Maps on page 356

1
2a
2b
3
4
5b
5a
6a
6b
7
8
9
10
11b
11a

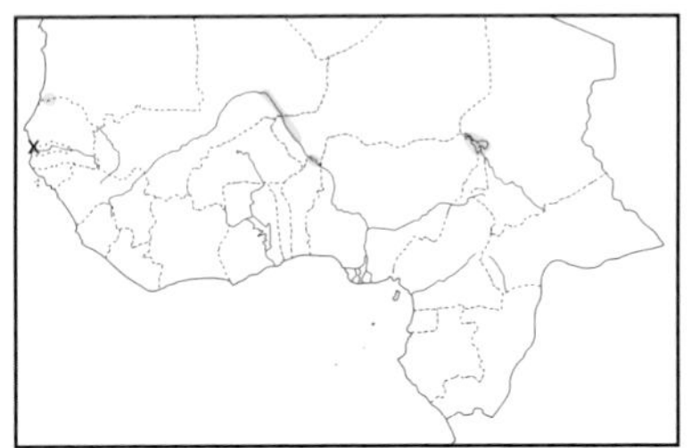

1. RIVER PRINIA

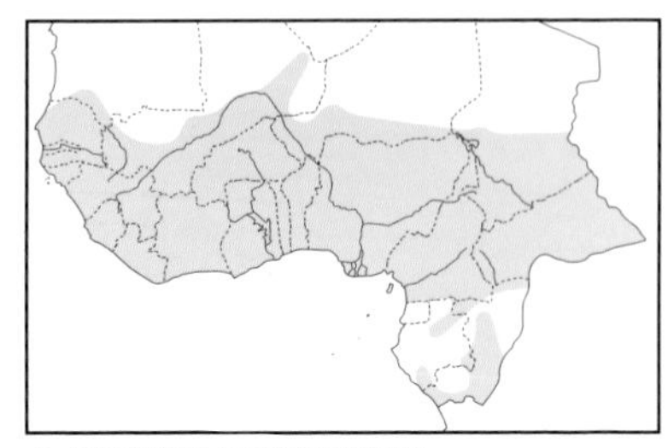

2. TAWNY-FLANKED PRINIA

3. WHITE-CHINNED PRINIA

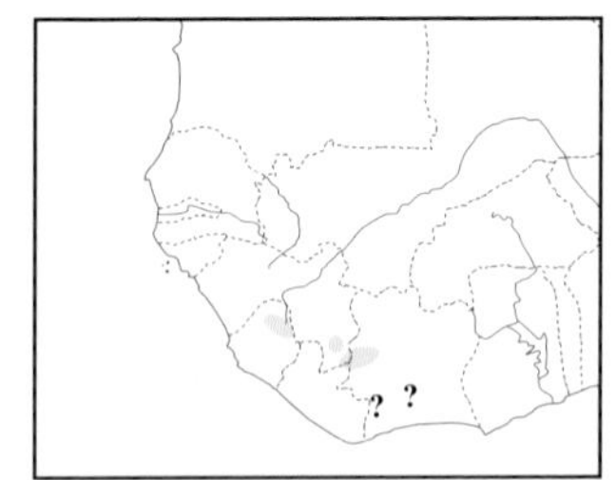

4. SIERRA LEONE PRINIA

5. BANDED PRINIA

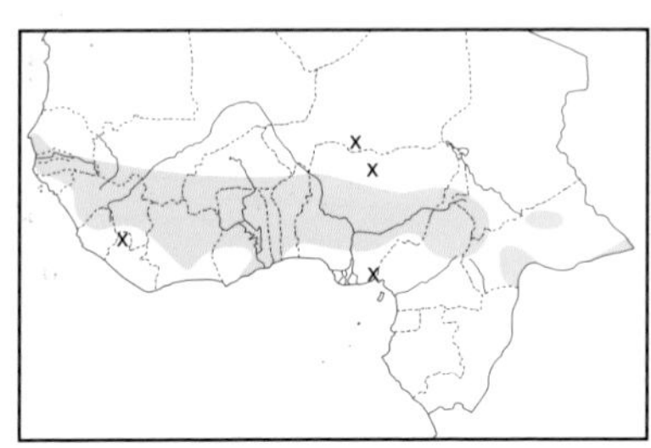

6. RED-WINGED WARBLER

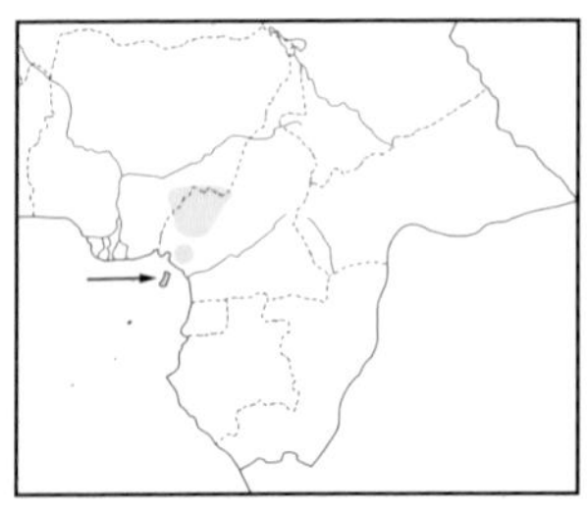

7. GREEN LONGTAIL

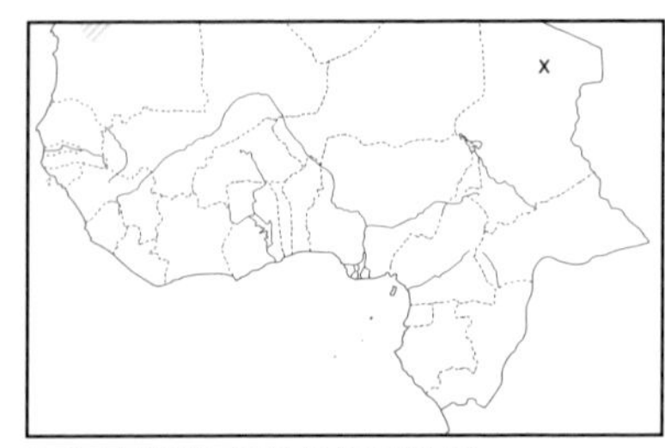

8. STREAKED SCRUB WARBLER

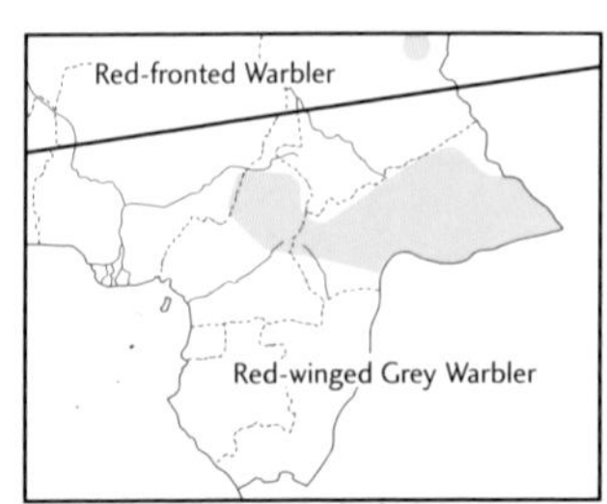

9. RED-WINGED GREY WARBLER
10. RED-FRONTED WARBLER

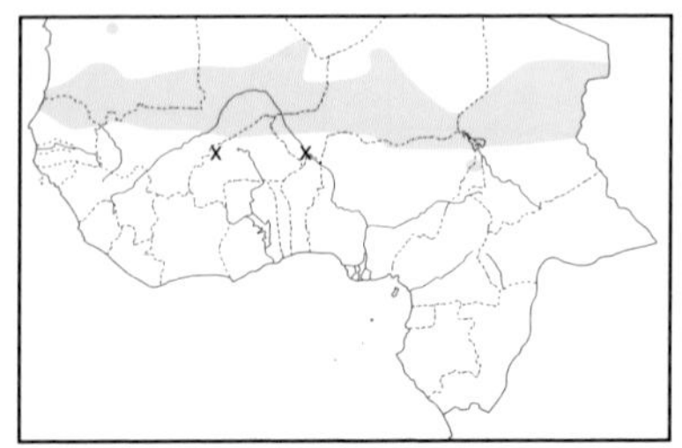

11. CRICKET WARBLER

Plate on page 354

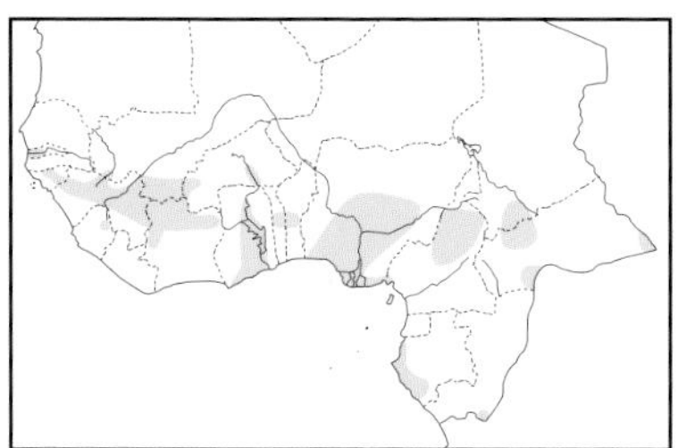
1. YELLOW-BREASTED APALIS

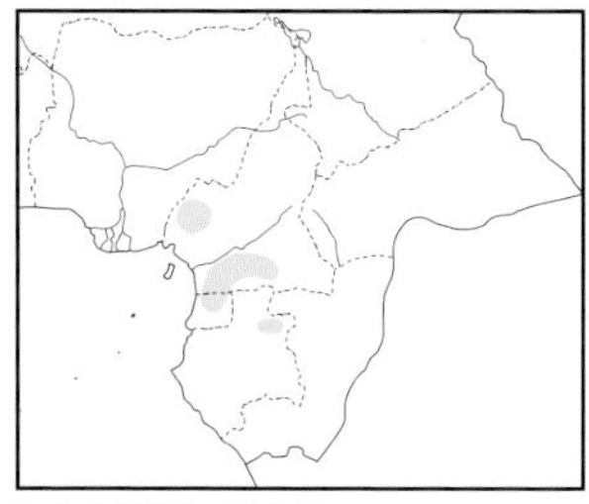
2. MASKED APALIS

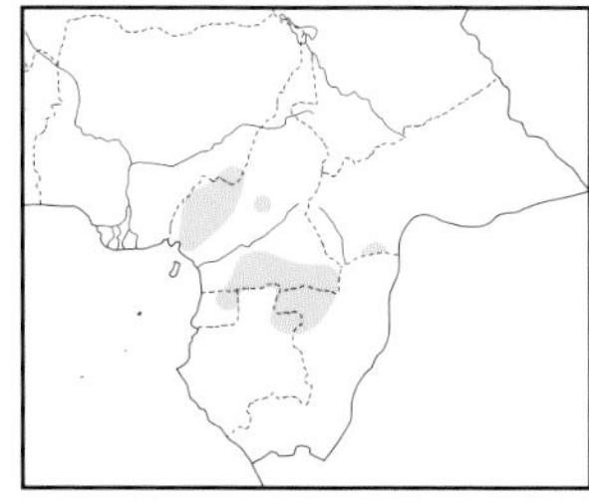
3. BLACK-THROATED APALIS

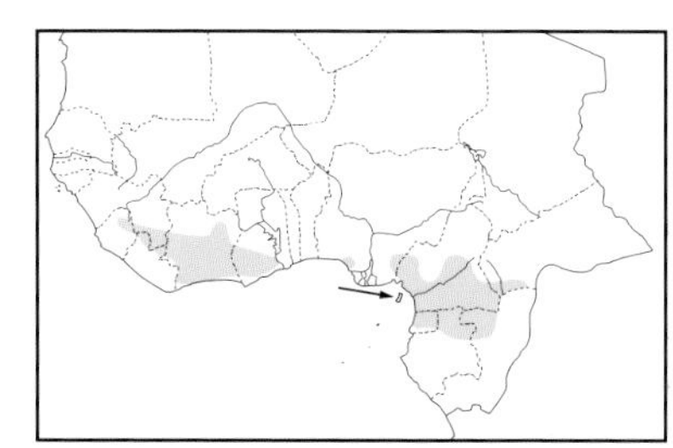
4. BLACK-CAPPED APALIS

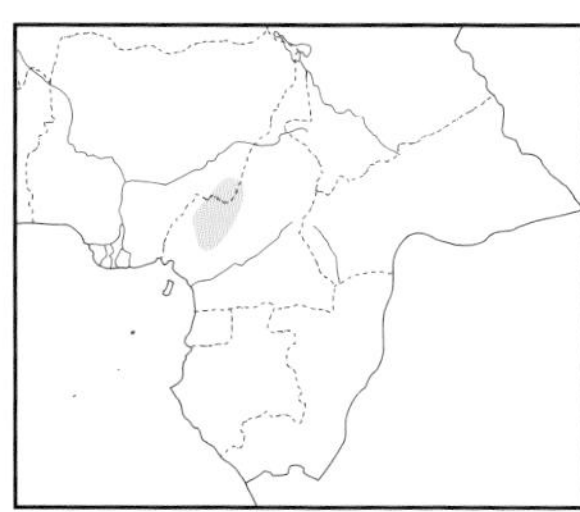
5. BLACK-COLLARED APALIS

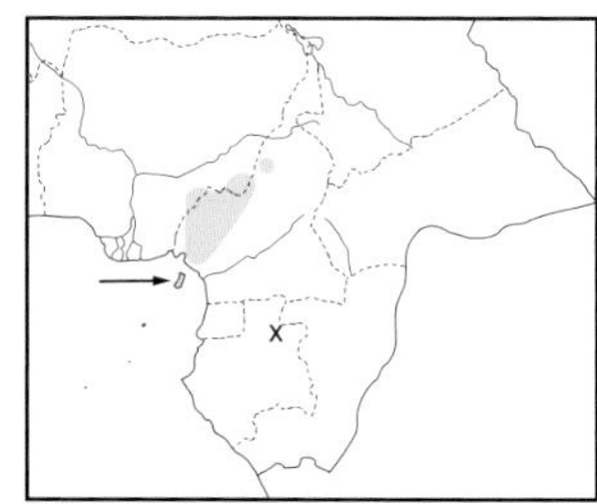

6. GREY APALIS

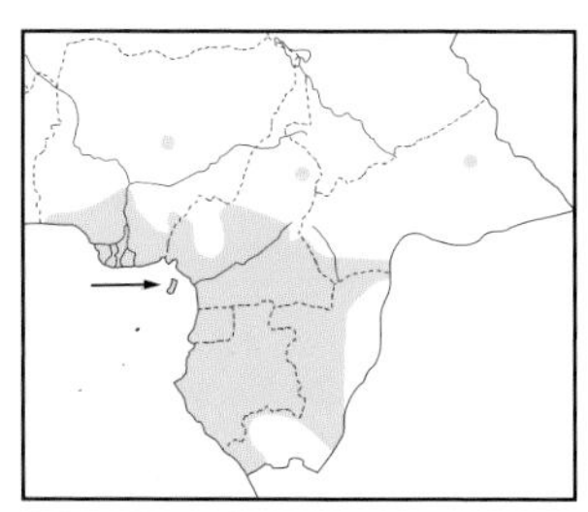
7. BUFF-THROATED APALIS

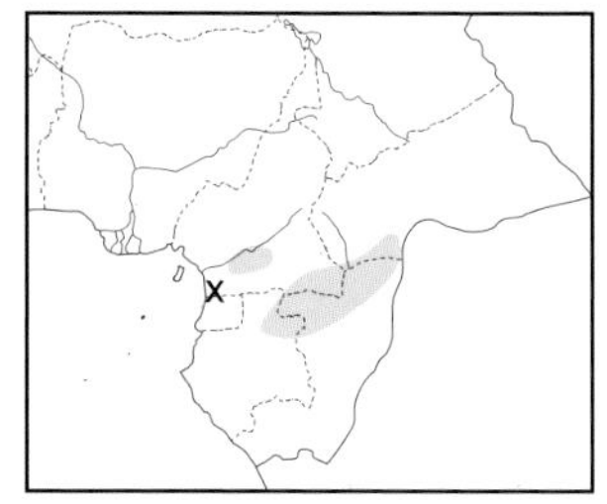

8. GOSLING'S APALIS

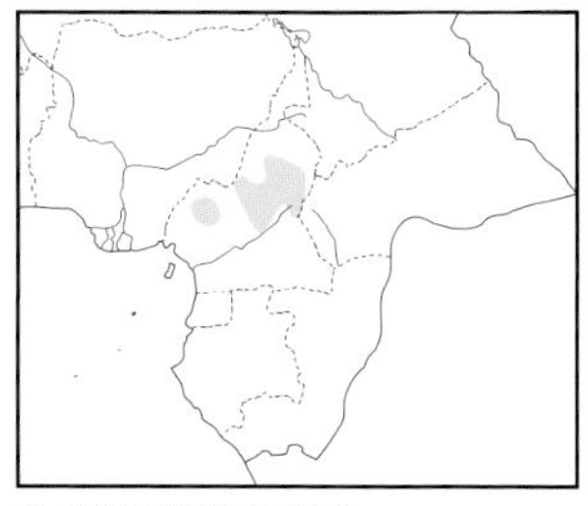
9. BAMENDA APALIS

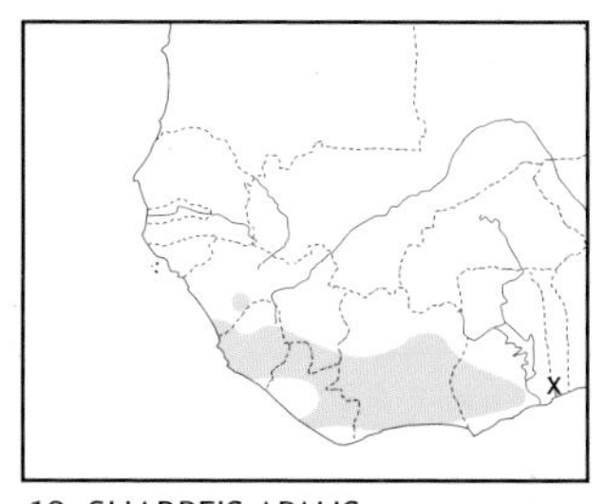

10. SHARPE'S APALIS

Plate on page 358

Genus *Apalis*. Slender; tail rather long and graduated. Mostly in forest. Active; often in mixed-species flocks. Song typically a rhythmic repetition of a single note.

1 YELLOW-BREASTED APALIS *Apalis flavida* 11.5 cm R lu/r
Grey head; green upperparts; yellow upper breast. Juvenile has green head, pale bill. ▲ Gallery forest, forest patches, bush, mangroves. ✤ Male song a rhythmic *churup-churup-churup-*... very similar to a song of Grey-backed Camaroptera. In duet, female utters sharp *kep-kep-kep-*.... [CD11:8]

2 MASKED APALIS *Apalis binotata* 10 cm R lf
2a Ad Dark slate-grey head; black throat; white malar stripe. **2b Juv** Greenish head and upperparts; greyish throat; underparts tinged yellow. ▲ Old clearings and second growth in lowland and submontane forest. ✤ Fast series of sharp, dry notes *tiree-tiree-tiree-*... or *trièk-trièk-trièk*...; with variations. Female often joins with fast *tatatatatata*... [CD11:9]

3 BLACK-THROATED APALIS *Apalis jacksoni* 11.5 cm R lc/r
3a Ad male Black head and throat; white moustachial stripe; bright yellow underparts. **3b Ad female** Grey head. ▲ Forest above *c.* 500 m. Canopy. ✤ Rapid *ku-kree ku-kree ku-kree*..., very different from other apalises and difficult to transcribe. Duets. [CD11:11]

4 BLACK-CAPPED APALIS *Apalis nigriceps* 11.5 cm R lc/r
4a Ad male Black head and breast-band; golden-yellow mantle; whitish underparts. **4b Ad female** Grey head and breast-band. ▲ Lowland and gallery forest. Canopy. ✤ Monotonous, rhythmic series of single note *tzrrrrr tzrrrr tzrrrr* or *turrrirrrp turrrirrrp*. Duets. [CD11:14]

5 BLACK-COLLARED APALIS *Apalis pulchra* 13 cm R lc
Slate-grey head and upperparts; black breast-band; rufous-chestnut flanks. ▲ Montane forest. Forages low; cocks tail. ✤ Repertoire includes rhythmic series of mewing *pew-pew-pew-pew-*..., rapid *kewkewkewkewkew*... and slower *pweet pweet pweet*... [CD11:20]

6 GREY APALIS *Apalis cinerea* 13 cm R lc
6a Ad Greyish-brown head; grey upperparts; creamy-white underparts. **6b Juv** Tinged olive above and yellow below. ▲ Montane and submontane forest. ✤ Monotonous series of single, rather croaking note *kwek-kwek-kwek-kwek-*...; often preceded by sweet, descending trill *pirrrrrrrr* (reminiscent of ringing telephone) or *trrrr-tik-tik*. Also duets, male (presumably) uttering *kwek* notes, female series of *pirrrrr* trills. [CD11:12]

7 BUFF-THROATED APALIS *Apalis rufogularis* 11.5 cm R f
7a Ad male Slate-grey head, throat and upperparts; white outer tail feathers. **7b Ad female** Cinnamon-rufous throat. **7c Juv** Olive-green above; pale yellow below. ▲ Lowland forest. Canopy. ✤ Fast, monotonous, rhythmic *truit-truit-truit-*... or *cheerk-cheerk-*... with variations. [CD11:16]

8 GOSLING'S APALIS *Apalis goslingi* 11.5 cm R lc
Slate-grey above; pale grey below; throat creamy-white. ▲ Along rivers in lowland forest. Mostly at mid-level. ✤ Short, fast, rhythmic series of single note *twit-twit-twit-twit-*... and faster *twitititititititit*, slightly reminiscent of Banded Prinia. [CD11:19]

9 BAMENDA APALIS *Apalis bamendae* 11.5 cm R f/lc
Dark brownish-grey above; grey below; forehead to throat rufous. ▲ Forest patches in highlands. Canopy. Endemic. ✤ Simple short note, monotonously repeated in relatively slow rhythm, resembling song of 8. Also a 3-note *tswee-tit-tit tswee-tit-tit* ... or *tsu-twit-twit tsu-twit-twit* ... and fast *tsutitititititititit*. [CD11:18]

10 SHARPE'S APALIS *Apalis sharpii* 11.5 cm R f/lc
10a Ad male All sooty-grey. **10b Ad female** Pale grey below; buff throat. **10c Juv** Greyish-olive above; pale lemon below; greyish breast-sides and flanks. ▲ Lowland forest. Mostly in canopy. Endemic. ✤ Monotonous, fast, rhythmic *cherit-cherit-cherit-*... or *tirrit-tirrit-tirrit-*... given throughout the day; very similar to 7. Duets. [CD11:15]

Maps on page 357

1
2b
2a
3b
3a
4b
4a
5
6b
6a
7c
7b
7a
8
10c
10b
10a
9

Streaked upperparts (warmer coloured in non-breeding); aerial displays; open grassy areas:

1 PECTORAL-PATCH CISTICOLA *Cisticola brunnescens* 10–11 cm R lf
1a Ad male breeding Top of head russet-brown; black lores; tail blackish above and below. **1b Ad male non-breeding/ad female** Top of head tawny broadly streaked black; faint loral mark. ✤ Rhythmic repetition of buzzing note, accelerating when diving to ground *tzit-tzit-tzit-tzit-... -tzitzitzitzitzit...*; accompanied by barely audible dry clicks. [CD10:69]

2 AYRES'S CISTICOLA *Cisticola ayresii* 9–10 cm R lf
Ad male *gabun* breeding (coastal Gabon, C Congo) Top of head and rump rufous-brown; very faint loral mark; very short tail plain blackish above and below. ▲ Status unclear. This race possibly refers to 1. ✤ Series of thin, high-pitched whistles, interspersed by dives accompanied by rapid *tiktiktik...* and occasional wing-snaps. [CD10:71]

3 PALE-CROWNED CISTICOLA *Cisticola cinnamomeus* 10–11 cm R lc
Ad male breeding Top of head plain rusty buff-brown; blackish around lores and forecrown; back heavily streaked. ✤ Short series of thin, high-pitched *eeeyip* followed by series of lower, slightly vibrating *rreee...* as bird dives. [CD10:70]

4 ZITTING CISTICOLA *Cisticola juncidis* 10 cm R lc/u
4a Ad male breeding Top of head brownish-buff heavily streaked black; rump dull rufous-brown; tail black above with black subterminal spots below. **4b Ad male non-breeding** More buff; streaking less heavy. ✤ Monotonous repetition of single harsh *tsip*. [CD4:36]

5 BLACK-BACKED CISTICOLA *Cisticola eximius* c. 10 cm R lf/r
5a Ad male breeding Top of head pale rufous-brown; rump bright rufous, contrasting with rest of upperparts; tail blackish above. **5b Ad male non-breeding** Top of head streaked black. ✤ Rather sharp, vibrant *tsree-tsree-tsree-...*, sometimes accompanied by rhythmic noise, sounding like wing-snapping, and dissonant *chereet-chereet.* [CD10:72]

6 DAMBO CISTICOLA *Cisticola dambo* 10–12 cm R lf
6a Ad male breeding Resembles 4, but more colourful; black better defined; unusually long tail plain black above and below. **6b Ad male non-breeding** Deeper buff; tail blackish edged rusty-buff. ✤ Rasping, rather piercing *hree-ep, hree-ep,...* with wing-snapping. [CD10:73]

7 DESERT CISTICOLA *Cisticola aridulus* 10–12 cm R u/f
Paler, more sandy coloured than 4. ✤ Rapid repetition of clear, high-pitched note *teehu*, interspersed with dry clicking bill-snaps (only audible at short range). [CD10:74]

Plain upperparts not contrasting with top of head; typically sing from perch; savanna:

8 FOXY CISTICOLA *Cisticola troglodytes* 10 cm R l
Upperparts bright russet. ▲ Local. Status and distribution inadequately known. ✤ Rapid series of similar *tsit* or *tsee* notes. [CD10:83]

9 RUFOUS CISTICOLA *Cisticola rufus* 10 cm R u/lf
Upperparts dull rust-brown. ▲ Endemic. ✤ Song 2–3 thin descending notes e.g. *see see hu, see see hu*, ... Also rapid, hesitant series *tsutitititsutititu....* [CD10:82]

10 SHORT-WINGED CISTICOLA *Cisticola brachypterus* c. 10 cm R f/lc
10a Ad breeding Upperparts dull brown. **10b Ad non-breeding** Upperparts warmer brown with faint dusky streaks. ✤ Continuously repeated short, rapid series of 2–3 high-pitched notes descending in pitch e.g. *see-se-swu see-su-swu ...* [CD10:81]

Continued on page 362

Maps on pages 362 and 363

1a
1b
2
3
4a
4b
5a
5b
6a
6b
7
8
9
10a
10b
11a
11b
12a
12b
13

Continued from page 360

Plain or streaked upperparts contrasting with rufous top of head; sing from perch; bush:

11 RED-PATE CISTICOLA *Cisticola ruficeps* 13 cm **R u/lc**
11a Ad breeding Top of head rusty-red; upperparts plain dark brown; undertail-coverts white. **11b Ad non-breeding** Upperparts heavily streaked black. ✤ Drawn-out, high-pitched note followed by 2 fast phrases of alternatively rising and falling notes. [CD10:79]

12 DORST'S CISTICOLA *Cisticola guinea* 13 cm **R lc DD**
12a Ad breeding Similar to 11a b ut undertail-coverts buff. Song different. **12b Ad non-breeding** Mantle indistinctly streaked. ▲ Range inadequately known. Endemic. ✤ Drawn-out trill of variable speed, often preceded by short vibrant note and followed by simple motif, repeated several times, e.g. *chirrrrrrr tsu-wheet tsu-wheet tsu wheet* .... [CD10:80]

13 PIPING CISTICOLA *Cisticola fulvicapillus* 10–11 cm **R lc/u**
Top of head dull rufous; upperparts plain earth-brown. ✤ Monotonous repetition of single, penetrating, but not very loud note *whee whee whee...* or *whip-whip-whip-...* [CD10:84]

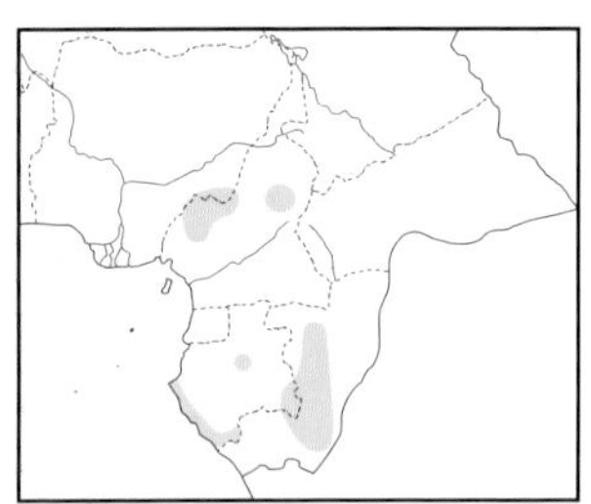

1. PECTORAL-PATCH CISTICOLA

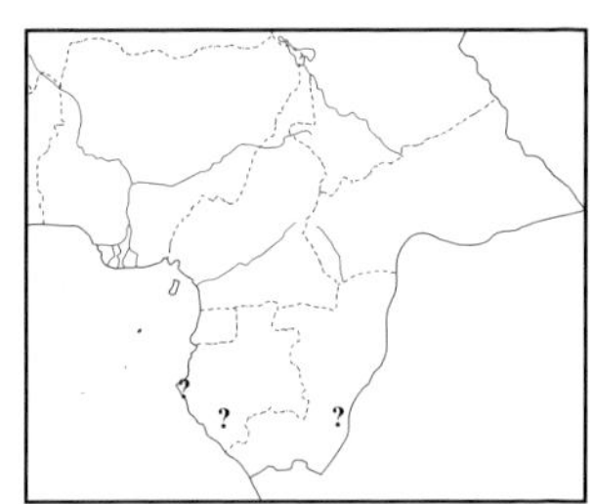

2. AYRES'S CISTICOLA

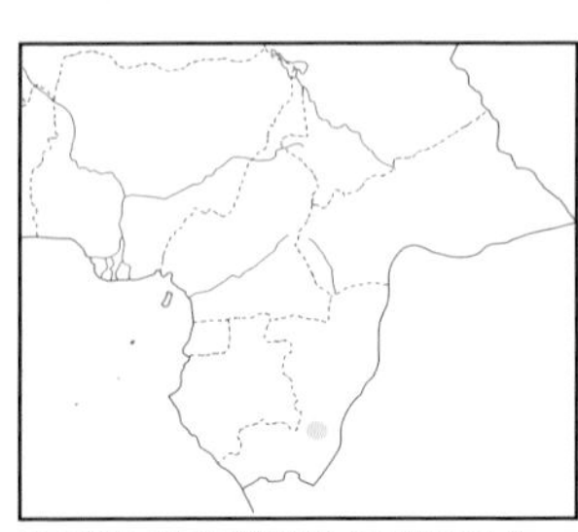

3. PALE-CROWNED CISTICOLA

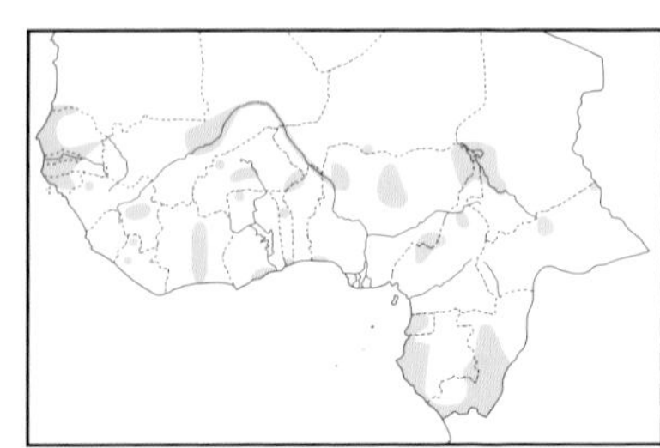

4. ZITTING CISTICOLA

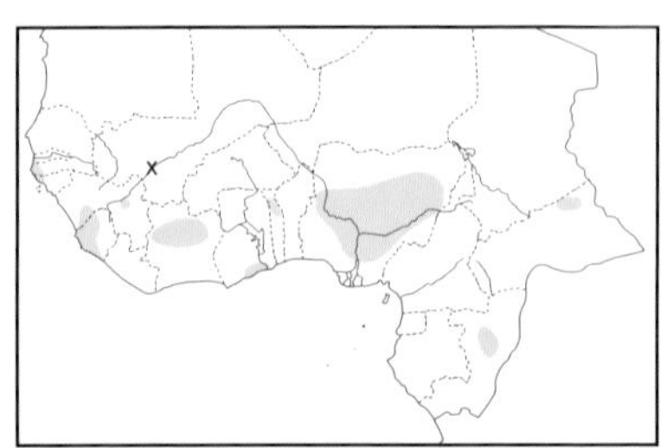

5. BLACK-BACKED CISTICOLA

6. DAMBO CISTICOLA

 Plate on page 360

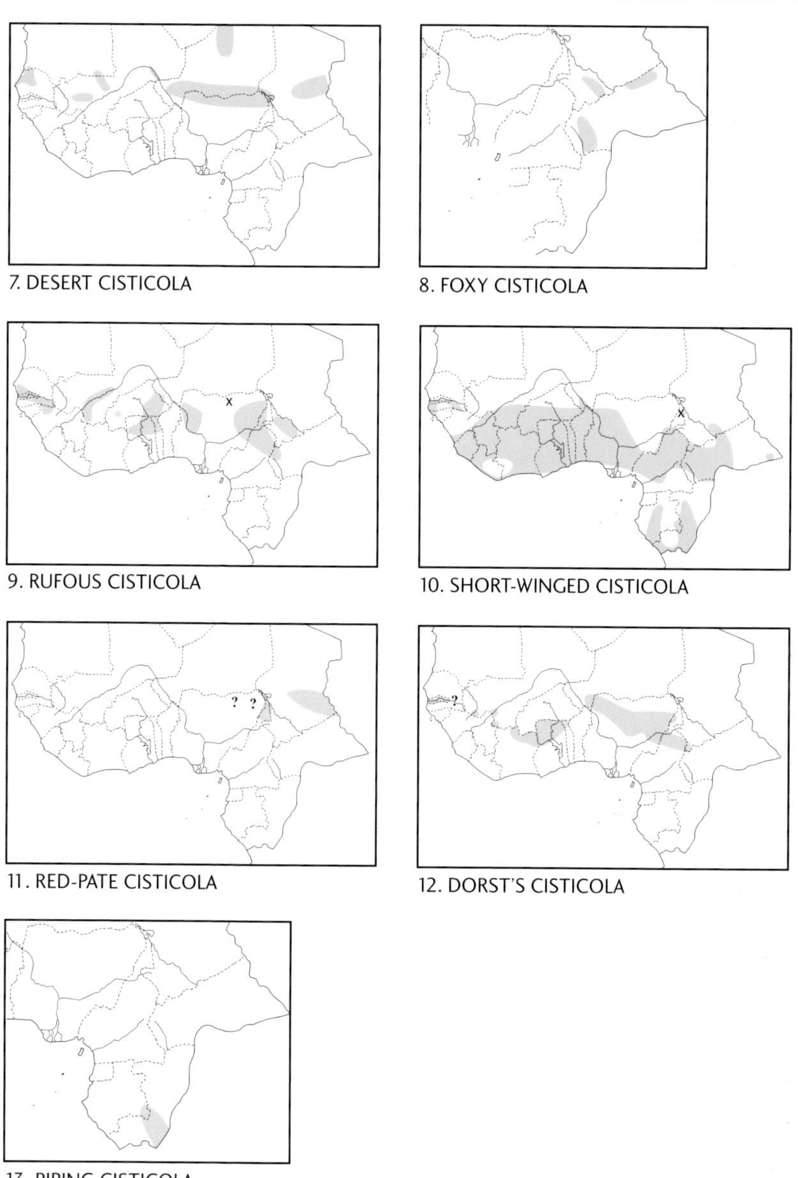

7. DESERT CISTICOLA

8. FOXY CISTICOLA

9. RUFOUS CISTICOLA

10. SHORT-WINGED CISTICOLA

11. RED-PATE CISTICOLA

12. DORST'S CISTICOLA

13. PIPING CISTICOLA

PLATE 109: MEDIUM-SIZED AND LARGE CISTICOLAS

Plain or faintly dappled upperparts (plumage warmer coloured in non-breeding):

1 ROCK-LOVING CISTICOLA *Cisticola aberrans* 13–15 cm **R lf**
Dull rufous top of head; rusty-buff supercilium; longish tail, black spots only visible from below. ▲ Rocky outcrops, savanna with boulders. ✤ Slow series of squeaky, metallic notes interspersed with short dry trills or rapid series of clicking notes. [CD10:86]

2 WHISTLING CISTICOLA *Cisticola lateralis* 12.5–14 cm **R c/f**
2a Sooty form Greyish-brown above; flight feathers edged rufous. **2b Foxy form** Duller rufous-brown. ▲ Woodland, derived savanna. ✤ Short, vigorous and melodious whistled phrase. East of Nigeria a more monotonous *thup thup thuthuthuthuthuw*. [CD10:87]

3 RED-FACED CISTICOLA *Cisticola erythrops* 12–14 cm **R c/u**
Rufous face. ▲ Various grassy and bushy habitats. ✤ Rhythmic succession of loud, varied notes, e.g. *ch-ch-ch trweet-trweet-trweet WEET-WEET-WEET plik-up plik-up plik-up WEET WEET WEET* ...; from low perch and usually in duet. [CD10:90]

4 CHUBB'S CISTICOLA *Cisticola chubbi* 14 cm **R lc/f**
C. c. discolor (Brown-backed Cisticola) Rusty top of head; black lores. ▲ Highlands. Dense herbage in forest clearings, edges, abandoned cultivation, bracken, etc. ✤ Loud, explosive duet of rapidly repeated phrases, e.g *switch-a-bee switch-a-bee* ... [CD10:92-93]

5 RATTLING CISTICOLA *Cisticola chiniana* 12–16 cm **R ls**
Dark rusty-brown top of head; dark brown upperparts; rufous-brown edges to flight feathers. ▲ Dry savanna woodland. ✤ 1–4 harsh notes followed by trill *chi-chi-chrrrr.* [CD10:78]

6 SINGING CISTICOLA *Cisticola cantans* 11.5–14.0 cm **R c/f**
Chestnut top of head and wing panel. ▲ Bush. ✤ Single, repeated, variable note, such as *kwleep krwleep krwleep*... or *p-lip p-lip* ... and *kwiplip kwiplip*... etc. [CD10:91]

7 BUBBLING CISTICOLA *Cisticola bulliens* 12.0–15.5 cm **?**
As 8 but top of head browner and head-sides paler. ▲ Grassy habitats. Status unclear; no certain records. Known range: Cabinda to Angola. ✤ Song described as three notes followed by bubbling flourish *di-di-di drrrreee*; apparently very similar to 8.

8 CHATTERING CISTICOLA *Cisticola anonymus* 12–15 cm **R lc**
As 5 but in different habitat; wings uniform with rest of upperparts. ▲ Grassy patches with bushes in forest clearings, low second growth, farmland etc. ✤ 2–3 harsh notes followed by bubbling trill *ch-ch-twurrrrlp* and fast, abrupt *tetete-tchrr*. Similar to 5. [CD10:89]

Streaked upperparts:

9 TINKLING CISTICOLA *Cisticola rufilatus* 13.0–14.5 cm **R s/lc**
Rufous top of head and tail; buff supercilium. ▲ Open grassland with scattered trees. ✤ Clear piping whistles *hweee-hweee-hweee-*... often followed by bill-snapping and trill. [CD10:77]

10 STOUT CISTICOLA *Cisticola robustus* 14.0–16.5 cm **R lc**
Rufous cap (streaked) and nape (unstreaked). ▲ Rank grass with shrubs. ✤ Short, piping ripple *tsri tsri tsrrrrrrr*, usually from low perch; occasionally in low display flight. [CD10:76]

11 WINDING CISTICOLA *Cisticola galactotes* 12–15 cm **R lc/f**
11a Ad breeding Forehead dull rust-brown; russet-edged flight feathers. **11b Ad non-breeding** Warmer coloured. ▲ Moist habitats. ✤ Drawn-out rasping note *zrrrreeeeeee*. [CD10:95]

12 CROAKING CISTICOLA *Cisticola natalensis* 12.5–16.0 cm **R c/s**
12a Ad breeding Large and bulky; upperparts dark earth-brown streaked dusky; almost plain when worn. **12b Ad non-breeding** Upperparts buffish-brown boldly streaked black. ▲ Grassy habitats. ✤ Loud *klink klunk* in display flight; and *kzeee klunk!* when perched [CD10:75]

Maps on page 366

1
2b
2a
3
4
5
6
7
8
9
10
12a
11b
12b
11a

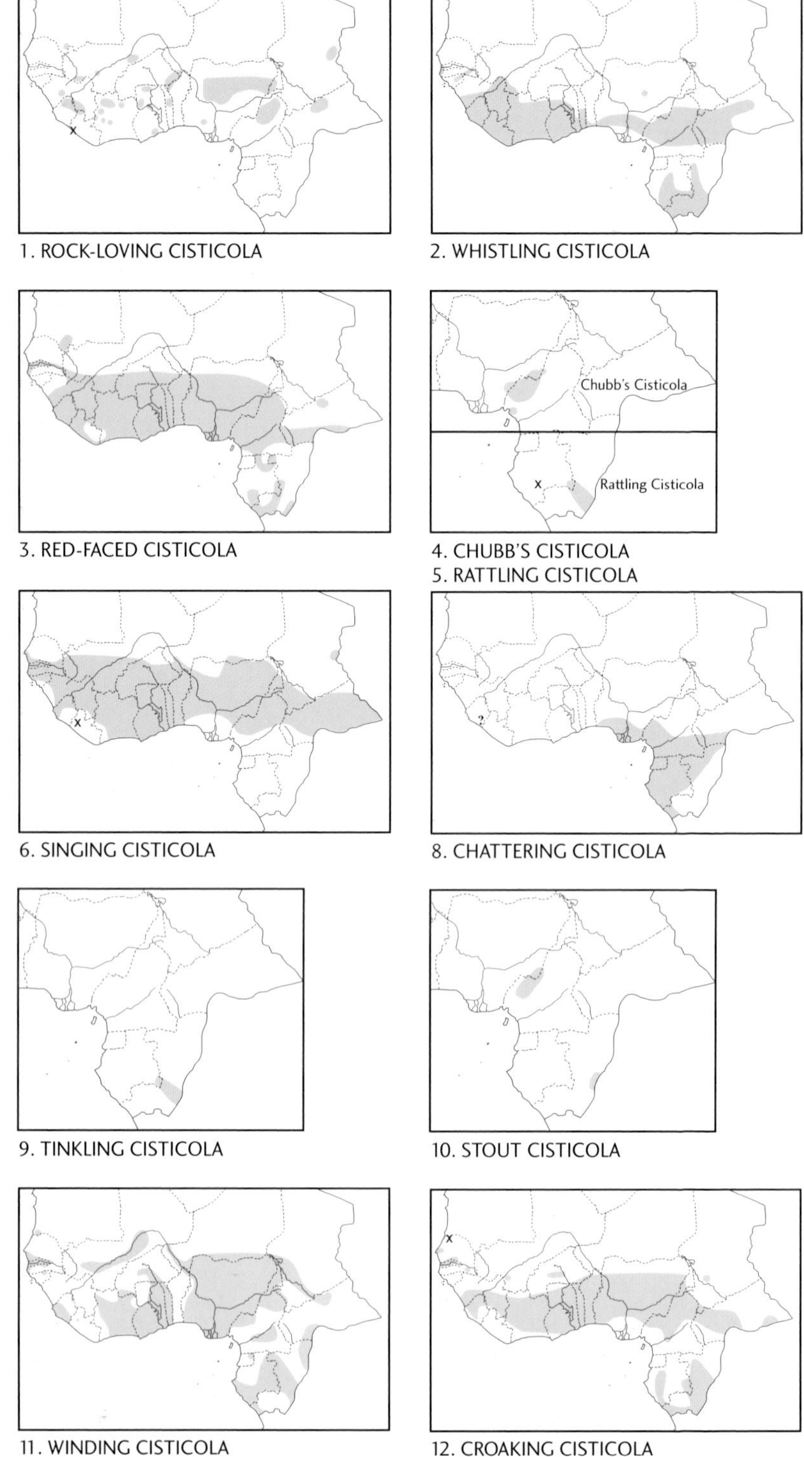

1. ROCK-LOVING CISTICOLA
2. WHISTLING CISTICOLA
3. RED-FACED CISTICOLA
4. CHUBB'S CISTICOLA
5. RATTLING CISTICOLA
6. SINGING CISTICOLA
8. CHATTERING CISTICOLA
9. TINKLING CISTICOLA
10. STOUT CISTICOLA
11. WINDING CISTICOLA
12. CROAKING CISTICOLA

Plate on page 364

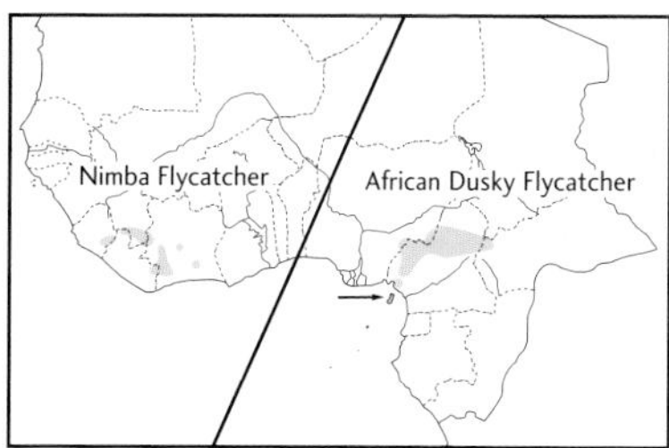

1. NIMBA FLYCATCHER
6. AFRICAN DUSKY FLYCATCHER

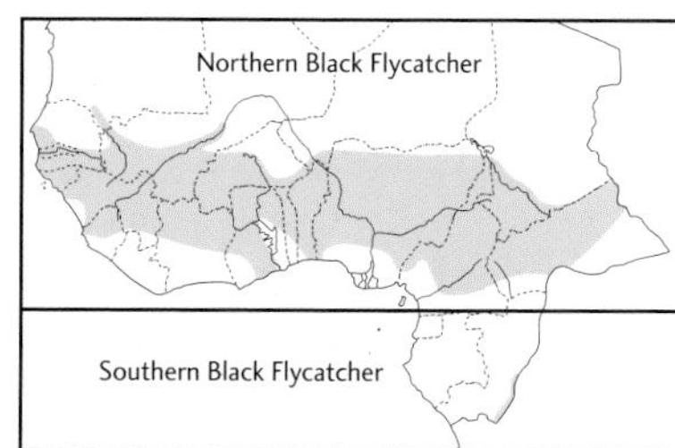

2. NORTHERN BLACK FLYCATCHER
3. SOUTHERN BLACK FLYCATCHER

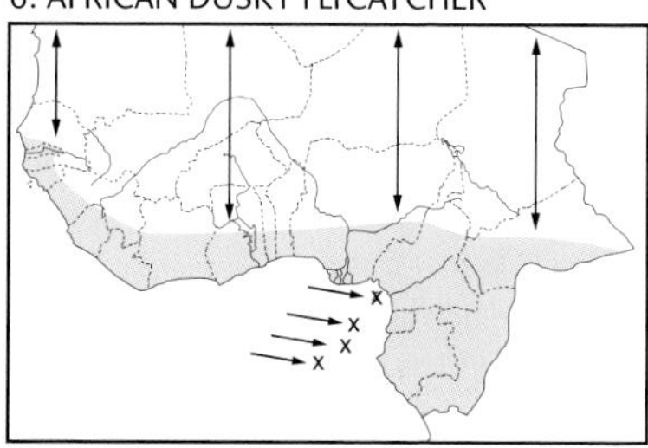

4. SPOTTED FLYCATCHER

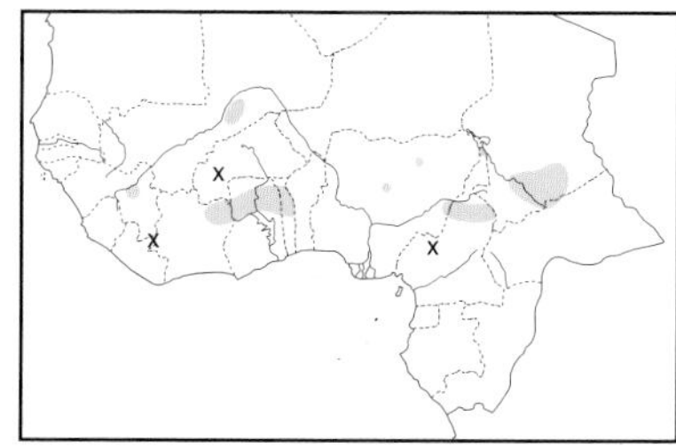

5. GAMBAGA FLYCATCHER

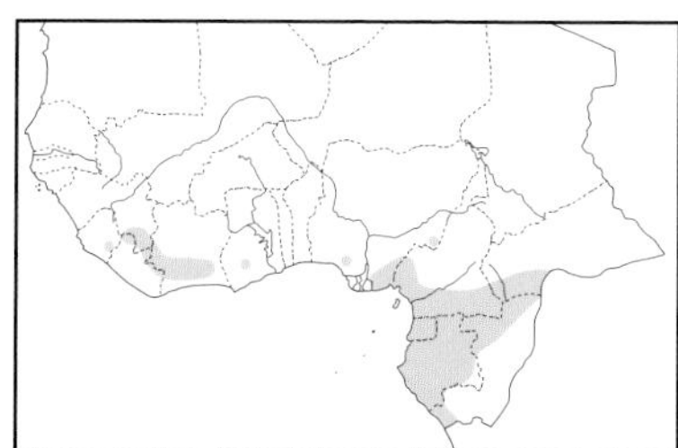

7. OLIVACEOUS FLYCATCHER

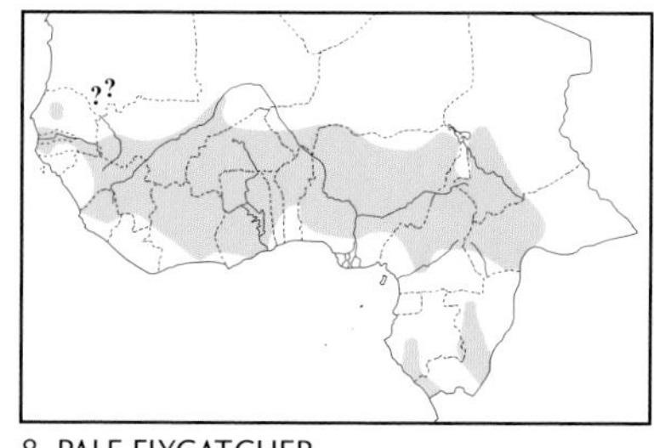

8. PALE FLYCATCHER

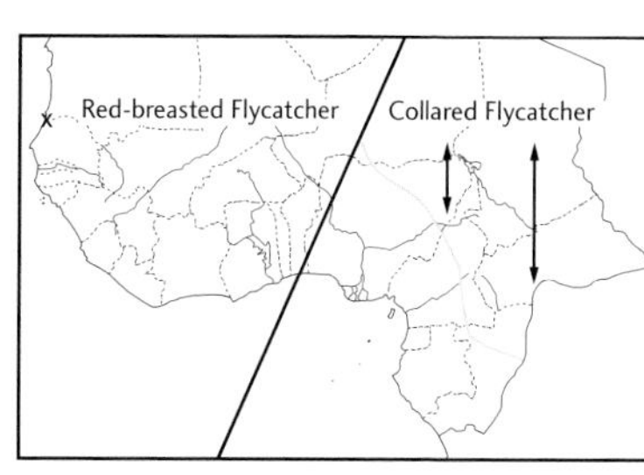

9. RED-BREASTED FLYCATCHER
13. COLLARED FLYCATCHER

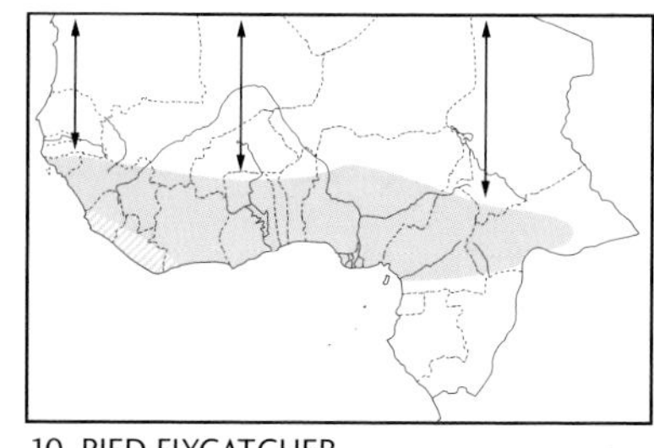

10. PIED FLYCATCHER

Continued from page 368

12 SEMI-COLLARED FLYCATCHER *Ficedula semitorquata* 13 cm **P**
No certain records. Almost intermediate between 10 and 13. **12a Ad male breeding** White half-collar (variable); second upper wingbar; more white in tail than 10 and 13. **12b Ad male non-breeding/ad female** As 13b but median coverts often white-tipped. ✤ [CD11:71]

13 COLLARED FLYCATCHER *Ficedula albicollis* 13 cm **P r/s**
13a Ad male breeding White collar; large white patch on forehead; whitish rump; all-black tail. **13b Ad male non-breeding/ad female** As 10b but primary patch larger. ▲ Distribution inadequately known. ✤ Mostly silent. Thin, sharp *seep*. [CD4:38]

Insectivorous and arboreal. Most have broad, flattened bills with broad gape and rictal bristles. Catch insects in short sallying flights from perch, by pounding on prey on ground or by picking it from foliage in warbler-like manner. Some join mixed-species flocks.

1 NIMBA FLYCATCHER *Melaenornis annamarulae* *c.* 19 cm R lr/s VU
Very dark plumbeous; robust. ▲ Forest. Endemic. ✤ Thin, soft *wheep-wheep* and harsh notes. Song: short, varied phrases of pleasant, melodious whistles. [CD11:67]

2 NORTHERN BLACK FLYCATCHER *Melaenornis edolioides* *c.* 20 cm R u/f
All black; long tail. ▲ Woodland. ✤ Rather silent. High-pitched *tseeeu*; also *tsik* and long churring note (alarm). Song a series of varied, mostly melodious phrases. [CD11:68]

3 SOUTHERN BLACK FLYCATCHER *Melaenornis pammelaina* 19–22 cm R lr
Similar to 2, but plumage glossed steel-blue. ▲ Wooded grassland. ✤ Rather silent. Song a series of thin notes forming short, varied, quite melodious phrases. [CD11:66]

4 SPOTTED FLYCATCHER *Muscicapa striata* 13.5–14.5 cm P/V+ c/f
Grey-brown; streaked crown and breast. ▲ Various wooded habitats. ✤ Mostly silent. Thin, scratchy *zreeht*; alarm *tik* or *tsee-tik*. [CD4:40]

5 GAMBAGA FLYCATCHER *Muscicapa gambagae* 12–13 cm R r/s
Resembles 4, but smaller; streaks very faint; wings shorter. ▲ Wooded grassland. ✤ Mostly silent. Call a sharp *tzik*. [CD11:72]

6 AFRICAN DUSKY FLYCATCHER *Muscicapa adusta* 10–11 cm R lf/r
6a Ad Smaller than 5; underparts darker; throat whitish. **6b Juv** Spotted above, mottled below. ▲ Various wooded habitats in highlands. ✤ Mostly silent. Thin, high-pitched *seeeeu*; sharp *trt-trt-trrtrrt*. Song an unhurried series of sharp notes *tsr tsit tsr tsr tsitit* ... [CD11:73]

7 OLIVACEOUS FLYCATCHER *Muscicapa olivascens* 14 cm R r/lf
Rather nondescript; mainly olivaceous-brown. ▲ Forest. ✤ Rather silent. Thin *seeeee* and *wit-wit*. Song: high-pitched whistles and jumble of buzzy notes and trills. [CD11:75]

8 PALE FLYCATCHER *Melaenornis pallidus* 15–17 cm R f
Medium-sized. Rather nondescript, brownish. ▲ Various types of woodland. ✤ Mostly silent. Soft *churr* and *see-see*. Song a jumble of harsh churring notes. [CD11:63]

9 RED-BREASTED FLYCATCHER *Ficedula parva* 11.5 cm V
Bold white sides to base of blackish tail (diagnostic). **9a Ad male breeding** Orange-red throat; grey-brown head. **9b Ad female** Creamy-buff throat; brown head. ▲ Palearctic vagrant. ✤ Dry rattle *tzrrr*, single *tzk*, soft, plaintive *tulee*, weak *shrrr*. [CD4:39]

The following four Palearctic migrants are very similar in male non-breeding/female plumages and often unidentifiable in the field. Note in particular wing and tail patterns, and rump colour.

10 PIED FLYCATCHER *Ficedula hypoleuca* 13 cm P c/s
10a Ad male breeding White wing patch; small white patch on forehead; very small primary patch; white edges to tail. **10b Ad male non-breeding/ad female** Brownish above; white wingbar narrower. ▲ Various wooded habitats. ✤ Sharp *whit*. [CD4:37]

11 ATLAS FLYCATCHER *Ficedula (hypoleuca) speculigera* 13 cm P
Ad male breeding As 10a but white forehead patch larger and more white in wing, thus similar to 13a, but lacks white collar (although some possess half-collar like 12a).

Continued on page 367

1
2
3
4
5
6b
6a
7
8
9b
9a
10a
10b
11
12a
12b
13a
13b

PLATE 112: MONARCHS

Diverse family of insectivorous, arboreal species with crests or incipient crests, and relatively long, graduated tails. Very active, gleaning most prey from branches in middle and lower levels, often brushing foliage with fanned tail and partially open wings to dislodge insects. Often in mixed-species flocks.

Genus *Terpsiphone* (paradise flycatchers). Distinctive. Some males have extremely elongated median tail feathers when breeding. Females as males but lack tail streamers. Some extremely variable and readily hybridise. Vocal and conspicuous. Call a rasping *zwhee-zwhèh*.

1 AFRICAN PARADISE FLYCATCHER *Terpsiphone viridis* 18 cm **R/M lc/r**
Glossy blue-black crested head. Adult males with very long, ribbon-like tail streamers (+10–18 cm). Upper- and underparts very variable; two extremes illustrated. Adult females have upperparts and tail always rufous; tail shorter. **1a Ad male rufous 'morph'**. **1b Ad male white 'morph'**. **1c Ad male *plumbeiceps*** (non-breeding visitor from south of equator, reaching S Cameroon) Duller than 1a; no white on wing. ▲ Savanna woodland, gallery forest, second growth, farmland. ✤ Song variable, typically loud and cheerful, e.g. *twee-twee-twee-twee twee-twee-twee-twee*. [CD12:28]

2 RED-BELLIED PARADISE FLYCATCHER *Terpsiphone rufiventer* 18 cm **R c**
Rufous underparts. Tail of male + max. 7 cm. **2a Ad male *rufiventer*** (Senegambia–W Guinea) Distinct crest, white wing panel, long tail streamers. **2b Ad male *nigriceps*** (Guinea–SW Benin) No crest, no wing panel, much shorter tail. *T. r. fagani* (Benin–SW Nigeria) and *ignea* (SE CAR) similar. In *schubotzi* (SW CAR–N Congo) and *mayombe* (S Congo) tail bluish-slate. **2c Ad male *neumanni*** (SE Nigeria–S Congo). Bluish-slate upperparts and tail. *T. r. tricolor* (Bioko) similar. ▲ Forest, second growth, cocoa plantations, thickets in savanna. ✤ Calls and songs typical of genus, including harsh, scolding *zwhee-zwhèh* and short, cheerful *tweedweedwee tweedweedwee*. [CD12:32]

ANNOBÓN PARADISE FLYCATCHER *Terpsiphone (rufiventer) smithii* See Plate 147:5

3 RUFOUS-VENTED PARADISE FLYCATCHER *Terpsiphone rufocinerea* 18 cm **R c**
Dark greyish-blue underparts; rufous undertail-coverts. Tail of male + 1–11 cm. ▲ Forest, woodland. ✤ Calls and songs typical of genus, including harsh, rasping *zwhee-zwhèh* and cheerful *thululululululu* or *twee-twee-twee-twee*. [CD12:30]

4 BATES'S PARADISE FLYCATCHER *Terpsiphone batesi* 18 cm **R r/c**
As 3 but less glossy and lacking crest. Tail of male + 1.0–10.5 cm. ▲ Forest. ✤ Song a cheerful *twee-twee-twee-twee twee-twee-twee-twee* similar to 2, but typically higher pitched and slightly faster than more variable songs of 1 and 3. [CD12:29]

5 AFRICAN BLUE FLYCATCHER *Elminia longicauda* 18 cm **R f**
Bright pale blue; short crest; long, graduated tail. ▲ Woodland. ✤ Various little call notes, including *dzp* and *tsluip*. Song a jumble of sibilant notes. [CD12:21]

6 CHESTNUT-CAPPED FLYCATCHER *Erythrocercus mccallii* 10 cm **R f**
Very small; rufous crown and tail. ▲ Lowland forest. Upper and middle strata. ✤ High-pitched little notes, frequently uttered. Song a fast, sibilant twittering. [CD12:20]

7 WHITE-BELLIED CRESTED FLYCATCHER *Elminia albiventris* 11 cm **R f**
Dark slate; dull black head; white belly. ▲ Montane forest. ✤ Striking, sharp *pink* or *slip*. Song a soft, rather hesitant, melodious simple warble. [CD12:23]

8 DUSKY CRESTED FLYCATCHER *Elminia nigromitrata* 11 cm **R u/lf**
Dark blue-slate; dull black crown. ▲ Lowland forest. ✤ Fast nasal sound, soft *tsep* or *tsi-ep*, *ptee-diew-diew* etc. Song fast, highly varied and sustained, including mimicry. [CD12:25]

9 BLUE-HEADED CRESTED FLYCATCHER *Trochocercus nitens* 15 cm **R s/f**
9a Ad male Glossy blackish-blue with clearly contrasting grey underparts; short glossy crest. **9b Ad female** Much greyer; only crest glossy blackish-blue; underparts wholly grey. ▲ Lowland and mid-elevation forest. ✤ Harsh 2-syllable *zwhee-zwheh*. Song a very fast, hollow, far-carrying *hohohohohohoho* often preceded by some harsh notes or rapid *tiktiktiktiktik*. [CD12:27]

Maps on page 376

1b
1a
2c
2b
1c
2a
3
4
5
6
7
9b
8
9a

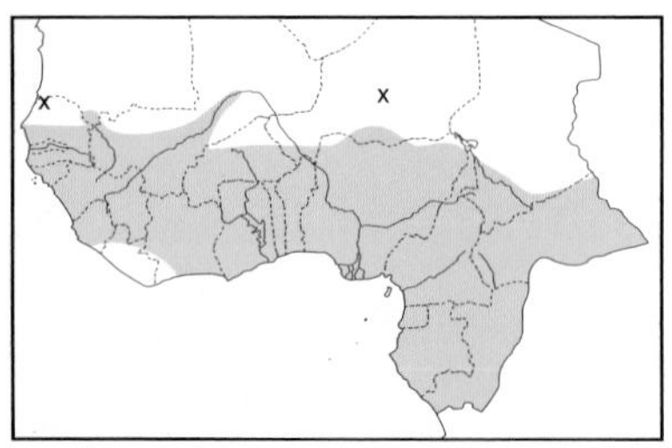
1. AFRICAN PARADISE FLYCATCHER

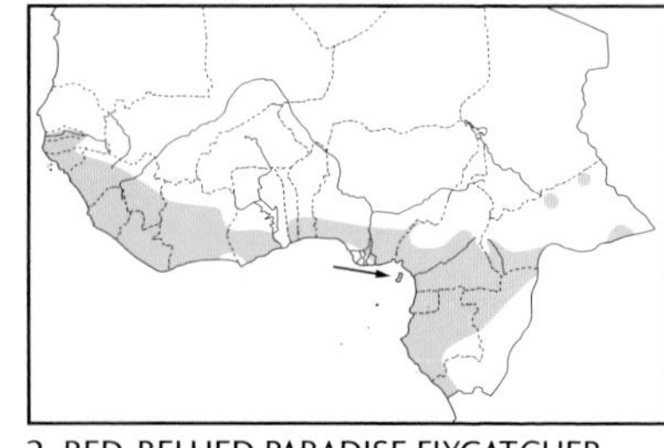
2. RED-BELLIED PARADISE FLYCATCHER

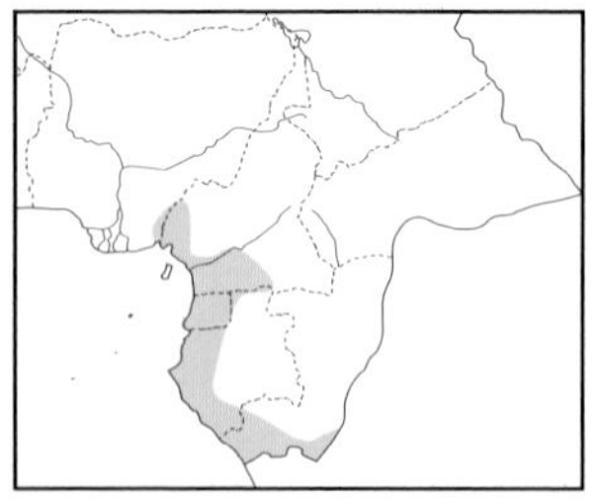
3. RUFOUS-VENTED PARADISE FLYCATCHER

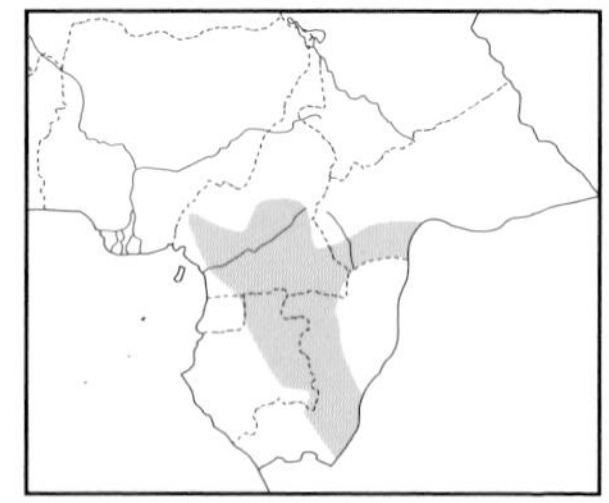
4. BATES'S PARADISE FLYCATCHER

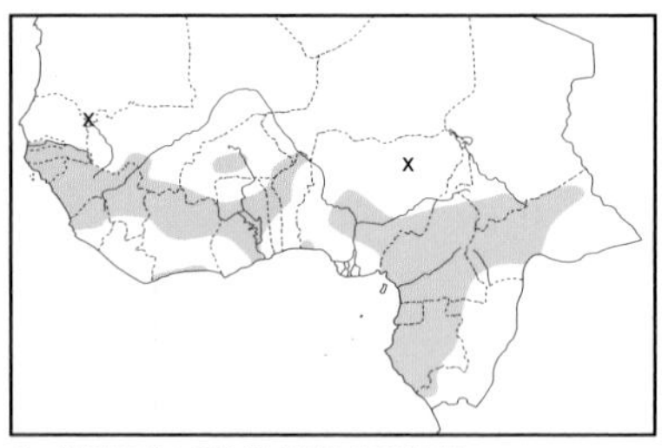
5. AFRICAN BLUE FLYCATCHER

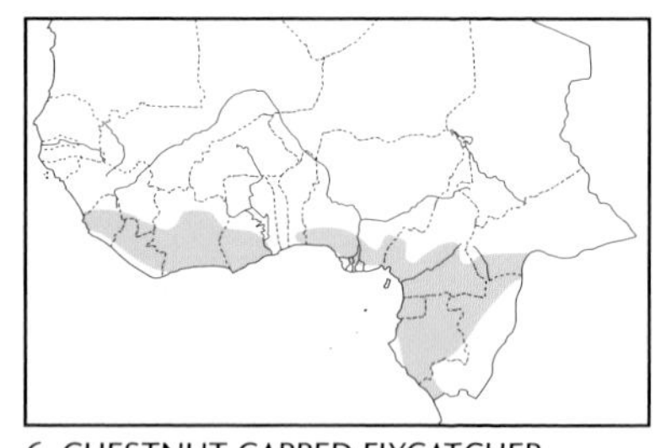
6. CHESTNUT-CAPPED FLYCATCHER

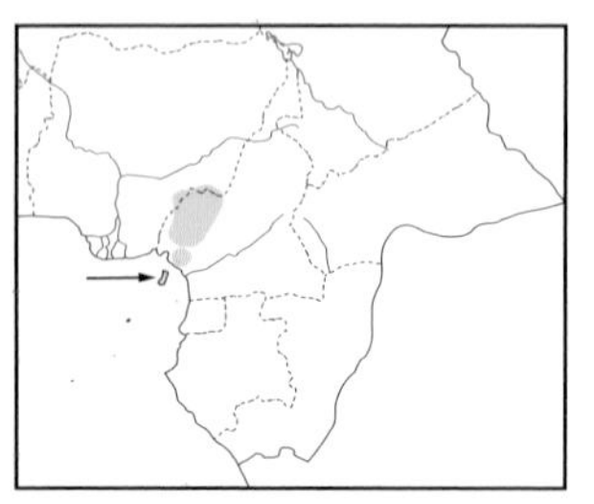
7. WHITE-BELLIED CRESTED FLYCATCHER

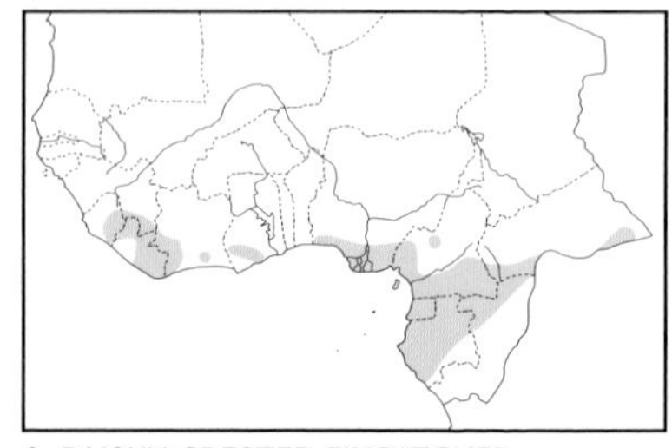
8. DUSKY CRESTED FLYCATCHER

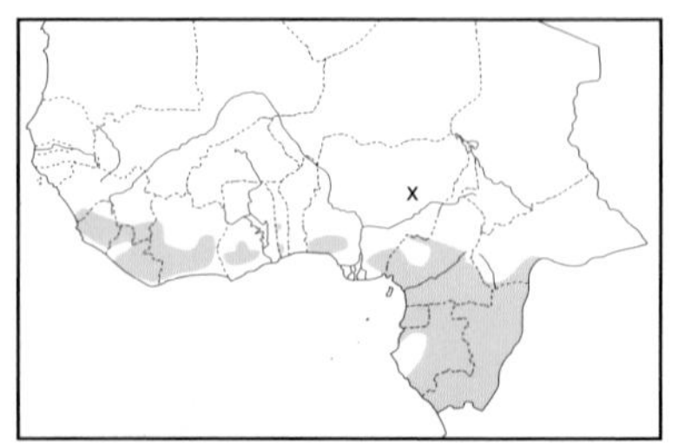
9. BLUE-HEADED CRESTED FLYCATCHER

Plate on page 374

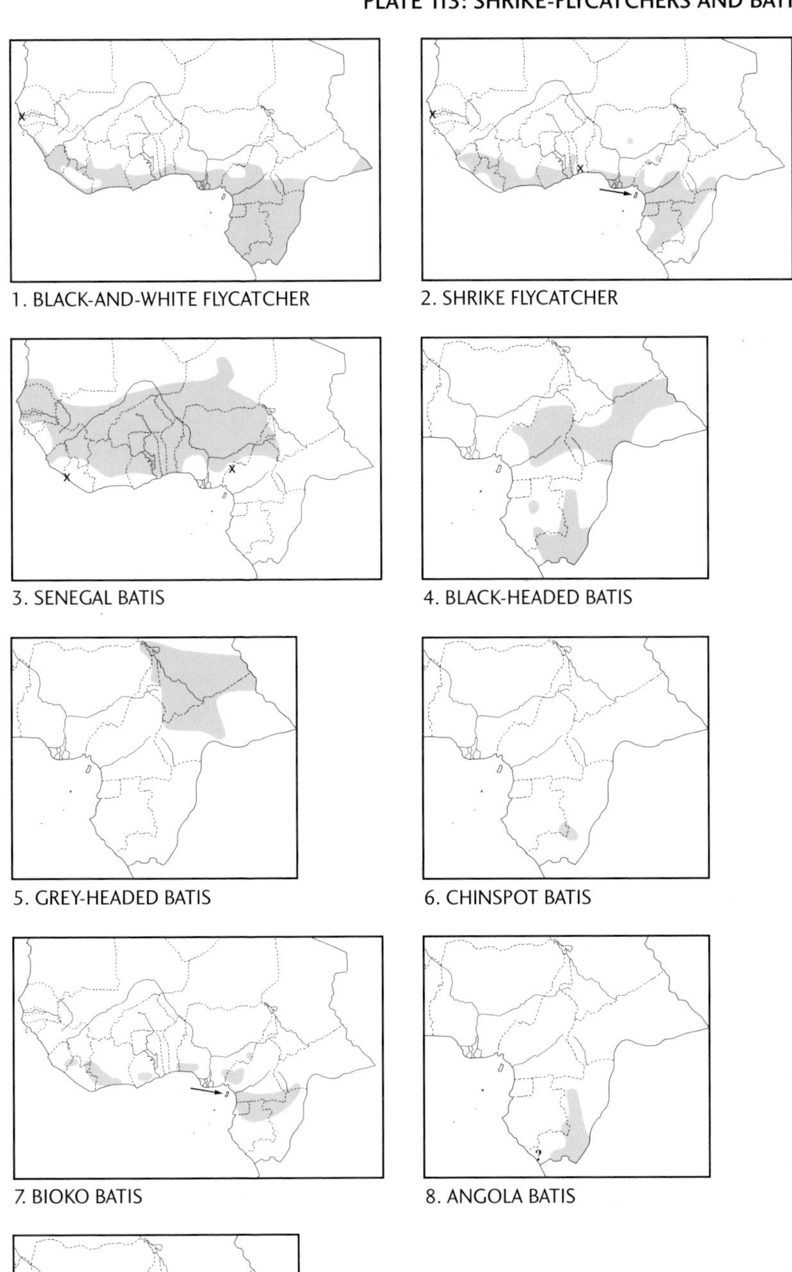

1. BLACK-AND-WHITE FLYCATCHER

2. SHRIKE FLYCATCHER

3. SENEGAL BATIS

4. BLACK-HEADED BATIS

5. GREY-HEADED BATIS

6. CHINSPOT BATIS

7. BIOKO BATIS

8. ANGOLA BATIS

9. VERREAUX'S BATIS

Plate on page 378

1 BLACK-AND-WHITE FLYCATCHER *Bias musicus* 16 cm R u
1a Ad male Glossy black and white; prominent crest; yellow eye. **1b Ad female** Mainly rufous-chestnut above. **1c Ad male in flight** Broad rounded wings with white patch; short tail. ▲ Forest edge. ✤ Vocal. Song, from perch or in flight, a loud, melodious *wheet-tee-tee-tiuw-tiuw* with variations. Call a harsh *tchèèèp* and *tk tk tk*. [CD12:2]

2 SHRIKE FLYCATCHER *Megabyas flammulatus* 15–16 cm R u
2a Ad male Glossy black above; white rump; white below. **2b Ad female** Earth-brown above; rufous rump; coarsely streaked brown below. ▲ Forest. Moves tail sideways. ✤ Song: characteristic *chewee-cheweeet* or *chuah-chuwheesiu*, short, rather melodious trill and fast *whee-whee-whee-chu*, alternating with bill snapping. [CD12:1]

Genus *Batis*. Small, active, arboreal species with relatively large heads and short tails. Flight swift and bouncing, often on whirring wings. Snap bill loudly when catching insects.

3 SENEGAL BATIS *Batis senegalensis* 10.5 cm R lc/f
3a Ad male Long broad supercilium. **3b Ad female** Supercilium and wingbar rusty-buff; pale chestnut breast-band. ▲ Open woodland. Endemic. ✤ Song a harsh buzzing note preceded or occasionally followed by 1–2 short notes *whut-tzeet* and *tzit-tzit-zheet* or *zheet-tit*, with variations. Calls include *tek-tek-tek-*..., and various buzzy and croaking notes. [CD12:6]

4 BLACK-HEADED BATIS *Batis minor* *c.* 11.5 cm R u/lc
4a Ad male As 5a, but crown darker grey; width of breast-band variable, usually slightly broader. **4b Ad female** As 5b, but breast-band usually slightly broader and paler chestnut. **4c Imm** As adult female, but with rusty-buff tinge above, esp. on supercilium and wingbar. Other immature batises have similar plumage. ▲ Open woodland, orchard bush. ✤ Song a series of loud, penetrating, identical whistles *heet-heet-heet-heet-*..., in flight with whirring wings and from perch; speed variable. Call a buzzing *dzip dzip*... [CD12:8]

5 GREY-HEADED BATIS *Batis orientalis* 10 cm R f/r
5a Ad male As 4a, but crown paler; width of breast-band variable, usually slightly narrower. **5b Ad female** As 4b, but breast-band usually slightly narrower and darker chestnut. ▲ Open woodland, thorn scrub. ✤ Song an endless series of clear, whistled notes *hee hee-hu hee-hu hee-hu-hu-hu hee-hu-hu-hu-hu-*... Call a buzzy *dzek-dzek*. [CD12:7]

6 CHINSPOT BATIS *Batis molitor* 12 cm R ls/r
6a Ad male Crown grey; narrow supercilium. Compare 4a (and 5a). **6b Ad female** Large chestnut throat patch. ▲ Wooded savanna. ✤ Song of male a vigorous, whistled, descending 2–3-note *hee-hu* or *hee-hee-hu* with variations; female joins with *wik* notes. Calls include *chik* (contact) and rapid *ch-ch-ch-ch* (alarm). Also snaps bill. [CD12:5]

7 BIOKO BATIS *Batis poensis* *c.* 12 cm R r/lf
7a Ad male *occulta* (West African Batis) Crown blackish; distinct supraloral spot; indistinct narrow supercilium. **7b Ad female *occulta*** Well-defined chestnut breast-band. ▲ Forest. Endemic. ✤ Song: *tee-tee-tee-trrruuu tu-tee-tee-tee*, second part often lacking; variation *trrruu-hu-hee-hee-hee* and *trrru-hu trrru-hu*...; rhythmic *trrip-trrip-trrip-*...; and long series of very high-pitched, penetrating *heet-heet-heet-*... Call a dry *trrr*. [CD12:11]. Bioko form ***poensis***: see Plate 148:11.

8 ANGOLA BATIS *Batis minulla* 10 cm R u/lc
8a Ad male No supercilium, broad breast-band. **8b Ad female** Bright chestnut breast-band broadening on flanks. ▲ Vestigial and fringing forest. ✤ Song a series of weak, high-pitched, sucking *heep heep heep*...; speed variable. Call: little buzzing notes. [CD12:9]

9 VERREAUX'S BATIS *Batis minima* 10 cm R ls NT
9a Ad male As 7a but supraloral spot indistinct, no supercilium, smaller size. **9b Ad female** Slate-grey breast-band. ▲ Forest. ✤ Song a series of high-pitched, identical whistles *heet-heet-heet* -...; speed variable. Various buzzing and clucking calls. [CD12:10]

Maps on page 377

1c
1b
1a
2b
2a
4c
3b
4b
5b
3a
4a
5a
6b
7b
8b
9b
6a
7a
8a
9a

Genus *Platysteira*. Distinguished by conspicuous scarlet eye-wattles. Behaviour similar to batises, but jizz different, due to longer tail and larger size. Snap bill and wings.

1 COMMON WATTLE-EYE *Platysteira cyanea* 13 cm R c
1a Ad male Scarlet eye-wattle, black breast-band, white wingbar. **1b Ad female** Throat and upper breast dark chestnut. **1c Imm** Olive-grey above; wingbar rusty-buff; buffish wash on breast and flanks. ▲ Various wooded habitats. ✤ Song a diagnostic, far-carrying series of clear, melodious whistles *hee-hu-huu-ho hee-hu-ho*; also variations. Various harsh low calls, often included in song. [CD12:12]

2 BANDED WATTLE-EYE *Platysteira laticincta* 13 cm R lf EN
2a Ad male As 1a but without wingbar. **2b Ad female** Throat and upper breast blue-black. ▲ Montane forest, W Cameroon (Bamenda Highlands). Endemic. ✤ Calls include soft, rather grating note.

Genus *Dyaphorophyia*. Small, colourful, very short-tailed species with blue, purplish or green eye-wattles. Active; often joining mixed-species flocks. Snap bill and wings.

3 WHITE-SPOTTED WATTLE-EYE *Dyaphorophyia tonsa* 9.5 cm R r/lf
3a Ad male As 4c but with short white superciliary streak (partly hidden by eye-wattle). **3b Ad female** As 4b but with glossy black crown, partly hidden superciliary streak, longer malar stripe. ▲ Upper levels of mature forest. ✤ Song: series of whistled notes *hu-hee-hu-hu-hu hu-hee-hu ...*, rhythmic series of identical notes *hu-hu-hu-...* or *ut ut ut ut ...* and *hee-hee-hee-...*, and series of *ptok-ut*. Calls a soft *yup* and *ptek*. [CD12:19]

4 CHESTNUT WATTLE-EYE *Dyaphorophyia castanea* 10 cm R c/u
4a Ad male *castanea* (Nigeria to S CAR–Congo; Bioko) Glossy blue-black above; white rump; broad breast-band; purplish eye-wattle. **4b Ad female *castanea*** Slate-grey head; chestnut upperparts, throat and breast. **4c Ad male *hormophora*** (Guinea–Benin) White neck collar. **4d Juv** Paler than adult female with broad, diffuse mottled band of grey and chestnut on upper breast. ▲ Mid-strata of forest. ✤ Varied. Song: rhythmic series of *whop pEEEE* (in west), sharp *ptik-ptik-ptik-...* and *klonk-klonk-klonk-...* (in east), *ptik-kwonk* or *whep-pleenk*, high-pitched *hit-hit-hit-...* etc. Call a low *kwonk*, soft *wop*, and nasal notes. [CD12:18]

5 BLACK-NECKED WATTLE-EYE *Dyaphorophyia chalybea* 9 cm R lf
Ad male Glossy greenish-black above; white underparts with pale yellow cast; emerald-green eye-wattle. Female similar but duller. ▲ Undergrowth in lowland and montane forest. ✤ Similar to 6, except advertising song, an excited *ptiukteehee* or *ptiuk-ti-hi hihihu* followed by 4–6 descending sweet notes, reminiscent of 1. Also long series of whistles when excited. Flight call *kwek-kwek-...* accelerating to *kwedekwedekwedek...* [CD12:17]

6 RED-CHEEKED WATTLE-EYE *Dyaphorophyia blissetti* 9 cm R u/lf
6a Ad male Glossy greenish-black above; chestnut patch on cheeks and neck-sides; greenish-blue eye-wattle. **6b Ad female** Duller. **6c Juv** Throat and upper breast tawny bordered chestnut. ▲ Undergrowth in lowland forest. Endemic. ✤ Song a far-carrying series of identical high-pitched notes *hee-hee-hee-hee-...*; speed of delivery and length variable. [CD12:15]

7 YELLOW-BELLIED WATTLE-EYE *Dyaphorophyia concreta* 10 cm R r/lc
7a Ad male *concreta* (Sierra Leone–SE Guinea to Ghana) Rich chestnut underparts; bright emerald-green eye-wattle. **7b Ad female *concreta*** Bright yellow underparts; throat and upper breast chestnut. **7c Ad male *graueri*** (SE Nigeria to SW CAR–Congo) Deep orange-yellow underparts. **7d Ad female *graueri*** Throat and upper breast have chestnut wash. **7e Juv** Greyish-olive above; dirty white washed olive and dull yellowish below. ▲ Undergrowth in lowland and montane forest. ✤ Song: rhythmic series of high-pitched, whistled note(s) followed by lower note(s) *hee-hee-hwot-hwot* or *hee-hu hee-hu... whot* and *heet whot*. Also fast series of identical notes *hwrit-hwrit-hwrit -...* Calls include various hoarse and guttural notes. [CD12:14]

Maps on page 382

1c
1b
1a
2b
2a
3b
3a
4d
4b
4c
4a
5
6c
6b
6a
7b
7e
7d
7a
7c

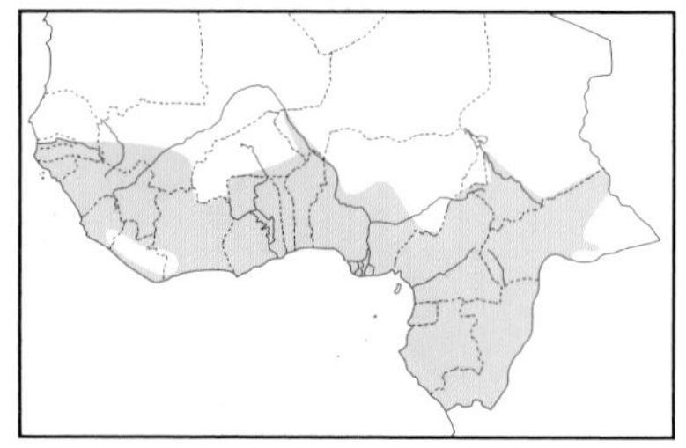
1. COMMON WATTLE-EYE

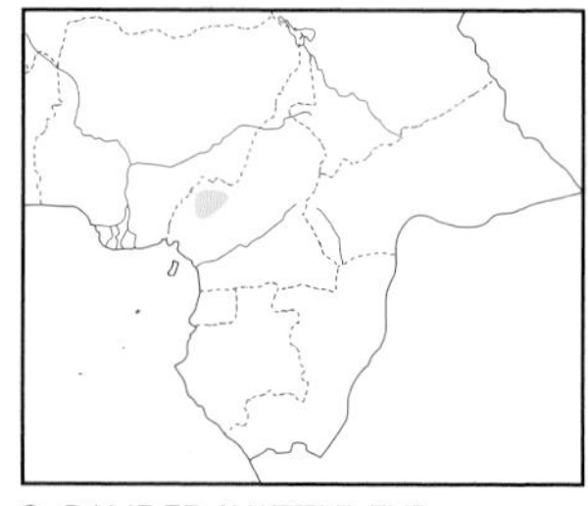
2. BANDED WATTLE-EYE

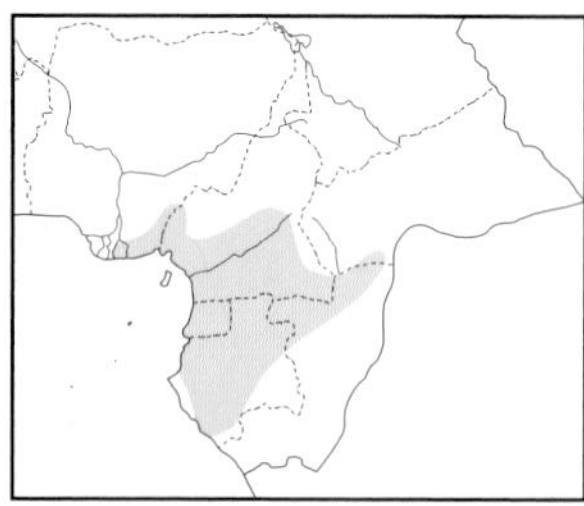
3. WHITE-SPOTTED WATTLE-EYE

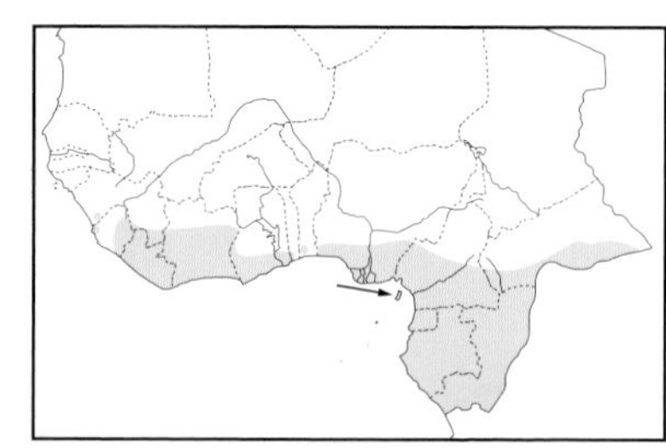
4 CHESTNUT WATTLE-EYE

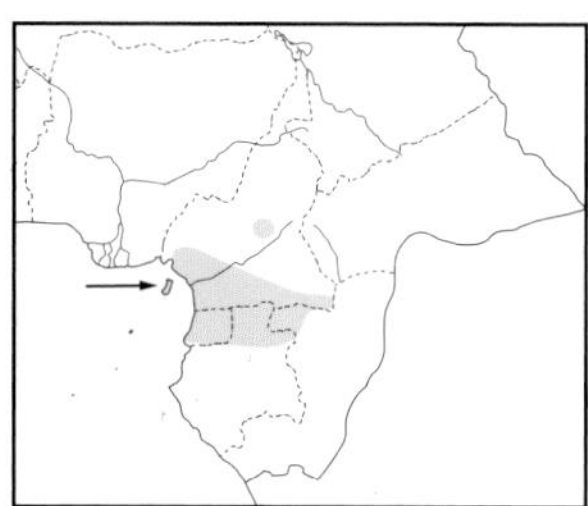
5. BLACK-NECKED WATTLE-EYE

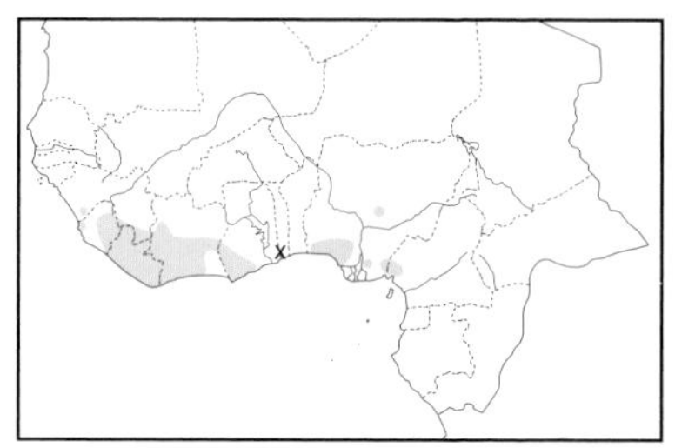
6. RED-CHEEKED WATTLE-EYE

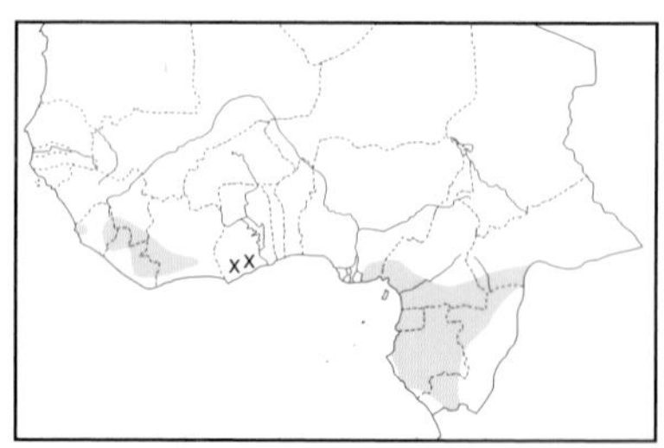
7. YELLOW-BELLIED WATTLE-EYE

Plate on page 380

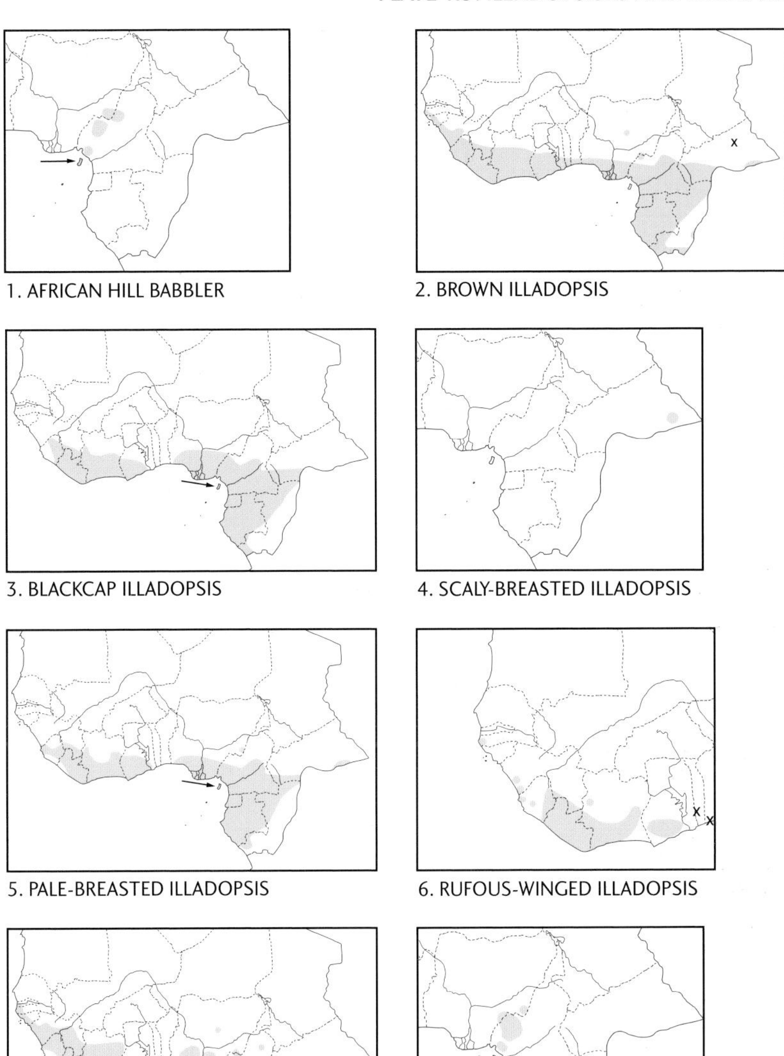

1. AFRICAN HILL BABBLER

2. BROWN ILLADOPSIS

3. BLACKCAP ILLADOPSIS

4. SCALY-BREASTED ILLADOPSIS

5. PALE-BREASTED ILLADOPSIS

6. RUFOUS-WINGED ILLADOPSIS

7. PUVEL'S ILLADOPSIS

8. GREY-CHESTED ILLADOPSIS

PLATE 115: ILLADOPSISES AND HILL BABBLER

Unobtrusive forest dwellers, mostly foraging low. Plumages often similar; voice key to identity. Songs, uttered in duet or in groups, consist of pure, far-carrying whistles that are hard to locate.

1 **AFRICAN HILL BABBLER** *Pseudoalcippe abyssinica* 14–15 cm **R lc**
1a *P. a. atriceps* (Ruwenzori Hill Babbler; E Nigeria–W Cameroon) Black head; dark chestnut above; slate-grey below. **1b *P. a. monachus*** (Mt Cameroon) Slate-grey head. *P. a. claudei* (Bioko) similar. ▲ Montane forest. ✤ Rich, clear, melodious warbling song, thrush-like and reminiscent of Blackcap or Garden Warbler. Also low, guttural chatter and disyllabic call. [CD12:41]

2 **BROWN ILLADOPSIS** *Illadopsis fulvescens* c. 16 cm **R c**
2a *I. f. fulvescens* (Cameroon to SW CAR–Congo) Dark fulvous-brown above; paler below; whitish throat. Compare 5. *I. f. gularis* (from Ghana west) and *ugandae* (SE CAR) similar. **2b *I. f. moloneyana*** (E Ghana–Togo) Wholly fulvous-brown below, darker than 2a. *I. f. iboensis* (S Nigeria–W Cameroon) similar. ▲ Forest, forest-scrub mosaic. ✤ Song a low, rather plaintive even whistle *wHEEEE* followed by pause and another, similar whistle on lower pitch, *wHUUUU*, the whole interspersed with a few short *huit* and *tew* notes (only audible at close range) by other group members. Another song type (in east of range only?) superimposes somewhat twanging notes resembling *dict-a-phone*. Drawn-out whistles easily imitated by human. [CD12:35]

3 **BLACKCAP ILLADOPSIS** *Illadopsis cleaveri* 15 cm **R f**
3a *I. c. cleaveri* (Ghana) Dull black cap; whitish or pale greyish lores and supercilium. *I. c. johnsoni* (west of 3a) darker; *batesi* (SE Nigeria to CAR–Congo) supercilium darker. **3b *I. c. marchanti*** (S Nigeria) Top of head olive-grey. *I. c. poensis* (Bioko) similar. ▲ Forest. Endemic. ✤ Song: 1–2 resonant whistles, usually introduced by 1–2 short notes (only audible at close range), *ptk whit wHU wHEEE*. Also faster, rising series of 3 whistles introduced by 1–2 short notes, *whit wHU-wHEE-wHIII*. Both forms endlessly repeated. Calls include dry *prrt* ... [CD12:37]

4 **SCALY-BREASTED ILLADOPSIS** *Illadopsis albipectus* 15 cm **R lc**
Breast feathers tipped olivaceous-brown giving slightly scaly appearance (hard to see). ▲ Forest. ✤ Similar to 3. Series of 2–3 pure, drawn-out whistles, usually preceded by 1–2 short notes e.g. *wHU-HEEE* or *whit tiuk wHUU wHEE wHIII*..., endlessly repeated. [CD12:38]

5 **PALE-BREASTED ILLADOPSIS** *Illadopsis rufipennis* 15 cm **R u/lc**
As 2a but upperparts slightly darker russet-brown; throat pure white; centre of belly whitish. ▲ Forest. ✤ Resembles 2, but whistled notes higher pitched (thus generally impossible to imitate by human). Two main song types: (1) pure whistle *wHEEE* repeated after short pause and accompanied by rapid series of clucking and hoarse notes e.g. *chuk-kuk-kuk whiz-whiz-wheez-*...; (2) short ascending series of 2–3 pure whistles, introduced and accompanied by short notes, e.g. *whit tew wHUUU wHEEE*, repeated constantly. Call a harsh *chuk* and *tzr*. [CD12:36]

6 **RUFOUS-WINGED ILLADOPSIS** *Illadopsis rufescens* 17 cm **R f/r NT**
Thrush-like. Dark russet-brown above; whitish below; flanks and undertail-coverts olivaceous-grey; strong, pale-coloured legs. Compare 7. ▲ Forest. ✤ Song a distinctive, far-carrying, rhythmic and fast *chk-chk-chk-HU-HU-HU* or shorter *chk-HU-HU*. [CD12:39]

7 **PUVEL'S ILLADOPSIS** *Illadopsis puveli* 18 cm **R u**
As 6 but paler, more rufous above. Voice different. Compare also Spotted Thrush Babbler (Plate 116:7). ▲ Forest edges, gallery forest, dense bush. ✤ Song far carrying, rhythmic and fast, resembling 6 in structure, but higher pitched and with slightly more varied notes and phrases, e.g. *chk-chk-whit HEE-HU-HUU* or *chk-whit-HE-HU-HU-HE-HU*. [CD12:40]

8 **GREY-CHESTED ILLADOPSIS** *Kakamega poliothorax* 17 cm **R r/lc**
Rich chestnut-brown above; grey below; throat and central belly white. ▲ Montane forest. ✤ Song: short series of distinctive, loud, clear, melodious whistles with almost oriole-like quality, e.g. *tchlee tlu tluweeo* or *chee-wee-woo, wee-woo*; also more rapid *trilutruleeo* and similar sounds. Thrush-like chirp call and harsh, chattering alarm. [CD12:33]

 Maps on page 383

1b
2b
2a
1a
3b
3a
4
5
6
7
8

PLATE 116: BABBLERS AND PICATHARTES

Babblers. Sturdy and thrush-like. Most species occur in small, noisy groups on or close to ground.

1 WHITE-THROATED MOUNTAIN BABBLER *Kupeornis gilberti* 21 cm R lc/f EN
Dark brown; face to breast white. ▲ Montane forest. In noisy, active groups in canopy and mid-stratum; occasionally lower. Often perches upside-down while searching for insects in moss, epiphytes and bark crevices. ✤ Harsh, explosive *chak*, usually singly, occasionally in short rapid series; soft *kiorr*. Groups often give harsh concerted chatter. [CD12:48]

2 CAPUCHIN BABBLER *Phyllanthus atripennis* 24 cm R u/f
2a *P. a. atripennis* (west of Ghana) Very dark chestnut; head to upper breast grey; greenish-yellow bill. **2b *P. a. haynesi*** (Ghana–Cameroon) Crown blackish; grey area reduced to head-sides and upper throat. *P. a. bohndorffi* (CAR) has chestnut throat. ▲ Forest edge, gallery forest, thickets. ✤ Short chuckling notes and loud, excited, raucous chattering similar to 4. Also drawn-out whistle *hu-wheew*. [CD12:50]

3 BLACKCAP BABBLER *Turdoides reinwardtii* 25 cm R u/lc
3a *T. r. reinwardtii* (west of Ivory Coast) Brown, with black head; contrasting creamy eye. **3b *T. r. stictilaema*** (from SE Mali–Ivory Coast east) Black of head not sharply defined; underparts darker than 3a. ▲ Bushy habitats. ✤ Harsh and scolding calls, similar to 4, but loud chorus of excited chattering interspersed with less harsh, rather nasal *ko-kwee ko-kwee...* [CD12:46]

4 BROWN BABBLER *Turdoides plebejus* 24 cm R c/f
Greyish-brown above; head-sides, rump and underparts paler; faint whitish squamations on breast. ▲ Bushy habitats. ✤ Harsh and scolding calls. Fast series of chuckling notes uttered simultaneously by several birds result in excited, raucous chattering. [CD12:43]

5 ARROW-MARKED BABBLER *Turdoides jardineii* 24 cm R lu
Slightly darker and better marked than 4. ▲ Bushy habitats. ✤ Similar to 4. [CD12:44]

6 DUSKY BABBLER *Turdoides tenebrosa* 24 cm
Dark brown; scaly forehead, throat and breast. ▲ Wooded savanna; usually near water. No certain records in W Africa. Occurs in Sudan. ✤ Hoarse *chow* and nasal *what-kow*.

7 SPOTTED THRUSH BABBLER *Ptyrticus turdinus* 20 cm R u/lf
Rufous above; white below spotted brown on breast; strong, pale legs. Compare Puvel's Illadopsis (Plate 115:7). ▲ Gallery forest, thickets. ✤ Song with loud, melodious, rather oriole-like whistles, *tiow* or *kiuw*, often preceded by some subdued notes. Recalls song of Grey-chested Illadopsis, but clearer and louder, and even more oriole-like. [CD12:42]

8 FULVOUS BABBLER *Turdoides fulva* 25 cm R lf/r
Wholly sandy-buff; long tail. ▲ Arid scrub country. ✤ Varied. Rather fast, slightly descending series of 6–9 whistled notes introduced by a more drawn-out one *peeeew peew-peew-peew-peew-peew-peew*. Also metallic trill *rirrrrrrrrrr*, short *pip* and clear *pee*. Alarm a sharp *pwit*. [CD4:41]

Picathartes. Strange-looking, slender forest birds with bare head and long, strong legs. Fast and agile, progressing in long springing hops. Dependent on caves or overhanging rocks for breeding (hence alternative name 'rockfowl'). Nest is a bowl of mud plastered to rock face. Often in small colonies. Remains close to nest-site year-round. Secretive but not shy.

9 YELLOW-HEADED PICATHARTES *Picathartes gymnocephalus* c. 38 cm R ls VU
Head yellow with large black patch; neck and underparts white. ▲ Forest with caves and rocky outcrops. Joins mixed-species flocks; attends ant-swarms. Endemic. ✤ Mostly silent. [CD12:53]

10 RED-HEADED PICATHARTES *Picathartes oreas* c. 38 cm R s/lf VU
Head blue, black and crimson; neck greyish; underparts yellowish-buff. ▲ As 9. Endemic. ✤ Usually silent. Alarm a muffled, hushing sound *kshhhhhhhhhhh*. [CD12:54]

Maps on page 388

1
2a
2b
3a
3b
4
5
6
7
8
9
10

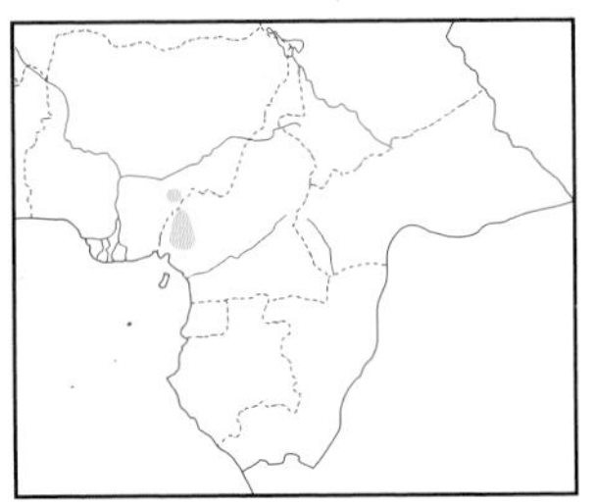
1. WHITE-THROATED MOUNTAIN BABBLER

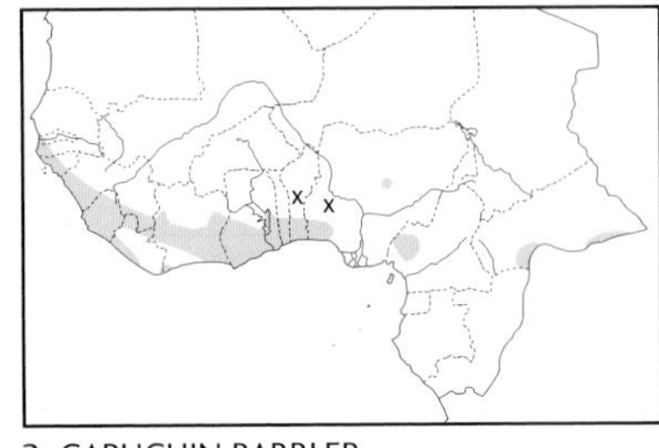
2. CAPUCHIN BABBLER

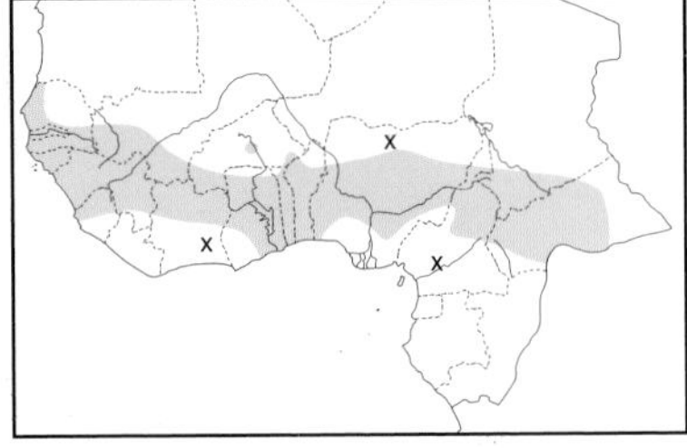
3. BLACKCAP BABBLER

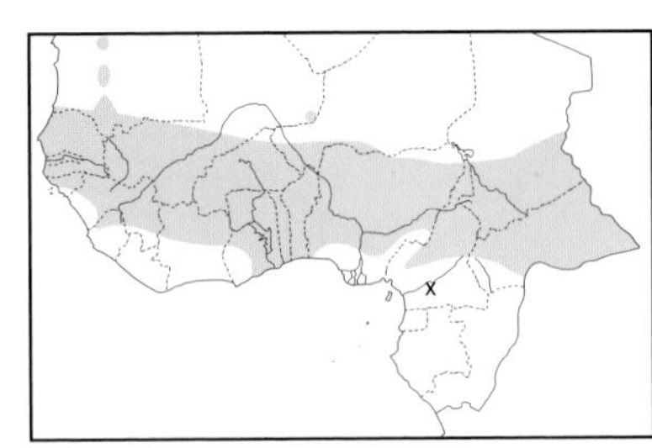
4. BROWN BABBLER

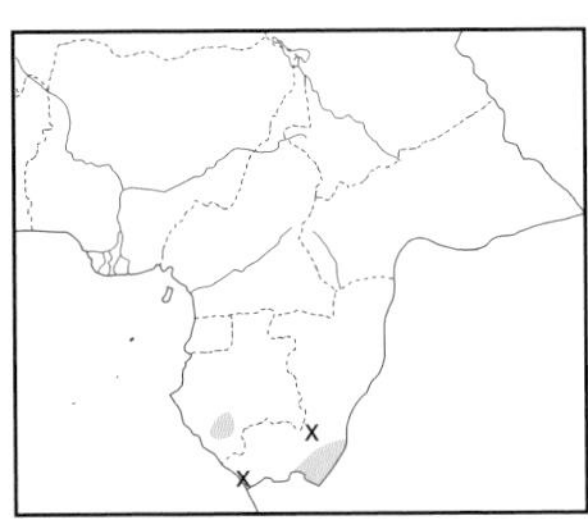
5. ARROW-MARKED BABBLER

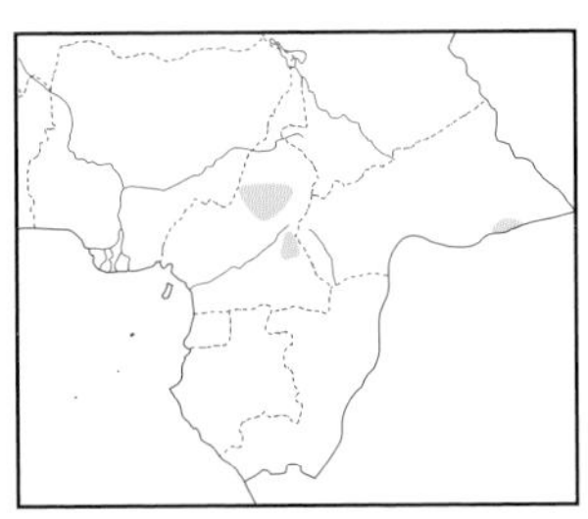
7. SPOTTED THRUSH BABBLER

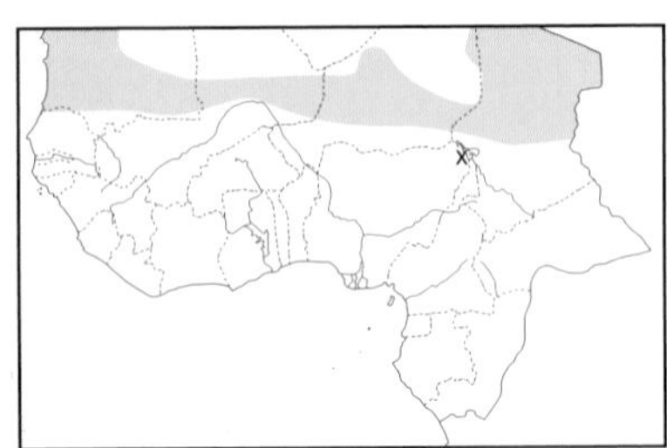
8. FULVOUS BABBLER

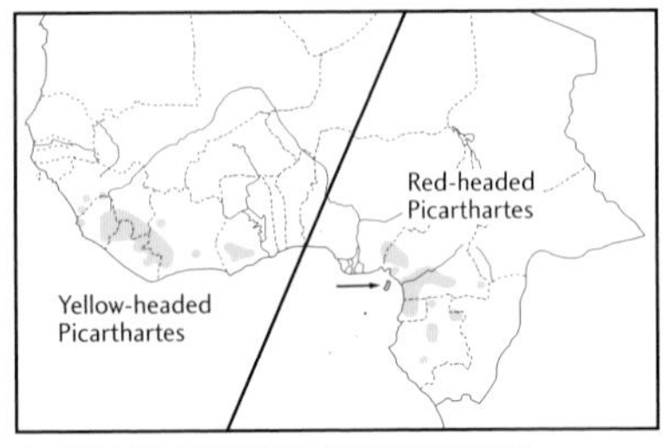

9. YELLOW-HEADED PICATHARTES
10. RED-HEADED PICATHARTES

Continued from page 390

13 **FOREST WHITE-EYE** *Zosterops (senegalensis) stenocricotus* 10-11 cm **R f**
Darker and greener than 12. ▲ Forest. ✤ Fast, vigorous *huhihuhichewchwchwchwchew*. [CD13:27]

14 **MOUNT CAMEROON SPEIROPS** *Speirops melanocephalus* 13 cm **R lc VU**
Olive-brown and grey; black cap; white throat; narrow white eye-ring and band on forehead. ▲ Forest, Mt Cameroon (1820–3000 m). Endemic. ✤ Hard, rattling *trrrr*. Song a fast *tsip-tsip-twrr-twrr tsip-tsip-twrr-twrr-tsee-ti-tew*. [CD13:25]

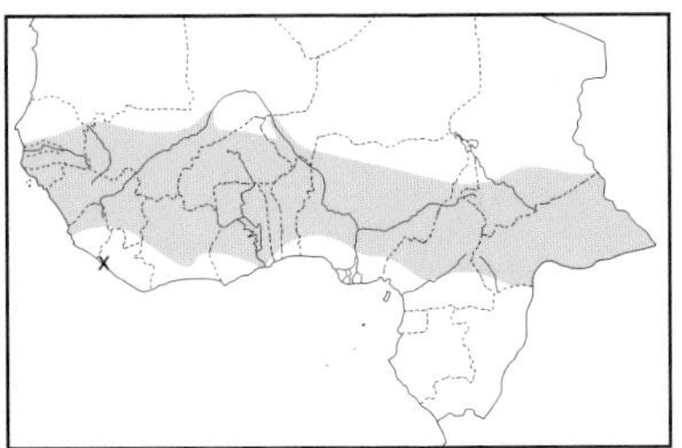

1. WHITE-SHOULDERED BLACK TIT

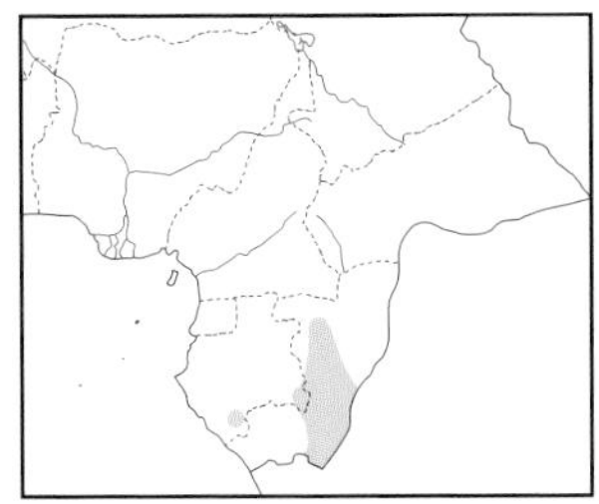

2. WHITE-WINGED BLACK TIT

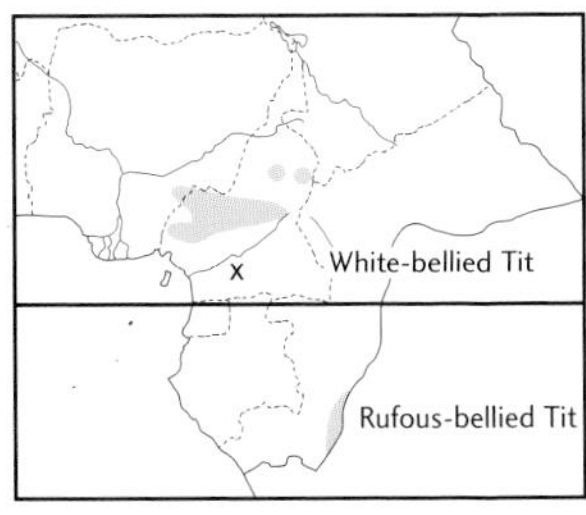

3. WHITE-BELLIED TIT
4. RUFOUS-BELLIED TIT

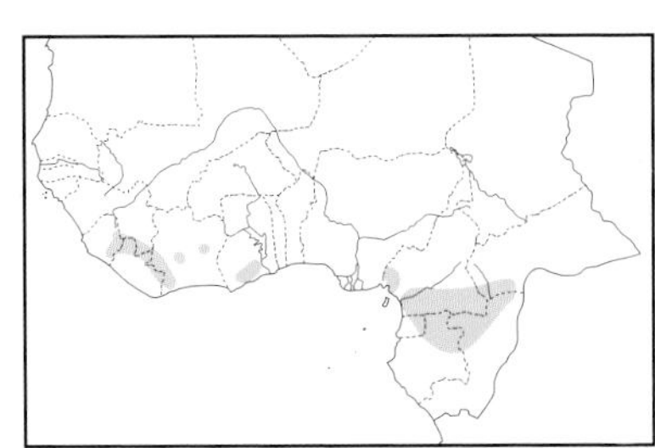

5. DUSKY TIT

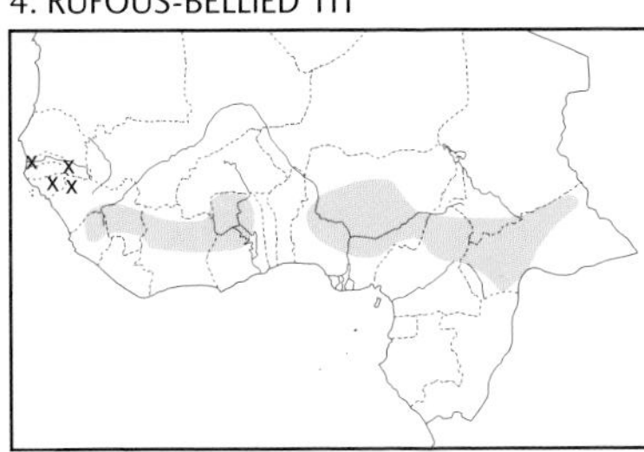

6. SPOTTED CREEPER

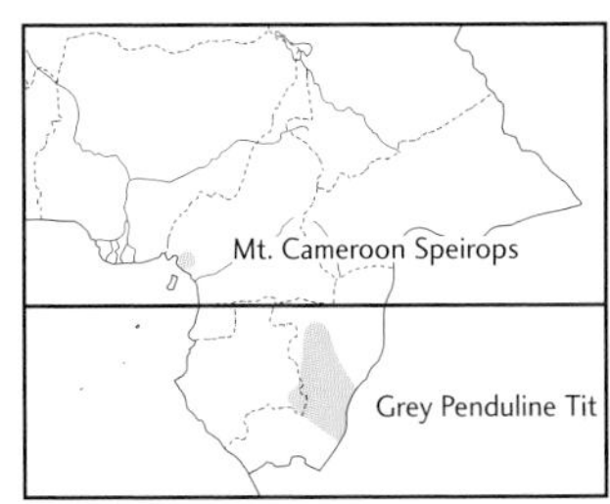

7. GREY PENDULINE TIT
14. MOUNT CAMEROON SPEIROPS

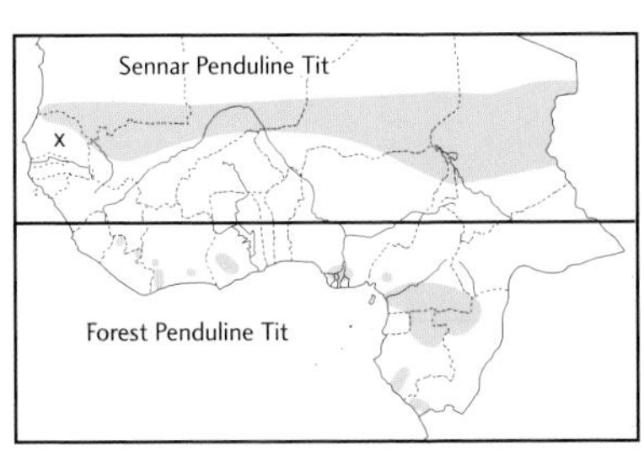

8. SENNAR PENDULINE TIT
9. FOREST PENDULINE TIT

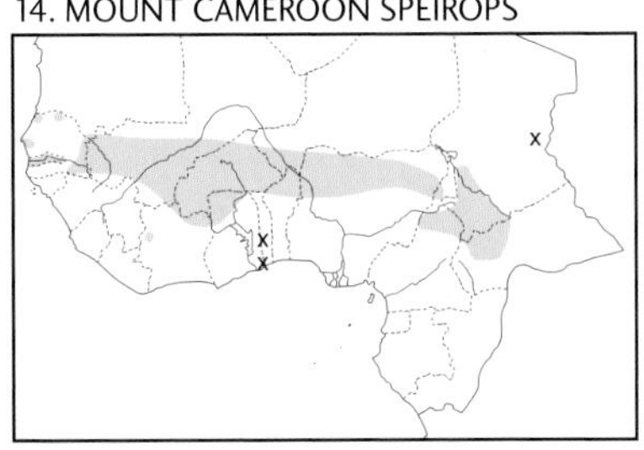

10. YELLOW PENDULINE TIT

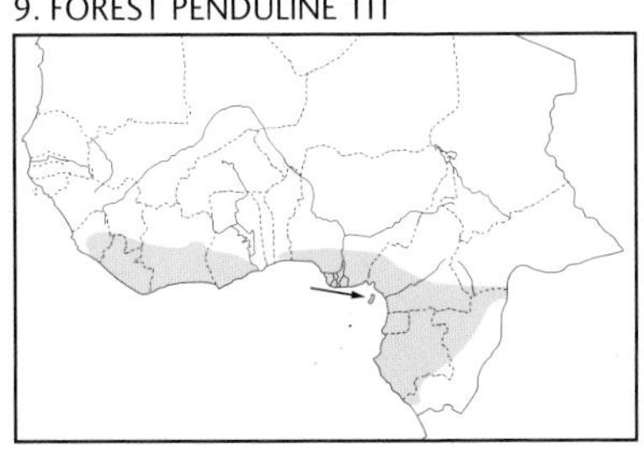

11. TIT-HYLIA

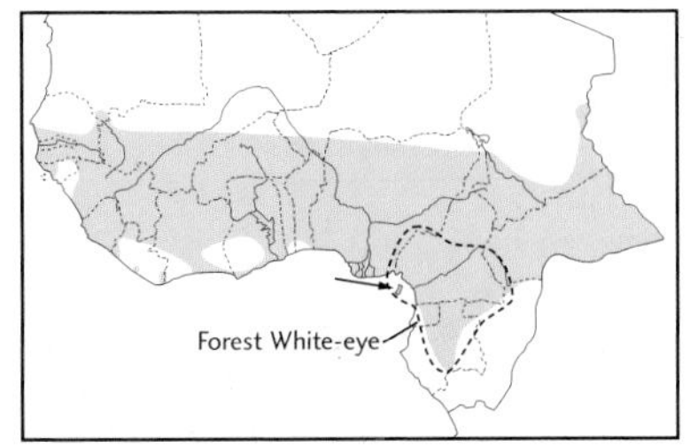

12. YELLOW WHITE-EYE
13. FOREST WHITE-EYE

Plate on page 390

Tits. Rather small, arboreal birds. Active and agile, often hanging upside-down when searching for insects, nuts and seeds. Join mixed-species flocks. Attract attention by rasping calls.

1 WHITE-SHOULDERED BLACK TIT *Parus (leucomelas) guineensis* 14 cm R f/u
Black; white wing patch; conspicuous yellow eye. ▲ Savanna. ✤ Varied. Calls harsh and buzzing. Songs include short, fast series of clear whistles and rolling notes. [CD12:59]

2 WHITE-WINGED BLACK TIT *Parus leucomelas* 15 cm R f/u
As 1 but eye brown; outer tail feathers narrowly edged and tipped white. ▲ Savanna. ✤ As 1.

3 WHITE-BELLIED TIT *Parus albiventris* 14–15 cm R f/r
Resembles 1 but lower breast and abdomen white; eye dark. ▲ Wooded savanna and forest edges in highlands. ✤ Fast *tzii-tzii chèr-chèr-chèr*. [CD12:60]

4 RUFOUS-BELLIED TIT *Parus rufiventris* 14–15 cm R r
Black head; cinnamon-rufous underparts; yellow eye. ▲ Savanna. ✤ Rasping *pzeet chrr chrr* and similar harsh notes. Also fast *whee-tee-uw whee-tee-uw*. [CD12:64]

5 DUSKY TIT *Parus funereus* 13–14 cm R r/s
Entirely black; red eye. ▲ Forest canopy. Noisy. ✤ Varied. Clear, whistled *tsee-tu* and *ptk-tsee-tu-tu* with variations; series of vibrant notes; also harsh, buzzing notes. [CD12:61]

Spotted Creeper. Unmistakable, arboreal bird with long, decurved bill and short, strong legs.

6 SPOTTED CREEPER *Salpornis spilonotus* 15 cm R lu/r
Cryptic brown plumage spotted and barred white; short tail. ▲ Savanna. Clings to tree trunks. ✤ High-pitched *tseee*. Song a short series of clear, high-pitched notes *tsip-tsee-tsu-tuwee*, with slight variations. Also faster *tsitsutsitsutsitsitsu*... [CD12:66]

Penduline tits. Tiny, mainly insectivorous, arboreal birds with short tails and short, sharp bills. Active but unobtrusive; often in small parties; some occasionally with other small insectivores.

7 GREY PENDULINE TIT *Anthoscopus caroli* 8.0–8.5 cm R s
Yellowish-olive above; off-white below; yellow forehead. ▲ Savanna. ✤ Thin squeaky *tseeep*, rhythmic *chisweep-chisweep-*... and rasping *chideZEE-chideZEE-*... [CD12:56]

8 SENNAR PENDULINE TIT *Anthoscopus punctifrons* 7.5–8.5 cm R u/lc
Yellowish-olive above; buffish-white below; yellowish forehead with blackish dots. ▲ Dry *Acacia* savanna. ✤ High-pitched *tsii* notes and churring and harsh sounds. [CD12:57]

9 FOREST PENDULINE TIT *Anthoscopus flavifrons* *c*. 9 cm R r/s
Olive-green above; olivaceous-grey washed yellow below; golden-yellow forehead (hard to see). ▲ Forest. Usually in canopy. ✤ High-pitched little calls. [CD12:58]

10 YELLOW PENDULINE TIT *Anthoscopus parvulus* 7.5–8.0 cm R lu/s
Olive-yellow above; bright yellow below; bright yellow forehead with blackish dots; variably distinct white wingbar. ▲ Wooded savanna, dry *Acacia* scrub. ✤ Fast *chipichipichipi*..., thin *si sli-li-liii*, high-pitched, slightly hoarse *bzee-bzee-*..., and rhythmical buzzing. [CD12:55]

11 TIT-HYLIA *Pholidornis rushiae* 7.5 cm R f
Grey head and breast finely streaked dusky; yellow rump. ▲ Forest. Usually in canopy, often also lower. ✤ Little *tsik*, *ptu* or *ptiu* notes and high-pitched *psee*. [CD11:62]

White-eyes. Small, arboreal, warbler-like birds. Most species have white eye-ring. Active.

12 YELLOW WHITE-EYE *Zosterops senegalensis* 10–11 cm R c/f
Yellowish-olive above; bright yellow below; white eye-ring. ▲ Wooded habitats. ✤ Quavering *ti-trrruutrrruu-ti-trrruut*i. Song a series of burred notes, including calls. [CD13:26]
Continued on page 388

1
2
3
4
5
6
7
10
9
8
11
12
13
14

Distinctive passerines with long, slender, sharply pointed, decurved bills (shortest and least decurved in *Anthreptes*, *Deleornis* and *Hedydipna*; this plate). Feed on nectar, insects and spiders. Active, restless and pugnacious, with rapid and dashing flight.

1 WESTERN VIOLET-BACKED SUNBIRD *Anthreptes longuemarei* 13–14 cm **R u/r**
1a Ad male Glossy violet above and on throat; white below. **1b Ad female** Brown above; white supercilium; yellow belly; tail glossed violet. ▲ Wooded savanna, gallery forest. ✤ Harsh *chep* and *tit*. Song a rapid, high-pitched jingle. [CD12:67]

2 BROWN SUNBIRD *Anthreptes gabonicus* 10 cm **R u/lc**
Grey-brown above; whitish below; narrow white lines above and below eye. Warbler-like. ▲ Mangroves, forested rivers. ✤ High-pitched little notes. [CD12:68]

3 VIOLET-TAILED SUNBIRD *Anthreptes aurantium* 13–14 cm **R lu**
3a Ad male Glossy violet-blue/green above and on throat; pale brownish-buff below. **3b Ad female** Glossy green-blue above; white supercilium; yellow lower underparts. ▲ Forested rivers; also mangroves. ✤ Hard *tsip*.

4 PYGMY SUNBIRD *Hedydipna platura* 9–10 cm **R/M c**
4a Ad male breeding Glossy coppery-green upperparts and throat; golden-yellow underparts; very long tail streamers (projecting up to 7 cm). **4b Ad male non-breeding** As 4c; some black on throat; some glossy green on wing-coverts. **4c Ad female** Grey-brown above; yellow below; faint yellowish supercilium. ▲ Wooded savanna, *Acacia* scrub, gardens. ✤ Call *cheep* or *twee*. Song a high-pitched jingle, including short trill. [CD12:78]

5 FRASER'S SUNBIRD *Deleornis fraseri* 11–13 cm **R c**
Warbler-like. Plain green plumage; pale, straight bill; pale eye-ring. Male has pectoral tufts. ▲ Forest. Often in mixed-species flocks. ✤ Thin, high-pitched notes. [CD12:70]

6 GREEN SUNBIRD *Anthreptes rectirostris* 10 cm **R u/lc**
6a Ad male *rectirostris* (Sierra Leone–Ghana) Yellow throat; glossy green breast-band; pale grey lower breast. Compare 8a. **6b Ad male *tephrolaemus*** (from Benin east) Grey throat. **6c Ad female** Warbler-like; olive-green above; mainly olive-yellow below. ▲ Forest edges, second growth. ✤ Vigorous *whseeet*; variable series of high-pitched notes. [CD12:69]

7 OLIVE SUNBIRD *Cyanomitra olivacea* 13–15 cm **R c**
Plain olive-green; non-glossy; long decurved bill. ▲ Forest, gallery forest. ✤ Vigorous *chip* or *cheep*, frequently and rapidly repeated. Song a series of clear, separate, piping notes *TSIT*, *tsee tut tsiu tsu tu tsu tsu-tsu-tututu*. [CD12:84]

8 COLLARED SUNBIRD *Hedydipna collaris* 10 cm **R c**
8a Ad male Bright glossy green upperparts and throat; yellow underparts. Compare 6a and Variable Sunbird (Plate 119:9a). **8b Ad female** Entire underparts yellow. **8c Juv** Olive above; greyish and lemon-yellow below; yellowish stripes above and below eye. ▲ Forest edge, gallery forest, thickets, woodland, farmbush, gardens. ✤ Vigorous *tsip* and *teew*, nasal *chee chee*. Song a series of high-pitched, separate notes. [CD12:77]

9 BATES'S SUNBIRD *Cinnyris batesi* 9.5 cm **R r/lc**
Very small, non-glossy. Bill decurved. Compare 7 and 10. ▲ Forest. ✤ High-pitched *tsip*. Song a series of *tsip* interspersed with *chep* and followed by subdued short trill. [CD13:17]

10 LITTLE GREEN SUNBIRD *Anthreptes seimundi* 9.5 cm **R u**
Very small, non-glossy. Bill rather straight. Compare 9. ▲ Forest. ✤ Soft *tsssip*. [CD12:71]

11 URSULA'S SUNBIRD *Cinnyris ursulae* 10 cm **R lc/r NT**
Very small. Olive above; greyish below. ▲ Montane and submontane forest. ✤ Soft *tsit-tsit*. Song a descending *tsee-see-see-see-see*. Also a fast, subdued jingle.
Continued on page 394

Maps on pages 394 and 395

1b
1a
2
4b
3b
3a
4c
4a
5
6c
6a
6b
7
8c
8b
8a
9
10
12c
11
12b
12a

Continued from page 392

12 BUFF-THROATED SUNBIRD *Chalcomitra adelberti* 11.5–12.0 cm **R lf**
12a Ad male Straw-coloured throat; black breast-band; rest of underparts chestnut. **12b Ad female** Olive-brown above; creamy streaked olive-brown below. **12c Imm male** Dark greyish-olive above; dark bib. ▲ Forest edges, plantations, etc. Endemic. ✤ Vigorous *che-pEEw*. Song: short phrases of 4–6 high-pitched *tsi* and *tsu* notes. [CD12:85]

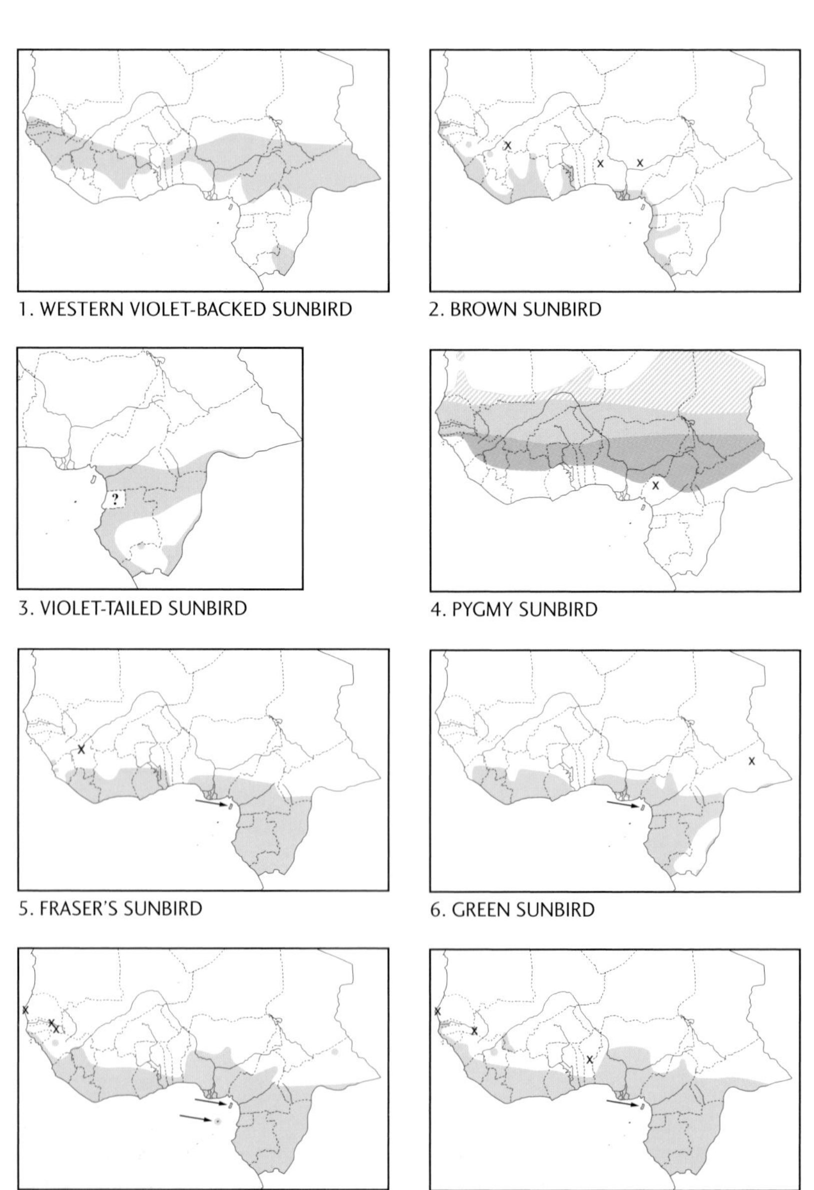

1. WESTERN VIOLET-BACKED SUNBIRD
2. BROWN SUNBIRD
3. VIOLET-TAILED SUNBIRD
4. PYGMY SUNBIRD
5. FRASER'S SUNBIRD
6. GREEN SUNBIRD
7. OLIVE SUNBIRD
8. COLLARED SUNBIRD

 Plate on page 392

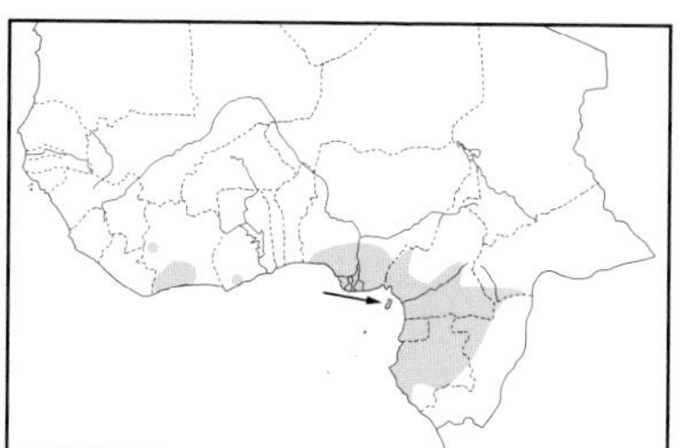
9. BATES'S SUNBIRD

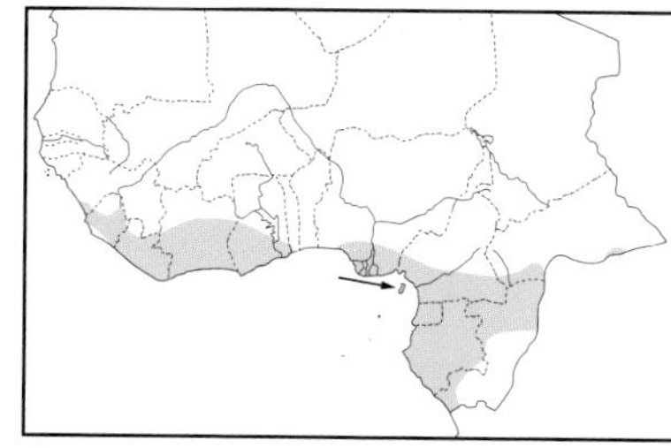
10. LITTLE GREEN SUNBIRD

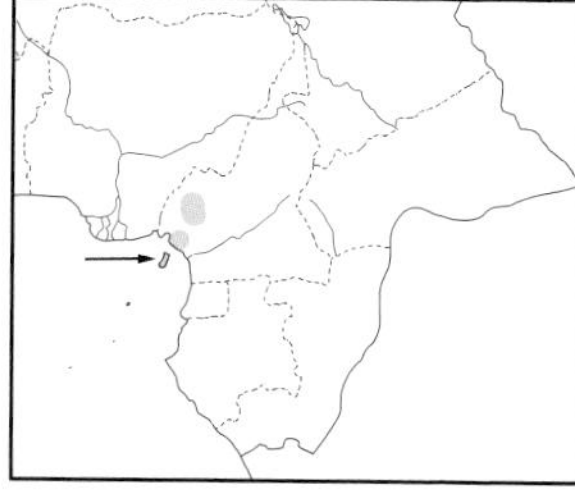
11. URSULA'S SUNBIRD

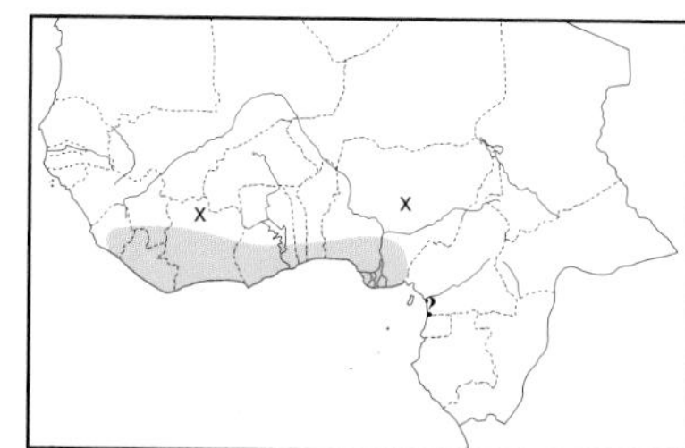

12. BUFF-THROATED SUNBIRD

Continued from page 396

11 OLIVE-BELLIED SUNBIRD *Cinnyris chloropygius* 10.5 cm **R c**
11a Ad male Glossy green above and on throat; bright red breast. Compare 10a. **11b Ad female** Dark olive above; yellowish supercilium; plain, dirty yellow below. ▲ Forest clearings, edges, thickets, shrubbery, gardens. ✤ Song: *tsup* or *tsip* notes stuttering into rapid, rising metallic jingle including many trills, descending at end. [CD13:1]

1 REICHENBACH'S SUNBIRD *Anabathmis reichenbachii* 13–14 cm **R u/lc**
1a Ad male Glossy blue head and throat; bright yellow lower abdomen; graduated, pale-tipped tail. Female similar. **1b Juv** Yellowish-green above; yellow lower abdomen. ▲ Coastal scrub, gardens. ✤ Call *chuwEE chuwEE*. Song a high-pitched jingle. [CD12:73]

2 GREEN-HEADED SUNBIRD *Cyanomitra verticalis* 13–14 cm **R lc**
2a Ad male Glossy blue-green head and throat; dusky-grey underparts. **2b Ad female** Wholly pale grey underparts. **2c Imm male** Forehead and throat blackish-grey. ▲ Wooded savanna, forest, gardens. ✤ Vigorous *chuwee* and *chi-ep*, harsh *chee* and high-pitched *chip*. Songs a harsh, accelerating series of *chip*s and high-pitched jingle. [CD12:79]

3 CAMEROON SUNBIRD *Cyanomitra oritis* 12–13 cm **R lc**
Ad male Glossy bluish-purple head and throat. Female similar but without lemon pectoral tufts. ▲ Montane and submontane forest. Endemic. ✤ Soft *tik tik tik*, descending *tsi-tsi-tsi-tsup*, series of *pee-tsu-pee-tsu-pee-tsu-*... and fast, high-pitched jingle. [CD12:82]

4 BLUE-THROATED BROWN SUNBIRD *Cyanomitra cyanolaema* 14–15 cm **R c**
4a Ad male Dark; glossy blue forecrown and throat; relatively long tail. **4b Ad female** White line above and below eye; whitish throat; mottled breast. ▲ Forest canopy. ✤ High-pitched trill *tsiiirrrrrrr* and descending *tsitsitsitsup*. Also *tsk-tsk* -... [CD12:81]

5 CARMELITE SUNBIRD *Chalcomitra fuliginosa* 13–14 cm **R lc/r**
5a Ad male (fresh) Dark brown; dark glossy blue forehead; glossy violet throat. Plumage bleaches with wear; then particularly conspicuous. **5b Ad female** Very pale; throat mottled brownish; breast indistinctly streaked. ▲ Mangroves, coastal and riverine scrub. ✤ Trill *srrrriiiirrrr*. Song: variable series of short notes, e.g. *chep chi chew chi chup*. [CD12:86]

6 AMETHYST SUNBIRD *Chalcomitra amethystina* 13.5–14.0 cm **R u/lc**
6a Ad male Usually appears black; glossy green crown; glossy violet throat. **6b Ad female** Olivaceous above; paler below; indistinctly streaked breast. **6c Imm male** Resembles 6b but throat glossy violet. ▲ Wooded savanna, forest edge, thickets, gardens. ✤ Sharp *chut* or *chyek*. Song a sustained, high-pitched twittering. [CD12:88]

7 GREEN-THROATED SUNBIRD *Chalcomitra rubescens* 12–13 cm **R c/f**
7a Ad male *rubescens* (Cameroon–S CAR and Congo) Usually appears black; glossy green forecrown and throat. **7b Ad male *crossensis*** (SE Nigeria–W Cameroon; rare) Lacks green throat. **7c Ad female** Brown above; pale supercilium; dirty yellowish below with streaks. ▲ Forest, gardens. ✤ Hard *tsik*. Song: short, rapid phrase of single notes, sometimes accelerating into trill, and series of clear separate notes, recalling 8. [CD12:87]

8 SCARLET-CHESTED SUNBIRD *Chalcomitra senegalensis* 14–15 cm **R/M c/f**
8a Ad male Glossy red lower throat and breast. **8b Ad female** Grey-brown above; broadly streaked below; dusky throat. ▲ Woodland, riverine forest, gardens. ✤ Vigorous *chuw*, scolding *chep*, and *tiup*. Song a simple series of clear separate notes. [CD12:89]

9 VARIABLE SUNBIRD *Cinnyris venustus* 10 cm **R c/u**
9a Ad male Violet forecrown, upper throat and breast; yellow belly. **9b Ad female** Grey-brown washed olive above; unstreaked, yellowish below. **9c Imm male** Blackish throat. ▲ Wooded savanna, clearings, mangroves, coastal scrub, farmbush, gardens. ✤ Song a series of *chuw* and *cheep* call notes accelerating into cheerful jingle. [CD13:13]

10 TINY SUNBIRD *Cinnyris minullus* 9–10 cm **R r/lf**
10a Ad male As 11a but with shorter, less decurved bill; rump bluish-green. **10b Ad female** As 11b but less yellow below; bill as 10a. ▲ Forest. ✤ Call *tsi-tsi-tsi-tsup*. Song: *tsup* or *tsip* notes stuttering into jingle; higher pitched, simpler than 11, and lacking trills. [CD13:2]

Continued on page 395

1b
1a
2c
2b
2a
3
4b
4a
5b
5a
6c
6b
6a
7b
7c
7a
8b
8a
9c
9b
9a
10b
10a
11b
11a

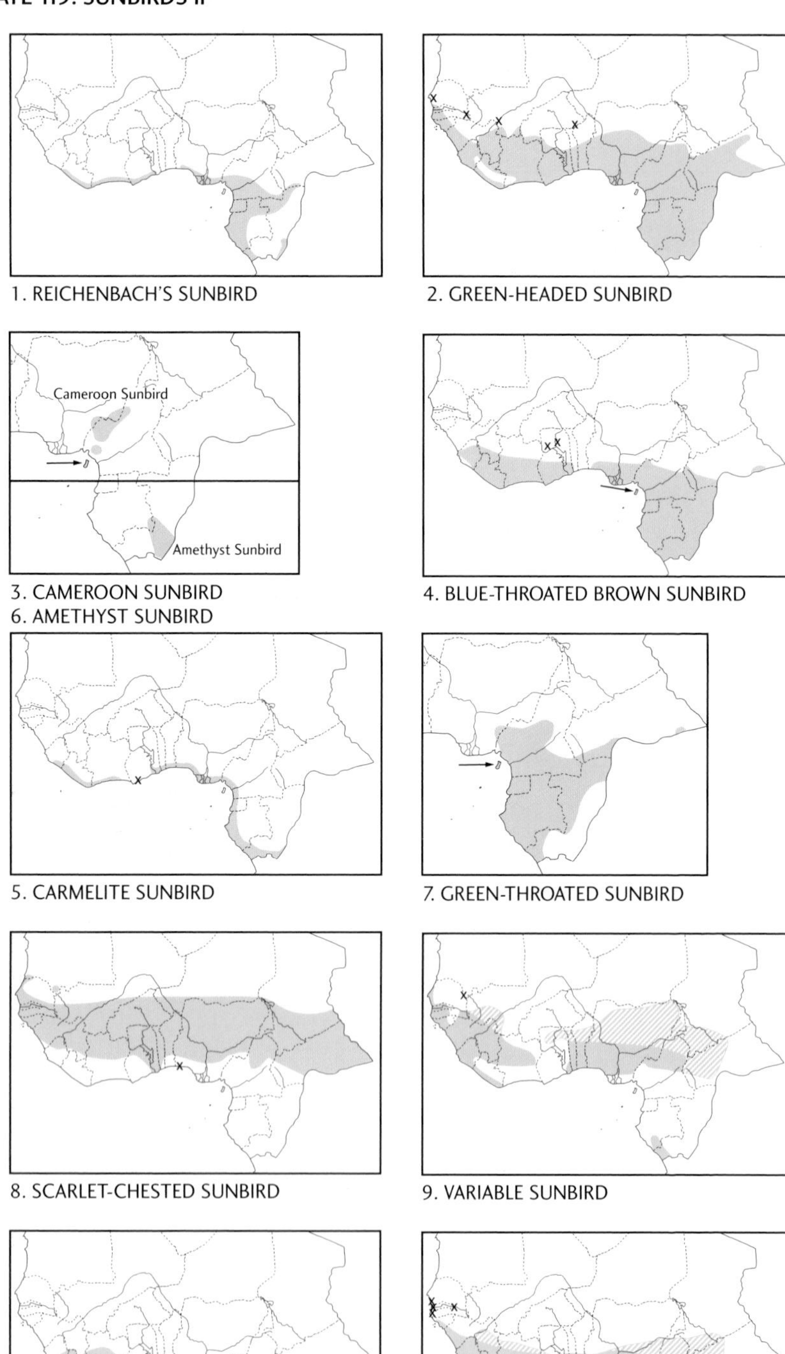

1. REICHENBACH'S SUNBIRD
2. GREEN-HEADED SUNBIRD
3. CAMEROON SUNBIRD
6. AMETHYST SUNBIRD
4. BLUE-THROATED BROWN SUNBIRD
5. CARMELITE SUNBIRD
7. GREEN-THROATED SUNBIRD
8. SCARLET-CHESTED SUNBIRD
9. VARIABLE SUNBIRD
10. TINY SUNBIRD
11. OLIVE-BELLIED SUNBIRD

Plate on page 396

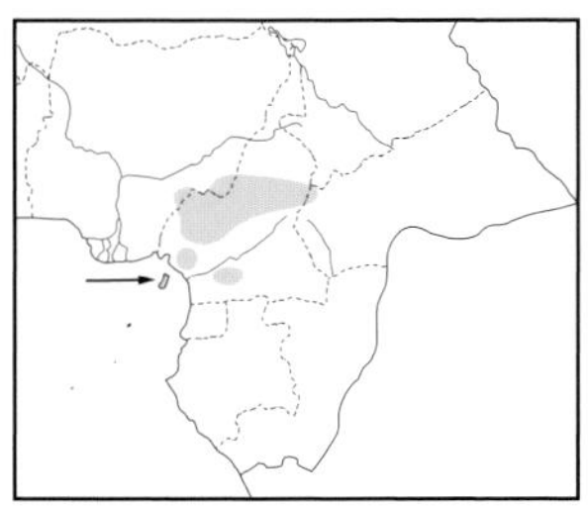

1. NORTHERN DOUBLE-COLLARED SUNBIRD

2. CONGO SUNBIRD

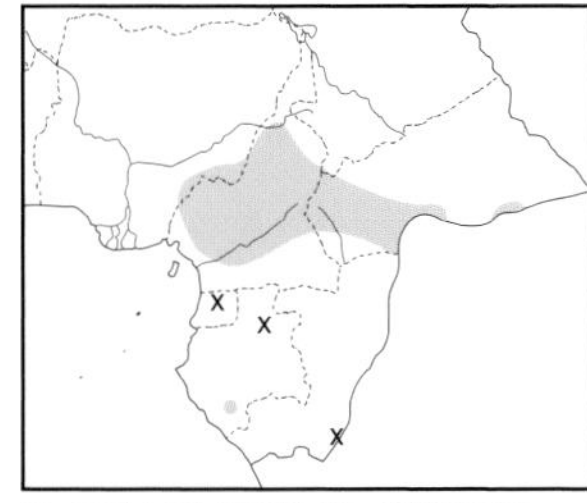

3. ORANGE-TUFTED SUNBIRD

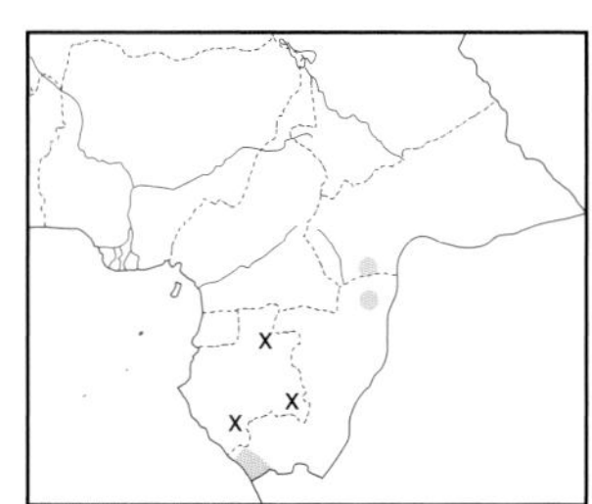

4. PURPLE-BANDED SUNBIRD

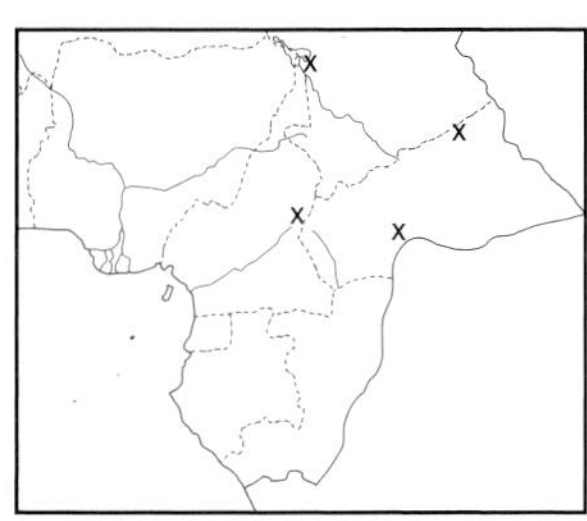

5. PALESTINE SUNBIRD

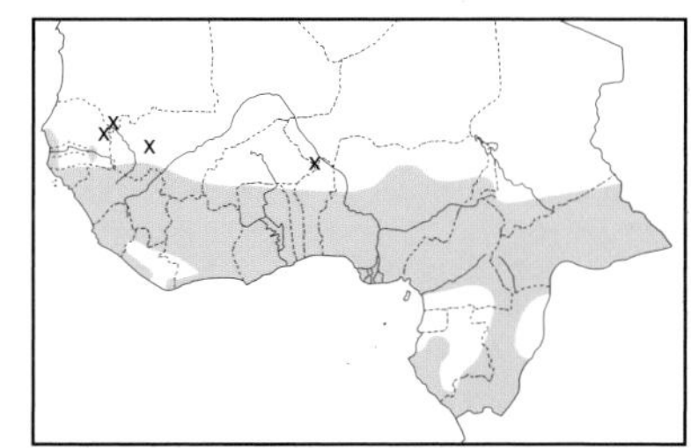

6. COPPER SUNBIRD

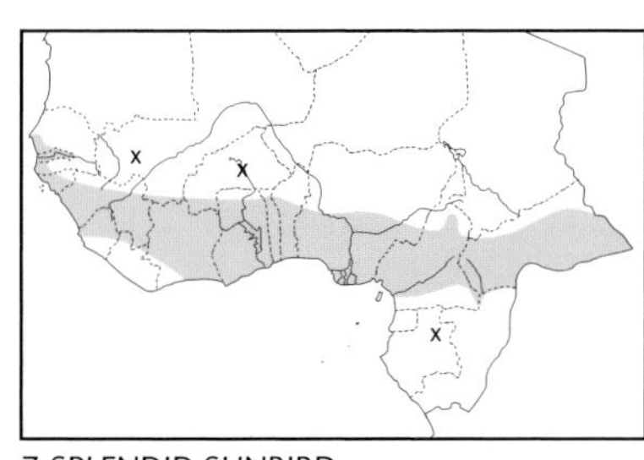

7. SPLENDID SUNBIRD

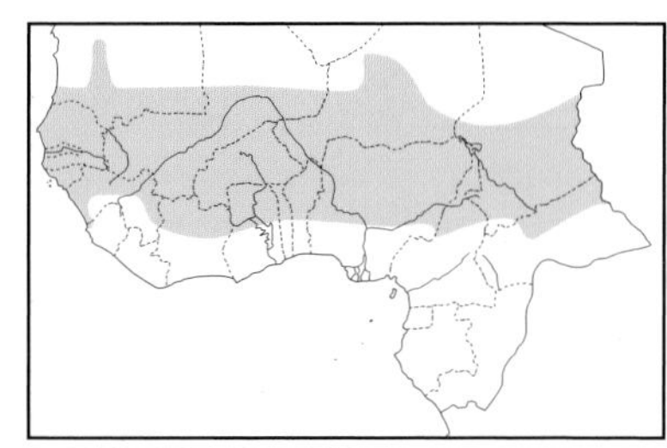

8. BEAUTIFUL SUNBIRD

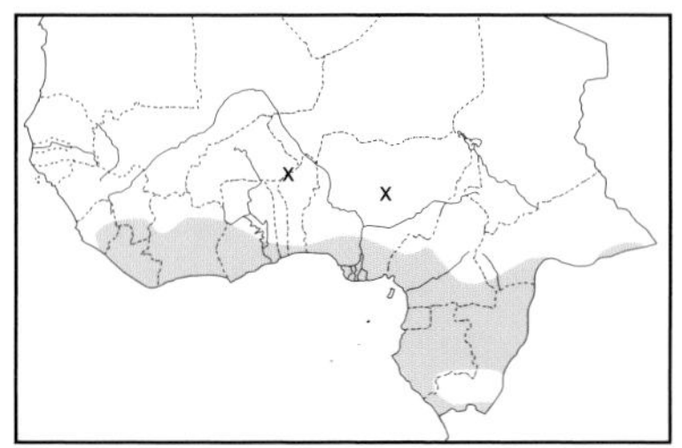

9. SUPERB SUNBIRD

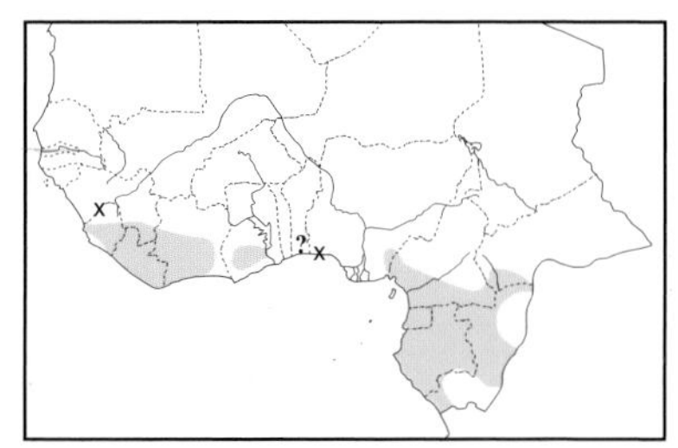

10. JOHANNA'S SUNBIRD

1 **NORTHERN DOUBLE-COLLARED SUNBIRD** *Cinnyris reichenowi* 11.5 cm R c
1a Ad male Resembles Olive-bellied Sunbird (Plate 119:11); glossy violet uppertail-coverts; darker belly. **1b Ad female** Dark grey-green above; paler below; no supercilium. ▲ Open montane forest, highland thickets, etc. ✤ Fast *chep-chep-chep-...* and high-pitched *siip* (alarm). Song a vigorous, rapid, harsh and buzzy jingle. [CD13:3]

2 **CONGO SUNBIRD** *Cinnyris congensis* 12–13 cm R lr
2a Ad male Very long tail streamers (projecting up to 7 cm); red breast-band; black belly. **2b Ad female** Grey-brown, paler on belly. ▲ Forested banks of large rivers. ✤ Unknown.

3 **ORANGE-TUFTED SUNBIRD** *Cinnyris bouvieri* 11.5–12.0 cm R lc/u
3a Ad male Glossy purple forehead; glossy purple and chestnut-red breast-bands. **3b Ad female** Resembles 4b but browner; throat dusky. ▲ Edges of montane and lowland forest. ✤ Low *cheep* and *chip-ip*; also *chew*. Song *tsik tsik tsik cheepa-cheepa-cheepa-cheepa-cheep*, or *tsit-tsit-tsit-chewchewchew*, also including hard churring *chrrrrrrr*. [CD13:9]

4 **PURPLE-BANDED SUNBIRD** *Cinnyris bifasciatus* 11–12 cm R u
4a Ad male Glossy blue and purple breast-bands; black belly. **4b Ad female** Olive-grey above; pale yellowish streaked dusky below. **4c Imm male** As 4b with black throat and blotched breast. ▲ Forest edges, thickets in savanna, coastal scrub. ✤ High-pitched *tsik-tsik-tsik* and buzzing *brrrzi*. Song a few *tsup* notes stuttering into twitter or trill. [CD13:11]

5 **PALESTINE SUNBIRD** *Cinnyris osea* 10.0–11.5 cm R/M lc/r
5a Ad male Iridescent green above; glossy violet-blue forecrown and throat. Appears all black in some lights. **5b Ad female** Greyish-brown above; dusky-white tinged yellowish below. ▲ Savanna. ✤ Short *chip* or *chip-ip-ip-ip*; also sharp *chuWEEp*. Song a rapid series of hard, rather metallic *cheep* or *chwing* notes, often preceded by call note. [CD13:10]

6 **COPPER SUNBIRD** *Cinnyris cupreus* 12–13 cm R c
6a Ad male Glossy coppery and purple. Appears all black in some lights. **6b Ad female** Olivaceous-green above; olive-yellow below; yellowish supercilium. ▲ Wooded savanna, coastal scrub, gardens. ✤ Hoarse *chip-chip*. Song a rapid series of *chip* notes followed by short metallic jingle. Also a rattling alarm call. [CD13:19]

7 **SPLENDID SUNBIRD** *Cinnyris coccinigastrus* 14 cm R c
7a Ad male Glossy purple head and throat; bright red breast; black belly. **7b Ad female** Brownish-olive above; pale yellow below; dusky streaks on breast. **7c Imm male** Black throat and upper breast. ▲ Wooded savanna, coastal thickets, gardens. ✤ Short *tsiup*, harsh *chew-chew-...*, and *hueet*. Song a series of 7–9 clear, separated notes *tip tiup tiup tiup* ... [CD13:15]

8 **BEAUTIFUL SUNBIRD** *Cinnyris pulchellus* 9–11 cm R/M c
8a Ad male breeding Long tail streamers (projecting up to 6 cm); glossy green plumage; red breast bordered yellow. Non-breeding male as female but retaining tail streamers and glossy green shoulders. **8b Ad female** Pale ashy-olive above; pale yellow below. ▲ Wooded savanna, gardens. ✤ Vigorous *chip* or *tut*. Song a rapid series of similar notes followed by jingle. [CD13:8]

9 **SUPERB SUNBIRD** *Cinnyris superbus* 16 cm R f/u
9a Ad male Large; conspicuously long bill; dark glossy blue throat and upper breast; dull dark red belly. **9b Ad female** Plain olive-yellow below; orange-red wash to undertail-coverts. ▲ Forest edge; also wooded savanna. ✤ Vigorous *cheep*. Also clear *wheet wheet...* and similar notes. Song a short series of single notes, occasionally followed by short jingle. [CD13:16]

10 **JOHANNA'S SUNBIRD** *Cinnyris johannae* 13–14 cm R r/lf
10a Ad male Resembles 9a but head and throat glossy green; red on underparts brighter; shape more compact. **10b Ad female** Heavily streaked below (diagnostic). **10c Imm.** ▲ Forest. ✤ Clear *tsik-peew* and series of loud, separate notes *pee pee pee pseew pee pee...* or *tseew tseew tseew* ... In flight a slight *pit pit*. Song a fast jingle. [CD13:14]

 Maps on page 399

1b
1a
2b
2a
3b
3a
4c
4b
4a
5b
5a
6b
6a
7c
7b
7a
8b
8a
9b
9a
10c
10b
10a

Small to medium-sized birds with strong, hooked bills and moderately long to very long tails. Bill and legs usually black(ish). Juveniles typically finely barred above and below; bill paler. Capture prey (insects, reptiles, young birds, small mammals) by pouncing from exposed perch. Some species impale prey on thorns. Flick, swing and fan tail when excited.

1 SOUSA'S SHRIKE *Lanius souzae* 17–18 cm R s
1a Ad male Pale grey crown and mantle; white scapulars; narrow tail. **1b Ad female** As male but flanks washed rufous. **1c Juv** Paler; narrow wavy bars above and (indistinctly) on breast. ▲ Savanna woodland. ✤ Muted whistle *peeeeht* (probably territorial call), harsh *tzzer* (contact) and *tzzjeht* (alarm). [CD13:34]

2 RED-BACKED SHRIKE *Lanius collurio* 17–18 cm P r
2a Ad male Blue-grey crown and rump; chestnut back; pinkish-white underparts. **2b Ad female** Rufous-brown above; underparts with dusky scaling. **2c Imm** Upperparts scaled; coverts and tertials tipped buffish with subterminal dark markings. ▲ Open habitats. Scattered records, mostly in east. ✤ Harsh *chak chak*. Subdued song, occasionally uttered in winter quarters, includes imitations of African and European birds. [CD4:52]

3 EMIN'S SHRIKE *Lanius gubernator* 15–16 cm R r/lu
3a Ad male As 2a but rump chestnut; underparts rusty; white wing patch. **3b Ad female** Duller; back and rump grey; underparts paler. **3c Juv** Barred dusky above and below. ▲ Savanna woodland. Perches on top of trees or other vantage point. Occasionally joins mixed-species flocks. ✤ Usually silent. Song consists of short, rapid series of a few simple, but varied, whistles, e.g. *tweet-u-wee-u-weet* and *trip-tu-trip-srtp*, interspersed by various low hoarse notes, e.g. *chweeeh*.

4 ISABELLINE SHRIKE *Lanius isabellinus* 17–18 cm P u/s
4a Ad male *phoenicuroides* Rather plain, greyish-brown; contrasting rufous rump and tail. *L. i. isabellinus* paler; greyer above; creamy below. Female duller; usually with indistinct scaling on breast and flanks; white patch at base of primaries reduced or absent. **4b Imm** Duller; mask generally poorly developed; underparts with indistinct scaling; coverts and tertials tipped buffish with subterminal dark markings. ▲ Open habitats. ✤ Nasal *kihet* and short *kzi-ek* or *tzea*. Subdued song, occasionally uttered in winter quarters, varied and sustained. [CD13:35]

5 WOODCHAT SHRIKE *Lanius senator* 18–19 cm P f/r
5a Ad male *senator* Chestnut crown and nape; white shoulder patch and rump. Pale fringes to wing feathers indicate fresh plumage. **5b Ad female *senator*** Duller and with more white on forehead and lores. **5c Imm *senator*** Variable amount of buff wash and indistinct crescents on scapulars, breast and flanks. **5d Ad male *badius*** No white on primaries; less white on shoulders; narrower black forehead. Worn plumage: no pale fringes to wing feathers. ▲ Open habitats. Generally frequent in west; rare in Cameroon–CAR. ✤ Hard *chak chak chak*. Mainly silent in winter quarters. [CD4:56]

6 MASKED SHRIKE *Lanius nubicus* 17–18 cm P u/r
6a Ad male White forehead; black rump; orange flanks. **6b Ad female** As male but duller. **6c Juv** Brown-grey and densely barred above; pale forehead and supercilium. ▲ Thorn scrub. ✤ Harsh *krret* and scolding *krrrr*. Mostly silent in winter quarters. [CD13:42]

Maps on page 404

1c
1b
1a
2c
2b
2a
3c
3b
3a
4b
4a
5d
5c
5b
5a
6c
6b
6a

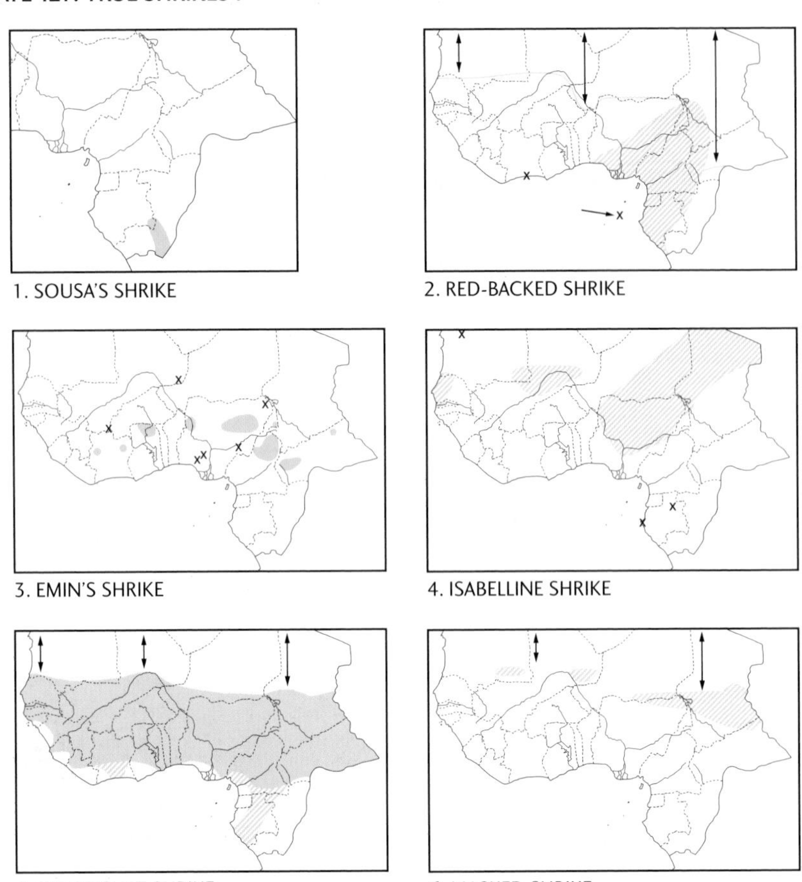

1. SOUSA'S SHRIKE

2. RED-BACKED SHRIKE

3. EMIN'S SHRIKE

4. ISABELLINE SHRIKE

5. WOODCHAT SHRIKE

6. MASKED SHRIKE

Plate on page 402

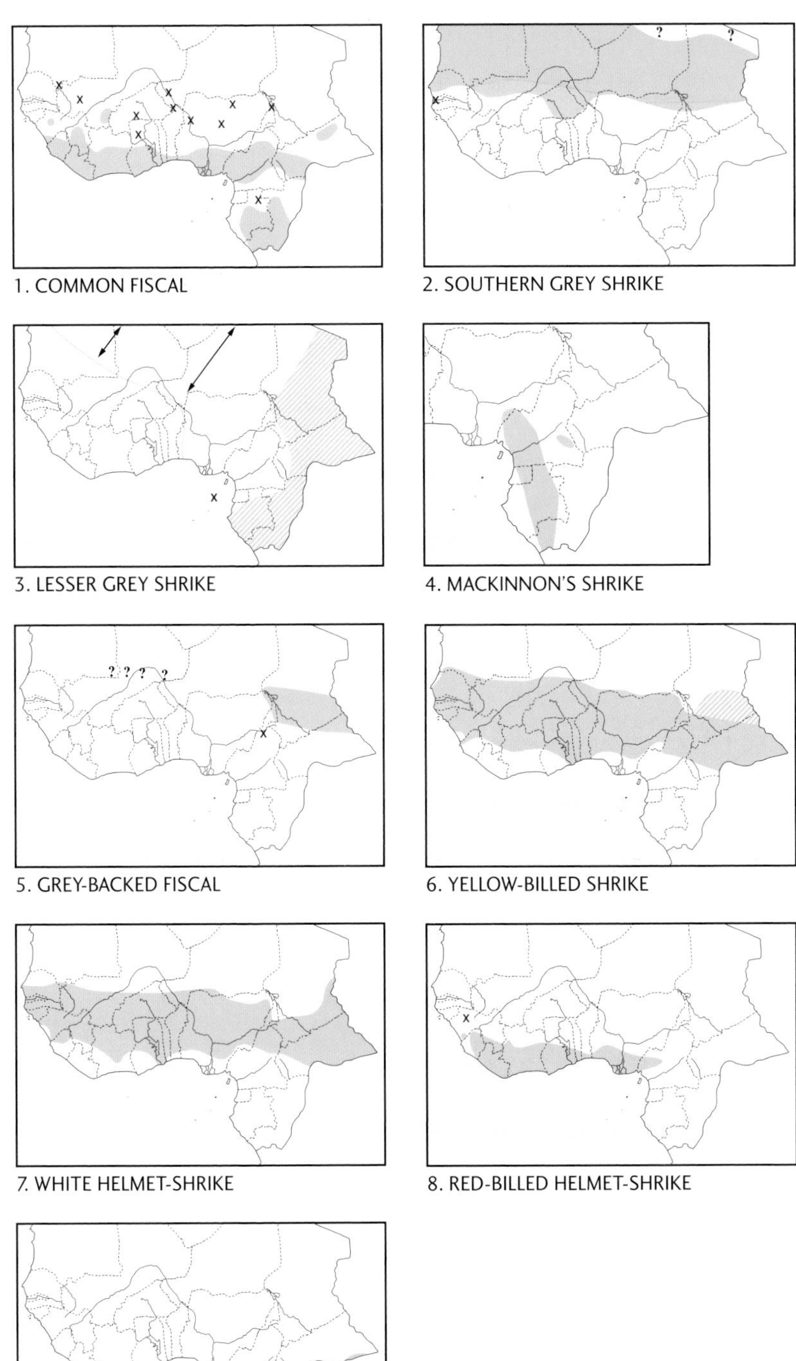

1. COMMON FISCAL

2. SOUTHERN GREY SHRIKE

3. LESSER GREY SHRIKE

4. MACKINNON'S SHRIKE

5. GREY-BACKED FISCAL

6. YELLOW-BILLED SHRIKE

7. WHITE HELMET-SHRIKE

8. RED-BILLED HELMET-SHRIKE

9. RUFOUS-BELLIED HELMET-SHRIKE

PLATE 122: TRUE SHRIKES II AND HELMET-SHRIKES

1 COMMON FISCAL *Lanius collaris* 21–23 cm **R c**
1a Ad Black and white; long, graduated tail. **1b Imm** Rufous-brown barred blackish above; variably barred dusky below. ▲ Inhabits variety of open habitats, incl. farmland, roadsides and vicinity of habitations. Perches conspicuously. ✤ Harsh *chaaa-chaaa* (alarm). Not vocal. [CD13:40]

2 SOUTHERN GREY SHRIKE *Lanius meridionalis* 24 cm **R f**
2a Ad *leucopygos* (Mauritania–Chad; semi-arid zone) Large; pale grey above; wings black and white; rump white or greyish-white; underparts white washed pale buff. *L. m. elegans* (Sahara) has rump greyer; underparts pure white. *L. m. algeriensis* (coastal Mauritania) darker grey above; no supercilium; white patch at base of primaries smaller; underparts washed pale grey. **2b Imm** Duller and tinged buffish. ▲ Thorn scrub. ✤ Dry *trr-trr* and harsh *zzeh-zzeh*; also a subdued, varied warble, including imitations. [CD4:54, 13:38]

3 LESSER GREY SHRIKE *Lanius minor* 20–21 cm **P/V r/lf**
3a Ad Black forehead; no white supercilium; long wings; variable pink wash on breast and flanks. **3b Imm** Duller; no black on forehead. ▲ Open habitats. ✤ Harsh *chak*. [CD13:37]

4 MACKINNON'S SHRIKE *Lanius mackinnoni* 20 cm **R lf**
4a Ad male All-black wings; white shoulder patch. Female has maroon flank patch. **4b Juv** Narrow dusky bars above and (indistinctly) below. ▲ Edges of lowland and montane forest. ✤ Usually silent. Song varied, sustained and melodious. [CD13:36]

5 GREY-BACKED FISCAL *Lanius excubitoroides* 25 cm **R r**
5a Ad Large and robust; black forehead; black scapulars. **5b Juv** Brownish, narrowly barred dusky above; smaller mask. ▲ Thorn scrub. Gregarious and noisy. ✤ Discordant, slightly hoarse and metallic chattering including liquid notes, often developing into excited chorus with entire group joining in. Alarm a harsh note. [CD13:39]

6 YELLOW-BILLED SHRIKE *Corvinella corvina* 30–33 cm **R lc**
Long brown tail; yellow bill. ▲ Open savanna woodland. Usually in small, vocal parties. ✤ Various rasping, chirping and chattering calls, usually in series uttered simultaneously by several members of a group. Calls include imitations of other birds. [CD13:43]

Helmet-shrikes. Gregarious species with rather stout, hooked bills, boldly patterned plumages and brightly coloured eye-wattles. Brush-like feathers on forehead give 'helmeted' appearance. Conspicuous and vocal, constantly chattering and often snapping bill.

7 WHITE HELMET-SHRIKE *Prionops plumatus* 19–23 cm **R f**
7a Ad *plumatus* (Senegambia–N Cameroon) Long white crest; yellow eye-wattle; black-and-white wings and tail. **7b Ad *concinnatus*** (C Cameroon–CAR) Crest shorter and more curly. **7c Juv** No crest; no eye-wattle; eye dark. **7d In flight** Broad, rounded wings with white bars. ▲ Savanna woodland. ✤ Various frequently uttered, hoarse growling sounds, e.g. *krrreew* and *kreepkrw*, often accompanied by bill-snapping. Alarm a high-pitched *tzzee-tzzee*. [CD14:2a]

8 RED-BILLED HELMET-SHRIKE *Prionops caniceps* 20 cm **R f**
8a Ad *caniceps* (west of Benin) Greyish-white crown; throat black; breast white merging with deep buff on lower underparts; red bill and orbital ring. *P. c. harterti* (Benin–W Cameroon) intermediate between 8a and 9a; head pattern (and vocalisations) more like latter (suggesting conspecificity?). **8b Juv** Blackish bill; no orbital ring; whitish throat. ▲ Forest. Endemic. ✤ Various indistinct hoarse, nasal sounds, e.g. *chok rrrr*, and hoarse, muffled *kwèh-kwèh*; also hard *chek*, and softer *tiuk*. Song a series of vigorous, ringing notes *kweeoo* or *whee-aw*. [CD14:3a]

9 RUFOUS-BELLIED HELMET-SHRIKE *Prionops rufiventris* 20 cm **R f**
9a Ad *rufiventris* (S Nigeria–Chad) Greyish-white extending onto head-sides and upper throat; underparts orange-chestnut from lower breast. *P. r. mentalis* (SE CAR) deeper coloured on crown and underparts. **9b In flight** Broad, rounded wings with white bar. ▲ Forest. ✤ Similar to 8. Also clear, melodious *huhu-huhu* and *kweekwee kweekwee*, and descending *chrrrrrrrr*. [CD14:3b]

Maps on page 405

1b
2b
3b
1a
2a
3a
4b
5b
6
4a
5a
9b
8b
7d
7c
9a
7a
8a
7b

Arboreal birds with stout, hooked bills. Many species often encountered in pairs. Most are highly vocal, with loud, ringing calls.

Bush-shrikes (*Malaconotus* and *Telophorus*). Brightly coloured birds of forest and savanna woodland. Forest species generally hard to observe but betray presence by melodious calls and readily respond to playback or whistled imitations.

1 MANY-COLOURED BUSH-SHRIKE *Malaconotus multicolor* 20 cm R u/s
1a Ad male scarlet-breasted form Black mask; grey crown; green upperparts; tail tip yellow and black. **1b Ad male orange-breasted form. 1c Ad male black-breasted form** Belly yellow to scarlet. **1d Ad female** Forehead and lores white; flanks washed green; tail green tipped yellow. ▲ Forest canopy. ✤ Melodious, resonant whistle *whoo-op*, often repeated and forming slow series. Also double *whop-wheeu* or *whu-whee*. Female a rasping note. [CD13:66]

2 PERRIN'S (GORGEOUS) BUSH-SHRIKE *Telophorus viridis* 19 cm R lf
2a Ad male Dark green; crimson throat; broad black breast-band; black tail. Female duller. **2b Imm** Olive-green above; yellow throat; rest of underparts greenish-yellow; some dark crescentic bars on breast; tail washed green. ▲ Thickets and shrubbery in wooded savanna. ✤ Characteristic, ringing *ko-ko-kwik ko-kwik*. [CD13:69]

3 SULPHUR-BREASTED BUSH-SHRIKE *Malaconotus sulfureopectus* 17–19 cm R f
3a Ad male Black mask (duller in female); yellow forehead and supercilium; orange-washed breast. **3b Imm** All-grey head. ▲ Savanna woodland. ✤ Series of far-carrying, clear whistles, varying in speed and motif, e.g. *hu-hu-hu-hweet* or *hu-wheet hu-wheet hu-wheet*. Call *tzzzzrr*, used with bill-snapping in alarm and occasionally also by female in duet; also *puwheet*. [CD13:65]

4 MOUNT KUPE BUSH-SHRIKE *Malaconotus kupeensis* 18–20 cm R lr EN
4a Ad Black mask; white throat; grey breast and belly; yellow vent. **4b Ad** An individual with maroon patch in centre of throat. ▲ Mid-levels of primary montane forest. Endemic. ✤ Short, babbler-like introductory chatter *thek-thek, kh-kh-kh* followed by a series of grating *tchraa* notes. Rarely a short, ascending series of three clearly detached whistles. [CD13:68]

5 GREY-HEADED BUSH-SHRIKE *Malaconotus blanchoti* 25 cm R lf/s
Large; grey head; upperparts and tail green; underparts bright yellow; breast washed orange; massive bill. ▲ Savanna woodland. ✤ Far-carrying, drawn-out, single hollow whistle *whoooop* (easily imitated by human), occasionally rising at end (*whoooo-up*). Also various harsh and rasping notes, bill clicks and shorter whistles. [CD13:74]

6 GREEN-BREASTED BUSH-SHRIKE *Malaconotus gladiator* 25–28 cm R s/lf VU
Large. Olive-green with grey head; massive bill. ▲ Montane forest. Endemic. ✤ Far-carrying, drawn-out, mournful whistle, repeated up to 10 times; similar to 5 and easily imitated by human. Also a harsh, grating call usually repeated with same timing as whistle, and a loud, unmusical, chattering alarm call. [CD13:73]

7 MONTEIRO'S BUSH-SHRIKE *Malaconotus monteiri* *c.* 25 cm R r DD
As 5 but white of lores extending above and below eye; underparts uniformly yellow. ▲ Montane forest. W Cameroon (Mt Cameroon, Mt Kupe). ✤ Similar to 6.

8 LAGDEN'S BUSH-SHRIKE *Malaconotus lagdeni* 23–25 cm R s NT
Large; deep yellow underparts; massive bill. ▲ Forest. ✤ Various melodious hoots and whistles; most distinctive are far-carrying, slow *hoot, hoot-hoot* followed, after pause, by two whistles *hweet-huuuu*. Also a harsh, grating *chrrrr*. [CD13:72]

9 FIERY-BREASTED BUSH-SHRIKE *Malaconotus cruentus* 25 cm R s/lf
9a Ad Large; grey head; underparts yellow and scarlet (variable in extent). **9b Ad yellow-breasted form** Rare. ▲ Forest, abandoned cultivation. ✤ Clear, hollow, whistled *whoop* or *whoo-p*, recalling 1, but lacking its slightly melancholic quality and more rapidly repeated, forming short series of *c.* 3–12 notes. Also various harsh calls. [CD13:71]

Maps on page 410

1c
2b
1b
2a
1d
3b
1a
3a
4a
4b
5
6
7
9b
8
9a

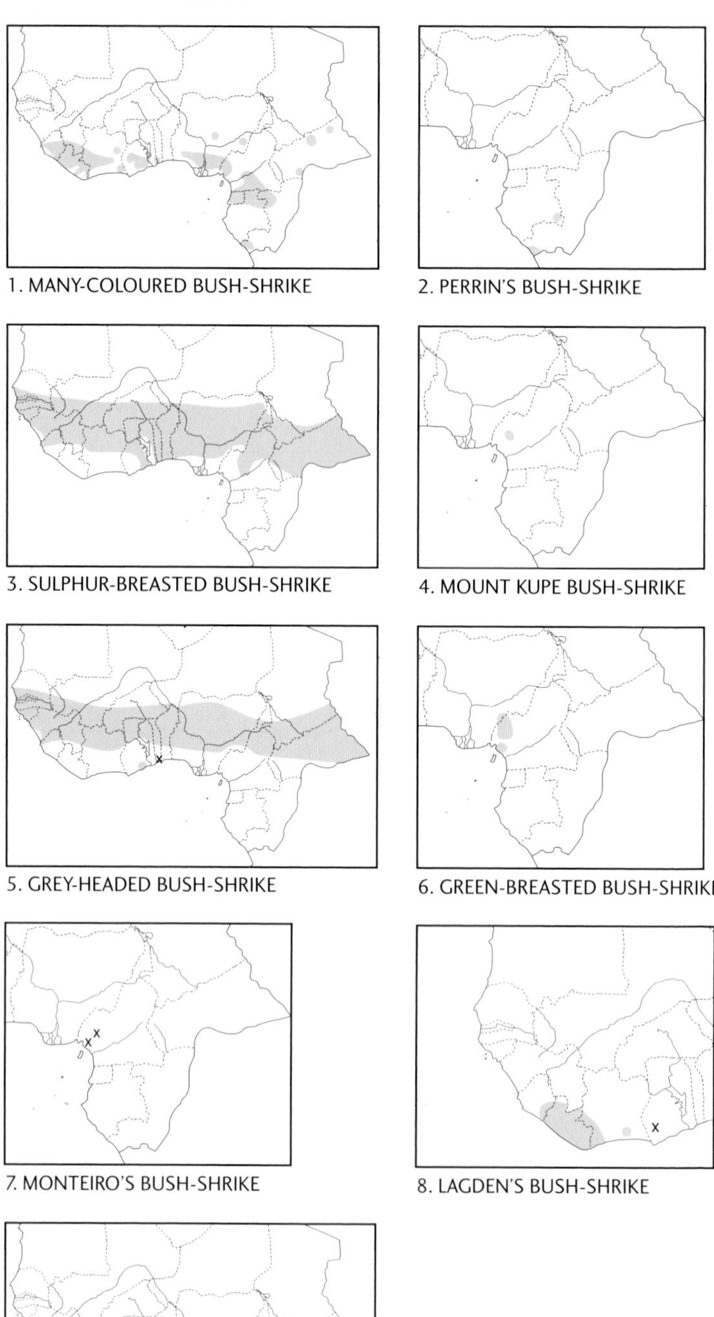

1. MANY-COLOURED BUSH-SHRIKE

2. PERRIN'S BUSH-SHRIKE

3. SULPHUR-BREASTED BUSH-SHRIKE

4. MOUNT KUPE BUSH-SHRIKE

5. GREY-HEADED BUSH-SHRIKE

6. GREEN-BREASTED BUSH-SHRIKE

7. MONTEIRO'S BUSH-SHRIKE

8. LAGDEN'S BUSH-SHRIKE

9. FIERY-BREASTED BUSH-SHRIKE

Plate on page 408

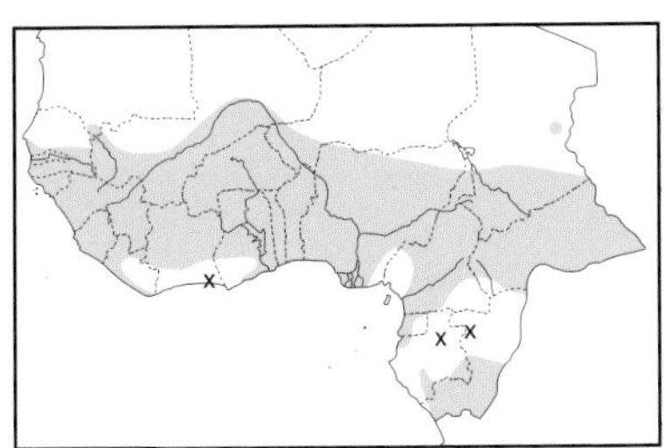

1. NORTHERN PUFFBACK

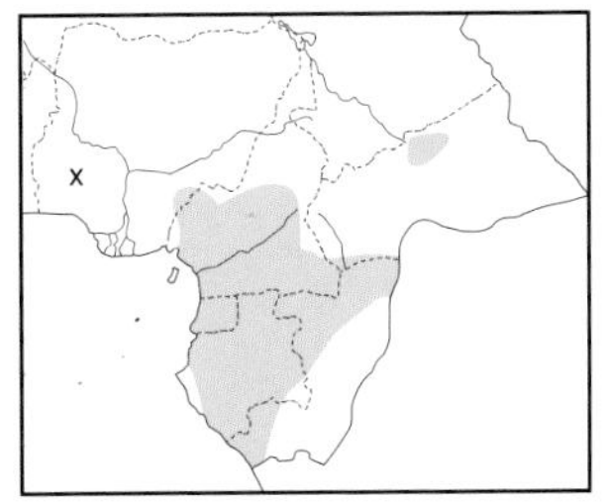

2. BLACK-SHOULDERED PUFFBACK

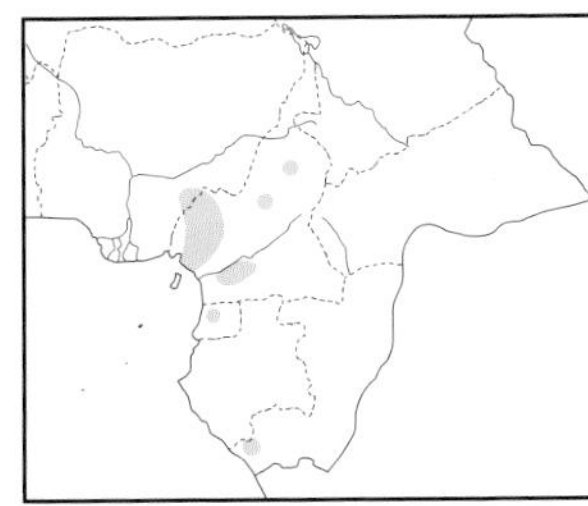

3. PINK-FOOTED PUFFBACK

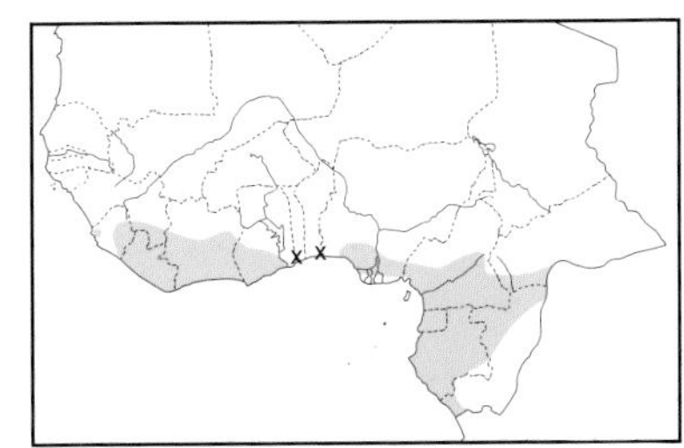

4. SABINE'S PUFFBACK

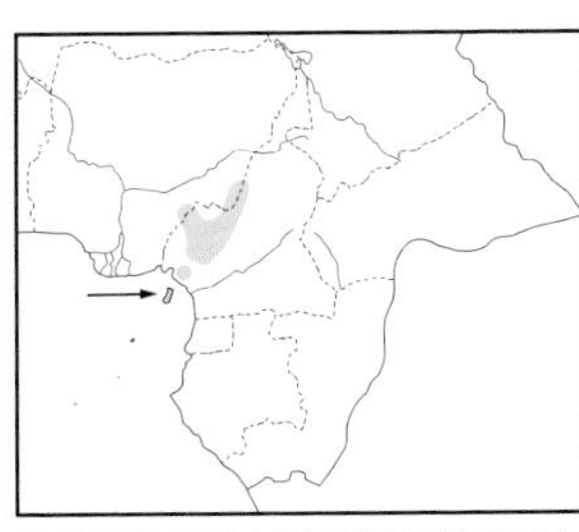

5. MOUNTAIN SOOTY BOUBOU

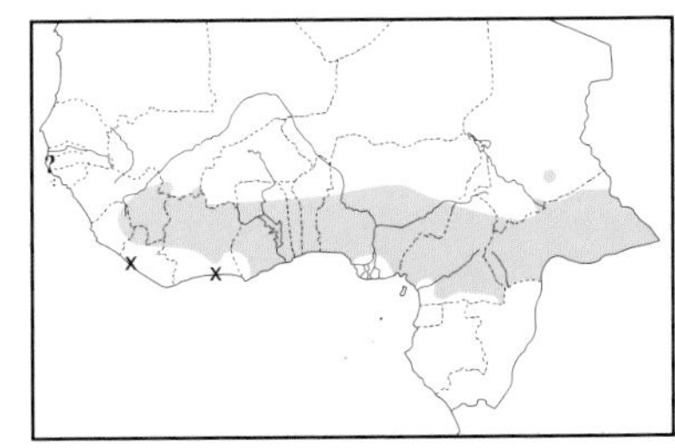

6. TROPICAL BOUBOU

7. SOOTY BOUBOU

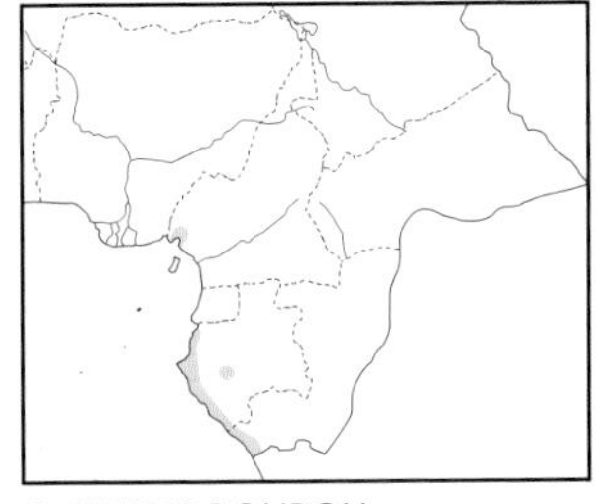

8. SWAMP BOUBOU

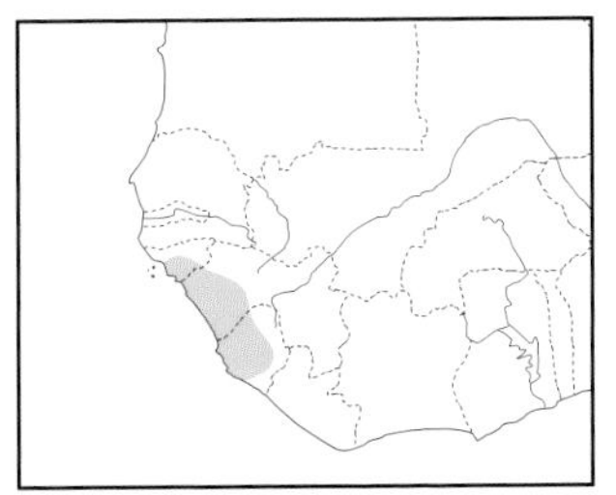

9. TURATI'S BOUBOU

Plate on page 412

PLATE 124: BUSH-SHRIKES II

Puffbacks (*Dryoscopus*). Males predominantly black and white with soft elongated feathers on lower back and rump, which are puffed out to produce spectacular white ball in display. Vocal.

1 NORTHERN PUFFBACK *Dryoscopus gambensis* 18–19 cm **R f/c**
1a Ad male The only puffback with whitish edges to wing feathers; orange-red eye. **1b Ad female *gambensis*** (most of range) Grey head; earth-brown upperparts; tawny-buff underparts. **1c Ad female *malzacii*** (CAR) Darker than 1b; head brown, as upperparts. ▲ Savanna woodland; also forest clearings, mangroves. ✤ Varied. Frequently repeated, loud, rasping *CHERP-CHERP-*... and harsh chattering, nasal and clicking sounds. [CD13:46]

2 BLACK-SHOULDERED PUFFBACK *Dryoscopus senegalensis* 16–17 cm **R r/lc**
2a Ad male Glossy black and pure white. Compare 1a and 4a. **2b Ad female** White supraloral streak. **2c Ad male displaying.** ▲ Forest edges, second growth. Joins mixed-species flocks. ✤ Male has loud *KYow! KYow! KYow!* ...; also harsh *kurrrWEERrr* and snapping sounds. Female responds with rasping, weaver-like note. [CD13:47]

3 PINK-FOOTED PUFFBACK *Dryoscopus angolensis* 15–17 cm **R lf/r**
3a Ad male Black head; grey upperparts; pinkish legs and feet. **3b Ad female** Resembles 1b, but more colourful. ▲ Montane and submontane forest. Often in mixed-species flocks. ✤ Harsh churring calls and rattles, and series of clicking *tik-tik-tik-tik-tik-*... [CD13:48]

4 SABINE'S PUFFBACK *Dryoscopus sabini* 18–19 cm **R u**
4a Ad male Larger than 2a, bill longer and heavier. **4b Ad female** Tawnier overall than 3b. ▲ Forest. Often in mixed-species flocks. ✤ Harsh calls; in flight *tok-tok-tok-*... Song a slowly descending series of clear whistles, *tsee tsu tsu tsu tsu tsu*... variable in length. [CD13:49]

Boubous (*Laniarius*). Either all black or with black upperparts and white or red/yellow underparts. Skulk in heavy shrubbery and forest edges. Vocal.

5 MOUNTAIN SOOTY BOUBOU *Laniarius poensis* 18 cm **R lc**
Entirely black. ▲ Dense undergrowth of montane forest edges and clearings. ✤ Very varied, including loud whistles, trills and rattles, squeals and grating notes. Most notable are duets of pair, *WHOO-EE* (male) followed by rasping *tchrerr* or *errgh* (female), with variations. Many calls not separable from Yellow-breasted Boubou's. [CD13:60–61]

6 TROPICAL BOUBOU *Laniarius aethiopicus* 23 cm **R f**
Black above; white below; long white wingbar. ▲ Various wooded habitats. ✤ Duet consists of ringing *HOOO* in short series of variable speed (male), answered by similar notes or grating *gkrzzz* (female). Call an explosive *KEK!*, often in long series. [CD13:53]

7 SOOTY BOUBOU *Laniarius leucorhynchus* 21.5 cm **R f/s**
7a Ad All black. **7b Juv** Bill whitish. ▲ Tangles, rank vegetation, thickets, overgrown cultivation in large clearings, dense regrowth at forest edges. Secretive but vocal. ✤ Male has clear, ringing *HOO-HOO*; also in long series of variable speed. Female answers with plaintive whistle *hweeeew*. Also repeated, hoarse *kchch* and *plt-plt-*... [CD13:63]

8 SWAMP BOUBOU *Laniarius bicolor* 23 cm **R lf**
Very similar to 6; whiter below; wingbar usually shorter. ▲ Savanna thickets, coastal scrub, mangroves. ✤ Duet consists of 2–3 ringing *HOOO* or *HIOO* whistles (male), answered by hard rattling *K-K-K-KKKKK*, shorter *K-K* or rasping *gha gha gha* (female). Also an explosive *KEK!* or *TUK!*, often in long series. [CD13:55]

9 TURATI'S BOUBOU *Laniarius turatii* 23 cm **R c**
As 6 but without wingbar. Variable pinkish tinge to underparts (often hard to see). ▲ Various wooded habitats. Endemic. ✤ Duet consists of series of ringing *HOO*'s (male) answered by a grating *gkrzzz* (female); similar to, but more stereotyped than 6. Also an explosive *KEK!*, often in long series. [CD13:54]

Maps on page 411

1c
1b
1a
2c
2b
2a
4b
4a
3b
3a
5
6
8
7b
7a
9

1 BLACK-HEADED GONOLEK *Laniarius erythrogaster* 23 cm R lc
Black above; crimson below. ▲ Bushy river banks. ✤ Male song a ringing whistle *tiu-WHEE-oo!* Female usually responds with harsh rasping. Call a hard *K-K-K-K.* [CD13:57]

2 YELLOW-CROWNED GONOLEK *Laniarius barbarus* 23 cm R lf/c
2a Ad Black above; crimson below; golden-yellow cap. **2b Juv** Feathers of upperparts tipped buff; below, yellowish-buff barred black. ▲ Thickets, thorn scrub, mangroves; often near watercourses. Endemic. ✤ Male song a ringing *WHEE-oo!*, likened to whip-lash, with variations; also quivering *whiiiir*. Female usually responds with *kik-kik*. Various other, harsh calls. [CD13:56]

3 YELLOW-BREASTED BOUBOU *Laniarius atroflavus* 18–19 cm R lf
Black above; deep yellow below. ▲ Dense undergrowth of clearings, secondary scrub, small forest remnants and bamboo in highlands. Endemic. ✤ Duets of pair *WHEEw!* (male) followed by *chek* or harsh note (female), with variations. Also a variety of loud whistles, swishing, rattling and grating notes, most inseparable from those of Mountain Sooty Boubou. [CD13:59]

4 BOCAGE'S BUSH-SHRIKE *Malaconotus bocagei* 16.5 cm R u
4a Ad Black and dark grey above; whitish below; long, white supercilium. **4b Juv** Upperparts washed greenish and speckled buff; some indistinct dusky barring below. ▲ Overgrown forest clearings and edges, gallery forest, wooded savanna with dense bushes. ✤ Varies geographically. Songs include rapid series of whistles *hu-hu-hu-hu-hu-huuu*, with variations, resembling Sulphur-breasted Bush-shrike; clear, double notes *tliuu-theeee*; rhythmic *tliu-tliu-tliu*; fast *tiuptiuptiuptiuptiup*; series of *kuli-kuli kuli-kuli* etc. May also duet, female answering with shorter whistles. Also harsh calls. [CD13:64]

5 LÜHDER'S BUSH-SHRIKE *Laniarius luehderi* 18–19 cm R u/f
5a Ad Black above; chestnut cap; deep cinnamon throat and breast. **5b Imm** Olivaceous-brown above; dirty yellowish below. ▲ Overgrown forest clearings, second growth, dense scrub, thickets. ✤ Varied. Guttural, quivering *krrrooh* or *keooow*, repeated at regular intervals; occasionally in duet by pair. Female also has dry *k-k-k-k-...* and sharp *gkssss*. Also hollow *hoo-up* similar to Many-coloured Bush-shrike, *k-krrrr* and harsh and churring notes. [CD13:52]

6 BRUBRU *Nilaus afer* 13–15 cm R f/lc
6a Ad male Small; mottled black and white above; chestnut on flanks. **6b Ad female** More brownish-grey above; flanks paler. **6c Juv** Speckled and mottled above; barred below. ▲ Savanna woodland. ✤ Male gives a far-carrying trill *brruuuu*. Female may respond with *wheeeu*. Also *tu* and *peep* (contact) and *chK-chK-...* (alarm). [CD13:44]

Tchagras (*Tchagra* and *Antichromus*). Mainly brown above with rufous wings and patterned head and tail. Melodious calls and characteristic display flights attract attention.

7 MARSH TCHAGRA *Antichromus minutus* 16–18 cm R lu
7a Ad male Jet-black cap; rufous-brown upperparts; black tail. **7b Ad female** Whitish supercilium. ▲ Rank herbage. ✤ Male song a few melodious whistles, usually in low, short display flight and preceded by wing-rattling. Female may respond with nasal *cherrruu*. Calls include various hard churrs and clucks. [CD13:50]

8 BLACK-CROWNED TCHAGRA *Tchagra senegalus* 20–22 cm R c
8a Black crown. **8b In flight** Rufous wings; black, white-tipped tail. ▲ Various wooded habitats (not forest). ✤ Male song a series of vigorous, far-carrying, melancholy whistles, in display flight or from perch. Female may respond with *trrrrrrrrrr*. Also various explosive, growling whistles, followed by rolling bubble and ending with *whu-heeuw*. [CD4:57]

9 BROWN-CROWNED TCHAGRA *Tchagra australis* 18–19 cm R f/u
Brown crown. Compare 8. ▲ Forest edge, thicket, scrub. ✤ Male song a jaunty, liquid, descending series *tree-tree-treeu-treeuu-treeuu...* usually in low display flight and preceded by wing-rattling. Female may respond with soft nasal *cheru-cheru*. Alarm *chuk-chuk-...* and *chrrr*. [CD13:51]

Maps on page 416

1
2b
2a
3
4b
4a
5b
5a
6c
6b
6a
7b
7a
8b
8a
9

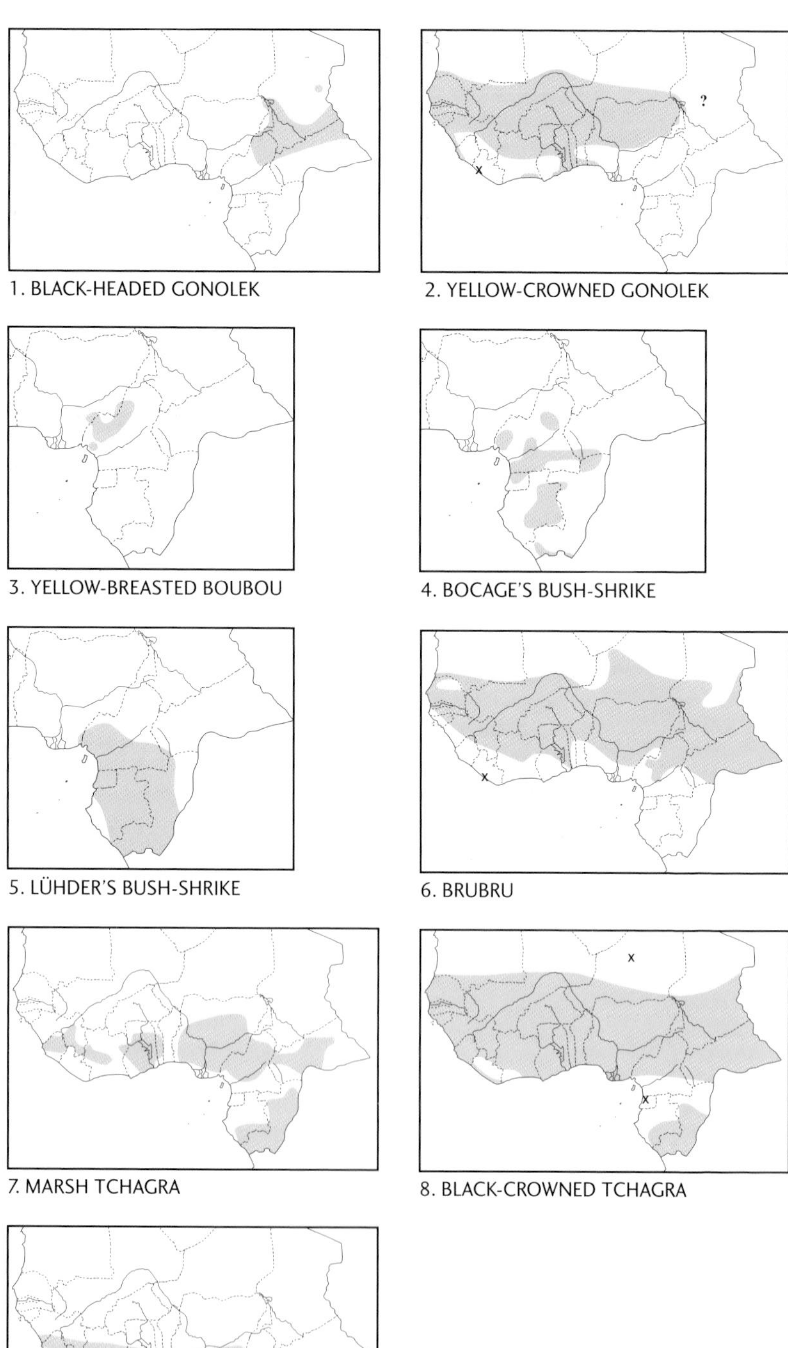

1. BLACK-HEADED GONOLEK

2. YELLOW-CROWNED GONOLEK

3. YELLOW-BREASTED BOUBOU

4. BOCAGE'S BUSH-SHRIKE

5. LÜHDER'S BUSH-SHRIKE

6. BRUBRU

7. MARSH TCHAGRA

8. BLACK-CROWNED TCHAGRA

9. BROWN-CROWNED TCHAGRA

Plate on page 414

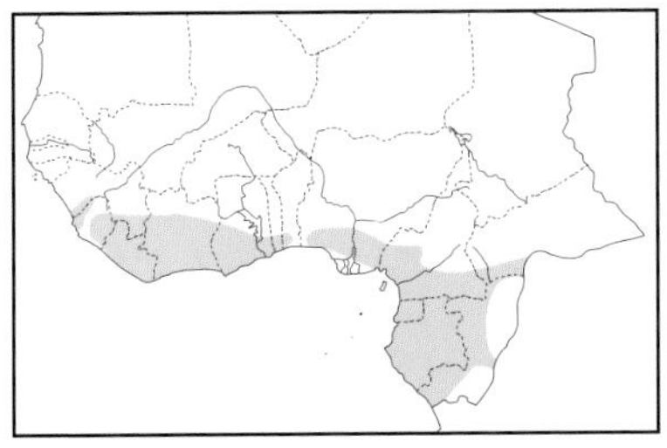

1. WESTERN BLACK-HEADED ORIOLE

3. BLACK-WINGED ORIOLE

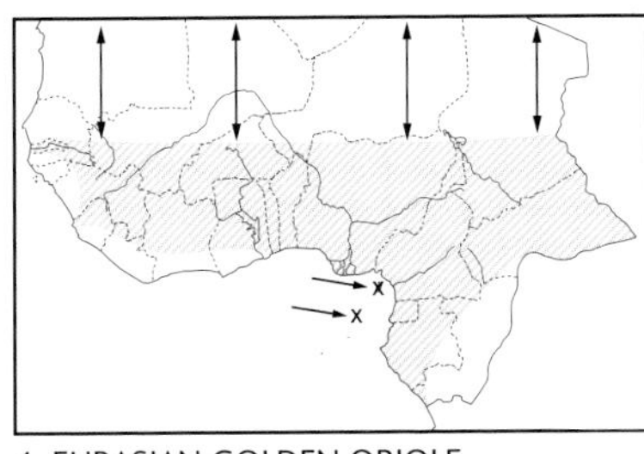

4. EURASIAN GOLDEN ORIOLE

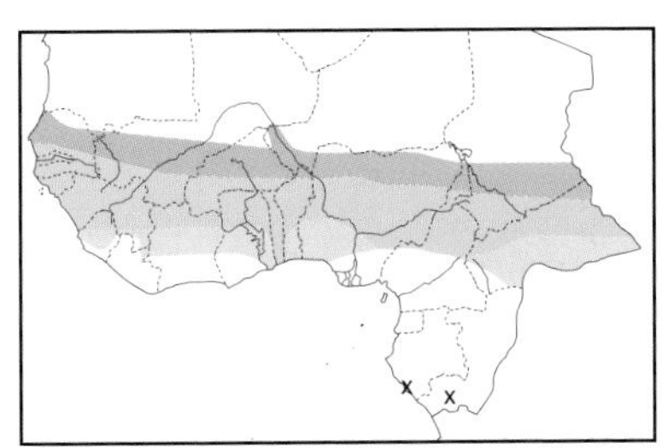

5. AFRICAN GOLDEN ORIOLE

PLATE 126: ORIOLES

Robust, arboreal birds. In most species occurring in Africa males are bright yellow and black, with black on head, wings and tail, and a strong, reddish bill. Feed on insects and fruit. Mostly in canopy, where often difficult to observe. Voice loud, fluty and melodious. Flight strong and undulating.

1 WESTERN BLACK-HEADED ORIOLE *Oriolus brachyrhynchus* *c.* 21 cm **R f/c**
1a Ad *brachyrhynchus* (Sierra Leone–SE Guinea to Benin) Black head; small white patch on edge of wing; green central tail feathers. **1b Ad *laetior*** (Nigeria to CAR–Congo) Broader yellow collar; yellower mantle. **1c Imm** Head and upperparts olive; throat olive streaked yellow; breast streaked black; bill dusky. ▲ Lowland and mid-elevation forest. ✤ Various fluty whistles with characteristic oriole quality e.g. *uoo-dleeo, uoo-uoo, tioolioo, whoolioo, whee-whooliu, too-too-tuloo* etc. Notes usually more detached and rather lower pitched than those of 3. Also a harsh *whit-chèèèw-chèèèw*. [CD13:29]

2 EASTERN BLACK-HEADED ORIOLE *Oriolus larvatus* *c.* 21 cm **?**
2a Ad As 1a, but flight feathers edged white; tertials edged yellow. **2b Imm** As 3b but with white wing patch. ▲ Open woodland. No certain records. Claims from SW CAR doubtful. ✤ Short phrases of melodious whistles. Also clear *kulEEw*, and harsh *whrreeaa*. [CD13:32]

3 BLACK-WINGED ORIOLE *Oriolus nigripennis* *c.* 20 cm **R f/c**
3a Ad Similar to 1a, but no white patch on edge of wing; central tail feathers black. **3b Imm** Throat black streaked yellow; breast streaked black. ▲ Lowland and montane forest. ✤ Whistles resemble 1 but even more melodious, very liquid and slightly higher pitched, e.g. *whuteluw* or *pteeuw-ee-ooleo* and variations. Also short *oo-ik* or *kloo-ik* and *tiup*, sometimes followed by descending *kirrrrrrr*, clear *kpi-uw* and *hiu* or *hoo-whEE* (very similar to human whistle), and harsh *whrrrèèèr* (distinctive). [CD13:30]

4 EURASIAN GOLDEN ORIOLE *Oriolus oriolus* *c.* 24 cm **P u/r**
4a Ad male Golden-yellow; black wings; black, yellow-tipped tail. **4b Ad female** More yellowish-olive above than male; streaked dusky below. **4c Imm** Greener above than adult female; whiter below with more distinct streaks. **4d Ad male in flight** Wings black. ▲ Wooded savanna. ✤ Melodious *weehla-weeoo*; also a harsh note. Mostly silent in W Africa. [CD4:51]

5 AFRICAN GOLDEN ORIOLE *Oriolus auratus* *c.* 24 cm **M lc**
5a Ad male Golden-yellow; black mask; black wings and tail broadly edged yellow. **5b Ad female** More yellowish-olive above than male; lightly streaked dusky below. **5c Imm** Yellowish-olive above; no mask; heavily streaked below. Compare 4c. **5d Ad male in flight** Black flight feathers broadly edged yellow. ▲ Wooded savanna. ✤ Various loud, melodious whistled phrases *wheetoliuw* and *tooleeoo* with longer variations; also harsh mewing *whrèèèh* or *mwaaarr*. [CD13:28]

Maps on page 417

1c
1a
1b
2b
3b
3a
2a
4d
4c
4b
4a
5d
5c
5b
5a

PLATE 127: DRONGOS AND CROWS

Drongos. Arboreal species with black, usually glossy, plumage and stout, slightly hooked bills. Eye red or orange in adults. Juveniles less glossy, with shorter tails. Hunt from perch, capturing insects on the wing like flycatchers. Conspicuous, bold, pugnacious and vocal. Voice varied and consisting of harsh, scolding notes interspersed by musical whistles.

1 FORK-TAILED DRONGO *Dicrurus adsimilis* 22.5–25.0 cm **R c/lf**
Wholly glossy blue-black; forked tail; red eye. ▲ Wooded savanna. Perches conspicuously and upright. ✤ Vigorous, discordant, grating and twanging notes. Song a loud medley of harsh, creaking, metallic notes and clear whistles. Sometimes includes imitations. [CD14:5b]

2 VELVET-MANTLED DRONGO *Dicrurus modestus* 24–27 cm **R c**
D. m. coracinus Unglossed velvety black mantle; deeply forked 'fish-tail'. ▲ Forest and forest edge. ✤ Similar to 1. [CD14:6a]
Príncipe form ***modestus***: see Plate 146:6.

3 SHINING DRONGO *Dicrurus atripennis* *c.* 21 cm **R u/c**
Strongly glossed blue-green; slightly forked tail; red eye. ▲ Mid-levels of forest interior. Often in mixed-species flocks. ✤ Short, rapid series of 4–6 explosive, ringing, discordant and harsh notes. Common phrase is fast *kwikwikwi-kwee-kwit* often preceded by rasping chatter. Also *whut-whut cheree* and *kzrr-kzrr tweet-tweet* and variations. Sometimes includes imitations. [CD14:5a]

4 SQUARE-TAILED DRONGO *Dicrurus ludwigii* 19 cm **R f/r**
Black with slight purplish-blue gloss; slightly notched tail; orange-red eye. ▲ Forest edge, second growth, thickets. ✤ Call *whit whit*. Song: series of short, explosive notes; also characteristic, fast, melodious *whididididid* followed by nasal, weaver-like *jeeeeezz*; hard *rrrwee rrwee*, rapid *chichichi* and *chi-rrrwee*, and rasping notes. [CD14:4b]

Crows. Medium-sized to large with stout bills and strong legs and feet. Plumage either wholly black, including bill and legs, or black with some white, brown or grey. Omnivorous; foraging mainly on ground. Calls mostly loud and harsh.

5 PIAPIAC *Ptilostomus afer* 35 cm **R c/u**
5a Ad Long (up to 28 cm), stiff, steeply graduated tail. **5b Imm** Pinkish bill with black tip. **5c In flight** Drab brown primaries (from above); ashy flight feathers (from below). ▲ Wooded savanna, where partial to palms. Not shy; often common in or near villages. Frequently associates with domestic stock. Usually in small flocks. ✤ Shrill, squeaking and scolding calls. [CD14:6b]

6 PIED CROW *Corvus albus* 46–50 cm **R c**
6a Black and white below; white collar on hindneck. **6b In flight** White on underparts visible from great distance. ▲ In all habitats except closed forest; typically near habitation and cultivation. ✤ Harsh *kwaar*, deep, guttural croak *kaarrh* or *kroh*, etc. [CD14:7a]

7 WESTERN JACKDAW *Corvus monedula* 33 cm **V**
Small. Black; grey neck; whitish eyes. ▲ Palearctic vagrant, N Mauritania. ✤ Diagnostic *kya* or *chak*. [CD4:62]

8 FAN-TAILED RAVEN *Corvus rhipidurus* 47 cm **R ls**
8a All black; wingtips project well beyond tail. **8b In flight** Very short tail; broad wings. ▲ Desert. ✤ High-pitched *kraah-kraah*, guttural *errrrow*, hollow *wok!* etc. [CD14:8a]

9 BROWN-NECKED RAVEN *Corvus ruficollis* 52–56 cm **R+ c/r**
9a Black; brown tinge to head and breast (often difficult to see). **9b In flight** Wedge-shaped tail. ▲ Desert, semi-desert. ✤ Harsh *karr*. [CD4:65]

Maps on page 422

1
2
3
4
5b
5a
6a
7
5c
6b
9b
8b
8a
9a

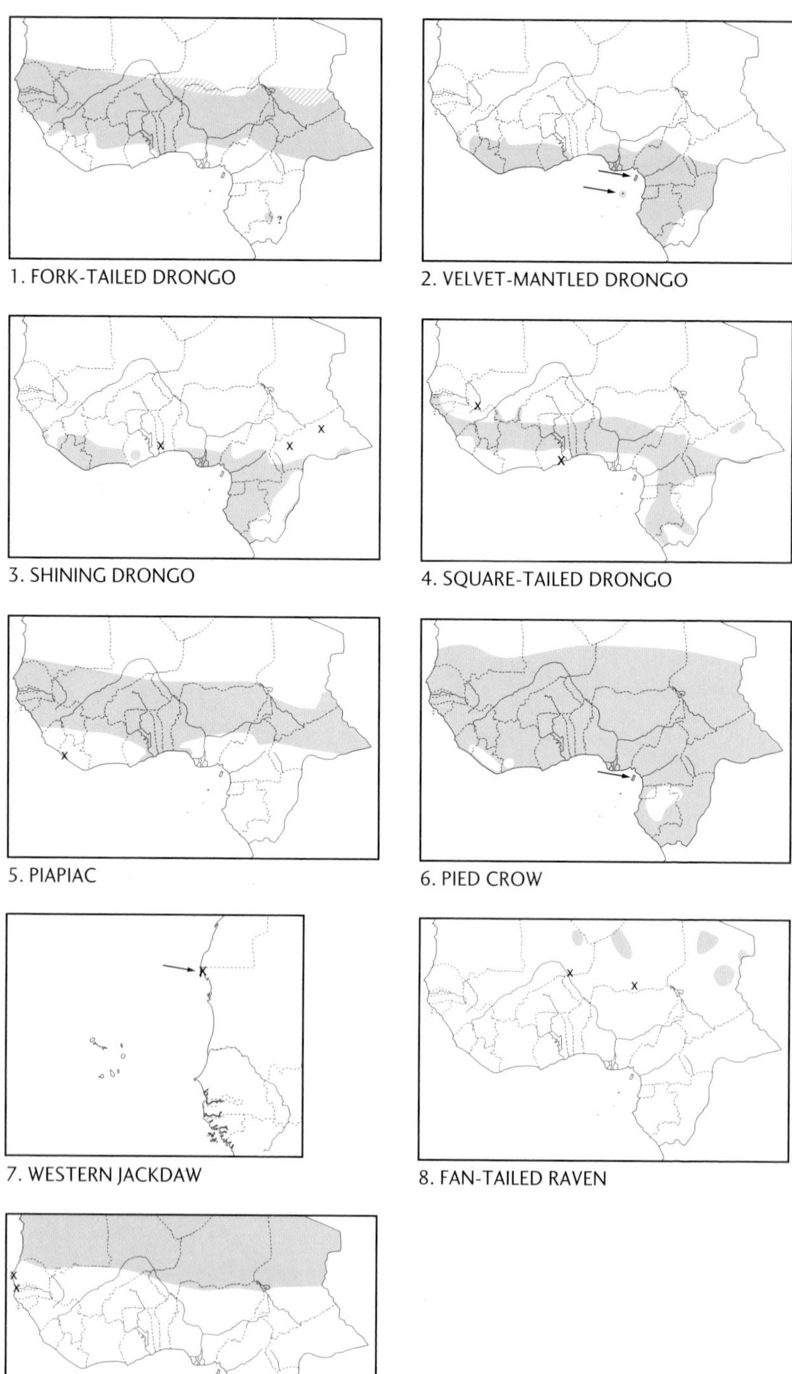

1. FORK-TAILED DRONGO

2. VELVET-MANTLED DRONGO

3. SHINING DRONGO

4. SQUARE-TAILED DRONGO

5. PIAPIAC

6. PIED CROW

7. WESTERN JACKDAW

8. FAN-TAILED RAVEN

9. BROWN-NECKED RAVEN

Plate on page 420

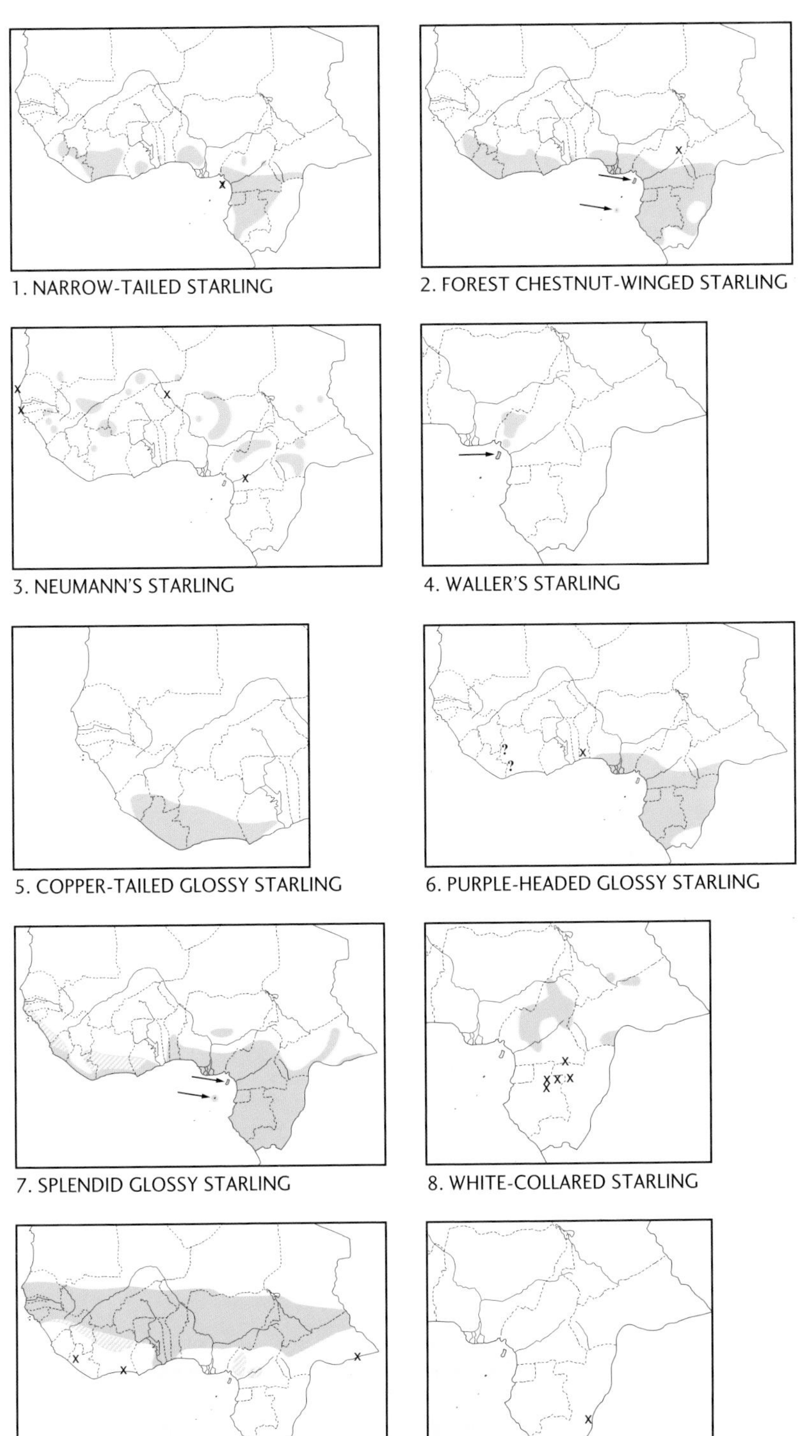

1. NARROW-TAILED STARLING

2. FOREST CHESTNUT-WINGED STARLING

3. NEUMANN'S STARLING

4. WALLER'S STARLING

5. COPPER-TAILED GLOSSY STARLING

6. PURPLE-HEADED GLOSSY STARLING

7. SPLENDID GLOSSY STARLING

8. WHITE-COLLARED STARLING

9. PURPLE GLOSSY STARLING

10. CAPE GLOSSY STARLING

Plate on page 424

PLATE 128: STARLINGS I

Mainly arboreal species with strong and pointed bills and sturdy legs. Juveniles a dull version of adult or different. Omnivorous, most feeding on fruit and insects. Most forage in flocks, roost communally and nest in holes. Calls mostly harsh and grating but also including pleasing sounds.

1 NARROW-TAILED STARLING *Poeoptera lugubris* 20–23 cm (tail 12 cm) R lf/s
1a Ad male Slender; very long, narrow, graduated tail; purple-black plumage; yellow eye. **1b Ad female** Much greyer. **1c Ad female in flight** Chestnut wing patch. ▲ Forest edges and clearings. ✤ Usually silent. Clear, whistled *wheew*; in flight a medley of shrill, cheeping notes. [CD14:9]

2 FOREST CHESTNUT-WINGED STARLING *Onychognathus fulgidus* 28–33 cm R f
Resembles 3 but occurs in different habitat. **2a Ad male *hartlaubii*** (Sierra Leone–S CAR, Bioko) Glossy purple-black; head with metallic green reflections. *O. f. intermedius* (Gabon–Congo) has reflections on head more bluish. Nominate (São Tomé) similar but usually substantially larger. **2b Ad female *hartlaubii*** Head and throat streaked ash-grey. **2c In flight** Chestnut wing patch smaller and darker than 3c. ▲ Forest and its outliers. ✤ Various resonant, rasping and melodious whistles. [CD14:13]

3 NEUMANN'S STARLING *Onychognathus neumanni* 28–33 cm R lf/s
3a Ad male Large; long, graduated tail; glossy purplish-black plumage. **3b Ad female** Head and throat ash-grey streaked blue-black. **3c In flight** Conspicuous chestnut wing patch. ▲ Crags, rocky outcrops. ✤ Clear loud whistles, e.g. *twee-lee-uw*, *whutcheerrleeo* and *peeeo* etc., melodious and oriole-like; from perch and in flight. [CD14:12]

4 WALLER'S STARLING *Onychognathus walleri* 23 cm R lc
4a Ad male Tail relatively short; chestnut wing patch; glossy purple-black plumage; head with metallic green reflections. **4b Ad female** Head with some grey streaking. ▲ Montane forest. ✤ Clear, loud whistles *teeeuw-tee-wheew*, with variations. Also melodious *preeti preeti*, clear *tewee tewee* and *wheet*, and soft nasal note. Alarm *chrrra*. [CD14:11]

5 COPPER-TAILED GLOSSY STARLING *Lamprotornis cupreocauda* 20 cm R lc/r NT
Glossy purple head to upper breast; rest of plumage glossy blue-black; eye yellow. Juvenile has eye initially dark. ▲ Forest. Endemic. ✤ Various harsh, squeaky notes. [CD14:15]

6 PURPLE-HEADED GLOSSY STARLING *Lamprotornis purpureiceps* 20 cm R u/lc
Resembles 5 but mantle and belly glossy blue-green, eye brown. ▲ Forest. ✤ Rather silent. Various chattering and whistling notes. In flight a characteristic, metallic *pleep!* or *twink!* [CD14:16]

7 SPLENDID GLOSSY STARLING *Lamprotornis splendidus* 27–30 cm R/M c/f
Large. Broad blackish band on wings and tail; eye whitish. Swishing wings in flight. ▲ Forest and its outliers. ✤ Vocal. Harsh, piping and rasping calls with nasal and metallic quality, e.g. *niyar-èh*; *spiyok!*; *kwank!*; explosive *KYAH!* Makes considerable noise at roost. [CD14:24]

8 WHITE-COLLARED STARLING *Grafisia torquata* 21.5–23.0 cm R/V u/lc
8a Ad male Purple-black; broad white breast-band. **8b Ad female** Greyish-brown slightly glossed blue-purple above; dull greyish below. ▲ Forest-savanna mosaic, grasslands. ✤ Reminiscent of *Lamprotornis*; includes chirruping notes and three short whistles.

9 PURPLE GLOSSY STARLING *Lamprotornis purpureus* 24 cm R c/lu
Glossy purple head, underparts and tail; relatively long bill and short tail; large yellow eye. ▲ Savanna. ✤ Various harsh, chattering and whistling notes. [CD14:17]

10 CAPE GLOSSY STARLING *Lamprotornis nitens* 23 cm R r
All glossy blue-green. ▲ Savanna. S Congo (Gabon?). ✤ Pleasant, slurred, rolling *churweelee*. Song a sustained series of varied syllables, incorporating call. [CD14:18]

Maps on page 423

1c
1a
2b
2a
1b
3c
2c
3b
4b
4a
3a
5
6
7
8b
8a
9
10

PLATE 129: STARLINGS II AND OXPECKER

1 YELLOW-BILLED OXPECKER *Buphagus africanus* 21–23 cm R lu/lf
1a Ad Yellow, red-tipped bill. **1b Juv** Dusky bill. **1c In flight** Long, pointed wings and tail; contrasting pale rump. ▲ Mainly wooded savanna and bush. Associated with large herbivores. ✤ Hard rasping and hissing, metallic *krrizs* or *pszrr* in flight or when perched; also rattling *kzsririririri*... [CD14:31]

2 WATTLED STARLING *Creatophora cinerea* 21.5 cm V/M lr
2a Ad male breeding Head with wattles and bare yellow-and-black skin. **2b Ad male non-breeding** Pale, drab grey; black wings and tail; greater coverts whitish; primary-coverts white. Female similar, but greater and primary-coverts blackish. Worn wings all brownish. Juvenile as female but browner; bill dusky. **2c In flight** Pointed wings; short tail; contrasting greyish-white rump. ▲ Savanna. ✤ Usually silent when not breeding. [CD14:30]

3 VIOLET-BACKED STARLING *Cinnyricinclus leucogaster* 16–18 cm M c
3a Ad male Brilliant violet; white breast and belly. **3b Ad female** Dark brown above; white streaked brown below. ▲ Various wooded habitats. Seasonally common; irregular in some places. ✤ Clear, high-pitched, drawn-out whistle *vfeeeee*. Also soft nasal call, usually on take-off. Song a short rapid series of whistling, twanging and clicking notes. [CD14:29]

4 CHESTNUT-BELLIED STARLING *Lamprotornis pulcher* 19 cm R lc/u
4a Ad Greyish-brown head; rufous-chestnut belly. **4b Juv** Throat and breast ashy-brown. **4c Ad in flight** Creamy wing patch. ▲ Dry bushy country and farmland, often near villages. ✤ Vigorous *whirrr*. Song a series of soft liquid notes. [CD14:27]

5 EMERALD STARLING *Lamprotornis iris* *c.* 20 cm R ls DD
Brilliant emerald-green; head-sides and abdomen glossy purple. ▲ Wooded savanna. Endemic. ✤ Various squeaky notes. A drawn-out *wheeezz* on take-off.

6 BRONZE-TAILED GLOSSY STARLING *Lamprotornis chalcurus* 21.5 cm R lc/u
Resembles 7 but tail distinctly shorter and purple; eye orange-yellow to reddish-orange. Also compare 9. ▲ Wooded savanna. ✤ Whining, drawn-out *weeaah*, abrupt *plip!* and various other nasal, chattering and whistling notes. [CD14:19]

7 GREATER BLUE-EARED STARLING *Lamprotornis chalybaeus* 21–24 cm R c/f
7a Ad Glossy metallic green; belly purple; ear-coverts and rump bluish; eye yellow. Compare 9a. **7b Juv** As adult but duller. Darker, more blackish than 9b. ▲ Thorn scrub with trees to desert edge. ✤ Nasal, whining *wèèh-aa-ah* or *skwee-aar*. Song a varied series of whistling, mewing, chirping and clicking notes, including call. [CD14:20]

8 LONG-TAILED GLOSSY STARLING *Lamprotornis caudatus* *c.* 51 cm R f/c
Very long (up to 33 cm), graduated, supple tail. ▲ Open wooded savanna, farmland, thorn scrub. ✤ Various shrill and harsh notes. [CD14:25]

9 LESSER BLUE-EARED STARLING *Lamprotornis chloropterus* 19–20 cm R c/f
9a Ad Very similar to 7a but ear-coverts more contrasting blue-black, tail relatively shorter, rump concolorous with upperparts, size smaller, eye yellow to orange-yellow, voice different. **9b Juv** Earth-brown head and underparts. Paler below than 7b. ▲ Wooded savanna. ✤ In flight a clear, distinctive *wirree-wirree*; from perch various rather similar calls, incl. *wreet* and *cherwee*. Song a rather simple, rhythmic series of 6–12 separated, chirruping and whistling syllables, e.g. *chirp chirp peelu chirp chrew whip*. [CD14:21]

10 COMMON STARLING *Sturnus vulgaris* 19–22 cm V+
Ad non-breeding Glossy black spotted buff above and white below; long pointed bill; short tail; triangular wings. ▲ Palearctic vagrant. ✤ Grating *cheerr*; also various other harsh whistling sounds. [CD4:67]

Maps on page 428

1b
1a
1c
2a
2b
2c
4c
4b
3b
3a
4a
6
5
8
7a
7b
10
9a
9b

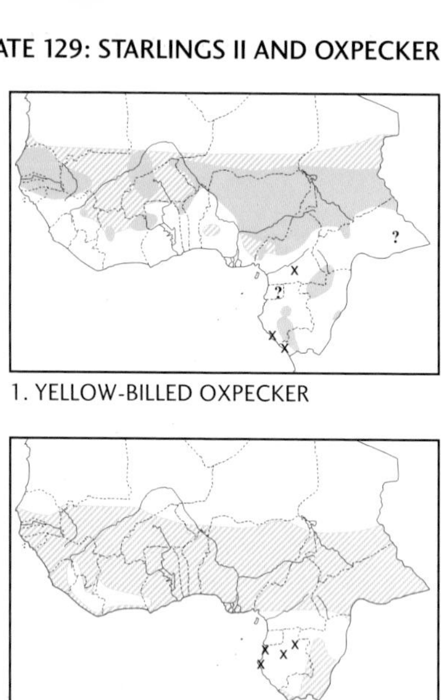

1. YELLOW-BILLED OXPECKER

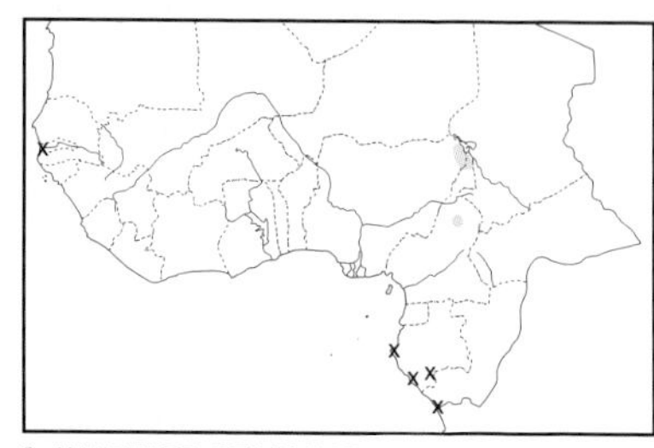

2. WATTLED STARLING

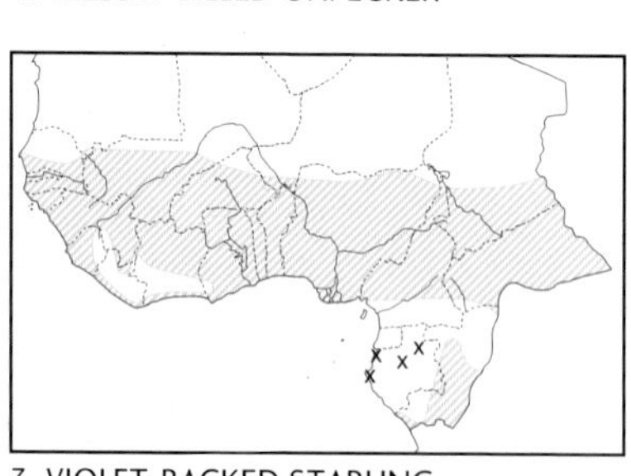

3. VIOLET-BACKED STARLING

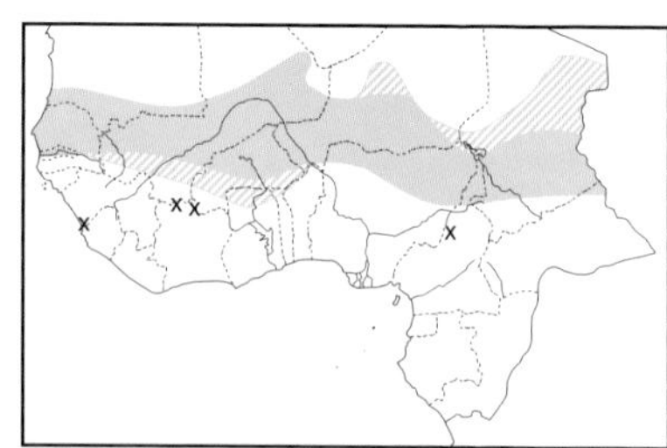

4. CHESTNUT-BELLIED STARLING

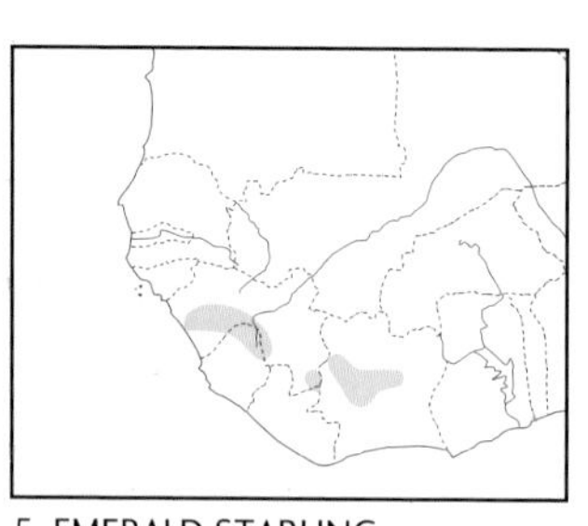

5. EMERALD STARLING

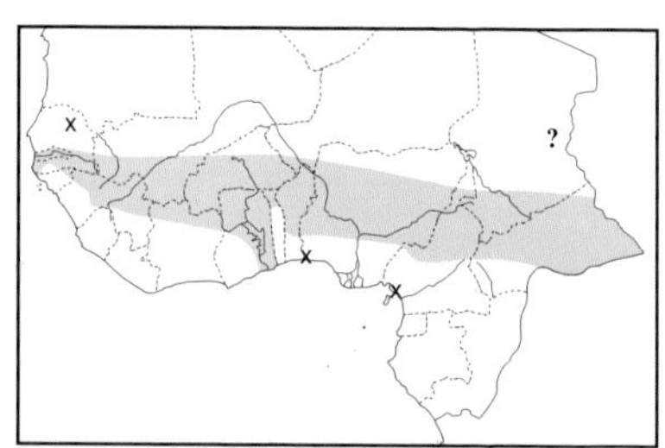

6. BRONZE-TAILED GLOSSY STARLING

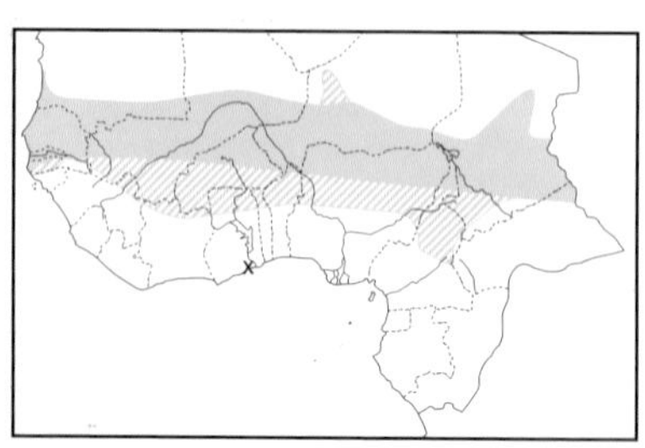

7. GREATER BLUE-EARED STARLING

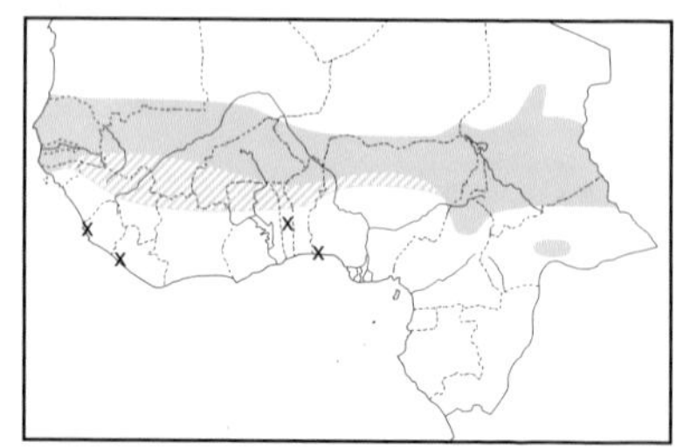

8. LONG-TAILED GLOSSY STARLING

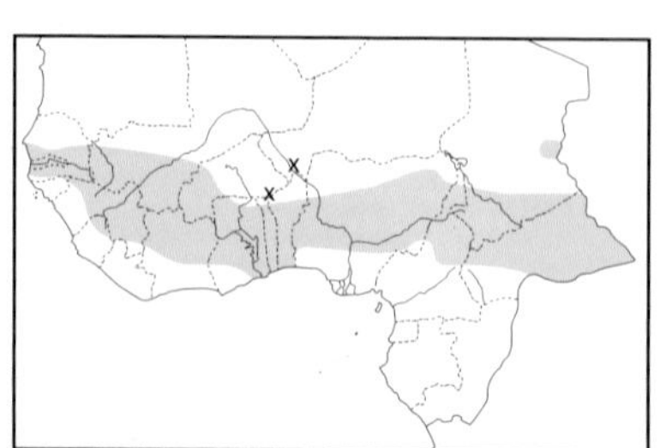

9. LESSER BLUE-EARED STARLING

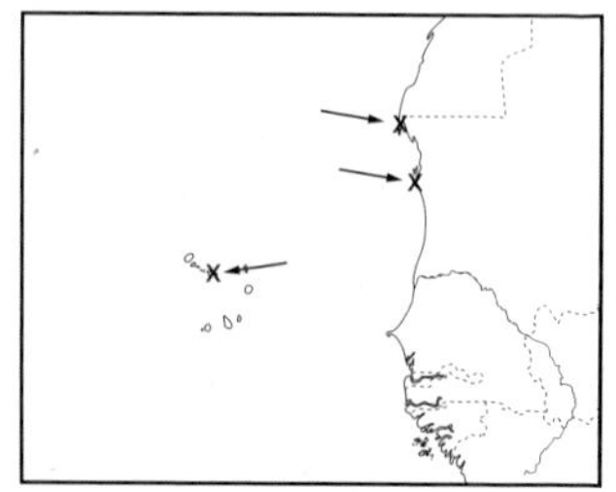

10. COMMON STARLING

Plate on page 426

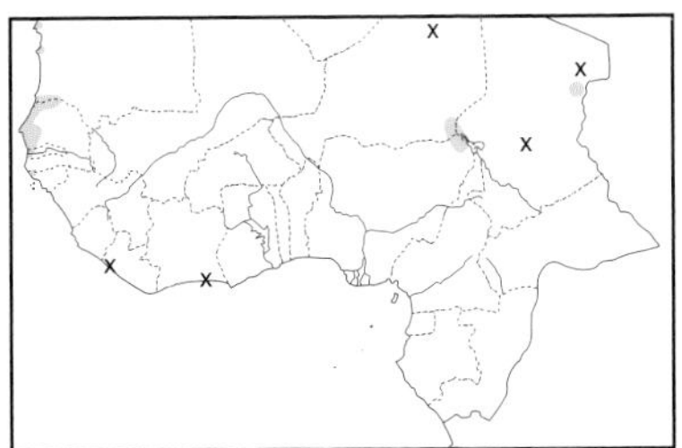

1. HOUSE SPARROW

2. EURASIAN TREE SPARROW
4. KORDOFAN RUFOUS SPARROW

3. SPANISH SPARROW

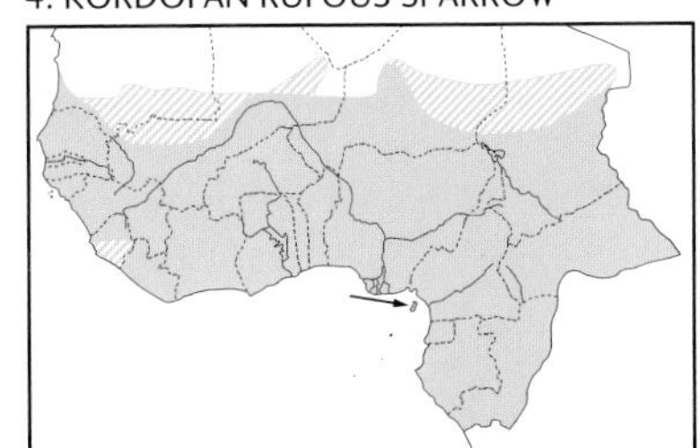

5. NORTHERN GREY-HEADED SPARROW

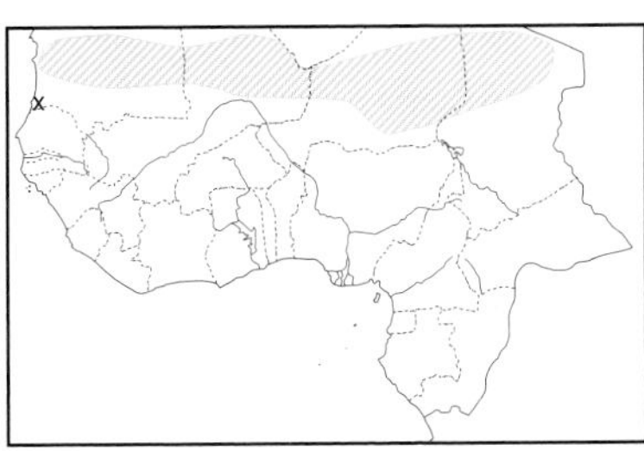

6. DESERT SPARROW

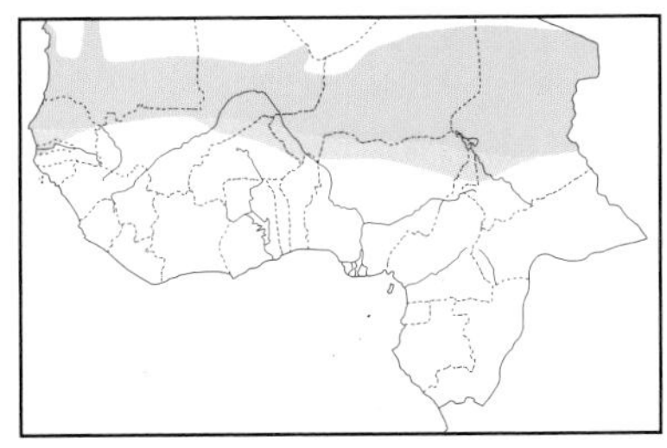

7. SUDAN GOLDEN SPARROW

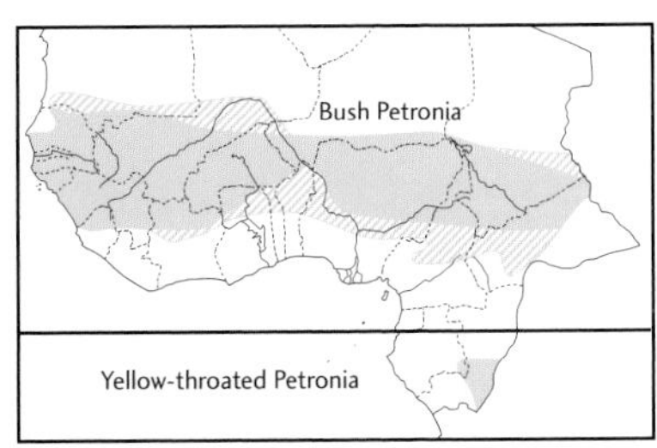

8. BUSH PETRONIA
9. YELLOW-THROATED PETRONIA

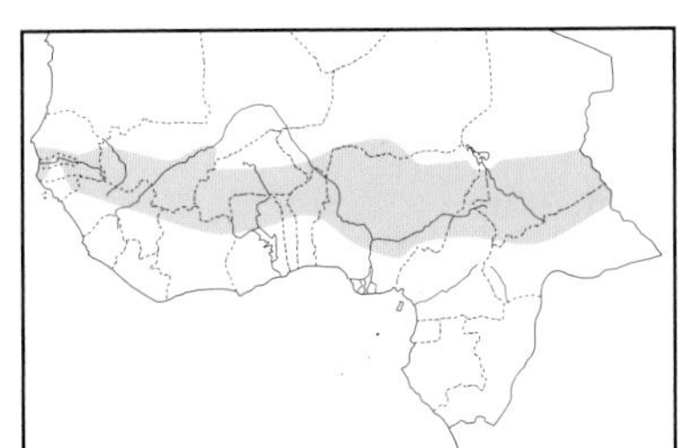

10. CHESTNUT-CROWNED SPARROW WEAVER

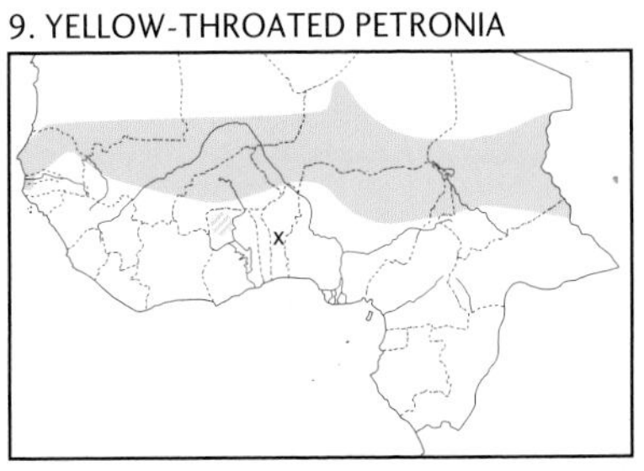

11. SPECKLE-FRONTED WEAVER

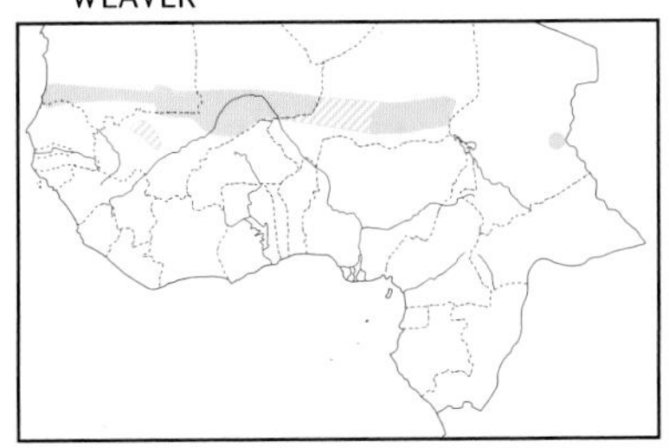

12. YELLOW-SPOTTED PETRONIA

Plate on page 430

Sparrows. Mostly brown and grey with short, conical bills. Bill of male typically turns from horn coloured to black in breeding season. Forage on or near the ground, feeding largely on seeds. Many species gregarious. Vocal, uttering a variety of rather harsh chirping calls.

1 HOUSE SPARROW *Passer domesticus* 14–15 cm R+ lc
1a Ad male breeding Grey crown and rump; black bib; whitish head-sides. **1b Ad female** Dull brown; buff supercilium; horn-coloured bill. ▲ Towns. Introduced. ✤ Various chirps; rattling *churrr-r-r-t-t*. Song a monotonous series of chirping notes. [CD4:69, 14:33]

2 EURASIAN TREE SPARROW *Passer montanus* 14 cm
Resembles 1a, but has chestnut crown and black ear-spot. ▲ Palearctic species claimed from Guinea (Conakry harbour). Ship-assisted or escape? ✤ [CD4:71]

3 SPANISH SPARROW *Passer hispaniolensis* 15 cm R*/V? c
3a Ad male breeding Chestnut crown; black bib extending onto breast; heavily streaked flanks. **3b Ad female** As 1b but more streaky. ▲ Towns, cultivated areas. ✤ Similar to 1. [CD4:70]

4 KORDOFAN RUFOUS SPARROW *Passer cordofanicus* 14 cm R lc
4a Ad male breeding Grey crown; crown-sides and upperparts rich chestnut; face and underparts whitish. **4b Ad female** Paler; bib dusky-grey. ▲ Arid scrub. ✤ Similar to 1. [CD14:34]

5 NORTHERN GREY-HEADED SPARROW *Passer griseus* 14 cm R c
Grey head and underparts; rich chestnut upperparts. ▲ Various habitats, but mainly towns and villages. ✤ Monotonous *chirp* or *cheerp*. Also rattling *churrr-r-r-t-t* when alarmed or excited. Song a monotonous series of chirps, including call. [CD14:35]

6 DESERT SPARROW *Passer simplex* 13.5 cm R lu
6a Ad male Pale greyish above; black mask, bib and bill. **6b Ad female** Pale sandy-buff; no face markings; bill pale horn. ▲ Desert. ✤ Chirps similar to 1. [CD4:73]

7 SUDAN GOLDEN SPARROW *Passer luteus* 14 cm R/M c
7a Ad male breeding Yellow head and underparts; chestnut upperparts; two white wing-bars. **7b Ad female** Buffish-brown head and upperparts; face and underparts washed yellow. ▲ Thorn scrub. Nomadic. Often in huge flocks. ✤ Chirping *chilp* or *chirrup*. In flight also fast *che-che-che-...* Song a monotonous series of chirps, including call. [CD14:36]

8 BUSH PETRONIA *Petronia dentata* 13 cm R/M u/lf
8a Ad male Chestnut supercilium; grey crown. **8b Ad female** Buff supercilium; brown crown. ▲ Wooded grassland. Most often in trees; rarely on ground. ✤ Sparrow-like chirps. [CD14:39]

9 YELLOW-THROATED PETRONIA *Petronia superciliaris* 15 cm R u/lc
9a Ad Broad whitish supercilium. **9b Juv** Warmer brown above; supercilium brownish-buff. ▲ Wooded grassland, cultivation. Forages in trees, rarely on ground. ✤ Series of 3 identical sparrow-like chirps *chreep-chreep-chreep*; also fast *chreechreechree*. [CD14:38]

10 CHESTNUT-CROWNED SPARROW WEAVER *Plocepasser superciliosus* 17 cm R u/s
Rufous-chestnut crown; white supercilium; black malar stripe. ▲ Thorn scrub, wooded grassland. ✤ Fast, metallic ticking trill *trrrrit* or *trrirr*, somewhat recalling *Euplectes*. Song a short, continuous phrase *witsweeweetsweeuseesweeuseeswee*. [CD14:42]

11 SPECKLE-FRONTED WEAVER *Sporopipes frontalis* 12 cm R u/lc
Small. Pale rufous nape; forehead and moustachial streak black speckled white. ▲ Dry savanna. ✤ Short, silvery trill *sriiii* or *tsrrrrk* and series of mainly high-pitched, short notes *tsisit chep tsisisit che ...*; rapid *tsip-tsip-tsip-tsip* when taking flight. [CD14:41]

12 YELLOW-SPOTTED PETRONIA *Petronia pyrgita* 15 cm R s/lu
Buffish-brown above; creamy-white below; small lemon patch on throat (difficult to see). ▲ Thorn scrub. ✤ Sparrow-like notes, *chirp*, *chillip* and *wurli*. Usually silent. [CD14:37]

 Maps on page 429

1b
1a
2
4b
4a
3b
3a
5
6b
6a
7b
8b
9b
7a
8a
9a
10
11
12

PLATE 131: TYPICAL WEAVERS I

Small to medium-sized birds with generally strong, conical bills. Many species gregarious, feeding and roosting in flocks and nesting in colonies. Vocalisations typically consist of characteristic, drawn-out, wheezy, buzzy, chirping and chattering notes.

Genus *Ploceus* (typical weavers). Males in breeding dress mostly bright yellow; females dull and streaky (sparrow-like). In non-breeding dress (if any) males similar to females.

1 LITTLE WEAVER *Ploceus luteolus* 11.5 cm **R c/f**

1a Ad male breeding Small; large black mask; mantle variably streaked dusky. **1b Ad male non-breeding** No mask; crown yellow-green; head-sides to breast buff; mantle buff streaked dusky. **1c Ad female breeding** No mask; head-sides to breast pale yellow; mantle yellow-green streaked dusky. Non-breeding female duller; bill horn. ▲ Dry wooded savanna. ✤ Soft *tsssp*, harsh *chep*, and jumble of chattering and chirping notes. [CD14:47]

2 SLENDER-BILLED WEAVER *Ploceus pelzelni* 11 cm **R lf**

2a Ad male As 1a but underparts deeper yellow; upperparts darker green, unstreaked. **2b Ad female** As 1b but brighter yellow; distinct yellow supercilium; unstreaked upperparts. ▲ Mangroves, marshes, along coastal lagoons and river banks. ✤ Jumble of subdued chattering and chirping notes. [CD14:46]

3 LOANGO WEAVER *Ploceus subpersonatus* 15 cm **R u VU**

3a Ad male As 2a but plumage duller, washed brownish, size larger. **3b Ad female** As 2b but darker and larger. ▲ Coastal scrub, mangroves, clumps of palm trees. ✤ Soft jumble of characteristic weaver notes; also a more melodious, subdued song. [CD14:48]

4 COMPACT WEAVER *Pachyphantes superciliosus* 15 cm **R u/lf**

4a Ad male breeding Black mask; short, stout bill; chestnut forehead; dark upperparts; dark eye. **4b Ad female breeding** As male, but forehead and centre of crown black. **4c Ad non-breeding** Brownish; blackish top of head and eye-stripe. ▲ Grassland. ✤ Crackling calls in flight. Also *cheee* and *pink*. Song: abrupt notes followed by typical weaver sizzling. [CD14:67]

5 VITELLINE MASKED WEAVER *Ploceus vitellinus* 14 cm **R c/f**

5a Ad male breeding Black mask (with narrow band on forehead); chestnut wash to crown and lower throat; red eye. Compare 6a and 8a. **5b Ad male non-breeding/ad female** No mask; brownish-olive above; legs pinkish. Compare 6b. ▲ Dry savanna. ✤ Jumble of chattering and wheezy notes. Also *tzik*. [CD14:58]

6 HEUGLIN'S MASKED WEAVER *Ploceus heuglini* 14 cm **R u/f**

6a Ad male breeding Black mask extending to upper breast; no chestnut wash on head and throat; eye yellow. **6b Ad female** As 5b but less distinct streaking on mantle. Bill brown to black (breeding) or horn (non-breeding). Non-breeding male similar but yellower below. ▲ Wooded savanna. ✤ Jumble of chattering and wheezy notes; more varied and including more melodious notes than most other *Ploceus*. [CD14:60]

7 ORANGE WEAVER *Ploceus aurantius* 15 cm **R lf**

7a Ad male Bright yellow-orange head and underparts. **7b Ad female** Olive above; white below; pale yellow head-sides to upper breast. ▲ Near water, coastal zone; inland along major rivers. Gregarious. ✤ Characteristic weaver chattering at colony. [CD14:54]

8 LESSER MASKED WEAVER *Ploceus intermedius* 14 cm **R lr/s**

8a Ad male breeding As 5a but mask larger; mantle indistinctly streaked; eye yellow. **8b Ad male non-breeding** No mask; similar to female. **8c Ad female** Yellowish-green above; head-sides yellow; legs bluish-grey. ▲ Near water. ✤ Jumble of chattering and wheezy notes. [CD14:57]

9 HOLUB'S GOLDEN WEAVER *Ploceus xanthops* 17–18 cm **R lr/s**

9a Ad male Large; mainly golden-yellow; heavy, black bill; yellow eye. Female is slightly duller. **9b Juv** Olive head; breast and flanks washed buff; horn-coloured bill. ▲ Open savanna. Not gregarious. ✤ Jumble of chattering and wheezy notes. Also sparrow-like chirp. [CD14:53]

Maps on page 434

1c
1b
2b
3b
4b
1a
2a
3a
4c
5b
6b
5a
4a
7b
8b
9b
6a
7a
8a
8c
9a

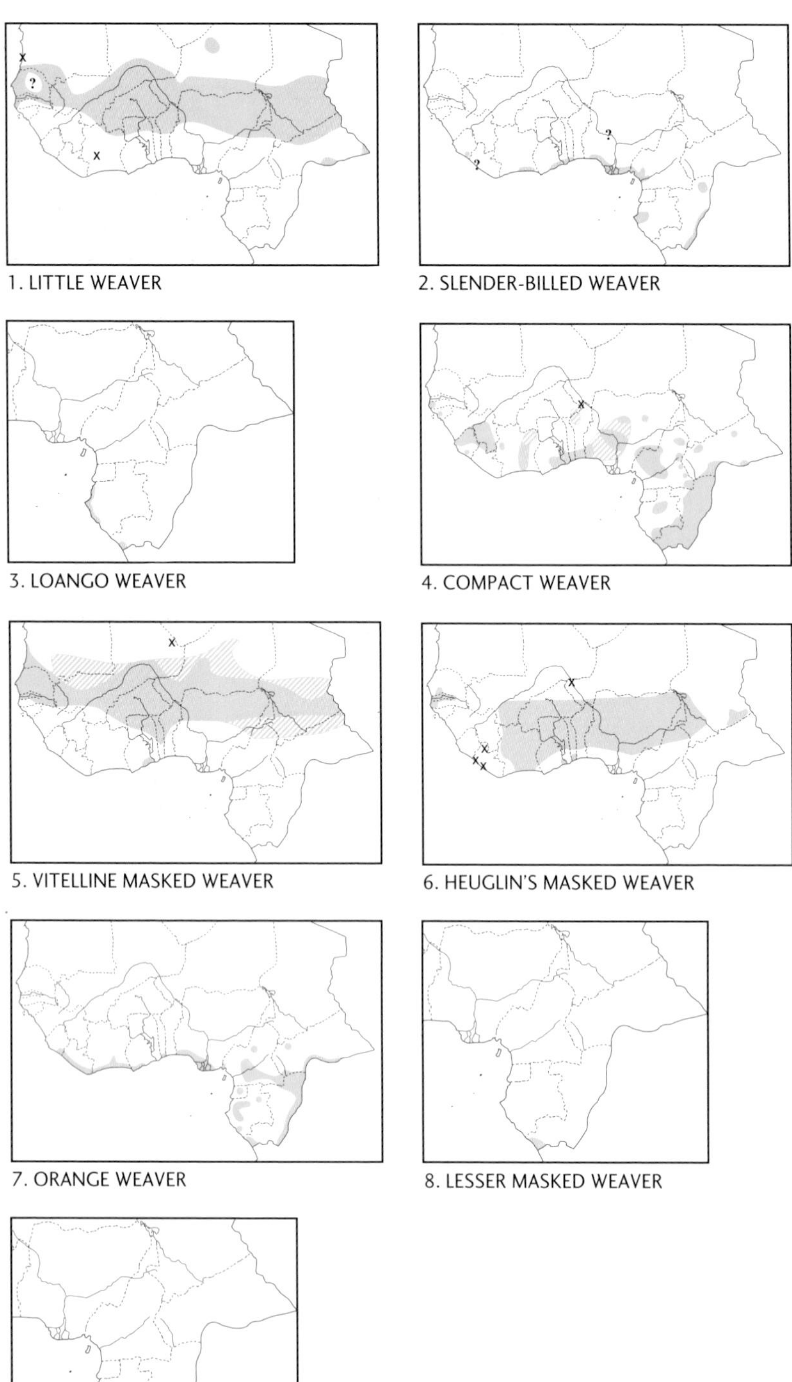

1. LITTLE WEAVER

2. SLENDER-BILLED WEAVER

3. LOANGO WEAVER

4. COMPACT WEAVER

5. VITELLINE MASKED WEAVER

6. HEUGLIN'S MASKED WEAVER

7. ORANGE WEAVER

8. LESSER MASKED WEAVER

9. HOLUB'S GOLDEN WEAVER

Plate on page 432

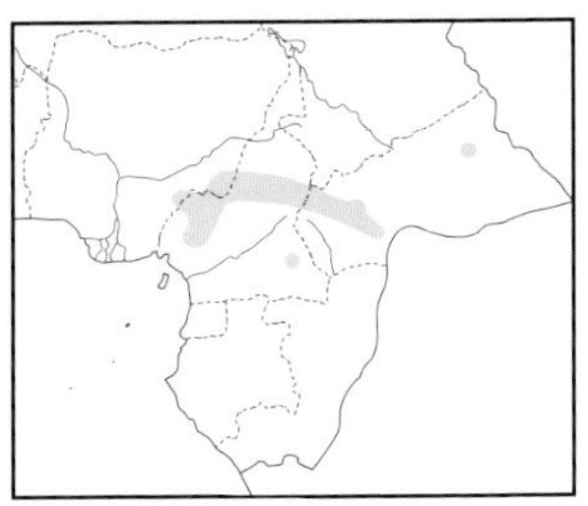
1. BAGLAFECHT WEAVER

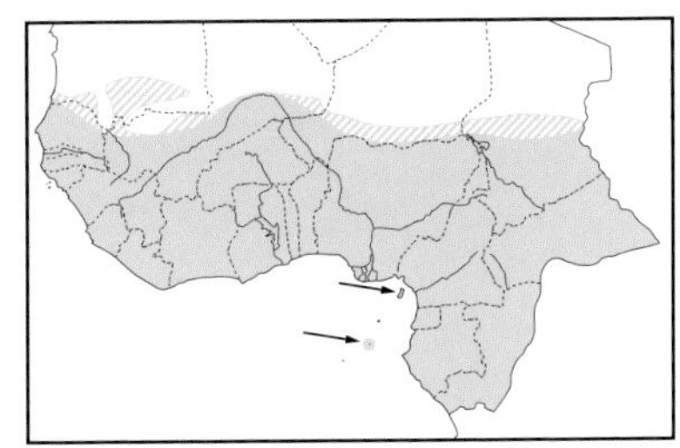
2. VILLAGE WEAVER

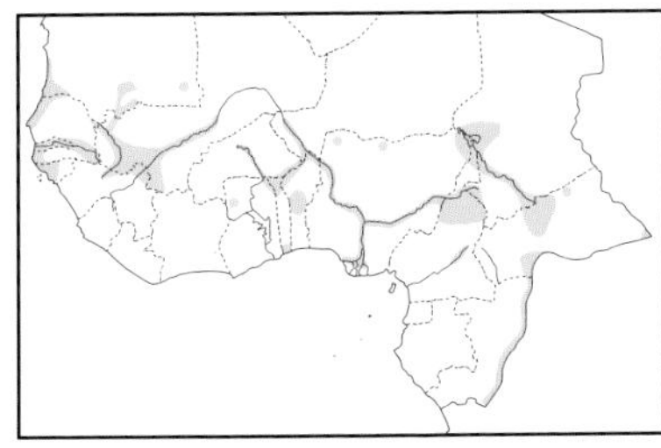
3. BLACK-HEADED WEAVER

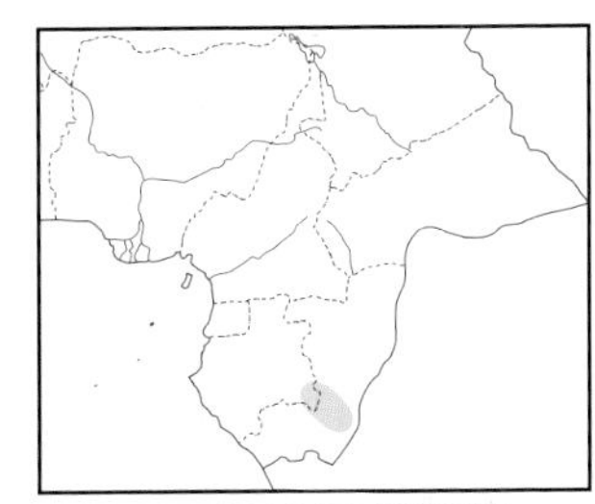
4. BLACK-CHINNED WEAVER

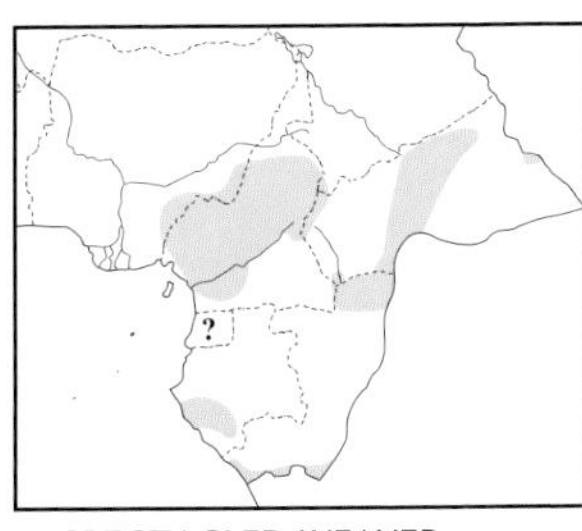

5. SPECTACLED WEAVER

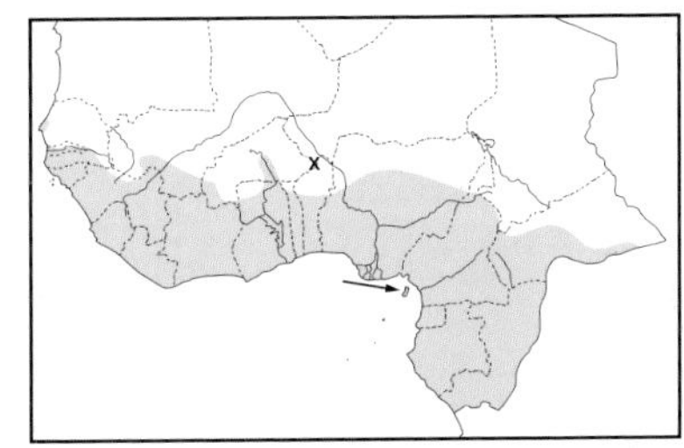

6. BLACK-NECKED WEAVER

Plate on page 436

1 BAGLAFECHT WEAVER *Ploceus baglafecht* 15 cm **R s/u**

1a Ad male breeding Bright yellow forehead; black mask with contrasting yellow eye; yellow-green mantle. **1b Ad female breeding** As male, but top of head yellow-green; dusky lores. **1c Ad non-breeding** Dusky area around eye; underparts whitish washed buff. ▲ Forest edge, montane (in west) and lowland (in east) areas. ✤ Shrill chattering and wheezy notes. Main call a distinctive, loud, repeated *zwenk*. [CD14:44]

2 VILLAGE WEAVER *Ploceus cucullatus* 15.0–17.5 cm **R c**

2a Ad male *cucullatus* breeding (throughout most of range) Black head; chestnut collar; black V bordering mantle. *P. c. bohndorffi* (Gabon–Congo) has less black on crown, more chestnut on breast. **2b Ad male *cucullatus* non-breeding** Lemon-yellow throat; underparts variable (yellow, or white with buffish breast and flanks). **2c Ad female breeding** Yellow-olive head contrasts with brownish-olive upperparts; lemon-yellow below; belly white. Non-breeding female duller; bill paler. **2d Ad male *collaris* breeding** (coastal Gabon–S Congo) No chestnut collar; no black V on upperparts; dark chestnut breast. ▲ The ubiquitous, well-known gregarious weaver, common in towns and villages. ✤ Harsh chattering and twittering notes at colony. Song a jumble of rasping, chattering notes ending in drawn-out rattling wheeze, *chit-chit chit-t-t-t-t shirrrzzzzwrrerr*. Alarm a sharp *zip*. In flight a short *chuk-chuk* or *chit-chit*. [CD14:63]

3 BLACK-HEADED WEAVER *Ploceus melanocephalus* 14–15 cm **R lc/f**

3a Ad male *melanocephalus* breeding (north of range) Black head; golden nuchal collar; yellowish-olive upperparts. **3b Ad male *capitalis* breeding** (south of range) Chestnut wash on breast. **3c Ad male non-breeding** Brownish-olive head; rest of plumage mainly brownish, buff and white. Female similar but smaller. ▲ Near water in semi-arid savanna. ✤ Discordant chattering and twittering; variety of calls, incl. hoarse, nasal *chè-èp*, *tshk* and *tseew*. [CD14:64]

4 BLACK-CHINNED WEAVER *Ploceus nigrimentus* *c.* 17 cm **R lf**

4a Ad male Black mask; black upperparts; wing feathers edged yellow. **4b Ad female** As male but head and nape black. ▲ Wooded grassland. Often perches atop trees. ✤ Harsh calls. Song a rather short, distinctive *whit-pui-pui-trrrrr-pui*.

5 SPECTACLED WEAVER *Ploceus ocularis* 16–17 cm **R f/s**

5a Ad male As 6a but head yellower; bill more slender. **5b Ad female** As male but without black on throat. ▲ Various wooded habitats. Often creeping through dense vegetation and hanging upside-down in search of insects. Joins mixed-species flocks. Rather shy and skulking, rarely emerging from cover. ✤ Distinctive. Fast, descending series of identical, resonant, piping notes *teeteeteeteeteetee*. Also *chirrrrdzrweew* and *chit*. [CD14:50]

6 BLACK-NECKED WEAVER *Ploceus nigricollis* 17 cm **R c**

6a Ad male *brachypterus* (from W Cameroon west) Golden-chestnut head; black eye-stripe and throat; yellowish-olive upperparts. *P. n. po* (Bioko) similar. **6b Ad female *brachypterus*** No black on throat; top of head yellowish-olive. **6c Ad male *nigricollis*** (from W Cameroon east) As 6a but upperparts blackish. **6d Ad female *nigricollis*** Top of head black; distinct yellow supercilium. ▲ Forest edges, wooded savanna. Occasionally joins mixed-species flocks. Forages for insects in leafy trees. ✤ Harsh *chet-chet* and a twittering *wheeze*. Usually rather silent. [CD14:49]

Maps on page 435

1c
1b
2d
2c
1a
2b
3c
2a
4b
3b
4a
3a
5b
6d
6b
5a
6c
6a

PLATE 133: TYPICAL WEAVERS III

1 **MAXWELL'S BLACK WEAVER** *Ploceus albinucha* 15 cm R u/f

1a Ad *albinucha* (west of range) All black except greyish nuchal patch; greyish-white eye. **1b Ad *holomelas*** (east of range) No nuchal patch. Nominate (Bioko) similar. **1c Juv** Dark sooty-grey, paler below. ▲ Canopy or mid-levels of mature lowland forest. In small groups. Searches branches for insects, often hanging upside-down. ✤ Jumble of chattering and wheezy notes at colony. [CD14:66]

2 **VIEILLOT'S BLACK WEAVER** *Ploceus nigerrimus* 17 cm R c

2a Ad male *nigerrimus* (Cameroon to CAR–Congo) All black. **2b Ad male *castaneofuscus*** (W Guinea–SE Nigeria) Black and chestnut; yellow eye. **2c Ad female *castaneofuscus*** Brownish above; buffish-yellow below with some rufous wash. ▲ Forest clearings and edges near villages, marshy habitats. ✤ Shrill jumble of chattering notes, esp. at colony. [CD14:61-62]

3 **BROWN-CAPPED WEAVER** *Ploceus insignis* 14 cm R lf

3a Ad male Golden-yellow; chestnut cap; black mask, wings and tail. **3b Ad female** All-black head. **3c Juv** Head olive mottled black; no black on throat; horn-coloured bill. ▲ Montane and submontane forest. Behaviour as 5. ✤ Usually silent. Short *twit* and wheezy sizzling. [CD14:69]

4 **YELLOW-MANTLED WEAVER** *Ploceus tricolor* 17 cm R f

4a Ad Broad yellow crescent on upper mantle; deep chestnut breast and belly. **4b Juv** Dull chestnut; wings and tail dull black. ▲ Lowland forest. ✤ Usually silent. Sharp *cherrit*; at nest, a typical weaver wheezing followed by short whistled note. [CD14:65]

5 **PREUSS'S (GOLDEN-BACKED) WEAVER** *Ploceus preussi* 14 cm R s/u

5a Ad male Golden-chestnut crown; yellow stripe from nape to uppertail-coverts. **5b Ad female** Black forehead; no chestnut wash on breast. **5c Juv** Pattern as adult female, but black on head replaced by yellow-olive; bill horn. ▲ Lowland forest. Climbs large branches and trunks of high trees in search of insects, often hanging upside-down. Joins mixed-species flocks. ✤ Usually silent. Harsh *chwep*. [CD14:70]

6 **DARK-BACKED WEAVER** *Ploceus bicolor* 15 cm R lf

Black head; blackish-slate upperparts; bright yellow underparts; blue-grey bill. Juvenile has mottled throat. ▲ Upper and mid-strata of montane and lowland forest. May cling to trunks like woodpecker when searching for insects. ✤ Song a short series of pleasant, melodious whistles ending in squeaky sizzling *who-he-who-he-hu kshshshrrr*, with variations. Call a high-pitched *pink-pink* and *zzrree*; alarm a rapid *tsi-tsi-tsi-*... [CD14:68]

7 **YELLOW-CAPPED WEAVER** *Ploceus dorsomaculatus* 14 cm R s

7a Ad male Similar to 5a but rump and uppertail-coverts black. **7b Ad female** Black head. ▲ Lowland forest. Searches foliage in search of insects; also flycatches in open canopy. Not known to forage on bark. Joins mixed-species flocks. ✤ Usually silent.

8 **BATES'S WEAVER** *Ploceus batesi* 14 cm R r EN

8a Ad male Chestnut head; black upper throat; yellowish-green upperparts and tail. **8b Ad female** As male but head black, throat yellow. **8c Juv** As adult female but head olive mottled black; bill horn coloured. ▲ Lowland forest. Endemic. ✤ Unknown.

9 **BLACK-BILLED WEAVER** *Ploceus melanogaster* 14 cm R lf

9a Ad male Black with yellow head; black throat bordered by yellow band. **9b Ad female** As male but throat yellow. **9c Juv** Pattern as adult female; head washed olive; underparts yellowish-olive. ▲ Montane forest. ✤ Song consists of clear, ringing notes ending in drawn-out rattling wheeze. Call a harsh *chet*. [CD14:51]

10 **BANNERMAN'S WEAVER** *Ploceus bannermani* 14 cm R lf VU

Black mask; yellowish-green upperparts and tail. ▲ Montane forest and scrub. Endemic. ✤ Sharp *pritt*. Song high pitched and with drawn-out wheeze, more pleasant than many other *Ploceus* weavers, *psi-psisi-trrir-si-psiuuuu-tsisisi-irrrr*, with variations. [CD14:45]

Maps on page 440

1b
1c
1a
2c
2a
2b
3b
3c
3a
4a
4b
5c
5b
5a
6
7b
7a
8c
8b
8a
9c
9b
9a
10

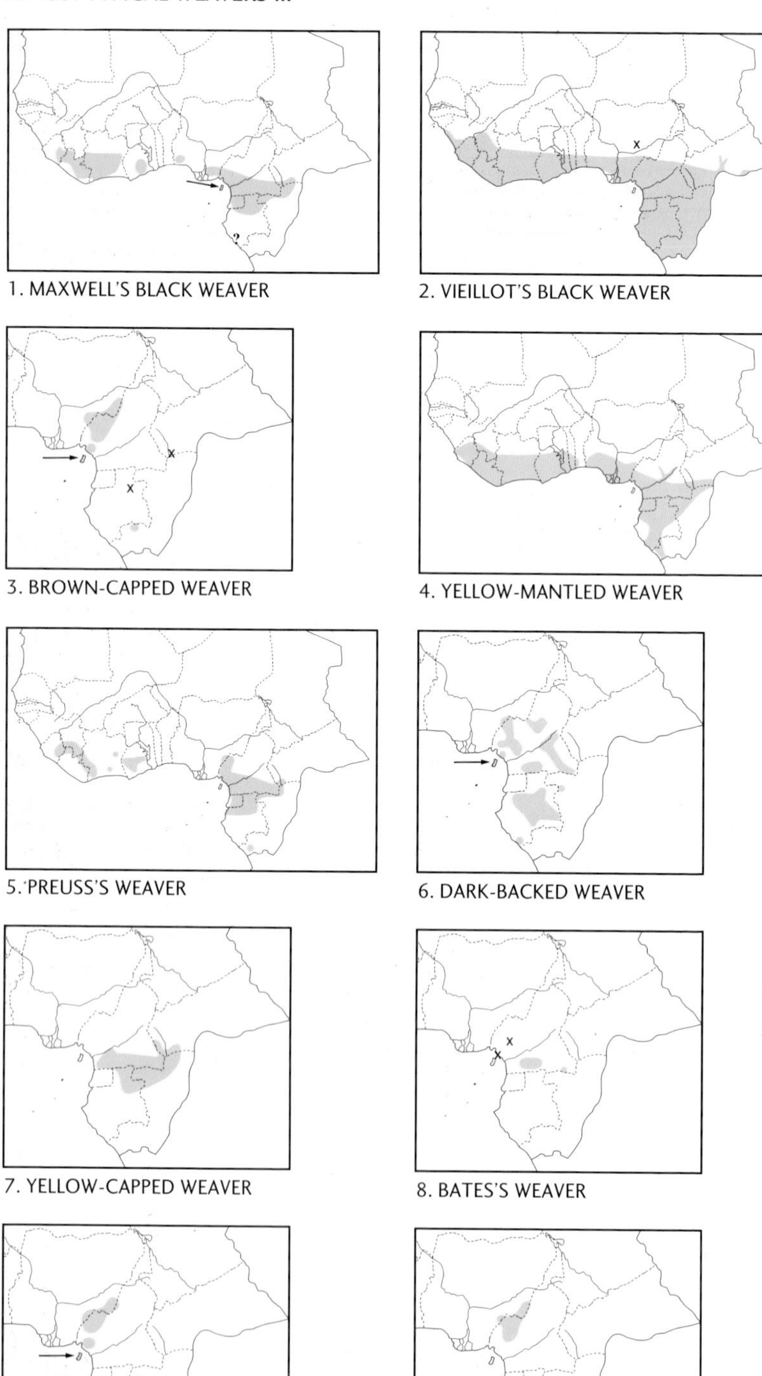

1. MAXWELL'S BLACK WEAVER

2. VIEILLOT'S BLACK WEAVER

3. BROWN-CAPPED WEAVER

4. YELLOW-MANTLED WEAVER

5. PREUSS'S WEAVER

6. DARK-BACKED WEAVER

7. YELLOW-CAPPED WEAVER

8. BATES'S WEAVER

9. BLACK-BILLED WEAVER

10. BANNERMAN'S WEAVER

Plate on page 438

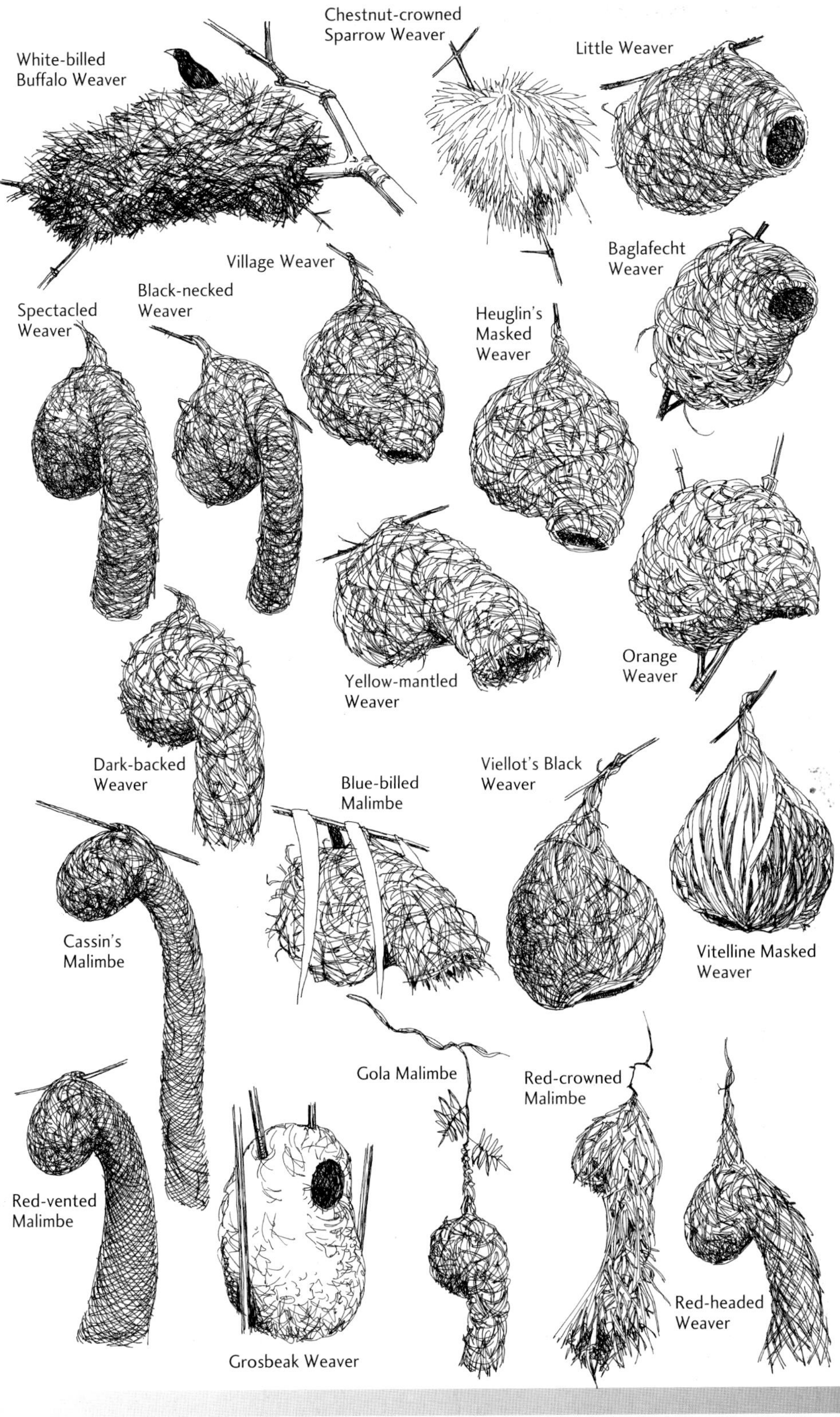
White-billed Buffalo Weaver
Chestnut-crowned Sparrow Weaver
Little Weaver
Baglafecht Weaver
Village Weaver
Black-necked Weaver
Spectacled Weaver
Heuglin's Masked Weaver
Yellow-mantled Weaver
Orange Weaver
Dark-backed Weaver
Blue-billed Malimbe
Viellot's Black Weaver
Vitelline Masked Weaver
Cassin's Malimbe
Gola Malimbe
Red-crowned Malimbe
Red-vented Malimbe
Grosbeak Weaver
Red-headed Weaver

PLATE 134: MALIMBES

Genus *Malimbus*. Distinctive black-and-red forest weavers. One species also has yellow in its plumage and another is black and yellow, lacking any red. Juveniles duller with horn-coloured bills variably tinged dusky. Most species join mixed-species flocks. Calls harsh and rasping.

1 BLUE-BILLED MALIMBE *Malimbus nitens* 17 cm R c
1a Ad Red patch from throat to upper breast; blue bill. **1b Juv** Dull red area extending onto throat, head-sides and forehead. ▲ Forest undergrowth, near water. ✤ Harsh *zheep*. Series of *zheet* followed by dissonant sounds and ending in weaver-like sizzling. [CD14:73]

2 CRESTED MALIMBE *Malimbus malimbicus* 17 cm R f/u
2a Ad male Red head and breast; small nuchal crest; black nape. **2b Ad female** No crest; red less extensive below. **2c Juv** Red area extending onto nape; throat blackish. ▲ Mid-levels of forest. ✤ Harsh *zheet*; other harsh chirps, also with more musical whistles. [CD14:74]

3 CASSIN'S MALIMBE *Malimbus cassini* 17 cm R c
3a Ad male Red head, neck and breast; black mask. **3b Ad female** Wholly black (as 10b). **3c Juv** Pale reddish on head, throat and breast. ▲ Forest canopy. ✤ Hard *tuk tuk tuk tuk*... followed by drawn-out nasal sizzling wheeze *szszszzuiiiiiin*. [CD14:75]

4 IBADAN MALIMBE *Malimbus ibadanensis* 17 cm R lu EN
4a Ad male As 3a, but red usually more extensive. **4b Ad female** As male, but red on underparts reduced. **4c Juv** Head to breast dull reddish; blackish mask. ▲ Forest edge, second growth. Endemic. ✤ Short series of clipped, jaunty notes ending in short, soft sizzle *chup-ti chup-ti tee-tu chupchupchupszszrrr*; rather melodious for a malimbe.

5 RED-VENTED MALIMBE *Malimbus scutatus* 17 cm R f/c
5a Ad male *scutatus* (west of range) As 3a but undertail-coverts red. **5b Ad female *scutatus*** Head black; underparts as male. **5c Ad female *scutopartitus*** (east of range) Red breast patch vertically separated. **5d Juv** As 5b, but red areas pale pinkish and extending onto throat and forehead. ▲ Forest edge, second growth, coastal gallery forest. In small, noisy groups in canopy. Endemic. ✤ Loud, very harsh *zee-zee-zee-zee* and abrupt *tuk tuk-tuk tuktuktuk tuk* ... [CD14:76]

6 RACHEL'S MALIMBE *Malimbus racheliae* 17 cm R lf
6a Ad male As 5a, but lower part of breast patch and undertail-coverts bright yellow. **6b Ad female** Head black; underparts as male. **6c Juv** As female, but breast patch extending onto throat. ▲ Forest canopy and mid-levels. Endemic. ✤ Fairly harsh *zhep-zhep*. Song short and ending in harsh, weaver-like sizzling, reminiscent of 1.

7 RED-HEADED MALIMBE *Malimbus rubricollis* 18 cm R f
7a Ad male Red top of head, nape and neck-sides. **7b Ad female** As male, but forehead and forecrown black. **7c Juv** Red area paler; some reddish feathers on face and breast. ▲ Upper levels of forest and outliers. Forages like woodpecker, examining trunks and large branches, adopting tit-like postures, often upside-down. ✤ Harsh *zheet*. Song a few rather soft, variable, dissonant whistles ending in short weaver-like sizzling. [CD14:77]

8 GOLA MALIMBE *Malimbus ballmanni* 17 cm R r/lc EN
8a Ad male Yellow nape, breast and undertail-coverts. **8b Ad female** As male, but nape black. **8c Juv male** As adult male, but top of head pale yellow to orange-brown; nape patch smaller; breast patch extending onto throat. ▲ Mid-levels of forest. Endemic. ✤ Series of discordant chattering sounds followed or not by a wheeze.

9 RED-BELLIED MALIMBE *Malimbus erythrogaster* 17 cm R f/r
9a Ad male Red head and underparts red; black mask. **9b Ad female** As male, but throat also red. **9c Juv** Head all red; indistinct mask; lower underparts greyish-brown. ▲ Forest canopy. ✤ Short, dry *ptsik*. Clicking notes in flight. [CD14:78]

Continued on page 445

 Maps on page 444

1b
1a
2c
2a
2b
3c
3a
4a
3b
4c
4b
5c
5d
5a
6c
6b
5b
7c
6a
8c
7b
7a
8b
8a
9c
9a
10c
10a
9b
10b

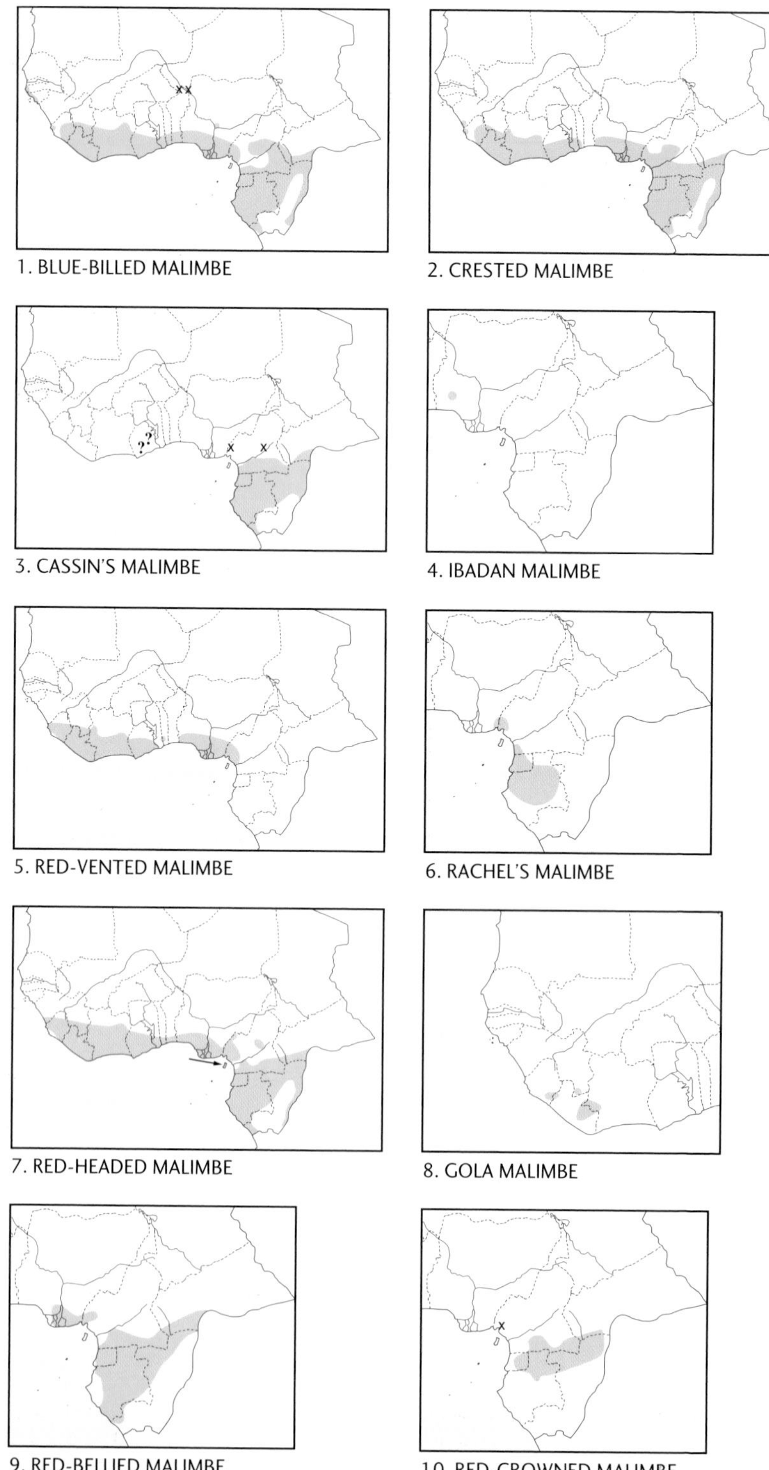

1. BLUE-BILLED MALIMBE

2. CRESTED MALIMBE

3. CASSIN'S MALIMBE

4. IBADAN MALIMBE

5. RED-VENTED MALIMBE

6. RACHEL'S MALIMBE

7. RED-HEADED MALIMBE

8. GOLA MALIMBE

9. RED-BELLIED MALIMBE

10. RED-CROWNED MALIMBE

Plate on page 442

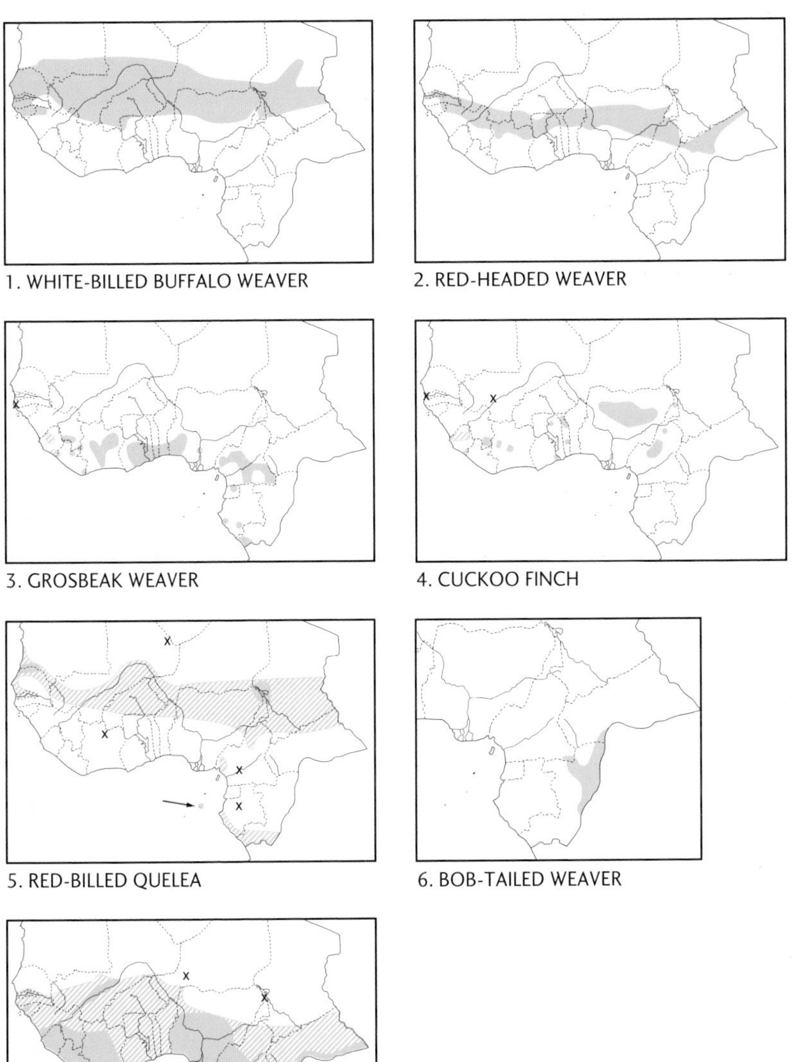

Continued from page 442

10 RED-CROWNED MALIMBE *Malimbus coronatus* 17 cm **R f/u**
10a Ad male Red crown patch. **10b Ad female** Wholly black (as 3b). **10c Juv** Rufous-brown crown patch. ▲ Forest canopy. Unobtrusive. ✣ Characteristic, drawn-out, wheezy note; also shorter calls. Song includes wheeze. [CD14:79]

1 WHITE-BILLED BUFFALO WEAVER *Bubalornis albirostris* 22–24 cm R c
1a Ad male breeding Very large; wholly black; massive whitish bill. **1b Ad male non-breeding** Bill black. Female similar. **1c Juv** Blackish-brown; underparts mottled white. ▲ Dry savanna with *Acacia* and thorn scrub. In small groups. Forages on ground. ✤ Various harsh, croaking and high-pitched, squeaky sounds, resulting in cackling chatter. [CD14:40]

2 RED-HEADED WEAVER *Anaplectes rubriceps* 14–15 cm R lu/r
2a Ad male Red head and bill; black mask. **2b Ad female** Greyish-brown head; red bill. ▲ Open woodland, scrub, *Acacia* savanna. Forages among foliage of trees and bushes, often hanging upside-down. Not gregarious. Joins mixed-species flocks. ✤ Usually silent. At nest a rapid sizzling, typical of weavers, but distinctly high pitched and squeaky. Call a sharp *tzik*. [CD14:80]

3 GROSBEAK WEAVER *Amblyospiza albifrons* 18 cm R lf
3a Ad male Heavy bill; white forehead and primary patch. **3b Ad female** Heavily streaked below; bill dark horn. **3c Juv** As adult female, but bill yellowish-horn. ▲ Usually in small scattered colonies in tall herbage. Forages in forest canopy. ✤ Chirping and twittering calls in flight and at breeding colony. Song a vigorous, fast jumble of hard, rasping, chattering and churring notes. [CD14:98]

4 CUCKOO FINCH (PARASITIC WEAVER) *Anomalospiza imberbis* 13 cm R lu/r
4a Ad male Small; bright yellow head and underparts; short tail; stubby, black bill. **4b Ad female** Buff-brown boldly streaked black above; buff below; bill horn. ▲ Wooded grassland. Parasitises cisticolas and prinias. Note: morphological and behavioural features support molecular evidence that Cuckoo Finch is a viduid finch, not a weaver. ✤ Song a fast series of squeaky, sibilant notes *tslee-tslee-tslee-...*, from perch or in display flight, occasionally followed by a drawn-out weaver-like wheeze. Flocks give soft chattering notes in flight. [CD14:97]

5 RED-BILLED QUELEA *Quelea quelea* 11–13 cm M lc/u
5a Ad male breeding Black mask (incl. frontal band); red bill. **5b Ad male breeding** Black-faced variant with wholly dark crown. **5c Ad male breeding** Black-faced variant without frontal band. **5d Ad male breeding** White-faced variant. **5e Ad female/ad male non-breeding** Small; streaky above; long off-white supercilium; greyish cheeks. Resembles 7b but with red bill. ▲ Highly gregarious, sometimes in very large flocks, in grasslands, ricefields, open savanna and dry *Acacia* country. ✤ Flocks utter rasping and squeaky chattering, esp. at breeding colony. [CD14:83]

6 BOB-TAILED WEAVER *Brachycope anomala* 11 cm R lf
6a Ad male Small; very short tail; black mask. **6b Ad female** Plain buffish-brown head and underparts. ▲ Along large rivers. ✤ Call *chk-chk-chk-...* Song a few call notes followed by a melodious, pleasant little trill, first rising in pitch, then descending. [CD14:84]

7 RED-HEADED QUELEA *Quelea erythrops* 11–12 cm M lc/u
7a Ad male breeding Red head; black bill. **7b Ad female/ad male non-breeding** Face washed with yellow; bill horn. ▲ In small or large flocks, sometimes with other small seed-eaters, in coastal thickets, forest clearings, moist grasslands and ricefields. ✤ Flocks utter a squeaky chattering, esp. at breeding colony. Song a slightly rising series of churring notes ending in drawn-out wheeze. [CD14:82]

 Maps on page 445

1a
1b
1c
2b
2a
3b
3a
5d
5c
5b
4b
4a
5e
5a
6b
6a
7b
7a

Genus *Euplectes*. Occur in open grassland and edges of marshy areas. Breeding males conspicuous in plumage and behaviour, and easily identified. In non-breeding season males resemble the nondescript, sparrow-like females, but are often noticeably larger and, in some species, may retain coloured shoulder patches. Females and many males in non-breeding plumage hard to identify; some not safely separable. Most breed during rains, forming mixed-species flocks in off-season.

1 RED-COLLARED WIDOWBIRD *Euplectes ardens* 12–14 cm R lc

1a Ad male breeding All black; very long, graduated tail (up to 20 cm). In south of range (S Gabon, Congo) some have yellow to red crescent-shaped collar on lower throat. **1b Ad male non-breeding** Black wing feathers. Compare 4b. **1c Ad female** Unstreaked below. Compare 4c. ▲ Grassy plains and hill sides. Display of male consists of slow, bouncing flight above territory with rapid wingbeats and depressed tail, uttering song; also short, low flight, with intermittent drops into grass. ✤ Rapid series of buzzes, chirps and ticks, in display flight and from perch. [CD14:94]

2 FAN-TAILED WIDOWBIRD *Euplectes axillaris* male *c.* 18 cm, female *c.* 14 cm R lf/s

2a Ad male breeding Orange-yellow shoulders above chestnut wing patch. **2b Ad male non-breeding** Retains orangey shoulders and black flight feathers. **2c Ad female** Lesser coverts black edged orangey. Compare 4c. ▲ Rank herbage. Male has slow flapping display flight ending with dive into grass. ✤ Series of weak *tseep* and similar notes, interspersed by rolling husky *twirrrlll*, in display flight and from perch. [CD14:91]

3 BLACK BISHOP *Euplectes gierowii* 15–16 cm R ls

3a Ad male breeding Large; orange hindcrown, neck and breast-band; orange-yellow mantle. **3b Ad female** As 4c but darker; more broadly streaked above; more washed yellowish-buff below; breast-sides more distinctly streaked. Non-breeding male similar to 4b but differs as 3b. ▲ Rank herbage. ✤ In display flight, male utters rapid series of thin, silvery notes followed by clearer *tee-ee-ee-ee-eee* and sizzling *see-zee see-zee see-zhe see-zhe SEE-ZHEE*, accelerating as volume increases.

4 BLACK-WINGED (RED) BISHOP *Euplectes hordeaceus* 13–14 cm R lc

4a Ad male breeding Scarlet crown; black wings; buff undertail-coverts. **4b Ad male non-breeding** Black flight feathers (conspicuous in flight). Compare 1b. **4c Ad female** Streaky above; yellowish supercilium; breast-sides faintly streaked. Compare 1c and 2c. ▲ Rank herbage. Male displays in slow, bouncing cruising flight with rapidly beating, rustling wings and puffed-out plumage, uttering wheezy song. ✤ Fast series of twittering, churring and buzzing notes, interspersed by occasional whistles. [CD14:86]

5 NORTHERN RED BISHOP *Euplectes franciscanus* 11–12 cm R c/f

5a Ad male breeding Black crown; brown wings; short tail concealed by long upper- and undertail-coverts. **5b Ad male moulting**. **5c Ad female/ad male non-breeding** Smaller than 4c. Compare Yellow-crowned Bishop (Plate 137:5b). ▲ Rank herbage. Male displays from perch or in slow, bouncing cruising flight with rapidly beating, rustling wings and puffed-out plumage, uttering song. ✤ Fast jumble of twittering, churring and wheezy notes. Call a sharp *chiz*. [CD14:88]

1b
1a
2b
1c
2c
2a
5b
4b
3b
5c
4c
3a
4a
5a

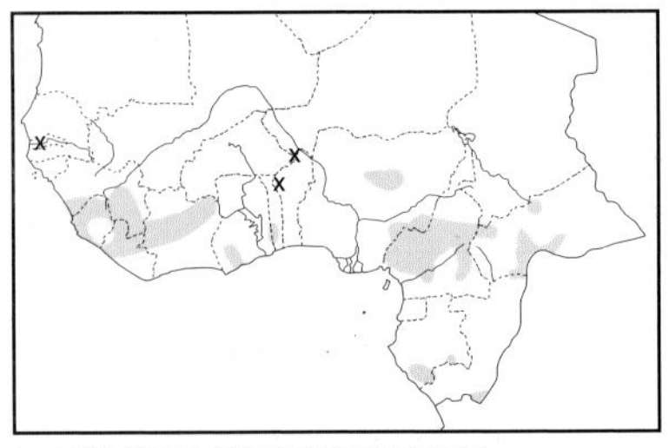

1. RED-COLLARED WIDOWBIRD

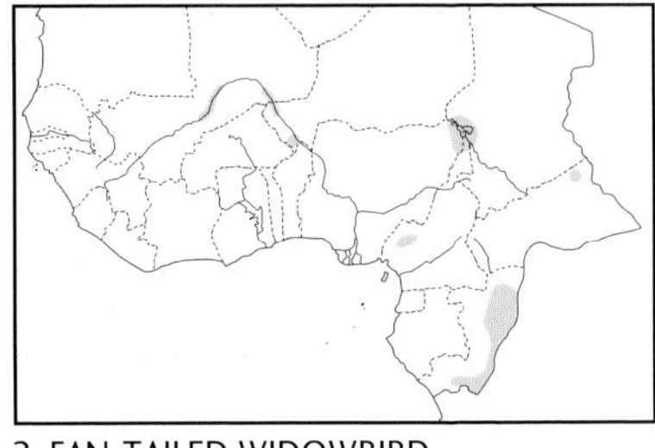

2. FAN-TAILED WIDOWBIRD

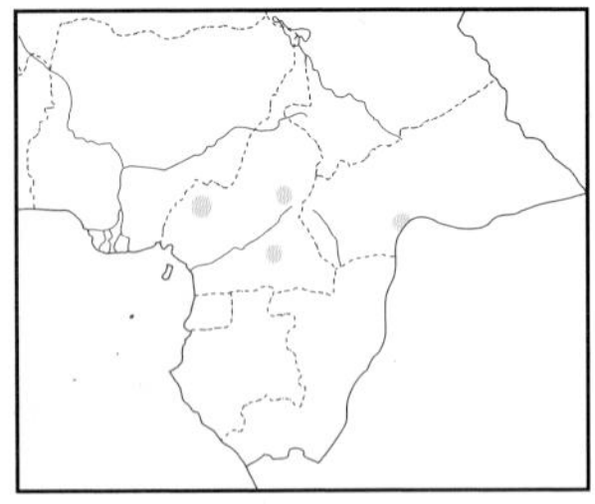

3. BLACK BISHOP

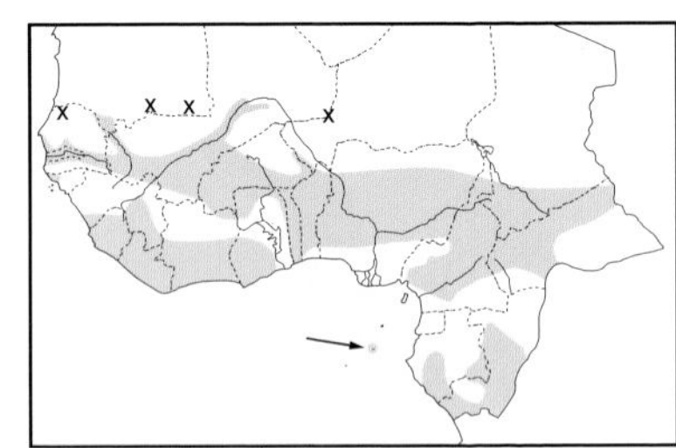

4. BLACK-WINGED BISHOP

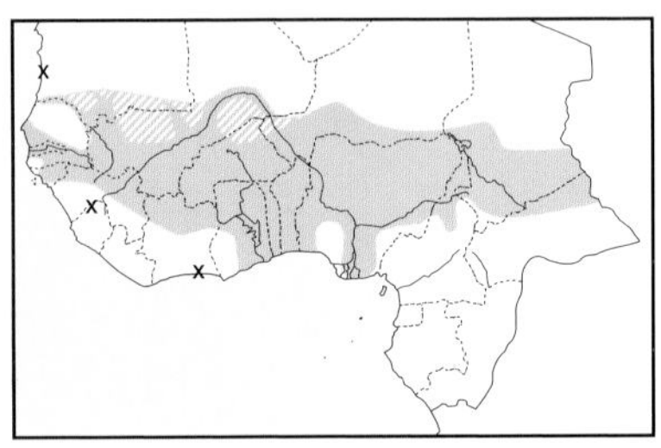

5. NORTHERN RED BISHOP

 Plate on page 448

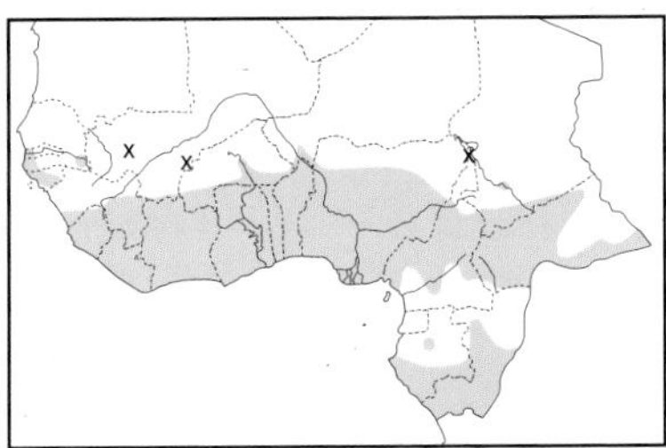

1. YELLOW-MANTLED WIDOWBIRD

2. HARTLAUB'S MARSH WIDOWBIRD

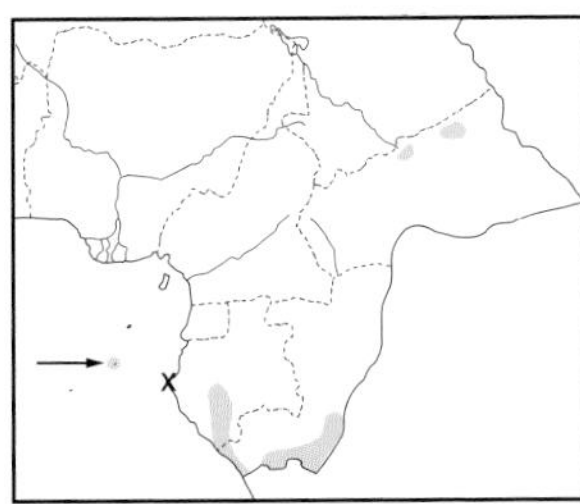

3. WHITE-WINGED WIDOWBIRD

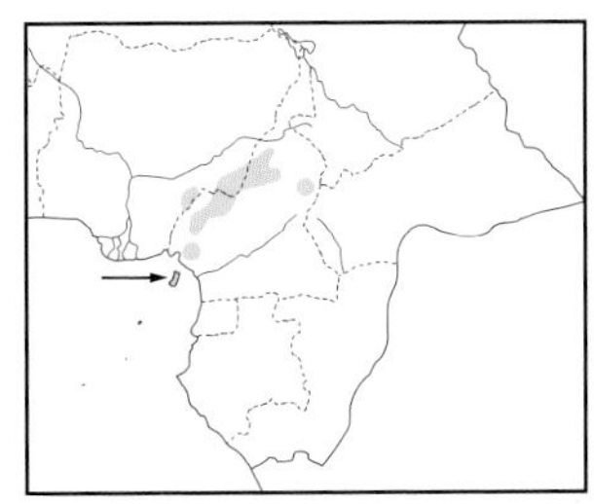

4. YELLOW BISHOP

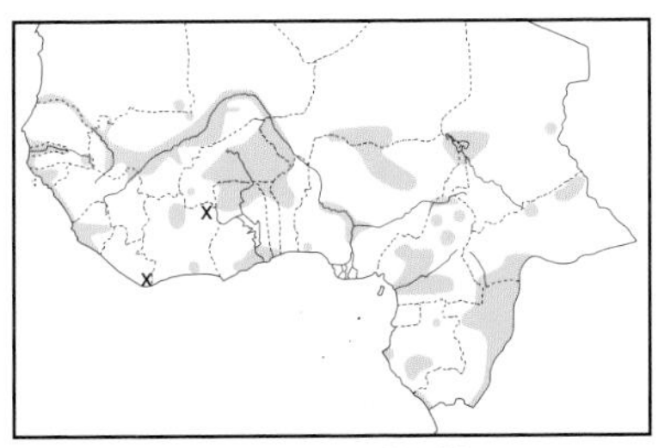

5. YELLOW-CROWNED BISHOP

Plate on page 452

1 YELLOW-MANTLED WIDOWBIRD *Euplectes macroura*

male 19–21 cm, female 13–14 cm **R c**

1a Ad male breeding Yellow mantle and shoulders; long black tail. **1b Ad male non-breeding** As 1c, but larger, with yellow shoulders. **1c Ad female** Lesser coverts tipped yellowish. ▲ Rank herbage. Male displays in low cruising flight above territory with ruffled neck and rump plumes, uttering song. ✤ Rhythmic series of thin, buzzing *zeet* and *tsweep* notes, in display flight and from perch. [CD14:92]

2 HARTLAUB'S MARSH WIDOWBIRD *Euplectes hartlaubi*

male 22 cm, female 16 cm **R lu/s**

2a Ad male breeding Orange-yellow shoulders; long, broad, black tail. **2b Ad female** Bulky; heavy bill. **2c Ad male non-breeding in flight** Orange-yellow shoulders; black flight feathers. ▲ Moist grasslands. Male displays in low cruising flight over territory with ruffled neck plumes, often starting from particular vantage point; frequently flicks tail when perched. ✤ Abrupt, metallic chirping or loud *yek!* followed by soft high-pitched sizzling or buzzy trill, in display. Male also gives dry *krrrt*. [CD14:96]

3 WHITE-WINGED WIDOWBIRD *Euplectes albonotatus*

male 14–15 cm, female 12–13 cm **R lu**

3a Ad male breeding Yellow shoulders; white primary patch (forming conspicuous wingbar in flight); long narrow tail (projecting up to 12 cm). **3b Ad male non-breeding** Sparrow-like but with yellow shoulders and white wing patch. **3c Ad female** Similar to 1c but with white underwing-coverts. ▲ Grasslands. Male displays from perch with fanned tail or in rather slow cruising flight on rapid wingbeats, uttering song. ✤ Usually rather silent. Rapid, rhythmic series of identical buzzing or chirping notes; also a dry, rustling papery sound. Twittering contact calls. [CD14:93]

4 YELLOW BISHOP *Euplectes capensis*

male 14 cm, female 11.0–12.5 cm **R lc**

4a Ad male breeding Black with yellow shoulders and rump. **4b Ad male non-breeding** Sparrow-like with yellow shoulders and rump. **4c Ad female** The only sparrow-like *Euplectes* with dull yellowish rump. ▲ Grasslands in montane and hilly areas. Male displays from perch or in cruising flight with rustling wings and ruffled neck and rump plumes. ✤ Thin *tseet* and *tsit*. In display flight loud wing rattling interspersed with short, fast series of high-pitched *tsip-tseep* or nasal *tzeep* notes. [CD14:90]

5 YELLOW-CROWNED BISHOP *Euplectes afer*

11 cm **R lc**

5a Ad male breeding Bright yellow and black; small. **5b Ad female breeding** Broad yellowish supercilium; dusky eye-stripe; sparse streaking on breast and flanks. Compare Northern Red Bishop (Plate 136:5c). Non-breeding male and female as 5b but more distinctly streaked on breast and flanks. ▲ Moist, rank vegetation, floodplains. Display of male consists of low, bouncing flight over territory with puffed-out feathers, uttering song. ✤ Series of high-pitched chirping and buzzing notes, e.g. *szit-szit-szit-*..., esp. in display flight. Call a sharp *tsip*. [CD14:85]

Maps on page 451

1b
2c
2b
2a
1c
1a
3b
3c
3a
4b
5b
4c
4a
5a

Small passerines with short, generally conical bills. Predominantly granivorous, but *Nesocharis* and forest-dwelling *Nigrita* and *Parmoptila* largely or entirely insectivorous. Forest species usually in pairs or small parties; open country estrildids may gather in flocks, especially outside breeding season.

1 BROWN TWINSPOT *Clytospiza monteiri* 13 cm R u/lf
1a Ad male Rich brown underparts densely spotted white; red throat stripe. **1b Ad female** As male but has white throat stripe. **1c Juv** Much duller; unspotted orange-brown underparts. ▲ Wooded grassland, forest edge. ✤ Sharp *chk*, frequently repeated, occasionally as rattling *chuk-chukchkrrrrrk*, and harsh, nasal *chèèp*. Song a fast, variable series of twitters and chirps. [CD15:20]

2 RED-FRONTED ANTPECKER *Parmoptila rubrifrons* 10–11 cm R s/lf
2a Ad male Bright red forehead and forecrown; dark above; rich brown below. **2b Ad female** Dark earth-brown above; whitish below, densely spotted. **2c Juv** Wholly rufous-brown. ▲ Forest. Endemic. ✤ Rather vigorous, hoarse *whseeeet*.

3 DYBOWSKI'S TWINSPOT *Euschistospiza dybowskii* 12 cm R lu/s
3a Ad male Crimson from mantle to uppertail-coverts; slate-grey from head to breast; white spots on black flanks. **3b Ad female** As male but duller. **3c Juv** Dull slate-grey underparts without spots; dull rusty upperparts. ▲ Wooded grassland. ✤ Call *chip-chip-*...; also *see* and *tswink-tswink-*... Song a medley of trills, melodious whistles and bubbling notes. [CD15:21]

4 WOODHOUSE'S (RED-HEADED) ANTPECKER *Parmoptila woodhousei* 11 cm R u/s
4a Ad male Orange-rufous face and throat; densely speckled below; some red on forehead. **4b Ad female** No red on forehead; less densely marked below. **4c Juv** As 2c, but slightly paler. ▲ Forest. ✤ Soft, thin *tseeu* and *tsee*. [CD15:1]

5 GREEN-WINGED PYTILIA *Pytilia melba* 12–13 cm R u/lf
5a Ad male *citerior* (most of range) Red face; olive-green upperparts; yellow upper breast; red bill. Tail longer than other pytilias. **5b Ad female *citerior*** Grey head; barred grey and white below. **5c Juv *citerior*** Dull olive upperparts; unbarred buffish-brown underparts. **5d Ad male *melba*** (SE Congo) Grey wedge from bill above eye. ▲ Thorn scrub, wooded grassland. ✤ Usually silent. Thin, penetrating *see-eh*; also short *wik*. Song a variable mix of buzzing and croaking sounds interspersed with call notes. [CD15:9]

6 RED-FACED CRIMSONWING *Cryptospiza reichenovii* 11–12 cm R lf/u
6a Ad male Red eye patch; deep red upperparts. **6b Ad female** Yellowish-buff eye patch. **6c Juv** Duller; no or indistinct eye patch. ▲ Dense undergrowth and edges of montane forest. ✤ High-pitched *tseet*. Song includes sharp, insect-like notes. [CD15:13]

7 YELLOW-WINGED PYTILIA *Pytilia hypogrammica* 12.5–13.0 cm R u/lf
7a Ad male Red face; mainly grey body; yellow wings; black bill. **7b Ad female** No red face; paler, browner grey. ▲ Wooded grassland. ✤ Frequently repeated *tsiee*. [CD15:11]

8 ORANGE-WINGED PYTILIA *Pytilia afra* 11 cm R lf
8a Ad male Red face; olive-green upperparts; orange wings; red bill. **8b Ad female** Pale grey head; underparts whitish barred olive. ▲ Wooded grassland. ✤ High-pitched, drawn-out *seee*, abrupt *pit pit* or *pwit pwit* and low *tiu*. [CD15:10]

9 GREEN TWINSPOT *Mandingoa nitidula* 10–11 cm R u/r
9a Ad male Bright green above; red face; lower breast and belly black spotted white; bill black with red cutting edges. **9b Ad female** Paler and duller; yellowish-buff face. **9c Juv** Olive and greyish-olive; pale buff face. ▲ Forest edge. ✤ Sharp *tsk*, very high-pitched *tseeet*, high-pitched trill *tsrrriii*. Song includes high-pitched whistles and trills. [CD15:19]

10 RED-WINGED PYTILIA *Pytilia phoenicoptera* 12.5–13.0 cm R u/lf
10a Ad male Grey with red wings and tail; black bill. **10b Ad female** As male but browner grey. ▲ Wooded grassland. ✤ Single *twink*. Song a short, variable phrase, consisting of some hard whistles followed by a rolling, croaking sound. [CD15:12]

Maps on page 456

1c
1b
2b
2c
1a
2a
3c
4c
3b
4b
3a
5d
4a
5c
5b
5a
6c
6b
7b
6a
7a
9c
8b
8a
9b
9a
10b
10a

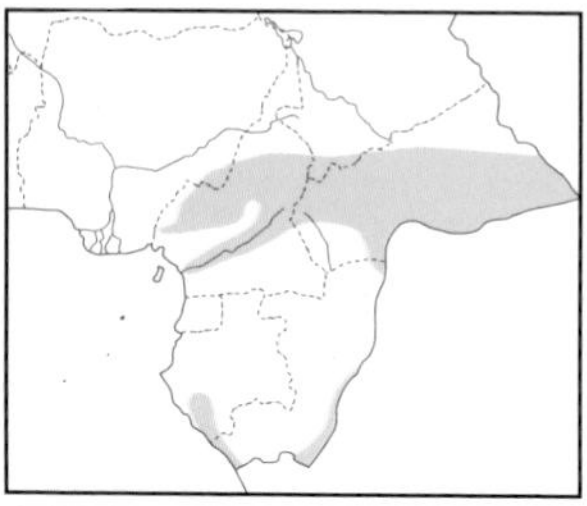

1. BROWN TWINSPOT

2. RED-FRONTED ANTPECKER

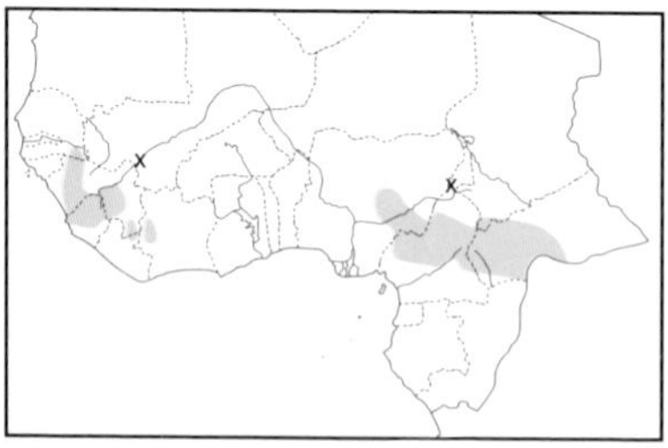

3. DYBOWSKI'S TWINSPOT

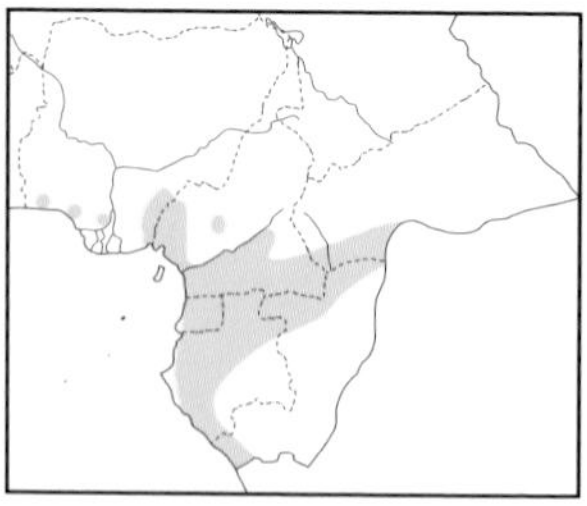

4. WOODHOUSE'S ANTPECKER

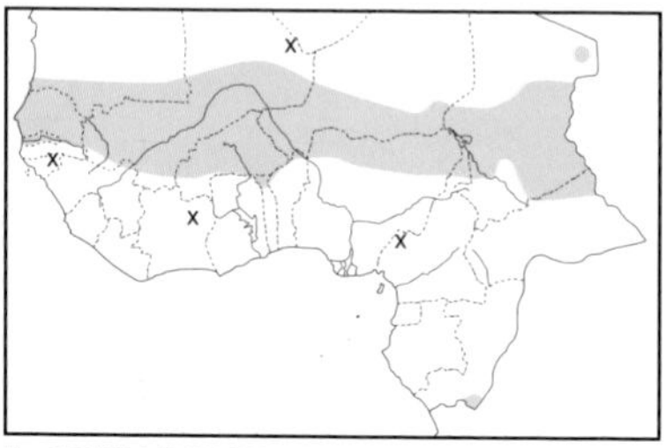

5. GREEN-WINGED PYTILIA

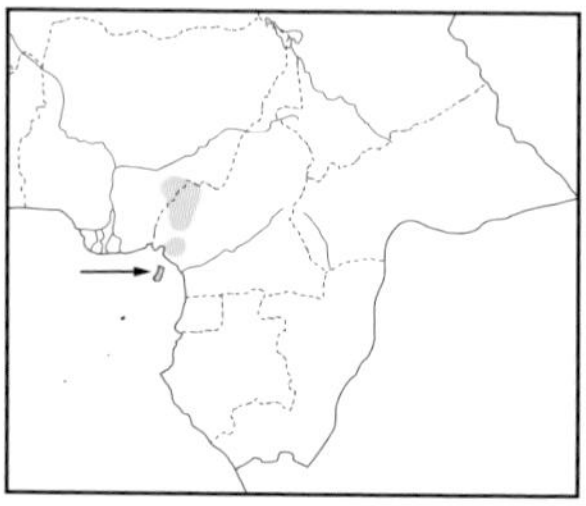

6. RED-FACED CRIMSONWING

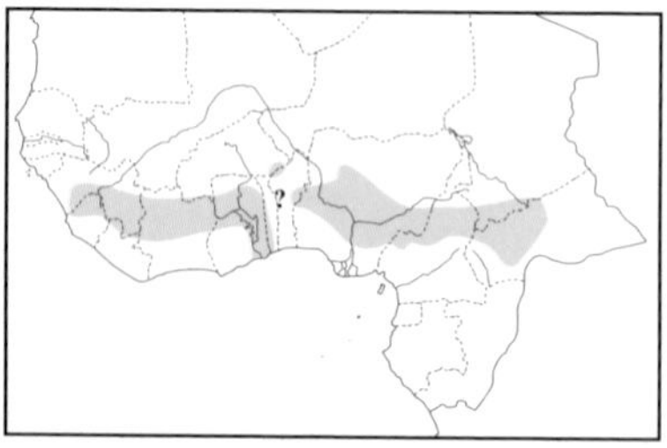

7. YELLOW-WINGED PYTILIA

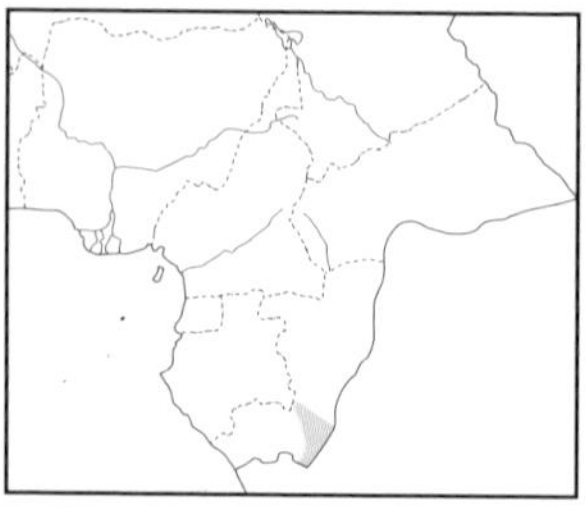

8. ORANGE-WINGED PYTILIA

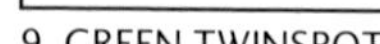

9. GREEN TWINSPOT

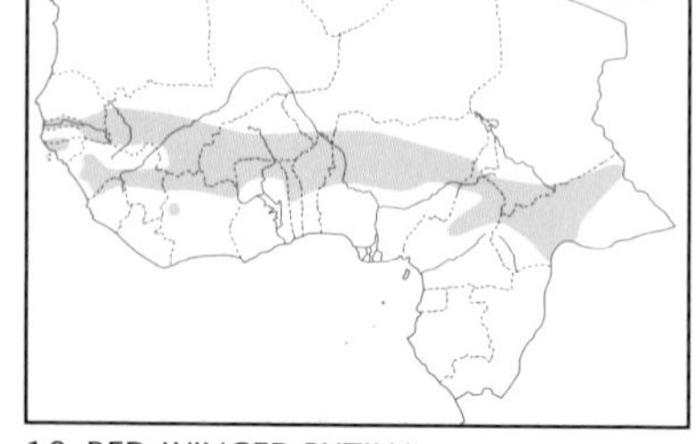

10. RED-WINGED PYTILIA

Plate on page 454

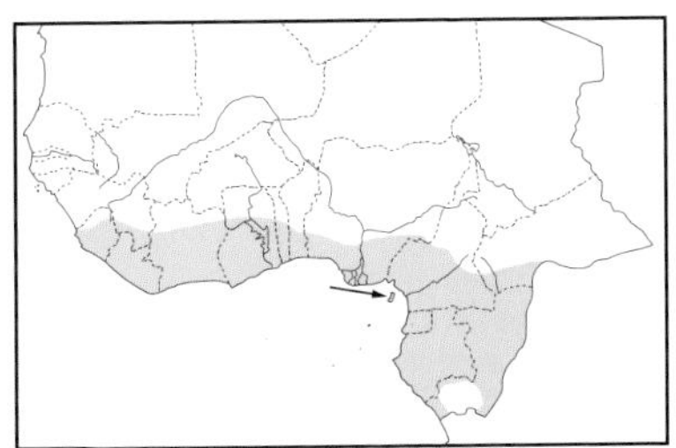

1. GREY-HEADED NEGROFINCH

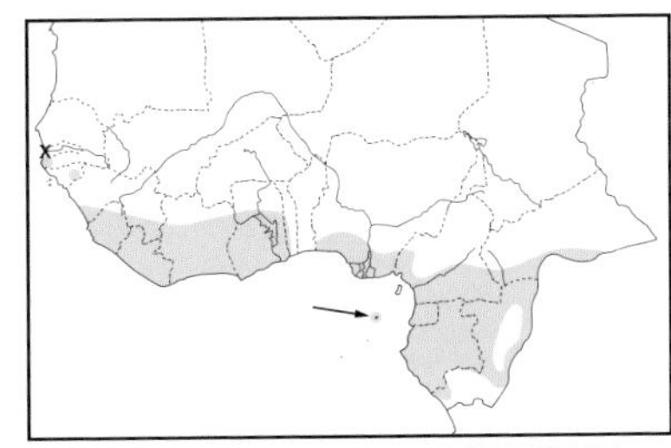

2. CHESTNUT-BREASTED NEGROFINCH

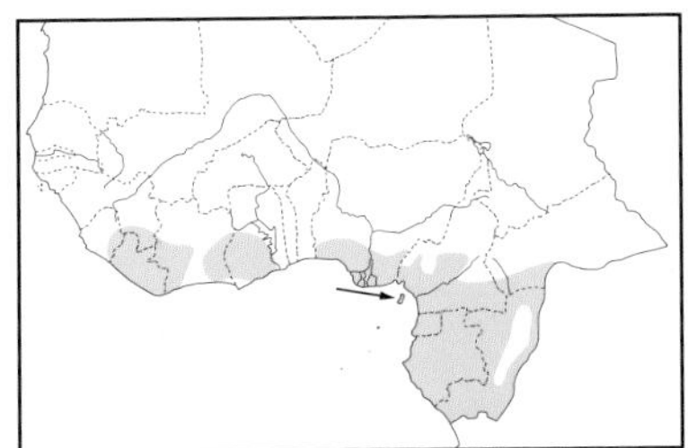

3. WHITE-BREASTED NEGROFINCH

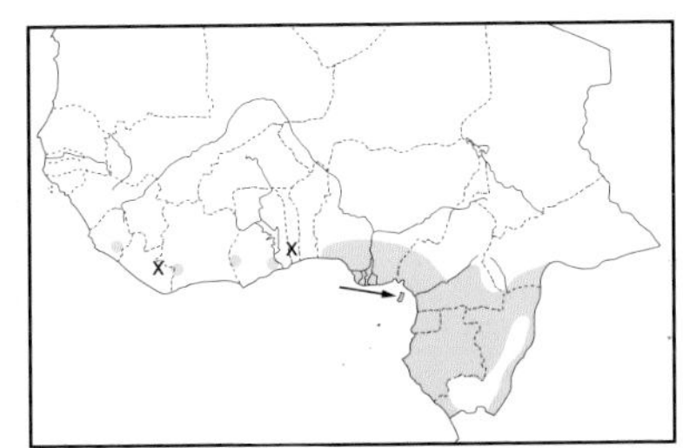

4. PALE-FRONTED NEGROFINCH

5. GREY-HEADED OLIVEBACK

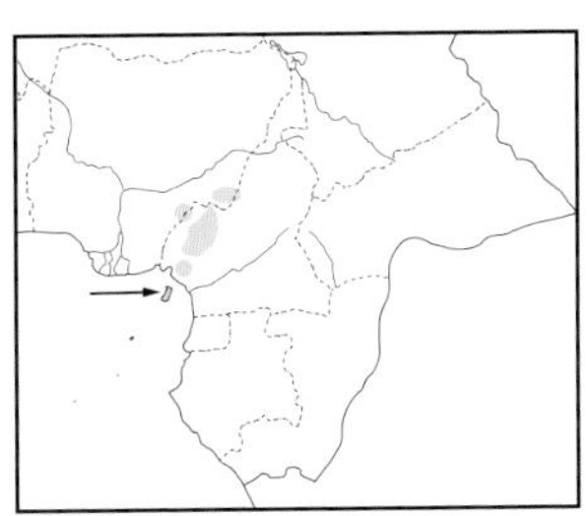

6. SHELLEY'S OLIVEBACK

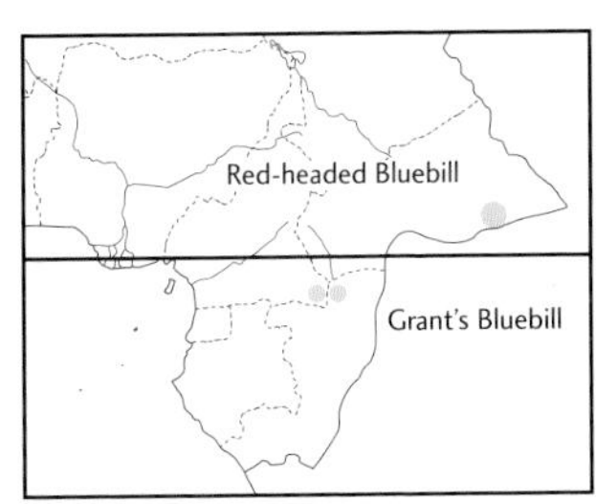

7. GRANT'S BLUEBILL
8. RED-HEADED BLUEBILL

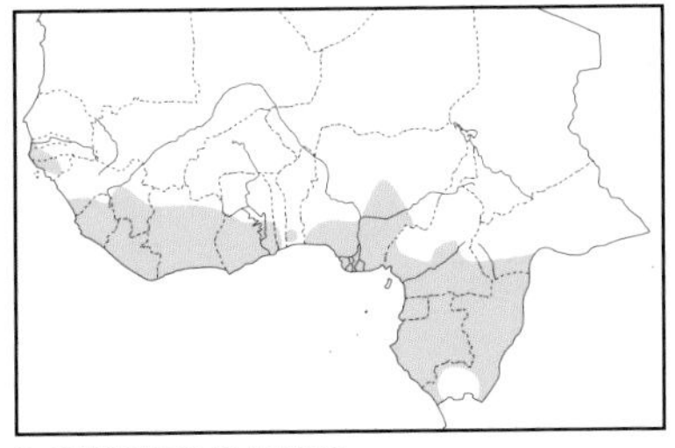

9. WESTERN BLUEBILL

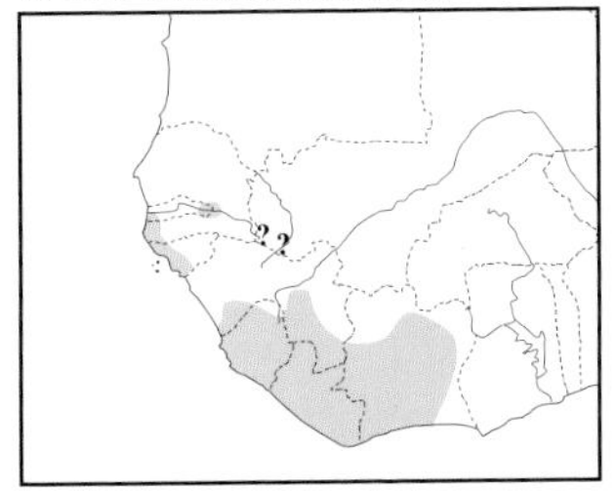

10. CRIMSON SEEDCRACKER

11. BLACK-BELLIED SEEDCRACKER

Plate on page 458

1 **GREY-HEADED (-CROWNED) NEGROFINCH** *Nigrita canicapillus* 15 cm **R c/u**
1a Ad *emiliae* (Upper Guinea forest) Mainly grey above; black face and underparts. **1b Ad *canicapillus*** (Lower Guinea forest) White line separating grey from black on head; white spots on wing-coverts; whitish-grey rump. **1c Juv** Wholly sooty-black. ▲ Forest and edges. Mostly in canopy. ✤ Short series of *c.* 2–7 distinctive, rather plaintive whistles. [CD15:2]

2 **CHESTNUT-BREASTED NEGROFINCH** *Nigrita bicolor* 11–12 cm **R lc/u**
2a Ad Slate-grey above; dark chestnut below. **2b Juv** Pale grey-brown above; rich buff-brown below. ▲ All levels of lowland forest and forest outliers in savanna. ✤ Series of cheerful whistles, e.g. *weet-weet-teeu weet-weet-teeu weet-weet-tiutiutiuhuu tiutiuhuu*. [CD15:4]

3 **WHITE-BREASTED NEGROFINCH** *Nigrita fusconotus* 10 cm **R u/lf**
Black head and tail; brown upperparts; white underparts. ▲ Canopy of mature lowland forest, forest edges, gallery forest. ✤ In west: clear whistled phrase *wheet-huw wheet heeew* and *tsip-tsip-rruuu tsee-uu*. In east: melodious descending series, *hwitwitwitwitwitwitwitwit* and trill, ending with some slower notes *trrrrrrrrititititsiutsiutsiutsiu*. [CD15:5]

4 **PALE-FRONTED NEGROFINCH** *Nigrita luteifrons* 11.5 cm **R lf/r**
4a Ad male Grey above; paler forehead tinged buffish. **4b Ad female** Wholly grey with small black mask. **4c Juv** Wholly dull grey. ▲ Forest, second growth, thickets in savanna. ✤ Simple, descending series of 4–6 whistles *wee-wee-wee-wee-wee-whuuuh*. [CD15:3]

5 **GREY-HEADED (WHITE-CHEEKED) OLIVEBACK** *Nesocharis capistrata* 12.5 cm **R lu/r**
5a Ad Grey head; white face; black throat; yellow flanks. **5b Juv** Lacks white cheeks; bill whitish. ▲ Moist grassland, riparian woodland. ✤ Usually silent. Thin, high-pitched *tsip*. Song a fast, descending series of high-pitched *tsi* notes. [CD15:8]

6 **SHELLEY'S (LITTLE) OLIVEBACK** *Nesocharis shelleyi* 8.0–8.5 cm **R lc/u**
6a Ad male Black head; grey nape; olive upperparts; white stripe on neck-sides; yellowish-olive breast. **6b Ad female** White stripe on neck-sides reduced or lacking; breast grey. ▲ Montane forest. Endemic. ✤ Very high-pitched twittering of *tsip* notes. [CD15:7]

7 **GRANT'S BLUEBILL** *Spermophaga poliogenys* 14 cm **R r**
7a Ad male As 8a but red restricted to face and forecrown. **7b Ad female** As 8b but head blackish; red restricted to throat and breast. ▲ Forest. ✤ Usually silent. Soft *tak* and *seeep*.

8 **RED-HEADED BLUEBILL** *Spermophaga ruficapilla* 15 cm **R c**
8a Ad male Entirely red head. Compare 7a. **8b Ad female** Entirely red head; red cutting edges to bill. ▲ Forest, thickets. ✤ Thin *spit-spit-spit-...* and *skwee*.

9 **WESTERN BLUEBILL** *Spermophaga haematina* 15 cm **R c**
9a Ad male *haematina* (Gambia–Ghana) Black with red throat, breast and flanks; heavy blue, red-tipped bill. **9b Ad female *haematina*** Face washed red; red uppertail-coverts; lower underparts densely spotted white. **9c Juv** Dark brownish-slate; dull reddish uppertail-coverts. **9d Ad male *pustulata*** (SE Nigeria to SW CAR–Congo) Red extends onto lower head-sides; red uppertail-coverts. *S. h. togoensis* (Togo–SW Nigeria) as 9d but has black head. ▲ Forest, thickets. ✤ Metallic *tswink-tswink -...*, short *tsik*, high-pitched *tsee*, sharp *tak*. Song a variable series of clear whistles, trills and/or rattles. [CD15:17]

10 **CRIMSON SEEDCRACKER** *Pyrenestes sanguineus* 13–14 cm **R u/lc**
10a Ad male Heavy, triangular, steel-blue bill; red head, breast and flanks; red rump and tail. Female as 11b. **10b Juv** Wholly brown except for dull red rump and tail. ▲ Forest edge, ricefields, farmbush. Endemic. ✤ Sharp *tsu-it*. Song a fast, short warble. [CD15:16]

11 **BLACK-BELLIED SEEDCRACKER** *Pyrenestes ostrinus* 15 cm **R u/lc**
11a Ad male As 10a but earth-brown areas replaced by blue-black. **11b Ad female** Red areas more restricted; rest of plumage warm brown. ▲ Forest edge, ricefields, farmbush. ✤ Soft, repeated *tak*, low metallic *peenk* and hard *trrr*. Song a very fast warble. [CD15:14]

 Maps on page 457

1c
1b
1a
4c
2b
3
4b
2a
4a
5b
5a
6b
7b
6a
7a
8b
9c
9b
8a
9a
9d
10b
11b
11a
10a

1 ORANGE-CHEEKED WAXBILL *Estrilda melpoda* 10 cm **R c**
1a Ad Bright orange face; red rump; red bill. **1b Juv** Duller and buffier; black bill. ▲ Various grassy habitats. ✤ Short trill *tsiririririt* and high-pitched *tsee-tsee*. [CD15:36]

2 FAWN-BREASTED WAXBILL *Estrilda paludicola* 11.5 cm **R lc**
2a Ad Greyish head; red rump; red bill. **2b Juv** Duller; black bill. ▲ Moist grassland, grassy clearings. ✤ Nasal *chyeek* and *chyep*; alarm *chyee-krr*.. [CD15:35]

3 ANAMBRA WAXBILL *Estrilda poliopareia* 11–12 cm **R s,l VU**
Pale, plain grey-brown head; red rump; pale yellowish-buff underparts; red bill. ▲ Rank grass along rivers, lagoon sand banks. Endemic. ✤ Various waxbill-type calls.

4 BLACK-CROWNED WAXBILL *Estrilda nonnula* 11 cm **R lc**
4a Ad male Black cap; white cheeks; red rump; white underparts washed pale grey. Female slightly duller. **4b Juv** Brownish above; buffish below; black bill. ▲ Grassland, forest regrowth, gardens. ✤ Thin *tsree* and *psit*. Song a few high-pitched notes. [CD15:39]

5 LAVENDER WAXBILL *Estrilda caerulescens* 10 cm **R c/s**
Mainly grey; red rump, tail and undertail-coverts. ▲ Wooded grassland. Endemic. ✤ Clear *tsee* and *tseeu*, dry *tik-tik*... and *skweeep*. Song a series of thin notes. [CD15:31]

6 BLACK-HEADED WAXBILL *Estrilda atricapilla* 10 cm **R lc**
6a Ad As 4a but darker; underparts smoke-grey; undertail-coverts black. **6b Juv** Duller and tinged dark brown; black bill. ▲ Grassy clearings, forest regrowth. ✤ As 4. [CD15:40]

7 GREY WAXBILL *Estrilda perreini* 11 cm **R u/s**
Mainly grey; red rump; black tail. ▲ Wooded grassland. ✤ High-pitched *psee*. [CD15:32]

8 BLACK-RUMPED WAXBILL *Estrilda troglodytes* 10 cm **R lc**
8a Ad male Red eye-stripe; black rump; white undertail. Female less pink below. **8b Juv** No red mask; black bill. ▲ Grassy habitats, scrub. ✤ Vigorous *cheeeu* and *chup*. [CD15:37]

9 COMMON WAXBILL *Estrilda astrild* 11 cm **R+ lc**
9a Ad *occidentalis* (Nigeria–CAR, Congo and Bioko) Resembles 8a but rump brown, undertail black. **9b Ad *rubriventris*** (coastal Gabon–Congo) Heavily tinged pinkish. **9c Juv** Duller; eye-stripe paler, smaller; bill blackish. ▲ Grassy habitats. ✤ Abrupt *pit*, nasal *pcher-pcher*, soft *chip*. [CD15:38]

10 SOUTHERN CORDON-BLEU *Uraeginthus angolensis* 12–13 cm **R u/lc**
As 11 but adult male lacks red on cheek. ▲ Gardens, towns, edges of cultivation. ✤ Calls as 11. Song a variable series of sibilant notes, dry rattles and harsh sounds. [CD15:41]

11 RED-CHEEKED CORDON-BLEU *Uraeginthus bengalus* 13 cm **R c**
11a Ad male Pale blue underparts, rump and tail; bright red cheek patch. **11b Ad female** Slightly duller; no red on cheek. ▲ Savanna, gardens, villages. ✤ Thin *sweep-sweep*; alarm *chchchrrrt*. Song a thin *tsee-tsee-tsu-tsuu*, with variations. [CD15:42]

12 BLACK-FACED (AFRICAN) QUAILFINCH *Ortygospiza atricollis* 9.5–10.0 cm **R lc**
12a Ad male Dark and dumpy. In flight short tail with white corners. **12b Ad female** Duller; head greyer without black mask. ▲ Open grassy habitats. ✤ Metallic *krreep-krreep*, in flight. Song a rapid series of *klik-klak-kloik-kluk*. [CD15:45]

13 ZEBRA WAXBILL *Sporaeginthus subflavus* 9–10 cm **R lc/s**
13a Ad male Bright orange below; red rump. **13b Ad female** Duller; no red supercilium. **13c Juv** Dull brownish above; rump washed orange; buffish below; blackish bill. ▲ Grassy habitats. ✤ Squeaky cheep; in flight *trip-trp-*... and soft metallic twittering. [CD15:43]

Continued on page 462

Maps on pages 462 and 463

1a
1b
2a
2b
3
4a
4b
5
6a
6b
7
8a
8b
9a
9b
9c
10
11a
11b
12a
12b
13a
13b
13c
14a
14b
15a
15b
15c

Continued from page 460

14 BLACK-CHINNED QUAILFINCH *Ortygospiza gabonensis* 9.5–10.0 cm **R lu/c**
14a Ad male As 12a but more streaky above, paler below; breast barred; bill bright red. **14b Ad female** Duller; no black mask. ▲ Open grassy habitats. ✤ Similar to 12. [CD15:46]

15 LOCUST FINCH *Paludipasser locustella* 9–10 cm **R lu NT**
15a Ad male Red face to breast; reddish wings (conspicuous in flight). **15b Ad female** Mainly greyish-brown with orange wings; barred flanks. **15c Juv** Browner than female; bill dark. ▲ Moist grassland. ✤ Mainly silent. Abrupt *chup-chup-...* in flight. [CD15:44]

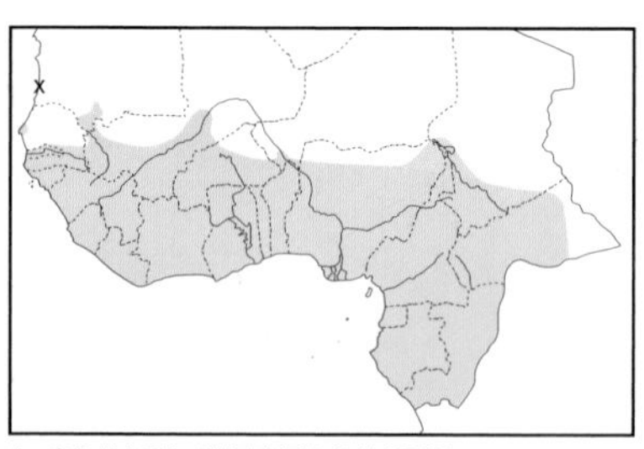

1. ORANGE-CHEEKED WAXBILL

2. FAWN-BREASTED WAXBILL

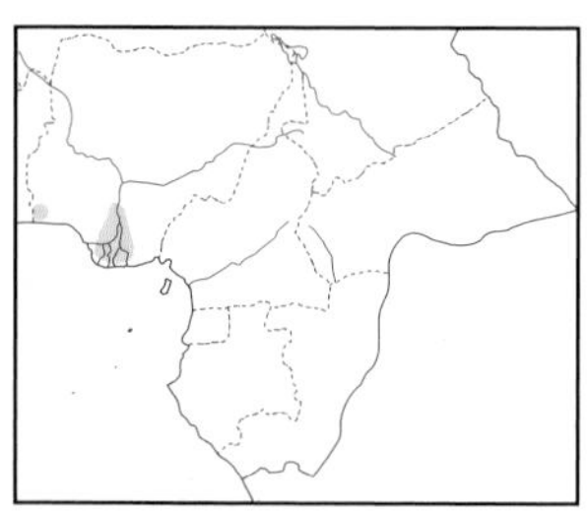

3. ANAMBRA WAXBILL

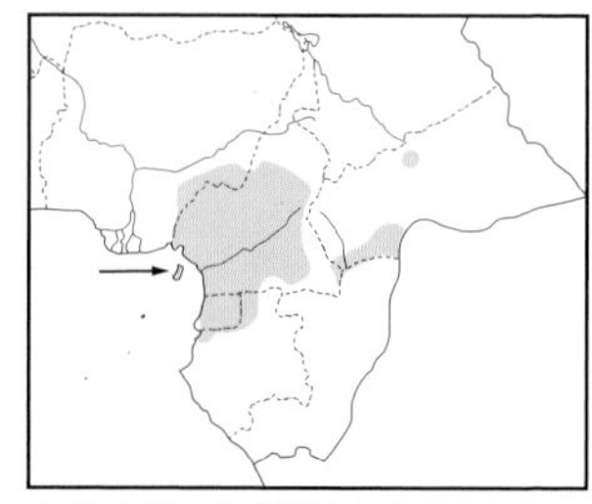

4. BLACK-CROWNED WAXBILL

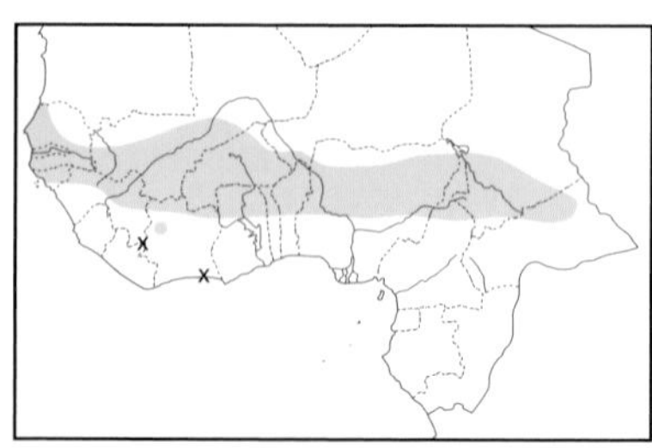

5. LAVENDER WAXBILL

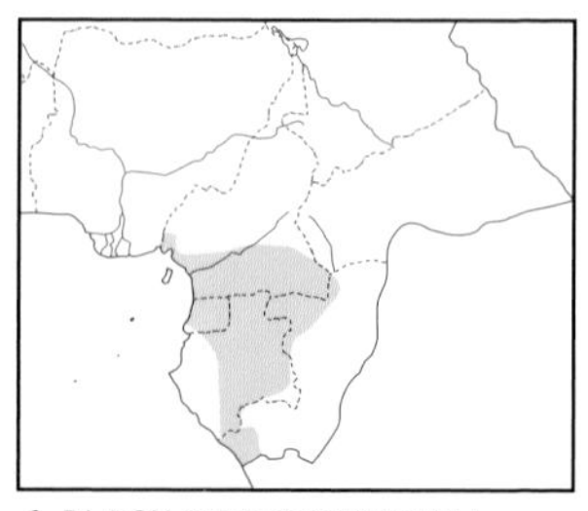

6. BLACK-HEADED WAXBILL

7. GREY WAXBILL

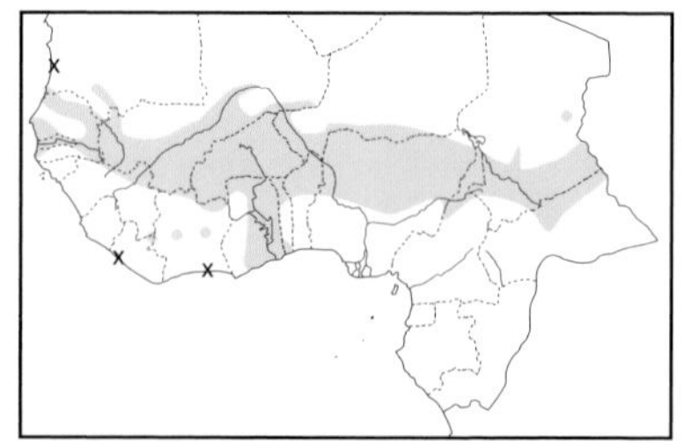

8. BLACK-RUMPED WAXBILL

Plate on page 460

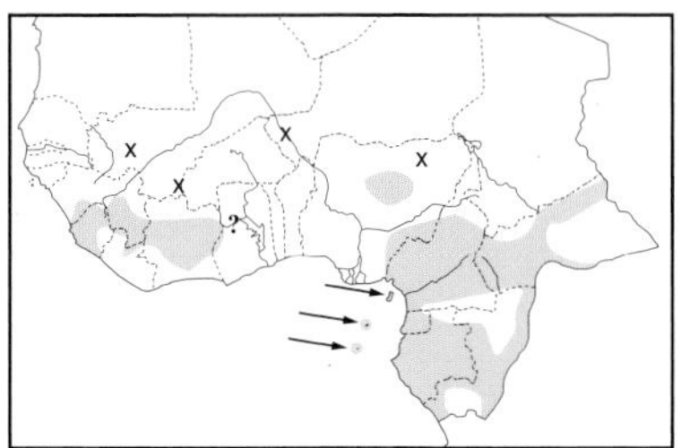

9. COMMON WAXBILL

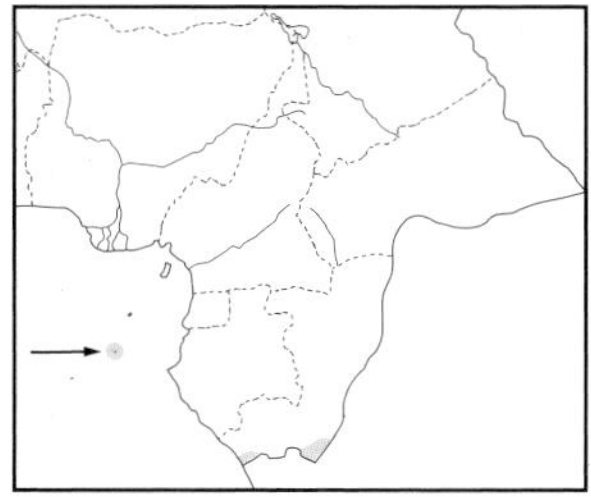

10. SOUTHERN CORDON-BLEU

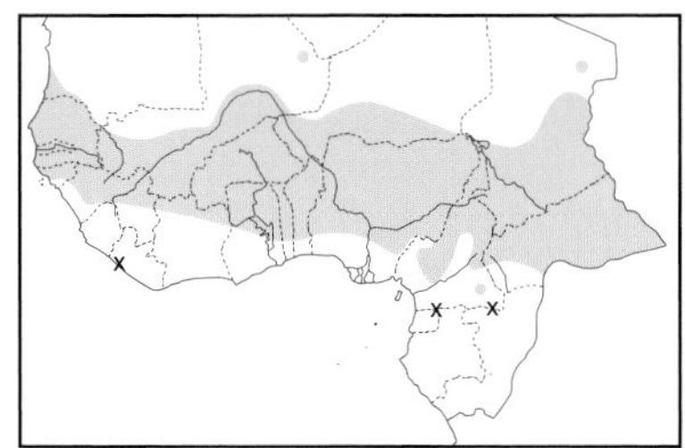

11. RED-CHEEKED CORDON-BLEU

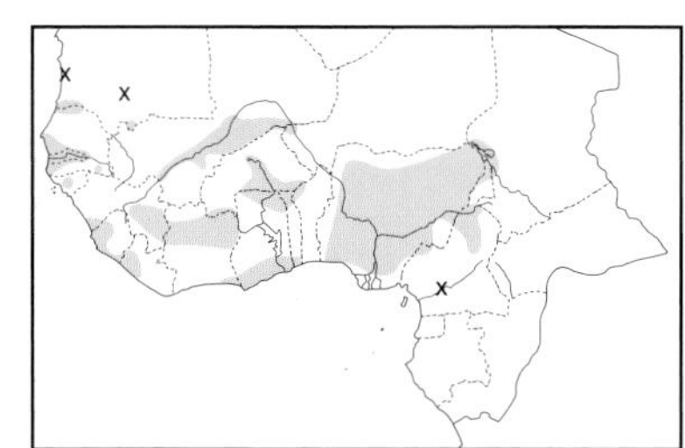

12. BLACK-FACED QUAILFINCH

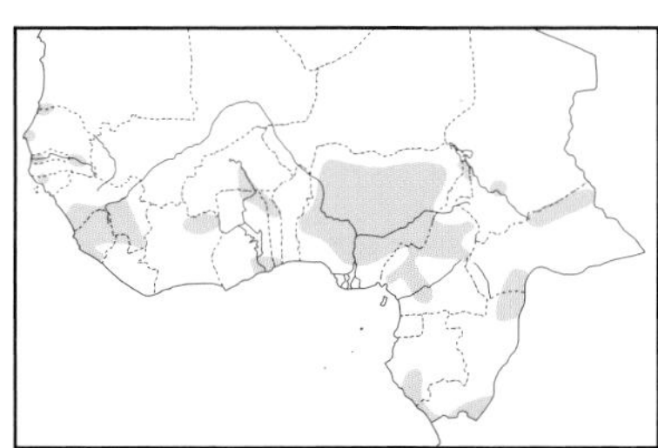

13. ZEBRA WAXBILL

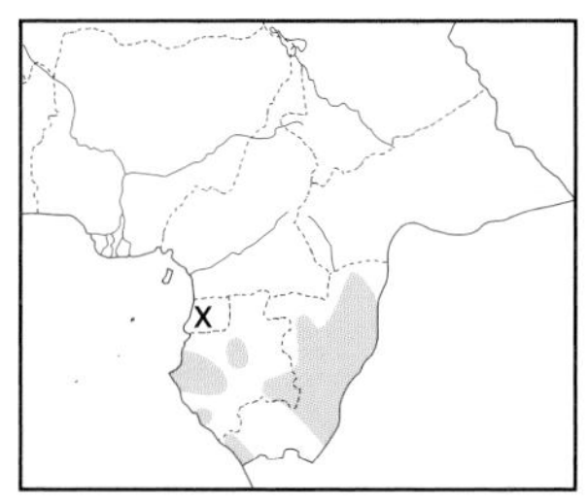

14. BLACK-CHINNED QUAILFINCH

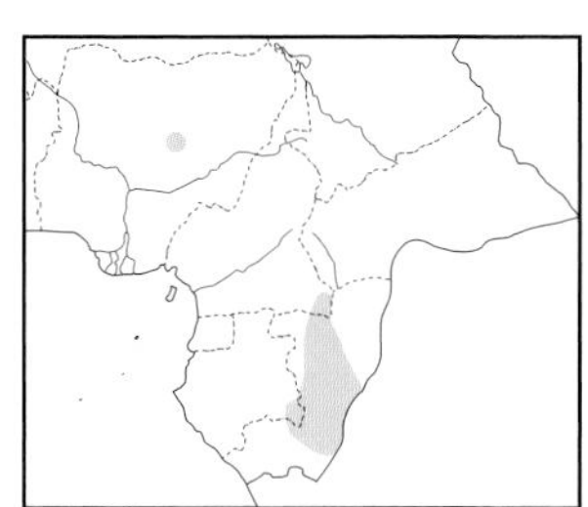

15. LOCUST FINCH

1 **BAR-BREASTED FIREFINCH** *Lagonosticta rufopicta* 11 cm R f
1a Ad Small white crescents on breast-sides and flanks; pinkish-red bill. **1b Juv** Dull grey-brown; face and breast tinged pinkish-red; dusky bill. ▲ Various open habitats. ✤ High-pitched *chip* or *pik*. Song a fast jumble of short, mostly clear notes. [CD15:22]

2 **RED-BILLED FIREFINCH** *Lagonosticta senegala* 10 cm R c
2a Ad male Mainly red; yellowish orbital ring; red bill. **2b Ad female** Mainly buff-brown above, paler below; red loral spot and rump; red bill. **2c Juv** As 2b but lacks red on lores; black bill. ▲ Villages, farmland, scrub, etc. ✤ Soft *dwee*; alarm *chep*. [CD15:24]

3 **ROCK FIREFINCH** *Lagonosticta sanguinodorsalis* c. 10 cm R lf
3a Ad male Grey crown; bright red mantle; two-toned bill (black and blue-grey). **3b Ad female** Bright reddish-brown mantle and back. ▲ Wooded grassland with rocky outcrops. Endemic. ✤ Trill *treeeee*, short *chew*, drawn-out whistles, and dry *pitpitpit*. [CD15:28]

4 **BLUE-BILLED FIREFINCH** *Lagonosticta rubricata* 10.0–11.5 cm R f/r
4a Ad male Top of head and upperparts grey-brown; black undertail-coverts; bluish bill. **4b Ad female** Paler than male. **4c Juv** Dull brownish, paler below; dark bill. ▲ Forest–savanna mosaic. ✤ Twittering *trrrrrrrr-t*, loud *chew-chew-chew*, dry *prttt* and *prrititit* (alarm) and rapid *wink-wink-wink* or *whitwhitwhitwhit*... [CD15:26]

5 **KULIKORO FIREFINCH** *Lagonosticta virata* 10–11 cm R lu/r
Ad male As 4a but slightly duller; fewer spots; bill with black tip, often all-black above. Female slightly paler below. ▲ Rocky areas with grass and bushes. Endemic. ✤ Nasal *chew*; rattling *chrrrrrrr* (alarm). Song has plaintive whistles *tseeeeeu*; also a trill. [CD15:29]

6 **BLACK-BELLIED FIREFINCH** *Lagonosticta rara* 10 cm R f/r
6a Ad male Deep wine-red; black up to lower breast; two-toned bill. **6b Ad female** Greyish head; black below as male. **6c Juv** Dull buff-brown; bill with pinkish base below. ▲ Wooded grassland. ✤ Short *pwit*, *pwit*, hard *chep*, nasal *cheeay*, short trill. [CD15:25]

7 **CHAD (REICHENOW'S) FIREFINCH** *Lagonosticta umbrinodorsalis* 10–11 cm R lf
Ad male Resembles 4a but top of head greyish. Female paler; red loral spot. ▲ Rocky hillsides with grassy areas. Endemic. ✤ Similar to 4. [CD15:27]

8 **BLACK-FACED FIREFINCH** *Lagonosticta larvata* 11.5 cm R lf
8a Ad male *vinacea* (west of range) Black face; pinkish-mauve above; bright pink below. **8b Ad female *vinacea*** Pale grey-brown head; buffish throat; pale pink below. **8c Juv *vinacea*** Grey-brown above; buff-brown below; no white specks on sides. **8d Ad male *togoensis*** (east of range) Grey above; paler grey below with variable pinkish wash. **8e Ad female *togoensis*** Grey-brown above, paler below; black bill. ▲ Wooded grassland. ✤ Piercing *seesee* and hard *twit-it-it*. Song a variable series of short notes. [CD15:30]

9 **CUT-THROAT FINCH** *Amadina fasciata* 11–12 cm R lc
9a Ad male Sandy coloured and scaly; red band on throat; chestnut on belly. **9b Ad female** No red on throat; face barred; no chestnut on belly. ▲ Dry woodland, thorn scrub, edges of cultivation. ✤ Abrupt *chilp*, nasal *dzèèp*. In flight, thin *eee-eee-*... [CD15:51]

10 **AFRICAN SILVERBILL** *Euodice cantans* 10 cm R s/lc
Pale sandy-brown; black rump; black, pointed tail; blue-grey bill. ▲ Dry savanna. ✤ Short *tsik*. Flocks give twittering *chip-chi-chi-chi-chip*... in flight.[CD15:47]

11 **BRONZE MANNIKIN** *Spermestes cucullatus* 9 cm R c
11a Ad Dark brown above; white below; barred rump and flanks. Small; dumpy. **11b Juv** Entirely buff-brown. ▲ Grassy habitats, bush, gardens. ✤ Rolling *trreep* in flight. [CD15:48]

Continued on page 466

Maps on pages 466 and 467

1a
1b
2a
2b
2c
3a
3b
4a
4b
4c
5
6a
6b
6c
7
8a
8b
8c
8d
8e
9b
9a
10
11b
11a
12b
12a
13b
13a

Continued from page 464

12 **MAGPIE MANNIKIN** *Spermestes fringilloides* 11.5–12.0 cm **R lu/r**
12a Ad Larger than 11a; bill heavier; black patch on breast-sides. **12b Juv** Dusky brown and buffish. ▲ Grassy habitats in forest zone. ✤ Fairly loud *teeoo*; alarm *tsek*. [CD15:50]

13 **BLACK-AND-WHITE MANNIKIN** *Spermestes bicolor* 9.5–10.5 cm **R c**
13a Ad Glossy black and pure white; heavy, grey-blue bill. **13b Juv** Dull brown above; pale buff-brown below; bill dark. ▲ Grassy habitats. ✤ Soft *kip* and *tseew*. [CD15:49]

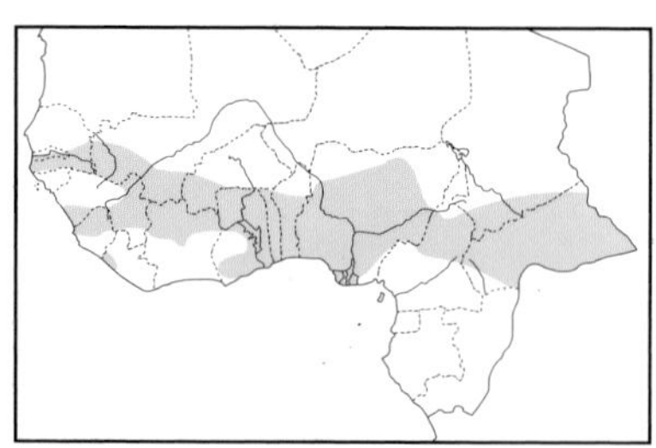

1. BAR-BREASTED FIREFINCH

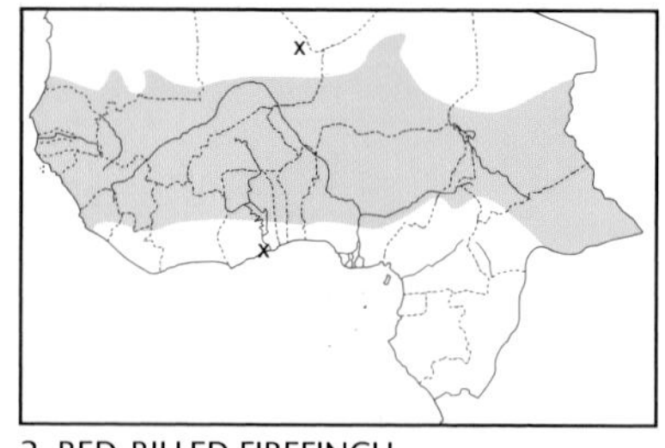

2. RED-BILLED FIREFINCH

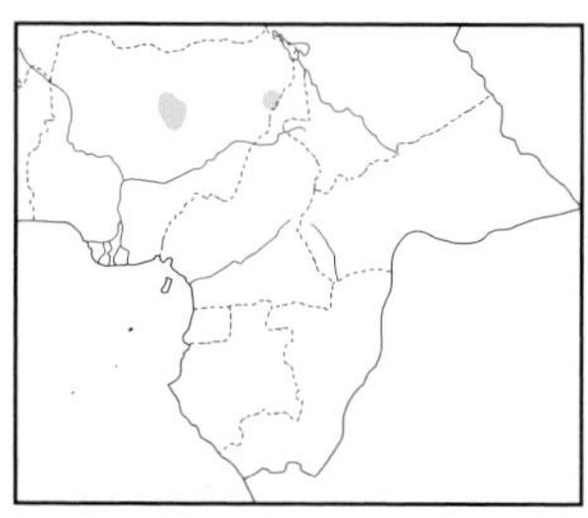

3. ROCK FIREFINCH

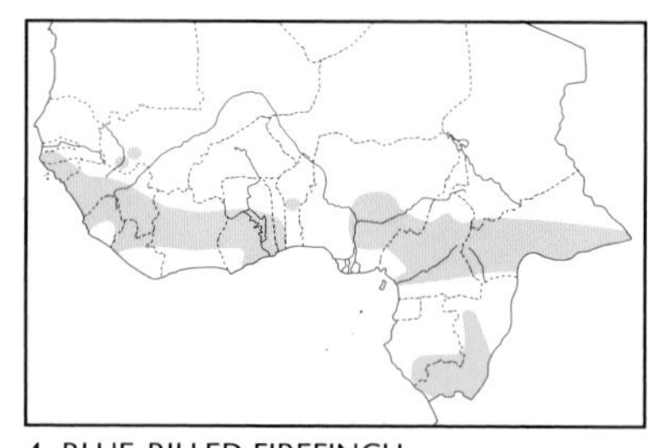

4. BLUE-BILLED FIREFINCH

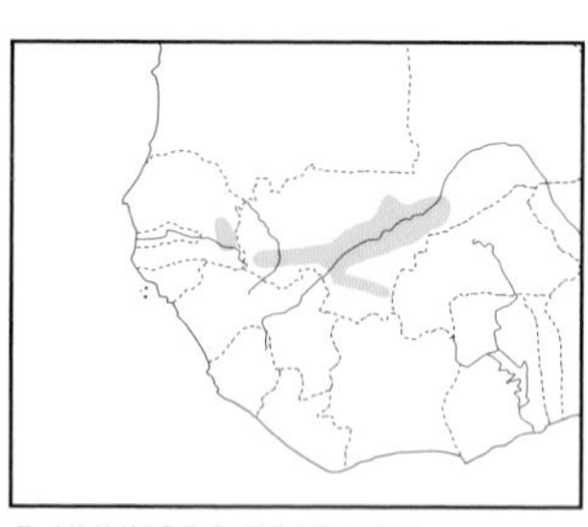

5. KULIKORO FIREFINCH

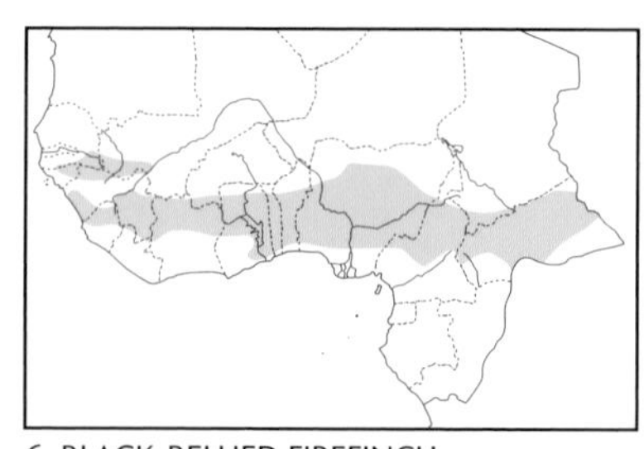

6. BLACK-BELLIED FIREFINCH

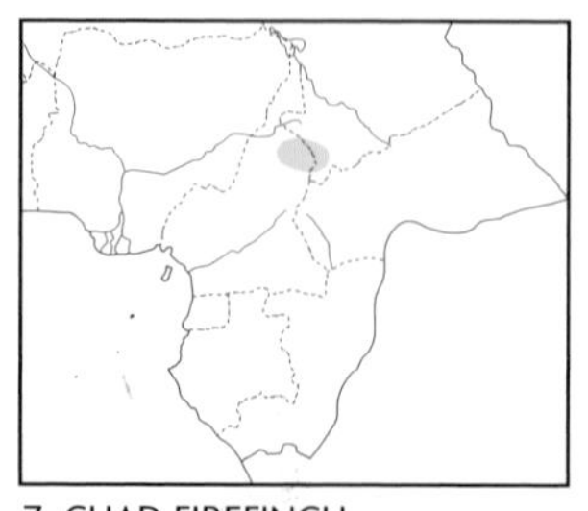

7. CHAD FIREFINCH

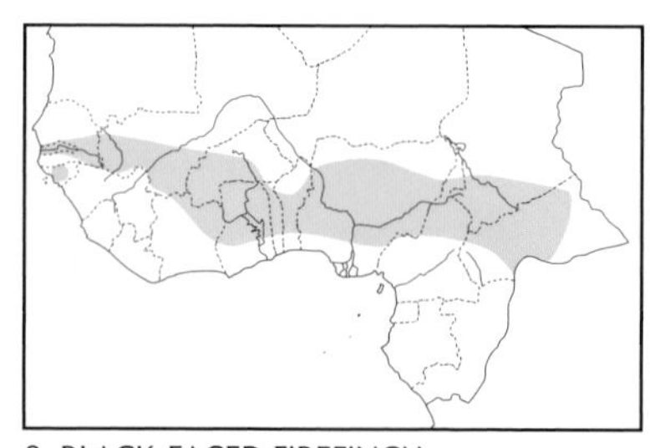

8. BLACK-FACED FIREFINCH

 Plate on page 464

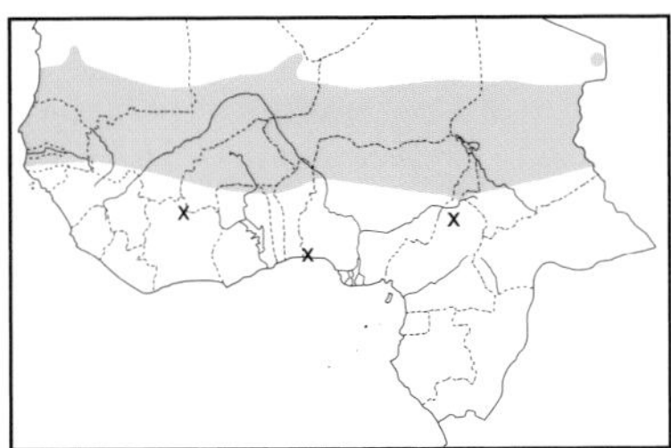

9. CUT-THROAT FINCH

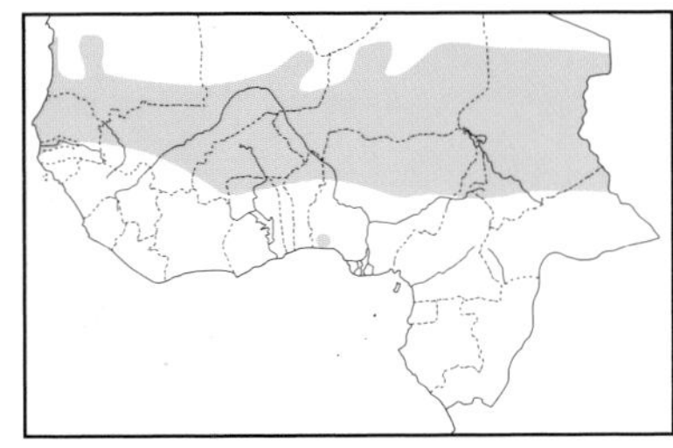

10. AFRICAN SILVERBILL

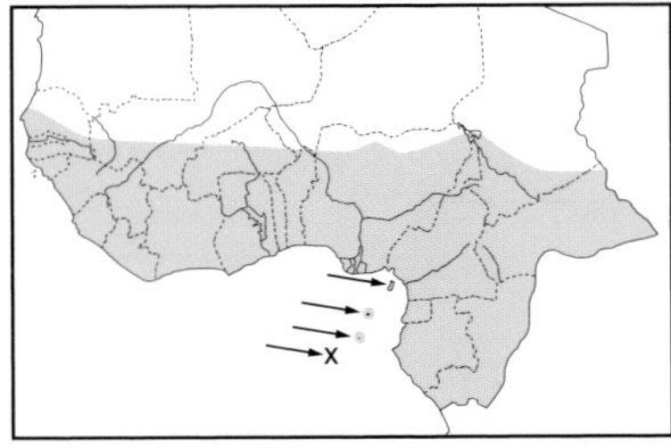

11. BRONZE MANNIKIN

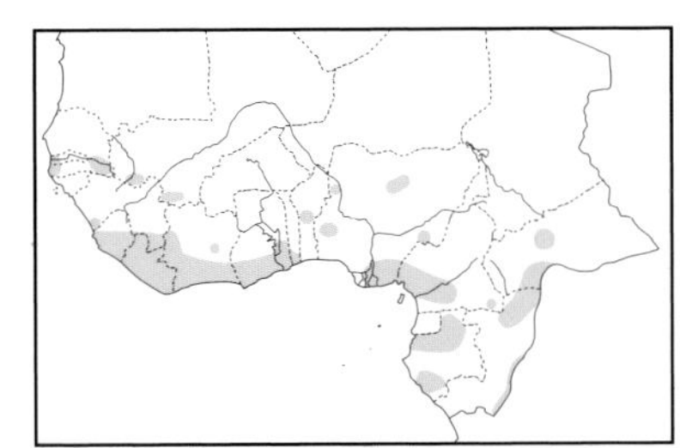

12. MAGPIE MANNIKIN

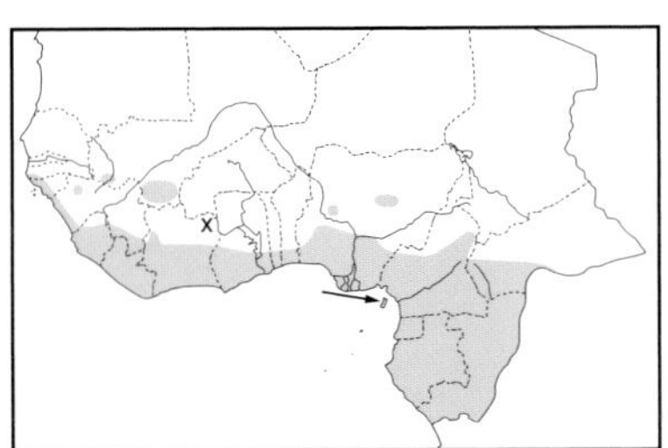

13. BLACK-AND-WHITE MANNIKIN

Small, seed-eating finches, occurring mainly in savanna. Forage on the ground. Polygamous, mostly species-specific brood parasites, laying eggs in nests of estrildids.
Whydahs. Males in breeding plumage have four greatly elongated central tail feathers. All paradise whydahs are very similar but differ mainly in length, width and shape of tail. Status and distribution inadequately known. Parasitise waxbills *Estrilda* and pytilias *Pytilia* species.

1 SAHEL PARADISE WHYDAH *Vidua orientalis* 12.5 cm (+tail of up to *c.* 24 cm) **R u**
1a Ad male breeding *aucupum* (S Mauritania–Nigeria) Very long broad tail streamers; chestnut collar. *V. o. orientalis* (N Cameroon–Chad) has collar more golden-buff. **1b Ad male non-breeding/ad female** Boldly striped head; dark bill and legs. **1c Juv** Plain earth-brown above; pale buff-brown below. ▲ Savanna. ✤ Sharp *chip*. Mimics Green-winged Pytilia. [CD15:62]

2 EXCLAMATORY PARADISE WHYDAH *Vidua interjecta* 12.5 cm (+tail of >27 cm) **R c**
2a Ad male breeding Tail streamers longer than 1a and broader than 3. **2b Ad male non-breeding/ad female** As 1b but with pale orange bill and pinkish-grey (male) or pinkish (female) legs. ▲ Savanna. ✤ Mimicry song includes *pik* calls and *to-wit-to-wit* song of Red-winged Pytilia. [CD15:63]

3 TOGO PARADISE WHYDAH *Vidua togoensis* 12.5 cm (+tail of >27 cm) **R u**
Ad male breeding Tail streamers narrower than 2a; underparts more uniformly yellow; nuchal collar paler. ▲ Savanna. Endemic. ✤ Mimics Yellow-winged Pytilia.

4 PIN-TAILED WHYDAH *Vidua macroura* 12.5 cm **R c**
4a Ad male breeding Black and white; long ribbon-like tail streamers of up to 20 cm. **4b Ad female breeding** Boldly striped head; mainly dark brown bill; dark legs. **4c Ad male/female non-breeding** As 4b but bill pinkish. **4d Juv** Plain pale brown above; pale buffish below; dusky bill. ▲ Various open habitats, incl. gardens. ✤ Call *tsip-tsip*. Song, uttered in display flight and at rest, a vigorous jerky series of high-pitched variations on the call note *tsip tsweep tswup tsweeu tswip* ... [CD15:61]

Indigobirds. Often only distinguishable in the field if song of foster species is known. Songs include non-mimetic chattering and clear, whistled notes mimicking those of hosts. Status and distribution imperfectly known; presence of host an indication of species possibly involved.

5 CAMEROON INDIGOBIRD *Vidua camerunensis* 11.5 cm **R**
Ad male breeding Blue gloss; pale purplish legs. Female and non-breeding male as 6b but with pale purplish legs. ✤ Mimics Blue-billed and Black-bellied Firefinches, Brown and Dybowski's Twinspots. [CD15:60]
The following species are very similar to 5:

JAMBANDU INDIGOBIRD *Vidua raricola* 11.5 cm **R**
Ad male breeding Blue to green gloss; brown wings. ✤ Mimics Zebra Waxbill. [CD15:54]

BARKA (BAKA) INDIGOBIRD *Vidua larvaticola* 11.5 cm **R**
Ad male breeding Greenish-blue gloss. ✤ Mimics Black-faced Firefinch. [CD15:55]

JOS PLATEAU INDIGOBIRD *Vidua maryae* 11.5 cm **R**
Ad male breeding Green gloss. Endemic to Jos Plateau. ✤ Mimics Rock Firefinch. [CD15:57]

QUAILFINCH INDIGOBIRD *Vidua nigeriae* 11.5 cm **R**
Ad male breeding Dull green. ✤ Mimics Black-faced Quailfinch. [CD15:59]

WILSON'S INDIGOBIRD *Vidua wilsoni* 11.5 cm **R**
Ad male breeding Purplish gloss; dark brown wings; purplish legs. ✤ Mimics Bar-breasted Firefinch. [CD15:58]

6 VILLAGE INDIGOBIRD *Vidua chalybeata* 11.5 cm **R**
6a Ad male breeding The only indigobird with orange to red legs. Gloss green (nominate) to blue (*neumanni*; E Mali–Chad); black wings. **6b Ad male non-breeding/ad female** Pale median crown-stripe; dark brown lateral crown-stripes; streaked upperparts; white bill. **6c Juv** Plain dusky-brown crown. ✤ Mimics Red-billed Firefinch. [CD15:52]

 Maps on page 470

1c
1a
2b
1b
3a
2a
3
4a
4c
4d
4b
6c
6b
5
6a

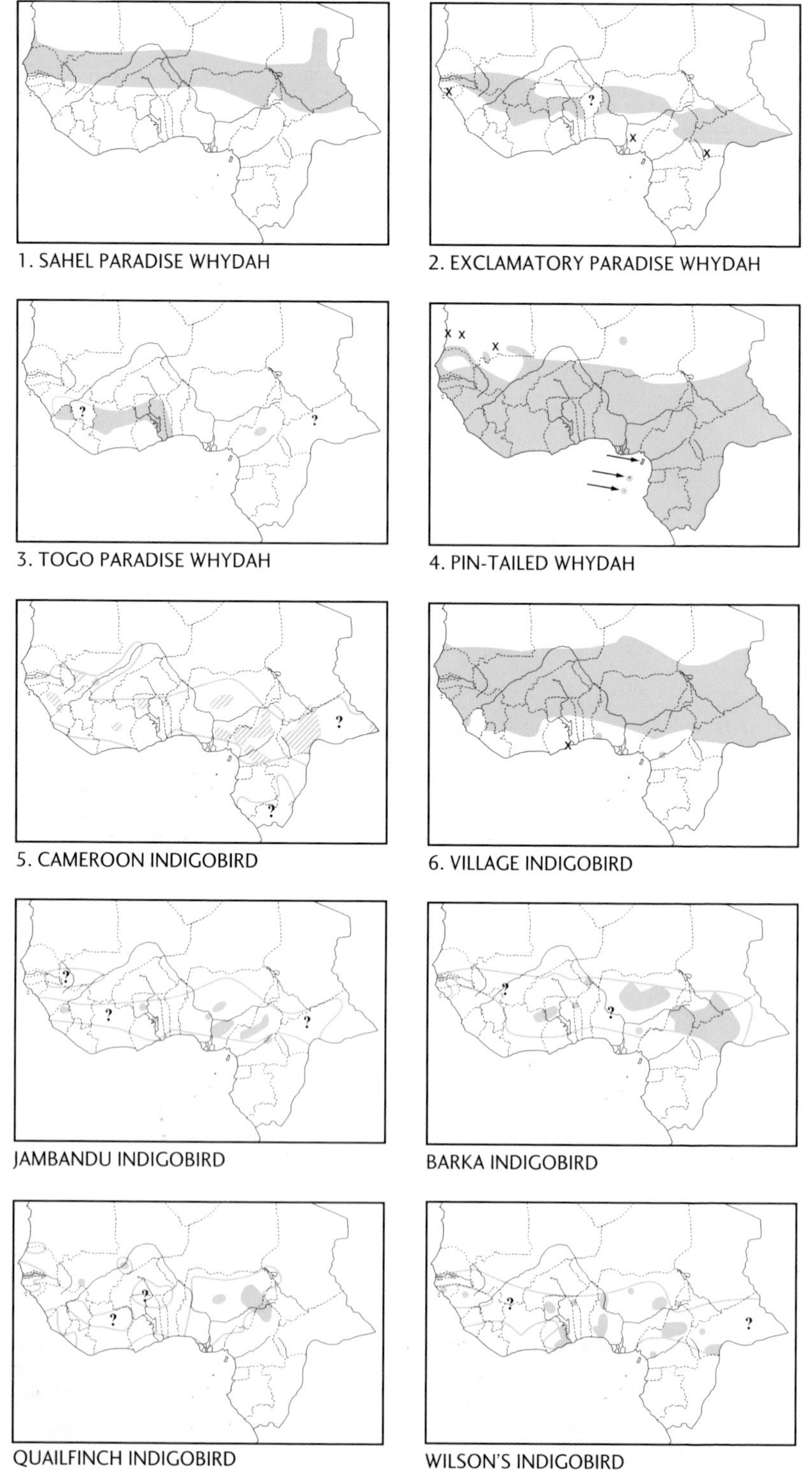

1. SAHEL PARADISE WHYDAH

2. EXCLAMATORY PARADISE WHYDAH

3. TOGO PARADISE WHYDAH

4. PIN-TAILED WHYDAH

5. CAMEROON INDIGOBIRD

6. VILLAGE INDIGOBIRD

JAMBANDU INDIGOBIRD

BARKA INDIGOBIRD

QUAILFINCH INDIGOBIRD

WILSON'S INDIGOBIRD

Plate on page 468

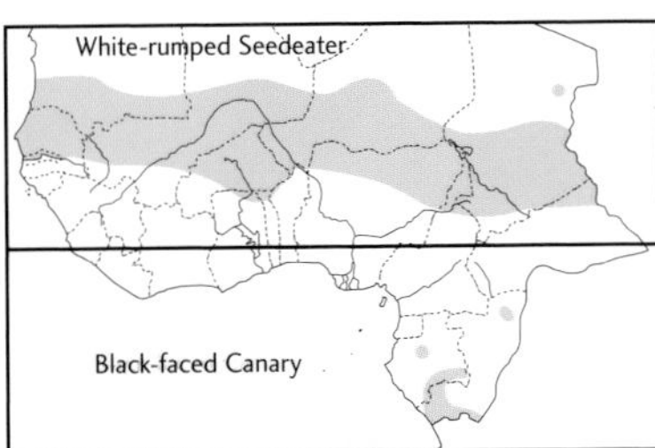

1. BLACK-FACED CANARY
2. WHITE-RUMPED SEEDEATER

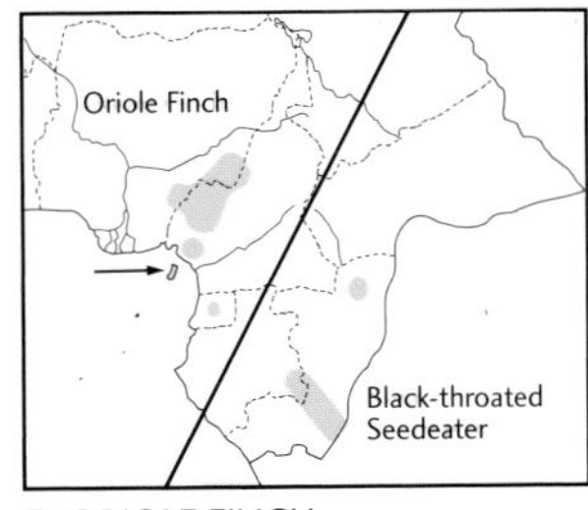

3. ORIOLE FINCH
4. BLACK-THROATED SEEDEATER

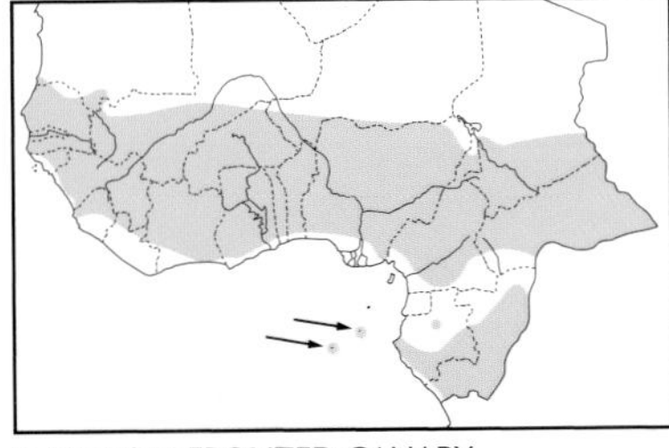

5. YELLOW-FRONTED CANARY

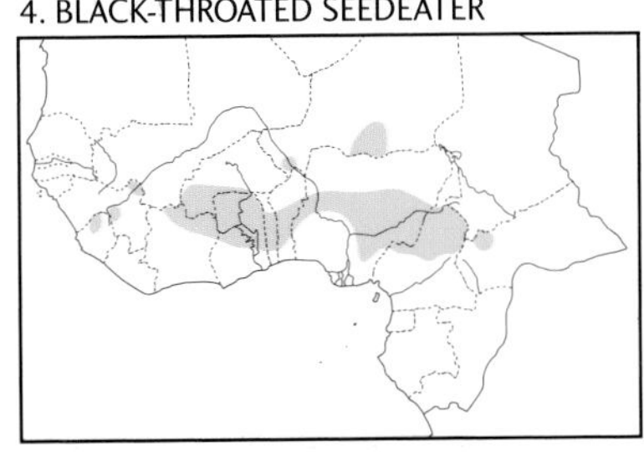

6. STREAKY-HEADED SEEDEATER

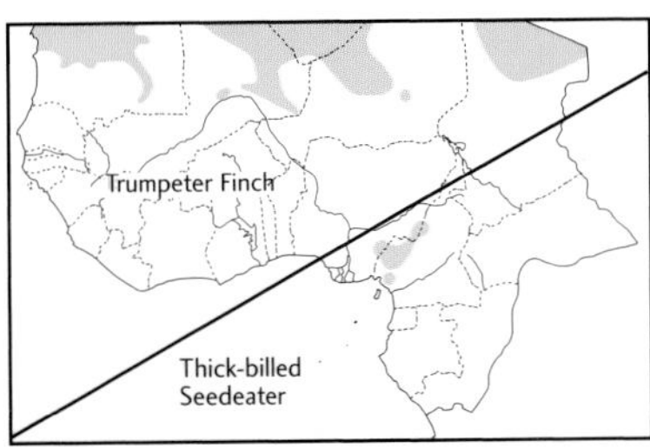

7. THICK-BILLED SEEDEATER
12. TRUMPETER FINCH

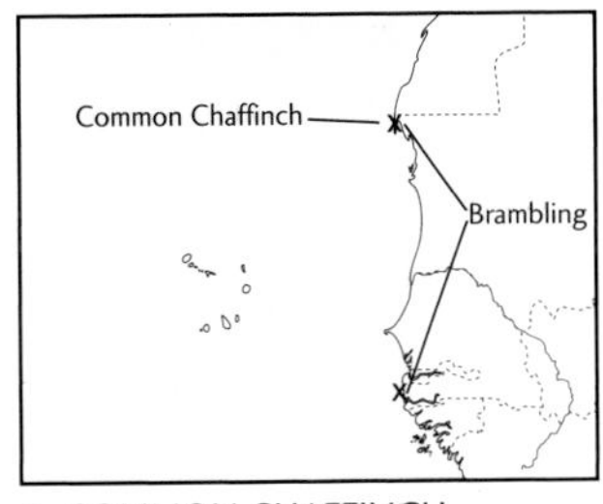

8. COMMON CHAFFINCH
9. BRAMBLING

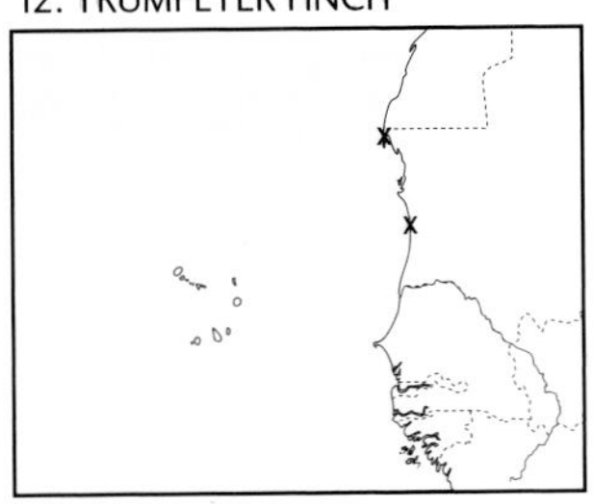

10. EUROPEAN GREENFINCH

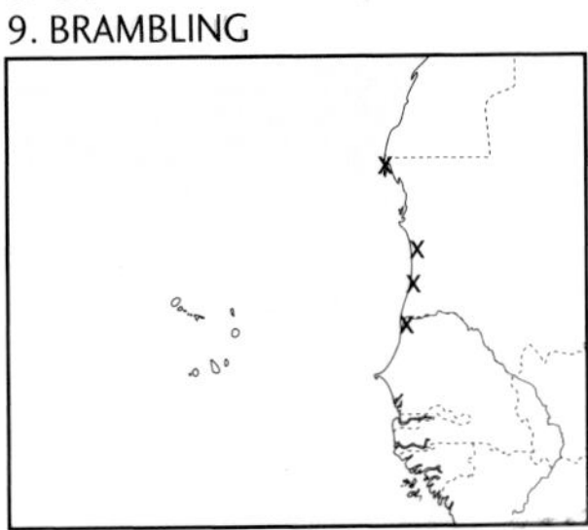

11. COMMON LINNET

Continued from page 472

12 TRUMPETER FINCH *Bucanetes githagineus* 12.5 cm **R lf**
12a Ad male breeding Grey head; distinct pink tinge esp. on underparts; orange-red bill. **12b Ad male non-breeding/ad female** Duller, sandy-grey; horn-coloured bill. ▲ Stony desert. ✤ Short *tip* or *chee-chup*; in flight a soft, nasal *veechp*. Song a distinctive drawn-out nasal wheeze. [CD4:85]

Small seed-eaters with stout, conical bills. Arboreal or terrestrial. Several have remarkable, melodious songs.

1 BLACK-FACED CANARY *Serinus capistratus* 11.5 cm R u/lc
1a Ad male Black mask; yellow underparts. **1b Ad female** No black mask; some streaks on throat, breast and flanks. ▲ Woodland, edges of cultivation, gardens. ✤ Various *chissik* notes. Song a jumble of whistles, trills and twittering phrases. [CD15:67]

2 WHITE-RUMPED SEEDEATER *Serinus leucopygius* 10.0–11.5 cm R lc/r
Small; pale greyish-brown; white rump. ▲ *Acacia* scrub, farmland, grassland near water, gardens. ✤ Call *twee-eet*. Song a varied series of clear, sweet whistled and twittering phrases, resembling (but superior to) 5. [CD15:69]

3 ORIOLE FINCH *Linurgus olivaceus* 13 cm R f
3a Ad male Black head; golden-yellow underparts; orange-yellow bill. **3b Ad female** Olive-green; paler bill. **3c Juv** As adult female but paler below with faint streaking; dusky bill. ▲ Montane forest. ✤ Mostly silent. Wheezy *tssit* or *twee*. Song a sustained, high-pitched twittering and soft churring followed by a melodious whistle. [CD15:79]

4 BLACK-THROATED SEEDEATER *Serinus atrogularis* 11–12 cm R lf
Small; grey-brown; bright yellow rump. ▲ Open woodland, cultivation. ✤ Rising *tweee*; in flight *chirrup*. Song a rapid, sustained jumble of trills and whistles, reminiscent of 5. [CD15:70]

5 YELLOW-FRONTED CANARY *Serinus mozambicus* 11–13 cm R lc
5a Ad male Distinctive head pattern; yellow rump. **5b Ad female** Duller; with or without necklace of dark spots. ▲ Open woodland, cultivation, gardens. ✤ Single or double *tseeu* or *tsssp* and *tuwu-tsilip* or *tuwu-tuwee*. Song a fairly simple but melodious, rapid jumble of twittering and whistled phrases. [CD15:71]

6 STREAKY-HEADED SEEDEATER *Serinus gularis* 15 cm R lf/s
6a Ad Mainly brown; broad white supercilium; dark face. **6b Juv** Duller; streaked underparts. ▲ Open woodland, cultivation, scrub. ✤ Nasal *shewee-uee*, short *chip*, *tseeee* and *chiririt*. Song a clear, rapid jumble of varied notes typical of canary. [CD15:74]

7 THICK-BILLED SEEDEATER *Serinus burtoni* 18 cm R s/lc
Robust; white forehead; heavy bill; double wingbar. ▲ Montane forest. ✤ Mostly silent. High-pitched *srip-sreep*. Song a soft jumble of tinkling notes, typical of canary. [CD15:77]

8 COMMON CHAFFINCH *Fringilla coelebs* 15 cm V
8a Ad male *africana* Bluish-grey head; olive-green above; pink below; broad white wingbars. **8b Ad female** Mainly grey-brown; white wingbars. **8c In flight** White wingbars; white outer tail feathers. ▲ Palearctic vagrant. ✤ Clear *pink*; in flight a low *yup*.

9 BRAMBLING *Fringilla montifringilla* 15.5 cm V
9a Ad male non-breeding Orangey breast and shoulders. **9b Ad female non-breeding** Duller. **9c In flight** White rump. ▲ Palearctic vagrant. ✤ Nasal *kèèhp*. [CD4:82]

10 EUROPEAN GREENFINCH *Carduelis chloris* 15 cm V
10a Ad male Robust; olive-green; yellow in wing and tail. Female similar but more grey-brown. **10b In flight** Yellow patches on primaries and sides of tail. ▲ Palearctic vagrant. ✤ Soft *dsooeet*; in flight a characteristic *djururrup*. [CD4:76]

11 COMMON LINNET *Carduelis cannabina* 14 cm V
11a Ad male Greyish head; chestnut upperparts. **11b Ad female** More heavily streaked. **11c In flight** White patches in wings and tail-sides. ▲ Palearctic vagrant. ✤ Dry stuttering *knutnutnutnut*, usually in flight. [CD4:79]

Continued on page 471

Maps on page 471

1b
1a
2
3c
3b
3a
4
5a
5b
6a
6b
7
8a
8b
8c
9a
9b
9c
10a
10b
11a
11b
11c
12a
12b

PLATE 144: BUNTINGS

Small, mainly seed-eating species with stout, conical bills. Predominantly terrestrial. Flight typically undulating. Usually singly or in pairs, but some gregarious outside breeding season. Vocalisations usually rather simple.

1 AFRICAN GOLDEN-BREASTED BUNTING *Emberiza flaviventris* 15.5 cm **R f/r**
1a Ad male *flavigaster* (Senegal–CAR) Bold black-and-white head pattern; white wingbars; grey rump. *E. f. flaviventris* (S Congo) has deeper coloured plumage. **1b Ad female *flavigaster*** Head markings browner. **1c Juv *flavigaster*** Duller and browner than adult; head pattern brown and buff. ▲ Dry grassy savanna, open woodland. ✤ Subdued *chruteeu*; in flight a soft *chup*. Song a rapid succession of variable series of 4–10 similar syllables. [CD15:81]

2 BROWN-RUMPED BUNTING *Emberiza affinis* 14 cm **R u/lf**
2a Ad male Head whiter than 1a; no white on wing; brown rump. Female has head markings browner. **2b Juv** Duller; head washed rusty-buff, stripes brown. ▲ Open wooded savanna, farmland. ✤ Liquid *pidrewdrlwi*; in flight a short *chip*. Song a rather short, slightly harsh warble. [CD15:83]

3 CABANIS'S BUNTING *Emberiza cabanisi* 16.5 cm **R u/lf**
3a Ad male *cabanisi* (Guinea–CAR) Blackish head; long white supercilium; black-and-grey upperparts. Female has head and upperparts browner; supercilium and stripe bordering head-sides tinged buffish. **3b Juv *cabanisi*** Much browner than adult; pale rufous supercilium. **3c Ad male *cognominata*** (S Gabon–Congo) Greyish median crown-stripe; buff-grey mantle with dark streaks. ▲ Wooded savanna. ✤ Soft *tureee*; in flight also *tsip*; alarm *seeee*. Song consists of rapid and variable series of 3–6 similar syllables, similar to those of 1. [CD15:82]

4 CINNAMON-BREASTED ROCK BUNTING *Emberiza tahapisi* 14 cm **R/M lc**
4a Ad male *goslingi* (north of range: Senegal–Chad and CAR) Black-and-white striped head; grey throat; rufous-brown body; rufous wing panel. **4b Ad female *goslingi*** Duller; head striped buffish-white and blackish-brown. **4c Ad male *tahapisi*** (Gabon–Congo) Black throat; less rufous in wings. ▲ Open savanna with rocky outcrops. ✤ Nasal *wee-eh*; alarm a thin *sweeee*. Song short and rapid, with emphasis on final note, e.g. *chrr-trr-erl-CHEEP*. [CD15:80]

5 CRETZSCHMAR'S BUNTING *Emberiza caesia* 15.5 cm **V**
5a Ad male Blue-grey head; rusty moustachial stripe and throat; creamy eye-ring. **5b Ad female** Duller and paler; finely streaked crown and breast. ▲ Palearctic vagrant. E Chad. ✤ Harsh *zee* or *chip*.

6 HOUSE BUNTING *Emberiza striolata* 14 cm **R/V u/lc**
6a Ad male Grey head streaked black; almost uniformly rufous body. **6b Ad female** Paler and duller; head more grey-brown; no streaks on breast. ▲ Rocky wadis, cultivation, villages. ✤ Nasal *chzwee* and thin *chik*. Song a rapid, simple and rather monotonous series of similar syllables. [CD4:92]

7 ORTOLAN BUNTING *Emberiza hortulana* 16.5 cm **P lr/u**
7a Ad male Olive-grey head; yellow moustache, throat and eye-ring; pink bill. **7b Ad female** Slightly duller; crown and breast streaked. **7c Imm** Duller than adult female; flanks streaked. ▲ Open upland habitats. ✤ Short *twit* and metallic *seee*. [CD4:91]

8 CORN BUNTING *Emberiza calandra* 18 cm **V**
Bulky and streaked; no white in tail; heavy bill. ▲ Palearctic vagrant. ✤ Dry, hard *kwit* and *kwitit* on take-off and in flight. [CD4:96]

Maps on page 476

1b
1a
1c
2a
2b
3a
3b
3c
4c
4d
4a
5b
5a
6b
6a
7c
7b
7a
8

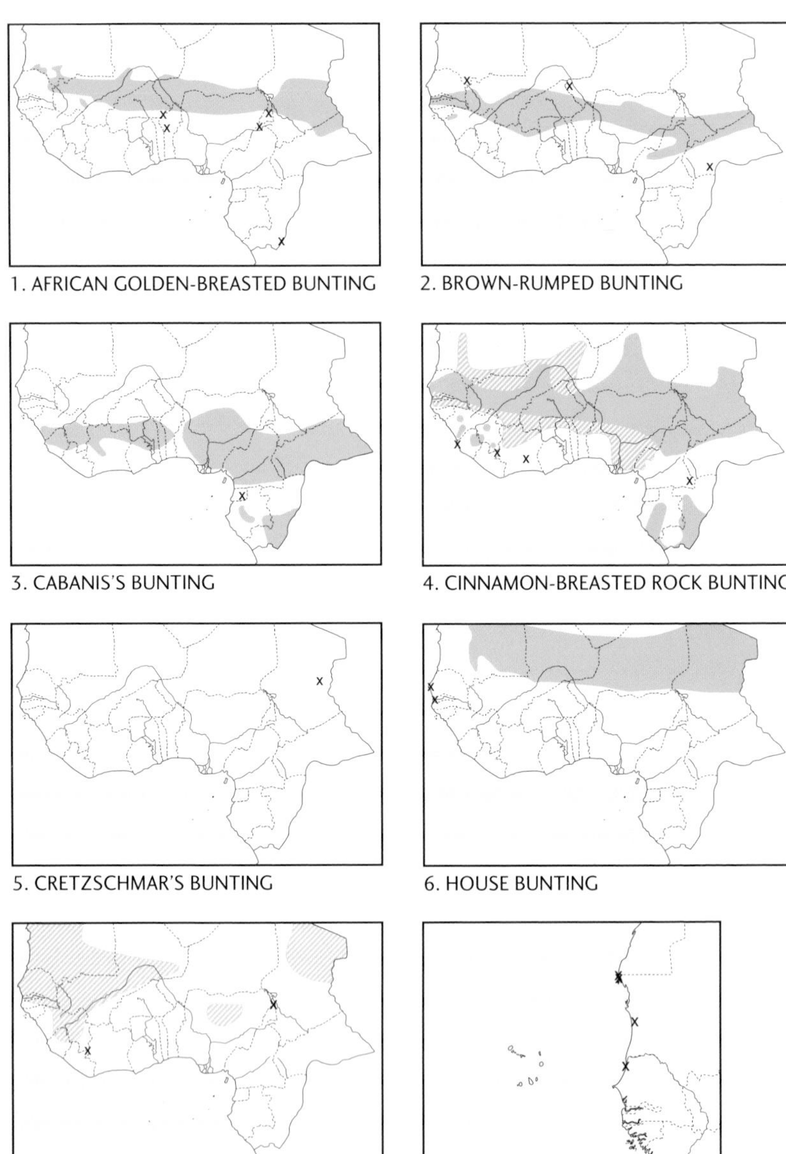

1. AFRICAN GOLDEN-BREASTED BUNTING

2. BROWN-RUMPED BUNTING

3. CABANIS'S BUNTING

4. CINNAMON-BREASTED ROCK BUNTING

5. CRETZSCHMAR'S BUNTING

6. HOUSE BUNTING

7. ORTOLAN BUNTING

8. CORN BUNTING

Plate on page 474

ISLAND FORMS

1 **CAPE VERDE WARBLER** *Acrocephalus brevipennis* 13.5 cm **R* lf EN**
Greyish head; grey-brown upperparts, rump brighter; short, faint pale supercilium; white underparts washed greyish on breast-sides and flanks. ▲ Santiago, São Nicolau. Well-vegetated valleys, esp. with patches of reeds, up to 500 m (mostly lower); also sugarcane and banana plantations, and gardens, usually near water. Secretive when not breeding. Occasionally in small parties, foraging in fruiting fig trees. Endemic. ✤ Song short and distinctive with harsh quality and including loud liquid bubbling and trills, reminiscent of Greater Swamp Warbler and/or Common Bulbul. Call a low, hard *kruk* or *kerr*. [CD4:10]

2 **RASO LARK** *Alauda razae* 12–13 cm **R* r CR**
Streaked; white throat; tail blackish with white outer feathers; rather long bill. ▲ Raso islet, where the only lark. Mainly on flat ground, sparsely vegetated with herbage and low bushes. Outside breeding season in flocks of up to 25. Not shy. Endemic. ✤ Rippling *trruwee*. Song, given in vertical display flight or from ground or perch, loud, clear and sustained, reminiscent of Eurasian Skylark but more repetitive and less varied. [CD3:50]

3 **IAGO SPARROW** *Passer iagoensis* 13 cm **R* c**
3a Ad male Blackish crown; rich chestnut from eye onto ear-coverts. **3b Ad female** Conspicuous pale, creamy-buff supercilium. ▲ All islands. Various arid habitats; also edges of cultivation, oases, villages and towns (esp. where Spanish Sparrow absent). In flocks when not breeding, occasionally with Spanish Sparrow. Tame. Endemic. ✤ Call and song consist of chirps, similar to House Sparrow's. [CD4:72]

4 **CAPE VERDE SWIFT** *Apus alexandri* 13 cm; WS 35 cm **R* c**
4a-b Small; relatively short wings; shallowly forked tail (other swifts recorded in Cape Verdes are larger and longer winged, with more deeply forked tails). ▲ Most islands. Over all habitats. Often in large flocks. Flight weak and fluttering. Endemic. ✤ As Common Swift but higher pitched and weaker.

5 **PEREGRINE FALCON** *Falco peregrinus* 33–50 cm; WS 80–115 cm **R r**
5a Ad *madens* (Cape Verde Peregrine) Crown and mantle tinged brown; underparts washed pink-buff. **5b In flight** Broad-based, pointed wings; relatively short tail. ▲ Probably on all islands. Coastal cliffs and mountain ranges in interior. Endemic.

6 **COMMON KESTREL** *Falco tinnunculus* 30–38 cm; WS 65–80 cm **R c/s**
6a Ad male *alexandri* (Alexander's Kestrel) Crown greyish streaked dark; upperparts and tail barred; tail grey. **6b Ad female *alexandri***. ▲ Santiago, Fogo, Brava, Sal, Boavista, Maio (c); also I. do Rombo. All habitats. Endemic. **6c Ad male *neglectus*** (Neglected Kestrel) Upperparts and tail heavily barred; head and tail with little or no grey. ▲ Santo Antão and São Nicolau (c), São Vicente (u), Santa Luzia (s), also Branco and Raso. All habitats. Endemic. **6d In flight**. See also plate 27:1a.

7 **CAPE VERDE PURPLE HERON** *Ardea (purpurea) bournei* 78–90 cm **R* r**
7a Ad Paler than Purple Heron (Plate 6:3), with less black on neck, breast and belly; centre of underparts white and pale chestnut. **7b Juv**. ▲ Santiago. Endemic. Feeds in wetlands and on dry hill slopes. Nests in tree crowns.

8 **CAPE VERDE KITE** *Milvus (milvus) fasciinucha* 55–72 cm; WS 140–180 cm **R* r**
8a As Red Kite (Plate 18:2), but tail less deeply forked and more barred; inner webs of primaries with darker grey marbling (not white). **8b In flight** Large whitish 'window' on primaries below. ▲ Santiago. Very rare endemic.

9 **CAPE VERDE BUZZARD** *Buteo (buteo) bannermani* 45–58 cm; WS 110–132 cm **R* r**
9a Similar to Common Buzzard (Plate 23:4) but has less individual variation. **9b In flight** Broad rounded wings and tail. ▲ Santiago, Santo Antão. Mountainous areas with steep rock faces, forested and cultivated hill slopes. Endemic.

1
2
3b
3a
4a
5a
6b
6a
6d
4b
5b
9b
7b
6c
8b
8a
7a
9a

1 **SÃO TOMÉ OLIVE PIGEON** *Columba thomensis* 37–40 cm R lc VU
Large and dark; yellow bill and feet. **1a Ad male** Mainly maroon; small white spots on wing-coverts. **1b Ad female** Duller; only slightly washed maroon. ▲ São Tomé. Forest. Endemic. ✤ Quavering series of very low, muffled coos. [CD6:66]

2 **SÃO TOMÉ SPINETAIL** *Zoonavena thomensis* 10 cm R c
2a–b Ad in flight Small. Broad white rump and white belly with dark streaks; square tail. ▲ São Tomé, Príncipe. Almost all habitats and altitudes. Endemic. ✤ Very high-pitched squeak.

3 **SÃO TOMÉ GREEN PIGEON** *Treron sanctithomae* 28 cm R c
Mainly olive-grey and olive-green. ▲ São Tomé, where the only green pigeon. Forest. Endemic. ✤ Similar to African Green Pigeon but even more complex: an accelerating rattle followed by fluty notes and ending with rasping, barking and growling sounds. [CD6:54]

4 **LEMON DOVE** *Aplopelia larvata* 24–25 cm R c/u
Ad male *simplex* Very similar to *inornata* (Plate 52:8) but slightly paler. Male *principalis* has more greyish face; darker belly, vinous paling to buffish. Female *principalis* and *simplex* as *inornata*. ▲ Bioko (*inornata*; f), São Tomé (*simplex*; c), Príncipe (*principalis*; c), Annobón (*inornata*; u). Forest. ✤ Monotonous series of 10–50 similar, low notes. [CD6:63]

5 **SÃO TOMÉ BRONZE-NAPED PIGEON** *Columba malherbii* 28 cm R c/u
Small; grey bill; reddish feet. **5a Ad male** Head and underparts grey; glossy nape; blackish upperparts; rufous undertail-coverts. **5b Ad female** Similar, but some ochre mottling below. ▲ São Tomé, Príncipe, Annobón. Forest. Endemic. ✤ 2–3 raucous notes followed by accelerating series of bouncing coos, ceasing abruptly. Call a low *krrreu*. [CD6:62]

6 **VELVET-MANTLED DRONGO** *Dicrurus modestus* 24–27 cm R c
D. m. modestus (Príncipe Drongo, NT) As *coracinus* (Plate 127:2); only slightly more glossy and usually larger. ▲ Príncipe. Edges of forest and cultivation. ✤ Commonest song a series of soft, melodious whistles *whee, hiu hiu hiuu*; also rhythmic *tiu-wee-tiuh tiu-wee-tiuh tiu-wee-tiuh tiu-weeeh*. Calls include various harsh, disharmonic and nasal sounds.

7 **GULF OF GUINEA THRUSH** *Turdus olivaceofuscus* 24 cm R c/r NT
7a Ad *olivaceofuscus* (São Tomé, where the only thrush; common) Dark olive-brown above; pale brown and white below with crescentic bars. **7b Ad *xanthorhynchus*** (Príncipe; extremely rare) Bill and eye-ring bright yellow. ▲ Forested habitats. Endemic. ✤ Song a few short whistles followed by sharper, higher pitched notes. Alarm, soft *whup*. [CD10:39]

8 **PRÍNCIPE GLOSSY STARLING** *Lamprotornis ornatus* 29 cm R c
Large; mainly bronzy plumage. Compare Splendid Glossy Starling (Plate 128:7). ▲ Príncipe. Forest, cultivation, forest regrowth. Endemic. ✤ Various nasal, twanging, wheezing and whistling notes. [CD14:23]

9 **BLUE-BREASTED KINGFISHER** *Halcyon malimbica* 25 cm R c
H. m. dryas As mainland races (Plate 67:5), but crown brown; bill larger. ▲ Príncipe. Forest, cultivation, gardens, etc. ✤ Song slower, more melancholic than mainland races. Alarm a raucous *chup, chup-chup-chup* or *kiah, kiah-kiah*. [CD7:79]

10 **SÃO TOMÉ ORIOLE** *Oriolus crassirostris* 23–24 cm R u/lc VU
10a Ad male Black head; whitish underparts; creamy collar; red bill. **10b Juv** Greyish head and upperparts; breast streaked blackish; dusky bill. ▲ São Tomé, where the only oriole. Forest. Endemic. ✤ Melodious fluting whistles, typically oriole-like in quality. Also a drawn-out *heeeew* or *hoo-heeew* and harsh *whrèèèh*. [CD13:33]

11 **DWARF OLIVE IBIS** *Bostrychia (olivacea) bocagei* c. 50 cm R r CR
As Olive Ibis (Plate 11:3) but much smaller. ▲ São Tomé, where the only ibis. Primary forest. Endemic. ✤ Raucous *kah-gah kah-gah kah-gah* or harsh *karh karh karh*...

1b
1a
2b
2a
3
5b
5a
4
6
7a
8
9
7b
11
10b
10a

1 **FERNANDO PO SPEIROPS** *Speirops brunneus* 13 cm **R lc VU**
Rusty-brown above with blackish cap; pale brown below with grey throat. ▲ Bioko. Montane forest, Pico Basilé (above 1900 m). Endemic. ✤ Long trill *trrrrrrrruuu* and rapid *trik-trik-trik*.

2 **ANNOBÓN WHITE-EYE** *Zosterops griseovirescens* 12 cm **R c VU**
Small. Greyish-olive above; whitish tinged olive-yellow below; white eye-ring. ▲ Annobón. Everywhere with bush and tree cover. Endemic. ✤ Similar to 3.

3 **PRÍNCIPE WHITE-EYE** *Zosterops ficedulinus* 10.5 cm **R u/r VU**
Z. f. feae (São Tomé; uncommon and local) Small; greyish-olive to olive-green above; greyish-white tinged yellow below; white eye-ring. *Z. f. ficedulinus* (Príncipe; rare) slightly paler below, with less grey wash. ▲ Forest regrowth, plantations. Endemic. ✤ Hard *trrrrr* and *pip pip pip...* Song a series of fairly loud, quavering notes.

4 **SÃO TOMÉ SPEIROPS** *Speirops lugubris* c. 14 cm **R c**
Mainly greyish-olive; black cap; white eye-ring; pale bill and legs. ▲ São Tomé. Wooded habitats. Endemic. ✤ Soft, quavering trill *rrriirrr*, hard *trrrr* and sharp *whseeeew*.

5 **PRÍNCIPE SPEIROPS** *Speirops leucophaeus* 13.5 cm **R c NT**
Head and underparts pale grey and white; upperparts brownish-grey. ▲ Príncipe. Forest and regrowth, plantations. Endemic. ✤ Soft little trill *rrrrrrrrr*, short *tiup tup* and fast, sibilant *whee-tsiu-tsiu-tseeu* and *tsee-tsitsiuu* or *tsiupti-ti-tiu*. [CD13:24]

6 **SÃO TOMÉ PRINIA** *Prinia molleri* 13 cm **R c**
Long, graduated tail; chestnut-brown head; grey upperparts. ▲ São Tomé. All habitats and altitudes, mainly disturbed and edge situations. Endemic. ✤ Nasal *dzik* or *dzi-dzi -dzi-...* Song a monotonous, rhythmic series of loud high-pitched single or double notes. [CD11:4]

7 **SÃO TOMÉ SHORT-TAIL** *Amaurocichla bocagei* c. 11 cm **R r VU**
Dark brown above; white upper throat; long straight bill; long legs. ▲ São Tomé. Forest. Endemic. ✤ Piercing *tsiiiii*; also *tsui-tsuii*. Song a longer version of call. [CD12:52]

8 **DOHRN'S THRUSH BABBLER** *Horizorhinus dohrni* c. 14 cm **R c**
Olivaceous-grey above; whitish below; olivaceous-grey breast-band and flanks; belly washed pale yellow. ▲ Príncipe. Forest regrowth, plantations. Endemic. ✤ Song melodious, vigorous and cheerful, starting with *tseeeeu-tu* then followed by fast, variable series of short notes. Calls a rapid *tyentyentyentyen...* and long trill *trrrrrrirrrrrr...* [CD12:51]

9 **PRÍNCIPE SUNBIRD** *Anabathmis hartlaubii* 13–14 cm **R c**
9a Ad male Dark glossy violet-blue throat and upper breast; dark olive above. **9b Ad female** Dark olive-green throat and upper breast. ▲ Príncipe. Forest, cultivation, gardens. Endemic. ✤ Clear *wheep*, fast *chrrrp-ch-chep* and various high-pitched notes. [CD12:74]

10 **NEWTON'S SUNBIRD** *Anabathmis newtonii* 10–11 cm **R c**
10a Ad male Dark glossy violet-blue throat and upper breast; rest of underparts yellow. **10b Ad female** Dark olive-green throat and upper breast. ▲ São Tomé. Forest, cultivation, gardens, dry savanna woodland. Endemic. ✤ Repeated *cheep* and harsh trill. Song a very rapid, high-pitched chattering jingle preceded by high-pitched *whseeeew*. [CD12:75]

11 **BIOKO BATIS** *Batis poensis* 12 cm **R f**
11a Ad male *poensis* As *B. p. occulta* (Plate 113:7) but supraloral spot generally smaller, supercilium indistinct or absent, breast-band broader and white edges to tertials and outer tail feathers narrower. **11b Ad female *poensis*** Chestnut breast-band. ▲ Bioko. Forest (to 1100 m). ✤ [CD12:11]

12 **GIANT SUNBIRD** *Dreptes thomensis* 18–23 cm **R r/lc VU**
Ad male Very large; blackish. Female similar but smaller. ▲ São Tomé. Forest. Endemic. ✤ Vigorous *cheep*, frequently in series. [CD12:76]

1
2
3
4
5
6
7
8
9b
9a
10b
10a
12
11b
11a

PLATE 148: ISLAND FORMS IV – Gulf of Guinea Islands

1 SÃO TOMÉ KINGFISHER *Alcedo (cristata) thomensis* 13.5 cm **R c**
1a Ad As Malachite Kingfisher (Plate 66:4) but darker; malar area barred dusky. **1b Juv** Blackish head-sides, mantle and breast. ▲ São Tomé, where the only kingfisher. Streams, beaches. Endemic. ✤ As Malachite Kingfisher.

2 PRÍNCIPE KINGFISHER *Alcedo (leucogaster) nais* 13 cm **R c**
As White-bellied Kingfisher (Plate 66:3) but no rufous supercilium and less white below. ▲ Príncipe, where the only small kingfisher. Streams, beaches; also forest. Endemic. ✤ High-pitched, sharp *(t)seee* or *tsee-eet*, in flight, and short *tsip*.

3 SÃO TOMÉ SCOPS OWL *Otus hartlaubi* 18 cm **R u VU**
Variable. **3a Pale morph** Mainly grey-brown and ochre. **3b Rufous morph** Much darker and rufous. ▲ São Tomé, where the only small owl. Most habitats with tall trees. Endemic. ✤ Soft *tuh* and *prr-u-u-p* or *pwu-huhu*, given at intervals of 15–20 seconds, reminiscent of African Scops Owl but more mellow and higher pitched. [CD7:29]

4 SÃO TOMÉ PARADISE FLYCATCHER *Terpsiphone atrochalybeia* *c.* 18 cm **R c**
4a Ad male Glossy blue-black with very long tail (extending 7.0–11.5 cm). **4b Ad female** Rufous upperparts and tail. ▲ São Tomé, where the only paradise flycatcher. Forest, edges, plantations. Endemic. ✤ Calls and songs include those typical of genus. [CD12:31]

5 ANNOBÓN PARADISE FLYCATCHER *Terpsiphone (rufiventer) smithii* 8 cm **R c VU**
Blue-black head; rich orange-rufous body; bluish-slate tail. Tail male + 0.5–1.0 cm. ▲ Annobón. Forest, cultivated areas. Endemic. ✤ As other paradise flycatchers.

6 SÃO TOMÉ FISCAL *Lanius newtoni* 23 cm **R r CR**
As Common Fiscal (Plate 122:1) but less white on scapulars; rump black; yellow wash to underparts. ▲ São Tomé, where the only shrike. Primary forest. Endemic. ✤ Series of *c.* 10 well-spaced, fluted *tiu* notes and more rapid, rhythmic *tsink-tsink-tsink* -... [CD13:41]

7 PRÍNCIPE GOLDEN WEAVER *Ploceus princeps* 18 cm **R c**
7a Ad male Orange-chestnut head; yellowish-green above; golden-yellow below. **7b Ad female** Yellowish-green top of head; white belly. ▲ Príncipe, where the only weaver. All habitats with trees; also villages. Endemic. ✤ Sharp *pzeep*; drawn-out wheeze. [CD14:55]

8 SÃO TOMÉ GROSBEAK *Neospiza concolor* 19–20 cm **R r CR**
Large; dark chestnut; massive bill. ▲ São Tomé. Primary forest. Endemic. ✤ Two-note whistle, second higher, recalling 11, but lower.

9 SOUTHERN MASKED WEAVER *Ploceus velatus* 14 cm **R c**
9a Ad male breeding Black mask extending in point onto lower throat; chestnut wash on crown and lower throat. **9b Ad female/ad male non-breeding** No mask; brownish-olive above. ▲ São Tomé. Open habitats, gardens. ✤ Typical weaver sounds. [CD14:59]

10 SÃO TOME WEAVER *Ploceus sanctithomae* 14 cm **R c**
10a Ad male Rusty-buff head-sides and underparts; blackish crown; double whitish wingbar; slender bill. **10b Ad female** Duller, paler. ▲ São Tomé. All forested habitats; also towns. Endemic. ✤ Abrupt *wik*, clear *psink*. Song ends in accelerating chatter. [CD14:72]

11 PRÍNCIPE SEEDEATER *Serinus rufobrunneus* 11.0–12.5 cm **R c/s**
11a Ad *thomensis* (São Tomé; c) Grey-brown; horn-coloured bill. **11b Ad *rufobrunneus*** (Príncipe; u/s) Rufous-brown. ▲ Forest, plantations, dry woodland, towns. ✤ Hard, tuneless *whsiiii*. Song a fast, melodious jumble of typical canary-like twitters and trills. [CD15:78]

Continued on page 486

1b
2
3b
1a
4a
3a
4b
5
7b
7a
6
9b
10b
10a
9a
8
11a
11b
12b
12a
13b
13a

Continued from page 484

12 GOLDEN-BACKED BISHOP *Euplectes aureus* 12 cm **R c**
12a Ad male breeding Black with orange-yellow mantle, back and rump. **12b Ad female/ ad male non-breeding** Rufous-buff upperparts heavily streaked black; lemon wash to face. ▲ São Tomé. Grassland near cultivation. ✤ Rasping *dzik*. [CD14:89]

13 GIANT WEAVER *Ploceus grandis* 22 cm **R c**
13a Ad male Very large; black head; chestnut collar; large bill. **13b Ad female** Above, greyish-olive streaked dusky; below, brownish-buff and white. ▲ São Tomé. Forest edge, plantations, etc. Endemic. ✤ Chattering notes ending in rather short and tuneless wheeze.

Plate on page 484

RECENT ADDITIONS

Since the publication of the first edition of this field guide, the following additional species have been reported from the region.

1 SNOWY EGRET *Egretta thula* 55–65 cm; WS 90–95 cm **V***
As Little Egret (Plate 7:7) but usually with rear of lower legs dull yellow (Little has all-dark legs); toes deeper yellow; lores bright yellow outside courtship period (in Little yellow during short courtship period only, otherwise grey-green). ▲ Aquatic habitats. N American vagrant (Cape Verde).

2 ORIENTAL HONEY BUZZARD *Pernis ptilorhynchus* 55–65 cm; WS 148–165 cm **V**
Very similar to European Honey Buzzard (Plate 23:1) but lacks dark carpal patches on underwings. Adult male has dark tail with broad pale band; female tail more like European Honey Buzzard but inner bar broader (hard to see). Also compare immature African Harrier Hawk (Plate 16:4). ▲ E Palearctic vagrant (Gabon).

3 WESTERN SANDPIPER *Calidris mauri* 14–17 cm **V**
Ad non-breeding Usually rather long bill with fine, slightly decurved tip combined with longish legs may recall miniature Dunlin (Plate 41:9). White supercilium prominent in front of eye. Compare Little Stint (Plate 41:3) and very similar Semipalmated Sandpiper (Plate 43:3) which also has half-webbed toes but different call (typically a short, harsh *churp*). ▲ N American vagrant (NW Senegal). ✤ Thin, high *jeet* or *cheet*.

4 EURASIAN COLLARED DOVE *Streptopelia decaocto* 31–34 cm **R? lc/s**
Very similar to African Collared Dove (Plate 53:3) but slightly larger; undertail-coverts grey (not white); tail slightly longer; voice different. ▲ Towns, villages. Palearctic species extending range southwards from Morocco (NW Mauritania). ✤ Trisyllabic *cu-cOOO-cu*. Nasal *kwèhr* on landing. [CD3:7]

5 WEYNS'S WEAVER *Ploceus weynsi* 15 cm **M? lc**
Ad male Pattern somewhat recalls Dark-backed Weaver (Plate 133:6) but throat and upper breast black, black wings edged yellow, flanks chestnut; eye pale yellow; bill black. **Ad female** Head and upperparts olive; wing feathers edged yellow; broad olivaceous breast-band; eye yellow; bill blackish. **Juv** As female, but duller. ▲ Forest, forest edge, farmbush. Usually in flocks. Nomadic (Congo). ✤ Song a high-pitched sizzling. Call a dry *chip*.

References

Buck, H. & Borrow, N. (in prep.) First record of Western Sandpiper *Calidris mauri* for Senegal and sub-Saharan Africa. *Bull. Afr. Bird Club.*

Clark, W.S. & Christy, P. (2006) First record of Oriental Honey Buzzard *Pernis ptilorhynchus* for Gabon and sub-Saharan Africa. *Bull. Afr. Bird Club* 13: 207–210.

Demey, R. (2004) Recent Reports. Mauritania. *Bull. Afr. Bird Club* 11: 176.

Demey, R. (2005) Recent Reports. Mauritania. *Bull. Afr. Bird Club* 12: 186.

Demey, R. (2006) Recent Reports. Cape Verde Islands. *Bull. Afr. Bird Club* 13: 100.

Rainey, H.J., Mokoko Ikonga, J., Vernon, R. & King, T. (in press) Birds new to the Republic of Congo. *Bull. Afr. Bird Club.*

REFERENCES

The references presented here are additions to those in Borrow & Demey (2001); only those that have been nominally cited in the present work (all belonging to part 1, except for two in 2) have been repeated. They are in four parts: (1) general and regional references, (2) country references, (3) family and species references and (4) acoustic reference.

1. GENERAL AND REGIONAL REFERENCES

Bannerman, D.A. (1930–1951) *The Birds of Tropical West Africa*, Vols. 1–8. Crown Agents, London.

Bannerman, D.A. (1953) *The Birds of West and Equatorial Africa*. 2 vols. Oliver & Boyd, Edinburgh & London.

Beaman, M. (1994) *Palearctic Birds: a Checklist of the Birds of Europe, North Africa and Asia north of the foothills of the Himalayas*. Harrier Publications, Stonyhurst.

BirdLife International (2000) *Threatened Birds of the World*. Lynx Edicions, Barcelona & BirdLife International, Cambridge.

BirdLife International (2004) *Threatened Birds of the World 2004*. CD-ROM. BirdLife International, Cambridge.

Borrow, N. & Demey, R. (2001) *Birds of Western Africa*. Christopher Helm, London.

Brown, L.H., Urban, E.K. & Newman, K. (1982) *The Birds of Africa*, Vol. 1. Academic Press, London.

David, N. & Gosselin, M. (2002a) Gender agreement of avian species names. *Bull. Brit. Orn. Club* 122: 14–49.

David, N. & Gosselin, M. (2002b) The grammatical gender of avian genera. *Bull. Brit. Orn. Club* 122: 257–282.

Dickinson, E.C. (ed.) (2003) *The Howard & Moore Complete Checklist of the Birds of the World*. Christopher Helm, London.

Dowsett, R.J. & Forbes-Watson, A.D. (1993) *Checklist of Birds of the Afrotropical and Malagasy Regions. Volume 1: Species limits and distribution*. Tauraco Press, Liège.

Fishpool, L.D.C. & Evans, M.I. (eds.) (2001) *Important Bird Areas in Africa and associated islands: Priority sites for conservation*. BirdLife Conservation Series No. 11. Pisces Publications & BirdLife International, Newbury & Cambridge, UK.

Fry, C.H., Keith, S. & Urban, E.K. (eds.) (1988, 2000) *The Birds of Africa*, Vols. 3 & 6. Academic Press, London.

del Hoyo, J., Elliot, A. & Sargatal, J. (eds.) (1992–2003) *Handbook of the Birds of the World*, Vols. 1–8. Lynx Edicions, Barcelona.

Inskipp, T., Lindsey, N. & Duckworth, W. (1996) *An Annotated Checklist of the Birds of the Oriental Region*. Oriental Bird Club, Sandy.

Keith, S., Urban, E.K. & Fry, C.H. (eds.) (1992) *The Birds of Africa*, Vol. 4. Academic Press, London.

Knox, A.G., Collinson, M., Helbig, A.J., Parkin, D.T. & Sangster, G. (2002) Taxonomic recommendations for British birds. *Ibis* 144: 707–710.

Mackworth-Praed, C.W. & Grant, C.H.B. (1970–1973) *Birds of West Central and Western Africa*. 2 vols. Longmans, London.

Sangster, G., Collinson, J.M., Helbig, A.J., Knox, A.G. & Parkin, D.T. (2004) Taxonomic recommendations for British birds: second report. *Ibis* 146: 153–157.

Sangster, G., Knox, A.G., Helbig, A.J. & Parkin, D.T. (2002) Taxonomic recommendations for European birds. *Ibis* 144: 153–159.

Urban, E.K., Fry, C.H. & Keith, S. (eds.) (1986, 1997) *The Birds of Africa*, Vols. 2 & 5. Academic Press, London.

2. COUNTRY REFERENCES

Benin

van den Akker, M. (2000) Red-tailed Greenbul *Criniger calurus* and Chestnut-breasted Negrofinch *Nigrita bicolor*, new to Bénin. *Bull. Afr. Bird Club* 7: 133.

van den Akker, M. (2003) Birds of Niaouli forest, southern Benin. *Bull. Afr. Bird Club* 10: 16–22.

van den Akker, M. (2003) First records for Benin of Yellow-whiskered Greenbul *Andropadus latirostris*, Western Bearded Greenbul *Criniger barbatus* and White-browed Forest Flycatcher *Fraseria cinerascens*. *Bull. Afr. Bird Club* 10: 122–124.

van den Akker, M. & Claffey, P. (2004) Further records from the remnant forests of Benin: White-tailed Ant Thrush *Neocossyphus poensis* and Bioko Batis *Batis poensis*. *Bull. Afr. Bird Club* 11: 32–33.

Anciaux, M.-R. (2000–2002) Approche de la phénologie de la migration des migrateurs intra-africains de l'intérieur des terres du Sud-Bénin (plateau d'Allada et sud de la dépression de la Lama). 1, 2a & 2b. *Alauda* 68: 311–320; 70: 203–211 & 413–419.

Claffey, P.M. (2003) Parasitic Weaver *Anomalospiza imberbis*, new to Benin. Bull. *Afr. Bird Club* 10: 49–50.

Burkina Faso

Portier, B. (2002) Red-necked Nightjar *Caprimulgus ruficollis*, new to Burkina Faso. *Bull. Afr. Bird Club* 9: 139–140.

Portier, B., Lungren, C. & Ouéda, G.H. (2002) Birding in Burkina Faso, more than just birdwatching. *Dutch Birding* 24: 127–141.

Cameroon

Bobo, S.K., Njabo, K.Y., Anye, D.N. & Languy, M. (2001) Status and distribution of the Bamenda Apalis *Apalis bamendae* in Cameroon, Central Africa. *Ostrich* Suppl. 15: 110–113.

Dowsett, R.J. & Dowsett-Lemaire, F. (2001) First records of Scarce Swift *Schoutedenapus myioptilus* and Grass Owl *Tyto capensis* from Mt Cameroon. *Malimbus* 23: 110-111.

Messemaker, R. (2004) First record of Sociable Lapwing *Vanellus gregarius* for Cameroon and western Africa. *Bull. Afr. Bird Club* 11: 34–35.

Mills, M., Hoff, R. & Myers, D. (2003) First breeding record of Ovambo Sparrowhawk *Accipiter ovampensis* in West Africa. *Malimbus* 25: 104–106.

Sinclair, I., Cassidy, R., Cope, A., Van Aswegen, H., Leslie, R. & Rose, B. (2003) The first Golden Nightjar *Caprimulgus eximius* in Cameroon. *Bull. Afr. Bird Club* 10: 124–125.

Cape Verde Islands

De Rouck, K. (2001) Bates' Swift in Cape Verde Islands? *Dutch Birding* 23: 24-25.

Hazevoet, C.J. (2003) Fifth report on birds from the Cape Verde Islands, including records of 15 taxa new to the archipelago. *Arquivos do Museu Bocage* Nova Série III(19): 503–528.

Congo

Herroelen, P. (2003) Reidentification of two Nyanza Swifts from the Congo basin. *Bull. Brit. Orn. Club* 123: 278–280.

Gabon

Dean, W.R.J., Walters, M.P. & Dowsett, R.J. (2003) Records of birds breeding collected by Dr W. Ansorge in Angola and Gabon. *Bull. Brit. Orn. Club* 123: 239–250.

The Gambia

Barlow, C.R. (2002) First nest record for Bronze-winged Courser *Cursorius chalcopterus* in Senegambia. *Bull. Afr. Bird Club* 9: 134–135.

Barlow, C.R. (2003) First conclusive evidence of breeding in Senegambia and parental behaviour of Black Coucal *Centropus grillii*. *Bull. Afr. Bird Club* 10: 53–54.

Clark, W.S. & Barlow, C. (2004) Status of Short-toed Snake Eagle *Circaetus gallicus* and Beaudouin's Snake Eagle *C. beaudouini* in The Gambia. *Bull. Afr. Bird Club* 11: 27–29.

Crewe, M.D. (2001) Selected observations from The Gambia, 1997–1999, with comments on the identification of a number of species. *Bull. Afr. Bird Club* 8: 113–116.

Crewe, M.D. & Small, B.J. (2002) Temminck's Horned Lark *Eremophila bilopha* – a new species for The Gambia. *Bull. Afr. Bird Club* 9: 137–138.

King, J.M.B. (2003) Baillon's Crake *Porzana pusilla*, new to The Gambia, with notes on seven other species. *Malimbus* 25: 59–61.

Kirk, G. & Barlow, C.R. (2002) Second confirmed record of Forbes's Plover *Charadrius forbesi* for The Gambia. *Bull. Afr. Bird Club* 9: 138–139.

Mikkola, A. & Mikkola, H. (2002) First record of Red-footed Falcon *Falco vespertinus* in The Gambia. *Bull. Afr. Bird Club* 9: 45. [Corrigendum 9: 93.]

Van Welie, L. (2003) Franklin's Gull *Larus pipixcan* in The Gambia. *Malimbus* 25: 97–99.

Guinea

Demey, R. & Rainey, H.J. (2004) The birds of the Pic the Fon Forest Reserve, Guinea: a preliminary survey. *Bull. Afr. Bird Club* 11: 126–138.

Demey, R. & Rainey, H.J. (in prep.) Surveys of the birds of Déré, Diécké and Mont Béro Forest Reserves, Guinea.

Trolliet, B. (2001) Première observation d'une *Zoothera* en Guinée. *Malimbus* 23: 113-115.

Guinea-Bissau

Catry, P. & Monteiro, H. (2003) House Sparrow *Passer domesticus* colonises Guinea-Bissau. *Malimbus* 25: 58–59.

Ivory Coast

Demey, R. & Rainey, H.J. (in prep.) The birds of Mont Sangbé National Park, Ivory Coast.

Rainey, H., Borrow, N., Demey, R. & Fishpool, L.D.C. (2003) First recordings of vocalisations of Yellow-footed Honeyguide *Melignomon eisentrauti* and confirmed records in Ivory Coast. *Malimbus* 25: 31–38.
Rainey, H. & Lachenaud, O. (2002) Recent bird observations from Ivory Coast. *Malimbus* 24: 23-37.
Rheindt, F.E., Grafe, U. & Linsenmair, K.E. (2002) New bird records in Comoé National Park, Ivory Coast. *Malimbus* 24: 38-40.

Mali

Clouet, M. & Goar, J.-L. (2003) L'avifaune de l'Adrar Tirharhar – Adrar des Iforas (Mali). *Alauda* 71: 469–474.
Lamarche, B. (1980–1981) Liste commentée des oiseaux du Mali. *Malimbus* 2: 121–158; 3: 73–102.
Moulin, S., Dobigny, G., Cornette, R. & Ag Sidiyene, E. (2001) Observations ornithologiques dans l'Adrar des Iforas (Mali). Alauda 69: 527-532. [+ Dowsett, R. (2002) *Alauda* 70: 236.]
Wyminga, E., Kone, B., van der Kamp, J. & Zwarts, L. (2002) *Delta Intérieur du Niger. Écologie et gestion durable des ressources naturelles.* Mali-PIN publication 2002-01. Wetlands International, Sévaré, RIZA, Rijkswaterstaat, Lelystad, and Altenburg & Wyminga, conseillers écologiques, Veenwouden.

Mauritania

Hamerlynck, O. & Messaoud, B. ould (2000) Suspected breeding of Lesser Flamingo *Phoeniconaias minor* in Mauritania. *Bull. Afr. Bird Club* 7: 109–110.
Lamarche, B. (1988) Liste commentée des oiseaux de Mauritanie. *Études Sahariennes et Ouest-Africaines* 1(4): 1–162. Privately published, Nouakchott & Paris.
Salewski, V., Altwegg, R., Liechti, F. & Peter, D. (2003) New records of Moussier's Redstart *Phoenicurus moussieri* and Lesser Striped Swallows *Hirundo abyssinica* from Mauritania. *Malimbus* 25: 103–104.

Niger

Ambagis, J., Brouwer, J. & Jameson, C. (2003) Seasonal waterbird and raptor fluctuations on the Niger and Mékrou Rivers in Niger. *Malimbus* 25: 39–51.
Crisler, T., Jameson, C. & Brouwer, J. (2003) An updated overview of the birds of W National Park, southwest Niger. *Malimbus* 25: 4–30.
Demey, R., Dowsett, R.J. & Fishpool, L.D.C. (2001) Comments on Black-throated Coucal *Centropus leucogaster*, claimed from Niger. *Malimbus* 23: 112-113.

Nigeria

de Bont, M. (2002) Avifauna of the Hwimo area, Nigeria. *Bull. Afr. Bird Club* 9: 101–106.
Demey, R., Dowsett-Lemaire, F. & Dowsett, R.J. (2003) Notable bird observations from Nigeria, including the first records of Spot-breasted Ibis *Bostrychia rara* and Yellow Longbill *Macrosphenus flavicans. Malimbus* 25: 85–94.
Dowsett, R.J. (2002) More on Boyd Alexander's types from Lake Chad. *Bull. Brit. Orn. Club* 122: 228–230.
Manu, S. (2001) Possible factors influencing the decline of Nigeria's rarest endemic bird, the Ibadan Malimbe *Malimbe ibadanensis. Ostrich* Suppl. 15: 119–121.
McGregor, R. (2004) New records of Ortolan Bunting *Emberiza hortulana* in Nigeria. *Bull. Afr. Bird Club* 11: 30–31.
McGregor, R. & Wilson, J.M. (2003) A major range extension of Locust Finch *Ortygospiza locustella* in West Africa. *Malimbus* 25: 99–101.
Ottoson, U., Bengtsson, D., Gustafsson, R., Hall, P., Hjort, C., Leventis, A.P., Neumann, R., Pettersson, J., Rhönnstad, P., Rumsey, S., Waldenström, J. & Velmala, W. (2002) New birds for Nigeria observed during the Lake Chad Bird Migration Project. *Bull. Afr. Bird Club* 9: 52-55.
Ottoson, U., Hjort, C., Hall, P., Velmala, W. & Wilson, J.M. (2003) On the occurrence of the Black Stork *Ciconia nigra* in Nigeria. *Malimbus* 25: 96–97.
Turk, A. (2000) Fishing owls at Agenebode, Nigeria. *Bull. Afr. Bird Club* 7: 107–108.
Velmala, W. & Gustafsson, R. (2003) Two new raptors for Nigeria and other raptor observations at Lake Chad. *Malimbus* 25: 52–55.
Wilson, J.M. (2002) First breeding record of Little Grey Woodpecker *Dendropicos elachus* in Nigeria. *Malimbus* 24: 42-43.
Wilson, J.M. & McGregor, R. (2002) House Sparrow *Passer domesticus* in NE Nigeria. *Malimbus* 24: 40-41.

São Tomé & Príncipe

Dallimer, M., King, T. & Atkinson, R. (2003) A high altitude sighting of the São Tomé Short-tail *Amaurocichla bocagei. Malimbus* 25: 101–103.
Dallimer, M., King, T. & Leitão, P. (2003) New records of the São Tomé Grosbeak *Neospiza concolor. Bull. Afr. Bird Club* 10: 23–25

Senegal

Brasseur, R.E. (2000) African Swallow-tailed Kite *Chelictinia riocourii* breeding in the Saloum Delta, Sénégal. *Bull. Afr. Bird Club* 7: 134.

Cruse, R. (2004) Yellow-browed Warbler in Senegal, December 2003. *Bull. Afr. Bird Club* 11: 147–148.

Haass, N. & Engelhardt, C. (2002) *Calonectris* shearwaters off West Africa. *Birding World* 14: 514.

Van der Have, T.M. & Van der Hoop, C. (2002) Spotted Sandpiper at Île de Gorée, Senegal, in February 2001. *Dutch Birding* 24: 156–157.

Togo

Selfe, G. (2003) Little Rush Warbler *Bradypterus baboecala*, new to Togo. *Bull. Afr. Bird Club* 10: 51.

3. FAMILY AND SPECIES REFERENCES

Alström, P. & Mild, K. (2003) *Pipits & Wagtails of Europe, Asia and North America*. Christopher Helm, London.

Beresford, P. (2003) Molecular systematics of *Alethe*, *Sheppardia* and some other African robins (Muscicapoidea). *Ostrich* 74: 58–73.

Blomdahl, A., Breife, B. & Holmström, N. (2003) *Flight Identification of European Seabirds*. Christopher Helm, London

Beyers, C., Olsson, U. & Curson, J. (1995) Buntings and Sparrows. *A Guide to the Buntings and North American Sparrows*. Pica Press, Mountfield.

Clarke, T. (1999) The Great Blue Heron on Tenerife. *Birding World* 12: 158-161.

Cleere, N. (2001) The validity of the genus *Veles* Bangs, 1918 (Caprimulgidae). *Bull. Brit. Orn. Club* 121: 278-279.

Clement, P. & Helbig, A.J. (1998) Taxonomy and identification of chiffchaffs in the Western Palearctic. *Brit. Birds* 91: 361–376.

Clouet, M. & Wink, M. (2000) The buzzards of Cape Verde *Buteo (buteo) bannermani* and Socotra *Buteo (buteo)* spp.: first results of a genetic analysis based on nucleotide sequences of the cytochrome b gene. *Alauda* 68: 55–58.

Craig, A. (2000) Identifying glossy starlings in the field. *Bull. Afr. Bird Club* 7: 119–127.

Diagana, C.H. (2003) Conservation action project for Black Storks in West Africa. *Fadama* 6: 6.

Dickerman, R. W. (1997) A substitute name for the Bioko race of *Pycnonotus virens*. *Bull. Brit. Orn. Club* 117: 75.

Dowsett-Lemaire, F., Borrow, N. & Dowsett, R.J. (2005) *Cisticola dorsti* (Dorst's Cisticola) and *C. ruficeps guinea* are conspecific. *Bull. Brit. Orn. Club* 125: 305–313.

Etherington, G. & Small, B. (2003) Taxonomy and identification of Atlas Flycatcher – a potential British vagrant. *Birding World* 16: 252–256.

Ferguson-Lees, J. & Christie, D.A. (2001) *Raptors of the World*. Christopher Helm, London.

Fontaine, B. (2003) Is Dja River Warbler *Bradypterus grandis* really globally threatened? *Bull. Afr. Bird Club* 10: 28–29.

Gantlett, S. (1998) Identification of Great Blue Heron and Grey Heron. *Birding World* 11: 12-20.

Grapputo, A., Pilastro, A., Baker, A.J. & Marin, G. (2001). Molecular evidence for phylogenetic relationships among buntings and American sparrows (Emberizidae). *J. Avian Biol.* 32: 95–101.

Gutiérrez, R. (1998) Flight identification of Cory's and Scopoli's Shearwaters. *Dutch Birding* 20: 216–225.

Harrop, A.H.J. (2004) The 'soft-plumaged petrel' complex: a review of the literature on taxonomy, identification and distribution. *Brit. Birds* 97: 6–15.

Hirschfeld, E., Roselaar, C.S. & Shrihai, H. (2000) Identification, taxonomy and distribution of Greater and Lesser Sand Plovers. *Brit. Birds* 93: 162–189.

Holyoak, D.T. (2001) *Nightjars and their Allies*. Oxford University Press, Oxford.

Jackson, H.D. (2002). Primary emargination as a guide to Afrotropical nightjar relationships. *Ostrich* 73: 69–71.

Jiguet, F., Jaramillo, A. & Sinclair, I. (2001) Identification of Kelp Gull. *Birding World* 14: 112-125.

Jones, P.J., Elliott, C.C.H. & Cheke, R.A. (2002) Methods for ageing juvenile Red-billed Queleas, *Quelea quelea*, and their potential for the detection of juvenile dispersal patterns. *Ostrich* 73: 43-48.

Lachenaud, O. (2003) On the plumages of Senegal Coucal *Centropus senegalensis* and a putative observation of Black-throated Coucal *C. leucogaster* in Niger. *Malimbus* 25: 55–56.

Lahti, D.C. & Payne, R.B. (2003) Morphological and behavioural evidence of relationships of the Cuckoo Finch *Anomalospiza imberbis*. *Bull. Brit. Orn. Club* 123: 112–125.

Lee, P.L.M., Richardson, L.J. & Bradbury, R.B. (2001) The phylogenetic status of the Corn Bunting *Miliaria calandra* based on mitochondrial control-region DNA sequences. *Ibis* 143: 299–303.

Louette, M. (2002) Relationship of the Red-thighed Sparrowhawk *Accipiter erythropus* and the African Little Sparrowhawk *A. minullus*. *Bull. Brit. Orn. Club* 122: 218–222.

McGeehan, A. & Gutiérrez, R. (1997) Dark secrets. Identification of Mediterranean Shearwater. *Birdwatch* 61: 26-30.

McGeehan, A. & Gutiérrez, R. (1998) Great dilemmas. Identification of Great and Cory's Shearwaters. *Birdwatch* 73: 32-36.

McGowan, R.Y. (2002) Racial identification of Pallid Swift. *Brit. Birds* 95: 454–455.

Nikolaus, G. (1982) Further notes on some birds new to South Sudan. *Scopus* 6: 1–4.

Nunn, G.B., Cooper, J., Jouventin, P., Robertson, C.J.R. & Robertson, G.G. (1996) Evolutionary relationships among extant albatrosses (Procellariiformes: Diomedeidae) established from complete cytochrome-b gene sequences. *Auk* 113: 784–801.

Parry, S.J., Clark, W.S. & Prakash, V. (2002) On the taxonomic status of the Indian Spotted Eagle *Aquila hastata*. *Ibis* 144: 665–675.

Sætre, G.-P., Borge, T. & Moum, T. (2001) A new bird species? The taxonomic status of 'the Atlas Flycatcher' assessed from DNA sequence analysis. *Ibis* 143: 494-497.

Salewski, V., Bairlein, F. & Leisler, B. (2002) Different wintering strategies of two Palearctic migrants in West Africa – a consequence of foraging startegies? *Ibis* 144: 85-93.

Salewski, V., Falk, K.H., Bairlein, F. & Leisler, B. (2003) Gambaga Flycatcher *Muscicapa gambagae*: evidence for migration in West Africa? *Bull. Brit. Orn. Club* 123: 48–51.

Salomon, M., Voisin, J.-F. & Bried, J. (2003) On the taxonomic status and denomination of the Iberian Chiffchaffs. *Ibis* 145: 87–97.

Sangster, G., Collinson, J.M., Helbig, A.J., Knox, A.G. & Parkin, D.T. (2002) The generic status of Black-browed Albatross and other albatrosses. *Brit. Birds* 95: 583–585.

Sangster, G., Collinson, J.M., Helbig, A.J., Knox, A.G. & Parkin, D. (2002) The specific status of Balearic and Yelkouan Shearwaters. *Brit. Birds* 95: 636–639.

Sangster, G., Knox, A.G., Helbig, A.J. & Parkin, D.T. (2002) Taxonomic recommendations for European birds. *Ibis* 144: 153–159.

Scott, M. (2002) A Brown Skua on the Isles of Scilly – the first for Europe? *Birding World* 15: 383–386.

Shirihai, H., Gargallo, G. & Helbig, A. J. (2001) *Sylvia Warblers: Identification, Taxonomy and Phylogeny of the Genus* Sylvia. Christopher Helm, London.

Shirihai, H. (2002) *A Complete Guide to Antarctic Wildlife*. Alula Press, Finland.

Short, L. L. & Horne, J.F.M. (2001) *Toucans, Barbets and Honeyguides*. Oxford University Press, Oxford.

Svensson, L. (2001) Identification of Western and Eastern Olivaceous, Booted and Sykes's Warblers. *Birding World* 14: 192-219.

Svensson, L. (2001) The correct name of the Iberian Chiffchaff *Phylloscopus ibericus* Ticehurst 1937, its identification and new evidence of its winter grounds. *Bull. Brit. Orn. Club* 121: 281-296.

Tickell, W.L.N. (2000) *Albatrosses*. Pica Press, Mountfield.

Todte, I. & Harz, M. (2003) Geschlechtsbestimmung, Mauser und Farbabweichungen beim Bienenfresser *Merops apiaster*. *Limicola* 17: 1–10.

Trolliet, B. & Fouquet, M. (2001) La population ouest-africaine du Flamant nain *Phoeniconaias minor*: effectifs, répartition et isolement. *Malimbus* 23: 87-92.

Ullman, M. (1994) Identification of Pied Wheatear and Eastern Black-eared Wheatear. *Dutch Birding* 16: 186–194.

Ullman, M. (2003) Separation of Western and Eastern Black-eared Wheatear. *Dutch Birding* 25: 77–97.

Urquhart, E. (2002) *Stonechats. A Guide to the Genus* Saxicola. Christopher Helm, London.

Walbridge, G., Small, B. & McGowan, R.Y. (2003) Ascension Frigatebird on Tiree – new to the Western Palearctic. *Brit. Birds* 96: 58–73.

Walsh, J.F. (2002) The status of Black Stork *Ciconia nigra* in West Africa. *Malimbus* 24: 41-42.

Waltert, M. & Mühlenberg, M. (2000) A nest of Grey-necked Picatharthes *Picathartes oreas* constructed on a tree. *Bull. Afr. Bird Club* 7: 132.

White, S.J. & Kehoe, C.V. (2001) Difficulties in determining the age of Common Terns in the field. *Brit. Birds* 94: 268-277.

Wilson, J.M. & Sallinen, P. (2003) First records of Didric Cuckoo *Chrysococcyx caprius* parasitizing Cricket Warbler *Spiloptila clamans*. *Malimbus* 25: 95–96.

Woodcock, M.°W. (2003) Systematics and confusion in the genus *Parmoptila*. *Bull. Brit. Orn. Club* 123: 274–277.

4. ACOUSTIC REFERENCE

Chappuis, C. (2000) *African Bird Sounds: Birds of North, West and Central Africa and Neighbouring Atlantic Islands*. 15 CDs. Société d'Études Ornithologiques de France, Paris.

SCIENTIFIC INDEX

ENGLISH INDEX

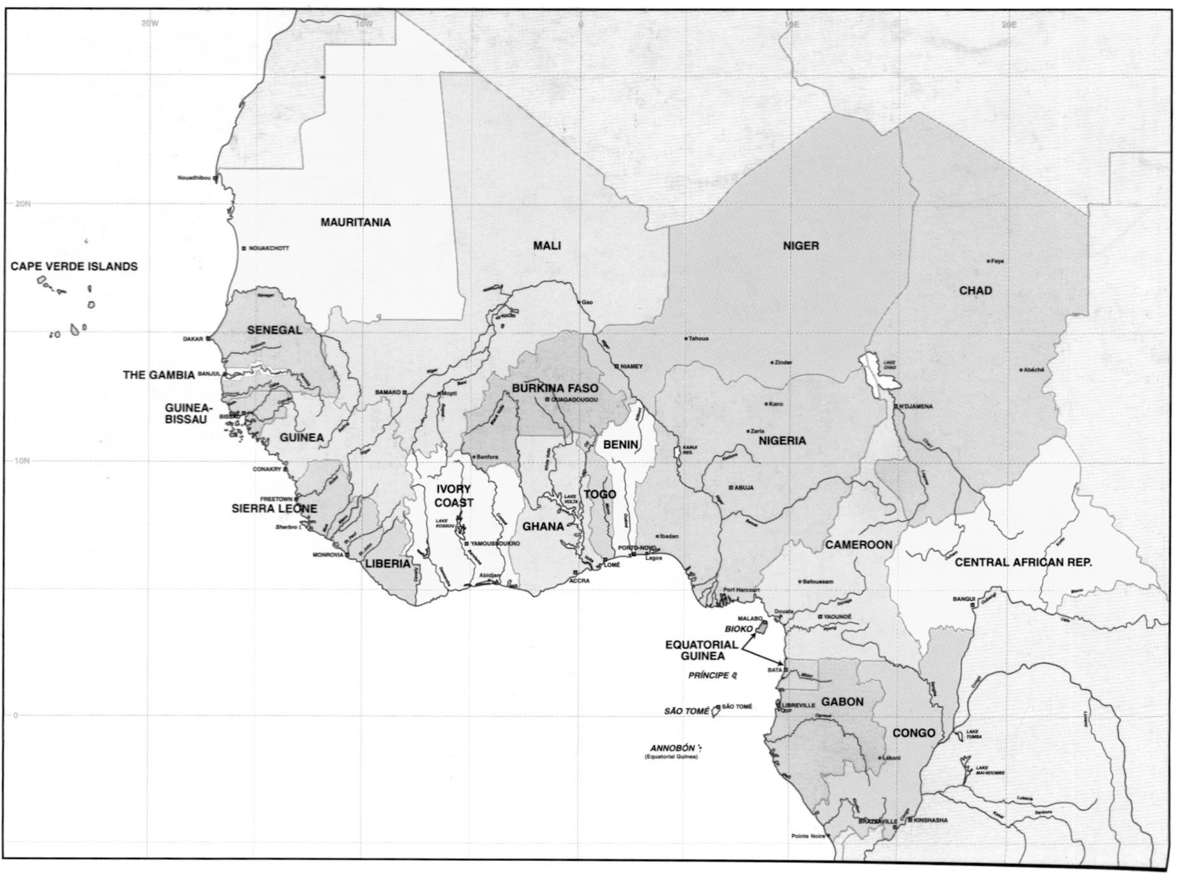
MAURITANIA
MALI
NIGER
CHAD
CAPE VERDE ISLANDS
SENEGAL
THE GAMBIA
GUINEA-BISSAU
GUINEA
SIERRA LEONE
LIBERIA
IVORY COAST
BURKINA FASO
GHANA
TOGO
BENIN
NIGERIA
CAMEROON
CENTRAL AFRICAN REP.
EQUATORIAL GUINEA
BIOKO
PRÍNCIPE
SÃO TOMÉ
ANNOBÓN
(Equatorial Guinea)
GABON
CONGO
Nouadhibou
NOUAKCHOTT
DAKAR
BANJUL
BAMAKO
Mopti
Gao
NIAMEY
Tahoua
Zinder
Kano
Zaria
ABUJA
OUAGADOUGOU
Banfora
CONAKRY
FREETOWN
MONROVIA
YAMOUSSOUKRO
Abidjan
ACCRA
LOMÉ
PORTO-NOVO
Lagos
Ibadan
Port Harcourt
MALABO
Douala
YAOUNDÉ
Bafoussam
BATA
LIBREVILLE
N'DJAMENA
Faya
Abéché
BANGUI
BRAZZAVILLE
KINSHASHA
Pointe Noire
Lékoni
20W
10W
0
10E
20E
20N
10N